Role of Natural Bioactive Compounds in the Rise and Fall of Cancers

Role of Natural Bioactive Compounds in the Rise and Fall of Cancers

Volume 1

Editor

Claudio Luparello

MDPI • Basel • Beijing • Wuhan • Barcelona • Belgrade • Manchester • Tokyo • Cluj • Tianjin

Editor
Claudio Luparello
Università di Palermo
Italy

Editorial Office
MDPI
St. Alban-Anlage 66
4052 Basel, Switzerland

This is a reprint of articles from the Special Issue published online in the open access journal *Cancers* (ISSN 2072-6694) (available at: https://www.mdpi.com/journal/cancers/special_issues/cancers-NBC).

For citation purposes, cite each article independently as indicated on the article page online and as indicated below:

LastName, A.A.; LastName, B.B.; LastName, C.C. Article Title. *Journal Name* **Year**, *Article Number*, Page Range.

Volume 1
ISBN 978-3-03943-254-7 (Pbk)
ISBN 978-3-03943-255-4 (PDF)

Volume 1-2
ISBN 978-3-03943-294-3 (Pbk)
ISBN 978-3-03943-295-0 (PDF)

Contents

About the Editor

Claudio Luparello, Ph.D. (Cell Biology) is a Professor of Cytology and Histology at the University of Palermo, Italy. He has authored more than 40 publications and 3 book chapters and was the Editor of the book "Novel aspects in PTHrP physiopatology" edited by Nova Sci Pub. He serves as the Editorial Board member of Biological Procedure Online International Journal of Breast Cancer and TheScientificWorldJournal domains Cell Biology and Oncology. He is a member of AICC.

Editorial

Role of Natural Bioactive Compounds in the Rise and Fall of Cancers

Claudio Luparello

Department of Biological, Chemical and Pharmaceutical Sciences and Technologies, University of Palermo, Viale delle Scienze, Edificio 16, 90128 Palermo, Italy; claudio.luparello@unipa.it; Tel.: +39-091-238-97405

Received: 31 August 2020; Accepted: 2 September 2020; Published: 3 September 2020

Recent years have seen the idea of a close association between nutrition and the modulation of cancer development/progression reinforced. In fact, an increasing number of experimental and epidemiological evidence has been produced, supporting the concept that many different bioactive components of food (e.g., polyphenols, mono- and polyunsaturated fatty acids, methyl-group donors …) may be implicated in either the promotion of or the protection against carcinogenesis. At the cellular level, such compounds can have an impact on different but sometimes intertwined processes, such as growth and differentiation, DNA repair, programmed cell death, and oxidative stress. In addition, compelling evidence is starting to build up of the existence of primary epigenetic targets of dietary compounds, such as oncogenic/oncosuppressor miRNAs or DNA-modifying enzymes, which in turn impair gene expression and function. This editorial aims to summarize the themes of the 31 papers (20 original articles and 11 reviews) published in the Special Issue "Role of Natural Bioactive Compounds in the Rise and Fall of Cancers" presenting the latest findings on the intracellular pathways and mechanisms affected by selected natural molecules influencing the fine-tuning of cancer phenotype.

Plant polyphenols have been among the most studied natural compounds by the contributors to this issue.

In the original article group, Polonio-Alcalà et al. [1] showed the additive and synergistic effects of the flavonoids (−)-epigallocatechin-3-gallate (EGCG) from green tea and its naphthalene derivative G28, which are fatty acid synthase inhibitors, in combination with epidermal growth factor receptor tyrosine kinase inhibitors on gefitinib-resistant lung adenocarcinoma models. Moreover, Wei et al. [2] examined the effect and mechanism of action of EGCG alone and in combination with current chemotherapeutics on pancreatic cancer cell growth, demonstrated the impairment of cell proliferation via the phosphofructokinase inhibition-mediated suppression of glycolysis in a ROS-dependent manner, and the additive enhancement of the anticancer effect of gemcitabine both in vitro and in pancreatic xenografts by the further inhibition of glycolysis and the impairment of cell kinetics. In the Review section, Farooqi et al. [3] analyzed the pleiotropic abilities of EGCG to regulate intracellular signalizations such as those related to JAK/STAT, Wnt/β-catenin, TGF/SMAD, SHH/GLI and NOTCH pathways, also commenting on the ability of EGCG to modulate non-coding RNAs and the methylation-associated machinery in different cancers.

Other natural phenolic compounds whose activity is discussed in the original articles of this issue are:

–   Gallic acid (3,4,5-trihydroxybenzoic acid), widely distributed in natural plants, fruits, and green tea, whose tumor-suppressive effect via the p53-mediated downregulation of the transmembrane protein PD-L1 was demonstrated by Kang et al. [4] in non-small-cell lung cancer models;

–   Oleuropein, the main bioactive phenolic component of *Olea europaea* L., whose presence in enriched extracts from olive leaves was proven to reduce the glycolytic rate of a wide range of solid and liquid tumor cells via the downregulation of the three key effectors of the glycolytic

pathway, GLUT-1, PKM2 and MCT4, likely resulting in a decreased glucose entrance and biomass production [5];
- Oleacein (3,4-dihydroxyphenylethanol), the main secoiridoid contained in extra virgin olive oil, able to elicit significant anti-tumor activity by promoting cell cycle arrest and apoptosis in multiple myeloma cells due to its histone deacetylase inhibitory properties [6];
- Dadzein (7,4'-dihydroxyisoflavone), present in soybeans, whose 4-sulphate metabolite produced by gut microbiota was found to exert an anti-estrogenic effect on ERα-positive breast cancer cells via the downregulation of the anti-apoptotic neuroglobin protein thus rendering cells more prone to the paclitaxel treatment [7];
- Gigantol, a bibenzyl compound from orchid species, whose ability to destabilize tumor integrity via the suppression of the PI3K/AKT/mTOR and JAK/STAT pathways was demonstrated by Losuwannarak et al. [8] in non-small-cell lung cancer models in vitro and in vivo;
- Lonchocarpin, a chalcone isolated from *Lonchocarpus sericeus*, proven to be a powerful inhibitor of the Wnt/β-catenin pathway able to selectively suppress the migration and proliferation of a panel of colorectal cancer cell lines in vitro and in a preclinical colorectal cancer mouse model [9];
- Isorhamnetin, (3'-methoxy-3,4',5,7-tetrahydroxyflavone), a flavonol aglycone found in some medicinal plants, able to exert an anti-proliferative effect on human bladder cancer cells via the induction of cell cycle arrest during the G2/M phase and apoptosis, accompanied by the activation of the AMPK signaling pathway and ROS overproduction [10];
- Erioquinol, eriopodol A and gibbilimbol B, derived from *Piper* genus plants, whose ability to inhibit XIAP protein, involved in the regulation of caspase-dependent/independent cell death pathways, was reported by Muñoz et al. [11] in breast cancer cell lines;
- Vatein, isolated from *Calocedrus formosana* Florin leaves extract, proven to interfere with cell cycle and microtubule dynamics in lung adenocarcinoma cells, also inhibiting tumor growth in a xenograft mouse model [12].

In the Review section, Perrone et al. [13] discussed the effects of polyphenols in preventing the progression of central and peripheral nervous system tumors underlining the beneficial effect of dietary compounds on the microbioma–intestine–brain axis. Barbosa and Martel [14] examined the role played by a wide variety of synthetic and natural substances, including polyphenols, on the impairment of glucose uptake by neoplastic breast cells thereby resulting in a tumor-restraining effect. Ong et al. [15] reported the broad-range in vitro/in vivo anticancer properties of the *Magnolia*-derived polyphenol honokiol based upon its ability to impair cell cycle progression, inhibit epithelial–mesenchymal transition, and suppress cell motility, invasion, metastasis and angiogenesis. Zhou et al. [16] summarized the late preclinical studies on the applications of bioactive polyphenols in lung cancer therapy, focusing on the molecular mechanisms at the basis of their therapeutic effects and also discussing the potential of the polyphenol-based combination therapy. Goh et al. [17] reviewed data on the anti-colon cancer properties of nobiletin, a polymethoxyflavone extracted from citrus peel, and its derivatives which are able to arrest the cell cycle, inhibit cell proliferation, induce apoptosis, prevent tumor formation, reduce inflammatory effects and limit angiogenesis, also exploring better drug delivery strategies due to the low oral bioavailability of the compounds. Ong et al. [18] focused their review on the pharmacological properties and therapeutic potential of formononetin [7-hydroxy-3-(4-methoxyphenyl)-4H-1-benzopyran-4-one], one of the main bioactive components of red clover, which regulates various transcription factor- and growth factor-mediated oncogenic pathways attenuating metastasis and tumor growth in vivo in multiple cancer cell models and also alleviating the possible causes of chronic inflammation that are linked to the cancer survival of neoplastic cells and their resistance against chemotherapy.

The other articles and reviews addressed further cancer-related issues relevant to types of compounds of a different nature, specifically:

- The methanolic extract of *Malva pseudolavatera* leaves, which showed a promising selective anti-proliferative and pro-apoptotic effect on acute myeloid leukemia cell lines, determining PARP cleavage, cytochrome-c release, Bax/Bcl-2 ratio increase and ROS overproduction [19];
- Eicosapentaenoic acid, an ω-3 polyunsaturated fatty acid, which played a protective role, both alone and in combination with angiotensin-converting enzyme inhibitors, in attenuating adipocyte-induced proinflammatory cytokine expression and the migration of breast cancer cells in an in vitro model of obesity-induced breast cancer [20];
- Fucoidan, a sulphated polysaccharide derived from brown seaweed, whose combination with gemcitabine determined an enhanced pro-apoptotic and cell cycle-inhibitory activity on selected uterine carcinosarcoma and stromal sarcoma cell lines [21];
- Nicotin, whose mechanisms underlying the promotion of melanoma cell proliferation and migration mediated through α9-nAChR-initiated carcinogenic signaling and PD-L1 expression were reported by Nguyen et al. [22];
- The ethyl acetate fraction of the crude extract of *Streptomyces* sp. MUM256, isolated from mangrove soil in Malaysia, and the cyclic dipeptides contained whose ability to induce cell cycle arrest and apoptosis was demonstrated by Tan et al. [23] in colon cancer cells;
- Manoalide, an antibiotic sesterterpenoid isolated from the marine sponge *Luffariella variabilis*, which preferentially inhibits the proliferation of oral cancer cells inducing apoptosis and DNA damages via oxidative stresses, such as intracellular ROS and MitoSOX/MitoMP [24];
- λ-carrageenan, a family of linear sulfated polysaccharides, proven to enhance the effect of radiotherapy by suppressing the survival and invasiveness of different cancer cell lines in vitro and in vivo through the Rac GTPase-activating protein 1 pathway [25];
- Ethanol, which was found to trigger a pro-survival autophagic response following the induction of oxidative and endoplasmic reticulum stress in colon cancer cells, and the activation of Nrf2 and HO-1 also leading to the acquisition of a more aggressive phenotype [26];
- Colchicine, an alkaloid present in the medicinal plant *Colchicum autumnale*, whose enhanced anticancer effects and reduced cytotoxicity on colon cancer cells when delivered in the nanoformulated form was reported by AbouAitah et al. [27].

In the Review section, Del Cornò et al. [28] discussed the modulatory effects of dietary β-glucans, present in diverse edible mushrooms, baker's yeast, cereals and seaweeds, on human innate immunity cells and their potential role in cancer control. Lee et al. [29] reviewed a large number of data on the role played by different cytokines, lipids and other natural molecules on the suppression of epithelial–mesenchymal transition in cancer progression. Ennour-Idrissi et al. [30] focused their attention on the bioaccumulation of persistent organic pollutants in the food chain and the association of exposure with breast cancer risk. Farooqi et al. [31] presented the current views about the ability of berberine, a natural alkaloid compound found in several medicinal plants, to target different signaling cascades in various cancers, also discussing the nanocarrier strategies developed to improve the delivery of the compound.

The number of manuscripts published in this Special Issue indicates an active interest in research about the molecular/pharmacological mechanisms used by natural products exerting anti-tumoral effects which deserve further and deeper studies. I wish to thank all the contributors of this issue for sharing with us their experimental or reviewed data which will surely attract readers' attention and encourage the publication of other high-quality papers in this field.

Funding: This research received no external funding.

Conflicts of Interest: The authors declare no conflict of interest.

References

1. Polonio-Alcalá, E.; Palomeras, S.; Torres-Oteros, D.; Relat, J.; Planas, M.; Feliu, L.; Ciurana, J.; Ruiz-Martínez, S.; Puig, T. Fatty acid synthase inhibitor G28 shows anticancer activity in EGFR tyrosine kinase inhibitor resistant lung adenocarcinoma models. *Cancers* **2020**, *12*, 1283. [CrossRef]

2. Wei, R.; Hackman, R.M.; Wang, Y.; Mackenzie, G.G. targeting glycolysis with epigallocatechin-3-gallate enhances the efficacy of chemotherapeutics in pancreatic cancer cells and xenografts. *Cancers* **2019**, *11*, 1496. [CrossRef]

3. Farooqi, A.A.; Pinheiro, M.; Granja, A.; Farabegoli, F.; Reis, S.; Attar, R.; Sabitaliyevich, U.Y.; Xu, B.; Ahmad, A. EGCG mediated targeting of deregulated signaling pathways and non-coding RNAs in different cancers: Focus on JAK/STAT, Wnt/β-Catenin, TGF/SMAD, NOTCH, SHH/GLI, and TRAIL mediated signaling pathways. *Cancers* **2020**, *12*, 951. [CrossRef]

4. Kang, D.Y.; Sp, N.; Jo, E.S.; Rugamba, A.; Hong, D.Y.; Lee, H.G.; Yoo, J.-S.; Liu, Q.; Jang, K.-J.; Yang, Y.M. The inhibitory mechanisms of tumor PD-L1 expression by natural bioactive gallic acid in Non-Small-Cell Lung Cancer (NSCLC) Cells. *Cancers* **2020**, *12*, 727. [CrossRef]

5. Ruzzolini, J.; Peppicelli, S.; Bianchini, F.; Andreucci, E.; Urciuoli, S.; Romani, A.; Tortora, K.; Caderni, G.; Nediani, C.; Calorini, L. Cancer glycolytic dependence as a new target of olive leaf extract. *Cancers* **2020**, *12*, 317. [CrossRef]

6. Juli, G.; Oliverio, M.; Bellizzi, D.; Gallo Cantafio, M.E.; Grillone, K.; Passarino, G.; Colica, C.; Nardi, M.; Rossi, M.; Procopio, A.; et al. Anti-tumor activity and epigenetic impact of the polyphenol oleacein in multiple myeloma. *Cancers* **2019**, *11*, 990. [CrossRef]

7. Montalesi, E.; Cipolletti, M.; Cracco, P.; Fiocchetti, M.; Marino, M. Divergent effects of daidzein and its metabolites on estrogen-induced survival of breast cancer cells. *Cancers* **2020**, *12*, 167. [CrossRef]

8. Losuwannarak, N.; Maiuthed, A.; Kitkumthorn, N.; Leelahavanichkul, A.; Roytrakul, S.; Chanvorachote, P. Gigantol targets cancer stem cells and destabilizes tumors via the suppression of the PI3K/AKT and JAK/STAT pathways in ectopic lung cancer xenografts. *Cancers* **2019**, *11*, 2032. [CrossRef]

9. Predes, D.; Oliveira, L.F.S.; Ferreira, L.S.S.; Maia, L.A.; Delou, J.M.A.; Faletti, A.; Oliveira, I.; Amado, N.G.; Reis, A.H.; Fraga, C.A.M.; et al. The chalcone lonchocarpin inhibits Wnt/β-Catenin signaling and suppresses colorectal cancer proliferation. *Cancers* **2019**, *11*, 1968. [CrossRef]

10. Park, C.; Cha, H.-J.; Choi, E.O.; Lee, H.; Hwang-Bo, H.; Ji, S.Y.; Kim, M.Y.; Kim, S.Y.; Hong, S.H.; Cheong, J.; et al. Isorhamnetin induces cell cycle arrest and apoptosis via reactive oxygen species-mediated AMP-activated protein kinase signaling pathway activation in human bladder cancer cells. *Cancers* **2019**, *11*, 1494. [CrossRef]

11. Muñoz, D.; Brucoli, M.; Zecchini, S.; Sandoval-Hernandez, A.; Arboleda, G.; Lopez-Vallejo, F.; Delgado, W.; Giovarelli, M.; Coazzoli, M.; Catalani, E.; et al. XIAP as a target of new small organic natural molecules inducing human cancer cell death. *Cancers* **2019**, *11*, 1336. [CrossRef]

12. Ho, S.-T.; Lin, C.-C.; Tung, Y.-T.; Wu, J.-H. Molecular mechanisms underlying yatein-induced cell-cycle arrest and microtubule destabilization in human lung adenocarcinoma cells. *Cancers* **2019**, *11*, 1384. [CrossRef] [PubMed]

13. Perrone, L.; Sampaolo, S.; Melone, M.A.B. Bioactive phenolic compounds in the modulation of central and peripheral nervous system cancers: Facts and misdeeds. *Cancers* **2020**, *12*, 454. [CrossRef]

14. Barbosa, A.M.; Martel, F. Targeting glucose transporters for breast cancer therapy: The effect of natural and synthetic compounds. *Cancers* **2020**, *12*, 154. [CrossRef]

15. Ong, C.P.; Lee, W.L.; Tang, Y.Q.; Yap, W.H. Honokiol: A review of its anticancer potential and mechanisms. *Cancers* **2020**, *12*, 48. [CrossRef]

16. Zhou, Q.; Pan, H.; Li, J. Molecular insights into potential contributions of natural polyphenols to lung cancer treatment. *Cancers* **2019**, *11*, 1565. [CrossRef]

17. Goh, J.X.H.; Tan, L.T.-H.; Goh, J.K.; Chan, K.G.; Pusparajah, P.; Lee, L.-H.; Goh, B.-H. Nobiletin and derivatives: Functional compounds from citrus fruit peel for colon cancer chemoprevention. *Cancers* **2019**, *11*, 867. [CrossRef]

18. Ong, S.K.L.; Shanmugam, M.K.; Fan, L.; Fraser, S.E.; Arfuso, F.; Ahn, K.S.; Sethi, G.; Bishayee, A. Focus on formononetin: Anticancer potential and molecular targets. *Cancers* **2019**, *11*, 611. [CrossRef]

19. El Khoury, M.; Haykal, T.; Hodroj, M.H.; Najem, S.A.; Sarkis, R.; Taleb, R.I.; Rizk, S. *Malva pseudolavatera* leaf extract promotes ROS induction leading to apoptosis in acute myeloid leukemia cells in vitro. *Cancers* **2020**, *12*, 435. [CrossRef]

20. Rasha, F.; Kahathuduwa, C.; Ramalingam, L.; Hernandez, A.; Moussa, H.; Moustaid-Moussa, N. Combined effects of eicosapentaenoic acid and adipocyte renin–angiotensin system inhibition on breast cancer cell inflammation and migration. *Cancers* **2020**, *12*, 220. [CrossRef]

21. Bobiński, M.; Okła, K.; Łuszczki, J.; Bednarek, W.; Wawruszak, A.; Moreno-Bueno, G.; Dmoszyńska-Graniczka, M.; Tarkowski, R.; Kotarski, J. Isobolographic analysis demonstrates the additive and synergistic effects of gemcitabine combined with fucoidan in uterine sarcomas and carcinosarcoma cells. *Cancers* **2020**, *12*, 107. [CrossRef] [PubMed]

22. Nguyen, H.D.; Liao, Y.-C.; Ho, Y.-S.; Chen, L.-C.; Chang, H.-W.; Cheng, T.-C.; Liu, D.; Lee, W.-R.; Shen, S.-C.; Wu, C.-H.; et al. The α9 nicotinic acetylcholine receptor mediates nicotine-induced PD-L1 expression and regulates melanoma cell proliferation and migration. *Cancers* **2019**, *11*, 1991. [CrossRef] [PubMed]

23. Tan, L.T.-H.; Chan, C.-K.; Chan, K.-G.; Pusparajah, P.; Khan, T.M.; Ser, H.-L.; Lee, L.-H.; Goh, B.-H. Streptomyces sp. MUM256: A source for apoptosis inducing and cell cycle-arresting bioactive compounds against colon cancer cells. *Cancers* **2019**, *11*, 1742. [CrossRef] [PubMed]

24. Wang, H.-R.; Tang, J.-Y.; Wang, Y.-Y.; Farooqi, A.A.; Yen, C.-Y.; Yuan, S.-S.F.; Huang, H.-W.; Chang, H.-W. Manoalide preferentially provides antiproliferation of oral cancer cells by oxidative stress-mediated apoptosis and DNA damage. *Cancers* **2019**, *11*, 1303. [CrossRef] [PubMed]

25. Wu, P.-H.; Onodera, Y.; Recuenco, F.C.; Giaccia, A.J.; Le, Q.-T.; Shimizu, S.; Shirato, H.; Nam, J.-M. Lambda-carrageenan enhances the effects of radiation therapy in cancer treatment by suppressing cancer cell invasion and metastasis through Racgap1 inhibition. *Cancers* **2019**, *11*, 1192. [CrossRef]

26. Cernigliaro, C.; D'Anneo, A.; Carlisi, D.; Giuliano, M.; Marino Gammazza, A.; Barone, R.; Longhitano, L.; Cappello, F.; Emanuele, S.; Distefano, A.; et al. Ethanol-mediated stress promotes autophagic survival and aggressiveness of colon cancer cells via activation of Nrf2/HO-1 pathway. *Cancers* **2019**, *11*, 505. [CrossRef]

27. AbouAitah, K.; Hassan, H.A.; Swiderska-Sroda, A.; Gohar, L.; Shaker, O.G.; Wojnarowicz, J.; Opalinska, A.; Smalc-Koziorowska, J.; Gierlotka, S.; Lojkowski, W. Targeted nano-drug delivery of colchicine against colon cancer cells by means of mesoporous silica nanoparticles. *Cancers* **2020**, *12*, 144. [CrossRef]

28. Del Cornò, M.; Gessani, S.; Conti, L. Shaping the innate immune response by dietary glucans: Any role in the control of cancer? *Cancers* **2020**, *12*, 155. [CrossRef]

29. Lee, C.H. Reversal of epithelial–mesenchymal transition by natural anti-inflammatory and pro-resolving lipids. *Cancers* **2019**, *11*, 1841. [CrossRef]

30. Ennour-Idrissi, K.; Ayotte, P.; Diorio, C. Persistent Organic pollutants and breast cancer: A systematic review and critical appraisal of the literature. *Cancers* **2019**, *11*, 1063. [CrossRef]

31. Farooqi, A.A.; Qureshi, M.Z.; Khalid, S.; Attar, R.; Martinelli, C.; Sabitaliyevich, U.Y.; Nurmurzayevich, S.B.; Taverna, S.; Poltronieri, P.; Xu, B. Regulation of cell signaling pathways by berberine in different cancers: Searching for missing pieces of an incomplete jig-saw puzzle for an effective cancer therapy. *Cancers* **2019**, *11*, 478. [CrossRef] [PubMed]

Article

Fatty Acid Synthase Inhibitor G28 Shows Anticancer Activity in EGFR Tyrosine Kinase Inhibitor Resistant Lung Adenocarcinoma Models

Emma Polonio-Alcalá [1,2,†], Sònia Palomeras [1,†], Daniel Torres-Oteros [3], Joana Relat [3,4,5], Marta Planas [6], Lidia Feliu [6], Joaquim Ciurana [2], Santiago Ruiz-Martínez [1,*] and Teresa Puig [1,*]

[1] New Therapeutic Targets Laboratory (TargetsLab)-Oncology Unit, Department of Medical Sciences, Faculty of Medicine, University of Girona, 17003 Girona, Spain; emma.polonio@udg.edu (E.P.-A.); sonia.palomeras@udg.edu (S.P.)

[2] Product, Process and Production Engineering Research Group (GREP), Department of Mechanical Engineering and Industrial Construction, University of Girona, 17003 Girona, Spain; quim.ciurana@udg.edu

[3] Department of Nutrition, Food Sciences and Gastronomy, School of Pharmacy and Food Sciences, Food Torribera Campus, University of Barcelona, E-08921 Santa Coloma de Gramanet, Spain; danytoot@hotmail.com (D.T.-O.); jrelat@ub.edu (J.R.)

[4] Institute of Nutrition and Food Safety of the University of Barcelona (INSA-UB), E-08921 Santa Coloma de Gramenet, Spain

[5] CIBER Physiopathology of Obesity and Nutrition (CIBER-OBN), Instituto de Salud Carlos III, E-28029 Madrid, Spain

[6] LIPPSO, Department of Chemistry, University of Girona, 17003 Girona, Spain; marta.planas@udg.edu (M.P.); lidia.feliu@udg.edu (L.F.)

[*] Correspondence: santiago.ruiz@udg.edu (S.R.-M.); teresa.puig@udg.edu (T.P.); Tel.: +34-972-419-548 (S.R.-M.); +34-972-419-628 (T.P.)

[†] These authors contributed equally to this work.

Received: 26 March 2020; Accepted: 16 May 2020; Published: 19 May 2020

Abstract: Epidermal growth factor receptor (EGFR) tyrosine kinases inhibitors (TKIs) are effective therapies for non-small cell lung cancer (NSCLC) patients whose tumors harbor an EGFR activating mutation. However, this treatment is not curative due to primary and secondary resistance such as T790M mutation in exon 20. Recently, activation of transducer and activator of transcription 3 (STAT3) in NSCLC appeared as an alternative resistance mechanism allowing cancer cells to elude the EGFR signaling. Overexpression of fatty acid synthase (FASN), a multifunctional enzyme essential for endogenous lipogenesis, has been related to resistance and the regulation of the EGFR/Jak2/STAT signaling pathways. Using EGFR mutated (EGFRm) NSCLC sensitive and EGFR TKIs' resistant models (Gefitinib Resistant, GR) we studied the role of the natural polyphenolic anti-FASN compound (−)-epigallocatechin-3-gallate (EGCG), and its derivative G28 to overcome EGFR TKIs' resistance. We show that G28's cytotoxicity is independent of TKIs' resistance mechanisms displaying synergistic effects in combination with gefitinib and osimertinib in the resistant T790M negative (T790M−) model and showing a reduction of activated EGFR and STAT3 in T790M positive (T790M+) models. Our results provide the bases for further investigation of G28 in combination with TKIs to overcome the EGFR TKI resistance in NSCLC.

Keywords: NSCLC; EGFR TKI; FASN inhibitors; resistance; STAT3; EGCG

---

1. Introduction

Lung cancer is the most incident and the leading cause of cancer death worldwide. Non-small cell lung cancer (NSCLC) subtype is the most common and it represents 80–85% of lung cancers diagnosed.

Early-stage NSCLC patients have long-term survival after surgery. However, approximately 75% of patients are diagnosed in advanced stages [1,2].

Gefitinib is a first generation epidermal growth factor receptor (EGFR) tyrosine kinase inhibitor (TKI). It was approved in 2003 by the Food and Drug Administration (FDA) for the treatment of patients whose tumors harbor an EGFR sensitizing/activating mutation (EGFRm) i.e., exon 19 deletion (ΔE746-A750) or the point mutation in exon 21 that leads to the amino acid substitution L858R [3–5]. Despite this therapy represents a breakthrough in the treatment of EGFRm NSCLC patients, in a median time of 9–16 months nearly all patients do not achieve a complete response. One of the most common resistance mechanisms described is the EGFR point mutation in exon 20 that leads to the replacement of threonine 790 by methionine (T790M), which normally derives to lethal disease progression [6]. Osimertinib is an irreversible third generation TKI effective in EGFRm T790M positive (T790M+) patients. However, the point mutation C797S in exon 20 has appeared as the main resistance mechanism to the latest FDA-approved TKI [6]. Other mechanisms for EGFR TKI resistance include Met amplification, phosphatidylinositol-4,5-bisphosphate 3-kinase catalytic subunit alpha (PI3KCA) mutations, appearance of stem-like properties as evidenced by increase in epithelial–mesenchymal transition (EMT) and histological transformation, epidermal growth factor receptor type 2 (ErbB2) gene amplification, increase of insulin-like growth factor 1 receptor (IGF-1R) signaling pathway through the loss of inhibitory insulin-like growth factor-binding protein (IGF-BP) and loss or reduction of phosphatase and tensin homolog (PTEN), activation of AXL tyrosine kinase receptor or B-Raf proto-oncogene, and serine/threonine kinase (BRAF) mutations [6–12].

Recently, the activation of a signal transducer and activator of transcription 3 (STAT3) has been described as an alternative resistance mechanism allowing cancer cells to escape the EGFR signaling or the TKI suppression [13]. STAT3 is involved in the transcription of many genes related to cell differentiation, proliferation, resistance to apoptosis, angiogenesis, metastasis, and immune response [14–16]. Besides being phosphorylated by EGFR [17], STAT3 can also be activated in response to different cytokines and growth factors such as interleukin 6 (IL-6), interferon-alpha (IFN-α) or epidermal growth factor (EGF), among others [18]. Approximately 60% of patients show STAT3 activation, which correlates with poorly differentiated tumors, the presence of metastasis, and the late clinical stage [19,20]. STAT3 phosphorylation has been related to disease progression in a small cohort of patients after EGFR TKI treatment [21]. Additionally, neither gefitinib nor osimertinib are able to inhibit STAT3 activation [22,23].

Energy metabolism deregulation has been described as a hallmark of cancer, allowing cell growth and proliferation [24]. Fatty acid synthase (FASN) is an essential enzyme for the de novo synthesis of long-chain fatty acids from acetyl-CoA, malonyl-CoA, and NADPH [25]. Unlike most normal cells, highly-proliferative cancer cells overexpress this lipogenic enzyme, being an interesting target in cancer therapy [26,27]. FASN is strongly associated to poor prognosis and resistance to treatment in different human tumors such as breast [28], bladder [29], pancreatic [30], or lung cancer [31]. Moreover, FASN overexpression has also been proposed as a multidrug resistance mechanism, protecting cells from drug-induced apoptosis through the overproduction of palmitic acid [32]. The natural compound (−)-epigallocatechin-3-gallate (EGCG) is a polyphenolic flavonoid from green tea that has been broadly studied for its cardiovascular, neuroprotective, anticancer, and antimicrobial properties [33,34]. EGCG has been reported to compete with NADPH to bind the β-ketoacyl reductase domain of FASN [35]. The ability of several FASN inhibitors to regulate the canonical EGFR/Jak2/STAT pathway has also been stated in the literature [36,37]. We and others have shown that FASN inhibition is mainly related to EGFR/HER2 signaling pathways, leading to cytotoxic effects in vitro and in vivo in a wide range of carcinomas, including breast and lung [38–42]. To date, many EGCG derivatives have been developed to improve efficacy and increase stability in physiological conditions. Among them, the naphthalene derivative G28 has shown interesting antiproliferative features against sensitive and resistant breast cancer cells [38,43,44].

The purpose of this work was to study the role of FASN inhibitors (EGCG and G28) to overcome TKI resistance in NSCLC. FASN inhibitors were tested alone and in combination with EGFR TKIs (gefitinib and osimertinib) in EGFRm NSCLC models resistant to EGFR TKIs (Gefitinib Resistant, GR). In addition, we also evaluated gene and protein expression changes of FASN, EGFR, and STAT3 resulting from treatments with FASN inhibitors and EGFR TKIs alone or in combination. We show that FASN inhibitor G28 cytotoxicity is independent of EGFR TKI resistance mechanisms. Interestingly, G28 compound exhibited a cytotoxic effect in combination with gefitinib showing changes in EGFR/STAT3 pathway in T790M positive (T790M+) GR models and strong synergism in combination with gefitinib or osimertinib in T790M negative (T790M−) GR model.

2. Results

2.1. EGFRm GR NSCLC Models Are Sensitive to FASN Inhibition

In order to study the role of FASN in the acquisition of EGFR TKI resistance in NSCLC, we used the sensitive PC9 cell line carrying the EGFR exon 19 deletion (ELREA) and three GR models, two T790M+ models (PC9-GR1 and PC9-GR4), and one T790M− model (PC9-GR3) [45].

2.1.1. EGFRm NSCLC Models Overexpress FASN

Firstly, we determined whether EGFRm NSCLC models express FASN enzyme. Hence, FASN protein (Figure 1a) and mRNA expression levels (Figure 1b) were analyzed by immunoblotting and quantitative real time PCR (qRT-PCR), respectively.

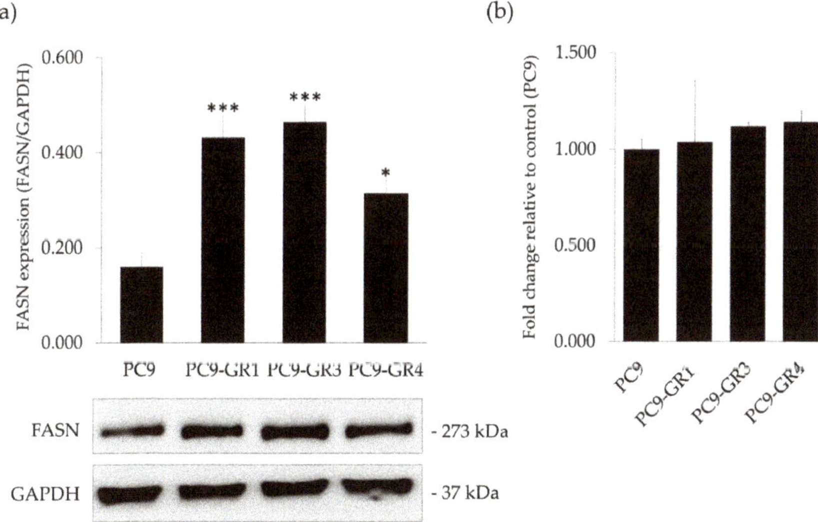

Figure 1. FASN protein and mRNA expression levels in sensitive (PC9) and Gefitinib Resistant (GR) models (PC9-GR1, PC9-GR3, and PC9-GR4). (**a**) Western blot analysis (quantification in upper panel and bands in lower panel) of FASN protein expression. GAPDH was used as a loading control. Results shown are representative from at least three independent experiments. (**b**) FASN endogenous mRNA levels were obtained by qRT-PCR and normalized against the GAPDH gene. FASN expression in the sensitive cells was normalized to 1 an expressed as a fold change, to which all other conditions were compared. Results shown are mean ± SE from three independent experiments. * $p < 0.050$, *** $p < 0.001$ indicate levels of statistically significance.

All models showed FASN protein and mRNA expression. Despite no differences in mRNA, GR models presented significantly higher protein expression levels (PC9-GR1 $p = 8.710 \times 10^{-4}$; PC9-GR3 $p = 3.160 \times 10^{-4}$, and PC9-GR4 $p = 0.049$) in comparison to PC9.

2.1.2. PC9-GR3 Model Is Resistant to Gefitinib and Osimertinib

We confirmed the resistance to EGFR TKIs in PC9 and GR models. For that, we measured the cytotoxic effect of gefitinib and osimertinib on all models by determining the half-maximal inhibitory concentration (IC_{50}) using the MTT assay (Figure 2).

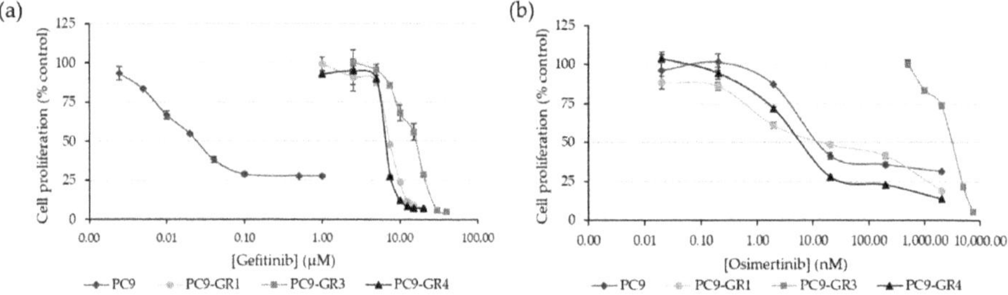

Figure 2. Cell proliferation inhibition of EGFR TKIs (gefitinib and osimertinib) in parental and Gefitinib Resistant (GR) models. Sensitive (PC9) and GR models (PC9-GR1, PC9-GR3, and PC9-GR4) were treated with increasing concentrations of (**a**) gefitinib (from 2.5×10^{-3} to 1 μM for PC9 and 1–40 μM for GR models) and (**b**) osimertinib (0.02–2000 nM for PC9, PC9-GR1, and PC9-GR4 and 500–7500 nM for PC9-GR3) for 72 h. Results shown are expressed as percentage of surviving cells after drug treatment (mean ± SE) and are representative from at least three independent experiments.

As expected, GR models were significantly more resistant to gefitinib with IC_{50} values in the micromolar range compared to the nanomolar IC_{50} found in the PC9 cell line (PC9-GR1 $p = 2.793 \times 10^{-7}$; PC9-GR3 $p = 1.631 \times 10^{-10}$, and PC9-GR4 $p = 1.000 \times 10^{-6}$). Although no significant differences were found in the IC_{50} value for gefitinib between the two T790M+ GR models, the IC_{50} value of the PC9-GR3 model for gefitinib was significantly greater than PC9-GR1 ($p = 7.953 \times 10^{-7}$) and PC9-GR4 ($p = 1.659 \times 10^{-7}$). PC9-GR3 model was also resistant to osimertinib compared to other models (PC9 $p = 2.799 \times 10^{-9}$; PC9-GR1 $p = 3.749 \times 10^{-8}$, and PC9-GR4 $p = 5.200 \times 10^{-9}$).

2.1.3. FASN Inhibitors Present Cytotoxic Effects in NSCLC Models

Cancer cells have been described to increase the de novo lipogenesis through the activation of FASN and its inhibition has proven to cause cell death. Therefore, this enzyme has become a promising candidate for the development of new anticancer therapies. Here we tested the cytotoxic activity of the two FASN inhibitors, EGCG and its derivative G28. MTT cell viability assays showed that the natural polyphenolic compound EGCG was cytotoxic for PC9 ($IC_{50} = 77.9 \pm 1.9$ μM), PC9-GR1 ($IC_{50} = 74.3 \pm 4.3$ μM), PC9-GR3 ($IC_{50} = 91.0 \pm 5.5$ μM), and PC9-GR4 ($IC_{50} = 75.6 \pm 2.4$ μM) NSCLC models with no significant differences ($p = 0.358$; Figure 3a).

The synthetic EGCG derivative G28 showed higher cytotoxicity in all tested models with IC_{50} of 12.8 ± 1.3 μM for PC9, 12.0 ± 0.8 μM for PC9-GR1, 17.8 ± 1.3 μM for PC9-GR3, and 11.2 ± 1.2 μM for PC9-GR4 (Figure 3b). Besides, only PC9-GR3 showed a significantly higher IC_{50} value compared to PC9 ($p = 0.030$), PC9-GR1 ($p = 0.005$), and PC9-GR4 ($p = 0.002$).

(a) (b)

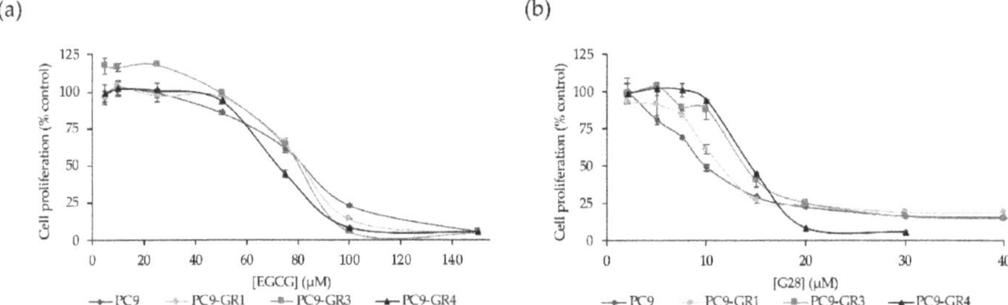

Figure 3. Cell proliferation inhibition of FASN inhibitors in parental and Gefitinib Resistant (GR) models. Sensitive (PC9) and GR models (PC9-GR1, PC9-GR3, and PC9-GR4) were treated with increasing concentrations of (**a**) EGCG (5–150 μM) and (**b**) G28 (2–40 μM) for 72 h. Results shown are expressed as the percentage of surviving cells after drug treatment (mean ± SE) and are representative from at least three independent experiments.

2.1.4. G28 Inhibits FASN in EGFRm NSCLC Models

The ability to internalize and inhibit FASN activity of EGCG and G28 after being exogenously added to the media was tested in sensitive and GR models (Figure 4). EGCG and G28 inhibited FASN in PC9 cells, resulting in a similar FASN activity reduction of roughly 80% ($p = 0.265$). Moreover, G28 significantly reduced FASN activity in all GR models in comparison with EGCG (PC9-GR1 $p = 3.000 \times 10^{-5}$; PC9-GR3 $p = 0.001$; PC9-GR4 $p = 0.008$) while the EGCG compound was not able to diminish FASN activity.

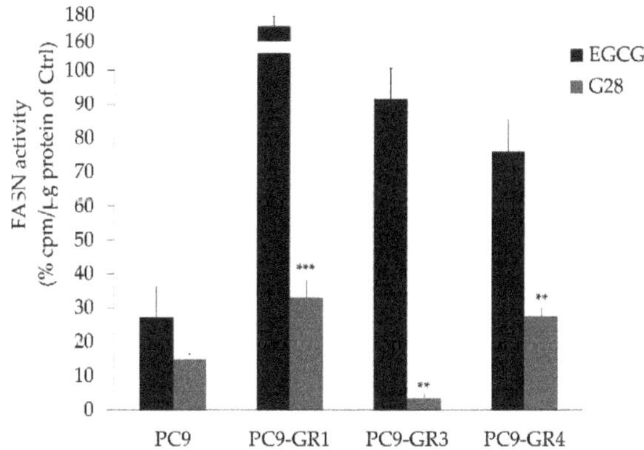

Figure 4. G28 compound inhibits FASN activity in Gefitinib Resistant (GR) models. Cells were treated for 72 h with EGCG or G28 at a concentration equal to their IC_{50} and with DMSO as control. FASN activity was assayed by counting radiolabeled fatty acids synthesized de novo. Bars represent the remaining activity as a percentage in treated versus untreated cells (control, Ctrl). Data are mean ± SE from at least three independent experiments. ** $p < 0.010$, *** $p < 0.001$ indicate levels of statistically significance.

2.1.5. Apoptosis Induction of FASN Inhibitors and EGFR TKIs Treatments

We also investigated whether the cell death caused by treatment with EGFR TKIs or FASN inhibitors was the result of apoptosis induction in both sensitive and GR models. Poly(ADP-ribose) polymerase (PARP) terminal proapoptotic protein activated after cleavage was used as an apoptosis marker. The effects of all compounds on PARP was determined by Western blot analysis in all models (Figure 5). The uncropped Western blots can be found in Figure S2a.

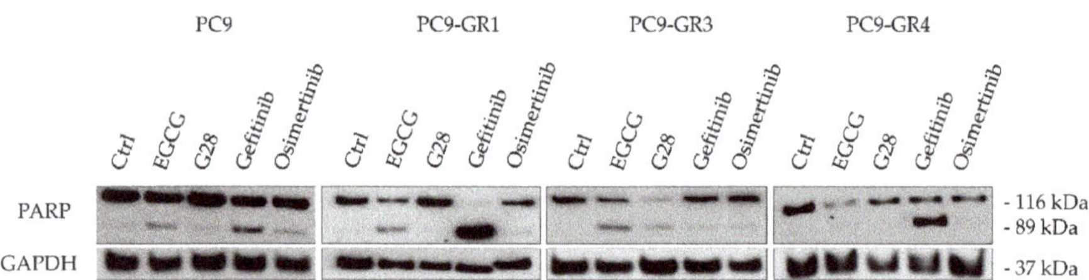

Figure 5. Effects of FASN inhibitors and EGFR TKIs on apoptosis determined by PARP cleavage. Sensitive (PC9) and Gefitinib Resistant (GR) models (PC9-GR1, PC9-GR3, and PC9-GR4) were treated for 72 h with a concentration equivalent to IC$_{50}$ of each drug. Untreated cells were used as an internal control (Ctrl) and GAPDH as a loading control. Results shown are representative from at least three independent experiments.

The cleaved form of PARP (89 kDa) appeared in either sensitive or resistant T790M− models after treatment with the IC$_{50}$ concentration of both EGFR TKIs (gefitinib and osimertinib) and FASN inhibitors (EGCG and G28), indicating the induction of apoptosis. Gefitinib treatment led to PARP cleavage in T790M+ GR models. Additionally, EGCG and osimertinib led to the formation of cleaved PARP in PC9-GR1 model.

2.2. G28 Increases EGFR Activation in EGFRm NSCLC Models

FASN has been previously related to AKT/ERK/EGFR signaling pathways [46] and the inhibition of the transcription factor STAT3 [36] in lung adenocarcinomas. Thus, differences in FASN, EGFR, and STAT3 protein and mRNA expression levels after FASN inhibitors or EGFR TKIs treatment were analyzed through immunoblotting (Figure 6a) and qRT-PCR (Figure 6b) in PC9 and GR models. The uncropped Western blots can be found in Figure S2b.

A reduction of FASN mRNA expression levels was observed in sensitive and GR models treated with FASN inhibitors, being statistically significant in the PC9-GR4 treated with G28 ($p = 0.004$) and PC9 cells treated with G28 ($p = 7.370 \times 10^{-4}$) and gefitinib ($p = 3.210 \times 10^{-4}$). T790M+ GR models presented a basal hyperactivation of EGFR that was inhibited after treatment with EGFR TKIs. Regarding FASN inhibitors, EGCG and G28 increased its activation in PC9 cells contrary to the PC9-GR1 model, while no changes were observed in PC9-GR3 and PC9-GR4 models.

EGCG reduced EGFR protein levels in sensitive and T790M+ GR models that did not correlate with mRNA expression levels. Contrary, G28 significantly increased EGFR mRNA expression in all models, without protein level modification. No changes in EGFR mRNA levels were observed after gefitinib treatment, however higher protein levels were detected in T790M− GR models. Osimertinib treatment did not lead to EGFR protein nor mRNA expression alteration. EGFR TKIs treatment increased STAT3 activation in all GR models. Gefitinib increased p-STAT3 levels in sensitive cells, while the highest STAT3 activation in GR models was found with osimertinib treatment. No STAT3 protein or mRNA expression differences were found in any of the models and treatments assayed.

Lung carcinomas are highly proliferative and resistance acquisition after EGFR TKI-based therapy is a major problem. Overproduction of palmitic acid by FASN emerges as a resistance mechanism, protecting cells from drug-induced apoptosis [47]. The combination of these drugs with a different mechanism of action may decrease resistance development and improve treatment response.

(a)

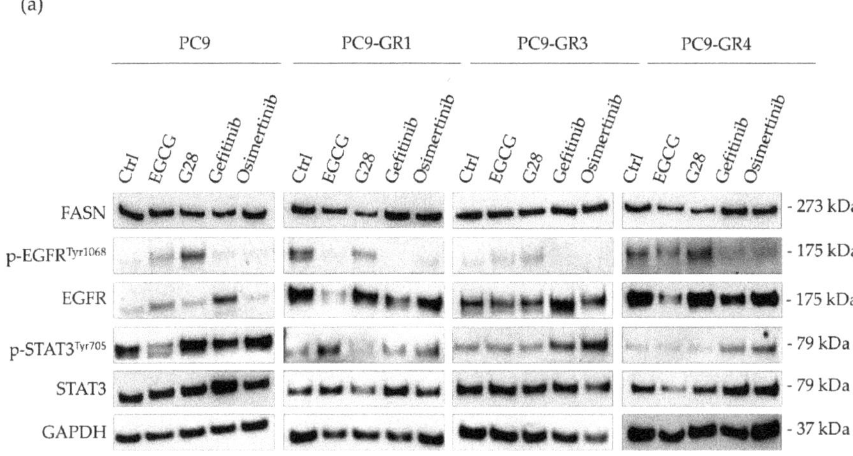

(b)

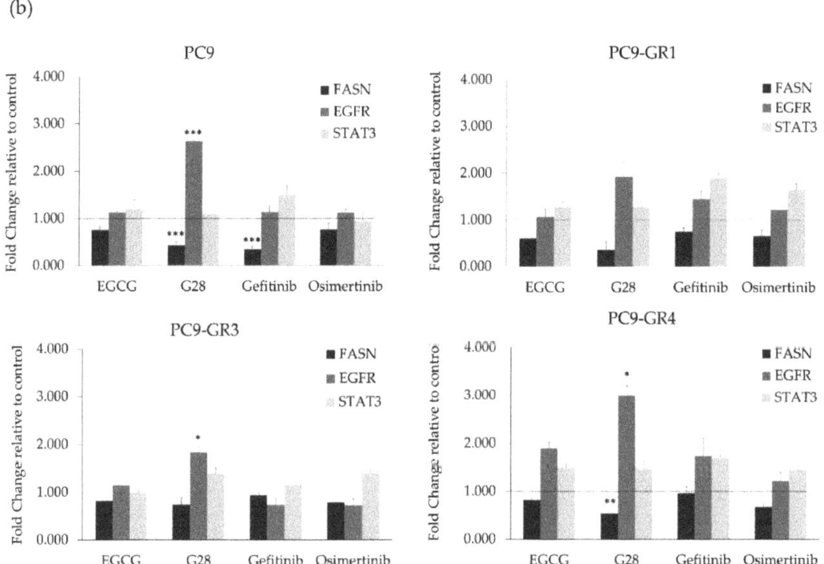

Figure 6. FASN, EGFR, and STAT3 protein and mRNA expression levels after FASN inhibitors (EGCG and G28) and EGFR TKIs (gefitinib and osimertinib) treatment in sensitive (PC9) and Gefitinib Resistant (GR) models (PC9-GR1, PC9-GR3, and PC9-GR4). (**a**) Western blot analysis of FASN, EGFR, and STAT3 protein expression after 72 h of FASN inhibitors (IC_{50}) and EGFR TKIs (IC_{50}) treatment in EGFR TKI sensitive and GR models. Untreated cells were used as an internal control (Ctrl) and GAPDH as a loading control. Results shown are representative from at least three independent experiments. (**b**) FASN, EGFR, and STAT3 mRNA levels after treatment with FASN inhibitors and EGFR TKIs in sensitive and GR models. mRNA levels were obtained by qRT-PCR and normalized against the GAPDH gene. All conditions were compared to control (untreated cells), which was normalized to 1 (indicated by the dotted line) and expressed as a fold change. Experiments were performed at least three times. * $p < 0.050$, ** $p < 0.010$, *** $p < 0.001$, indicate levels of statistically significance.2.3. G28 Combined with EGFR TKIs Outcomes in Synergistic Effects.

Once the effects of FASN inhibitors and EGFR TKIs alone were analyzed in all models, the combinatorial treatment between FASN inhibitors and EGFR TKIs was also studied in order to evaluate the possible synergistic interactions (Figure 7 and Figure S1). Despite the fact that EGCG did not reduce FASN activity in GR models, no IC_{50} differences were found in comparison with sensitive cells. Therefore, GR models were treated with EGCG or G28 in combination with the EGFR TKI to which they were resistant. Three EGFR TKIs concentrations were chosen based on MTT assays (vide supra). The combination of EGCG with either gefitinib or osimertinib resulted in mostly additive effect as shown by the combination index (CI; Figure 7a). The combination of G28 with gefitinib generally led to additivism in T790M+ GR models, with some synergism found in G28 concentrations ranging from 10 to 20 µM. Remarkably, T790M− GR model treated with G28 combined with gefitinib or osimertinib showed greater synergistic effects (Figure 7b).

2.3. G28 Reduces STAT3 Activation in T790M+ GR Models Alone or in Combination with Gefitinib

In order to discern whether G28 is able to reduce the STAT3 activation produced by EGFR TKIs, a combinatorial analysis was performed (Figure 8). As before, FASN, EGFR, and STAT3 protein and mRNA levels were analyzed using Western blot (Figure 8a) and qRT-PCR (Figure 8b) in GR models treated with synergistic drug concentrations (all of them under the IC_{50} value). The uncropped Western blots can be found in Figure S2c. Therefore, GR models were treated with G28 at 15 µM in combination with 1 µM gefitinib, which is the highest clinically achievable plasma concentration [5]. Osimertinib-resistant PC9-GR3 model was co-treated with 15 µM G28 and 0.5 µM osimertinib. All concentrations were also used in single-treatment to elucidate the effects produced by the combination.

T790M+ GR models treated with G28 alone and in combination with gefitinib showed both FASN protein and mRNA expression decrease. In the PC9-GR3 model, the FASN protein was slightly diminished in mono- and co-treatments. This decrease is in accordance to FASN mRNA expression, which was significantly reduced in combination with osimertinib. Although G28 alone showed more activated EGFR compared to control in all GR models, the combination of G28 with both EGFR TKIs decreased p-EGFR levels. Regarding the total EGFR, co-treatment resulted in higher protein levels compared to monotreatment in all GR models. However, only T790M+ GR models treated with G28 in combination with gefitinib presented significantly higher EGFR mRNA expression (PC9-GR1 $p = 0.036$ and PC9-GR4 $p = 0.040$). Moreover, G28 both alone and in combination with gefitinib reduced STAT3 activation in T790M+ GR models. No changes in STAT3 activation were seen in PC9-GR3 in any treatment. None of the models analyzed showed alterations in the STAT3 protein or mRNA expression levels.

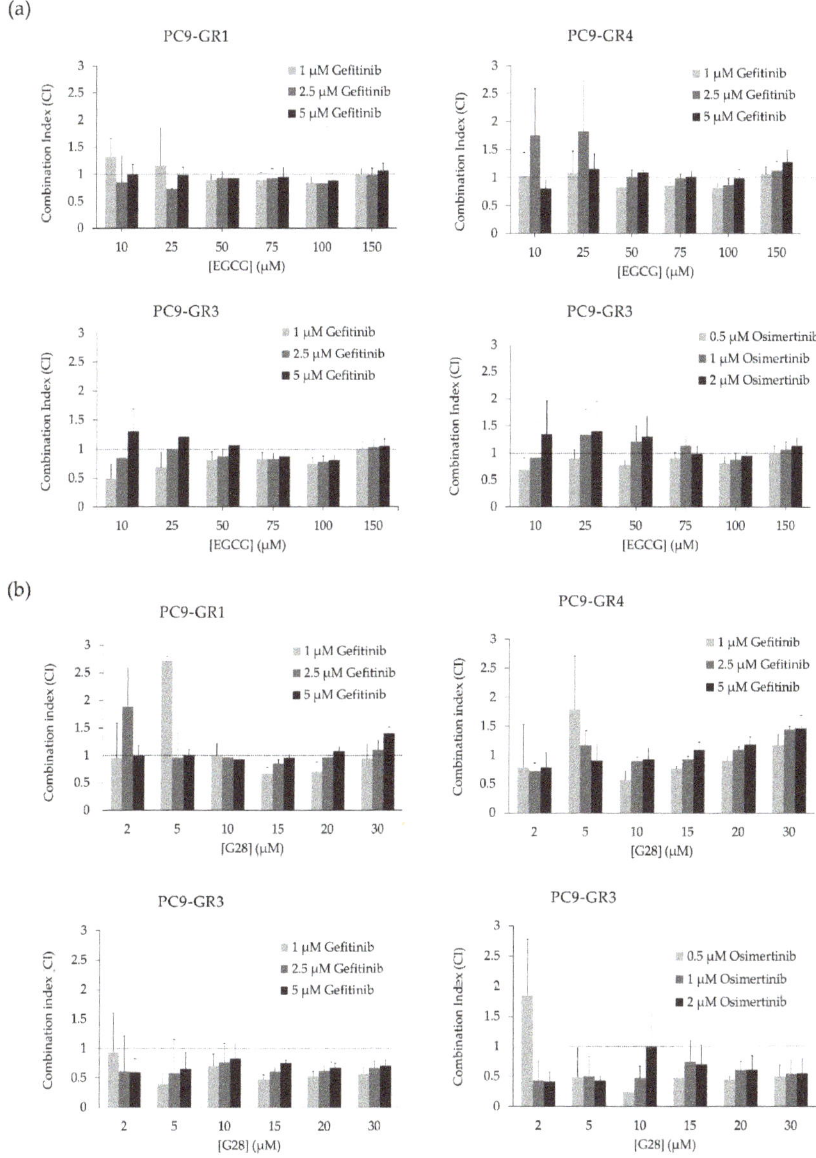

Figure 7. Combination index (CI) of FASN inhibitors (EGCG and G28) and EGFR TKIs (gefitinib and osimertinib) treatment in Gefitinib Resistant (GR) models (PC9-GR1, PC9-GR3, and PC9-GR4). (**a**) CI of EGCG and gefitinib in PC9-GR1, PC9-GR3, and PC9-GR4 models or osimertinib in PC9-GR3. (**b**) CI of G28 and gefitinib in PC9-GR1, PC9-GR3, and PC9-GR4 models or osimertinib in PC9-GR3. PC9-GR1, PC9-GR3, and PC9-GR4 models were treated with EGCG (10–150 μM) or G28 (2-30 μM) in combination with gefitinib (1, 2.5, and 5 μM) for 72 h. PC9-GR3 cells were also treated with EGCG (10–150 μM) or G28 (2–30 μM) in combination with osimertinib (0.5, 1, and 2 μM) for 72 h. Results were determined using the MTT assay and are expressed as the CI based on the Chou and Talalay method [48]. The dotted line indicates additivism (CI approximately equal to 1). CI > 1 designates antagonistic effects and CI < 1 synergistic effects. Experiments were performed at least three times. Results shown are mean ± SE.

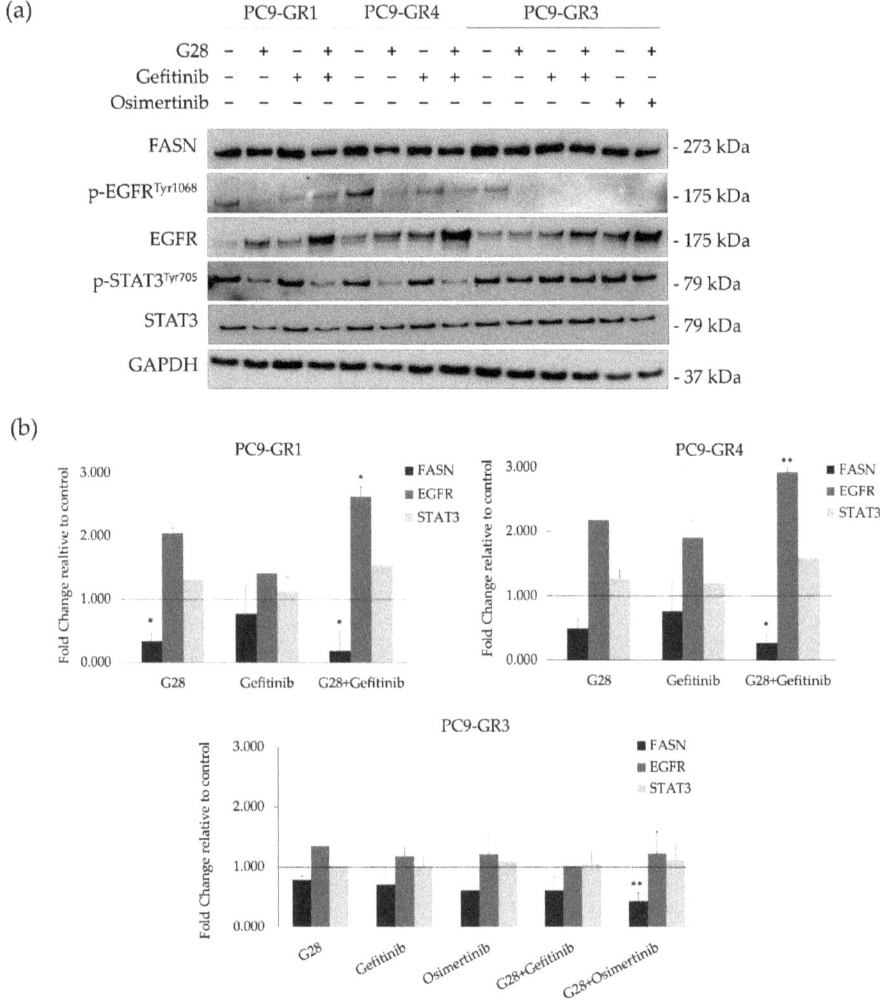

Figure 8. FASN, EGFR, and STAT3 protein and mRNA expression in sensitive (PC9) and Gefitinib Resistant (GR) models (PC9-GR1, PC9-GR3, and PC9-GR4) treated with FASN inhibitors (EGCG and G28) in combination with EGFR TKIs (gefitinib and osimertinib). (**a**) Western blot analysis of FASN, EGFR, and STAT3 in PC9-GR1, PC9-GR3, and PC9-GR4 models treated with G28 alone and in combination with gefitinib, and PC9-GR3 treated with G28 and osimertinib for 72 h. Results shown are representative from three independent experiments. Untreated cells are used as internal control (Ctrl) and GAPDH as a loading control. (**b**) FASN, EGFR, and STAT3 mRNA levels analysis in PC9-GR1, PC9-GR3, and PC9-GR4 models treated with G28 in combination with gefitinib, PC9-GR3 model treated with the combination G28 and osimertinib for 72 h. mRNA levels were obtained by qRT-PCR and normalized against the GAPDH gene. All conditions were compared to the control (untreated cells), which was normalized to 1 (indicated by the dotted line) and expressed as a fold change. Experiments were performed at least three times. * $p < 0.050$, ** $p < 0.010$, *** $p < 0.001$ indicate levels of statistical significance.

3. Discussion

Despite significant advances in EGFRm NSCLC treatment, current therapy is still ineffective to many patients due to the late stage of diagnosis and acquisition of resistance to EGFR TKIs [2,6]. Hence, several efforts have been made on the identification of EGFR TKIs resistance mechanisms to develop an effective treatment for these patients. Some authors pointed out the important role of FASN in drug resistance due to its capacity to allow fast synthesis of new phospholipids for membrane remodeling and plasticity [49,50]. Although the relationship between EGFR and FASN remains unclear, it has recently been described that EGFR upregulates FASN in TKI-resistant EGFRm NSCLC [41]. In addition, FASN inhibition showed cytotoxic effects in lung cancer [40] and resensitized cells to chemotherapy, anti-EGFR and anti-HER2 therapies in breast cancer [38,51]. Therefore, the FASN enzyme may become a promising target for anticancer therapy in EGFRm NSCLC.

Here we studied the effects of natural polyphenolic compound EGCG and its derivative G28 to overcome EGFR TKI resistance in sensitive and GR models. The higher FASN protein levels observed in EGFR TKI resistant models (Figure 1) demonstrated its potential involvement in EGFR TKI resistance acquisition. FASN inhibitors showed similar cytotoxic effects between sensitive and resistant models with IC_{50} values ranging from 75 to 90 µM for EGCG and from 12 to 18 µM for G28 (Figure 3). As determined by K. Jacobsen and coworkers, PC9-GR1 is a T790M+ GR model that also presents MET and EphA2 activation, the PC9-GR3 model exhibits AXL overexpression, and the T790M+ PC9-GR4 model shows EphA2 activation and AXL overexpression [45]. In correlation with this, the higher G28 IC_{50} found in PC9-GR3 model and the significantly similar IC_{50} values in the two T790M+ GR models indicate that none of the known resistance mechanisms described are related to FASN inhibition.

Both compounds have the ability to internalize and inhibit FASN activity as observed in parental PC9 cells (Figure 4), however G28 was of average 5.5 times more effective than EGCG. We have previously shown the ability of the natural compound EGCG to inhibit FASN activity in wild type EGFR NSCLC cells and different breast cancer subtypes [40,43]. Despite their cytotoxicity, only the synthetic compound significantly reduced FASN activity in GR models (Figure 4). Other studies proved the ability of G28 to inhibit FASN activity in triple-negative breast cancer (TNBC) cells [43]. Nevertheless, our study demonstrated that EGCG anticancer activity was independent of FASN inhibition in GR models. It has been extensively described that EGCG has multiple targets and it is involved in multiple signaling pathways and transcription factors, membrane-associated receptors tyrosine kinase (RTKs), or DNA methylation [33]. Some authors exposed that the mechanism underlying EGCG antitumor potency is due to the suppression of the EGFR signaling pathway in NSCLC [52]. Others observed a very stable complex between EGCG and EGFRm (exon 19 ELREA deletion) that was lost with the T790M substitution [53]. Kozue and coworkers showed that EGCG reduced stemness and immunogenicity in EGFR wild type NSCLC cells in vitro and in vivo through the inhibition of AXL [54]. AXL is a tyrosine kinase receptor that has been related to drug resistance and the induction of malignant properties [55] and is overexpressed in GR models [45].

The apoptosis induction was verified by PARP cleavage for all treatments in sensitive and T790M− GR models (Figure 5). However, no PARP cleavage was observed through Western blot analysis in T790M+ GR models after G28 treatment. These results suggest that G28 might cause an apoptosis-independent proliferation reduction in T790M+ GR models as found by others in NSCLC wild-type EGFR models treated with the natural plant polyphenol resveratrol (3,5,40-trihydroxystilben) [56]. The anticancer activity of polyphenols has been shown to be mediated by numerous mechanisms including cell cycle arrest. EGCG, among other natural compounds, showed down-regulation of cyclin-dependent kinases (CDK) and modulation in CDK inhibitors in different human carcinomas [33,57]. The lack of PARP cleavage in some of the treatments and models could be due to the activation of other programmed cell death mechanisms such as autophagy [58].

Alteration in EGFR expression was observed after treatment with FASN inhibitors. EGCG produced a decrease in EGFR protein levels in T790M+ GR models, which is in agreement with Ali et al., who observed an EGFR decrease in sensitive and resistant PC9 cells treated with Orlistat, a FDA-approved FASN inhibitor for obesity management [41]. The synthetic G28 compound increased EGFR mRNA levels in all models, indicating that EGFR overexpression could be an alternative pathway to FASN inhibition. The EGFR activation after FASN gene overexpression in epithelial breast cells has been previously shown [42]. Furthermore, our results suggest that the EGFR pathway could be implicated in FASN regulation at a transcriptional or translational level as exposed before [40]. Despite all models increased EGFR mRNA expression after G28 treatment, only PC9 cells seemed to compensate the effect of G28 by increasing EGFR activation. Therefore, cell proliferation reduction found in all models after FASN inhibition (Figure 3) could be due to the lack of post-translational palmitoylation substrate [41].

It seems increasingly clear that persistent STAT3 activation is related to EGFR-based therapies resistance [13]. STAT3 is under the control of different cytokines and growth factors playing an important role in metastasis, proliferation, survival, invasion, migration, and angiogenesis [14–16]. Natural polyphenols, normally multitarget inhibitors, are now emerging as promising STAT3 inhibitors or its upstream signaling molecules Src, gp130, or NFκB [18]. Among them, EGCG has been reported to reduce STAT3 phosphorylation in head and neck carcinomas [59] and pancreatic cancer [60]. Based on the above, we hypothesized that EGCG and its derivative G28 would diminish the STAT3 activation produced by EGFR TKIs in NSCLC cells. Both compounds reduced p-STAT3 levels in a dose-dependent manner when used alone in comparison to control samples. In agreement with previous studies, STAT3 was activated by EGFR TKIs (Figure 6) [22].

Combinatorial treatments of EGCG and G28 with EGFR TKIs were performed in order to study their effects on GR models (Figure S1). The combination of gefitinib and osimertinib with EGCG showed additivism whereas synergistic effects were identified in combination with G28 (Figure 7). Other authors have previously reported synergistic outcomes after co-treatment with a STAT3 inhibitor (TPCA-1) and EGFR TKI in EGFRm NSCLC models [61]. Previous results from our group demonstrated that G28 compound had a synergistic interaction with pertuzumab and temsirolimus in HER2+ breast cancer cells [38] and EGCG with cetuximab in TNBC [51]. Interestingly, G28 in combination with gefitinib decreased the activation of STAT3 to the same extent as when used alone in T790M+ GR models. Thus, other mechanisms must be involved in the synergistic effects found. On the other hand, no differences in p-STAT3 were observed in PC9-GR3. This could be, in part, due to the multiple pathways that can be altered in the acquisition of gefitinib and osimertinib double-resistance in PC9-GR3 model such as mutations in EGFR/STAT3 or other related up- or downstream signaling molecules, leading to the constitutive activation of STAT3. The analysis of the main genes regulated by the STAT3 transcription factor could provide relevant information to identify some pathways alterations after FASN inhibitors treatment. The synergistic effects found in GR models co-treated with FASN inhibitor G28 and EGFR TKIs supports the idea that EGFR palmitoylation mediated by FASN leads to TKI resistance acquisition in EGFRm NSCLC [41]. However, the specific G28 mechanism of action and possible targets are still unknown and further studies are needed. Wu et al. suggested that FASN inhibition may cause an imbalance in the membrane lipids levels, which may produce a membrane localization decrease of IGF-1R, and the inactivation of the downstream STAT3 signaling pathway [36]. Furthermore, the IGF-1R is a transmembrane tyrosine kinase linked to MAPK and PI3K/AKT pathways, shared with EGFR, which could explain the effects found only in T790M+ GR models [62].

Taken together, our observations suggest that FASN has a key role in acquired TKI-resistant EGFRm NSCLC. The inhibition of this enzyme resensitizes cells to EGFR TKIs treatments. These results encourage for further studies to analyze the combinatorial treatment of FASN inhibitors and EGFR TKIs to overcome the EGFR TKI resistance in NSCLC.

4. Materials and Methods

4.1. Cell Lines and Culture Conditions

Human adenocarcinoma PC9 cells and its gefitinib resistant derivatives PC9-GR1, PC9-GR3, and PC9-GR4 models [45] were kindly provided by Dr. R. Rosell and Dr. M. A. Molina (Barcelona, Spain). All cells were routinely grown in RPMI-1640 medium (Lonza, Basel, Switzerland), supplemented with 10% fetal bovine serum (FBS; HyClone Laboratories, GE Healthcare, Chicago, IL, USA), 50 U/mL penicillin, and 50 μg/mL streptomycin (Lonza). In all cases, the cells were used immediately after resuscitation and were maintained at 37 °C in a humidified atmosphere with 5% CO_2, propagated following established protocols, remaining free of mycoplasma throughout the experiments.

4.2. Cell Proliferation Assays

To investigate the effect of EGFR TKIs or FASN inhibitors cells were seeded into 96-well plates at the appropriate density in their growth medium. Gefitinib and osimertinib were kindly provided by AstraZeneca. EGCG was purchased from Sigma (USA) and G28 was synthesized as described elsewhere [43]. For monotreatment assays, after 24 h cells were treated with different concentrations of each drug for 72 h. Cell viability was determined by the 3-(4,5-dimethyl-2-thiazolyl)-2,5-diphenyl-tetrazolium bromide (MTT) assay (Sigma) as described elsewhere [44]. For drug combination experiments, cells were treated with three fixed concentrations of gefitinib (1, 2.5, and 5 μM) or osimertinib (0.5, 1, and 2 μM) in combination of a series of increasing concentrations of EGCG or G28 for 72 h. Following treatment, cell proliferation was measured using the standard colorimetric MTT assay. Using the multi-well plate reader Benchmark Plus (Bio-Rad Laboratories, Inc., CA, USA), absorbance was determined at 570 nm. Combinatorial effects were evaluated using the combination index (CI) based on the Chou and Talalay method [48] using the CompuSyn™ software (Biosoft, MO, USA). CompuSyn™ calculates the CI values; if the value equals 1 the effect is considered additive, if above 1 antagonistic and below 1 synergistic. Data presented are from three separate wells per assay, and the assay was performed at least three times.

4.3. Western Blot Analysis of Cell Lysates

Gefitinib-sensitive and -resistant models were treated with EGFR TKI, FASN inhibitor, or the combination of both drugs for 72 h. Afterwards, attached and floating cells were harvested and lysed in ice-cold lysis buffer (Cell Signaling Technology Inc., MA, USA) containing 100 μg/mL phenylmethylsulfonylfluoride (PMSF; Sigma) by vortexing every 5 min for 30 min. Protein concentration was determined by the Lowry method (DC Protein Assay, Bio-Rad Laboratories, Inc.). Equal amounts of protein were heated in lithium dodecyl sulfate (LDS) sample buffer with a sample reducing agent (Invitrogen, CA, USA) for 10 min at 70 °C, separated by SDS-PAGE and transferred to nitrocellulose membranes (ThermoFisher Scientific Inc., MA, USA). Membranes were incubated for 1 h at room temperature in blocking buffer (5% skim milk powder in Tris-buffered saline (TBS)) 0.05% Tween (TBS-T) and overnight at 4 °C with the following primary antibodies diluted in blocking buffer: FASN (Cell Signaling Technology Inc.; #3180), p-STAT3^{Tyr705} (Cell Signaling Technology Inc.; #9131), p-EGFR^{Tyr1068} (Cell Signaling Technology Inc.; #2234), PARP (Cell Signaling Technology Inc.; #9542), EGFR (Cell Signaling Technology Inc.; #2232), STAT3 (Cell Signaling Technology Inc.; #4904), and GAPDH (Proteintech, Manchester, UK; #60004-1-IG). Specific horseradish peroxidase (HRP)-conjugated secondary antibodies were incubated for 1 h at room temperature. The immune complexes were detected using chemiluminescent HRP substrate Clarity™ Western ECL Substrate (Bio-Rad Laboratories, Inc.) or West Femto Maximum Sensitivity Substrate (ThermoFisher Scientific Inc.) in a Bio-Rad ChemiDoc™ MP Imaging System (Bio-Rad Laboratories, Inc.). Western blot analyses were repeated at least three times and representative results are shown.

4.4. Quantitative Real-Time PCR (qRT-PCR) Analysis

Cells were treated with EGFR TKIs or FASN inhibitors as a single agent or in combination for 72 h. Then, cells were PBS washed and resuspended in 1 mL of Qiazol (Qiagen, Hilden, Germany). GeneJET RNA Purification Kit (ThermoFisher Scientific Inc.) was used to isolate total RNA following the manufacturer's instructions. RNA samples were quantified using a Nano-Drop 2000 Spectophotometer (ThermoFisher Scientific). Total RNA was reverse-transcribed into complementary DNA (cDNA) using a High Capacity cDNA Archive Kit (Applied Biosystems, CA, USA). Different gene expression levels were determined using QuantStudio3 Real-time PCR System (ThermoFisher Scientific Inc.) with qPCRBIO SyGreen Mix Lo-Rox real-time PCR (PCR Biosystems Inc., PA, USA), following manufacture instructions. Primers used are shown in Table 1. qRT-PCR analyses were performed at least four times and each gene was run in triplicate. Gene expression levels were quantified using the standard formula 2^{-dCT} and normalized to the housekeeping GAPDH (2^{-dCT}).

Table 1. Primer design.

FASN	Forward	CAGGCACACACGATGGAC
	Reverse	CGGAGTGAATCTGGGTTGAT
STAT3	Forward	CACCTTCAGGATGTCCGGAA
	Reverse	ATCCTGGAGATTCTCTACCACTTTCA
EGFR7	Forward	CATGTCGATGGACTTCCAGA
	Reverse	GGGACAGCTTGGATCACACT
GAPDH	Forward	TCTTCCAGGAGCGAGATC
	Reverse	CAGAGATGATGACCCTTTTG

4.5. Inhibition of Fatty Acid Synthase Activity

Cells were seeded in 24-well plates at a density of 3×10^4 cells/well in RPMI supplemented with 10% fetal bovine serum (FBS). 24 h later, the maintenance medium was replaced by the treatment medium (RPMI-1640 supplemented with 1% lipoprotein-deficient FBS (Sigma-Aldrich)) along with the assayed compounds or vehicle (dimethyl sulfoxide, DMSO, Sigma-Aldrich). Cells were treated with a concentration corresponding to the previously determined IC_{50} for 72 h. For the last 6 h (1,2-14C) acetic acid sodium salt (53.9 mCi/mmol, Perkin Elmer Biosciences, Waltham, MA, USA) was added to the medium (0.5 µCi/mL). The lipid extraction was performed as previously described [43]. Cells were washed twice with phosphate-buffered saline (PBS, HyClone Laboratories, Logan, UT, USA) and once with MeOH/PBS (2:3). Cell pellets were resuspended in 0.2 M NaCl and lysed with freeze (liquid N_2)—thaw (37 °C) cycles. Then, lipids were extracted with $CHCl_3$/MeOH (2:1) and 0.1 M KOH. The organic phase was washed with $CHCl_3$/MeOH/H_2O (3:48:47) and the solvents were evaporated under vacuum conditions. Finally, pellets were resuspended in EtOH and counted by scintillation. The total protein content was quantified by the Bradford assay (Sigma-Aldrich).

4.6. Statistical Analysis

Parametric data were analyzed by the Student's *t* test when comparing two groups or the one-way analysis of variance (ANOVA) followed by Bonferroni or Tamhane's T2 post hoc test for multiple comparisons. The non-parametric data were analyzed with the Mann–Whitney U tests for non-normally independent variables; otherwise, the Kruskal–Wallis test was used for more than two groups. All data are expressed as mean ± SE. Levels of significance were set at $p < 0.050$ and are represented by asterisks, as follows: $p < 0.050$ (denoted as *), $p < 0.010$ (denoted as **), and $p < 0.001$ (denoted as ***). The statistical analysis was performed using the IBM SPSS software (Version 21.0; SPSS Inc., IL, USA).

5. Conclusions

The need to find more effective and less toxic therapeutic treatments for EGFRm NSCLC patients is one of the major challenges in lung cancer research. Natural compounds are emerging as potential anticancer candidates for their safety and multitarget intrinsic features. Improvement of the properties of natural compounds through the design of synthetic derivatives is aimed at maintaining these features while increasing, efficacy, bioavailability, and stability in physiological conditions.

Here we show additive and synergistic effects of the polyphenolic plant-derived EGCG compound and its derivative G28, respectively, in combination with EGFR TKIs in GR models. Despite the exact mechanism by which these compounds are cytotoxic is still unknown, our results shed light on their ability to modulate FASN/EGFR/STAT3 pathways. The capacity to affect multiple pathways might prove useful in overcoming other drug resistances. Further analyses are required to completely understand the mechanism of action.

Taken together, this paper supports the inhibition of the metabolic enzyme FASN by G28 compound in combination with EGFR TKIs as a new potential strategy for resistant EGFRm NSCLC.

Supplementary Materials: The following are available online at http://www.mdpi.com/2072-6694/12/5/1283/s1, Figure S1: Proliferative curves of FASN inhibitors (EGCG and G28) alone and in combination with EGFR TKIs (gefitinib and osimertinib) in EGFR TKIs resistant models, Figure S2. Whole Western blot figures showing PARP, FASN, EGFR, STAT3 and GAPDH protein bands with molecular weight markers (merge of colorimetric and chemiluminescence). (a) Western blot images from Figure 5. (b) Western blot images from Figure 6. (c) Western blot images from Figure 8.

Author Contributions: Conceptualization, T.P. and S.R.-M.; methodology, E.P.-A., S.P., S.R.-M., J.R., D.T.-O., M.P., L.F.; validation, E.P.-A., S.P., S.R.-M.; resources, T.P., and J.C.; writing—original draft preparation, E.P.-A., S.P., S.R.-M.; writing—review and editing, E.P.-A., S.P., S.R.-M., T.P. and J.C.; supervision, T.P. and S.R.-M.; funding acquisition, T.P. and J.C. All authors have read and agreed to the published version of the manuscript.

Funding: The authors acknowledge the financial support from AstraZeneca, the E. P.-A pre-doctoral grant (2019FI_B01011), and the support of the Catalan Government (2017SGR00385).

Acknowledgments: Authors thank R. Rosell and M. A. Molina from laboratory of Oncology Pangaea (Barcelona, Spain) for kindly provided PC9 models and AstraZeneca (London, UK) for supplying gefitinib and osimertinib.

Conflicts of Interest: The authors declare no conflict of interest. The funders had no role in the design of the study; in the collection, analyses, or interpretation of data; in the writing of the manuscript, or in the decision to publish the results.

References

1. Testa, U.; Castelli, G.; Pelosi, E. Lung Cancers: Molecular Characterization, Clonal Heterogeneity and Evolution, and Cancer Stem Cells. *Cancers* **2018**, *10*. [CrossRef] [PubMed]

2. Sun, W.; Song, L.; Ai, T.; Zhang, Y.; Gao, Y.; Cui, J. Prognostic value of MET, cyclin D1 and MET gene copy number in non-small cell lung cancer. *J. Biomed. Res.* **2013**, *27*, 220. [CrossRef] [PubMed]

3. Choo, J.R.E.; Tan, C.S.; Soo, R.A. Treatment of EGFR T790M−Positive Non-Small Cell Lung Cancer. *Target. Oncol.* **2018**, *13*, 141–156. [CrossRef] [PubMed]

4. Lynch, T.J.; Bell, D.W.; Sordella, R.; Gurubhagavatula, S.; Okimoto, R.A.; Brannigan, B.W.; Harris, P.L.; Haserlat, S.M.; Supko, J.G.; Haluska, F.G.; et al. Activating Mutations in the Epidermal Growth Factor Receptor Underlying Responsiveness of Non−Small-Cell Lung Cancer to Gefitinib. *N. Engl. J. Med.* **2004**, *350*, 2129–2139. [CrossRef] [PubMed]

5. Cohen, M.H.; Williams, G.A.; Sridhara, R.; Chen, G.; McGuinn, W.D.; Morse, D.; Abraham, S.; Rahman, A.; Liang, C.; Lostritto, R.; et al. United States Food and Drug Administration Drug Approval summary: Gefitinib (ZD1839; Iressa) tablets. *Clin. Cancer Res.* **2004**, *10*, 1212–1218. [CrossRef]

6. Barnes, T.A.; O'Kane, G.M.; Vincent, M.D.; Leighl, N.B. Third-Generation Tyrosine Kinase Inhibitors Targeting Epidermal Growth Factor Receptor Mutations in Non-Small Cell Lung Cancer. *Front. Oncol.* **2017**, *7*, 113. [CrossRef]

7. Bean, J.; Brennan, C.; Shih, J.-Y.; Riely, G.; Viale, A.; Wang, L.; Chitale, D.; Motoi, N.; Szoke, J.; Broderick, S.; et al. MET amplification occurs with or without T790M mutations in EGFR mutant lung tumors with acquired resistance to gefitinib or erlotinib. *Proc. Natl. Acad. Sci. USA* **2007**, *104*, 20932–20937. [CrossRef]

8. Marks, J.L.; Gong, Y.; Chitale, D.; Golas, B.; McLellan, M.D.; Kasai, Y.; Ding, L.; Mardis, E.R.; Wilson, R.K.; Solit, D.; et al. Novel MEK1 mutation identified by mutational analysis of epidermal growth factor receptor signaling pathway genes in lung adenocarcinoma. *Cancer Res.* **2008**, *68*, 5524–5528. [CrossRef]

9. Shien, K.; Toyooka, S.; Yamamoto, H.; Soh, J.; Jida, M.; Thu, K.L.; Hashida, S.; Maki, Y.; Ichihara, E.; Asano, H.; et al. Acquired Resistance to EGFR Inhibitors Is Associated with a Manifestation of Stem cell–like Properties in Cancer Cells. *Cancer Res.* **2013**, *73*, 3051. [CrossRef]

10. Byers, L.A.; Diao, L.; Wang, J.; Saintigny, P.; Girard, L.; Peyton, M.; Shen, L.; Fan, Y.; Giri, U.; Tumula, P.K.; et al. An Epithelial-Mesenchymal Transition Gene Signature Predicts Resistance to EGFR and PI3K Inhibitors and Identifies Axl as a Therapeutic Target for Overcoming EGFR Inhibitor Resistance. *Clin. Cancer Res.* **2013**, *19*, 279–290. [CrossRef]

11. Takezawa, K.; Pirazzoli, V.; Arcila, M.E.; Nebhan, C.A.; Song, X.; de Stanchina, E.; Ohashi, K.; Janjigian, Y.Y.; Spitzler, P.J.; Melnick, M.A.; et al. HER2 amplification: A potential mechanism of acquired resistance to EGFR inhibition in EGFR-mutant lung cancers that lack the second-site EGFR T790M mutation. *Cancer Discov.* **2012**, *2*, 922–933. [CrossRef] [PubMed]

12. Ohashi, K.; Sequist, L.V.; Arcila, M.E.; Moran, T.; Chmielecki, J.; Lin, Y.-L.; Pan, Y.; Wang, L.; de Stanchina, E.; Shien, K.; et al. Lung cancers with acquired resistance to EGFR inhibitors occasionally harbor BRAF gene mutations but lack mutations in KRAS, NRAS, or MEK1. *Proc. Natl. Acad. Sci. USA* **2012**, *109*, E2127–E2133. [CrossRef] [PubMed]

13. Zulkifli, A.A.; Tan, F.H.; Putoczki, T.L.; Stylli, S.S.; Luwor, R.B. STAT3 signaling mediates tumour resistance to EGFR targeted therapeutics. *Mol. Cell. Endocrinol.* **2017**, *451*, 15–23. [CrossRef] [PubMed]

14. Germain, D.; Frank, D.A. Targeting the Cytoplasmic and Nuclear Functions of Signal Transducers and Activators of Transcription 3 for Cancer Therapy. *Clin. Cancer Res.* **2007**, *13*, 5665–5669. [CrossRef]

15. Frank, D.A. STAT3 as a central mediator of neoplastic cellular transformation. *Cancer Lett.* **2007**, *251*, 199–210. [CrossRef]

16. Aggarwal, B.B.; Kunnumakkara, A.B.; Harikumar, K.B.; Gupta, S.R.; Tharakan, S.T.; Koca, C.; Dey, S.; Sung, B. Signal transducer and activator of transcription-3, inflammation, and cancer: How intimate is the relationship? *Ann. N. Y. Acad. Sci.* **2009**, *1171*, 59–76. [CrossRef]

17. Rosell, R.; Karachaliou, N. Large-scale screening for somatic mutations in lung cancer. *Lancet* **2016**, *387*, 1354–1356. [CrossRef]

18. Levy, D.E.; Darnell, J.E. Stats: Transcriptional control and biological impact. *Nat. Rev. Mol. Cell Biol.* **2002**, *3*, 651–662. [CrossRef]

19. Yu, Y.; Zhao, Q.; Wang, Z.; Liu, X.Y. Activated STAT3 correlates with prognosis of non-small cell lung cancer and indicates new anticancer strategies. *Cancer Chemother. Pharmacol.* **2015**, *75*, 917–922. [CrossRef]

20. Xu, Y.H.; Lu, S. A meta-analysis of STAT3 and phospho-STAT3 expression and survival of patients with non-small-cell lung cancer. *Eur. J. Surg. Oncol.* **2014**, *40*, 311–317. [CrossRef]

21. Lee, H.J.; Zhuang, G.; Cao, Y.; Du, P.; Kim, H.J.; Settleman, J. Drug resistance via feedback activation of stat3 in oncogene-addicted cancer cells. *Cancer Cell* **2014**, *26*, 207–221. [CrossRef]

22. Ninomiya, T.; Takigawa, N.; Ichihara, E.; Ochi, N.; Murakami, T.; Honda, Y.; Kubo, T.; Minami, D.; Kudo, K.; Tanimoto, M.; et al. Afatinib Prolongs Survival Compared with Gefitinib in an Epidermal Growth Factor Receptor-Driven Lung Cancer Model. *Mol. Cancer Ther.* **2013**, *12*, 589–597. [CrossRef] [PubMed]

23. Chaib, I.; Karachaliou, N.; Pilotto, S.; Servat, J.C.; Cai, X.; Li, X.; Drozdowskyj, A.; Servat, C.C.; Yang, J.; Hu, C.; et al. Co-activation of STAT3 and YES-Associated Protein 1 (YAP1) Pathway in EGFR-Mutant NSCLC. *J. Natl. Cancer Inst.* **2017**, *109*, 1–12. [CrossRef] [PubMed]

24. Hanahan, D.; Weinberg, R.A.A. Hallmarks of cancer: The next generation. *Cell* **2011**, *144*, 646–674. [CrossRef] [PubMed]

25. Jayakumar, A.; Tai, M.H.; Huang, W.Y.; al-Feel, W.; Hsu, M.; Abu-Elheiga, L.; Chirala, S.S.; Wakil, S.J. Human fatty acid synthase: Properties and molecular cloning. *Proc. Natl. Acad. Sci. USA* **1995**, *92*, 8695–8699. [CrossRef] [PubMed]

26. Kuhajda, F.P. Fatty acid synthase and cancer: New application of an old pathway. *Cancer Res.* **2006**, *66*, 5977–5980. [CrossRef] [PubMed]

27. Liu, H.; Liu, J.-Y.; Wu, X.; Zhang, J.-T. Biochemistry, molecular biology, and pharmacology of fatty acid synthase, an emerging therapeutic target and diagnosis/prognosis marker. *Int. J. Biochem. Mol. Biol.* **2010**, *1*, 69–89.

28. Puig, T.; Vázquez-Martín, A.; Relat, J.; Pétriz, J.; Menéndez, J.A.; Porta, R.; Casals, G.; Marrero, P.F.; Haro, D.; Brunet, J.; et al. Fatty acid metabolism in breast cancer cells: Differential inhibitory effects of epigallocatechin gallate (EGCG) and C75. *Breast Cancer Res. Treat.* **2008**, *109*, 471–479. [CrossRef]

29. Jiang, B.; Li, E.-H.; Lu, Y.-Y.; Jiang, Q.; Cui, D.; Jing, Y.-F.; Xia, S.-J. Inhibition of Fatty-acid Synthase Suppresses P-AKT and Induces Apoptosis in Bladder Cancer. *Urology* **2012**, *80*, 484.e9–484.e15. [CrossRef]

30. Yang, Y.; Liu, H.; Li, Z.; Zhao, Z.; Yip-Schneider, M.; Fan, Q.; Schmidt, C.M.; Chiorean, E.G.; Xie, J.; Cheng, L.; et al. Role of fatty acid synthase in gemcitabine and radiation resistance of pancreatic cancers. *Int. J. Biochem. Mol. Biol.* **2011**, *2*, 89–98.

31. Zhan, N.; Li, B.; Xu, X.; Xu, J.; Hu, S. Inhibition of FASN expression enhances radiosensitivity in human non-small cell lung cancer. *Oncol. Lett.* **2018**, *15*, 4578–4584. [CrossRef] [PubMed]

32. Flavin, R.; Peluso, S.; Nguyen, P.L.; Loda, M. Fatty acid synthase as a potential therapeutic target in cancer. *Futur. Oncol.* **2010**, *6*, 551–562. [CrossRef] [PubMed]

33. Colomer, R.; Sarrats, A.; Lupu, R.; Puig, T. Natural Polyphenols and their Synthetic Analogs as Emerging Anticancer Agents. *Curr. Drug Targets* **2016**, *18*, 147–159. [CrossRef] [PubMed]

34. Chakrawarti, L.; Agrawal, R.; Dang, S.; Gupta, S.; Gabrani, R. Therapeutic effects of EGCG: A patent review. *Expert Opin. Ther. Pat.* **2016**, *26*, 907–916. [CrossRef]

35. Wang, X.; Tian, W. Green Tea Epigallocatechin Gallate: A Natural Inhibitor of Fatty-Acid Synthase. *Biochem. Biophys. Res. Commun.* **2001**, *288*, 1200–1206. [CrossRef]

36. Wu, J.; Du, J.; Fu, X.; Liu, B.; Cao, H.; Li, T.; Su, T.; Xu, J.; Tse, A.K.-W.; Yu, Z.-L. Icaritin, a novel FASN inhibitor, exerts anti-melanoma activities through IGF-1R / STAT3 signaling. *Oncotarget* **2016**, *7*, 51251–51269. [CrossRef]

37. Ou, Y.-C.; Li, J.-R.; Kuan, Y.-H.; Raung, S.-L.; Wang, C.-C.; Hung, Y.-Y.; Pan, P.-H.; Lu, H.-C.; Chen, C.-J. Luteolin sensitizes human 786-O renal cell carcinoma cells to TRAIL-induced apoptosis. *Life Sci.* **2014**, *100*, 110–117. [CrossRef]

38. Blancafort, A.; Giró-Perafita, A.; Oliveras, G.; Palomeras, S.; Turrado, C.; Campuzano, Ò.; Carrión-Salip, D.; Massaguer, A.; Brugada, R.; Palafox, M.; et al. Dual fatty acid synthase and HER2 signaling blockade shows marked antitumor activity against breast cancer models resistant to anti-HER2 drugs. *PLoS ONE* **2015**, *10*, e0131241. [CrossRef]

39. Giró-Perafita, A.; Rabionet, M.; Puig, T.; Ciurana, J. Optimization of Poli(ε-caprolactone) Scaffolds Suitable for 3D Cancer Cell Culture. *Procedia CIRP* **2016**, *49*, 61–66. [CrossRef]

40. Relat, J.; Blancafort, A.; Oliveras, G.; Cufí, S.; Haro, D.; Marrero, P.F.; Puig, T. Different fatty acid metabolism effects of (-)-epigallocatechin-3-gallate and C75 in adenocarcinoma lung cancer. *BMC Cancer* **2012**, *12*, 280. [CrossRef]

41. Ali, A.; Levantini, E.; Teo, J.T.; Goggi, J.; Clohessy, J.G.; Wu, C.S.; Chen, L.; Yang, H.; Krishnan, I.; Kocher, O.; et al. Fatty acid synthase mediates EGFR palmitoylation in EGFR mutated non-small cell lung cancer. *EMBO Mol. Med.* **2018**, *10*. [CrossRef] [PubMed]

42. Vazquez-Martin, A.; Colomer, R.; Brunet, J.; Lupu, R.; Menendez, J.A. Overexpression of fatty acid synthase gene activates HER1/HER2 tyrosine kinase receptors in human breast epithelial cells. *Cell Prolif.* **2008**, *41*, 59–85. [CrossRef] [PubMed]

43. Crous-Masó, J.; Palomeras, S.; Relat, J.; Camó, C.; Martínez-Garza, Ú.; Planas, M.; Feliu, L.; Puig, T. (−)-Epigallocatechin 3-gallate synthetic analogues inhibit fatty acid synthase and show anticancer activity in triple negative breast cancer. *Molecules* **2018**, *23*. [CrossRef]

44. Giró-Perafita, A.; Rabionet, M.; Planas, M.; Feliu, L.; Ciurana, J.; Ruiz-Martínez, S.; Puig, T. EGCG-Derivative G28 Shows High Efficacy Inhibiting the Mammosphere-Forming Capacity of Sensitive and Resistant TNBC Models. *Molecules* **2019**, *24*, 1027. [CrossRef]

45. Jacobsen, K.; Bertran-Alamillo, J.; Molina, M.A.; Teixidó, C.; Karachaliou, N.; Pedersen, M.H.; Castellví, J.; Garzón, M.; Codony-Servat, C.; Codony-Servat, J.; et al. Convergent Akt activation drives acquired EGFR inhibitor resistance in lung cancer. *Nat. Commun.* **2017**, *8*. [CrossRef] [PubMed]

46. Chang, L.; Fang, S.; Chen, Y.; Yang, Z.; Yuan, Y.; Zhang, J.; Ye, L.; Gu, W. Inhibition of FASN suppresses the malignant biological behavior of non-small cell lung cancer cells via deregulating glucose metabolism and AKT/ERK pathway. *Lipids Health Dis.* **2019**, *18*, 118. [CrossRef]

47. Liu, H.; Liu, Y.; Zhang, J.-T. A new mechanism of drug resistance in breast cancer cells: Fatty acid synthase overexpression-mediated palmitate overproduction. *Mol. Cancer Ther.* **2008**, *7*, 263–270. [CrossRef]

48. Chou, T.C.; Talalay, P. Quantitative analysis of dose-effect relationships: The combined effects of multiple drugs or enzyme inhibitors. *Adv. Enzym. Regul.* **1984**, *22*, 27–55. [CrossRef]

49. Wu, X.; Qin, L.; Fako, V.; Zhang, J.-T. Molecular mechanisms of fatty acid synthase (FASN)-mediated resistance to anti-cancer treatments. *Adv. Biol. Regul.* **2014**, *54*, 214–221. [CrossRef]

50. Liu, H.; Wu, X.; Dong, Z.; Luo, Z.; Zhao, Z.; Xu, Y.; Zhang, J.-T. Fatty acid synthase causes drug resistance by inhibiting TNF-α and ceramide production. *J. Lipid Res.* **2013**, *54*, 776–785. [CrossRef]

51. Giró-Perafita, A.; Palomeras, S.; Lum, D.H.; Blancafort, A.; Viñas, G.; Oliveras, G.; Pérez-Bueno, F.; Sarrats, A.; Welm, A.L.; Puig, T. Preclinical Evaluation of Fatty Acid Synthase and EGFR Inhibition in Triple Negative Breast Cancer. *Clin. Cancer Res.* **2016**, *22*, 4687–4697. [CrossRef] [PubMed]

52. Ma, Y.C.; Li, C.; Gao, F.; Xu, Y.; Jiang, Z.B.; Liu, J.X.; Jin, L.Y. Epigallocatechin gallate inhibits the growth of human lung cancer by directly targeting the EGFR signaling pathway. *Oncol. Rep.* **2014**, *31*, 1343–1349. [CrossRef] [PubMed]

53. Minnelli, C.; Laudadio, E.; Mobbili, G.; Galeazzi, R. Conformational Insight on WT- and Mutated-EGFR Receptor Activation and Inhibition by Epigallocatechin-3-Gallate: Over a Rational Basis for the Design of Selective Non-Small-Cell Lung Anticancer Agents. *Int. J. Mol. Sci.* **2020**, *21*, 1721. [CrossRef] [PubMed]

54. Namiki, K.; Wongsirisin, P.; Yokoyama, S.; Sato, M.; Rawangkan, A.; Sakai, R.; Iida, K.; Suganuma, M. (−)-Epigallocatechin gallate inhibits stemness and tumourigenicity stimulated by AXL receptor tyrosine kinase in human lung cancer cells. *Sci. Rep.* **2020**, *10*, 1–11. [CrossRef] [PubMed]

55. Zhu, C.; Wei, Y.; Wei, X. AXL receptor tyrosine kinase as a promising anti-cancer approach: Functions, molecular mechanisms and clinical applications. *Mol. Cancer* **2019**, *18*. [CrossRef] [PubMed]

56. Yousef, M.; Vlachogiannis, I.; Tsiani, E. Effects of Resveratrol against Lung Cancer: In Vitro and In Vivo Studies. *Nutrients* **2017**, *9*, 1231. [CrossRef]

57. Bailon-Moscoso, N.; Cevallos-Solorzano, G.; Romero-Benavides, J.C.; Orellana, M.I.R. Natural Compounds as Modulators of Cell Cycle Arrest: Application for Anticancer Chemotherapies. *Curr. Genom.* **2017**, *18*, 106–131. [CrossRef]

58. Hasima, N.; Ozpolat, B. Regulation of autophagy by polyphenolic compounds as a potential therapeutic strategy for cancer. *Cell Death Dis.* **2014**, *5*, e1509. [CrossRef]

59. Lin, H.-Y.; Hou, S.-C.; Chen, S.-C.; Kao, M.-C.; Yu, C.-C.; Funayama, S.; Ho, C.-T.; Way, T.-D. (−)-Epigallocatechin Gallate Induces Fas/CD95-Mediated Apoptosis through Inhibiting Constitutive and IL-6-Induced JAK/STAT3 Signaling in Head and Neck Squamous Cell Carcinoma Cells. *J. Agric. Food Chem.* **2012**, *60*, 2480–2489. [CrossRef]

60. Tang, S.-N.; Fu, J.; Shankar, S.; Srivastava, R.K. EGCG Enhances the Therapeutic Potential of Gemcitabine and CP690550 by Inhibiting STAT3 Signaling Pathway in Human Pancreatic Cancer. *PLoS ONE* **2012**, *7*, e31067. [CrossRef]

61. Codony-servat, C.; Codony-servat, J.; Karachaliou, N.; Angel, M.; Chaib, I.; Ramirez, J.L.; Gil, M.D.L.L.; Solca, F. Activation of signal transducer and activator of transcription 3 (STAT3) signaling in EGFR mutant non-small-cell lung cancer (NSCLC). *Oncotarget* **2017**, *8*, 47305–47316. [CrossRef] [PubMed]

62. Steelman, L.S.; Chappell, W.H.; Abrams, S.L.; Kempf, C.R.; Long, J.; Laidler, P.; Mijatovic, S.; Maksimovic-Ivanic, D.; Stivala, F.; Mazzarino, M.C.; et al. Roles of the Raf/MEK/ERK and PI3K/PTEN/Akt/mTOR pathways in controlling growth and sensitivity to therapy-implications for cancer and aging. *Aging (Albany. N.Y.)* **2011**, *3*, 192–222. [CrossRef] [PubMed]

Article

Targeting Glycolysis with Epigallocatechin-3-Gallate Enhances the Efficacy of Chemotherapeutics in Pancreatic Cancer Cells and Xenografts

Ran Wei [1,2], **Robert M. Hackman** [2], **Yuefei Wang** [1] **and Gerardo G. Mackenzie** [2,3,*]

[1] Tea Science Institute, Zhejiang University, Hangzhou 310058, China; wran@ucdavis.edu (R.W.); zdcy@zju.edu.cn (Y.W.)

[2] Department of Nutrition, University of California, Davis, CA 95616, USA; rmhackman@ucdavis.edu

[3] Davis Comprehensive Cancer Center, University of California, Sacramento, CA 95817, USA

* Correspondence: ggmackenzie@ucdavis.edu; Tel.: +1-530-752-2140

Received: 22 August 2019; Accepted: 3 October 2019; Published: 5 October 2019

Abstract: Pancreatic cancer is a complex disease, in need of new therapeutic approaches. In this study, we explored the effect and mechanism of action of epigallocatechin-3-gallate (EGCG), a major polyphenol in green tea, alone and in combination with current chemotherapeutics on pancreatic cancer cell growth, focusing on glycolysis metabolism. Moreover, we investigated whether EGCG's effect is dependent on its ability to induce reactive oxygen species (ROS). EGCG reduced pancreatic cancer cell growth in a concentration-dependent manner and the growth inhibition effect was further enhanced under glucose deprivation conditions. Mechanistically, EGCG induced ROS levels concentration-dependently. EGCG affected glycolysis by suppressing the extracellular acidification rate through the reduction of the activity and levels of the glycolytic enzymes phosphofructokinase and pyruvate kinase. Cotreatment with catalase abrogated EGCG's effect on phosphofructokinase and pyruvate kinase. Furthermore, EGCG sensitized gemcitabine to inhibit pancreatic cancer cell growth in vitro and in vivo. EGCG and gemcitabine, given alone, reduced pancreatic tumor xenograft growth by 40% and 52%, respectively, whereas the EGCG/gemcitabine combination reduced tumor growth by 67%. EGCG enhanced gemcitabine's effect on apoptosis, cell proliferation, cell cycle and further suppressed phosphofructokinase and pyruvate kinase levels. In conclusion, EGCG is a strong combination partner of gemcitabine reducing pancreatic cancer cell growth by suppressing glycolysis.

Keywords: pancreatic cancer; epigallocatechin-3-gallate (EGCG); gemcitabine; glycolysis; ROS; phosphofructokinase

---

1. Introduction

Pancreatic cancer is one of the top five causes of cancer-related death in the United States, with an overall five-year survival rate of approximately eight percent [1]. While surgery, which offers the only realistic hope, is a viable option in a limited number of patients, current radiation therapy and chemotherapy regimens offer no significant clinical benefit [2]. For example, gemcitabine is an antimetabolite recognized as the current standard of care for unresectable locally advanced or metastatic pancreatic cancer, but its therapeutic effect for patients is limited [3]. Thus, an urgent need exists for new ways to combat pancreatic cancer, with the exploration of novel therapeutic strategies being a critical component.

Deregulated energy metabolism is recognized as one of the hallmarks of cancer [4]. Unlike normal cells, tumor cells prefer to rely on aerobic glycolysis to produce energy and to obtain intermediate metabolites that can then be directed to other biosynthetic pathways that promote tumor growth [5,6]. Moreover, the higher glycolytic rate in cancer cells has been shown to correlate to chemotherapeutic resistance [7]. Therefore, targeting tumor glycolysis remains attractive for therapeutic interventions, not only as a main target, but also for its ability to sensitize cancer cells to chemotherapeutics.

Epigallocatechin-3-gallate (EGCG), the most active polyphenol component found in green tea, presents several anticancer properties [8–10]. Accumulating evidence, including a previous study by our group, has shown that EGCG can inhibit cancer cell growth by modulating metabolic pathways [11–13]. An important consideration when exploring EGCG's mechanism of action is that in the presence of oxygen, EGCG undergoes auto-oxidation to generate reactive oxygen species (ROS) [14]. Since cancer cells are more susceptible to an increase in ROS levels [15], this has been proposed as a potential mechanism of action for the inhibitory cell growth effect by EGCG. However, to date, whether the glycolytic pathway is modulated by EGCG in pancreatic cancer, and whether it is ROS-dependent is not completely elucidated.

In the present study, we explored the effect and mechanism of action of EGCG alone and in combination with current chemotherapeutics on pancreatic cancer cell growth, focusing on glycolysis metabolism involved in pancreatic cancer cell growth, and investigated whether it is dependent on the ability of EGCG to induce ROS. Our results show that EGCG strongly reduced pancreatic cancer cell growth by suppressing glycolysis in a ROS-dependent manner. Moreover, EGCG enhanced the anticancer effect of gemcitabine in vitro and in vivo by inhibiting glycolysis and affecting cell kinetics.

2. Results

2.1. EGCG Affects Glycolysis in Pancreatic Cancer Cells

We initially evaluated the efficacy of EGCG on pancreatic cancer cells in culture. For this purpose, we treated multiple human pancreatic cancer cells with or without increasing concentrations of EGCG (20–100 μM) for 72 h. In parallel, we also compared the effect of EGCG in human pancreatic cancer cells to that of human pancreatic normal epithelial cells (HPNE). EGCG reduced human pancreatic cancer cell growth in a concentration-dependent manner. Of note, EGCG reduced cell growth more potently in human pancreatic cancer cells compared to HPNE cells. For instance, at 72 h, EGCG 80 μM reduced MIA PaCa-2 and Panc-1 cell growth by 84% and 64% ($p < 0.05$), respectively. In contrast, EGCG at 80 μM for 72 h had significantly less effect on the HPNE cells, reducing cell growth by only 24% (Figure 1A).

Given that alterations in cancer cell metabolism can lead to an inhibition of cell growth [16], and because EGCG has been shown to affect glycolysis in other tumor types [11–13], we evaluated the effect of EGCG on glucose metabolism in two human pancreatic cancer cells and one murine pancreatic cancer cell line (KPC). We assessed the impact of EGCG on glycolysis by measuring the extracellular acidification rate (ECAR) in the MIA PaCa-2, Panc-1, and KPC cell lines. After 24 h of treatment, EGCG strongly reduced ECAR in all three cell lines, revealing an inhibitory effect of EGCG on glycolysis (Figure 1B and Figure S1).

Next, we determined the effect of EGCG on pancreatic cancer cell growth under the conditions of either glucose deprivation or treatment with 2-Deoxy-D-glucose (2-DG), a glucose analog. As shown in Figure 1C, the reduction in cell growth induced by EGCG was enhanced under glucose deprivation or 2-DG treatment. For example, 2-DG alone reduced the cell growth rate by 5%, 27%, and 8% in Panc-1, MIA PaCa-2, and KPC cells, respectively, whereas when combined with EGCG, the cell growth was reduced by 71%, 58%, and 89%, respectively ($p < 0.01$; Figure 1C). Moreover, EGCG treatment reduced ATP levels concentration-dependently (Figure 1D). Treatment with EGCG at 40 μM for 24 h reduced ATP levels in Panc-1 and MIA PaCa-2 cells by 35% and 32%, respectively ($p < 0.01$ for both, Figure 1D).

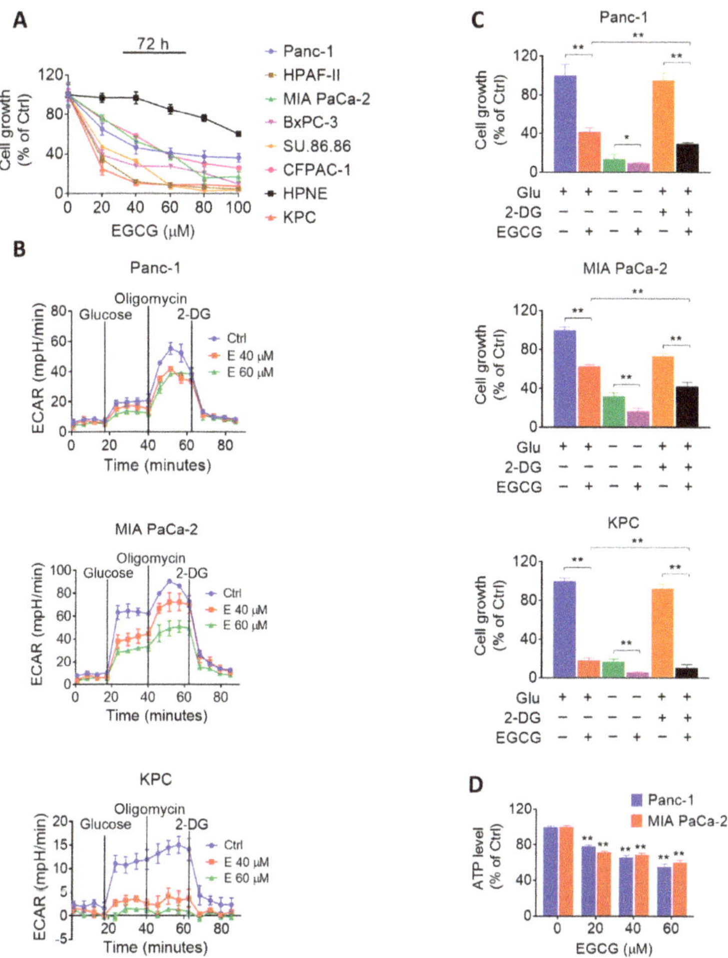

Figure 1. Epigallocatechin-3-gallate (EGCG) inhibits pancreatic cancer cell growth through glycolysis suppression. (**A**) EGCG inhibits human pancreatic cancer cell growth in a concentration-dependent manner. Cell growth was determined in Panc-1, MIA PaCa-2, HPAF-II, BxPC-3, SU-86.86, CFPAC-1, and KPC pancreatic cancer cells, and in the human pancreatic normal epithelial (HPNE) cells after treatment with increasing EGCG concentrations for 72 h. Results are expressed as a percentage of control. (**B**) EGCG suppresses glycolysis capacity in Panc-1, MIA PaCa-2, and KPC cells after 24 h. Glucose (25 mM), Oligomycin (1 μM) and 2-Deoxy-D-glucose (2-DG) (75 mM) were injected and the extracellular acidification rate (ECAR) of live cells was monitored during the experimental period. Results are presented as the mean ± SD of ECAR. (**C**) Cell growth was measured in Panc-1, MIA PaCa-2, and KPC cells treated with or without EGCG (40 μM) under glucose deprivation or 2-DG (10 mM) treatment condition. Results are expressed as a percentage of control. * $p < 0.05$, ** $p < 0.01$ vs. control. (**D**) EGCG reduced ATP levels in Panc-1 and MIA PaCa-2 cells after 24 h. Results are expressed as a percentage of control. * $p < 0.05$, ** $p < 0.01$ vs. control.

2.2. EGCG Inhibits Glycolysis through Suppressing Rate-Limiting Enzymes

Given the effect of EGCG on glycolysis, we evaluated whether EGCG could affect any particular step in the glycolytic pathway by measuring the activity and levels of glycolytic enzymes. EGCG treatment reduced both the activity and expression levels of phosphofructokinase (PFK) and pyruvate kinase (PK) in Panc-1 and MIA PaCa-2 cells, having a stronger effect on PFK (Figure 2A–C, Figure S5). For instance, EGCG at 40 μM reduced the levels of platelet-type phosphofructokinase (PFKP) and the pyruvate kinase M2 (PKM2), an isoform of PK, by 65% and 49%, respectively, in Panc-1 cells, and by 57% and 34%, respectively in MIA PaCa-2 cells (Figure 2B, Figure S4). However, EGCG failed to reduce hexokinases II (HK2) and lactate dehydrogenase A (LDHA) protein expression levels (Figure S2, Figure S12). In agreement with the in vitro results, EGCG reduced the levels of PFKP and PKM2 (p < 0.01 for both) in pancreatic tumor xenograft homogenates, obtained from mice treated with EGCG (Figure 2D, Figure S6).

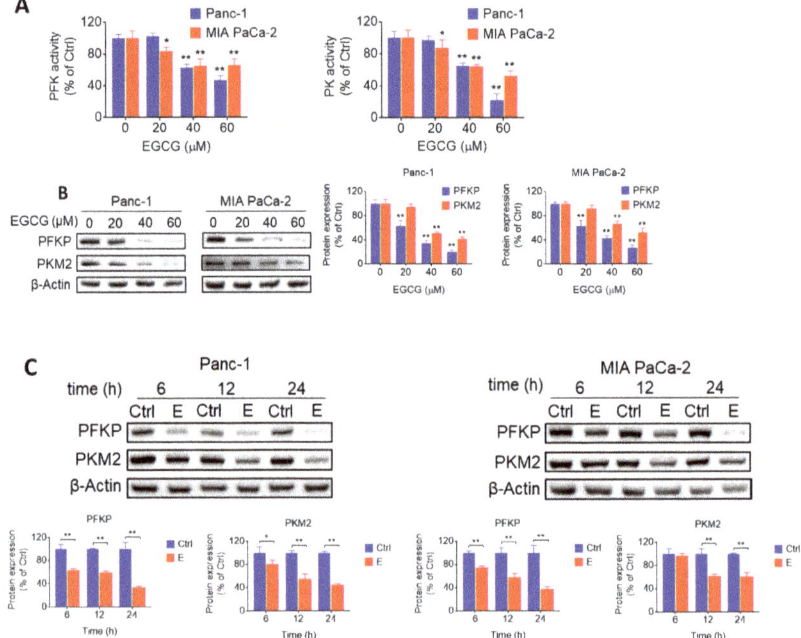

Figure 2. *Cont.*

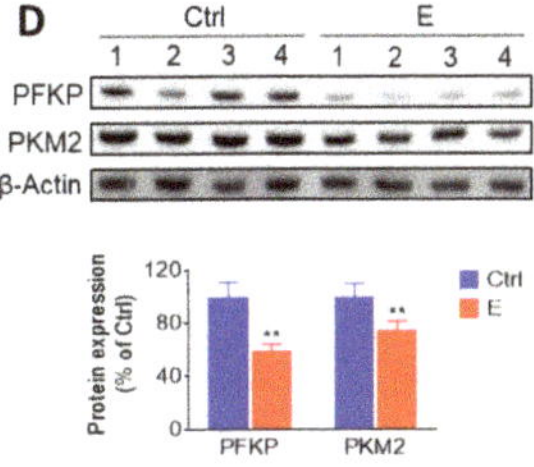
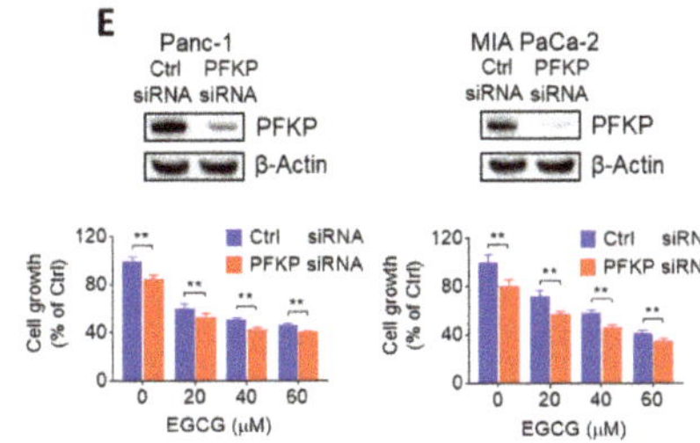

Figure 2. EGCG inhibits glycolysis through suppressing rate-limiting enzyme activity and expression. (**A**) Phosphofructokinase (PFK) and pyruvate kinase (PK) activities were determined in Panc-1 and MIA PaCa-2 cells after treatment with EGCG for 24 h. Results are expressed as a percentage of control. * $p < 0.05$, ** $p < 0.01$ vs. control. (**B**) Immunoblots for platelet-type phosphofructokinase (PFKP) and pyruvate kinase M2 (PKM2) in total cell protein extracts from Panc-1 and MIA PaCa-2 cells treated with escalating concentrations of EGCG, as indicated, for 24 h. Loading control: β-Actin. Bands were quantified and results are expressed as a percentage of control. * $p < 0.05$, ** $p < 0.01$ vs. control. (**C**) EGCG (40 μM) inhibited PFKP and PKM2 protein expression in a time-dependent manner in Panc-1 and MIA PaCa-2 cells. Results are expressed as percentage of control and presented as the mean ± SD. * $p < 0.05$, ** $p < 0.01$ vs. control. (**D**) Immunoblots of PFKP, PKM2 expression on tumor tissue from control- and EGCG-treated (10mg/kg/d) mice. Results are expressed as a percentage of control. * $p < 0.05$, ** $p < 0.01$ vs. control. (**E**) Effect of silencing PFKP on EGCG-induced cell growth reduction. Panc-1 and MIA PaCa-2 cells were transfected with either control or PFKP siRNA. After transfection, cells were treated with EGCG for 72 h and cell growth was evaluated. Results are expressed as percentage of control; * $p < 0.05$, ** $p < 0.01$ vs. control. Immunoblots to verify PFKP silencing were performed on whole cell extracts obtained from these cells (top panel).

To confirm the role of PFKP on EGCG-induced cell growth inhibition, we silenced PFKP in Panc-1 and MIA PaCa-2 cells. Knocking-down PFKP reduced Panc-1 and MIA PaCa-2 cell growth by 15% and 19%, respectively. In Panc-1 and MIA PaCa-2 cells transfected with nonspecific siRNA, EGCG 40 μM decreased the number of viable cells by 49% and 42%, whereas in PFKP-silenced cells, EGCG reduced cell growth by 57% and 54% in Panc-1 and MIA PaCa-2 cells, respectively (Figure 2E, Figure S6). Taken together, these findings indicate that silencing PFKP function has a slight additive effect on the growth inhibitory response of EGCG in both cell lines, and suggest that regulating glycolysis represents an important mechanism of EGCG in inhibiting pancreatic cancer cell growth.

2.3. EGCG Affects the Glycolytic Pathway in Pancreatic Cancer Cells through a ROS-Dependent Manner

Because EGCG and other polyphenols have been shown to undergo rapid oxidization to generate free radicals in the presence of oxygen [17], we evaluated the ability of EGCG to induce reactive oxygen species (ROS) in pancreatic cancer cells. Compared to the control, treatment with EGCG significantly increased ROS levels concentration-dependently, as determined with the general probe 2′,7′-Dichlorodihydrofluorescein diacetate (H2DCFDA). For example, EGCG at 40 μM increased ROS levels by 1.4- and 1.6-fold in Panc-1 and MIA PaCa-2 cells, respectively ($p < 0.01$ for both; Figure 3A).

Next, we determined the levels of hydrogen peroxide (H_2O_2) using the Amplex™ Red indicator probe. EGCG strongly increased H_2O_2 levels in pancreatic cancer cells. Compared to controls, EGCG at 40 μM increased H_2O_2 levels by 1.5- and 1.9-fold in Panc-1 and MIA PaCa-2 cells, respectively (Figure 3B). In both cell lines, cotreatment with catalase significantly abrogated the increase in H_2O_2 levels induced by EGCG (Figure 3B). Moreover, EGCG increased the levels of mitochondria superoxide anion in a concentration and time-dependent manner in both Panc-1 and MIA PaCa-2 cells (Figure 3C).

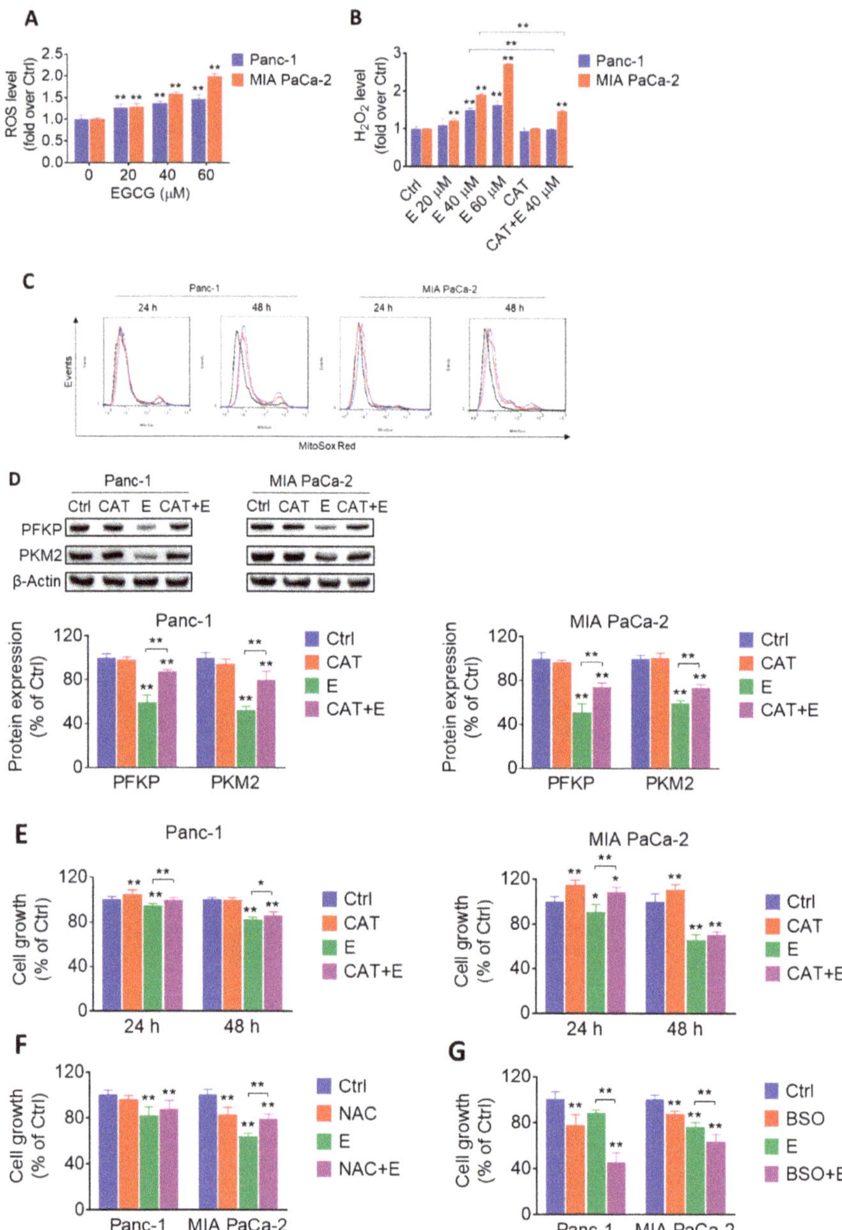

Figure 3. EGCG affects the glycolytic pathway in pancreatic cancer through a reactive oxygen species (ROS)-dependent manner. (**A**) 2′,7′-Dichlorodihydrofluorescein diacetate (H2DCFDA) fluorescence was measured in Panc-1 and MIA PaCa-2 cells treated without (control) or with EGCG at various concentration for 24 h. Results are expressed as fold change of control. * $p < 0.05$, ** $p < 0.01$ vs. control. (**B**) Amplex^TM Red was used to evaluate the levels of hydrogen peroxide level released by Panc-1 and MIA PaCa-2 cells. Results are expressed as fold change of control. * $p < 0.05$, ** $p < 0.01$ vs. control.

(**C**) EGCG induces a concentration- and time-dependent increase in mitochondrial superoxide anion levels. The levels of superoxide anion in the mitochondria were determined by flow cytometry using the MitoSOX-Red fluorescent probe. Black, red and blue lines represent control, EGCG 40 µM, and EGCG 60 µM groups, respectively. (**D**) The reduction in the expression levels of PFKP and PKM2 triggered by EGCG (E) was reversed by catalase (CAT). Immunoblots for PFKP and PKM2 in total cell protein extracts from Panc-1 and MIA PaCa-2 cells treated with 40 µM EGCG (E), 1500 U/mL catalase (CAT) or both, for 12 h. Loading control: β-Actin. Bands were quantified and results are expressed as a percentage of control. * $p < 0.05$, ** $p < 0.01$ vs. control. (**E**) Catalase (CAT) partly ameliorated the cell growth inhibitory effect induced by EGCG (E). Cells were treated with 40 µM EGCG (E), 1500 U/mL CAT or both, for 24 h or 48 h. Results are expressed as a percentage of control. * $p < 0.05$, ** $p < 0.01$ vs. control. (**F**) N-Acetyl-L-Cysteine (NAC) partly reversed cell inhibition effect of EGCG after 48 h. Cells were treated with 40 µM EGCG (E), 5 mM NAC or both for 48 h. Results are expressed as percentage of control. * $p < 0.05$, ** $p < 0.01$ vs. control. (**G**) L-Buthionine Sulfoximine (BSO) further aggravated cell inhibition effect of EGCG after 24 h. Cells were treated with 40 µM EGCG (E), 1 mM BSO or both for 24 h. Results are expressed as a percentage of control. * $p < 0.05$, ** $p < 0.01$ vs. control.

To elucidate whether the effect of EGCG on glycolysis is mediated through the increase in ROS, Panc-1 and MIA PaCa-2 cells were incubated with or without EGCG at 40 µM in the presence or absence of catalase for 12 h. In both cell lines, catalase mostly abrogated the inhibitory effect of EGCG on PFKP and PKM2, suggesting that the effect of EGCG on glycolytic enzymes is dependent on the ability of EGCG to induce ROS (Figure 3D, Figure S7).

However, catalase could only, in part, prevent the reduction in pancreatic cancer cell growth induced by EGCG (Figure 3E). Somewhat similar results were obtained when cotreating with N-Acetyl-L-Cysteine (NAC) (Figure 3F). In contrast, treatment with L-Buthionine Sulfoximine (BSO), an inhibitor of glutamylcysteine synthetase, strongly enhanced the reduction in cell growth induced by EGCG in both cell lines (Figure 3G). For example, in Panc-1 cells at 24 h, BSO and EGCG alone reduced cell growth by 22% and 12%, respectively, and by 54% when combined. These results suggest that EGCG reduces pancreatic cancer growth, partly, through an ROS-dependent mechanism.

2.4. EGCG Sensitizes Pancreatic Cancer Cells to Gemcitabine In Vitro

Because chemotherapeutics often display limited effects, show resistance, and cause side-effects, we evaluated whether EGCG would represent a useful partner in combination with commonly used drugs. For this purpose, we treated Panc-1 and MIA PaCa-2 cells with EGCG alone or in combination with gemcitabine, Abraxane®, 5-Fluorouracil, irinotecan, or oxaliplatin, five commonly used chemotherapeutics in pancreatic cancer patients, for 72 h and analyzed their combination index (CI) by means of the Chou–Talalay method. In both Panc-1 and MIA PaCa-2 cells, EGCG strongly synergized with gemcitabine (Figure 4 and Table S1). For example, in Panc-1 cells, the CI of all tested groups, except one, showed synergistic effects. In MIA PaCa-2 cells, the CI effects between EGCG and gemcitabine were also indicative of a strong additive effect (Figure 4). The combination effect of EGCG with the other four chemotherapeutics tested presented effects that were more variable with some showing an additive effect (Figure 4 and Table S1).

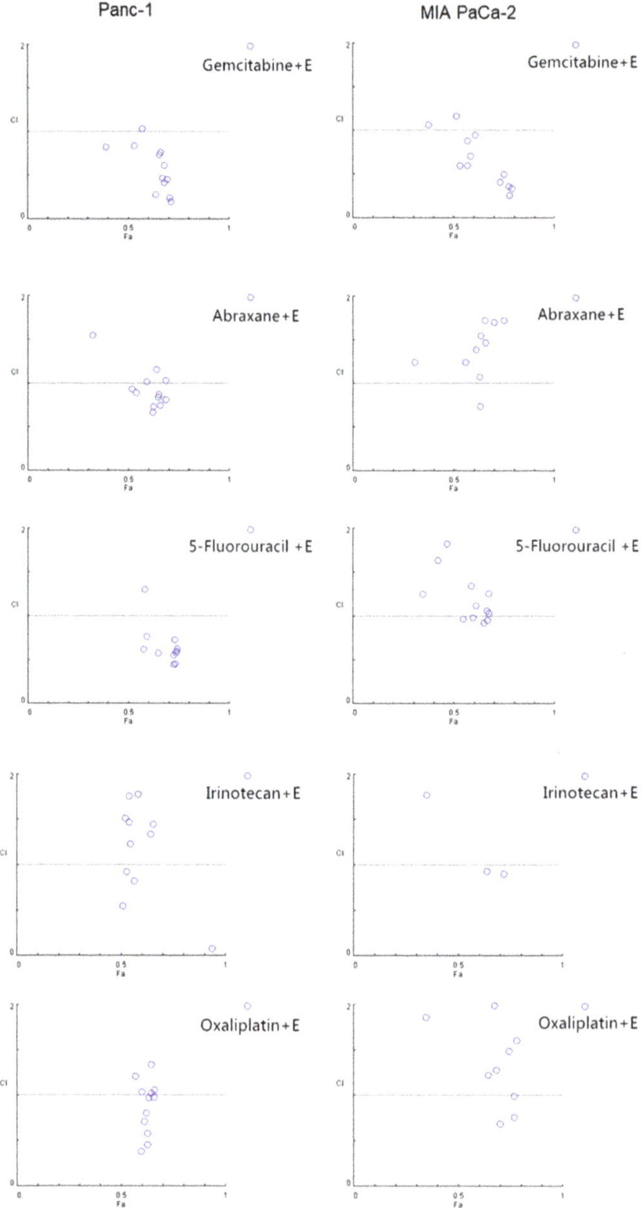

Figure 4. EGCG enhances the cell growth inhibitory effect of gemcitabine in pancreatic cancer cells. Combination index (CI) plots of EGCG (20, 40, 60 µM) in combination with gemcitabine (1, 10, 20 and 40 nM), abraxane (1, 10, 20 and 40 nM), 5-Fluorouracil (1, 10, 20 and 40 µM), irrinotecan (1, 10, 20 and 40 µM), and oxialiplatin (1, 10, 20 and 40 µM) in Panc-1 (left) and MIA PaCa-2 (right) cells. Drug interactions were quantitatively determined using the Chou–Talalay method, and CI < 1, =1, and >1 indicates synergism, additive and antagonism effect, respectively. Of note, the CI value dots depicted are based on concentrations of EGCG and the chemotherapeutic drugs shown on Table S1.

2.5. EGCG Enhances the Anticancer Effect of Gemcitabine in Pancreatic Cancer Xenografts through a Strong Cytokinetic Effect

Next, we evaluated whether EGCG could enhance the chemotherapeutic effect of gemcitabine in vivo. For this purpose, murine pancreatic cancer KPC cells were injected subcutaneously into immunocompetent mice, which gave rise to exponentially growing tumors. Once the tumors reached ~300 mm^3, the mice were treated either with EGCG (10 mg/kg) suspended in phosphate buffered saline (PBS) pH 7.4 given intraperitoneally (I.P.) once daily, gemcitabine (100 mg/kg) given I.P. twice per week, or both drugs. On day 16 of treatment, EGCG and gemcitabine, given as single agents, reduced tumor weight by 40% and 52%, respectively, compared to vehicle-treated controls ($p < 0.05$ and $p < 0.01$). In combination, EGCG plus gemcitabine reduced tumor weight by 67%, compared to controls ($p < 0.01$). Of note, the effect of the EGCG plus gemcitabine combination was significantly different from that of gemcitabine alone ($p < 0.05$; Figure 5A). Importantly, the drug combination was well tolerated by the mice, as indicated by the comparable mean body weight of the experimental groups throughout the treatment period (Figure 5B).

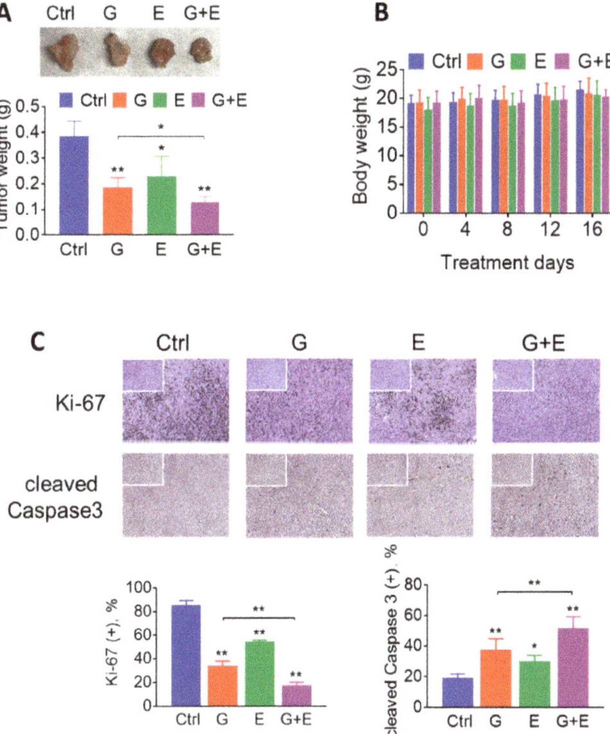

Figure 5. EGCG enhances the anticancer effect of gemcitabine in pancreatic cancer xenografts. (**A**) EGCG (E) promoted tumor weight lost with gemcitabine (G). Results are presented as the mean ± SD. * $p < 0.05$, ** $p < 0.01$ vs. control. (**B**) Mice body weight during treatment days for control, gemcitabine-treated groups (G), EGCG-treated groups (E), and the combination (G + E). Results are presented as the mean ± SD. (**C**) Immunostaining of the cell proliferation marker Ki-67 and of cleaved Caspase 3 were performed on KPC tumor sections and photographs were taken at 100× magnification. Representative images are shown. The consecutive section was stained with isotype Immunoglobulin G (IgG) as negative staining control and it is shown in the upper left corner. Results were expressed as percent of positive (+) cells ± SD per 100x field. * $p < 0.05$, ** $p < 0.01$ vs. control.

Because EGCG has been shown to be hepatotoxic at higher doses [18], and to further evaluate the safety of EGCG plus gemcitabine, we performed an acute toxicity study and determined the levels of multiple liver enzymes and kidney markers. After 16 days of EGCG plus gemcitabine treatment, liver and kidney function markers (including activity of Alanine Transaminase, Alkaline Phosphatase, Aspartate Transaminase, and the levels of blood urea nitrogen, creatinine, etc.) were in the normal range (Table S2). Consistent with the efficacy study, the mean body weights were comparable between the groups (Figure S3).

To investigate the mechanism by which EGCG plus gemcitabine reduced tumor growth, we determined cell proliferation (Ki-67 expression), and apoptosis (cleaved Caspase 3 expression) levels by immunohistochemistry in tumor tissue sections from control and EGCG plus/minus gemcitabine-treated mice (Figure 5C). Compared to controls, EGCG alone and EGCG plus gemcitabine inhibited cell proliferation (Ki-67 positive cells) by 45% and 83% in the KPC xenografts, respectively ($p < 0.01$, Figure 5C). In addition, EGCG increased the percentage of cleaved Caspase 3 positive cells by 1.6-fold when given alone and by 2.7-fold when given in combination ($p < 0.01$, Figure 5C).

In vitro, we determined the effect of EGCG and gemcitabine on cell cycle progression and cell death by apoptosis. As expected, gemcitabine, an inhibitor of DNA synthesis [19], strongly induced Synthesis/Gap2 (S/G2) phase arrest after 48 h of treatment. While EGCG at 40 µM only slightly arrested cell cycle progression, it strongly enhanced the effect of gemcitabine, with the combination further blocking the cell cycle at S/G2 in Panc-1 and MIA PaCa-2 cells (Figure 6A). Concomitant with cell cycle arrest, expression levels of S/G2 checkpoint proteins, including phosphorylated checkpoint kinases 1 (p-Chk1), phosphorylated tumor protein p53 (p-p53), and cyclin-dependent kinase (cdk) inhibitor p21 Waf1/Cip1 (p21) further increased, while cdc2 and CyclinB1 were further reduced in the EGCG plus gemcitabine-treated cells compared to those treated with gemcitabine alone (Figure 6B, Figure S8).

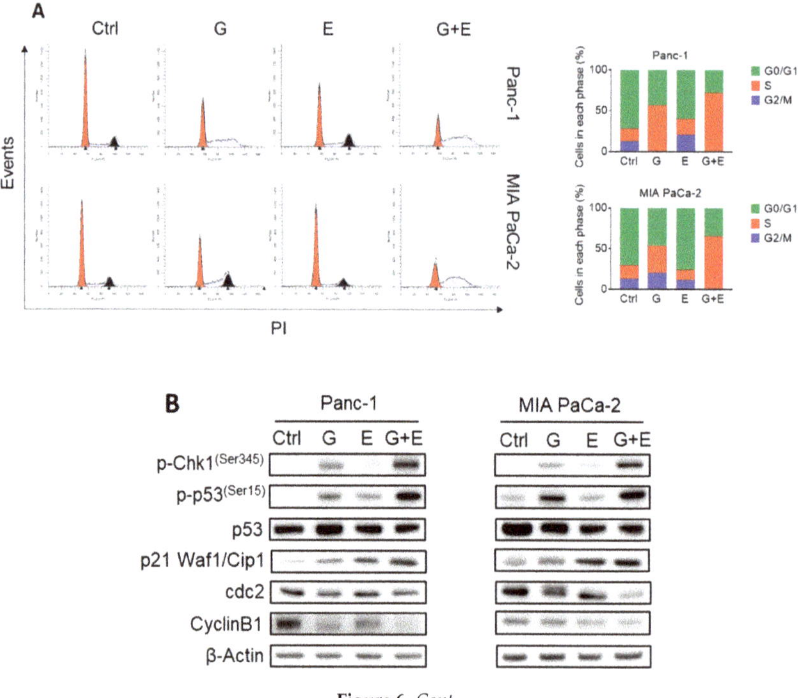

Figure 6. *Cont.*

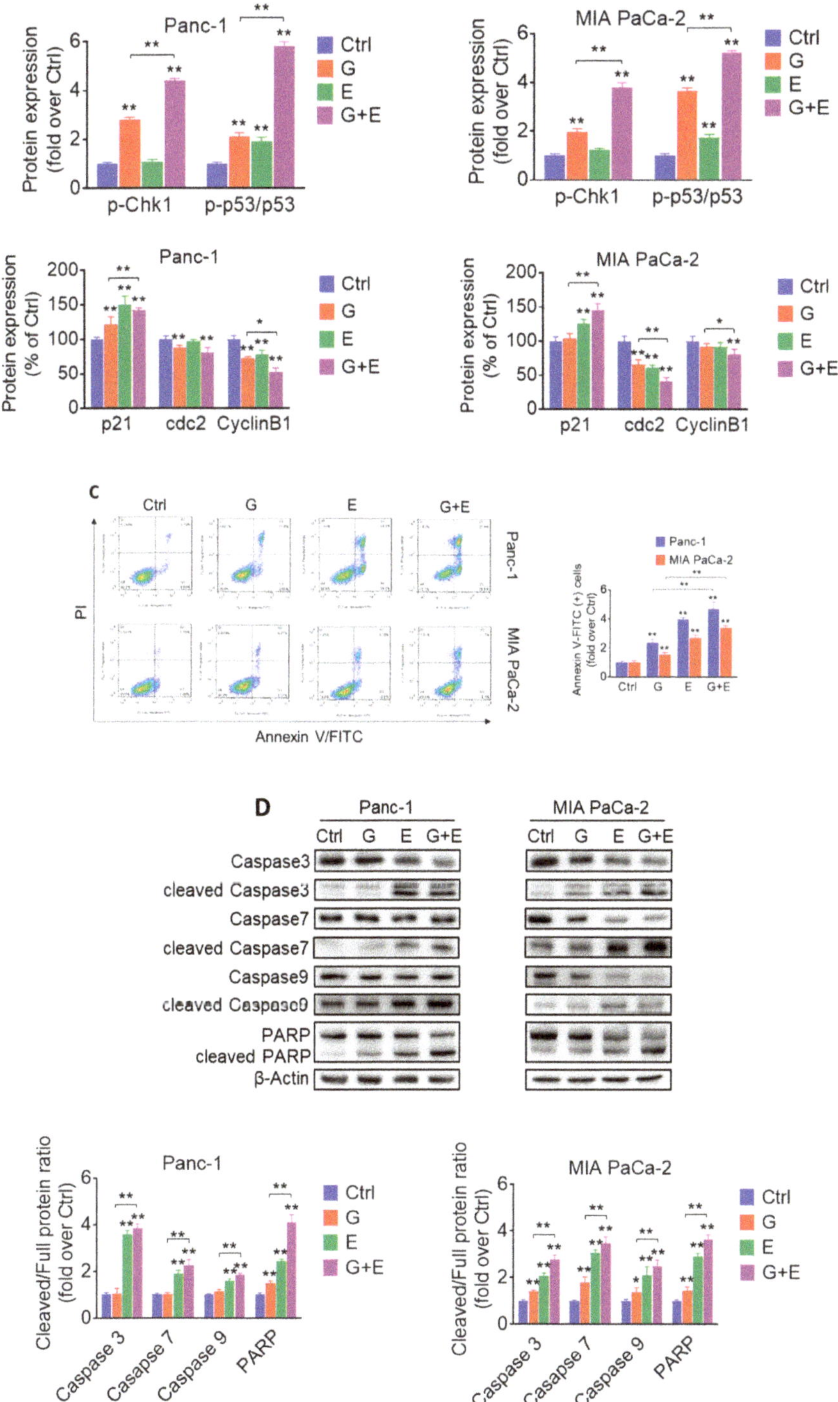

Figure 6. *Cont.*

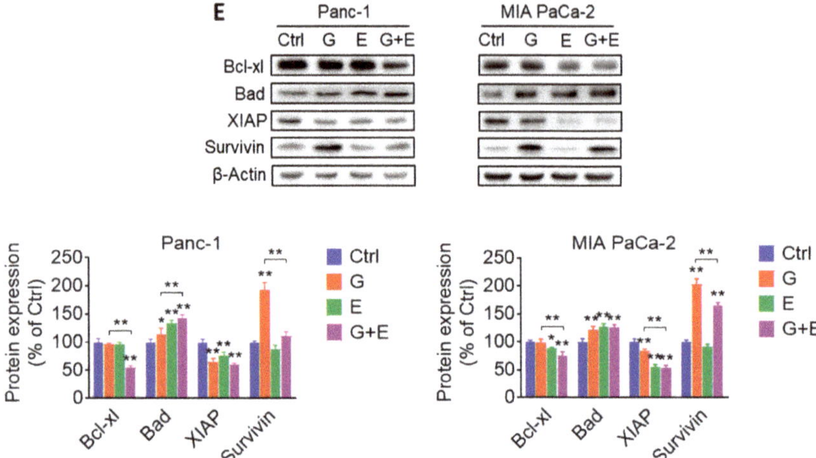

Figure 6. EGCG and gemcitabine together further affected cell kinetic in pancreatic cancer cells.
(**A**) EGCG (E) enhanced the effect of gemcitabine (G) on the cell cycle. Following treatment with 40 μM
EGCG (E), 20 nM gemcitabine (G), or both (G + E) for 48 h, cells were stained with propidium iodide
(PI) and the number of cells in each phase of the cell cycle was measured by flow cytometry. Results
are expressed as a percentage of control. (**B**) EGCG (E) and gemcitabine (G) modulated S/G2 phase
protein expression. Immunoblots for phosphorylated checkpoint kinases 1 (p-Chk1), phosphorylated
and total tumor protein p53 (p53), cyclin-dependent kinase (cdk) inhibitor p21 Waf1/Cip1 (p21), cell
division cycle 2 (cdc2) and Cyclin B1 in total cell protein extracts from Panc-1 and MIA PaCa-2 cells
treated with EGCG (E), gemcitabine (G), or both (G + E), for 48 h. Loading control: β-Actin. Bands
were quantified and results are expressed as a percentage of control. * $p < 0.05$, ** $p < 0.01$ vs. control.
(**C**) Panc-1 and MIA PaCa-2 cells were treated with EGCG (E), gemcitabine (G), or both (G + E) for
48 h, and the percentage of apoptotic cells were determined by flow cytometry using dual staining
(Annexin V and propipium iodide). The percentages of Annexin V (+) cells was calculated, and results
are expressed as the fold-increase over control. Co-treatment with EGCG (E) further increased the
apoptosis rate induced by gemcitabine (G) alone after 48 h. Results are expressed as percentage of
control. * $p < 0.05$, ** $p < 0.01$ vs. control. (**D**) Immunoblots for full length and cleaved Caspases 3,
7 and 9 as well as full length and cleaved poly (ADP-ribose) polymerase (PARP) in total cell protein
extracts from Panc-1 and MIA PaCa-2 cells treated with EGCG (E), gemcitabine (G), or both (G + E)
for 48 h. Loading control: β-Actin. Bands were quantified and results are shown as the ratio between
the cleaved/full length protein; * $p < 0.05$, ** $p < 0.01$ vs. control. (**E**) EGCG sensitized gemcitabine on
apoptosis induction by regulating B-cell lymphoma 2 (Bcl-2) family and Inhibitors of apoptosis proteins
(IAP) family protein expression in Panc-1 and MIA PaCa-2 cells after 48 h. Results are expressed as
percentage of control. * $p < 0.05$, ** $p < 0.01$ vs. control.

Besides blocking cell cycle progression and consistent with the in vivo results, gemcitabine
also induced apoptosis in pancreatic cancer cells in culture. Compared to controls, treatment with
gemcitabine for 48 h resulted in a 2.4- and 1.6-fold increase in apoptosis in Panc-1 and MIA PaCa-2
cells, respectively (Figure 6C). While EGCG at 40 μM induced apoptosis by 3.9- and 2.7-fold, the effect
of EGCG plus gemcitabine was enhanced 4.7- and 3.4-fold over control in Panc-1 and MIA PaCa-2
cells, a response that was approximately two times that of gemcitabine alone ($p < 0.01$).

We then determined the expression of proteins involved in the mechanism of apoptosis by Western
blot. Compared to controls, EGCG plus gemcitabine significantly affected the expression of cleaved
Caspase 3, 7, 9, cleaved poly (ADP-ribose) polymerase (PARP), proapoptotic member of the Bcl-2
family protein Bad, antiapoptotic member of the Bcl-2 family protein Bcl-xl, and X-linked inhibitor of
apoptosis protein (XIAP) levels, but not survivin (Figure 6D,E and Figure S9). Of note, no additive

effect was observed on the expression of cleaved Caspase 3, 7, 9, Bad, Bcl-xl, or XIAP between EGCG plus gemcitabine and EGCG alone (Figure 6E, Figure S10), suggesting that the effect of the combination is most likely being driven by EGCG.

2.6. EGCG Plus Gemcitabine Further Inhibits Glycolysis

Given that EGCG strongly affected the glycolytic pathway, we evaluated whether combining EGCG with gemcitabine would lead to any additional glycolysis inhibitory effect. For this purpose, we determined the activity and levels of glycolytic enzymes in Panc-1 and MIA PaCa-2 cells, treated with EGCG and gemcitabine alone or in combination. The activity and levels of PFK and PK were significantly reduced compared to controls and gemcitabine alone groups. However, the effect of the combination group was not significant compared to the EGCG group (Figure 7A,B and Figure S11).

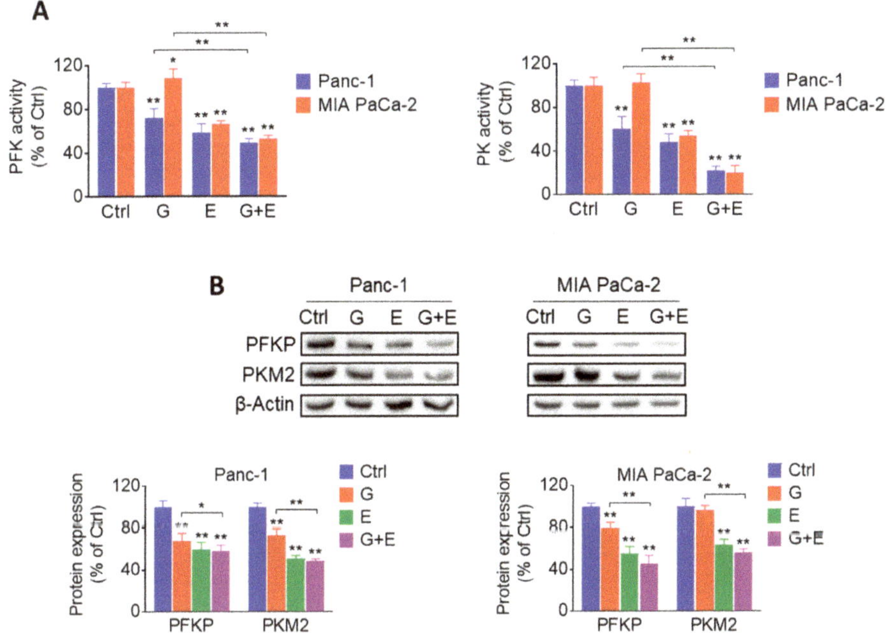

Figure 7. EGCG plus gemcitabine inhibit glycolysis. (**A**) The activity of PFK and PK was determined in Panc-1 and MIA PaCa-2 cells treated with 40 μM EGCG (E), 20 nM gemcitabine (G), or both (G + E) for 24 h. Results are expressed as a percentage of control. * $p < 0.05$, ** $p < 0.01$ vs. control. (**B**) Immunoblots for PFKP and PKM2 in total cell protein extracts from Panc-1 and MIA PaCa-2 cells treated with EGCG (E), gemcitabine (G), or both (G + E) for 12 h. Loading control: β-Actin. Bands were quantified and results are expressed as a percentage of control. * $p < 0.05$, ** $p < 0.01$ vs. control.

3. Discussion

Pancreatic cancer continues to be a significant health problem around the world. Given the lack of effective treatments that can meaningfully prolong a patient's life, an active search for agents that have additive or synergistic effect with chemotherapeutic drugs is critical in order to enhance their efficacy. In this study, we show that EGCG sensitizes pancreatic cancer cells to gemcitabine by suppressing glycolysis.

Our work identified the glycolytic pathway as one of the major signaling mechanisms involved in eliciting the growth inhibitory effect of EGCG. Glycolysis supports cell growth by rapidly generating ATP and metabolic intermediates for other biosynthetic pathways. A high rate of glycolysis is characteristic of cancer cells, even in the presence of sufficient oxygen [20]. EGCG strongly inhibited the glycolytic rate in pancreatic cancer cells by affecting the activity and expression of PFK and PK, two essential rate-limiting enzymes involved in glycolysis. The finding that glucose deprivation or 2-DG treatment enhances the growth inhibition effect of EGCG indicates that glycolysis is related to pancreatic cancer cell growth.

PFK catalyzes the first irreversible reaction of glycolysis and is usually highly expressed in tumor tissues. The platelet isoform of PFK, PFKP, functions as an important mediator in cancer cell proliferation and metastasis [21,22]. On the other hand, PKM2, an isoform of PK, is involved in regulating cell growth and metastasis [23]. Our results showed that EGCG decreased the enzyme activity and protein expression of PFK and PK in vitro and in vivo, but not HK2 or LDHA. Of note, silencing PFKP had a slight additive effect on the growth inhibitory effect of EGCG, suggesting that regulating glycolysis represents an important mechanism of EGCG in inhibiting pancreatic cancer cell growth. These results are consistent with other studies that also report inhibition of growth through the suppression of glycolysis by EGCG, such as in breast and hepatocellular cancer cell models [11,12].

The induction of oxidative stress plays a significant role in the effect of many anticancer agents [24]. Compared with normal cells, cancer cells exhibit higher levels of ROS and antioxidant levels to maintain redox homeostasis. Cancer cells are thus more susceptible to oxidative stress [15], which precedes the induction of apoptotic cell death [25]. EGCG induced ROS levels in pancreatic cancer cells in culture, consistent with its known pro-oxidant activity [26]. Interestingly, the inhibitory effect of EGCG on PFKP and PKM2 levels was mostly reversed by catalase, suggesting that the effect of EGCG on glycolytic processes is ROS-dependent. However, catalase only partly prevented the growth inhibition effect of EGCG, indicating that the ROS produced by EGCG could only explain part of the growth inhibitory effect by EGCG. Interestingly, we have recently shown that EGCG inhibits the protein kinase B (PKB, also known as Akt) pathway though a ROS-independent effect [27]. Thus, the effect of EGCG on the growth of pancreatic cancer cells appears to be the result of the sum of EGCG's ROS-dependent and independent effects. Moreover, the significance of oxidative stress in the reduction of cell growth can be further evidenced by two manipulations of the system affecting the levels of glutathione. Pretreatment with BSO, which depletes intracellular glutathione, strongly enhanced the growth inhibitory effect of EGCG. Alternatively, supplementing the cells with NAC attenuated the growth inhibitory effect of EGCG.

A common practice in the clinic is to administer multiple drugs concomitantly to cancer patients to help enhance a beneficial effect at lower doses while reducing side effects. The combination of gemcitabine with Abraxane® is the most widely used regimen for patients with newly diagnosed pancreatic cancer [28]. Another option is the combined therapy of leucovorin-modulated 5-Fluorouracil (5-FU), irinotecan, and oxaliplatin (FOLFIRINOX). Unfortunately, this regimen is given only to patients that can tolerate its toxicity, which includes a higher degree of neutropenia, diarrhea, and sensory neuropathy [29]. In this work, we explored whether EGCG could enhance the anticancer efficacy of the Food and Drug Administration (FDA)-approved drugs mentioned above. EGCG successfully enhanced the cell growth inhibition effect of gemcitabine in vitro and in vivo. Importantly, EGCG plus gemcitabine, at their effective doses, appeared to be safe to mice, being well tolerated and showing no signs of liver toxicity. As an analog of deoxycytidine, gemcitabine enters the cells, embeds into DNA, inhibits DNA synthesis, and induces cell cycle arrest. EGCG enhanced gemcitabine's suppression of cell growth through the inhibition of cell proliferation, the arrest of the cell cycle, and the induction of apoptosis through activation of execution Caspases [30], modulation of Bcl-2, and inhibition of apoptosis protein families. Mechanistically, the two agents together showed an additive effect of inhibiting glycolysis in pancreatic cancer cells. Of note, additional studies in complementary preclinical pancreatic cancer models are needed to validate the anti-tumor effect of EGCG with gemcitabine and advance the preclinical development of this drug combination. In summary, these results suggest the possibility to utilize EGCG as a useful adjuvant drug with gemcitabine to inhibit pancreatic cancer by further modulating cell kinetics and suppressing glycolysis.

4. Materials and Methods

4.1. Chemicals and Reagents

EGCG (≥98% purity) was purchased from Tocris (Minneapolis, MN, USA). 3-(4,5-dimethylthiazol-2-yl)-2,5-diphenyltetrazolium bromide (MTT) (≥97.5%), D-(+)-Glucose (≥99.5%), Oligomycin (≥90%), 2-Deoxy-D-glucose (≥99%), Irinotecan hydrochloride, Oxaliplatin, Catalase, *N*-Acetyl-L-cysteine (≥99%), DL-Buthionine-sulfoximine (≥99%), RIPA lysis buffer, Halt Protease Inhibitor Cocktail, Phosphatase Inhibitor Cocktail, and the following kits: Phosphofructokinase Activity Colorimetric assay and Pyruvate Kinase Activity assay were purchased from MilliporeSigma (St. Louis, MO, USA). Seahorse XF24 Extracellular Flux assay kits were purchased from Agilent (Santa Clara, CA, USA). CellTiter-Glo® reagent and rATP were purchased from Promega (Madison, WI, USA). PFKP siRNA plasmid was purchased from Santa Cruz Biotechnology (Dallas, TX, USA). Gemcitabine-HCL (>99%) was purchased from BIOTANG (Waltham, MA, USA). 5-Fluorouracil (≥99%) was purchased from Alfa Aesar (Haverhill, MA, USA). Annexin V-FITC conjugate, propidium iodide (PI), 2′,7′-dichlorodihydrofluorescein diacetate (H2DCFDA), Amplex® Red Hydrogen Peroxide kit, MitoSOX™ Red Mitochondrial Superoxide Indicator, Lipofectamine™ 3000 Transfection reagent, and SuperSignal™ West Dura Extended Duration Substrate were purchased from ThermoFisher Scientific (Waltham, MA, USA). Antibodies for Western blot were purchased from Cell Signaling Technology (Danvers, MA, USA). Bradford protein assay reagent, 30% (*w/v*) Acrylamide/Bis Solution, 4xLaemmli sample buffer, and Immun-Blot® PVDF Membranes were purchased from Bio-Rad (Hercules, CA, USA).

4.2. Cell Culture

Human pancreatic cancer cell lines (BxPC-3, HPAF-II, CFPAC-1, Su.86.86, Panc-1 and MIA PaCa-2), and human pancreatic normal epithelial (HPNE) cells were sourced from the American Type Culture Collection (Manassas, VA, USA). FC1245 cells (KPC, murine pancreatic cancer cells) were a gift from Dr. David Tuveson (Cold Spring Harbor Laboratory, Cold Spring Harbor, NY, USA). All cell lines were grown as monolayers in a specific medium under conditions suggested by the vendor. Although these cells lines were not authenticated in our lab, they were characterized by cell morphology and growth rate, and cultured in our laboratory less than six months after being received. We also routinely test for mycoplasma contamination in all cell lines every three months.

4.3. Cell Viability

Following the treatment with EGCG or the various chemotherapeutic drugs for 72 h, the reduction of 3-(4,5-dimethylthiazol-2-yl)-2,5-diphenyltetrazolium bromide dye (MTT) was determined according to the manufacture's protocol (MilliporeSigma, St. Louis, MO, USA).

4.4. Cellular Glycolytic Rate Measurements

The cellular glycolytic rate, represented as the extracellular acidification rate (ECAR), was measured using a Seahorse XF24 Extracellular Flux Analyzer (Agilent, Santa Clara, CA, USA). Briefly, Panc-1 and MIA PaCa-2 cells, plated on XF24 cell culture plates, were incubated with the agents for 24 h, and then assayed with a glycolytic stress test kit, following the manufacturer's instruction (Agilent, Santa Clara, CA, USA).

4.5. Cellular ATP Levels

Cells, plated in white 96-well plates, were treated with test drugs for 24 h. After treatment, ATP levels were measured with the CellTiter-Glo® reagent, following the manufacturer's protocol (Promega, Madison, WI, USA). A standard curve was generated using escalating concentrations of ATP.

4.6. Glycolysis-Related Enzymes Activity

Cells were seeded into the 6-well plates overnight and treated with EGCG for 24 h. Following the incubation, cells were collected, washed with cold PBS, and homogenized. After centrifugation, the supernatant was collected and used to measure the phosphofructokinase (PFK) and pyruvate kinase (PK) enzyme activities following the manufacturer's protocol (MilliporeSigma, Saint Louis, MO, USA). Protein concentration was determined using the Bradford protein assay.

4.7. Gene Silencing

Cells were plated in 6-well plates overnight, and transiently transfected with PFKP siRNA or nonspecific control siRNA for several hours using Lipofectamine™ 3000 reagent according to the manufacturer's instructions (ThermoFisher Scientific, Waltham, MA, USA). Following transfection, cells were replated and treated with EGCG for up to 72 h and cell viability was tested. The gene silencing efficiency was determined by immunoblotting.

4.8. Determination of ROS Levels

After treatment with EGCG for 24 h, cells were incubated with 10 µM H2DCFDA for 30 min at 37 °C and their fluorescence intensity was analyzed using a Synergy H1 microplate reader (Biotek, Winooski, VT, USA). Hydrogen peroxide (H_2O_2) levels were detected using an Amplex™ Red hydrogen peroxide kit according to the manufacturer's instruction (ThermoFisher Scientific, Waltham, MA, USA).

4.9. Mitochondrial Superoxide Level Analysis

After treatment with EGCG for 24 or 48 h, cells were collected and incubated with 5 µM MitoSOX™ Red mitochondrial superoxide probe at 37 °C for 30 min. The fluorescence intensity was determined by FACScan flow cytometry (Becton Dickinson, San Jose, CA, USA) and results were analyzed with FlowJo software (v7.6, Tree Star, Inc., Ashland, OR, USA).

4.10. Cell Apoptosis

After cells were treated with the test agents in 6-well plates, they were trypsinized and stained with Annexin V-FITC (100× dilution) and PI (0.5 µg/mL) for 15 min. Annexin V-FITC and PI fluorescence intensities were analyzed by FACScan (Becton Dickinson, San Jose, CA, USA). Annexin V (+)/PI (-) cells are apoptotic cells, Annexin V (+)/ PI (+) cells have undergone secondary necrosis, and Annexin V (-)/ PI (+) cells are necrotic cells. Results were analyzed by using FlowJo software.

4.11. Cell Cycle Analysis

Cells were seeded in 6-well plates and treated for 48 h. After each treatment, cells were trypsinized and fixed in 70% (*v/v*) ethanol overnight, stained with PI (50 µg/mL) and RNase A (10 mg/mL) for 15 min, and subjected to flow cytometric analysis (Becton Dickinson, San Jose, CA, USA).

4.12. Western Blot

Whole cell protein lysates were prepared, and electrophoresis and electroblotting were performed as previously described [31]. Membranes were probed overnight with the following primary antibodies (1:1000 dilution) from Cell Signaling Technology (Danvers, MA, USA): PFKP (Cat #8164), PKM2 (Cat #4053), HK2 (Cat #2867), LDHA (Cat #3582), phospho-Chk1 (Ser345) (Cat #2348), phospho-p53 (Ser15) (Cat #9286), p53 (Cat #2527), p21 Waf1/Cip1 (Cat #2947), cdc2 (Cat #28439), Cyclin B1 (Cat #12231), Caspase-3 (Cat #14220), Caspase-7 (Cat #12827), Caspase-9 (Cat #9508), and PARP (Cat #9542). β-Actin (Cat #8457) was used at the same time as a loading control. After incubation for 60 min at room temperature in the presence of the secondary antibody (HRP-conjugated; 1:5000 dilution), the conjugates were developed and visualized using a Molecular Imager FX™ System (BioRad; Hercules, CA, USA).

4.13. Animal Studies

All animal studies were approved by the Institutional Animal Care and Use Committee at the University of California, Davis (protocol # 20716; approved on September 6, 2018). For the efficacy study, C57BL/6J mice (4–6 weeks) were bilaterally, s.c injected with 0.3×10^6 KPC cells/tumor suspended in 0.1 mL sterile PBS. When KPC cells reached palpable tumor size (~300 mm^3), mice (n = 5/group) were divided randomly into four groups. Mice were either given vehicle, EGCG 10 mg/kg, 7x/week by intraperitoneal injection (I.P.) injections, gemcitabine 100 mg/kg, 2x/week by i.p injections, or EGCG in combination with gemcitabine at the above doses. The dose of EGCG was based on our previous studies [11,27]. Mice were treated for 16 days. Tumor size and body weight were measured every two days, and tumor size was determined by the equation length × width × (length + width/2) × 0.56, in millimeters [32]. At the end of the study, animals were euthanized by carbon dioxide asphyxiation, and tumor weights measured. Tumor and liver tissues were collected for analysis. For the drug combination toxicity study, C57BL/6J mice (n = 4/group) were treated either with PBS (vehicle control), or EGCG 10 mg/kg, 7x/week by i.p injections, in combination with gemcitabine 100 mg/kg, 2x/week by i.p injections. On day 16, mice were euthanized, blood was drawn, serum was collected, and a liver-kidney function panel was performed.

4.14. Immunohistochemistry

Immunohistochemical staining for for Ki-67 (Cat #12202) and cleaved Caspase-3 (Cat #9661, both from Cell Signaling Technology, Danvers, MA, USA) was performed as previously described [33]. Briefly, paraffin-embedded sections (5 μm thick) were deparaffinized and rehydrated, followed by antigen retrieval performed by microwave-heating in 0.01 M citrate buffer (pH 6.0). 3% H_2O_2 was used to block endogenous peroxidase activity for 10 min at room temperature. Slides were blocked for 60 min with serum, and incubated with primary antibody overnight at 4 °C. The following morning, slides were washed thrice with PBS, and then incubated with the biotinylated secondary antibody and the streptavidin-biotin complex (Invitrogen, Carlsbad, CA, USA) for 1 h of each at room temperature. After washing with PBS three times, slides were stained with 3,3′-Diaminobenzidine tetrahydrochloride hydrate (DAB) solution, and then counterstained with hematoxylin. Images were taken at 100× magnification. At least five fields per sample were scored and analyzed using Image J software (v1.46, NIH, Bethesda, MD, USA).

4.15. Statistical Analysis

Data were obtained from at least three independent biological experiments and results expressed as mean ± SD. One-way analysis of variance (ANOVA) and the Duncan test were used to analyze differences among multiple groups. *T*-tests were performed to compare the difference between two groups. $p < 0.05$ was regarded as being statistically significant.

5. Conclusions

EGCG strongly suppresses glycolysis through the inhibition of PFK, an effect that is ROS-dependent. In addition, EGCG presents a strong additive effect when combined with gemcitabine in pancreatic xenografts by further inhibiting glycolysis and affecting cell kinetics. Although additional studies to validate the above findings in complementary preclinical models of pancreatic cancer are warranted, our results suggest that EGCG is a useful combination partner of gemcitabine in pancreatic cancer treatment.

Supplementary Materials: The following are available online at http://www.mdpi.com/2072-6694/11/10/1496/s1, Figure S1: EGCG reduced cell glycolysis, glycolytic capacity and glycolytic reserve in Panc-1, MIA PaCa-2 and KPC cells. Figure S2: Effect of EGCG on HK2 and LDHA protein expression. Table S1: Cell growth combination effects of EGCG with various chemotherapeutics in MIA PaCa-2 and Panc-1 cells. Table S2: Serum levels of multiple biochemical enzymes and markers of liver and kidney function for control and EGCG plus gemcitabine the end of the treatment period. Figure S3: Mice body weight progression for control and gemcitabine+EGCG treated groups. Figure S4. Western blot images with molecular weight for PFKP and PKM2 shown in Figure 2B. Figure S5. Western blot images with molecular weight for PFKP and PKM2 shown in Figure 2C. Figure S6. Western blot images with molecular weight for PFKP and PKM2 shown in Figures 2D and 2E. Figure S7. Western blot images with molecular weight for PFKP and PKM2 shown in Figure 3D. Figure S8. Western blot images with molecular weight for p-Chk1, p53, p21, cdc2 and CyclinB1 shown in Figure 6B. Figure S9. Western blot images with molecular weight for Caspase 3, Caspase 7, Caspase 9 and PARP shown in Figure 6D. Figure S10. Western blot images with molecular weight for Bcl-xl, Bad, XIAP, Survivin shown in Figure 6E. Figure S11. Western blot images with molecular weight for PFKP and PKM2 shown in Figure 7B. Figure S12. Western blot images with molecular weight for HK2 and LDHA shown in Figure S2.

Author Contributions: Conceptualization, R.W. and G.G.M.; methodology, R.W. and G.G.M.; validation, R.W.; formal analysis, R.W.; investigation, R.W.; resources, G.G.M.; data curation, R.W.; writing—original draft preparation, R.W.; writing—review and editing, R.W., R.M.H., Y.W. and G.G.M.; supervision, G.G.M.; project administration, G.G.M.; funding acquisition, G.G.M.

Funding: This study was supported by the funds from the University of California, Davis and NIFA-USDA (CA-D-XXX-2397-H) to GGM. Ran Wei is sponsored by a China Scholarship Council fellowship. Flow cytometry experiments were funded in part by the UC Davis Comprehensive Cancer Center Support Grant (CCSG) (NCI P30CA093373).

Conflicts of Interest: The authors declare no conflict of interest.

References

1. Aier, I.; Semwal, R.; Sharma, A.; Varadwaj, P.K. A systematic assessment of statistics, risk factors, and underlying features involved in pancreatic cancer. *Cancer Epidemiol.* **2019**, *58*, 104–110. [CrossRef] [PubMed]

2. Kleeff, J.; Korc, M.; Apte, M.; La Vecchia, C.; Johnson, C.D.; Biankin, A.V.; Neale, R.E.; Tempero, M.; Tuveson, D.A.; Hruban, R.H.; et al. Pancreatic cancer. *Nat. Rev. Dis. Primers* **2016**, *2*. [CrossRef] [PubMed]

3. Von Hoff, D.D.; Ervin, T.; Arena, F.P.; Chiorean, E.G.; Infante, J.; Moore, M.; Seay, T.; Tjulandin, S.A.; Ma, W.W.; Saleh, M.N.; et al. Increased survival in pancreatic cancer with nab-paclitaxel plus gemcitabine. *N. Engl. J. Med.* **2013**, *369*, 1691–1703. [CrossRef] [PubMed]

4. Li, Z.; Zhang, H. Reprogramming of glucose, fatty acid and amino acid metabolism for cancer progression. *Cell Mol. Life Sci.* **2016**, *73*, 377–392. [CrossRef] [PubMed]

5. Dhup, S.; Dadhich, R.K.; Porporato, P.E.; Sonveaux, P. Multiple biological activities of lactic acid in cancer: Influences on tumor growth, angiogenesis and metastasis. *Curr. Pharm. Des.* **2012**, *18*, 1319–1330. [CrossRef] [PubMed]

6. Liberti, M.V.; Locasale, J.W. The warburg effect: How does it benefit cancer cells? *Trends Biochem. Sci.* **2016**, *41*, 211–218. [CrossRef] [PubMed]

7. Bhattacharya, B.; Mohd Omar, M.F.; Soong, R. The warburg effect and drug resistance. *Br. J. Pharmacol.* **2016**, *173*, 970–979. [CrossRef]

8. Lewis, K.A.; Jordan, H.R.; Tollefsbol, T.O. Effects of SAHA and EGCG on growth potentiation of triple-negative breast cancer cells. *Cancers* **2018**, *11*, 23. [CrossRef]

9. Du, G.J.; Zhang, Z.Y.; Wen, X.D.; Yu, C.H.; Calway, T.; Yuan, C.S.; Wang, C.Z. Epigallocatechin Gallate (EGCG) Is the most effective cancer chemopreventive polyphenol in green tea. *Nutrients* **2012**, *4*, 1679–1691. [CrossRef]

10. Wang, Y.K.; Chen, W.C.; Lai, Y.H.; Chen, Y.H.; Wu, M.T.; Kuo, C.T.; Wang, Y.Y.; Yuan, S.S.F.; Liu, Y.P.; Wu, I.C. Influence of tea consumption on the development of second esophageal neoplasm in patients with head and neck cancer. *Cancers* **2019**, *11*, 387. [CrossRef]

11. Wei, R.; Mao, L.; Xu, P.; Zheng, X.; Hackman, R.M.; Mackenzie, G.G.; Wang, Y. Suppressing glucose metabolism with epigallocatechin-3-gallate (EGCG) reduces breast cancer cell growth in preclinical models. *Food Funct.* **2018**, *9*, 5682–5696. [CrossRef] [PubMed]

12. Li, S.; Wu, L.; Feng, J.; Li, J.; Liu, T.; Zhang, R.; Xu, S.; Cheng, K.; Zhou, Y.; Zhou, S.; et al. In vitro and in vivo study of epigallocatechin-3-gallate-induced apoptosis in aerobic glycolytic hepatocellular carcinoma cells involving inhibition of phosphofructokinase activity. *Sci. Rep.* **2016**, *6*, e28479. [CrossRef] [PubMed]

13. Gao, F.; Li, M.; Liu, W.B.; Zhou, Z.S.; Zhang, R.; Li, J.L.; Zhou, K.C. Epigallocatechin gallate inhibits human tongue carcinoma cells via HK2mediated glycolysis. *Oncol. Rep.* **2015**, *33*, 1533–1539. [CrossRef] [PubMed]

14. Long, L.H.; Clement, M.V.; Halliwell, B. Artifacts in cell culture: Rapid generation of hydrogen peroxide on addition of (-)-epigallocatechin, (-)-epigallocatechin gallate, (+)-catechin, and quercetin to commonly used cell culture media. *Biochem. Biophys. Res. Commun.* **2000**, *273*, 50–53. [CrossRef]

15. Sullivan, L.B.; Chandel, N.S. Mitochondrial reactive oxygen species and cancer. *Cancer Metab.* **2014**, *2*, e17. [CrossRef]

16. Villa, E.; Ricci, J.E. How does metabolism affect cell death in cancer? *Febs. J.* **2016**, *283*, 2653–2660. [CrossRef]

17. Wei, Y.; Chen, P.; Ling, T.; Wang, Y.; Dong, R.; Zhang, C.; Zhang, L.; Han, M.; Wang, D.; Wan, X.; et al. Certain (-)-epigallocatechin-3-gallate (EGCG) auto-oxidation products (EAOPs) retain the cytotoxic activities of EGCG. *Food Chem.* **2016**, *204*, 218–226. [CrossRef]

18. Lambert, J.D.; Kennett, M.J.; Sang, S.; Reuhl, K.R.; Ju, J.; Yang, C.S. Hepatotoxicity of high oral dose (-)-epigallocatechin-3-gallate in mice. *Food Chem. Toxicol.* **2010**, *48*, 409–416. [CrossRef]

19. Mini, E.; Nobili, S.; Caciagli, B.; Landini, I.; Mazzei, T. Cellular pharmacology of gemcitabine. *Ann. Oncol.* **2006**, *17*, 7–12. [CrossRef]

20. Yu, L.; Chen, X.; Wang, L.; Chen, S. The sweet trap in tumors: Aerobic glycolysis and potential targets for therapy. *Oncotarget* **2016**, *7*, 38908–38926. [CrossRef]

21. Lee, J.H.; Liu, R.; Li, J.; Zhang, C.; Wang, Y.; Cai, Q.; Qian, X.; Xia, Y.; Zheng, Y.; Piao, Y.; et al. Stabilization of phosphofructokinase 1 platelet isoform by AKT promotes tumorigenesis. *Nat. Commun.* **2017**, *8*, e949. [CrossRef] [PubMed]

22. Lang, L.; Chemmalakuzhy, R.; Shay, C.; Teng, Y. PFKP signaling at a glance: An emerging mediator of cancer cell metabolism. *Adv. Exp. Med. Biol.* **2019**, *1134*, 243–258. [CrossRef] [PubMed]

23. Azoitei, N.; Becher, A.; Steinestel, K.; Rouhi, A.; Diepold, K.; Genze, F.; Simmet, T.; Seufferlein, T. PKM2 promotes tumor angiogenesis by regulating HIF-1alpha through NF-kappaB activation. *Mol. Cancer* **2016**, *15*, e3. [CrossRef] [PubMed]

24. Sun, Y.; Huang, L.Q.; Mackenzie, G.G.; Rigas, B. Oxidative stress mediates through apoptosis the anticancer effect of phospho-nonsteroidal anti-Inflammatory drugs: Implications for the role of oxidative stress in the action of anticancer agents. *J. Pharmacol. Exp. Ther.* **2011**, *338*, 775–783. [CrossRef] [PubMed]

25. Hegedus, C.; Kovacs, K.; Polgar, Z.; Regdon, Z.; Szabo, E.; Robaszkiewicz, A.; Forman, H.J.; Martner, A.; Virag, L. Redox control of cancer cell destruction. *Redox. Biol.* **2018**, *16*, 59–74. [CrossRef] [PubMed]

26. Qanungo, S.; Das, M.; Haldar, S.; Basu, A. Epigallocatechin-3-gallate induces mitochondrial membrane depolarization and caspase-dependent apoptosis in pancreatic cancer cells. *Carcinogenesis* **2005**, *26*, 958–967. [CrossRef] [PubMed]

27. Wei, R.; Penso, N.E.C.; Hackman, R.M.; Wang, Y.; Mackenzie, G.G. Epigallocatechin-3-gallate (EGCG) suppresses pancreatic cancer cell growth, invasion, and migration partly through the inhibition of Akt pathway and epithelial-mesenchymal transition: Enhanced efficacy when combined with gemcitabine. *Nutrients* **2019**, *11*, 1856. [CrossRef]

28. Frese, K.K.; Neesse, A.; Cook, N.; Bapiro, T.E.; Lolkema, M.P.; Jodrell, D.I.; Tuveson, D.A. nab-Paclitaxel potentiates gemcitabine activity by reducing cytidine deaminase levels in a mouse model of pancreatic cancer. *Cancer Discov.* **2012**, *2*, 260–269. [CrossRef]

29. Conroy, T.; Desseigne, F.; Ychou, M.; Bouche, O.; Guimbaud, R.; Becouarn, Y.; Adenis, A.; Raoul, J.L.; Gourgou-Bourgade, S.; de la Fouchardiere, C.; et al. FOLFIRINOX versus gemcitabine for metastatic pancreatic cancer. *N. Engl. J. Med.* **2011**, *364*, 1817–1825. [CrossRef]

30. Shi, Y. Mechanisms of caspase activation and inhibition during apoptosis. *Mol. Cell* **2002**, *9*, 459–470. [CrossRef]

31. Mallangada, N.A.; Vargas, J.M.; Thomas, S.; DiGiovanni, M.G.; Vaeth, B.M.; Nemesure, M.D.; Wang, R.X.; LaComb, J.F.; Williams, J.L.; Golub, L.M.; et al. A novel tricarbonylmethane agent (CMC2.24) reduces human pancreatic tumor growth in mice by targeting Ras. *Mol. Carcinogen.* **2018**, *57*, 1130–1143. [CrossRef] [PubMed]

32. Mackenzie, G.G.; Huang, L.; Alston, N.; Ouyang, N.; Vrankova, K.; Mattheolabakis, G.; Constantinides, P.P.; Rigas, B. Targeting mitochondrial STAT3 with the novel phospho-valproic acid (MDC-1112) inhibits pancreatic cancer growth in mice. *PLoS ONE* **2013**, *8*, e61532. [CrossRef] [PubMed]

33. Bartels, L.E.; Mattheolabakis, G.; Vaeth, B.M.; LaComb, J.F.; Wang, R.; Zhi, J.; Komninou, D.; Rigas, B.; Mackenzie, G.G. The novel agent phospho-glycerol-ibuprofen-amide (MDC-330) inhibits glioblastoma growth in mice: An effect mediated by cyclin D1. *Carcinogenesis* **2016**, *37*, 420–429. [CrossRef] [PubMed]

Review

EGCG Mediated Targeting of Deregulated Signaling Pathways and Non-Coding RNAs in Different Cancers: Focus on JAK/STAT, Wnt/β-Catenin, TGF/SMAD, NOTCH, SHH/GLI, and TRAIL Mediated Signaling Pathways

Ammad Ahmad Farooqi [1], Marina Pinheiro [2,*], Andreia Granja [2], Fulvia Farabegoli [3], Salette Reis [2], Rukset Attar [4], Uteuliyev Yerzhan Sabitaliyevich [5], Baojun Xu [6] and Aamir Ahmad [7]

[1] Institute of Biomedical and Genetic Engineering (IBGE), Islamabad 54000, Pakistan; ammadfarooqi@rlmclahore.com

[2] LAQV, REQUIMTE, Departamento de Ciências Químicas, Faculdade de Farmácia, Universidade do Porto, 4050-313 Porto, Portugal; andreia26293@gmail.com (A.G.); shreis@ff.up.pt (S.R.)

[3] Department of Pharmacy and Biotechnology (FaBiT), University of Bologna, 40126 Bologna, Italy; fulvia.farabegoli@unibo.it

[4] Department of Obstetrics and Gynecology, Yeditepe University, Ataşehir/İstanbul 34755, Turkey; ruksetattar@hotmail.com

[5] Department of Health Policy and Health Care Development, Kazakh Medical University of Continuing Education, Almaty 050004, Kazakhstan; e.uteuliyev@ksph.kz

[6] Food Science and Technology Program, Beijing Normal University-Hong Kong Baptist University United International College, Zhuhai 519087, China; baojunxu@uic.edu.hk

[7] Department of Medicine, University of Alabama at Birmingham, Birmingham, AL 35205, USA; aamirahmad100@gmail.com

* Correspondence: mpinheiro@ff.up.pt

Received: 16 March 2020; Accepted: 4 April 2020; Published: 12 April 2020

Abstract: Decades of research have enabled us to develop a better and sharper understanding of multifaceted nature of cancer. Next-generation sequencing technologies have leveraged our existing knowledge related to intra- and inter-tumor heterogeneity to the next level. Functional genomics have opened new horizons to explore deregulated signaling pathways in different cancers. Therapeutic targeting of deregulated oncogenic signaling cascades by products obtained from natural sources has shown promising results. Epigallocatechin-3-gallate (EGCG) has emerged as a distinguished chemopreventive product because of its ability to regulate a myriad of oncogenic signaling pathways. Based on its scientifically approved anticancer activity and encouraging results obtained from preclinical trials, it is also being tested in various phases of clinical trials. A series of clinical trials associated with green tea extracts and EGCG are providing clues about significant potential of EGCG to mechanistically modulate wide ranging signal transduction cascades. In this review, we comprehensively analyzed regulation of JAK/STAT, Wnt/β-catenin, TGF/SMAD, SHH/GLI, NOTCH pathways by EGCG. We also discussed most recent evidence related to the ability of EGCG to modulate non-coding RNAs in different cancers. Methylation of the genome is also a widely studied mechanism and EGCG has been shown to modulate DNA methyltransferases (DNMTs) and protein enhancer of zeste-2 (EZH2) in multiple cancers. Moreover, the use of nanoformulations to increase the bioavailability and thus efficacy of EGCG will be also addressed. Better understanding of the pleiotropic abilities of EGCG to modulate intracellular pathways along with the development of effective EGCG delivery vehicles will be helpful in getting a step closer to individualized medicines.

Keywords: EGCG; signaling pathways; non-coding RNAs; anti-cancer drug

---

1. Introduction

Genomic approaches such as whole genome sequencing and genetic mapping have helped considerably in the identification of many genetic variants in multiple components of cell signaling pathways. Moreover, advancements in functional genomics have marked a new frontier in molecular oncology. Epigallocatechin-3-gallate (EGCG) is a phenolic compound present in tea and has captivated tremendous attention in the past two decades because of its premium pharmacological properties. There is a wide variety of reviews published with reference to EGCG mediated anticancer effects [1–4]. However, in this review we focused on EGCG mediated modulation of deregulation cell signaling pathways in different cancers. We partitioned this multi-component review into different sections. We will open the review by critical analysis of layered regulation of the JAK-STAT pathway by EGCG.

2. Targeting of JAK/STAT Signaling

The JAK-STAT pathway constitutes a rapid membrane-to-nucleus signaling module that has been shown to play fundamental role in cancer development and progression (shown in Figure 1). In this section, we will discuss in detail how EGCG modulated JAK/STAT signaling. EGCG has been shown to interfere with the JAK/STAT pathway at different steps, which includes inhibition of STAT phosphorylation and restriction of nuclear transportation of STAT proteins.

EGCG remarkably reduced tyrosine and serine phosphorylation of signal transducer and activator of transcription 1 (STAT1) [5]. Moreover, phosphorylation of protein kinase C delta PKC-delta, Janus kinase 1 (JAK1), and Janus kinase 2 (JAK2), which are the upstream activators of STAT1 are also inhibited by EGCG in interferon gamma (IFNγ)-stimulated oral cancer cells (shown in Figure 1) [5]. EGCG-mono-palmitate (EGCG-MP), a highly active derivative of EGCG effectively activated Src homology 2 domain-containing tyrosine phosphatase-1 (SHP-1) which consequentially resulted in reduction of phosphorylated levels of BCR-ABL and signal transducer and activator of transcription 3 (STAT3) in human chronic myeloid leukemia (CML) cells (shown in Figure 1) [6]. EGCG-MP treatment more efficiently induced regression of tumor growth in BALB/c athymic nude mice [6]. EGCG potently inhibited BCR/ABL oncoprotein and the JAK2/STAT3/AKT pathway in BCR/ABL+ CML cell lines [7]. Curcumin worked synchronously with EGCG and considerably interfered with tumor conditioned media-induced transition of normal endothelial cells toward tumor endothelial cells by inhibition of the JAK/STAT3 signaling pathway [8].

EGCG significantly reduced phosphorylation of STAT3 on the 705th tyrosine residue and improved sensitivity of cisplatin-resistant oral cancer cells [9]. Fundamental role-play of STAT signaling had previously been studied in invasive breast cancers and matched lymph nodes using quantitative immunofluorescence [10]. STAT proteins were analyzed in lymph nodes and paired primary breast cancer tissues. There was higher expression of cytoplasmic STAT1, p-STAT3 (Ser727), STAT5, and nuclear p-STAT3 (Ser727) in the nodes [10]. c-Myb overexpression induced activation of NF-κB and STAT3 signaling to enhance proliferation, invasion, and resistance against cisplatin [11]. However, c-Myb silencing inhibited proliferation, invasive potential, and sensitized ovarian cancer cells to cisplatin. EGCG completely inhibited c-Myb-mediated proliferative and invasive abilities of ovarian cancer cells [11].

EGCG dose-dependently reduced phosphorylated levels of STAT1 and STAT3 [12]. Quercetin and EGCG worked synergistically and exerted inhibitory effects on cytokine-mediated upregulation of iNOS (inducible nitric oxide synthase) and ICAM-1 (intercellular adhesion molecule-1) via JAK/STAT cascade in cholangiocarcinoma cells (Figure 1) [12].

Indoleamine 2,3-dioxygenase (IDO) is a tryptophan catabolic enzyme. IDO mechanistically regulates immunological response and enables tumor cells to evade the immune system [13].

IFN-γ increased mRNA and protein levels of IDO in HT29 and SW837 colorectal cancer cells. EGCG dose-dependently decreased IFN-γ-induced expression of IDO in SW837 cells. Increase in p-STAT1 level induced by IFN-γ was also found to be markedly repressed by EGCG. Data obtained from reporter assays clearly revealed that EGCG inhibited the transcriptional activity of IDO promoter and blocked binding of p-STAT1 to gamma-activated sequence (GAS) sites on the promoters of target genes (Figure 1) [13].

Toxicological analysis of EGCG highlighted its efficacy and minimum off-target effects. Orally administered EGCG mitigated cisplatin-induced hearing loss along with a marked reduction in the loss of outer hair cells in the basal cochlear region. Importantly, chemotherapeutic drug-induced toxicity was also reduced mainly though suppression of apoptotic markers and oxidative stress [14].

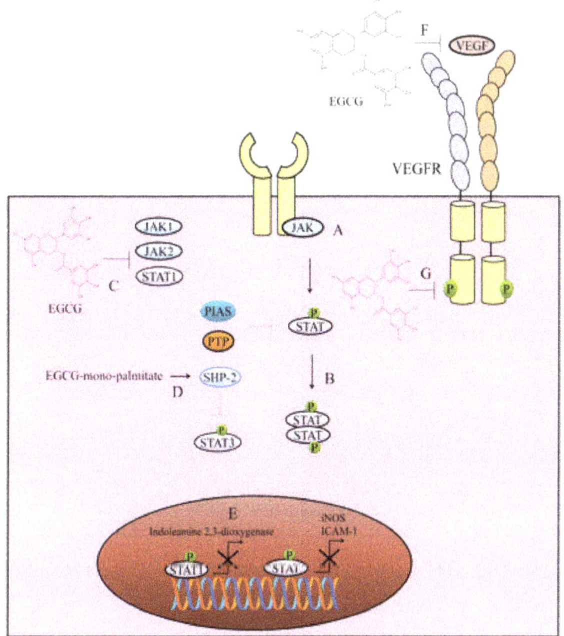

Figure 1. Regulation of the JAK/STAT pathway by epigallocatechin-3-gallate (EGCG). (**A,B**) Janus kinase (JAK) mediated phosphorylation of STAT proteins promoted their accumulation in nucleus to regulate expression of a plethora of genes. (**C–E**) EGCG showcased remarkable ability to shut down the JAK/STAT pathway by inhibition of Janus kinase 1 (JAK1), Janus kinase 2 (JAK2), signal transducer and activator of transcription 1 (STAT1), signal transducer and activator of transcription 3 (STAT3). EGCG also activated negative regulators of STAT-driven signaling. Activation of Src homology 2 domain-containing tyrosine phosphatase-1 (SHP-2) was effective in inhibition of JAK/STAT signaling. Different oncogenes particularly, inducible nitric oxide synthase (iNOS), intercellular adhesion molecule-1 (ICAM-1), and indoleamine 2,3-dioxygenase have been shown to be under direct control of STAT signaling. (**F,G**) Vascular endothelial growth factor vascular endothelial growth factor receptor (VEGF/VEGFR) signaling is also regulated by EGCG. EGCG interacted with VEGF. Additionally, EGCG inhibited phosphorylation of VEGFR.

It has recently been reported that IFNγ-mediated PD-L1 levels were noted to be downregulated after treatment with green tea extracts and EGCG mainly through inhibition of JAK2/STAT1 signaling in A549 cells [15]. Likewise, EGF-stimulated PD-L1 upregulation was reduced in EGCG-treated Lu99 cells by inactivation of EGFR/AKT transduction cascade. Additionally, green tea extracts notably reduced average number of tumors and percentage of PD-L1$^+$ cells in lungs of A/J mice intraperitoneally injected with a cigarette smoke toxin. EGCG reduced mRNA levels of PD-L1 in F10-OVA cells and enhanced expression of interleukin-2 in tumor-specific CD3$^+$ T cells [15]. Collectively these findings suggested that green tea catechin acted as a useful immunological checkpoint inhibitor.

Confluence of information suggested central role of JAK/STAT signaling in different cancers. EGCG mediated inhibition of JAK/STAT signaling via activation of negative regulators (SHP-2) and inactivation of positive regulators (JAK1, JAK2) has gradually gained appreciation. Additionally, different fusion oncoproteins (BCR-ABL) are also exclusively targeted by EGCG.

3. VEGF/VEGFR Signaling

EGCG and silibinin worked synergistically and inhibited vascular endothelial growth factor/vascular endothelial growth factor receptor (VEGF/VEGFR) signaling. EGCG and Silibinin also reduced migratory potential of A549 cells [16]. EGCG interacted with VEGF mainly through hydrophobic interactions and induced a change in the secondary structure of the protein (Figure 1) [17].

Vandetanib (ZD6474), a VEGFR inhibitor was co-loaded with EGCG in mesoporous Silica-Gold nanoclusters for effective targeting of tamoxifen-resistant breast cancer cells [18]. Vandetanib and EGCG effectively reduced phosphorylated levels of EGFR2 and VEGFR2 in drug-resistant breast cancer cells [18]. EGCG also worked with superior efficacy when used in combination with tamoxifen. Tamoxifen worked powerfully with EGCG and reduced the levels of EGFR1, VEGF, and VEGFR1 in breast cancer cells [19]. SU5416 (Semaxanib) also worked remarkably with EGCG and induced apoptosis in malignant neuroblastoma SK-N-BE2 and SH-SY5Y cells [20]. SU5416 and EGCG also inhibited VEGFR2 expression [20].

EGCG dose-dependently decreased levels of VEGFR2 and p-VEGFR2 in HCC and colorectal cancer cells (Figure 1) [21,22]. EGCG induced regression of tumors in mice xenografted with either HuH7 or SW837 cells. EGCG decreased total and phosphorylated levels of VEGFR2 in these xnografts [21,22].

Detailed mechanistic insights revealed that p-STAT1 and p-STAT3 formed complexes with VEGFR1 and VEGFR2 in chronic lymphocytic leukemia (CLL) cells [23]. VEGF induced nuclear accumulation of p-STAT3 in primary CLL B cells. VEGF/VEGFR complex facilitated shuttling of STAT3 from the plasma membrane to perinuclear regions. VEGF induced co-localization of STAT3, VEGFR1 and VEGFR2 to the same perinuclear regions. Collectively these findings provided clear evidence that the VEGF/VEGFR pathway "switched on" STAT proteins which induced resistance against apoptosis. EGCG decreased levels of p-STAT3 [23]. EGCG also remarkably reduced phosphorylated levels of VEGFR1 and VEGFR2 in B-cell chronic lymphocytic leukemia cells [24].

4. Regulation of Methylation-Associated Machinery

PRC2 (Polycomb repressive complex-2) is a transcriptional repressive complex that consists of three essential proteins: EZH2 (enhancer of zeste-2), EED (embryonic ectoderm development), and SUZ12 (suppressor of zeste 12). A series of structural studies have shown that EZH2 context-dependent trimethylates lysine 27 on histone 3 (H3K27) to promote transcriptional inactivation of target genes (shown in Figure 2).

EZH2-mediated trimethylation of H3K27 induced transcriptional repression of TIMP3 (tissue inhibitor of metalloproteinases-3). However, EGCG demonstrated remarkable ability to inhibit EZH2-mediated trimethylation. There was a considerable reduction in the levels of enhancers of zeste homolog 2 (EZH2) and H3K27me3 repressive marks at the promoter region of TIMP-3. Additionally, there was an evident increase in histone H3K9/18 acetylation [25]. Essentially, green tea polyphenols and EGCG treatment significantly reduced class I histone deacetylases (HDAC) activity/expression in prostate cancer cells. Furthermore, levels of EZH2 and H3K27me3 were also found to be reduced in prostate cancer cells [25]. Data clearly suggested that EGCG efficiently demonstrated multi-layered regulation of HDACs and EZH2.

Due to the fundamental role of EZH2 in cancer progression, different inhibitors of EZH2 have been designed and tested for evaluation of efficacy. EGCG and GSK343 (EZH2 inhibitor) exerted inhibitory effects on the proliferation, invasive and migratory potential of the cells, and suppressed EZH2-mediated trimethylation of H3K27 [26].

Recent advancements in the biochemical characterization of polycomb-group (PcG) complexes have revealed a broad range of new proteins, which assemble to form multi-protein complexes. All PRC1 complexes contain Ring1B, which has the E3 ubiquitin ligase activity of the complex. Complexes also include PCGF4/BMI-1 in association with Ring1B to regulate epigenetic modifications [27]. EGCG reduced BMI-1 and EZH2 levels in SCC-13 cells [28].

PML–RARα homodimers worked synchronously with co-repressors and histone deacetylases (HDACs) and consequentially enhanced DNA methylation [29]. EGCG reduced the levels of HDAC1 and PML/RARα in leukemic cells (Figure 2) [30].

Groundbreaking discoveries in biology of epigenome have enabled us to develop a sharp comprehension of highly intricate and well-coordinated interplay of HDACs, histone methyltransferases, and DNA methyltransferases. EGCG has emerged as a master-regulator of epigenetic-associated machinery.

Chromatin immunoprecipitation (ChIP) analyses revealed that EGCG enhanced hyperacetylated H4 and acetylated H3K14 histones within promoter regions of p27, PCAF, C/EBP and reduced binding of PRC2 core component genes EZH2, SUZ12, and EED [31].

EGCG significantly reduced enzymatic activities of DNA methyltransferase (DNMT) and HDAC in HeLa cells [32]. Moreover, EGCG also reduced expression level of DNMT3B whereas expression levels of HDAC1 remained unchanged [32]. GTP/EGCG-promoted acetylation of p53 and enhanced its binding to the promoters of Bax and p21/waf1. Treatment of cells with GTPs and EGCG dose- and time-dependently inhibited class I HDACs [33].

Am80 is a structurally different synthetic retinoid from all-trans-retinoic acid. EGCG and Am80 increased acetylated-p53 and acetylated-α-tubulin through suppression of HDAC activity. Use of specific inhibitors against HDAC4 and HDAC5 strongly induced p21waf1 gene expression. Additionally, HDAC6 inhibition induced upregulation of GADD153 and p21waf1 [34].

UHRF1 (ubiquitin-like containing PHD and Ring finger 1) contributed to inactivation of tumor suppressor genes by directing the binding of DNA methyltransferase 1 (DNMT1) to hemi-methylated promoters [35]. EGCG downregulated DNMT1 and UHRF1 expression and consequently upregulated p73 and p16 (INK4A) in Jurkat cells. UHRF1 downregulation was dependent upon the generation of ROS by EGCG. Upregulation of p16 (INK4A) correlated strongly with reduction in the binding of UHRF1 to the promoter region. UHRF1 overexpression counteracted EGCG-induced apoptosis and upregulation of p73 and p16 (INK4A) [35].

EGCG effectively reduced 5-methylcytosine, DNMT activity, mRNA and protein levels of DNMT1, DNMT3a, and DNMT3b [36]. EGCG decreased HDAC activity and increased levels of acetylated H3K9 and H3K14, H4K5, H4K12, and H4K16 but decreased levels of methylated H3-Lys 9. Collectively, because of inhibition of DNMTs and HDACs, EGCG induced re-expression of p16INK4a and Cip1/p21 [36].

Gazing through a molecular lens clearly highlighted contextual push and pull between various versatile regulators associated with methylation. Substantial fraction of information gathered through high-quality research has unraveled that a broad range of tumor suppressors are epigenetically silenced during cancer progression. Selective targeting of DNMTs and HDACs specifically in cancer cells is very challenging and needs to be comprehensively investigated in EGCG-treated preclinical models. In the upcoming section we will analyze how EGCG modulated deregulated TGF/SMAD signaling.

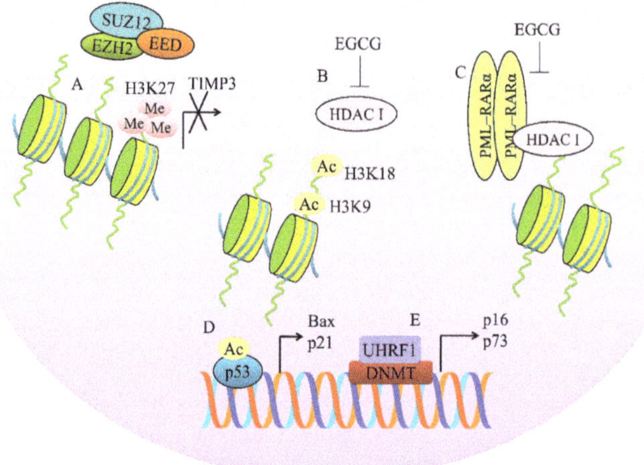

Figure 2. Interconnected and orchestrated interplay among various regulators of epigenetic modifying machinery. (**A**) Protein enhancer of zeste-2 (EZH2), embryonic ectoderm development (EED), and suppressor of zeste 12 (SUZ12) worked synchronously to trimethylate H3K27 and transcriptionally repressed tissue inhibitor of metalloproteinases-3 (TIMP-3). (**B**) Class 1 histone deacetylases (HDACs) were inhibited by EGCG to increase acetylation at H3K9 and H3K18. (**C**) PML–RARα homodimers worked collaboratively with HDAC to regulate expression of target genes. However, EGCG effectively inhibited PML–RARα and HDAC. (**D**) Acetylation of proteins has also been investigated. Acetylated p53 stimulated expression of Bax and p21. (**E**) Ubiquitin-like containing PHD and Ring finger 1 (UHRF1) and DNA methyltransferase (DNMT) also notably downregulated p16 and p73.

5. TGF/SMAD Signaling

Binding of TGFβ superfamily ligands to a type II receptor facilitated closer positioning of type I receptor and phosphorylated it [37]. More importantly, type-I receptor mediated phosphorylation of receptor-regulated SMADs (R-SMADs), which promoted formation of a complex with common mediator SMAD (co-SMAD) (shown in Figure 3). Structural studies had shown that the R-SMAD/co-SMAD complex accumulated in the nucleus to transcriptionally modulate the expression of target genes [38]. Epithelial to mesenchymal transition (EMT) is a highly complex mechanism induced by TGF/SMAD signaling. SMAD2/3 proteins have been shown to stimulate the expression of Snail and Slug in different cancers [39].

In this section, we will provide an overview of multi-layered regulation of TGF/SMAD signaling by EGCG in different cancers. Inhibition of phosphorylation of R-SMADs will inhibit TGF/SMAD signaling. Consequentially, TGF/SMAD signaling inhibition will result in repression of EMT-associated markers.

EGCG effectively reduced p-SMAD3, Snail, and Slug levels in ovarian cancer cells [40]. EGCG considerably suppressed EMT, invasive and migratory capacity of anaplastic thyroid carcinoma (ATC) 8505C cells by regulation of the TGFβ/SMAD pathway [41]. EGCG exerted inhibitory effects on TGFβ1-induced expression of EMT markers (vimentin) in 8505C cells. EGCG was noted to completely block the phosphorylation of SMAD2/3 and nuclear accumulation of SMAD4 [41].

Apart from phosphorylation, acetylation of SMAD proteins is also an intricate mechanism. p300/CBP, a histone acetyltransferase, has been shown to post-translationally modify SMAD proteins. TGFβ1-driven activation of p300/CBP mediated EMT mainly through acetylation of SMAD2 and SMAD3 [42]. EGCG inhibited p300/CBP activity in lung cancer cells. EGCG strongly repressed

TGFβ1-induced EMT and reversed the upregulation of different target genes associated with EMT. EGCG inhibited TGFβ1-mediated activation of p300/CBP. EGCG inhibited TGFβ1-mediated EMT by interfering with the acetylated state of SMAD2 and SMAD3 in lung cancer cells [42].

TGFβ potently induced epithelial–mesenchymal transition (EMT) in NSCLC cells but EGCG reversed TGFβ-induced morphological alterations [43]. EGCG upregulated the expression of E-cadherin and downregulated the expression of vimentin. Data obtained through immunofluorescence also provided clear clues that E-cadherin was upregulated, and vimentin was downregulated by EGCG [43]. Moreover, EGCG effectively inhibited TGFβ-induced migratory and invasive potential of NSCLC cells. EGCG inhibited TGFβ-induced EMT at the transcriptional level. Expectedly, EGCG reduced phosphorylated levels of ERK1/2 (extracellular signal-regulated protein kinases 1/2) and SMAD2 and also inhibited the accumulation of SMAD2 in the nucleus. EGCG repressed the expression of transcriptional factors Slug, Snail, Twist, and ZEB1 and upregulated E-cadherin expression (Figure 3) [43].

Interestingly, different peptide aptamers have been designed to effectively inhibit interaction of SMAD2 and SMAD3 with SMAD4. Therefore, it might be advantageous to combine EGCG with different TGFβ signaling inhibitors to inhibit tumor growth in xenografted mice. More importantly, it will also be exciting to evaluate EGCG-mediated regulation of negative regulators (SMURFs and NEDDs) of the TGF/SMAD pathway.

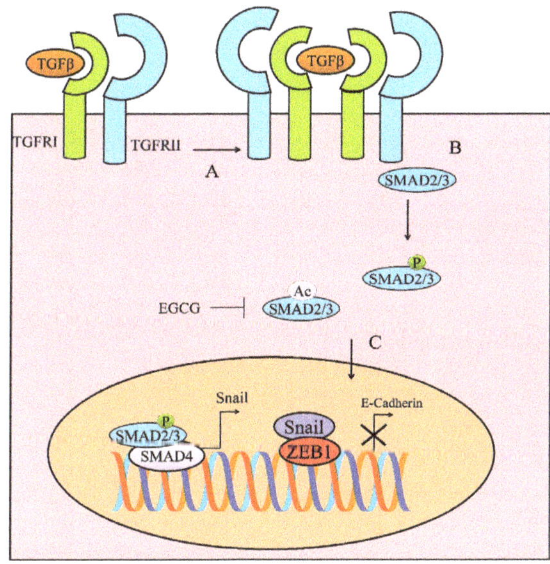

Figure 3. (**A**,**B**) Binding of TGFβ superfamily ligands to a type II receptor induced juxtapositioning of type I receptor. Phosphorylation of SMAD2/3 promoted its accumulation in the nucleus. SMAD2/3 have been shown to stimulate expression of Snail and Slug. Apart from phosphorylation, additional post-translational modifications, particularly acetylation, have also been observed in TGF/SMAD signaling. EGCG inhibited acetylation of SMAD proteins.

6. Regulation of Wnt/β-Catenin Pathway

Detailed mechanistic insights revealed that in the absence of Wnt signal, β-catenin was phosphorylated by APC (adenomatous polyposis coli)/Axin/GSK3β complex and degraded by proteasome [44]. However, activation of the membrane receptors by Wnt signal resulted in the phosphorylation and degradation of GSK3β. EGCG inhibited phosphorylation of GSK3β, upregulated GSK3β expression, and decreased the levels of β-catenin in colorectal cancer cells [44].

O^6-methylguanine (O6-meG) DNA-methyltransferase (MGMT) is a versatile mediator of temozolomide resistance in glioblastomas. TCF/LEF-binding sites within the promoter region of the MGMT gene have previously been identified [45]. Intriguingly, there is evidence of regulation of MGMT by WNT/β-catenin signaling. EGCG not only prevented translocation of β-catenin into the nucleus but also reduced the levels of transcriptional factors TCF1 and LEF1 [46]. Overall these findings clearly suggested that EGCG repressed MGMT expression via inhibition of the β-catenin-driven pathway.

EGCG not only reduced mRNA levels and transcriptional activities of β-catenin in p53 wild-type-expressing KB cells but also promoted ubiquitylation and degradation of β-catenin [47]. EGCG dose-dependently suppressed β-catenin expression in MDA-MB-231 cells [48]. EGCG worked synergistically with gemcitabine and exerted stronger inhibitory effects on β-catenin and N-cadherin in pancreatic cancer cells [49].

Clinical trials of CWP232291 (NCT01398462) and PRI-724 (NCT01302405; NCT01764477) are providing important clinically relevant information and it will be interesting to combine these agents with EGCG for evaluation of robust inhibition of β-catenin-driven signaling and tumor growth inhibitory effects in xenografted mice.

7. Regulation of Notch Pathway

The Notch signaling pathway consists of the Notch receptor, Notch ligand, DNA-binding protein, and Notch regulatory molecules. Notch is a transmembrane protein that mediates communication between neighboring cells. Binding of the ligands to a Notch receptor promoted proteolytic cleavage of NICD (Notch intracellular domain) and its consequential nuclear translocation where it complexed with CSL and formed NICD/CSL transcriptional activation assembly for stimulation of HES and HEY.

EGCG decreased mRNA levels of Notch1, Hey1, and Hes1 [50]. Western blot assay clearly indicated that EGCG dose-dependently reduced protein levels of Notch1 in cancer stem cells (CSCs) of head and neck squamous carcinoma (HNSC). Additionally, EGCG dose-dependently decreased the Notch promoter activity [50].

Tumor growth was significantly reduced in HuCC-T1 tumor-bearing mice subcutaneously injected with EGCG. Notch1 was found to be markedly reduced by EGCG treatment [51].

Expression levels of Hes1 and Notch2 were observed to be considerably reduced in EGCG treated colorectal cancer cells [52]. More importantly, EGCG inhibited Notch1 and cleaved-Notch1 in 5-fluorouracil-resistant colorectal cancer cells [53].

8. Regulation of TRAIL Mediated Apoptosis

Increasingly it is being realized that cancer cells harbor highly complex signaling networks that resist apoptotic programming. High-quality research works related to the TRAIL-mediated pathway in the past two decades have ignited the field of molecular oncology and yielded a stream of preclinical and clinical insights that have reshaped current knowledge of apoptotic cell death.

GCG and TRAIL synergistically reduced Bcl-XL, Bcl-2, and FLIP. Whereas, combinatorial treatment activated capase-8, -9, and -3 in nasopharyngeal carcinoma cells [54].

EGCG and TRAIL also worked effectively against renal cell carcinoma and pancreatic cancer cells [55,56].

EGCG restored sensitivity of HCC cells to TRAIL-induced apoptosis [57]. EGCG upregulated caspase-3 activity and simultaneously downregulated Bcl-2 levels. EGCG also induced upregulation of DR4 and DR5 [57]. EGCG and TRAIL robustly enhanced DR4 levels. Furthermore, FLIP levels were reduced in prostate cancer cells treated in combination with EGCG and TRAIL [58]. Collectively these findings suggested that EGCG might be helpful in increasing the protein levels as well as cell surface expression of death receptors. There is sufficient experimental evidence related to reduction in the cell surface levels of death receptors. Death receptors are internalized and degraded in various cancers. Therefore, EGCG might play its role in stabilizing the levels of death receptors.

PEA15 (phosphoprotein-enriched in astrocytes) is an oncoprotein [59]. It has previously been reported that AKT-induced PEA15 phosphorylation and increased its stability. EGCG downregulated PEA levels mainly through inactivation of AKT. However, overexpression of PEA15 severely impaired apoptotic cell death induced by EGCG and TRAIL [59].

Certain hints had emerged which highlighted that EGCG inhibited TRAIL-induced apoptosis and activated autophagic flux in colon cancer cells. Inhibition of autophagic flux induced death receptor-driven apoptosis in colon cancer cells [60].

These scientific findings are intriguing and future research must converge on identification of additional protein targets in the TRAIL-driven pathway. Essentially, the TRAIL mediated pathway is regulated by long non-coding RNAs as well. Therefore, it will be paramount to unravel underlying mechanisms of TRAIL resistance and identification of proteins, which can be targeted to restore apoptosis in TRAIL-resistant cancers. Keeping in view the fact that TRAIL-based therapeutics and death receptor-targeting agonistic antibodies have entered into various phases of clinical trials, any progress in improving the efficacy of TRAIL-based therapeutics will be advantageous.

9. Regulation of Non-Coding RNAs by EGCG in Different Cancers

Discovery of non-coding RNAs has revolutionized our current understanding of the mechanisms that regulate post-transcriptional processes. The field of non-coding RNA has been extensively explored and researchers have witnessed groundbreaking advancements in disentangling the complicated web ranging from biogenesis of non-coding RNAs to post-transcriptional regulation of a myriad of target mRNAs.

A wealth of information has unveiled a fundamental role of non-coding RNAs in different cancers and researchers have experimentally verified the effects of natural and synthetic products on wide-ranging microRNAs and long non-coding RNAs in different cancers.

9.1. Tumor Suppressor miRNAs

miR-485, a tumor suppressor microRNA, has been found to be frequently downregulated in various cancers. CD44 was directly targeted by miR-485 in A549-cisplatin resistant lung cancer cells. CD44 was overexpressed in A549-cisplatin resistant lung cancer cells but EGCG treatment exerted repressive effects on CD44 levels by enhancing miR-485-mediated targeting of CD44 [61]. EGCG also induced regression of tumors in mice xenografted with A549-cisplatin resistant lung cancer cells.

miR-485-5p directly targeted RXRα in drug-resistant lung cancer cells. EGCG repressed CSC-like properties via modulation of the miR-485-5p/RXRα axis [62]. miR-155 is a tumor suppressor miRNA reportedly involved in enhancing drug sensitivity of cancer cells [63]. EGCG promoted NF-κB mediated upregulation of miR 155. Resultantly, miR 155 enhanced drug sensitivity of colorectal cancer cells by directly targeting MDR1 [63]. IGF2BP1 and IGF2BP3 are direct targets of miR-1275 in different cancers [64]. EGCG stimulated the expression of miR-1275 and potentiated targeting of IGF2BP1 and IGF2BP3 by miR-1275 in HCC cells [65].

9.2. Oncogenic miRNAs

miR-221/222 played a central role in drug resistance. Knockdown of miR-221/222 inhibited proliferation of drug-resistant breast cancer cells [66]

Suberoylanilide hydroxamic acid (SAHA), an HDAC inhibitor worked effectively with EGCG and markedly reduced expression of miR-221/222 in triple-negative breast cancer cells [67].

9.3. Targeting of Oncogenic LncRNAs

SOX2OT variant 7 effectively promoted Notch3/DLL3 signaling in osteosarcoma stem cells (OSCs) [68]. NOTCH target genes HEY1 and HES1 were found to be notably enhanced in variant 7-expressing OSCs. EGCG efficiently inhibited the levels of HEY1 and HES1 in OSCs. However, EGCG mediated inhibitory effects were noted to be impaired in variant 7-expressing cells [68]. EGCG mediated tumor regression was not observed in mice xenografted with variant 7-expressing OSCs. However, EGCG treatment and NOTCH3 knockdown induced reduction in tumor growth in mice inoculated with variant 7-expressing OSCs [68].

EGCG also downregulated lncRNA-AF085935 in HCC cells. It was suggested that lncRNA-AF085935 promoted proliferation of HCC cells [69]. However, the study did not clearly provide a link between lncRNA-AF085935 and its targets and how lncRNA-AF085935 regulated apoptosis and proliferation in HCC cells.

9.4. Tumor Suppressor LncRNAs

EGCG had been shown to induce the expression of cisplatin transporter CTR1 (copper transporter 1) in cancer cells [70]. EGCG upregulated CTR1 and enhanced accumulation of intracellular platinum in NSCLC cells. hsa-miR-98-5p suppressed CTR1, whereas NEAT1 (nuclear enriched abundant transcript 1) enhanced its expression. hsa-miR-98-5p had specific complementary binding sites for NEAT1. Essentially, NEAT1 acted as a competitive endogenous RNA and upregulated EGCG-triggered CTR1 by sponging away hsa-miR-98-5p in NSCLC cells [70].

It seems surprising to note that available scientific proof related to regulation of non-coding RNAs by EGCG is limited. Keeping in view the wealth of information about remarkable pharmacological properties of EGCG, it is paramount to uncover how EGCG modulated different miRNAs, lncRNA, and circular RNAs in different cancers. Identification of the list of tumor suppressor and oncogenic non-coding RNAs regulated by EGCG in different cancers will be highly valuable in combinatorial treatments.

10. Nanotechnological Approaches for Effective Delivery of EGCG to the Target Sites

Despite the ability of EGCG to modulate several cancer-related mechanisms there are still major hurdles for the establishment of EGCG in clinical settings. The therapeutic concentrations of EGCG (between 1 and 10 μM) in the majority of the studies are much higher than the concentrations monitored in the plasma or tissues of animals or in human plasma (usually lower than 1 μM) after tea ingestion. In fact, even after the consumption of 7–9 cups of tea the EGCG concentration in plasma was still lower than 1 μM [71] and for that reason the use of nanotechnology, particularly the development of nanoparticles (NPs) as drug delivery systems, represent a promising approach to increase the bioavailability of EGCG. Nanotechnology corresponds to the science that studies and creates materials with dimensions between 1 and 1000 nm. NPs have at least one of the dimensions in the nanoscale range [72]. There are several types of NPs and for more comprehensive and detailed information the reader can consult the following revisions [73–76]. The different properties of the NPs can be used for medical purposes. Due to their small scale, NPs are excellent drug carriers, and since they can be modified in various factors such as size, chemical composition, outer layer, and others they are very versatile [77]. Furthermore, NPs can modify the pharmacokinetics and the stability of the carrier compound, being, for that reason, a promising strategy to improve EGCG bioavailability profile [78]. Another interesting characteristic of NPs is the possibility to enhance the cellular uptake or even the cellular targeting by modifying the outer layer with different ligands expressed in the target cells to assign particular characteristics in a strategy known by active targeting [79]. This is a useful strategy to improve the bioavailability and stability of EGCG even further, enhancing the utilization options and ultimately enhancing the anti-cancer properties of EGCG. The main types of NPs used for the delivery of EGCG reported in the literature are gold, polymeric, lipid-based, and inorganic NPs (shown in Figure 4). The majority of the NPs are designed to be at the range of approximately 200 nm since this size allows the administration of the NPs by the oral and intravenous routes. Other types of NPs were also used for the encapsulation of EGCG for the purpose of cancer therapy, including carbohydrates, transition metals, and inorganic materials [80–82]. The use of targeting ligands further increased cancer cell specificity and improved the anti-tumor effects of EGCG and, for that reason, folic acid has been used frequently to functionalize the NPs, since the folic acid receptor is overexpressed in tumor cells.

However, other ligands can also be used, including antibodies, carbohydrates, or polysaccharides and other molecules [83]. A summary of the studies using different EGCG nanocarriers for cancer management and carried out in cell lines and in animals is depicted in Table 1.

Table 1. Different types of EGCG nanocarriers for cancer management.

Type of Nanoparticles	Route of Administration	Target Organ	Outcome	Ref.
Gold	Oral, intra-tumoral and intra-peritoneal	Bladder	Tumor volume reduction in a bladder xenograft model	[84]
Gold	Intra-tumoral	Skin	Tumor volume reduction in a melanoma cells in a mouse model	[85]
Gold	N/A	Autonomic nervous system	Induction of apoptosis in neuroblastoma cells	[86]
Gold	N/A	Liver	Toxicity in tumor cells and protection of normal mouse hepatocytes	[87]
Polymeric	N/A	Prostate	Toxicity in prostate cancer cell line	[88]
Polymeric	N/A	Colon and rectum	DNA damage levels in samples of lymphocytes from colorectal cancer patients	[89]
Polymeric	N/A	Breast	Toxicity in breast cancer cell line and patient-derived cells	[90]
Polymeric	Intra-tumoral	Prostate	Tumor size reduction in mice model of prostate cancer	[91]
Polymeric	Oral	Prostate	Tumor size reduction in mice model of prostate cancer	[92]
Polymeric	Oral	Skin	Toxicity in human melanoma cells	[93]
Polymeric	N/A	Stomach and intestine	Anti-tumoral activity in gastrointestinal cancer cell line	[94]
Polymeric	N/A	Breast	Inhibition of breast cancer cell line viability	[95]
Lipid-based	Topic and intra-tumoral	Skin	Accumulation of EGCG in the tissues in a mice model of basal cell carcinoma	[96]
Lipid-based	Intra-tumoral	Skin	Apoptosis in a mice model of basal cell carcinoma	[97]
Lipid-based	N/A	Breast	Anti-proliferative and pro-apoptotic effect in a breast cancer cell line	[98]
Lipid-based	N/A	Breast	Cell apoptosis and cell invasion inhibition in a breast cancer cell line	[99]
Sugar-based	N/A	Prostate	Cell viability inhibition in a prostate cancer cell line	[80]
Inorganic	Intra-tumoral	Liver	Tumor growth reduction in a mouse model of liver cancer	[81]
Inorganic	N/A	Prostate	Anti-tumoral activity in prostate cell line	[82]

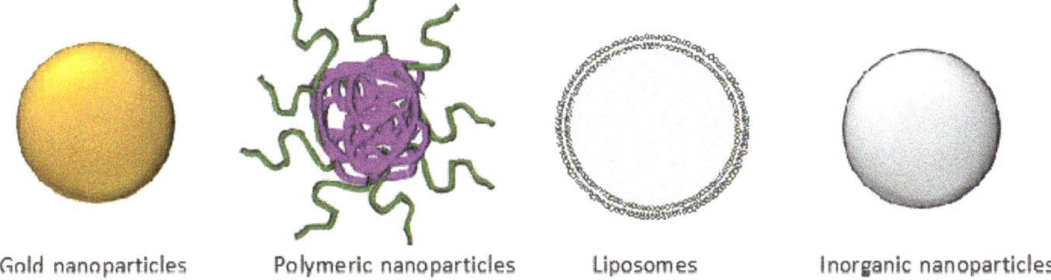

Figure 4. Main types of nanoparticles (NPs) used for the delivery of EGCG.

Gold NPs as EGCG delivery systems have been exploited in several types of cancer since gold has anti-cancer properties per se [85,100]. Several reports have described the in vitro and in vivo efficacy of gold NPs in conjugation with EGCG for cancer treatment, including for the bladder, melanoma, neuroblastoma, and hepatocarcinoma [84–87]. These nanocarriers also demonstrated a high biocompatibility, inducing low damage to human red blood cells and therefore no toxicity for the dose tested was observed. The NPs made of polymers approved and recognized as safe by the US Food and Drug Administration (FDA) are also suitable for cancer applications [89,101]. Several groups have already encapsulated EGCG into different polymeric NPs for cancer therapy, including for the treatment of prostate cancer, colorectal, breast cancer, melanoma, and gastrointestinal cancer [88–95]. Despite the high toxicity towards cancer cells these NPs demonstrated absence of toxicity for normal cells. Liposomes and lipid NPs are lipid-based NPs in composition and for that reason are biodegradable and present minimal levels of toxicity [97]. There are some studies reporting the use of lipid-nanocarriers for the delivery of EGCG to cancer cells [96–99]. All of the studies were used for the treatment of breast cancer with results that demonstrated efficacy and security in vitro and in vivo, including in the MDA-MB-231 cell line, which is a model of the triple-negative cancer and considered more aggressive and associated with poorer outcome than other types of breast cancer.

11. Potential Clinical Applications

EGCG drug delivery systems based in NPs might represent an extraordinary resource to improve the application of EGCG in chemoprevention or to introduce the use of EGCG in the therapy of cancer. The idea of using drug delivery systems, such as NPs for loading EGCG, preserving its structure, and allowing to circumvent the limitations of the low bioavailability associated with the oral administration of free EGCG has a tremendous potentiality since increasing the amount of EGCG inside the cells will potentialize the effect of EGCG in the molecular targets and the effect of deregulated oncogenic signaling cascades and, therefore, determine better cancer outcomes in comparison with free EGCG. For instance, EGCG loaded in polylactic acid–polyethylene glycol NPs preserved the biological activity and efficacy on molecular targets in vitro and in xenograft tumors with over 10-fold dose advantage in comparison with EGCG alone [91]. Indeed, in vitro and in vivo studies are mandatory to verify whether EGCG loaded in NPs maintain EGCG mechanism of action and to understand if the efficacy on molecular targets is at least retained or increased. In view of a safe application, the toxicity of engineered NPs associated with EGCG needs to be fully investigated. For instance, transition metal oxide NPs have been found to increase oxidative stress, disturb calcium homeostasis, and deregulate cell cycle [102]. The activation of the immune system, specifically macrophage activation and cytokine release has been also reported [103]. Thus, lipid-based NPs show higher level of biocompatibility and bioavailability, emerging as the best candidates for pharmaceutical and clinical applications. In this context, EGCG loaded solid lipid NPs as an oral delivery system did not show any toxicity in rats [104]. Different nanoformulations, including EGCG, also showed great biocompatibility with no or very modest toxicity in animal models [105,106]. All these findings encourage the efforts to invest in biocompatible EGCG NPs to be used on humans, as interventional studies in pre-cancerous lesions, including prostate, breast, colon, and Barret's Esophagus [107–110] demonstrated EGCG efficacy despite the poor bioavailability and low plasma concentrations. Therefore, EGCG NPs are expected to improve the chemopreventive effects and to widen the applications in pre-neoplastic lesions, where the results were unclear or incomplete. In addition, EGCG mechanism of action can be improved by the association with anti-cancer drugs already used in cancer treatment since numerous drugs used in cancer therapy, including doxorubicin, 5-flurouracile, cisplatin, paclitaxel, act synergistically with EGCG [111], the best combinations being predictable on the basis of in vitro and in vivo studies. Lastly, active targeting also represents a strategy to preferentially address NPs to cancer cells. Nanomedicine-based therapy is at the beginning, but in the context of cancer chemoprevention and therapy, EGCG NPs might become a powerful strategy over the conventional chemotherapy approach.

Clinical Trials Evaluating EGCG

Given the promising reports from preclinical studies, EGCG has been tested in various clinical studies. Postmenopausal women are at high risk of developing breast cancer and, therefore, EGCG safety clinical trials have been conducted targeting this population. EGCG can afford benefit in terms of regulating LDL-cholesterol as well as glucose and insulin, as reported by a double-blind, randomized, placebo-controlled intervention study in healthy postmenopausal women [112]. A subsequent ancillary study of a double-blind, randomized, placebo-controlled, parallel-arm trial further confirmed the benefit of EGCG but reported the total cholesterol levels reductions only in women with elevated baseline levels [113]. In postmenopausal women, a daily dose of 843 mg EGCG has been reported to be generally well-tolerated with only a small fraction (6.7%) of women reporting adverse events [114]. This dose of 843 mg EGCG, when administered for a year, can reduce mammographic density in relatively younger women (50–55 years) but not in postmenopausal women, as suggested by phase II trial [115]. Not only in breast cancer patients or women at high breast cancer risk, EGCG is well-tolerated by chronic lymphocytic leukemia (CLL) patients as well [116]. Further, EGCG, at a daily dose as low as 44.9 mg for 4 weeks prior to surgery, has been reported to result in increased bioavailability, including accumulations in breast tumor tissue, in early stage breast cancer patients [108].

A randomized trial reported no reduction in likelihood of prostate cancer in men with high-grade prostatic intraepithelial neoplasia, compared to placebo, after a year on 400 mg EGCG dose per day [117]. It is possible that this might be related to the dose tested in this study as a previous study which tested the effects of 800 mg EGCG administered to 26 patients with positive prostate biopsies reported significant reductions in PSA, HGF, and VEGF, with no associated liver toxicity [118]. Similarly, a phase II pharmacodynamic prevention trial in bladder cancer patients indicated a possible reduction in PCNA and clusterin levels upon 2–4 weeks administration of EGCG prior to transurethral resection of bladder tumor or cystectomy [119].

EGCG has been tested in cancer clinical trials not just for the direct anticancer effects, but also for possible effects on co-morbidities. In lung cancer patients with an unresectable stage III disease, a phase I study was conducted to evaluate the efficacy of EGCG against chemotherapy related esophagitis [120]. Patients, divided in six cohorts receiving six different doses of EGCG, were administered EGCG once grade 2 esophagitis occurred. The study reported dramatic regression of esophagitis to grade 0/1 in 22 of 24 patients (91.7% cases), thus underlying the effectiveness of EGCG. On similar lines, a prospective phase II trial confirmed that EGCG can be effective against acute radiation-induced esophagitis as well [121]. Topical administration of EGCG to the radiation field, post-mastectomy and radiotherapy, can resolve radiation dermatitis, as revealed in a phase I study [122].

12. Concluding Remarks

Recent breakthroughs in novel single-cell profiling and spatial transcriptomics have leveraged our understanding to a new level and helped us to find new answers to a critical question of how cancers move through space and time. Importantly, with rapidly increasing sensitivity of detection methods, we also require novel approaches to conceptually analyze single-cell data with observations at the tissue and organ level.

We have developed a near to complete understanding of VEGF/VEGFR signaling pathways. Studies have shown that relative abundance of the cell surface expression of various VEGFRs and their affiliations for specific VEGF ligands play a fundamental role in the initial set of dimeric constellations. Deeper knowledge of this multifaceted signaling web is key to result-oriented therapeutic targeting. Likewise, EGCG mediated targeting of Wnt/β-catenin has been explored and it needs to be tested comprehensively in different types of cancers. Henceforth xenografted mice bearing β-catenin-overexpressing cancer cells will be helpful in uncovering the true potential of EGCG. Likewise, there is a need to unveil if EGCG inhibited β-catenin activation by functionalization of negative regulators of Wnt signaling. Accordingly, TGF/SMAD signaling regulation by EGCG needs to be addressed more conceptually. Inhibition of SMAD phosphorylation by EGCG is a single dimension of this highly intricate mechanism. Available evidence enlightens involvement of SMURFs and NEDDs in inhibition of TGF/SMAD signaling. Therefore, additional key players of TGF/SMAD signaling also need in-depth research. Regulation of Notch signaling by EGCG seems to be sparsely studied. Therefore, we still have incomplete information about targeting of proteolytically cleaved segment of Notch-intracellular domain (NICD) in regulation of the target gene network. Does EGCG inhibit NICD nuclear accumulation or whether it also interferes with repressor/co-repressor and activator/co-activator machinery needs more answers. On a similar note, SHH/Gli pathway regulation by EGCG requires initial cellular studies. Furthermore, Gli-overexpressing cancers have to be treated with EGCG and combinatorial treatments.

Despite the absence of clinical trials, the NPs loaded with EGCG might be an efficient and safe strategy for the treatment of several cancers, especially breast and prostate cancer. Thus, clinical trials should be conducted to establish the clinical potential of the NPs loaded with EGCG alone or in addition with the conventional anti-cancer drugs.

Author Contributions: Conceptualization, A.A.F. and M.P.; Writing Original Draft Preparation, A.A.F., M.P., A.G., F.F., S.R., R.A., U.Y.S., B.X., A.A.; Writing Review & Editing, A.A.F., M.P., A.G., F.F., S.R., R.A., U.Y.S., B.X., A.A.; Visualization, A.A.F., M.P., A.G., F.F., S.R., R.A., U.Y.S., B.X., A.A.; Supervision, A.A.F. and M.P.; Funding and Acquisition. M.P. and S.R. All authors have read and agreed to the published version of the manuscript.

Acknowledgments: This work received financial support from the European Union (FEDER funds through COMPETE POCI-01-0145-FEDER-30624) and National Funds (FCT, Fundação para a Ciência e Tecnologia) through project PTDC/BTM-MAT/30624/2017. Marina Pinheiro thanks FCT for funding through program DL 57/2016—Norma transitória. Andreia Granja thanks FCT and POPH (Programa Operacional Potencial Humano) for the PhD grant (SFRH/BD/130147/2017). This work was also supported by FCT through the FCT PhD Programs, specifically by the BiotechHealth Program (Doctoral Program on Cellular and Molecular Biotechnology Applied to Health Sciences).

Conflicts of Interest: The authors declare no conflict of interest.

References

1. Yang, C.S.; Wang, X.; Lu, G.; Picinich, S.C. Cancer prevention by tea: Animal studies, molecular mechanisms and human relevance. *Nat. Rev. Cancer* **2009**, *9*, 429–439. [CrossRef] [PubMed]

2. Tauber, A.L.; Schweiker, S.S.; Levonis, S.M. From tea to treatment; epigallocatechin gallate and its potential involvement in minimizing the metabolic changes in cancer. *Nutr. Res.* **2020**, *74*, 23–36. [CrossRef] [PubMed]

3. Saeki, K.; Hayakawa, S.; Nakano, S.; Ito, S.; Oishi, Y.; Suzuki, Y.; Isemura, M. In Vitro and In Silico Studies of the Molecular Interactions of Epigallocatechin-3-O-gallate (EGCG) with Proteins That Explain the Health Benefits of Green Tea. *Molecules* **2018**, *23*, 1295. [CrossRef] [PubMed]

4. Negri, A.; Naponelli, V.; Rizzi, F.; Bettuzzi, S. Molecular Targets of Epigallocatechin-Gallate (EGCG): A Special Focus on Signal Transduction and Cancer. *Nutrients* **2018**, *10*, 1936. [CrossRef] [PubMed]

5. Cheng, C.W.; Shieh, P.C.; Lin, Y.C.; Chen, Y.J.; Lin, Y.H.; Kuo, D.H.; Liu, J.Y.; Kao, J.Y.; Kao, M.C.; Way, T.D. Indoleamine 2,3-dioxygenase, an immunomodulatory protein, is suppressed by (-)-epigallocatechin-3-gallate via blocking of gamma-interferon-induced JAK-PKC-delta-STAT1 signaling in human oral cancer cells. *J. Agric. Food Chem.* **2010**, *58*, 887–894. [CrossRef]

6. Jung, J.H.; Yun, M.; Choo, E.J.; Kim, S.H.; Jeong, M.S.; Jung, D.B.; Lee, H.; Kim, E.O.; Kato, N.; Kim, B.; et al. A derivative of epigallocatechin-3-gallate induces apoptosis via SHP-1-mediated suppression of BCR-ABL and STAT3 signalling in chronic myelogenous leukaemia. *Br. J. Pharmacol.* **2015**, *172*, 3565–3578. [CrossRef]

7. Xiao, X.; Jiang, K.; Xu, Y.; Peng, H.; Wang, Z.; Liu, S.; Zhang, G. (-)-Epigallocatechin-3-gallate induces cell apoptosis in chronic myeloid leukaemia by regulating Bcr/Abl-mediated p38-MAPK/JNK and JAK2/STAT3/AKT signalling pathways. *Clin. Exp. Pharmacol. Physiol.* **2019**, *46*, 126–136. [CrossRef]

8. Jin, G.; Yang, Y.; Liu, K.; Zhao, J.; Chen, X.; Liu, H.; Bai, R.; Li, X.; Jiang, Y.; Zhang, X.; et al. Combination curcumin and (-)-epigallocatechin-3-gallate inhibits colorectal carcinoma microenvironment-induced angiogenesis by JAK/STAT3/IL-8 pathway. *Oncogenesis* **2017**, *6*, e384. [CrossRef]

9. Yuan, C.H.; Horng, C.T.; Lee, C.F.; Chiang, N.N.; Tsai, F.J.; Lu, C.C.; Chiang, J.H.; Hsu, Y.M.; Yang, J.S.; Chen, F.A. Epigallocatechin gallate sensitizes cisplatin-resistant oral cancer CAR cell apoptosis and autophagy through stimulating AKT/STAT3 pathway and suppressing multidrug resistance 1 signaling. *Environ. Toxicol.* **2017**, *32*, 845–855. [CrossRef]

10. Huang, R.; Faratian, D.; Sims, A.H.; Wilson, D.; Thomas, J.S.; Harrison, D.J.; Langdon, S.P. Increased STAT1 signaling in endocrine-resistant breast cancer. *PLoS ONE* **2014**, *9*, e94226. [CrossRef]

11. Tian, M.; Tian, D.; Qiao, X.; Li, J.; Zhang, L. Modulation of Myb-induced NF-kB-STAT3 signaling and resulting cisplatin resistance in ovarian cancer by dietary factors. *J. Cell. Physiol.* **2019**, *234*, 21126–21134. [CrossRef] [PubMed]

12. Senggunprai, L.; Kukongviriyapan, V.; Prawan, A.; Kukongviriyapan, U. Quercetin and EGCG exhibit chemopreventive effects in cholangiocarcinoma cells via suppression of JAK/STAT signaling pathway. *Phytother. Res. PTR* **2014**, *28*, 841–848. [CrossRef] [PubMed]

13. Ogawa, K.; Hara, T.; Shimizu, M.; Nagano, J.; Ohno, T.; Hoshi, M.; Ito, H.; Tsurumi, H.; Saito, K.; Seishima, M.; et al. (-)-Epigallocatechin gallate inhibits the expression of indoleamine 2,3-dioxygenase in human colorectal cancer cells. *Oncol. Lett.* **2012**, *4*, 546–550. [CrossRef] [PubMed]

14. Borse, V.; Al Aameri, R.F.H.; Sheehan, K.; Sheth, S.; Kaur, T.; Mukherjea, D.; Tupal, S.; Lowy, M.; Ghosh, S.; Dhukhwa, A.; et al. Epigallocatechin-3-gallate, a prototypic chemopreventative agent for protection against cisplatin-based ototoxicity. *Cell Death Dis.* **2017**, *8*, e2921. [CrossRef] [PubMed]

15. Rawangkan, A.; Wongsirisin, P.; Namiki, K.; Iida, K.; Kobayashi, Y.; Shimizu, Y.; Fujiki, H.; Suganuma, M. Green Tea Catechin Is an Alternative Immune Checkpoint Inhibitor that Inhibits PD-L1 Expression and Lung Tumor Growth. *Molecules* **2018**, *23*, 2071. [CrossRef] [PubMed]

16. Mirzaaghaei, S.; Foroughmand, A.M.; Saki, G.; Shafiei, M. Combination of Epigallocatechin-3-gallate and Silibinin: A Novel Approach for Targeting Both Tumor and Endothelial Cells. *ACS Omega* **2019**, *4*, 8421–8430. [CrossRef]

17. Perez-Moral, N.; Needs, P.W.; Moyle, C.W.A.; Kroon, P.A. Hydrophobic Interactions Drive Binding between Vascular Endothelial Growth Factor-A (VEGFA) and Polypheolic Inhibitors. *Molecules* **2019**, *24*, 2785. [CrossRef]

18. Kumar, B.N.P.; Puvvada, N.; Rajput, S.; Sarkar, S.; Mahto, M.K.; Yallapu, M.M.; Pathak, A.; Emdad, L.; Das, S.K.; Reis, R.L.; et al. Targeting of EGFR, VEGFR2, and Akt by Engineered Dual Drug Encapsulated Mesoporous Silica-Gold Nanoclusters Sensitizes Tamoxifen-Resistant Breast Cancer. *Mol. Pharm.* **2018**, *15*, 2698–2713. [CrossRef]

19. Scandlyn, M.; Stuart, E.; Somers-Edgar, T.; Menzies, A.; Rosengren, R. A new role for tamoxifen in oestrogen receptor-negative breast cancer when it is combined with epigallocatechin gallate. *Brit. J. Cancer* **2008**, *99*, 1056–1063. [CrossRef]

20. Mohan, N.; Karmakar, S.; Banik, N.L.; Ray, S.K. SU5416 and EGCG work synergistically and inhibit angiogenic and survival factors and induce cell cycle arrest to promote apoptosis in human malignant neuroblastoma SH SY5Y and SK N BE2 cells. *Neurochem. Res.* **2011**, *36*, 1383–1396. [CrossRef]

21. Shirakami, Y.; Shimizu, M.; Adachi, S.; Sakai, H.; Nakagawa, T.; Yasuda, Y.; Tsurumi, H.; Hara, Y.; Moriwaki, H. (-)-Epigallocatechin gallate suppresses the growth of human hepatocellular carcinoma cells by inhibiting activation of the vascular endothelial growth factor-vascular endothelial growth factor receptor axis. *Cancer Sci.* **2009**, *100*, 1957–1962. [CrossRef] [PubMed]

22. Shimizu, M.; Shirakami, Y.; Sakai, H.; Yasuda, Y.; Kubota, M.; Adachi, S.; Tsurumi, H.; Hara, Y.; Moriwaki, H. (-)-Epigallocatechin gallate inhibits growth and activation of the VEGF/VEGFR axis in human colorectal cancer cells. *Chem. Biol. Interact.* **2010**, *185*, 247–252. [CrossRef] [PubMed]

23. Lee, Y.K.; Shanafelt, T.D.; Bone, N.D.; Strege, A.K.; Jelinek, D.F.; Kay, N.E. VEGF receptors on chronic lymphocytic leukemia (CLL) B cells interact with STAT 1 and 3: Implication for apoptosis resistance. *Leukemia* **2005**, *19*, 513–523. [CrossRef] [PubMed]

24. Lee, Y.K.; Bone, N.D.; Strege, A.K.; Shanafelt, T.D.; Jelinek, D.F.; Kay, N.E. VEGF receptor phosphorylation status and apoptosis is modulated by a green tea component, epigallocatechin-3-gallate (EGCG), in B-cell chronic lymphocytic leukemia. *Blood* **2004**, *104*, 788–794. [CrossRef]

25. Deb, G.; Shankar, E.; Thakur, V.S.; Ponsky, L.E.; Bodner, D.R.; Fu, P.; Gupta, S. Green tea-induced epigenetic reactivation of tissue inhibitor of matrix metalloproteinase-3 suppresses prostate cancer progression through histone-modifying enzymes. *Mol. Carcinog.* **2019**, *58*, 1194–1207. [CrossRef]

26. Ying, L.; Yan, F.; Williams, B.R.; Xu, P.; Li, X.; Zhao, Y.; Hu, Y.; Wang, Y.; Xu, D.; Dai, J. (-)-Epigallocatechin-3-gallate and EZH2 inhibitor GSK343 have similar inhibitory effects and mechanisms of action on colorectal cancer cells. *Clin. Exp. Pharmacol. Physiol.* **2018**, *45*, 58–67. [CrossRef]

27. Schwartz, Y.B.; Pirrotta, V. A new world of Polycombs: Unexpected partnerships and emerging functions. *Nat. Rev. Genet.* **2013**, *14*, 853–864. [CrossRef]

28. Balasubramanian, S.; Adhikary, G.; Eckert, R.L. The Bmi-1 polycomb protein antagonizes the (-)-epigallocatechin-3-gallate-dependent suppression of skin cancer cell survival. *Carcinogenesis* **2010**, *31*, 496–503. [CrossRef]

29. De The, H.; Chen, Z. Acute promyelocytic leukaemia: Novel insights into the mechanisms of cure. *Nat. Rev. Cancer* **2010**, *10*, 775–783. [CrossRef]

30. Moradzadeh, M.; Roustazadeh, A.; Tabarraei, A.; Erfanian, S.; Sahebkar, A. Epigallocatechin-3-gallate enhances differentiation of acute promyelocytic leukemia cells via inhibition of PML-RARalpha and HDAC1. *Phytother. Res. PTR* **2018**, *32*, 471–479. [CrossRef]

31. Borutinskaite, V.; Virksaite, A.; Gudelyte, G.; Navakauskiene, R. Green tea polyphenol EGCG causes anti-cancerous epigenetic modulations in acute promyelocytic leukemia cells. *Leuk. Lymphoma* **2018**, *59*, 469–478. [CrossRef] [PubMed]

32. Khan, M.A.; Hussain, A.; Sundaram, M.K.; Alalami, U.; Gunasekera, D.; Ramesh, L.; Hamza, A.; Quraishi, U. (-)-Epigallocatechin-3-gallate reverses the expression of various tumor-suppressor genes by inhibiting DNA methyltransferases and histone deacetylases in human cervical cancer cells. *Oncol. Rep.* **2015**, *33*, 1976–1984. [CrossRef] [PubMed]

33. Thakur, V.S.; Gupta, K.; Gupta, S. Green tea polyphenols increase p53 transcriptional activity and acetylation by suppressing class I histone deacetylases. *Int. J. Oncol.* **2012**, *41*, 353–361. [CrossRef] [PubMed]

34. Oya, Y.; Mondal, A.; Rawangkan, A.; Umsumarng, S.; Iida, K.; Watanabe, T.; Kanno, M.; Suzuki, K.; Li, Z.; Kagechika, H.; et al. Down-regulation of histone deacetylase 4, -5 and -6 as a mechanism of synergistic enhancement of apoptosis in human lung cancer cells treated with the combination of a synthetic retinoid, Am80 and green tea catechin. *J. Nutr. Biochem.* **2017**, *42*, 7–16. [CrossRef]

35. Achour, M.; Mousli, M.; Alhosin, M.; Ibrahim, A.; Peluso, J.; Muller, C.D.; Schini-Kerth, V.B.; Hamiche, A.; Dhe-Paganon, S.; Bronner, C. Epigallocatechin-3-gallate up-regulates tumor suppressor gene expression via a reactive oxygen species-dependent down-regulation of UHRF1. *Biochem. Biophys. Res. Commun.* **2013**, *430*, 208–212. [CrossRef]

36. Nandakumar, V.; Vaid, M.; Katiyar, S.K. (-)-Epigallocatechin-3-gallate reactivates silenced tumor suppressor genes, Cip1/p21 and p16INK4a, by reducing DNA methylation and increasing histones acetylation in human skin cancer cells. *Carcinogenesis* **2011**, *32*, 537–544. [CrossRef]

37. Massague, J. How cells read TGF-beta signals. *Nat. Rev. Mol. Cell Biol.* **2000**, *1*, 169–178. [CrossRef]

38. Ikushima, H.; Miyazono, K. TGFbeta signalling: A complex web in cancer progression. *Nat. Rev. Cancer* **2010**, *10*, 415–424. [CrossRef]

39. Brandl, M.; Seidler, B.; Haller, F.; Adamski, J.; Schmid, R.M.; Saur, D.; Schneider, G. IKK(alpha) controls canonical TGF(ss)-SMAD signaling to regulate genes expressing SNAIL and SLUG during EMT in panc1 cells. *J. Cell Sci.* **2010**, *123*, 4231–4239. [CrossRef]

40. Sicard, A.A.; Dao, T.; Suarez, N.G.; Annabi, B. Diet-Derived Gallated Catechins Prevent TGF-beta-Mediated Epithelial-Mesenchymal Transition, Cell Migration and Vasculogenic Mimicry in Chemosensitive ES-2 Ovarian Cancer Cells. *Nutr. Cancer* **2020**, 1–12. [CrossRef]

41. Li, T.; Zhao, N.; Lu, J.; Zhu, Q.; Liu, X.; Hao, F.; Jiao, X. Epigallocatechin gallate (EGCG) suppresses epithelial-Mesenchymal transition (EMT) and invasion in anaplastic thyroid carcinoma cells through blocking of TGF-beta1/Smad signaling pathways. *Bioengineered* **2019**, *10*, 282–291. [CrossRef] [PubMed]

42. Ko, H.; So, Y.; Jeon, H.; Jeong, M.H.; Choi, H.K.; Ryu, S.H.; Lee, S.W.; Yoon, H.G.; Choi, K.C. TGF-beta1-induced epithelial-mesenchymal transition and acetylation of Smad2 and Smad3 are negatively regulated by EGCG in human A549 lung cancer cells. *Cancer Lett.* **2013**, *335*, 205–213. [CrossRef] [PubMed]

43. Liu, L.C.; Tsao, T.C.; Hsu, S.R.; Wang, H.C.; Tsai, T.C.; Kao, J.Y.; Way, T.D. EGCG inhibits transforming growth factor-beta-mediated epithelial-to-mesenchymal transition via the inhibition of Smad2 and Erk1/2 signaling pathways in nonsmall cell lung cancer cells. *J. Agric. Food Chem.* **2012**, *60*, 9863–9873. [CrossRef] [PubMed]

44. Chen, Y.; Wang, X.Q.; Zhang, Q.; Zhu, J.Y.; Li, Y.; Xie, C.F.; Li, X.T.; Wu, J.S.; Geng, S.S.; Zhong, C.Y.; et al. (-)-Epigallocatechin-3-Gallate Inhibits Colorectal Cancer Stem Cells by Suppressing Wnt/beta-Catenin Pathway. *Nutrients* **2017**, *9*, 572. [CrossRef]

45. Wickstrom, M.; Dyberg, C.; Milosevic, J.; Einvik, C.; Calero, R.; Sveinbjornsson, B.; Sanden, E.; Darabi, A.; Siesjo, P.; Kool, M.; et al. Wnt/beta-catenin pathway regulates MGMT gene expression in cancer and inhibition of Wnt signalling prevents chemoresistance. *Nat. Commun.* **2015**, *6*, 8904. [CrossRef]

46. Xie, C.R.; You, C.G.; Zhang, N.; Sheng, H.S.; Zheng, X.S. Epigallocatechin Gallate Preferentially Inhibits O(6)-Methylguanine DNA-Methyltransferase Expression in Glioblastoma Cells Rather than in Nontumor Glial Cells. *Nutr. Cancer* **2018**, *70*, 1339–1347. [CrossRef]

47. Shin, Y.S.; Kang, S.U.; Park, J.K.; Kim, Y.E.; Kim, Y.S.; Baek, S.J.; Lee, S.H.; Kim, C.H. Anti-cancer effect of (-)-epigallocatechin-3-gallate (EGCG) in head and neck cancer through repression of transactivation and enhanced degradation of beta-catenin. *Phytomedicine* **2016**, *23*, 1344–1355. [CrossRef]

48. Hong, O.-Y.; Noh, E.-M.; Jang, H.-Y.; Lee, Y.-R.; Lee, B.K.; Jung, S.H.; Kim, J.-S.; Youn, H.J. Epigallocatechin gallate inhibits the growth of MDA-MB-231 breast cancer cells via inactivation of the β-catenin signaling pathway. *Oncol. Lett.* **2017**, *14*, 441–446. [CrossRef]

49. Wei, R.; Penso, N.E.C.; Hackman, R.M.; Wang, Y.; Mackenzie, G.G. Epigallocatechin-3-Gallate (EGCG) Suppresses Pancreatic Cancer Cell Growth, Invasion, and Migration partly through the Inhibition of Akt Pathway and Epithelial-Mesenchymal Transition: Enhanced Efficacy when Combined with Gemcitabine. *Nutrients* **2019**, *11*, 1856. [CrossRef]

50. Lee, S.H.; Nam, H.J.; Kang, H.J.; Kwon, H.W.; Lim, Y.C. Epigallocatechin-3-gallate attenuates head and neck cancer stem cell traits through suppression of Notch pathway. *Eur. J. Cancer* **2013**, *49*, 3210–3218. [CrossRef]

51. Kwak, T.W.; Park, S.B.; Kim, H.-J.; Jeong, Y.-I.; Kang, D.H. Anticancer activities of epigallocatechin-3-gallate against cholangiocarcinoma cells. *Oncol. Targets Ther.* **2016**, *10*, 137–144. [CrossRef] [PubMed]

52. Jin, H.; Gong, W.; Zhang, C.; Wang, S. Epigallocatechin gallate inhibits the proliferation of colorectal cancer cells by regulating Notch signaling. *Oncol. Targets Ther.* **2013**, *6*, 145–153. [CrossRef] [PubMed]

53. Toden, S.; Tran, H.M.; Tovar-Camargo, O.A.; Okugawa, Y.; Goel, A. Epigallocatechin-3-gallate targets cancer stem-like cells and enhances 5-fluorouracil chemosensitivity in colorectal cancer. *Oncotarget* **2016**. [CrossRef] [PubMed]

54. Li, P.; Li, S.; Yin, D.; Li, J.; Wang, L.; Huang, C.; Yang, X. EGCG sensitizes human nasopharyngeal carcinoma cells to TRAIL-mediated apoptosis by activation NF-kappaB. *Neoplasma* **2017**, *64*, 74–80. [CrossRef]

55. Wei, R.; Zhu, G.; Jia, N.; Yang, W. Epigallocatechin-3-gallate Sensitizes Human 786-O Renal Cell Carcinoma Cells to TRAIL-Induced Apoptosis. *Cell Biochem. Biophys.* **2015**, *72*, 157–164. [CrossRef]

56. Basu, A.; Haldar, S. Combinatorial effect of epigallocatechin-3-gallate and TRAIL on pancreatic cancer cell death. *Int. J. Oncol.* **2009**, *34*, 281–286. [CrossRef]

57. Abou El Naga, R.N.; Azab, S.S.; El-Demerdash, E.; Shaarawy, S.; El-Merzabani, M.; Ammar el, S.M. Sensitization of TRAIL-induced apoptosis in human hepatocellular carcinoma HepG2 cells by phytochemicals. *Life Sci.* **2013**, *92*, 555–561. [CrossRef]

58. Siddiqui, I.A.; Malik, A.; Adhami, V.M.; Asim, M.; Hafeez, B.B.; Sarfaraz, S.; Mukhtar, H. Green tea polyphenol EGCG sensitizes human prostate carcinoma LNCaP cells to TRAIL-mediated apoptosis and synergistically inhibits biomarkers associated with angiogenesis and metastasis. *Oncogene* **2008**, *27*, 2055–2063. [CrossRef]

59. Siegelin, M.D.; Habel, A.; Gaiser, T. Epigallocatechin-3-gallate (EGCG) downregulates PEA15 and thereby augments TRAIL-mediated apoptosis in malignant glioma. *Neurosci. Lett.* **2008**, *448*, 161–165. [CrossRef]

60. Kim, S.W.; Moon, J.H.; Park, S.Y. Activation of autophagic flux by epigallocatechin gallate mitigates TRAIL-induced tumor cell apoptosis via down-regulation of death receptors. *Oncotarget* **2016**, *7*, 65660–65668. [CrossRef]

61. Jiang, P.; Xu, C.; Chen, L.; Chen, A.; Wu, X.; Zhou, M.; Haq, I.U.; Mariyam, Z.; Feng, Q. EGCG inhibits CSC-like properties through targeting miR-485/CD44 axis in A549-cisplatin resistant cells. *Mol. Carcinog.* **2018**, *57*, 1835–1844. [CrossRef] [PubMed]

62. Jiang, P.; Xu, C.; Chen, L.; Chen, A.; Wu, X.; Zhou, M.; Haq, I.U.; Mariyam, Z.; Feng, Q. Epigallocatechin-3-gallate inhibited cancer stem cell-like properties by targeting hsa-mir-485-5p/RXRalpha in lung cancer. *J. Cell Biochem.* **2018**, *119*, 8623–8635. [CrossRef] [PubMed]

63. La, X.; Zhang, L.; Li, Z.; Li, H.; Yang, Y. (-)-Epigallocatechin Gallate (EGCG) Enhances the Sensitivity of Colorectal Cancer Cells to 5-FU by Inhibiting GRP78/NF-kappaB/miR-155-5p/MDR1 Pathway. *J. Agric. Food Chem.* **2019**, *67*, 2510–2518. [CrossRef] [PubMed]

64. Fawzy, I.O.; Hamza, M.T.; Hosny, K.A.; Esmat, G.; El Tayebi, H.M.; Abdelaziz, A.I. miR-1275: A single microRNA that targets the three IGF2-mRNA-binding proteins hindering tumor growth in hepatocellular carcinoma. *FEBS Lett.* **2015**, *589*, 2257–2265. [CrossRef]

65. Shaalan, Y.M.; Handoussa, H.; Youness, R.A.; Assal, R.A.; El-Khatib, A.H.; Linscheid, M.W.; El Tayebi, H.M.; Abdelaziz, A.I. Destabilizing the interplay between miR-1275 and IGF2BPs by Tamarix articulata and quercetin in hepatocellular carcinoma. *Nat. Prod. Res.* **2018**, *32*, 2217–2220. [CrossRef]

66. Rao, X.; Di Leva, G.; Li, M.; Fang, F.; Devlin, C.; Hartman-Frey, C.; Burow, M.E.; Ivan, M.; Croce, C.M.; Nephew, K.P. MicroRNA-221/222 confers breast cancer fulvestrant resistance by regulating multiple signaling pathways. *Oncogene* **2011**, *30*, 1082–1097. [CrossRef]

67. Lewis, K.A.; Jordan, H.R.; Tollefsbol, T.O. Effects of SAHA and EGCG on Growth Potentiation of Triple-Negative Breast Cancer Cells. *Cancers* **2018**, *11*, 23. [CrossRef]

68. Wang, W.; Chen, D.; Zhu, K. SOX2OT variant 7 contributes to the synergistic interaction between EGCG and Doxorubicin to kill osteosarcoma via autophagy and stemness inhibition. *J. Exp. Clin. Cancer Res.* **2018**, *37*, 37. [CrossRef]

69. Sabry, D.; Abdelaleem, O.O.; El Amin Ali, A.M.; Mohammed, R.A.; Abdel-Hameed, N.D.; Hassouna, A.; Khalifa, W.A. Anti-proliferative and anti-apoptotic potential effects of epigallocatechin-3-gallate and/or metformin on hepatocellular carcinoma cells: In vitro study. *Mol. Biol. Rep.* **2019**, *46*, 2039–2047. [CrossRef]

70. Jiang, P.; Wu, X.; Wang, X.; Huang, W.; Feng, Q. NEAT1 upregulates EGCG-induced CTR1 to enhance cisplatin sensitivity in lung cancer cells. *Oncotarget* **2016**, *7*, 43337–43351. [CrossRef]

71. Peter, B.; Bosze, S.; Horvath, R. Biophysical characteristics of proteins and living cells exposed to the green tea polyphenol epigallocatechin-3-gallate (EGCg): Review of recent advances from molecular mechanisms to nanomedicine and clinical trials. *Eur. Biophys. J. EBJ* **2017**, *46*, 1–24. [CrossRef] [PubMed]

72. Nikalje, A.P. Nanotechnology and its applications in medicine. *Med. Chem.* **2015**, *5*, 081–089. [CrossRef]

73. Bhatia, S. Nanoparticles types, classification, characterization, fabrication methods and drug delivery applications. In *Natural Polymer Drug Delivery Systems*; Springer: Cham, Switzerland, 2016; pp. 33–93.

74. Senapati, S.; Mahanta, A.K.; Kumar, S.; Maiti, P. Controlled drug delivery vehicles for cancer treatment and their performance. *Signal. Transduct. Target. Ther.* **2018**, *3*, 7. [CrossRef] [PubMed]

75. Liu, W.T. Nanoparticles and their biological and environmental applications. *J. Biosci. Bioeng.* **2006**, *102*, 1–7. [CrossRef] [PubMed]

76. Yih, T.C.; Al-Fandi, M. Engineered nanoparticles as precise drug delivery systems. *J. Cell Biochem.* **2006**, *97*, 1184–1190. [CrossRef] [PubMed]

77. Sanvicens, N.; Marco, M.P. Multifunctional nanoparticles—Properties and prospects for their use in human medicine. *Trends Biotechnol.* **2008**, *26*, 425–433. [CrossRef]

78. Granja, A.; Frias, I.; Neves, A.R.; Pinheiro, M.; Reis, S. Therapeutic Potential of Epigallocatechin Gallate Nanodelivery Systems. *BioMed Res. Int.* **2017**, *2017*, 5813793. [CrossRef]

79. Cho, K.; Wang, X.U.; Nie, S.; Shin, D.M. Therapeutic nanoparticles for drug delivery in cancer. *Clin. Cancer Res.* **2008**, *14*, 1310–1316. [CrossRef]

80. Rocha, S.; Generalov, R.; Peres, I.; Juzenas, P. Epigallocatechin gallate-loaded polysaccharide nanoparticles for prostate cancer chemoprevention. *Nanomedicine* **2011**, *6*, 79–87. [CrossRef]

81. Zhou, Y.; Yu, Q.; Qin, X.; Bhavsar, D.; Yang, L.; Chen, Q.; Zheng, W.; Chen, L.; Liu, J. Improving the anticancer efficacy of laminin receptor-specific therapeutic ruthenium nanoparticles (RuBB-loaded EGCG-RuNPs) via ROS-dependent apoptosis in SMMC-7721 cells. *ACS Appl. Mater. Interf.* **2016**, *8*, 15000–15012. [CrossRef]

82. Shafiei, S.S.; Solati-Hashjin, M.; Samadikuchaksaraei, A.; Kalantarinejad, R.; Asadi-Eydivand, M.; Abu Osman, N.A. Epigallocatechin Gallate/Layered Double Hydroxide Nanohybrids: Preparation, Characterization, and In Vitro Anti-Tumor Study. *PLoS ONE* **2015**, *10*, e0136530. [CrossRef] [PubMed]

83. Muhamad, N.; Plengsuriyakarn, T.; Na-Bangchang, K. Application of active targeting nanoparticle delivery system for chemotherapeutic drugs and traditional/herbal medicines in cancer therapy: A systematic review. *Int. J. Nanomed.* **2018**, *13*, 3921–3935. [CrossRef] [PubMed]

84. Hsieh, D.-S.; Wang, H.; Tan, S.-W.; Huang, Y.-H.; Tsai, C.-Y.; Yeh, M.-K.; Wu, C.-J. The treatment of bladder cancer in a mouse model by epigallocatechin-3-gallate-gold nanoparticles. *Biomaterials* **2011**, *32*, 7633–7640. [CrossRef] [PubMed]

85. Chen, C.C.; Hsieh, D.S.; Huang, K.J.; Chan, Y.L.; Hong, P.D.; Yeh, M.K.; Wu, C.J. Improving anticancer efficacy of (-)-epigallocatechin-3-gallate gold nanoparticles in murine B16F10 melanoma cells. *Drug Des. Dev. Ther.* **2014**, *8*, 459–473. [CrossRef]

86. Sanna, V.; Pala, N.; Dessi, G.; Manconi, P.; Mariani, A.; Dedola, S.; Rassu, M.; Crosio, C.; Iaccarino, C.; Sechi, M. Single-step green synthesis and characterization of gold-conjugated polyphenol nanoparticles with antioxidant and biological activities. *Int. J. Nanomed.* **2014**, *9*, 4935–4951. [CrossRef]

87. Mukherjee, S.; Ghosh, S.; Das, D.K.; Chakraborty, P.; Choudhury, S.; Gupta, P.; Adhikary, A.; Dey, S.; Chattopadhyay, S. Gold-conjugated green tea nanoparticles for enhanced anti-tumor activities and hepatoprotection—Synthesis, characterization and in vitro evaluation. *J. Nutr. Biochem.* **2015**. [CrossRef]

88. Sanna, V.; Pintus, G.; Roggio, A.M.; Punzoni, S.; Posadino, A.M.; Arca, A.; Marceddu, S.; Bandiera, P.; Uzzau, S.; Sechi, M. Targeted biocompatible nanoparticles for the delivery of (-)-epigallocatechin 3-gallate to prostate cancer cells. *J. Med. Chem.* **2011**, *54*, 1321–1332. [CrossRef]

89. Alotaibi, A.; Bhatnagar, P.; Najafzadeh, M.; Gupta, K.C.; Anderson, D. Tea phenols in bulk and nanoparticle form modify DNA damage in human lymphocytes from colon cancer patients and healthy individuals treated in vitro with platinum based-chemotherapeutic drugs. *Nanomedicine* **2012**, *4*, 2012. [CrossRef]

90. Narayanan, S.; Mony, U.; Vijaykumar, D.K.; Koyakutty, M.; Paul-Prasanth, B.; Menon, D. Sequential release of epigallocatechin gallate and paclitaxel from PLGA-casein core/shell nanoparticles sensitizes drug-resistant breast cancer cells. *Nanomed. Nanotechnol. Biol. Med.* **2015**, *11*, 1399–1406. [CrossRef]

91. Siddiqui, I.A.; Adhami, V.M.; Bharali, D.J.; Hafeez, B.B.; Asim, M.; Khwaja, S.I.; Ahmad, N.; Cui, H.; Mousa, S.A.; Mukhtar, H. Introducing Nanochemoprevention as a Novel Approach for Cancer Control: Proof of Principle with Green Tea Polyphenol Epigallocatechin-3-Gallate. *Cancer Res.* **2009**, *69*, 1712–1716. [CrossRef]

92. Khan, N.; Bharali, D.J.; Adhami, V.M.; Siddiqui, I.A.; Cui, H.; Shabana, S.M.; Mousa, S.A.; Mukhtar, H. Oral administration of naturally occurring chitosan-based nanoformulated green tea polyphenol EGCG effectively inhibits prostate cancer cell growth in a xenograft model. *Carcinogenesis* **2014**, *35*, 415–423. [CrossRef] [PubMed]

93. Siddiqui, I.A.; Bharali, D.J.; Nihal, M.; Adhami, V.M.; Khan, N.; Chamcheu, J.C.; Khan, M.I.; Shabana, S.; Mousa, S.A.; Mukhtar, H. Excellent anti-proliferative and pro-apoptotic effects of (-)-epigallocatechin-3-gallate encapsulated in chitosan nanoparticles on human melanoma cell growth both in vitro and in vivo. *Nanomed. Nanotechnol. Biol. Med.* **2014**, *10*, 1619–1626. [CrossRef] [PubMed]

94. Hu, B.; Xie, M.; Zhang, C.; Zeng, X. Genipin-structured peptide-polysaccharide nanoparticles with significantly improved resistance to harsh gastrointestinal environments and their potential for oral delivery of polyphenols. *J. Agric. Food Chem.* **2014**, *62*, 12443–12452. [CrossRef] [PubMed]

95. Shutava, T.G.; Balkundi, S.S.; Vangala, P.; Steffan, J.J.; Bigelow, R.L.; Cardelli, J.a.; O'Neal, D.P.; Lvov, Y.M. Layer-by-layer-coated gelatin nanoparticles as a vehicle for delivery of natural polyphenols. *ACS Nano* **2009**, *3*, 1877–1885. [CrossRef]

96. Fang, J.-Y.; Hung, C.-F.; Hwang, T.-L.; Huang, Y.-L. Physicochemical characteristics and in vivo deposition of liposome-encapsulated tea catechins by topical and intratumor administrations. *J. Drug Target.* **2005**, *13*, 19–27. [CrossRef]

97. Fang, J.-Y.; Lee, W.-R.; Shen, S.-C.; Huang, Y.-L. Effect of liposome encapsulation of tea catechins on their accumulation in basal cell carcinomas. *J. Dermatol. Sci.* **2006**, *42*, 101–109. [CrossRef]

98. De Pace, R.C.C.; Liu, X.; Sun, M.; Nie, S.; Zhang, J.; Cai, Q.; Gao, W.; Pan, X.; Fan, Z.; Wang, S. Anticancer activities of (−)-epigallocatechin-3-gallate encapsulated nanoliposomes in MCF7 breast cancer cells. *J. Liposome Res.* **2013**, *23*, 187–196. [CrossRef]

99. Ramadass, S.K.; Anantharaman, N.V.; Subramanian, S.; Sivasubramanian, S.; Madhan, B. Paclitaxel/epigallocatechin gallate coloaded liposome: A synergistic delivery to control the invasiveness of MDA-MB-231 breast cancer cells. *Coll. Surf. B Biointerfaces* **2015**, *125*, 65–72. [CrossRef]

100. Jain, S.; Hirst, D.G.; O'Sullivan, J.M. Gold nanoparticles as novel agents for cancer therapy. *Brit. J. Radiol.* **2012**, *85*, 101–113. [CrossRef]

101. Elsabahy, M.; Wooley, K.L. Design of polymeric nanoparticles for biomedical delivery applications. *Chem. Soc. Rev.* **2012**, *41*, 2545–2561. [CrossRef]

102. Huang, Y.-W.; Cambre, M.; Lee, H.-J. The Toxicity of Nanoparticles Depends on Multiple Molecular and Physicochemical Mechanisms. *Int. J. Mol. Sci.* **2017**, *18*, 2702. [CrossRef] [PubMed]

103. Halamoda-Kenzaoui, B.; Bremer-Hoffmann, S. Main trends of immune effects triggered by nanomedicines in preclinical studies. *Int. J. Nanomed.* **2018**, *13*, 5419–5431. [CrossRef] [PubMed]

104. Ramesh, N.; Mandal, A.K.A. Pharmacokinetic, toxicokinetic, and bioavailability studies of epigallocatechin-3-gallate loaded solid lipid nanoparticle in rat model. *Drug Dev. Ind. Pharm.* **2019**, *45*, 1506–1514. [CrossRef] [PubMed]

105. Sanna, V.; Singh, C.K.; Jashari, R.; Adhami, V.M.; Chamcheu, J.C.; Rady, I.; Sechi, M.; Mukhtar, H.; Siddiqui, I.A. Targeted nanoparticles encapsulating (−)-epigallocatechin-3-gallate for prostate cancer prevention and therapy. *Sci. Rep.* **2017**, *7*. [CrossRef] [PubMed]

106. Ding, J.; Liang, T.; Min, Q.; Jiang, L.; Zhu, J.J. "Stealth and Fully-Laden" Drug Carriers: Self-Assembled Nanogels Encapsulated with Epigallocatechin Gallate and siRNA for Drug-Resistant Breast Cancer Therapy. *ACS Appl. Mater. Interfaces* **2018**, *10*, 9938–9948. [CrossRef]

107. Bettuzzi, S.; Brausi, M.; Rizzi, F.; Castagnetti, G.; Peracchia, G.; Corti, A. Chemoprevention of human prostate cancer by oral administration of green tea catechins in volunteers with high-grade prostate intraepithelial neoplasia: A preliminary report from a one-year proof-of-principle study. *Cancer Res.* **2006**, *66*, 1234–1240. [CrossRef]

108. Lazzeroni, M.; Guerrieri-Gonzaga, A.; Gandini, S.; Johansson, H.; Serrano, D.; Cazzaniga, M.; Aristarco, V.; Macis, D.; Mora, S.; Caldarella, P.; et al. A Presurgical Study of Lecithin Formulation of Green Tea Extract in Women with Early Breast Cancer. *Cancer Prev. Res.* **2017**, *10*, 363–370. [CrossRef]

109. Shin, C.M.; Lee, D.H.; Seo, A.Y.; Lee, H.J.; Kim, S.B.; Son, W.C.; Kim, Y.K.; Lee, S.J.; Park, S.H.; Kim, N.; et al. Green tea extracts for the prevention of metachronous colorectal polyps among patients who underwent endoscopic removal of colorectal adenomas: A randomized clinical trial. *Clin. Nutr.* **2018**, *37*, 452–458. [CrossRef]

110. Joe, A.K.; Schnoll-Sussman, F.; Bresalier, R.S.; Abrams, J.A.; Hibshoosh, H.; Cheung, K.; Friedman, R.A.; Yang, C.S.; Milne, G.L.; Liu, D.D.; et al. Phase Ib Randomized, Double-Blinded, Placebo-Controlled, Dose Escalation Study of Polyphenon E in Patients with Barrett's Esophagus. *Cancer Prev. Res.* **2015**, *8*, 1131–1137. [CrossRef]

111. Cao, J.; Han, J.; Xiao, H.; Qiao, J.; Han, M. Effect of Tea Polyphenol Compounds on Anticancer Drugs in Terms of Anti-Tumor Activity, Toxicology, and Pharmacokinetics. *Nutrients* **2016**, *8*, 762. [CrossRef]

112. Wu, A.H.; Spicer, D.; Stanczyk, F.Z.; Tseng, C.C.; Yang, C.S.; Pike, M.C. Effect of 2-month controlled green tea intervention on lipoprotein cholesterol, glucose, and hormone levels in healthy postmenopausal women. *Cancer Prev. Res.* **2012**, *5*, 393–402. [CrossRef] [PubMed]

113. Samavat, H.; Newman, A.R.; Wang, R.; Yuan, J.M.; Wu, A.H.; Kurzer, M.S. Effects of green tea catechin extract on serum lipids in postmenopausal women: A randomized, placebo-controlled clinical trial. *Am. J. Clin. Nutr.* **2016**, *104*, 1671–1682. [CrossRef] [PubMed]

114. Dostal, A.M.; Samavat, H.; Bedell, S.; Torkelson, C.; Wang, R.; Swenson, K.; Le, C.; Wu, A.H.; Ursin, G.; Yuan, J.M.; et al. The safety of green tea extract supplementation in postmenopausal women at risk for breast cancer: Results of the Minnesota Green Tea Trial. *Food Chem. Toxicol.* **2015**, *83*, 26–35. [CrossRef] [PubMed]

115. Samavat, H.; Ursin, G.; Emory, T.H.; Lee, E.; Wang, R.; Torkelson, C.J.; Dostal, A.M.; Swenson, K.; Le, C.T.; Yang, C.S.; et al. A Randomized Controlled Trial of Green Tea Extract Supplementation and Mammographic Density in Postmenopausal Women at Increased Risk of Breast Cancer. *Cancer Prev. Res.* **2017**, *10*, 710–718. [CrossRef] [PubMed]

116. Shanafelt, T.D.; Call, T.G.; Zent, C.S.; Leis, J.F.; LaPlant, B.; Bowen, D.A.; Roos, M.; Laumann, K.; Ghosh, A.K.; Lesnick, C.; et al. Phase 2 trial of daily, oral Polyphenon E in patients with asymptomatic, Rai stage 0 to II chronic lymphocytic leukemia. *Cancer* **2013**, *119*, 363–370. [CrossRef] [PubMed]

117. Kumar, N.B.; Pow-Sang, J.; Egan, K.M.; Spiess, P.E.; Dickinson, S.; Salup, R.; Helal, M.; McLarty, J.; Williams, C.R.; Schreiber, F.; et al. Randomized, Placebo-Controlled Trial of Green Tea Catechins for Prostate Cancer Prevention. *Cancer Prev. Res.* **2015**, *8*, 879–887. [CrossRef]

118. McLarty, J.; Bigelow, R.L.; Smith, M.; Elmajian, D.; Ankem, M.; Cardelli, J.A. Tea polyphenols decrease serum levels of prostate-specific antigen, hepatocyte growth factor, and vascular endothelial growth factor in prostate cancer patients and inhibit production of hepatocyte growth factor and vascular endothelial growth factor in vitro. *Cancer Prev. Res.* **2009**, *2*, 673–682. [CrossRef]
119. Gee, J.R.; Saltzstein, D.R.; Kim, K.; Kolesar, J.; Huang, W.; Havighurst, T.C.; Wollmer, B.W.; Stublaski, J.; Downs, T.; Mukhtar, H.; et al. A Phase II Randomized, Double-blind, Presurgical Trial of Polyphenon E in Bladder Cancer Patients to Evaluate Pharmacodynamics and Bladder Tissue Biomarkers. *Cancer Prev. Res.* **2017**, *10*, 298–307. [CrossRef]
120. Zhao, H.; Zhu, W.; Xie, P.; Li, H.; Zhang, X.; Sun, X.; Yu, J.; Xing, L. A phase I study of concurrent chemotherapy and thoracic radiotherapy with oral epigallocatechin-3-gallate protection in patients with locally advanced stage III non-small-cell lung cancer. *Radiother. Oncol.* **2014**, *110*, 132–136. [CrossRef]
121. Zhao, H.; Xie, P.; Li, X.; Zhu, W.; Sun, X.; Sun, X.; Chen, X.; Xing, L.; Yu, J. A prospective phase II trial of EGCG in treatment of acute radiation-induced esophagitis for stage III lung cancer. *Radiother. Oncol.* **2015**, *114*, 351–356. [CrossRef]
122. Zhao, H.; Zhu, W.; Jia, L.; Sun, X.; Chen, G.; Zhao, X.; Li, X.; Meng, X.; Kong, L.; Xing, L.; et al. Phase I study of topical epigallocatechin-3-gallate (EGCG) in patients with breast cancer receiving adjuvant radiotherapy. *Brit. J. Radiol.* **2016**, *89*, 20150665. [CrossRef] [PubMed]

Article

The Inhibitory Mechanisms of Tumor PD-L1 Expression by Natural Bioactive Gallic Acid in Non-Small-Cell Lung Cancer (NSCLC) Cells

Dong Young Kang [1,†], Nipin Sp [1,†], Eun Seong Jo [1], Alexis Rugamba [1], Dae Young Hong [2], Hong Ghi Lee [3], Ji-Seung Yoo [4], Qing Liu [5], Kyoung-Jin Jang [1,*] and Young Mok Yang [1,*]

[1]   Department of Pathology, School of Medicine, Institute of Biomedical Science and Technology, Konkuk University, Seoul 05029, Korea; kdy6459@naver.com (D.Y.K.); nipinsp@gmail.com (N.S.); eses0706@naver.com (E.S.J.); rugambalex@gmail.com (A.R.)

[2]   Department of Emergency Medicine, School of Medicine, Konkuk University, Seoul 05029, Korea; kuhemhdy@gmail.com

[3]   Division of Hematology-Oncology, Department of Internal Medicine, Konkuk University Medical Center, Seoul 05029, Korea; mlee@kuh.ac.kr

[4]   Department of Immunology, Hokkaido University Graduate School of Medicine, Sapporo 060-0808, Japan; jiseungy@pop.med.hokudai.ac.jp

[5]   Jilin Green Food Engineering Research Institute, Changchun 130000, Jilin, China; liuqing0523@hotmail.com

*   Correspondence: jangkj@konkuk.ac.kr (K.-J.J.); ymyang@kku.ac.kr (Y.M.Y.); Tel.: +82-2-2030-7839 (K-J.J.); +82-2-2030-7812 (Y.M.Y.)

†   These authors contribute equally to this paper.

Received: 9 March 2020; Accepted: 16 March 2020; Published: 19 March 2020

Abstract: Non-small-cell lung cancer (NSCLC) is the most common lung cancer subtype and accounts for more than 80% of all lung cancer cases. Epidermal growth factor receptor (EGFR) phosphorylation by binding growth factors such as EGF activates downstream prooncogenic signaling pathways including KRAS-ERK, JAK-STAT, and PI3K-AKT. These pathways promote the tumor progression of NSCLC by inducing uncontrolled cell cycle, proliferation, migration, and programmed death-ligand 1 (PD-L1) expression. New cytotoxic drugs have facilitated considerable progress in NSCLC treatment, but side effects are still a significant cause of mortality. Gallic acid (3,4,5-trihydroxybenzoic acid; GA) is a phenolic natural compound, isolated from plant derivatives, that has been reported to show anticancer effects. We demonstrated the tumor-suppressive effect of GA, which induced the decrease of PD-L1 expression through binding to EGFR in NSCLC. This binding inhibited the phosphorylation of EGFR, subsequently inducing the inhibition of PI3K and AKT phosphorylation, which triggered the activation of p53. The p53-dependent upregulation of miR-34a induced PD-L1 downregulation. Further, we revealed the combination effect of GA and anti-PD-1 monoclonal antibody in an NSCLC-cell and peripheral blood mononuclear–cell coculture system. We propose a novel therapeutic application of GA for immunotherapy and chemotherapy in NSCLC.

Keywords: natural bioactive compound; gallic acid; EGFR signaling; p53; PD-L1; immunotherapy

---

1. Introduction

Lung cancer, constituting 18% of all global cancer deaths, is one of the representative causes of death globally [1]. It is classified into two main groups: small-cell lung cancer (SCLC; 15% of all lung cancers) and non-SCLC (NSCLC; 85% of all lung cancers) [2,3]. NSCLC can be further subcategorized into three subtypes: adenocarcinoma, squamous cell carcinoma, and large cell carcinoma [4]. Despite many efforts to treat NSCLC, the overall survival rate is only 15.9% within five years [5]. Further, many patients receiving NSCLC chemotherapies struggle with adverse reactions, drug resistance, and the necessary target specificity of some types of drugs. Thus, unsatisfactory outcomes in NSCLC treatment have motivated researchers to identify novel agents such as natural compounds [6–8]. Moreover, the main advantage of such drugs having fewer side effects relative to non-natural drugs has spurred scientists to reveal their molecular mechanisms.

Gallic acid (3,4,5-trihydroxybenzoic acid; GA), a natural phenolic compound, is widely distributed in natural plants, fruits, and green tea [9,10]. Many studies have reported that GA exhibits anti oxidative, anti mutagenic, anti carcinogenic, antibacterial, antiviral, and anti-inflammatory effects [11–15]. However, the main interest in GA and its derivatives surrounds its anticancer activity. Previous studies have revealed that GA effectively induces selective apoptosis in various cancer cells, including HeLa, HCT-15, SH-SY5Y, and NSCLC cells [16–19] and inhibits proliferation and migration via regulating fatty acid synthase in TSGH-8301 cells [20]. Recent studies also revealed potential anticancer effects of GA is due to its ability to inhibit cell proliferation and to induce apoptosis *in vivo* [21,22].

Programmed cell death ligand-1 (PD-L1), also known as CD274 and B7-H1, is a transmembrane protein expressed on the surface of antigen-presenting cells such as dendritic cells, macrophages, and B-cells. It is also overexpressed and found in various types of cancer [23–26]. PD-L1 specifically binds to programmed cell death-1 (PD-1), which is an important inhibitory receptor expressed on the surface of immune-related lymphocytes like T-cells, B-cells, and myeloid cells [27]. The binding of PD-L1 to PD-1 inhibits the proliferation, cytokine generation and release, and cytotoxicity of T-cells. Thus, the binding leads to an immunosuppressive effect and allows cancer cells to escape immune eradication via assistance from tumor-specific T-cells [28]. PD-L1 overexpression in cancer cells promotes cancer progression and leads cancer cells to malignancy. Moreover, the intrinsic signal transduction by PD-L1 enhances the survival of cancer cells through increasing the resistance toward proapoptotic stimuli such as interferons [29]. PD-L1 expression at the transcriptional level is regulated individually or cooperatively by many oncogenic transcription factors such as MYC, AP-1, STAT, IRF1, HIF, and NF-κB. Some studies have demonstrated that there is a tendency toward higher PD-L1 expression in *TP53*-mutated and low p53-expression cancer cells, which imply that PD-L1 expression is considerably related to p53 status in cancer cells [30–33]. In addition, tumor suppressor gene *PTEN*, one of the most frequently mutated genes in human cancers, downregulates PD-L1 expression, which signifies that tumor suppressors play an important role in controlling PD-L1 expression [34–36].

Previous research has demonstrated that apigenin, a kind of flavonoid, induces growth-suppressive and proapoptotic effects in melanoma cells. Additionally, such significantly inhibits the expression of interferon (IFN)-γ-induced PD-L1, which may indicate the existence of an immunosuppressive effect shown by a natural compound [37]. In this study, we examined the anticancer and immunosuppressive activities of natural bioactive GA in NSCLC A549 and H292 cells (wild-type p53 and epidermal growth factor receptor (EGFR)), which are kinds of NSCLC cells. The binding of GA to EGFR inhibited the EGFR phosphorylation, leading to the promotion of p53 expression in both A549 and H292 cells. Furthermore, highly expressed p53 decreased PD-L1 protein expression through enhancing the miR-34a related to PD-L1 downregulation at the transcriptional level. As such, our results suggest an immunosuppressive effect of GA toward NSCLC cells, which might imply a potential possibility for clinical application in NSCLC treatment.

2. Results

2.1. GA Downregulates the PD-L1 Expression in NSCLC Cells

To determine whether GA inhibits the cell proliferation of A549 and H292 cell lines, GA-treated cells were compared with non-treated control cells. The results of MTT assay showed that the cell growth of GA-treated cells is significantly inhibited in a time- and concentration-dependent manner (Figure S1A). This result was confirmed with crystal violet assay by treating GA in A549 and H292 cells for 48 h (Figure S1B). From this data, we respectively identified an IC$_{50}$ dosage of 400 μM in A549 and 100 μM in H292 cell lines at 48 hours, respectively, information which was used for further studies. We checked the same concentration in non-cancerous cells (HUVEC cell line) and found that

400 µM GA inducing around 8% cell death which indicated that this concentration does not make much toxicity in normal cells (Figure S2). Recently, cancer immunotherapy based on PD-1/PD-L1 blockade has shown clinical efficacy in the treatment of multiple cancers [26,28]. In addition, a study of the drug-induced inhibition of PD-L1 expression in cancer cells has been conducted [37]. To investigate whether GA influences PD-L1 expression, we assessed the expression levels of PD-L1 by the impact of GA in NSCLC cells. As shown in Figure 1A, results from western blotting suggested that GA strongly decreases the expression levels of PD-L1 protein in A549 and H292 NSCLC cells. In addition, GA showed a greater than 70% inhibitory effect as compared with in non-treated control cells among A549 cells (Figure 1B). Subsequently, we performed a real time PCR experiment to examine the influence of GA on the messenger RNA (mRNA) expression of PD-L1 in A549 and H292 cell lines. In accordance with data from real time PCR, GA also downregulated the expression of PD-L1 mRNA in a concentration-dependent manner in both A549 and H292 cells (Figure 1C). These results suggest that bioactive natural GA has a significant inhibitory effect on PD-L1 expression (both protein and mRNA) in A549 and H292 cell lines, which imply the potential of using it as an immune anticancer agent.

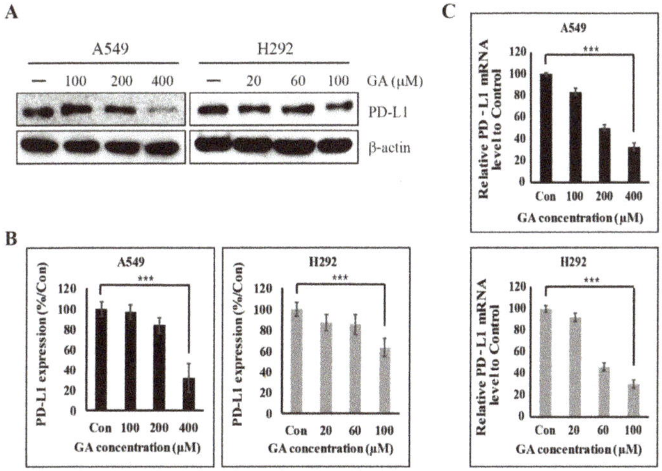

Figure 1. Gallic acid (GA) reduces the programmed death-ligand 1 (PD-L1) expression in non-small-cell lung cancer (NSCLC) cells. (**A**) The expression levels of PD-L1 protein in A549 and H292 cells were detected after GA treatment in concentrations indicated for 48 hours. (**B**) The relative expression levels of PD-L1 protein were determined by densitometry and normalized to β actin. Data are representative of three independent experiments. *** $p < 0.001$ (*t*-test). (**C**) The expression levels of PD-L1 mRNA in A549 and H292 cells were detected after GA treatment in concentrations indicated for 48 hours. The relative expression levels of PD-L1 mRNA were determined by real time qPCR and normalized to GAPDH mRNA. Data are representative of three independent experiments. *** $p < 0.001$ (*t*-test).

2.2. GA Binds to EGFR and Then Inhibits its Phosphorylation

EGFR phosphorylation induces various oncogenic signaling pathways for cell proliferation, invasion, and metabolic reprogramming in many cancer cells [38]. Therefore, to inhibit EGFR phosphorylation in cancer cells, many clinical applications have been stimulated to develop EGFR tyrosine kinase inhibitors (TKIs) such as erlotinib, gefitinib, and lapatinib [39,40]. In previous studies, we found that the binding of natural compounds to EGFR, causing a decrease in EGFR phosphorylation, inhibited the proliferation, migration, invasion, and angiogenesis of human breast adenocarcinoma cells [41,42]. To understand the impact of GA for EGFR phosphorylation, we identified the binding ability of GA to EGFR. Molecular docking was performed with an AutoDock Vina platform (Oleg Trott, The Scripps Research Institute, La Jolla, CA, USA). We found that GA is docked in the ATP binding site of EGFR, and this result may imply the direct binding of GA to EGFR (Figure 2A). Subsequently, we further performed an immunoblot analysis for understanding whether the GA/EGFR binding influences the phosphorylation of EGFR and found that, GA significantly downregulated the phosphorylation of EGFR in both A549 and H292 cells (EGFR wild-type NSCLC cells) (Figure 2B,C). However, this treatment did not affect the expression levels of total EGFR mRNA (Figure 2D). These results may imply that GA could influence the inhibition of EGFR signal transduction in two NSCLC cells. Moreover, these results led us to investigate the binding specificity of GA to EGFR, where we conducted a competitive binding experiment of GA and EGF (25 ng/mL pre-treatment for 15 min) versus EGFR. Here, GA significantly inhibited EGF-induced EGFR phosphorylation in both A549 and H292 cells (Figure 2E,F). This result suggests that GA binds to EGFR as compared with the natural ligand (EGF) for EGFR, and this act of binding clearly induces the inhibition of EGFR phosphorylation.

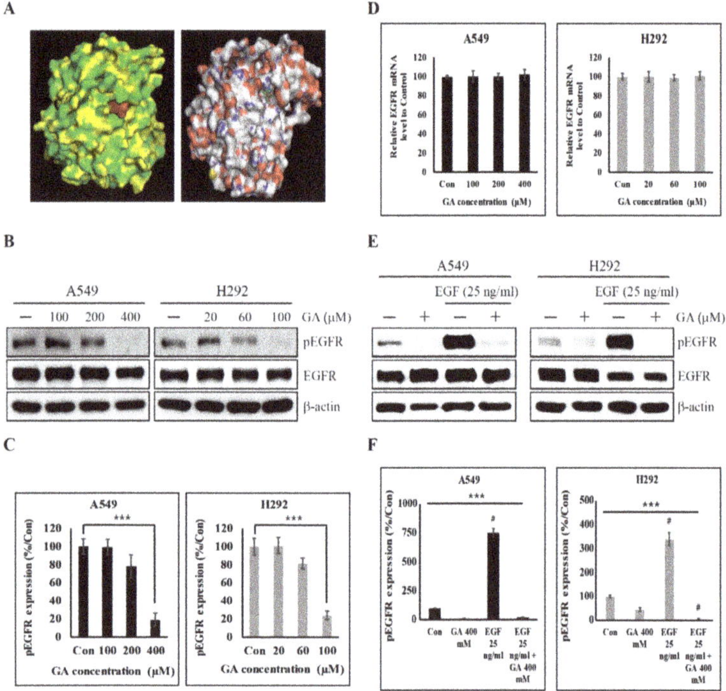

Figure 2. GA binding to epidermal growth factor receptor (EGFR) inhibits the phosphorylation of EGFR. (**A**) Binding of GA (PubChem CID: 370) to the ATP-binding domain of EGFR (Protein Data Bank ID: 4LQM) as determined by molecular docking using AutoDock Vina. (**B**) The expression levels of total EGFR and phosphorylated EGFR (pEGFR) protein in A549 and H292 cells were detected after GA treatment in concentrations indicated for 48 hours. (**C**) The relative levels of pEGFR protein were determined by densitometry and normalized to β-actin. Data are representative of three independent experiments. *** $p < 0.001$ (*t*-test). (**D**) The expression levels of EGFR mRNA in A549 and H292 cells were detected by real time PCR after GA treatment in concentrations indicated for 48 hours. The relative levels of EGFR mRNA were determined and normalized to GAPDH mRNA. Data are representative of three independent experiments. (**E**) A549 and H292 cells for detecting the expression levels of total EGFR and pEGFR protein were treated with or without 25 ng/mL EGF for 15 minutes and then further treated with GA (A549: 400 µM; H292: 100 µM) for 48 hours. (**F**) The relative levels of pEGFR protein were determined by densitometry and normalized to β-actin. Data are representative of three independent experiments. *** $p < 0.001$ (*t*-test). # $p < 0.001$ vs. control.

2.3. GA Reduces the Phosphorylation of PI3K/AKT That is One of the Downstream Targets of EGFR Signaling

EGF/EGFR signal transduction has been known to lead to the constitutive activation of downstream signaling pathways associated with MAPKs, STAT3, and PI3K for regulating PD-L1 expression in various cancer cells [43]. A previous study found that the PD-L1 expression of EGFR–mutated PC-9 cells was significantly higher than those of EGFR wild-type LU-99, A549, and PC-14 cells. In EGFR inhibitor experiments, the EGFR TKI gefitinib induced a lower expression of phosphorylated AKT and STAT3, which prompted the downregulation of PD-L1 expression [44]. To determine a key EGFR-downstream pathway associated with PD-L1 expression, we used an immunoblot analysis. As shown in Figure S3, GA treatment did not inhibit the phosphorylation of JAK2/STAT3, which is one of the main pathways. However, GA efficiently controlled the PI3K/AKT pathway by inhibiting their phosphorylation but not total protein level (Figure 3A,B). These results clearly show that the regulation of PI3K/AKT phosphorylation by GA could be responsible for PD-L1 expression in both A549 and H292 cells. Moreover, the downregulation of PI3K/AKT phosphorylation by GA may indicate a beneficial effect in terms of controlling various oncogenic signals, such as cellular proliferation, invasion, angiogenesis, and metastasis.

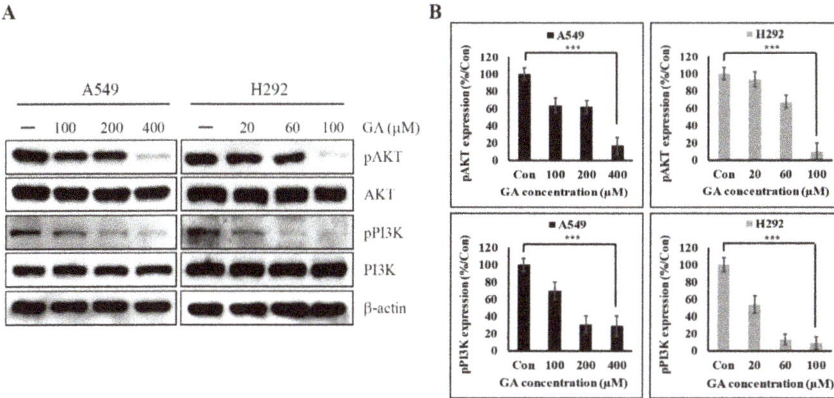

Figure 3. GA inhibits the phosphorylation of AKT and PI3K protein in a GA concentration-dependent manner. (**A**) The expression levels of pAKT and pPI3K protein in A549 and H292 cells were detected after GA treatment in concentrations indicated for 48 hours. (**B**) The relative expression levels of pAKT and pPI3K protein were determined by densitometry and normalized to β-actin. Data are representative of three independent experiments. *** $p < 0.001$ (*t* test).

2.4. GA Activates the Expression of Tumor Suppressor Factor p53 for Inhibiting the Expression of PD-L1

The tumor suppressor factor p53 plays an important role in cell-cycle arrest and apoptosis induction in response to oncogenic or other stresses for the prevention of cancer development. However, it is downregulated or mutated in an inactive form in almost all human cancer cells. A previous study found that p53 is led into Mdm2-mediated ubiquitination and degradation by PI3K/AKT signal transduction in breast cancer MCF-7 cells but not p53 mRNA [45]. Furthermore, p53-regulated IFN-γ induced PD-L1 expression in melanoma cells [32]. To investigate the effect on p53 by GA, we checked the protein levels of p53 with or without GA treatment in A549 and H292 cells and found that GA upregulates the expression levels of p53 protein in a concentration-dependent manner (Figure 4A,B). In addition, the expression levels of p53 protein were nearly doubled in A549 cells. Further, the mRNA levels of p53 identified by real time PCR showed a significant increase in a GA concentration-dependent manner in H292 cells as well as in A549 cells (Figure 4C). From these data, although a previous study revealed that PI3K/AKT signaling induced by their phosphorylation regulates only p53 protein levels, the PI3K/AKT signaling controlled by GA plays a key role in regulating both protein and mRNA levels of p53. These results additional imply that GA regulates p53 from mRNA levels through the downregulation of PI3K/AKT phosphorylation. To further understand the role of p53 in PD-L1 regulation, we used GA with or without p53 siRNA and determined whether specific gene silencing influences PD-L1 expression in A549 and H292 cells. As shown in Figure 4D,E, the gene silencing of p53 significantly affected the increase in PD-L1 proteins compared to non-treated control, which was decreased by GA treatment. In contrast, the effect by GA regulated the protein levels of p53 and PD-L1 in two NSCLC cells. These results imply that the regulation of PD-L1 by GA is indirectly controlled by way of inducing an increase in p53 protein level. In addition, the upregulation of p53 by GA may induce various p53-mediated anti-oncogenic factors such as the regulation of miRNA.

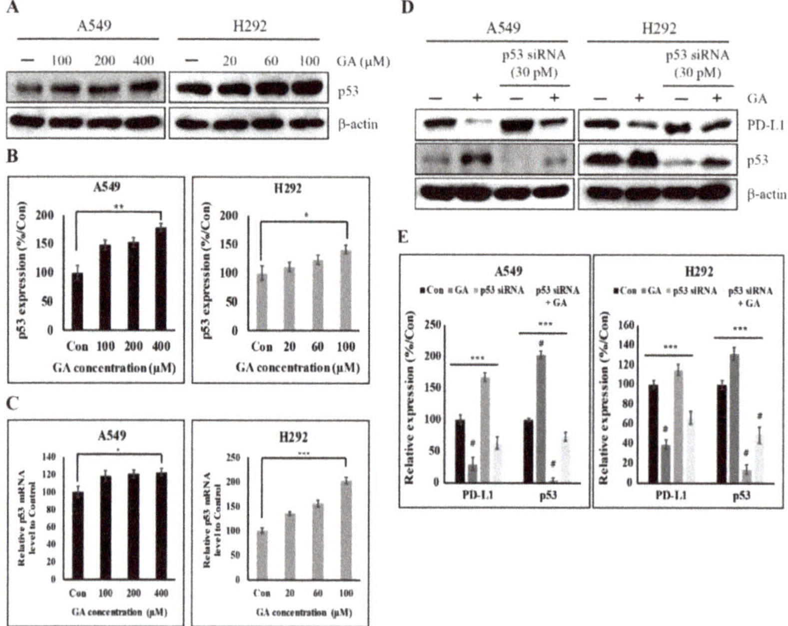

Figure 4. GA increased p53 expression. (**A**) The expression levels of p53 protein in A549 and H292 cells were detected after GA treatment in concentrations indicated for 48 hours. (**B**) The relative expression levels of p53 protein were determined by densitometry and normalized to β-actin. Data are representative of three independent experiments. * $p < 0.05$ and ** $p < 0.01$ (*t*-test). (**C**) The expression levels of p53 mRNA in A549 and H292 cells were detected by real time qPCR after GA treatment in concentrations indicated for 48 hours. The relative expression levels of p53 mRNA were determined and normalized to GAPDH mRNA. Data are representative of three independent experiments. * $p < 0.05$ and *** $p < 0.001$ (*t*-test). (**D**) The expression levels of p53 and PD-L1 protein in A549 and H292 cells, treated with GA (A549: 400 μM; H292: 100 μM) or 30 pM p53 siRNA, were detected by western blotting at 48 hours. (**E**) The relative expression levels of p53 and PD-L1 protein were determined using densitometry and normalized to β-actin. Data are representative of three independent experiments. *** $p < 0.001$ (ANOVA); # $p < 0.001$ vs. control.

2.5. GA Upregulates p53-Dependent MiR-34a for Inhibiting the Expression of PD-L1

miRNAs, a family of small noncoding RNAs, regulate wide biological processes including carcinogenesis, which severely is dysregulated in many cancer cells. Some miRNAs such as miR-513 and miR-570 directly target PD-L1 [46,47]. However, p53 indirectly regulates the expression levels of PD-L1 through inducing miR-34a in cancer cells [33]. Although many studies have shown results for the regulation of PD-L1 expression directly by miRNA, detailed studies of the actions brought on indirectly by p53 via drugs including natural compounds is poorly understood. To understand the expression level of miR-34a by GA, we performed a real time PCR experiment because miR-34a is a well-known molecule transcriptionally induced by p53. As shown in Figure 5A, we found that it was significantly increased in a time- and GA concentration-dependent manner in both A549 and H292 cells. To further investigate miR-34a regulation by GA via p53, we additionally used p53 siRNA. The expression levels of miR-34a were decreased by p53 siRNA, but their expression levels were slightly increased by additional GA (Figure 5B). These results clearly suggest that miR-34a expression is regulated by GA-dependent p53. Additionally, we used a miR-34a inhibitor with or without GA to determine a more detailed interrelation analysis in the regulation of PD-L1 expression. In this experiment, we demonstrated that the inhibition of miR-34a function by its inhibitor is induced into an increase of PD-L1 protein which reversed by GA, but not p53 (Figure 5C,D). These results support that the expression of PD-L1 is regulated via miR-34a-induction through GA-dependent p53 in A549 and H292 cells.

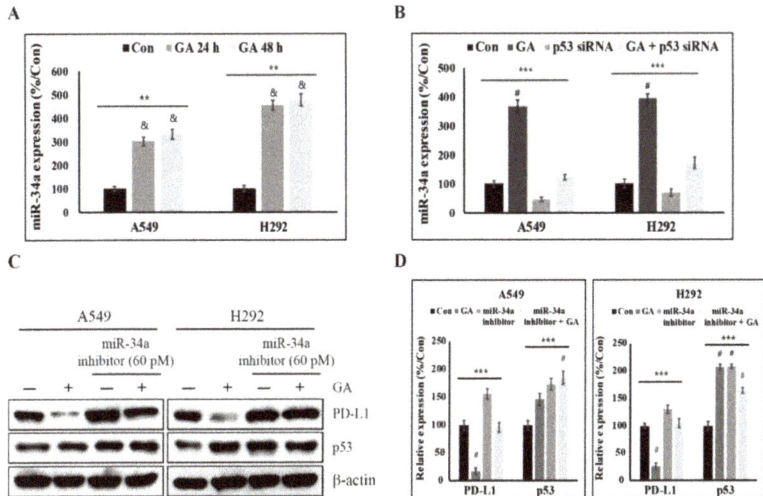

Figure 5. GA upregulates the PD-L1 expression by p53 via miR-34a. (**A**) Relative expression levels of miR34a after treatment of GA (A549: 400 µM; H292: 100 µM) for 24 and 48 hours. Data are representative of three independent experiments. ** $p < 0.01$ (ANOVA); & $p < 0.01$ vs. control. (**B**) The expression levels of miR-34a in A549 and H292 cells, treated with GA (A549: 400 µM; H292: 100 µM), p53 siRNA (60 pM) or GA plus p53 siRNA were detected by RT-PCR at 48 hours. The relative expression levels of miR-34a were determined using densitometry and normalized to U6. Data are representative of three independent experiments. *** $p < 0.001$ (ANOVA); # $p < 0.001$ vs. control. (**C**) The expression levels of p53 and PD-L1 protein in A549 and H292 cells, treated with GA (A549: 400 µM; H292: 100 µM) or 30 pM miR-34a inhibitor, were detected by western blotting at 48 hours. (**D**) The relative expression levels of p53 and PD-L1 protein were determined by densitometry and normalized to β-actin. Data are representative of three independent experiments. *** $p < 0.001$ (ANOVA); # $p < 0.001$ vs. control.

2.6. The Downregulation of PD-L1 Expression by GA Induces the Combination Effect with PD-1 Blockade

To test the combination effect of PD-1 blockade and GA on antitumor activity, we evaluated cytotoxicity in an NSCLC-cell and peripheral blood mononuclear-cell (PBMC) coculture system in the presence of the anti-PD-1 monoclonal antibody (mAb) nivolumab, GA, or both. We observed a considerable apoptotic effect in the presence of both PD-1 mAb and GA in A549 and H292 cells (Figure 6A). Further, GA reduced the viability of cancer cells more effectively in comparison with a single blockade of PD-1 with PD-1 mAb. These results may indicate that the decrease of PD-L1 expression by GA regulates not only reducing survival signals of PD-L1 downstream but also activates the T-cell-mediated immune response. To further investigate the combination effect on PBMC cytokine expression, we performed an IFN-γ analysis by enzyme-linked immunosorbent assay (ELISA). As shown in Figure 6B, GA treatment was observed to slightly increase the IFN-γ level more so than a single blockade of PD-1 with PD-1 mAb. In addition, treatment with both GA and PD-1 mAb considerably enhanced the IFN-γ production in the supernatant of the NSCLC-cell and PBMC coculture system. These results suggest that the decrease in PD-L1 expression brought about by GA enhances the effect observed with PD-1 mAb in the production of IFN-γ. Figure 7 is a graphical abstract which gave the conclusion of all these results. We checked the effect of this combination in a non-cancerous cell (HUVEC cell line) and found that these combination does not induce much cell toxicity in non-cancerous cells (Figure S3).

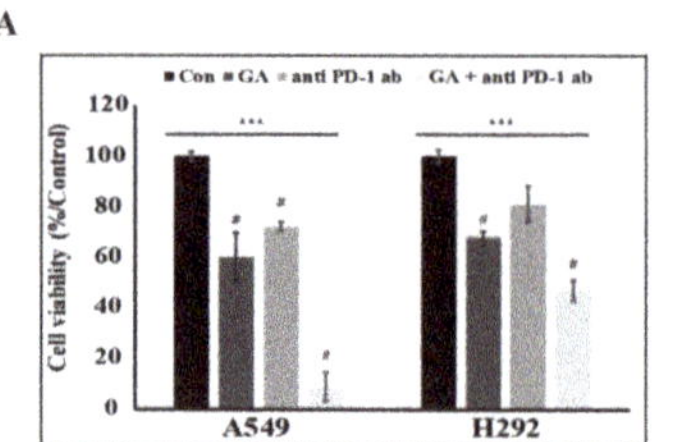
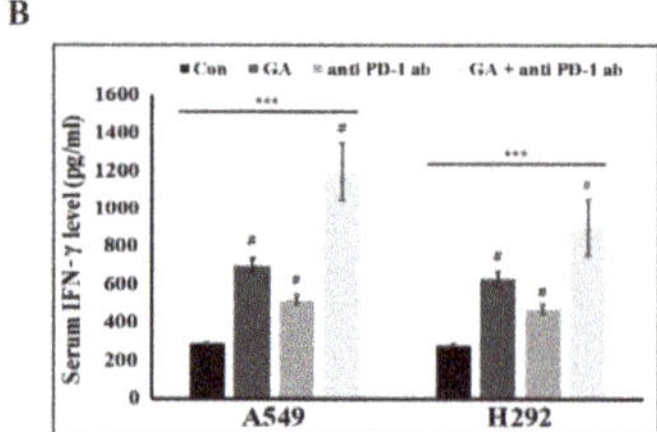

Figure 6. GA enhances the combination effect with anti-PD-1 mAb in an NSCLC-cell and peripheral blood mononuclear-cell (PBMC) coculture system. (**A**) The cell survival rates of NSCLC cells cocultured with PBMCs were examined after treatment with mock, GA (A549: 200 μM; H292: 50 μM), anti-PD-1 mAb (5 μg/mL), or both GA and anti-PD-1 mAb for 48 hours. *** $p < 0.001$ (ANOVA); # $p < 0.001$ vs. control. (**B**) The levels of interferon (IFN)-γ from the supernatants of the coculture system were measured by ELISA also following treatment with mock, GA (A549: 200 μM; H292: 50 μM), anti-PD-1 mAb (5 μg/mL), or both GA and anti-PD-1 mAb for 48 hours. *** $p < 0.001$ (ANOVA); # $p < 0.001$ vs. control.

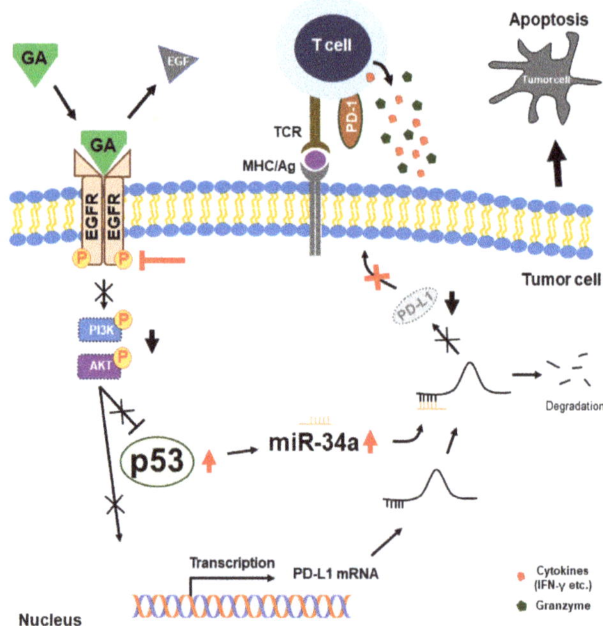

Figure 7. Molecular regulatory mechanism of programmed death-ligand 1 (PD-L1) by natural bioactive gallic acid in NSCLC cells and proposed combination effect for NSCLC immunotherapy.

3. Discussion

An important concept in cancer treatment is that the cancerous cells should ideally be removed without influencing normal cells. Chemotherapy is the most common type of treatment, where chemicals or drugs to destroy cancer cells and cancer microenvironments are applied. Genomic studies such as those on *TP53*, *BCL2*, and *c-MYC* have accelerated the effective application of chemotherapy for developing anticancer drugs and reagents in cancer treatment [48–50]. Anticancer drugs, according to their mechanisms of action, are generally classified as either alkylating agents for damaging cancer cell DNA, antimetabolites for replacing the normal building blocks of RNA and DNA, or antibiotics for interfering with the enzymes involved in DNA replication [51–53].

Although observed therapeutic issues for NSCLC are still deemed to be unsatisfactory because of multidrug resistance and adverse effects [54,55], chemical drugs such as vinorelbine and cisplatin have been tested in NSCLC treatment [56,57]. To overcome these problems, the combined effects of two chemotherapy drugs including cisplatin or carboplatin plus one other drug have often been deployed to treat early-stage NSCLC. Despite many efforts, these chemotherapy-based regimens seem to have reached a therapeutic limit. Recently, many studies have reported the potential possibility of applying natural compounds in the treatment or control of various cancerous diseases. In previous studies, we demonstrated various anticancer effects of natural compounds [41,58,59]. Moreover, combination treatment using a chemotherapy drug and naturally derived drugs showed more effective anticancer effects, which imply that such might reduce the burden of adverse effects brought on by chemotherapy drugs alone [60]. However, a therapeutic strategy using natural compounds is difficult to apply without knowing the specific targets, which is one of the disadvantages of use. Thus, a targeted study focused on using natural compounds is essential to achieve more effective anticancer treatment. Many studies have investigated a phenolic natural compound, gallic acid, that acts as an anticancer agent against various cancers [16,17,19,20]. Nevertheless, these studies did not identify where the target position of GA is against various cancer cells or did they reveal detailed molecular mechanisms underlying the anticancer effects of GA in cancer cell death. In this study, we demonstrated that GA influences cancer cell viability and specifically binds to the tyrosine kinase receptor, EGFR in NSCLC cell lines.

EGFR is a cell-surface protein that binds with epidermal growth factor (EGF) [61]. EGFR often is mutated and/or overexpressed in several types of human cancers, including lung, ovary, breast, head, and neck cancer, and it serves to modulate the growth, differentiation, signaling, adhesion, migration, and survival of cancer cells. Usually, EGF-mediated EGFR phosphorylation induces three main signal transductions including JAK-STAT, KRAS-ERK, and PI3K-AKT-mTOR. These pathways are known to be involved in the growth, proliferation, inhibition of apoptosis, angiogenesis, and invasion of cancer cells [62,63]. For this reason, EGFR has been regarded as an attractive candidate for anticancer treatment because of its multifunctional role in tumorigenesis [38]. To date, two monoclonal antibodies, cetuximab and panitumumab, capable of inhibiting EGF or growth factor-mediated signaling pathways have been used for cancer therapy [64]. In addition, several TKIs such as erlotinib and gefitinib have been employed for the inhibition of EGFR phosphorylation. In this study, we demonstrated that GA inhibits EGFR phosphorylation by binding to EGFR in two NSCLC cells. Moreover, GA showed binding specificity and inhibited EGFR phosphorylation despite EGF-binding. These results may imply that GA is a selective and potent inhibitor against EGFR phosphorylation. Furthermore, the inhibition of EGFR phosphorylation by GA induced the downregulation of phosphorylated PI3K and AKT. Previous studies revealed that the inhibition of EGFR TKI-mediated EGFR phosphorylation induces the downregulation of phospho-PI3K and AKT [65,66]. Therefore, GA, which showed a similar effect to that of TKIs, may be a useful drug candidate for NSCLC treatment.

The tumor suppressor p53 is a transcription factor and plays a pivotal role in cell-cycle, DNA repair, senescence, and apoptosis [67–69]. Under various stresses such as DNA damage, p53 is phosphorylated and acetylated via posttranslational modification and then it is translocated to the nucleus for trans-activating numerous target genes that direct processes including cell-cycle arrest and/or apoptosis. Mutations of *TP53* have been discovered in more than 50% of human cancers and p53 mutation leads to not only the loss of cancer suppressive functions but also the acquisition of additional oncogenic functions such as growth and survival [70]. Wild-type p53 proteins (WTp53) are frequently downregulated because of their function of tumor suppression in many cancer cells. Previous studies have identified that the downregulation of WTp53 is associated with EGFR signal-mediated PI3K/AKT pathway activation in cancer cells [65,66,71,72]. As mentioned above, we demonstrated that the downregulation of EGFR phosphorylation by GA leads to the inhibition of PI3K and AKT phosphorylation. The decrease in their phosphorylation by GA induced the upregulation of WTp53 protein and mRNA in A549 and H292 cells. Furthermore, the competitive activity of GA in an EGF-dependent condition suggested that binding of GA to EGFR, may associated with the upregulation of p53 through inhibiting EGFR/PI3K/AKT4 phosphorylation. These results also suggest that natural bioactive GA may have a potential role as a chemotherapeutic drug for NSCLC treatment. Although many studies have revealed that p53 is related to some immune responses including IFN signaling [73,74], the expression of inflammatory cytokines and Toll-like receptors [75–77], and the activation of T- and natural killer cells [78], the correlation of p53 and tumor immune evasion is poorly understood. Recently, some studies reported that p53 interacts with the apoptotic pathway by regulating miRNAs in cancer cells [79,80]. Furthermore, the effect on p53 was augmented in miR-34a expression, which leads to decreased expression levels of PD-L1 in NSCLC cells [33]. Interestingly, we found that bioactive GA decreased the protein and mRNA levels of PD-L1 as compared with the control experiment, and the expression levels of p53 and miR-34a were upregulated by GA in NSCLC cells. These results propose that GA controls the expression of PD-L1 by regulating the p53–miR-34a pathway.

Finally, studies have revealed that PD-L1 expression in cancer cells enhances cell proliferation and resistance toward pro-apoptotic stimuli [29,81]. Furthermore, PD-L1 expression in cancer also enhances PD-L1-mediated tumor immune resistance from cytotoxic cluster of differentiation (CD)8 T-cells through the PD-1/PD-L1 blockade [82]. Thus, inhibition of PD-L1 expression will activate cytotoxic CD8 T-cell responses to various cancers. This approach has been labeled as PD-1/PD-L1 based-immunotherapy. Recently, many clinical approaches and successes are emerging through PD-1/PD-L1 blockade therapy. As mentioned above, we found that GA decreases PD-L1 expression in A549 and H292 cells. In combination experiments with a human monoclonal anti-PD-1 mAb (nivolumab), GA exhibited a more effective effect regarding cancer cell viability. In accordance with the decreased expression levels of PD-L1 by GA, the experimental condition involving anti-PD-1 mAb decreased NSCLC cell viability and oppositely increased the level of IFN-γ in the NSCLC-cell and PBMC coculture system.

4. Materials and Methods

4.1. Cell Lines and Cell Culture

Roswell Park Memorial Institute (RPMI) 1640 media and a penicillin–streptomycin solution was purchased from Gibco (Gaithersburg, MD, USA). Fetal bovine serum (FBS) was purchased from Sigma-Aldrich (St. Louis, MO, USA). Trypsin ethylenediaminetetraacetic acid (0.05%) was obtained from Gibco (Gaithersburg, MD, USA). The human NSCLC cell lines H292 (no. 21848; Korean Cell Line Bank, Seoul, South Korea) and A549 (CCL-185; American Type Culture Collection, Manassas, VA, USA) were cultured in RPMI-1640 supplemented with 10% FBS and antibiotics (1% penicillin–streptomycin) at 37 °C with 5% CO_2. For each experiment, at 70% to 80% confluence, cells were gently washed twice with phosphate-buffered saline. Unless otherwise specified, cells were treated with 100 µM of GA (in H292 cells) or 400 µM of GA (in A549 cells) for 48 hours at 37°C under an atmosphere of 5% CO_2.

4.2. Immunoblotting

Whole-cell lysates were prepared by radioimmunoprecipitation assay buffer (EMD Millipore, Burlington, MA, USA) containing phosphatase and protease inhibitors. Antibodies specific for β-actin (sc-47778), p53 (sc-126) and secondary antibodies (antimouse (sc-516102), and antirabbit (sc-2357) antibody) were obtained from Santa Cruz Biotechnology, Inc. (Dallas, TX, USA). pEGFR (#2234), EGFR (#3776s), pAKT (#4060), AKT (#4691), pPI3K (#4228), and PI3K (#4257) antibodies were purchased from Cell Signaling Technology Inc. (Danvers, MA, USA). PD-L1 (R30949) antibody was purchased from NSJ Bioreagents (San Diego, CA, USA). Recombinant human EGF (AF-100-15) was purchased from PeproTech Inc. (Rocky Hill, NJ, USA).

4.3. Real Time Quantitative PCR (qPCR)

Total RNA was isolated with the RNeasy Mini Kit (Qiagen GmbH, Hilden, Germany) according to the manufacturer's protocol. The extracted RNA was quantified spectrophotometrically at 260 nm, and cDNA was synthesized at 42 °C for 1 h and 95 °C for 5 min with a first-strand cDNA synthesis kit (K-2041; Bioneer Corporation, Daejeon, Korea) and oligo d(T) primers. Real-time qPCR was conducted in a thermal cycler (C1000 Thermal Cycler, Bio-Rad, Hercules, CA) as follows: 2 μL diluted cDNA was added to diluted forward and reverse primers (1 μL each, 100 pM) and 10 μL TB Green Advantage Premix (Takara Bio, Japan) according to the manufacturer's instructions. We used the following primers for EGFR, sense 5'- TGGCAGTGTCTTAGCTGGTTGT -3' and anti-sense 5'- TGCACTCAGAGAGCTCAGGA -3', for PD-L1, sense 5'- TGCCAGGCATTGAATCTACA -3' and anti-sense 5'- GGCCTATTTCCTCCTCTTGG -3', for p53, sense 5'- AGGCCTTGGAACTCAAGGAT -3' and anti-sense 5'- CAGTCTGAGTCAGGCCCTTC -3', and for miR34a analysis: for miR34a, sense 5'- TGGCAGTGTCTTAGCTGGTTGT -3'. The measurement was carried out in triplicate. The relative expression of target genes was normalized to GAPDH or U6 snRNA.

4.4. Transfections of siRNA and miRNA

Lung cancer cells (1×10^5 cells) were seeded in six-well plates and grown to 60% confluence. The cells were then transfected with p53 siRNA (sc-29435; Santa Cruz Biotechnology, Dallas, TX, USA) or miR-34a inhibitor (AM 17000; Thermo Fisher Scientific, Inc., Waltham, MA, USA) using Lipofectamine transfection reagent (Thermo Fisher Scientific, Inc., Waltham, MA, USA). After 48 hours, transfected cells additionally were cultured with/without GA for an additional 48 hours under a cell culture condition.

4.5. NSCLC-Cell and PBMC Co-Culture Experiments

Lung cancer cells (5×10^4 cells) were seeded in 24-well plate until 70% to 80% confluence under a cell culture condition. Human PBMCs were isolated by Ficoll Paque density centrifugation from peripheral blood donated by healthy volunteers using Lymphoprep™ and SepMate™-50 (Stemcell Technologies, Vancouver, Canada). Then, the acquired PBMCs were added into each coculture system at a PBMCs/attached NSCLC cells ratio of 5:1. Some cocultured wells were treated with 5 μg/mL of anti-PD-1 mAb (nivolumab, #A1307; BioVision, Milpitas, CA, USA) and/or GA (A549: 200 μM and H292: 50 μM) and cultured for 48 hours. After 48 hours of co-culture, the culture supernatant was used to analyze the human IFN-γ level, while the viability of attached NSCLC cells was analyzed by MTT assay. The human IFN-γ level in cell-free supernatant was determined using an ELISA kit (#430104; BioLegend, San Diego, CA, USA) according to the manufacturer's protocol.

4.6. Statistical Analyses

All experiments were performed at least three times. Results are expressed as means ± standard errors of the mean. Statistical analyses were conducted using a one-way analysis of variance (ANOVA) or the Student's t-test. The one-way ANOVA was performed with Duncan's multiple-range test as a post-hoc test. Analyses were performed using the SAS 9.3 program (SAS Institute, Inc., Cary, NC, USA). A *p*-value of less than 0.05 was taken to indicate a statistically significant difference.

5. Conclusions

In summary, our results constitute the first study to disclose the detailed mechanism of PD-L1 downregulation, which could be mediated by bioactive natural GA in NSCLC cells. Moreover, we demonstrated that GA might not only directly inhibit cancer cell survival through the upregulation of tumor suppressor p53 but also indirectly enhance antitumor immunity through the downregulation of PD-L1. Thus, our findings additionally pave the way for further research on bioactive natural compounds to study its efficiency in combinations with immune checkpoint-based therapies and chemotherapeutic agents.

Supplementary Materials: The following are available online at http://www.mdpi.com/2072-6694/12/3/727/s1, Figure S1: Effects of GA on A549 and H292 cell viability. (A) MTT-based evaluation of A549 and H292 cell viability in different time- and GA concentration-dependent manners. Data are representative of three independent experiments. *** $p < 0.001$ (*t*-test). (B) A549 and H292 cell lines were treated with increasing concentration of GA for 48 h and the viability was checked using the crystal violet method. Data are representative of three independent experiments. *** $p < 0.001$ (*t*-test). Figure S2: Effects of GA on Huvec cell viability. The cell survival rates of Huvec cells were examined after treatment with 400 μM GA, anti-PD-1 mAb (5 μg/mL), or both GA and anti-PD-1 mAb for 48 hours. Figure S3: Effect of GA in Jak2/STAT3 pathway. (A) The expression levels of pSTAT3 and pJAK2 protein in A549 and H292 cells were detected after GA treatment in concentrations indicated for 48 h. Data are representative of three independent experiments. * $p < 0.05$ (*t*-test). (B) Relative protein levels pJak2 and pSTAT3 were determined by densitometry analysis and normalized to β-actin. Data are representative of three independent experiments.

Author Contributions: K.-J.J. and Y.M.Y. designed the experiments. D.Y.K. and N.S. performed the most of experiments. E.S.J., A.R., and D.Y.H. helped on some of the experiments. K.-J.J., Y.M.Y., D.Y.K., N.S., H.G.L., J.-S.Y., and Q.L. analyzed the data, and K.-J.J. wrote the manuscript. All authors have read and agreed to the published version of the manuscript.

Funding: This work was funded by the National Research Foundation of Korea (NRF) grant funded by the Korea government (MSIT) (No. 2018R1C1B6006146) and by Basic Science Research Program through the National Research Foundation of Korea (NRF) funded by the Ministry of Education (No. 2019R1I1A1A01060399 and 2019R1I1A1A01060537).

Conflicts of Interest: The authors declare no conflict of interest.

References

1. Reck, M.; Heigener, D.F.; Mok, T.; Soria, J.C.; Rabe, K.F. Management of non-small-cell lung cancer: Recent developments. *Lancet* **2013**, *382*, 709–719. [CrossRef]
2. Bender, E. Epidemiology: The dominant malignancy. *Nature* **2014**, *513*, S2–S3. [CrossRef] [PubMed]
3. Inamura, K. Lung Cancer: Understanding its molecular pathology and the 2015 WHO classification. *Front. Oncol.* **2017**, *7*, 193. [CrossRef] [PubMed]
4. Brambilla, E.; Travis, W.D.; Colby, T.V.; Corrin, B.; Shimosato, Y. The new World Health Organization classification of lung tumours. *Eur. Respir. J.* **2001**, *18*, 1059–1068. [CrossRef]
5. Ettinger, D.S.; Akerley, W.; Borghaei, H.; Chang, A.C.; Cheney, R.T.; Chirieac, L.R.; D'Amico, T.A.; Demmy, T.L.; Govindan, R.; Grannis, F.W.; et al. Non-Small Cell Lung Cancer, Version 2.2013 Featured Updates to the NCCN Guidelines. *J. Natl. Compr. Cancer Netw.* **2013**, *11*, 645–653. [CrossRef]
6. Di, S.Y.; Fan, C.X.; Yang, Y.; Jiang, S.; Liang, M.M.; Wu, G.L.; Wang, B.D.; Xin, Z.L.; Hu, W.; Zhu, Y.F.; et al. Activation of endoplasmic reticulum stress is involved in the activity of icariin against human lung adenocarcinoma cells. *Apoptosis* **2015**, *20*, 1229–1241. [CrossRef]
7. Ma, Z.Q.; Yang, Y.; Fan, C.X.; Han, J.; Wang, D.J.; Di, S.Y.; Hu, W.; Liu, D.; Li, X.F.; Reiter, R.J.; et al. Melatonin as a potential anticarcinogen for non-small-cell lung cancer. *Oncotarget* **2016**, *7*, 46768–46784. [CrossRef]
8. Aung, T.N.; Qu, Z.P.; Kortschak, R.D.; Adelson, D.L. Understanding the Effectiveness of Natural Compound Mixtures in Cancer through Their Molecular Mode of Action. *Int. J. Mol. Sci.* **2017**, *18*, 656. [CrossRef]
9. Shahrzad, S.; Aoyagi, K.; Winter, A.; Koyama, A.; Bitsch, I. Pharmacokinetics of gallic acid and its relative bioavailability from tea in healthy humans. *J. Nutr.* **2001**, *131*, 1207–1210. [CrossRef]

10. Abdelwahed, A.; Bouhlel, I.; Skandrani, I.; Valenti, K.; Kadri, M.; Guiraud, P.; Steiman, R.; Mariotte, A.M.; Ghedira, K.; Laporte, F.; et al. Study of antimutagenic and antioxidant activities of gallic acid and 1,2,3,4,6-pentagalloylglucose from Pistacia lentiscus. Confirmation by microarray expression profiling. *Chem. Biol. Interact.* **2007**, *165*, 1–13. [CrossRef]

11. Velderrain-Rodriguez, G.R.; Torres-Moreno, H.; Villegas-Ochoa, M.A.; Ayala-Zavala, J.F.; Robles-Zepeda, R.E.; Wall-Medrano, A.; Gonzalez-Aguilar, G.A. Gallic Acid Content and an Antioxidant Mechanism Are Responsible for the Antiproliferative Activity of 'Ataulfo' Mango Peel on LS180 Cells. *Molecules* **2018**, *23*, 695. [CrossRef] [PubMed]

12. Kim, S.W.; Han, Y.W.; Lee, S.T.; Jeong, H.J.; Kim, S.H.; Kim, I.H.; Lee, S.O.; Kim, D.G.; Kim, S.Z.; Park, W.H. A superoxide anion generator, pyrogallol, inhibits the growth of HeLa cells via cell cycle arrest and apoptosis. *Mol. Carcinog.* **2008**, *47*, 114–125. [CrossRef] [PubMed]

13. Sorrentino, E.; Succi, M.; Tipaldi, L.; Pannella, G.; Maiuro, L.; Sturchio, M.; Coppola, R.; Tremonte, P. Antimicrobial activity of gallic acid against food-related Pseudomonas strains and its use as biocontrol tool to improve the shelf life of fresh black truffles. *Int. J. Food Microbiol.* **2018**, *266*, 183–189. [CrossRef]

14. Lee, J.H.; Oh, M.; Seok, J.H.; Kim, S.; Lee, D.B.; Bae, G.; Bae, H.I.; Bae, S.Y.; Hong, Y.M.; Kwon, S.O.; et al. Antiviral Effects of Black Raspberry (Rubus coreanus) Seed and Its Gallic Acid against Influenza Virus Infection. *Viruses-Basel* **2016**, *8*, 157. [CrossRef] [PubMed]

15. Dludla, P.V.; Nkambule, B.B.; Jack, B.; Mkandla, Z.; Mutize, T.; Silvestri, S.; Orlando, P.; Tiano, L.; Louw, J.; Mazibuko-Mbeje, S.E. Inflammation and Oxidative Stress in an Obese State and the Protective Effects of Gallic Acid. *Nutrients* **2019**, *11*, 23. [CrossRef] [PubMed]

16. You, B.R.; Moon, H.J.; Han, Y.H.; Park, W.H. Gallic acid inhibits the growth of HeLa cervical cancer cells via apoptosis and/or necrosis. *Food Chem. Toxicol.* **2010**, *48*, 1334–1340. [CrossRef]

17. Subramanian, A.P.; Jaganathan, S.K.; Mandal, M.; Supriyanto, E.; Muhamad, I.I. Gallic acid induced apoptotic events in HCT-15 colon cancer cells. *World J. Gastroentero.* **2016**, *22*, 3952–3961. [CrossRef]

18. Tang, H.M.; Cheung, P.C.K. Gallic Acid Triggers Iron-Dependent Cell Death with Apoptotic, Ferroptotic, and Necroptotic Features. *Toxins* **2019**, *11*, 492. [CrossRef]

19. Phan, A.N.; Hua, T.N.; Kim, M.K.; Vo, V.T.; Choi, J.W.; Kim, H.W.; Rho, J.K.; Kim, K.W.; Jeong, Y. Gallic acid inhibition of Src-Stat3 signaling overcomes acquired resistance to EGF receptor tyrosine kinase inhibitors in advanced non-small cell lung cancer. *Oncotarget* **2016**, *7*, 54702–54713. [CrossRef]

20. Liao, C.C.; Chen, S.C.; Huang, H.P.; Wang, C.J. Gallic acid inhibits bladder cancer cell proliferation and migration via regulating fatty acid synthase (FAS). *J. Food Drug Anal.* **2018**, *26*, 620–627. [CrossRef]

21. Bhattacharya, S.; Muhammad, N.; Steele, R.; Peng, G.; Ray, R.B. Immunomodulatory role of bitter melon extract in inhibition of head and neck squamous cell carcinoma growth. *Oncotarget* **2016**, *7*, 33202–33209. [CrossRef] [PubMed]

22. Raina, K.; Rajamanickam, S.; Deep, G.; Singh, M.; Agarwal, R.; Agarwal, C. Chemopreventive effects of oral gallic acid feeding on tumor growth and progression in TRAMP mice. *Mol. Cancer Ther.* **2008**, *7*, 1258–1267. [CrossRef] [PubMed]

23. Baitsch, L.; Baumgaertner, P.; Devevre, E.; Raghav, S.K.; Legat, A.; Barba, L.; Wieckowski, S.; Bouzourene, H.; Deplancke, B.; Romero, P.; et al. Exhaustion of tumor-specific CD8(+) T cells in metastases from melanoma patients. *J. Clin. Investig.* **2011**, *121*, 2350–2360. [CrossRef] [PubMed]

24. Pauken, K.E.; Wherry, E.J. Overcoming T cell exhaustion in infection and cancer. *Trends Immunol.* **2015**, *36*, 265–276. [CrossRef]

25. Sharma, P.; Allison, J.P. The future of immune checkpoint therapy. *Science* **2015**, *348*, 56–61. [CrossRef]

26. Zou, W.; Wolchok, J.D.; Chen, L. PD-L1 (B7-H1) and PD-1 pathway blockade for cancer therapy: Mechanisms, response biomarkers, and combinations. *Sci. Transl. Med.* **2016**, *8*, 328rv4–328rv4. [CrossRef]

27. Havel, J.J.; Chowell, D.; Chan, T.A. The evolving landscape of biomarkers for checkpoint inhibitor immunotherapy. *Nat. Rev. Cancer* **2019**, *19*, 133–150. [CrossRef]

28. Swaika, A.; Hammond, W.A.; Joseph, R.W. Current state of anti-PD-L1 and anti-PD-1 agents in cancer therapy. *Mol. Immunol.* **2015**, *67*, 4–17. [CrossRef]

29. Escors, D.; Gato-Canas, M.; Zuazo, M.; Arasanz, H.; Garcia-Granda, M.J.; Vera, R.; Kochan, G. The intracellular signalosome of PD-L1 in cancer cells. *Signal Transduct. Target. Ther.* **2018**, *3*, 26. [CrossRef]

30. Pascual, M.; Mena-Varas, M.; Robles, E.F.; Garcia-Barchino, M.J.; Panizo, C.; Hervas-Stubbs, S.; Alignani, D.; Sagardoy, A.; Martinez-Ferrandis, J.I.; Bunting, K.L.; et al. PD-1/PD-L1 immune checkpoint and p53 loss facilitate tumor progression in activated B-cell diffuse large B-cell lymphomas. *Blood* **2019**, *133*, 2401–2412. [CrossRef]

31. Cha, Y.J.; Kim, H.R.; Lee, C.Y.; Cho, B.C.; Shim, H.S. Clinicopathological and prognostic significance of programmed cell death ligand-1 expression in lung adenocarcinoma and its relationship with p53 status. *Lung Cancer* **2016**, *97*, 73–80. [CrossRef] [PubMed]

32. Thiem, A.; Hesbacher, S.; Kneitz, H.; di Primio, T.; Heppt, M.V.; Hermanns, H.M.; Goebeler, M.; Meierjohann, S.; Houben, R.; Schrama, D. IFN-gamma-induced PD-L1 expression in melanoma depends on p53 expression. *J. Exp. Clin. Cancer Res.* **2019**, *38*, 397. [CrossRef] [PubMed]

33. Cortez, M.A.; Ivan, C.; Valdecanas, D.; Wang, X.; Peltier, H.J.; Ye, Y.; Araujo, L.; Carbone, D.P.; Shilo, K.; Giri, D.K.; et al. PDL1 Regulation by p53 via miR-34. *J. Natl. Cancer Inst.* **2016**, *108*. [CrossRef] [PubMed]

34. Wang, X.; Cao, X.; Sun, R.; Tang, C.; Tzankov, A.; Zhang, J.; Manyam, G.C.; Xiao, M.; Miao, Y.; Jabbar, K.; et al. Clinical Significance of PTEN Deletion, Mutation, and Loss of PTEN Expression in De Novo Diffuse Large B-Cell Lymphoma. *Neoplasia* **2018**, *20*, 574–593. [CrossRef] [PubMed]

35. Xu, C.; Fillmore, C.M.; Koyama, S.; Wu, H.; Zhao, Y.; Chen, Z.; Herter-Sprie, G.S.; Akbay, E.A.; Tchaicha, J.H.; Altabef, A.; et al. Loss of Lkb1 and Pten leads to lung squamous cell carcinoma with elevated PD-L1 expression. *Cancer cell* **2014**, *25*, 590–604. [CrossRef]

36. Buchakjian, M.R.; Merritt, N.M.; Moose, D.L.; Dupuy, A.J.; Tanas, M.R.; Henry, M.D. A Trp53fl/flPtenfl/fl mouse model of undifferentiated pleomorphic sarcoma mediated by adeno-Cre injection and in vivo bioluminescence imaging. *PLoS ONE* **2017**, *12*, e0183469. [CrossRef]

37. Xu, L.; Zhang, Y.; Tian, K.; Chen, X.; Zhang, R.; Mu, X.; Wu, Y.; Wang, D.; Wang, S.; Liu, F.; et al. Apigenin suppresses PD-L1 expression in melanoma and host dendritic cells to elicit synergistic therapeutic effects. *J. Exp. Clin. Cancer Res.* **2018**, *37*, 261. [CrossRef]

38. Sigismund, S.; Avanzato, D.; Lanzetti, L. Emerging functions of the EGFR in cancer. *Mol. Oncol.* **2018**, *12*, 3–20. [CrossRef]

39. Bell, D.W.; Lynch, T.J.; Haserlat, S.M.; Harris, P.L.; Okimoto, R.A.; Brannigan, B.W.; Sgroi, D.C.; Muir, B.; Riemenschneider, M.J.; Iacona, R.B.; et al. Epidermal growth factor receptor mutations and gene amplification in non-small-cell lung cancer: Molecular analysis of the IDEAL/INTACT gefitinib trials. *J. Clin. Oncol.* **2005**, *23*, 8081–8092. [CrossRef]

40. Erlichman, C.; Hidalgo, M.; Boni, J.P.; Martins, P.; Quinn, S.E.; Zacharchuk, C.; Amorusi, P.; Adjei, A.A.; Rowinsky, E.K. Phase I study of EKB-569, an irreversible inhibitor of the epidermal growth factor receptor, in patients with advanced solid tumors. *J. Clin. Oncol.* **2006**, *24*, 2252–2260. [CrossRef]

41. Sp, N.; Kang, D.Y.; Joung, Y.H.; Park, J.H.; Kim, W.S.; Lee, H.K.; Song, K.D.; Park, Y.M.; Yang, Y.M. Nobiletin Inhibits Angiogenesis by Regulating Src/FAK/STAT3-Mediated Signaling through PXN in ER+ Breast Cancer Cells. *Int. J. Mol. Sci.* **2017**, *18*. [CrossRef] [PubMed]

42. Kang, D.Y.; Sp, N.; Kim, D.H.; Joung, Y.H.; Lee, H.G.; Park, Y.M.; Yang, Y.M. Salidroside inhibits migration, invasion and angiogenesis of MDAMB 231 TNBC cells by regulating EGFR/Jak2/STAT3 signaling via MMP2. *Int. J. Oncol.* **2018**, *53*, 877–885. [PubMed]

43. Chen, J.; Jiang, C.C.; Jin, L.; Zhang, X.D. Regulation of PD-L1: A novel role of pro-survival signalling in cancer. *Ann. Oncol.* **2016**, *27*, 409–416. [CrossRef] [PubMed]

44. Abdelhamed, S.; Ogura, K.; Yokoyama, S.; Saiki, I.; Hayakawa, Y. AKT-STAT3 Pathway as a Downstream Target of EGFR Signaling to Regulate PD-L1 Expression on NSCLC cells. *J. Cancer* **2016**, *7*, 1579–1586. [CrossRef] [PubMed]

45. Ogawara, Y.; Kishishita, S.; Obata, T.; Isazawa, Y.; Suzuki, T.; Tanaka, K.; Masuyama, N.; Gotoh, Y. Akt enhances Mdm2-mediated ubiquitination and degradation of p53. *J. Biol. Chem.* **2002**, *277*, 21843–21850. [CrossRef] [PubMed]

46. Gong, A.Y.; Zhou, R.; Hu, G.; Li, X.; Splinter, P.L.; O'Hara, S.P.; LaRusso, N.F.; Soukup, G.A.; Dong, H.; Chen, X.M. MicroRNA-513 regulates B7-H1 translation and is involved in IFN-gamma-induced B7-H1 expression in cholangiocytes. *J. Immunol.* **2009**, *182*, 1325–1333. [CrossRef]

47. Guo, W.; Tan, W.; Liu, S.; Huang, X.; Lin, J.; Liang, R.; Su, L.; Su, Q.; Wang, C. MiR-570 inhibited the cell proliferation and invasion through directly targeting B7-H1 in hepatocellular carcinoma. *Tumour Biol.* **2015**, *36*, 9049–9057. [CrossRef]

48. Lowe, S.W.; Bodis, S.; Mcclatchey, A.; Remington, L.; Ruley, H.E.; Fisher, D.E.; Housman, D.E.; Jacks, T. P53 Status and the Efficacy of Cancer-Therapy in-Vivo. *Science* **1994**, *266*, 807–810. [CrossRef]

49. Santoro, A.; Vlachou, T.; Luzi, L.; Melloni, G.; Mazzarella, L.; D'Elia, E.; Aobuli, X.; Pasi, C.E.; Reavie, L.; Bonetti, P.; et al. p53 Loss in Breast Cancer Leads to Myc Activation, Increased Cell Plasticity, and Expression of a Mitotic Signature with Prognostic Value. *Cell Rep.* **2019**, *26*, 624. [CrossRef]

50. Chiarugi, V.; Ruggiero, M. Role of three cancer "master genes" p53, bcl2 and c-myc on the apoptotic process. *Tumori* **1996**, *82*, 205–209.

51. Espinosa, E.; Zamora, P.; Feliu, J.; Baron, M.G. Classification of anticancer drugs - a new system based on therapeutic targets. *Cancer Treat. Rev.* **2003**, *29*, 515–523. [CrossRef]

52. Sun, J.C.; Wei, Q.; Zhou, Y.B.; Wang, J.Q.; Liu, Q.; Xu, H. A systematic analysis of FDA-approved anticancer drugs. *BMC Syst. Biol.* **2017**, *11*, 27–43. [CrossRef]

53. Xu, X.L.; Gu, H.; Wang, Y.; Wang, J.; Qin, P. Autoencoder Based Feature Selection Method for Classification of Anticancer Drug Response. *Front Genet.* **2019**, *10*, 233. [CrossRef] [PubMed]

54. Schiller, J.H.; Harrington, D.; Belani, C.P.; Langer, C.; Sandler, A.; Krook, J.; Zhu, J.M.; Johnson, D.H.; Grp, E.C.O. Comparison of four chemotherapy regimens for advanced non-small-cell lung cancer. *New Engl. J. Med.* **2002**, *346*, 92–98. [CrossRef]

55. Hanna, N.; Shepherd, F.A.; Fossella, F.V.; Pereira, J.R.; De Marinis, F.; von Pawel, J.; Gatzemeier, U.; Tsao, T.C.Y.; Pless, M.; Muller, T.; et al. Randomized phase III trial of pemetrexed versus docetaxel in patients with non-small-cell lung cancer previously treated with chemotherapy. *J. Clin. Oncol.* **2004**, *22*, 1589–1597. [CrossRef] [PubMed]

56. Winton, T.; Livingston, R.; Johnson, D.; Rigas, J.; Johnston, M.; Butts, C.; Cormier, Y.; Goss, G.; Inculet, R.; Vallieres, E.; et al. Vinorelbine plus cisplatin vs. observation in resected non-small-cell lung cancer. *New Engl. J. Med.* **2005**, *352*, 2589–2597. [CrossRef]

57. Le Chevalier, T.; Arriagada, R.; Le Pechoux, C.; Grunenwald, D.; Dunant, A.; Pignon, J.P.; Tarayre, M.; Abratt, R.; Arriagada, R.; Bergman, B.; et al. Cisplatin-based adjuvant chemotherapy in patients with completely resected non-small-cell lung cancer. *New Engl. J. Med.* **2004**, *350*, 351–360.

58. Sp, N.; Kang, D.Y.; Kim, D.H.; Park, J.H.; Lee, H.G.; Kim, H.J.; Darvin, P.; Park, Y.M.; Yang, Y.M. Nobiletin Inhibits CD36-Dependent Tumor Angiogenesis, Migration, Invasion, and Sphere Formation Through the Cd36/Stat3/Nf-Kb Signaling Axis. *Nutrients* **2018**, *10*, 772. [CrossRef]

59. Kang, D.Y.; Darvin, P.; Yoo, Y.B.; Joung, Y.H.; Sp, N.; Byun, H.J.; Yang, Y.M. Methylsulfonylmethane inhibits HER2 expression through STAT5b in breast cancer cells. *Int. J. Oncol.* **2016**, *48*, 836–842. [CrossRef]

60. Sp, N.; Darvin, P.; Yoo, Y.B.; Joung, Y.H.; Kang, D.Y.; Kim, D.N.; Hwang, T.S.; Kim, S.Y.; Kim, W.S.; Lee, H.K.; et al. The combination of methylsulfonylmethane and tamoxifen inhibits the Jak2/STAT5b pathway and synergistically inhibits tumor growth and metastasis in ER-positive breast cancer xenografts. *BMC Cancer* **2015**, *15*, 474. [CrossRef]

61. Schlessinger, J. Receptor Tyrosine Kinases: Legacy of the First Two Decades. *Cold Spring Harbor Perspect. Biol.* **2014**, *6*, a008912. [CrossRef] [PubMed]

62. Hynes, N.E.; Lane, H.A. ERBB receptors and cancer: The complexity of targeted inhibitors. *Nat. Rev. Cancer* **2005**, *5*, 341–354. [CrossRef] [PubMed]

63. Yarden, Y.; Sliwkowski, M.X. Untangling the ErbB signalling network. *Nat. Rev. Mol. Cell Biol.* **2001**, *2*, 127–137. [CrossRef] [PubMed]

64. Sforza, V.; Martinelli, E.; Ciardiello, F.; Gambardella, V.; Napolitano, S.; Martini, G.; Della Corte, C.; Cardone, C.; Ferrara, M.L.; Reginelli, A.; et al. Mechanisms of resistance to anti-epidermal growth factor receptor inhibitors in metastatic colorectal cancer. *World J. Gastroenterol.* **2016**, *22*, 6345–6361. [CrossRef] [PubMed]

65. Li, M.; Yang, J.Y.; Zhou, W.L.; Ren, Y.; Wang, X.X.; Chen, H.P.; Zhang, J.Y.; Chen, J.L.; Sun, Y.H.; Cui, L.J.; et al. Activation of an AKT/FOXM1/STMN1 pathway drives resistance to tyrosine kinase inhibitors in lung cancer. *Brit. J. Cancer* **2017**, *117*, 974–983. [CrossRef]

66. Jacobsen, K.; Bertran-Alamillo, J.; Molina, M.A.; Teixido, C.; Karachaliou, N.; Pedersen, M.H.; Castellvi, J.; Garzon, M.; Codony-Servat, C.; Codony-Servat, J.; et al. Convergent Akt activation drives acquired EGFR inhibitor resistance in lung cancer. *Nat. Commun.* **2017**, *8*, 1–14. [CrossRef]

67. Sionov, R.V.; Haupt, Y. The cellular response to p53: The decision between life and death. *Oncogene* **1999**, *18*, 6145–6157. [CrossRef]

68. Burns, T.F.; El-Deiry, W.S. The p53 pathway and apoptosis. *J. Cell Physiol.* **1999**, *181*, 231–239. [CrossRef]
69. Matsuda, K.; Tanikawa, C. The transcriptional landscape of p53 signaling pathway. *Cancer Res.* **2017**, *77*, 109–119.
70. Blandino, G.; Di Agostino, S. New therapeutic strategies to treat human cancers expressing mutant p53 proteins. *J. Exp. Clin. Cancer Res.* **2018**, *37*, 30. [CrossRef]
71. Zheng, L.; Ren, J.Q.; Ll, H.; Kong, Z.L.; Zhu, H.G. Downregulation of wild-type p53 protein by HER-2/neu mediated PI3K pathway activation in human breast cancer cells: Its effect on cell proliferation and implication for therapy. *Cell Res.* **2004**, *14*, 497–506. [CrossRef] [PubMed]
72. Zhang, Y.Z.; Han, C.Y.; Duan, F.G.; Fan, X.X.; Yao, X.J.; Parks, R.J.; Tang, Y.J.; Wang, M.F.; Liu, L.; Tsang, B.K.; et al. p53 sensitizes chemoresistant non-small cell lung cancer via elevation of reactive oxygen species and suppression of EGFR/PI3K/AKT signaling. *Cancer Cell Int.* **2019**, *19*, 188. [CrossRef] [PubMed]
73. Porta, C.; Hadj-Slimane, R.; Nejmeddine, M.; Pampin, M.; Tovey, M.G.; Espert, L.; Alvarez, S.; Chelbi-Alix, M.K. Interferons alpha and gamma induce p53-dependent and p53-independent apoptosis, respectively. *Oncogene* **2005**, *24*, 605–615. [CrossRef] [PubMed]
74. Takaoka, A.; Hayakawa, S.; Yanai, H.; Stoiber, D.; Negishi, H.; Kikuchi, H.; Sasaki, S.; Imai, K.; Shibue, T.; Honda, K.; et al. Integration of interferon-alpha/beta signalling to p53 responses in tumour suppression and antiviral defence. *Nature* **2003**, *424*, 516–523. [CrossRef] [PubMed]
75. Xue, W.; Zender, L.; Miething, C.; Dickins, R.A.; Hernando, E.; Krizhanovsky, V.; Cordon-Cardo, C.; Lowe, S.W. Senescence and tumour clearance is triggered by p53 restoration in murine liver carcinomas. *Nature* **2007**, *445*, 656–660. [CrossRef]
76. Menendez, D.; Shatz, M.; Azzam, K.; Garantziotis, S.; Fessler, M.B.; Resnick, M.A. The Toll-Like Receptor Gene Family Is Integrated into Human DNA Damage and p53 Networks. *PLos Genet.* **2011**, *7*, e1001360. [CrossRef]
77. Shatz, M.; Menendez, D.; Resnick, M.A. The Human TLR Innate Immune Gene Family Is Differentially Influenced by DNA Stress and p53 Status in Cancer Cells. *Cancer Res.* **2012**, *72*, 3948–3957. [CrossRef]
78. Textor, S.; Fiegler, N.; Arnold, A.; Porgador, A.; Hofmann, T.G.; Cerwenka, A. Human NK Cells Are Alerted to Induction of p53 in Cancer Cells by Upregulation of the NKG2D Ligands ULBP1 and ULBP2. *Cancer Res.* **2011**, *71*, 5998–6009. [CrossRef]
79. Quan, X.W.; Li, X.L.; Yin, Z.H.; Ren, Y.W.; Zhou, B.S. p53/miR-30a-5p/SOX4 feedback loop mediates cellular proliferation, apoptosis, and migration of non-small-cell lung cancer. *J. Cell Physiol.* **2019**, *234*, 22884–22895. [CrossRef]
80. Biamonte, F.; Battaglia, A.M.; Zolea, F.; Oliveira, D.M.; Aversa, I.; Santamaria, G.; Giovannone, E.D.; Rocco, G.; Viglietto, G.; Costanzo, F. Ferritin heavy subunit enhances apoptosis of non-small cell lung cancer cells through modulation of miR-125b/p53 axis. *Cell Death Dis.* **2018**, *9*. [CrossRef]
81. Wu, Y.L.; Chen, W.Y.; Xu, Z.P.; Gu, W.Y. PD-L1 Distribution and Perspective for Cancer Immunotherapy-Blockade, Knockdown, or Inhibition. *Front. Immunol.* **2019**, *10*, 2022. [CrossRef] [PubMed]
82. Pardoll, D.M. The blockade of immune checkpoints in cancer immunotherapy. *Nat. Rev. Cancer* **2012**, *12*, 252–264. [CrossRef] [PubMed]

Article

Cancer Glycolytic Dependence as a New Target of Olive Leaf Extract

Jessica Ruzzolini [1], Silvia Peppicelli [1], Francesca Bianchini [1], Elena Andreucci [1], Silvia Urciuoli [2], Annalisa Romani [2], Katia Tortora [3], Giovanna Caderni [3], Chiara Nediani [1,*] and Lido Calorini [1,4,*]

[1] Department of Experimental and Clinical Biomedical Sciences "Mario Serio", University of Florence, 50134 Florence, Italy; jessica.ruzzolini@unifi.it (J.R.); silvia.peppicelli@unifi.it (S.P.); francesca.bianchini@unifi.it (F.B.); e.andreucci@unifi.it (E.A.)

[2] PHYTOLAB (Pharmaceutical, Cosmetic, Food Supplement Technology and Analysis)-DiSIA, Scientific and Technological Pole, University of Florence, 50019 Sesto Fiorentino (Florence), Italy; silvia.urciuoli@gmail.com (S.U.); annalisa.romani@unifi.it (A.R.)

[3] NEUROFARBA Department, Pharmacology and Toxicology Section, University of Florence, 50139 Florence, Italy; katia.tortora@unifi.it (K.T.); giovanna.caderni@unifi.it (G.C.)

[4] Center of Excellence for Research, Transfer and High Education DenoTHE University of Florence, 50134 Florence, Italy

* Correspondence: chiara.nediani@unifi.it (C.N.); lido.calorini@unifi.it (L.C.); Tel.: +39-0552751203 (C.N.); +39-0552751286 (L.C.)

Received: 19 December 2019; Accepted: 26 January 2020; Published: 29 January 2020

Abstract: Oleuropein (Ole), the main bioactive phenolic component of *Olea europaea* L. has recently attracted the scientific attention for its several beneficial properties, including its anticancer effects. This study is intended to investigate whether an olive leaf extract enriched in Ole (OLEO) may counteract the aerobic glycolysis exploited by tumor cells. We found that OLEO decreased melanoma cell proliferation and motility. OLEO was also able to reduce the rate of glycolysis of human melanoma cells without affecting oxidative phosphorylation. This reduction was associated with a significant decrease of glucose transporter-1, protein kinase isoform M2 and monocarboxylate transporter-4 expression, possible drivers of such glycolysis inhibition. Extending the study to other tumor histotypes, we observed that the metabolic effects of OLEO are not confined to melanoma, but also confirmed in colon carcinoma, breast cancer and chronic myeloid leukemia. In conclusion, OLEO represents a natural product effective in reducing the glycolytic metabolism of different tumor types, revealing an extended metabolic inhibitory activity that may be well suited in a complementary anti-cancer therapy.

Keywords: olive leaf extract; oleuropein; Seahorse analysis; cancer metabolism; glycolytic markers

---

1. Introduction

The Pasteur effect describes the inhibition of glycolysis by oxygen, through the inhibition of phosphofructokinase-1, the most important controlling enzyme of glycolysis, by ATP and citrate [1]. Otto Warburg, for the first time, showed that cancer cells do not follow this principle, since even under normoxic condition they prefer to exploit the glycolytic pathway, producing lactic acid from glucose. Indeed, while normal cells in the presence of oxygen use the oxidative phosphorylation, most cancer cells prefer the glycolysis, a phenomenon termed "Warburg effect" or aerobic glycolysis [2]. This is of a special importance for proliferating cancer cells which may regenerate NAD^+, increase the availability of glycolytic biosynthetic intermediates and lactate production. Lactate may contribute to sustain proliferation either by stimulating the production of vascular endothelial growth factor or by promoting cellular motility, two favorable aspects for proliferating cancer cells, e.g., generation of new blood vessels and expansion in neighboring tissues [3]. Lactic acid production and its release into the tumor microenvironment helps reduce the local extracellular pH, which might be instrumental for tumor progression, by promoting the invasive abilities of cancer cells [4], their resistance to apoptotic stimuli as well as chemo- and target therapies [5], by inducing anoikis resistance thus favouring tumor cell survival into the circulatory system [6], and importantly, by inhibiting the immune response supporting tumor cell escape [7]. The Warburg phenotype is regulated by numerous oncogenes, e.g., MYC transcription factor has been found to activate lactate dehydrogenase (LDH)A [8], and promote the switch from pyruvate kinase muscle isozyme 1 (PKM1) to 2 (PKM2), a limiting glycolytic enzyme of the final step of glycolysis, involved in the pyruvate and ATP production from phosphoenolpyruvate [9]. PKM2, in its less active dimeric form, reduces ATP generation leading to the production of lactate and several glycolytic intermediates, used as building blocks for the biosynthesis of cellular macromolecules, such as amino acids, lipids and nucleotides. In addition, mammalian target of rapamycin (mTOR) was demonstrated to be a key activator of the Warburg effect, as it induces under normoxic conditions several glycolytic enzymes, including PKM2 [10].

Recently, plant-derived compounds have drawn the attention of the scientific community for their several beneficial properties. In particular, polyphenols have been subjected to numerous studies and they showed anti-oxidant, anti-inflammatory, cardio- and neuro-protective functions as well as anti-cancer activity [11–15]. Moreover, their anti-cancer activity has been proved in a broad range of cancer models, so that some of these natural compounds have been included in clinical trials [16,17], as they showed promising effects in terms of promoting the anti-cancer response and decreasing at the same time the toxicity of conventional therapies [18–22].

Oleuropein (Ole) is the main bioactive phenolic compound of *Olea europaea* L. that has attracted great interest in the prevention and therapy of several non–communicable diseases, including cancer [23]. As to its anti-cancer properties, Ole affects and modulates multiple different biochemical processes and pathways involved in carcinogenesis. Indeed, Ole exerts an inhibitory effect on cancer cell proliferation, tumor growth and angiogenesis; it reduces inflammation and induces apoptosis [23–25]. In our previous study we found that Ole affects both the proliferation and the viability of A375 BRAF melanoma cells and potentiates their therapy response through pAKT/mTOR pathway [26]. In addition, we observed that an olive leaf extract enriched in Ole (OLEO), used at equimolar Ole concentration, was more effective to potentiate the cytotoxic effect, co-administered with conventional chemotherapeutic agents, compared to Ole alone [26]. Following this line of research, we decided to investigate if OLEO could be able to inhibit the metabolism of BRAF melanoma cells, that are usually glycolysis-addicted. The existence of a strong link between tumor-specific signalling pathways and metabolic adaptations is well known. Therefore, interfering with metabolic processes and metabolic enzymes may be a key strategy for cancer therapy. In this context, significant efforts have been recently done to elucidate how plant-derived natural compounds may act as modulators of tumor cell metabolism and, in this way, exert their anti-cancer activity [27].

Gerhauser, revising the knowledge on tumor metabolism and epigenetic variation of glycolytic genes, discovered that several of these processes are influenced by natural compounds [28]. Then, Gao and Chen underlined how several natural compounds may regulate HIF-1α-dependent anaerobic glycolysis of tumor cells: this actually represents a great contribution underlining the ability of natural products to inhibit one of the most critical transcription factors, i.e., HIF-1α, in cancer progression [29]. In this study, we proved that OLEO is able to reduce the glycolytic rate of both primary and metastatic melanoma cells, reducing the expression levels of critical glucose and lactate transporters (glucose transporter-1 (GLUT1) and monocarboxylate transporter-4 (MCT4), respectively) and enzymes, such as PKM2. Extending the study to other tumor types, we observed that OLEO is able to inhibit the glycolytic metabolism also in colorectal, breast and chronic myeloid leukemia cancer cells.

2. Results

In a previous work, with the aim to verify whether Ole might potentiate drug efficiency on BRAF mutant melanoma cells, we decided to use a non-toxic 250 μM dose able to reduce cell proliferation rate without affecting cancer cell viability and apoptosis. We found that Ole potentiates the cytotoxic effect of everolimus against BRAF melanoma cells inhibiting pAKT/mTOR pathway, as measured by the decrease of pAKT/S6. This effect was also demonstrated using an olive leaf extract enriched in an equimolar concentration of Ole [26]. Here, we confirmed that a similar OLEO, at a 200 μM dose, reduces the viability of A375 melanoma cells in a very limited amount (see the 48 and 72 h of treatment), as cell proliferation without modifying cell cycle phase distribution (Figure 1A–C). The same concentration of the extract does not modify viability of human mesenchymal stem cells at each time point of the experiments (see Figure S1). Further, the OLEO, at a 200 μM dose, significantly reduced the closure of a wound (Figure 1D), which was used as an assay of cell motility. The reduced closure of wounds of OLEO-treated melanoma cells discloses the ability of this natural product to inhibit cell motility. These findings prompted to investigate effects of OLEO on melanoma metabolism. We know that V600E mutant BRAF melanoma cells are strictly addicted to glycolysis, the so-called Warburg effect, thus it was possible that a reduction of the glycolytic pathway may have a role in the decreased proliferation and motility of OLEO-treated melanoma cells.

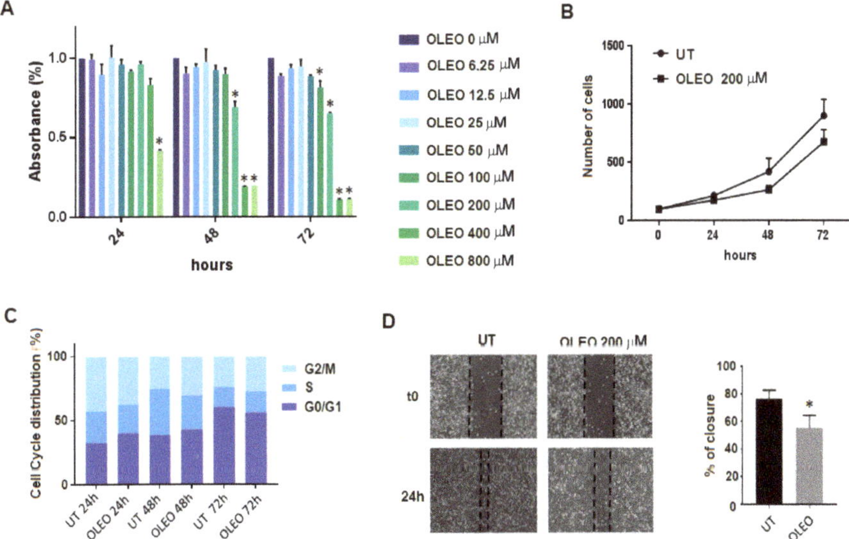

Figure 1. Effects of Ole-enriched leaf extract (OLEO) on A375 melanoma cells. (**A**) Dose-time response evaluated by MTT assay. Significance is indicated with *; (**B**) Cell growth of A375 human melanoma cells treated with OLEO 200 μM; (**C**) Cell cycle distribution analyzed using FACS; (**D**) Effect of OLEO 200 μM on the motility of A375 cells evaluated by scratch wound healing assay. Significance is indicated with *.

We evaluated the metabolic profile of melanoma cells after 200 μM OLEO administration through Seahorse Bioanalyzer XF96 analysis, thereafter studying the metabolic markers through real time PCR and Western Blot analysis.

We first tested the effect of OLEO extract on the glycolytic activity of A375 melanoma cells using a glycostress standard assay. Overall, OLEO impairs glycolysis rate without modifying glycolytic capacity and reserve of melanoma cells (Figure 2A). On the other hand, Mito stress analysis indicates that OLEO does not modify the respiration of melanoma cells.

To add information on OLEO-driven glycolysis inhibition, we exploited the Seahorse Bioanalyzer XF96 to check the metabolism of A375 melanoma cells exposed for 24 h to an equimolar concentration of pure Oleuropein, and we found an equivalent reducing effect on the glycolysis rate (Figure 2B), without any modification of respiration of these cells.

Further, to sustain the inhibitory role on the glycolytic metabolism of melanoma cells by OLEO, we tested its glycolysis inhibitory effect on a metastatic clone of A375 melanoma cells, called A375-M6, isolated from the lung of immunodeficient animals. In line with the previous results, we observed that OLEO exerts the same inhibitory effect on glycolysis of these metastatic cells, as showed in the glycostress analysis (Figure 2C). Overall, OLEO is able to repress the aerobic glycolysis of primary and metastatic melanoma cells.

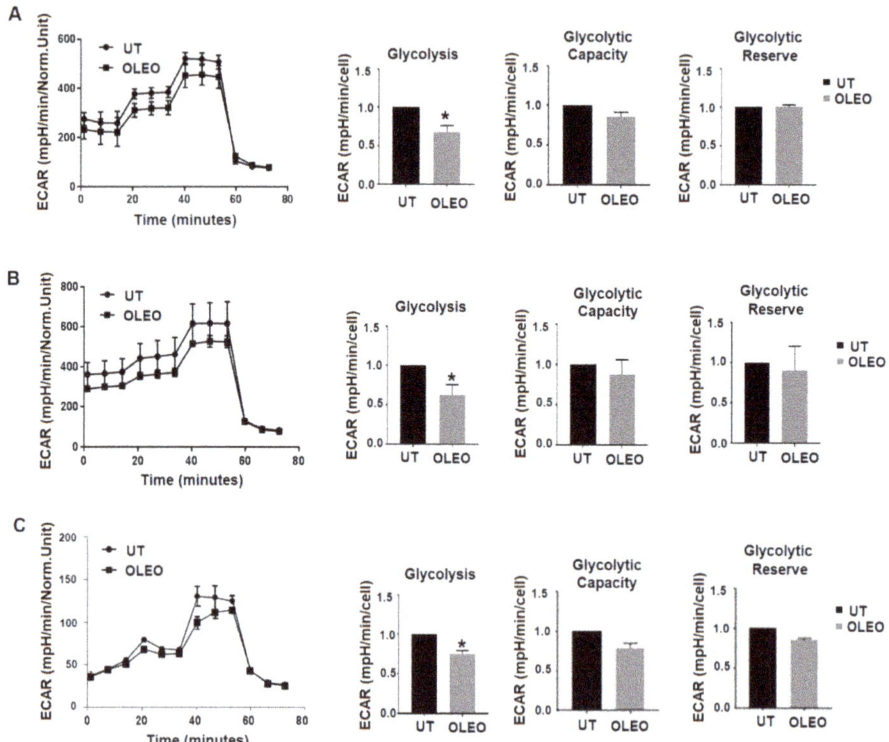

Figure 2. Representative results of a glucose stress test of A375 melanoma cells treated with 200 μM OLEO (**A**) or 200 μM Ole (**B**) for 24 h, and of A375-M6 melanoma cells treated with 200 μM OLEO for 24 h (**C**). Plots on the right represent glycolysis, glycolytic reserve and glycolytic capacity extracted from glycolysis stress assay results obtained using the Seahorse Analyzer. Significance is indicated with *.

Along with the dynamic investigation of metabolism expressed by A375 and A375-M6 melanoma cells following OLEO treatment, we identified a series of glycolytic biomarkers down regulated by our nutraceutical product. Testing both mRNA and protein levels, we observed that glucose transporter isoform 1 (GLUT1), pyruvate kinase isozymes M2 (PKM2) and monocarboxylate transporter 4 (MCT4) of OLEO-treated A375 melanoma cells are reduced by a 50% compared to control (i.e., untreated melanoma cells) (Figure 3). To underline the importance of these three key glycolytic biomarkers inhibited by OLEO: (1) GLUT1 is the major glucose transporter in cancer cells; (2) PKM2 is a modulator of glucose metabolism sustaining building block generation needed for cell proliferation; (3) MCT4 exports lactate and protons produced by glycolysis, preventing the inhibition of glycolytic enzymes such as phosphofructokinase activity, that is reduced by intracellular acidification.

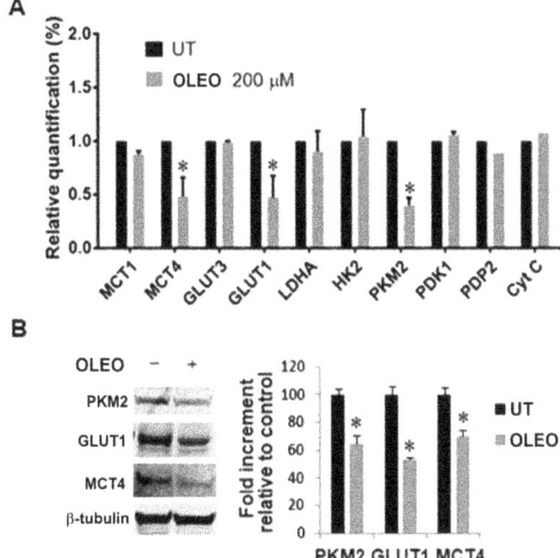

Figure 3. Change in metabolic markers of A375 melanoma cells treated with OLEO 200 µM for 24 h. (**A**) Evaluation by quantitative real-time PCR of genes involved in metabolism; (**B**) Representative Western blot panels of PKM2, GLUT1 and MCT4 protein levels. Each band in the Western blot was quantified by densitometric analysis and the corresponding histogram was constructed by normalizing the density of each band to that of β-tubulin. Values presented are means ± SEM of three independent experiments. Significance is indicated with *.

To extend our investigation on OLEO metabolic inhibition in cancer cells, we also tested HCT116 (a human colorectal carcinoma cell line), MDA-MB-231 (an undifferentiated triple-negative breast cancer cell line) and K562 cells (a chronic myeloid leukemia cell line) through the Seahorse Bioanalyzer XF96. OLEO does not modify substantially number and viability of colorectal, breast and leukemia cancer cells (Figure 4A,C,E), but was effective in reducing glycolysis rate of all these type of cancer cells; a higher dose of OLEO was needed to inhibit glycolysis of K562 cells. Overall, these results demonstrate a metabolic inhibitory activity of OLEO on a wide array of cancer histotypes, including that of malignant cells of a clonal disorder of hematopoietic stem cells (Figure 4B,D,F).

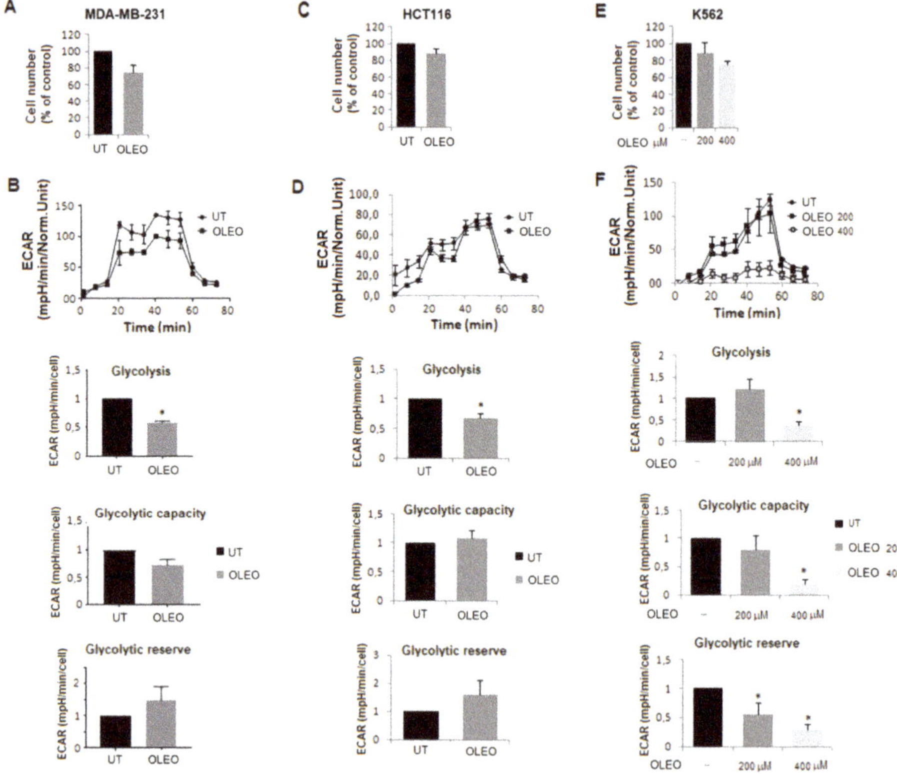

Figure 4. Effect of OLEO on glycolytic metabolism of breast cancer, colon carcinoma and myeloid leukemia cells. Cell number (**A,C,E**) and representative results of a glucose stress test of MDA-MB-231 (**B**), HCT116 (**D**), and K562 (**F**) cells, treated with 200 or 400 μM OLEO for 48 h. Plots on the right represent glycolysis, glycolytic reserve and glycolytic capacity extracted from glycolysis stress assay results obtained using the Seahorse Analyzer. Significance is indicated with *.

3. Discussion

It is well known that cancer cells, compared to normal tissues, are characterized by a high rate of glycolytic metabolism. They indeed prefer to use glycolysis even in the presence of enough oxygen to sustain the oxidative phosphorylation (the so-called "Warburg effect" or aerobic glycolysis, to be distinguished from the anaerobic glycolysis exploited under hypoxic conditions). The higher glycolytic rate of cancer cells ensures them an adequate amount of energy and an ample availability of intermediate macromolecules useful to sustain a rapid cell proliferation and tumor mass expansion [3]. Nowadays, the deregulated metabolism is considered a hallmark of cancer and the identification of new compounds able to modulate tumor metabolism is under intense investigation. For this reason, natural agents can be a great importance, in particular because they demonstrated to interfere with most of the activities of cancer cells, at the same time showing, very low toxic effects on normal cells [11,18,30]. To sum up, several authors have underlined how plant-derived natural products interfere with tumor metabolism [22].

Here we show that OLEO is able to exert a significant inhibitory effect on cancer cell glycolysis. In particular, by a dynamic evaluation of cancer cell metabolism through the Seahorse Bioanalyzer XF96 platform, we observed that OLEO reduces the glycolytic rate of primary and metastatic melanoma cells, but also of colorectal, breast and chronic myeloid leukemia cancer cells. In line with our results, Sharma and colleagues showed that morin and/or esculetin impaired glycolysis and glutaminolysis preventing colon carcinogenesis [31]. Moreover, Gomez de Cedron and colleagues identified a new series of polyphenols characterized by a galloyl based "head" and a hydrophobic N-acyl "tail", able to inhibit glycolysis and mitochondrial respiration in colon cancer cells [32].

We observed that the glycolytic reduction exerted by our OLEO in melanoma cells is associated with a decreased GLUT1 expression, at both the mRNA and protein levels. GLUT1 is an important target in cancer treatment, being over-expressed by a wide range of tumor cells. Cancer cells may indeed take great advantages of the GLUT1 rapid response and its high affinity for glucose, in order to overcome the several stress conditions encountered in the host microenvironment and continue the progression towards malignancy. The K_M value of GLUT1 for glucose is near 1 mM, a significantly less amount compared to the normal glucose level found in serum, allowing a relentless glucose transport into the cells. Of interest, GLUT1 represents the predominant glucose transporter isoform of fetus tissues, which exhibit a higher growth rate than adult ones, at comparable levels to those observed in tumor cells, requiring an increased supply of energy-producing substrates [32]. After birth, GLUT1 expression levels decrease and, even though the reasons behind its decline are not yet clear, it could occur a possible switch form a carbohydrate to a fat source of fuel that may induce this change in some organs [33]. For all these reasons, the development of new clinical strategies involving natural GLUT1 inhibitors such as OLEO, in combination with conventional anticancer agents, deserves the attention of the scientific community, sounding as promising combined therapeutic strategy, as recently reported for other natural compounds [34,35].

Along with GLUT1, we showed that OLEO is also able to down regulate PKM2, one of the four pyruvate kinase isoforms which is highly expressed in rapidly proliferating tissues including cancer. This metabolic enzyme is regulated by oncogenic tyrosine kinases which usually lead to an increase glycolytic rate in tumor cells. Despite, tyrosine phosphorylation of glycolytic enzymes usually increases the activities of a majority of glycolytic enzymes, the tyrosine phosphorylation of PKM2 paradoxically results in a decreased PKM2 activity that in turn promotes the Warburg effect [36]. It is possible that the OLEO-driven PKM2 reduction may reduce its glycolytic promotion [37]. PKM2 overexpression was observed in melanoma human samples compared to naevi, showing a gradient of increased expression from radial growth phase to metastatic melanoma. Furthermore, recent studies have shown that PKM2 is also able to act as a protein kinase using phosphoenolpyruvate as a substrate to promote tumorigenesis [36]. Then, Zhang and colleagues found that miR-625-5p regulates PKM2 expression at both mRNA and protein levels in melanoma cells, disclosing a miR/PKM2 role in glucose metabolism of melanoma cells [38]. PKM2 expression has been shown to be also reduced by other natural products such as resveratrol and curcumin. In particular, resveratrol inhibits aerobic glycolysis and PKM2 enzyme in HeLa (human cervical cancer), HepG2 (human liver cancer) and MCF-7 (human breast cancer) cancer cells through the inhibition of mTOR signaling [39]. Resveratrol was also demonstrated to impair hexokinase-2 enzyme in human non-small cell lung cancer cells inhibiting Akt signaling pathway [40], and pyruvate dehydrogenase complex in colon cancer cells [41]. Curcumin, a further well-known phytopolyphenolic compound, has been shown to decrease glucose uptake and lactate production in several cancer cells (lung, breast, cervical, prostate and embryonic kidney cancer cell lines) down-regulating PKM2 expression, interfering with the mTOR-HIF-1α axis [42].

MCT4 is the other glycolytic marker that we found to be inhibited by OLEO in melanoma cells. This monocarboxylate transporter acquires a key role in the metabolic activity of glycolytic cells through the proton-coupled transport of monocarboxylates, such as L-lactate, ketone bodies and pyruvate. An immunohistochemical study of the expression of MCT4 in 356 melanoma-bearing patients revealed that this glycolytic marker is significantly increased in metastatic lesions and associated with a poor prognosis [43].

To conclude, in this study we demonstrated that OLEO is able to reduce the high glycolytic activity of various solid tumors, like melanoma, colorectal and breast cancer, but also of chronic myeloid leukemia cells, suggesting a possible usage of this natural product in combination with conventional therapy for a wide range of malignancy.

4. Materials and Methods

4.1. Olive Leaf Extract's Preparation and Toxicity

The OLEO used to treat normal and cancer cells was prepared and characterized as previously described ([26], see Figures S3 and S4). *Olea europaea* L. (cultivar Leccino), organic green leaves, were collected in April 2018 in Tuscany (Vinci, Florence, Italy) and immediately processed. The extraction using 15% of *Olea* leaves (45 g leaves/300 g double-distilled and purified water), was performed in water at a temperature of 50 °C for 60 min and at room temperature over the night (12 h) [44]. The final powder is obtained by lyophilization with a LYOVAC GT 2 system (Leybold GmbH, Cologne, Germany), freeze-drying yield 1.85%. The identity of the phenolic compounds of *Olea* dry extract powder and the composition of the solution used for the test in vitro, enriched in oleuropein, was ascertained using data from the HPLC/DAD and HPLC/mass spectrometry analyses, in accordance with a previous paper [45].

All the solvents (HPLC grade) and formic acid (ACS reagent) were purchased from Aldrich Chemical Company Inc. (Milwaukee, WI, USA). Tyrosol, luteolin 7-O-glucoside, chlorogenic and Ole were obtained from Extrasynthese S.A. (Genay, France). The HPLC-grade water was obtained via double-distillation and purification with a Labconco Water Pro PS polishing station (Labconco Corporation, Kansas City, MO, USA).

The OLEO used in our study has been tested in a sub-acute test of toxicity (7 days) on female F344 rats fed a diet containing 2.7 g of extract/kg of diet (corresponding to a dosage of 100 mg of extract/kg of b.w.) without inducing any change of body weight.

This lack of toxicity is in agreement with a recent study of Guex et al., which showed that an olive leaf extract, in vivo tested up to 2000 mg/kg, had no toxic or unwanted effects on rats [46]. Moreover, Sepporta et al. [47]'s paper demonstrated absence of toxicity in mice administered 125 mg of Ole/kg (b.w.). The standard dose of Ole used in vivo animal model, from 10 to 125 mg/kg, did not induce toxic effects, evaluated in terms of viability of the animals [48] or liver biomarkers, such as alanine and aspartate aminotransferase activities [49].

4.2. Cell Lines and Culture Conditions

In this study we used A375 human melanoma cell lines, obtained from American Type Culture Collection (ATCC, Rockville, MD, USA), A375M6, isolated in our laboratory from lung metastasis of SCID bg/bg mice i.v. injected with A375 [5]; human colorectal carcinoma cell line HT116, a kind gift of Dr. Matteo Lulli (Department of Clinical and Experimental Biomedical Sciences, University of Florence, Italy); human breast carcinoma cells MDA-MB-231, obtained from American Type Culture Collection (ATCC); human leukemia cells K562, a kind gift of Prof. Persio Dello Sbarba (Department of Clinical and Experimental Biomedical Sciences, University of Florence) and human mesenchymal stem cells (MSC) obtained from bone marrow aspirates of donors which signed informed consent [50]. A375, A375M6, HCT116 and MDA-MB-231 were cultured in Dulbecco's Modified Eagle Medium high glucose (DMEM 4500, EuroClone, Milan, Italy) supplemented with 10% fetal bovine serum (FBS, EuroClone); K562 were cultured in Roswell Park Memorial Institute 1640 medium (RPMI, EuroClone) supplemented with 10% FBS; MSC were expanded in Dulbecco's modified Eagle's medium with low glucose (DMEM 1000; Gibco, Life Technologies, Monza, Italy) supplemented with 20% FBS. Cells were maintained at 37 °C in humidified atmosphere containing 90% air and 10% CO_2 and they were harvested from subconfluent cultures by incubation with a trypsin-EDTA solution (EuroClone), and propagated every three days. Viability of the cells was determined by trypan blue exclusion test. Cultures were periodically monitored for mycoplasma contamination using Chen's fluorochrome test. Cells were treated with OLEO for 24–72 h.

4.3. MTT Assay

Cell viability was assessed using MTT (3-(4,5-dimethylthiazol-2-yl)-2,5-diphenyltetrazolium bromide) tetrazolium reduction assay (Sigma Aldrich, Milan, Italy) as described in [26]. Cells were plated into 96-multiwell plates in complete medium without red phenol. The treatment was added to the medium colture at different dose and times, according to the experiment. Then the MTT reagent was added to the medium and plates were incubated at 37 °C. After 2 h, MTT was removed and the blue MTT–formazan product was solubilized with dimethyl sulfoxide (DMSO, Sigma Aldrich). The absorbance of the formazan solution was read at 595 nm using the microplate reader (Bio-Rad, Milan, Italy).

4.4. Cell Cycle Analysis

Cell cycle distribution was analyzed by the DNA content using propidium iodide (PI) staining method. Cells were centrifugated and stained with a mixture of 50 µg/mL PI (Sigma-Aldrich, St. Louis, MO, USA), 0.1% trisodium citrate and 0.1% NP40 (or triton x-100) in the dark at 4°C for 30 min. The stained cells were analyzed by flow cytometry (BD-FACS Canto, BD-Biosciences, San Jose, CA, USA) using red propidium-DNA fluorescence as previously described [5].

4.5. Wound Healing Assay

Cell migration was evaluated by an in vitro wound healing assay as previously described [6]. Cells were treated for 24 h with the extract, then cells have been detached and sown in 35 mm dishes at high confluence; cell monolayer was wounded with a sterile 200 mL pipette tip, washed with PBS and incubated in 1% FBS culture medium. Wound was analyzed following a 24-h incubation and photographed using phase contrast microscopy.

4.6. Seahorse Analysis

Seahorse analysis has been performed as previously described [51]. The extracellular acidification rate (ECAR) and the Oxygen Consumption Rate (OCR) were determined using the Seahorse XF96 Extracellular Flux Analyzer (Seahorse Bioscience, Billerica, MA, USA) through Seahorse XF Glycolysis Stress Test Kit (Agilent Technologies, Santa Clara, CA, USA), measuring preferentially the glycolytic function in cells, or Seahorse XF Mito Stress Test Kit (Agilent Technologies), measuring the dependence of cells on the oxidative metabolism. Cells were counted and seeded in XF96 Seahorse® microplates precoated with poly-D-lysine (ThermoFisher Scientific, Waltham, MA, USA). Cells were suspended in XF Assay Medium supplemented with 1 mM glutamine (from EuroClone, Paington, UK) in order to assess ECAR, in XF Assay Medium supplemented with 2 mM glutamine in order to assess OCR. Cells were left to adhere for a minimum of 30 min at 37 °C. The plate was left to equilibrate in a CO_2-free incubator before being transferred to the Seahorse XF96 analyzer. The pre-hydrated cartridge was filled with the indicated compounds and calibrated for 30 min in the Seahorse Analyzer. All the experiments were performed at 37 °C. Normalization to protein content was performed after each experiment. The Seahorse XF Report Generator automatically calculated the parameters from Wave data that have been exported to Excel or Graphpad.

4.7. RNA Isolation and Quantitative PCR (qPCR)

Total RNA was isolated from cells by using TRI Reagent (Sigma, Milan, Italy). The amount and purity of RNA were determined spectrophotometrically. cDNAwas obtained by incubating 2 µg of total

RNA with 4 U/µL of M-MLV reverse transcriptase (Promega, San Luis Obispo, CA, USA) according to the manufacturer's instructions. Quantitative real time PCR (qPCR) was performed as reported in [51] using the GoTaq® Probe Systems (Promega). The qPCR analysis was carried out in triplicate using an Applied Biosystems 7500 Sequence Detector with the default PCR setting: 40 cycles at 95 °C for 15 s, 60 °C for 60 s. mRNA was quantified with the DDCt method as described [52]. mRNA levels were normalized to β2 microglobulin as an endogenous control. The primer sequences used are listed in Table 1.

Table 1. Primer sequences for PCR.

Gene	FW	RV
MCT1	5′-GTGGCTCAGCTCCGTATTGT-3′	5′-GAGCCGACCTAAAAGTGGTG-3′
MCT4	5′-CAGTTCGAGGTGCTCATGG-3′	5′-ATGTAGAGGTGGGTCGCATC-3′
GLUT1	5′-CGGGCCAAGAGTGTGCTAAA-3′	5′-TGACGATACCGGAGCCAATG-3′
GLUT3	5′-CGAACTTCCTAGTCGGATTG-3′	5′-AGGAGGCACGACTTAGACAT-3′
LDHA	5′-AGCCCGATTCCGTTACCT-3′	5′-CACCAGCAACATTCATTCCA-3′
PKM2	5′-CAGAGGCTGCCATCTACCAC-3′	5′-CCAGACTTGGTGAGGACGAT-3′
PDK1	5′-CCAAGACCTCGTGTTGAGACC-3′	5′-AATACAGCTTCAGGTCTCCTTGG-3′
HK2	5′- CAAAGTGACAGTGGGTGTGG-3′	5′- GCCAGGTCCTTCACTGTCTC-3′
18s	5′-CGCCGCTAGAGGTGAAATTCT-3′	5′-CGAACCTCCGA CTTTCGTTCT-3′
PDP2	5′-ACCACCTCCGTGTCTATTGG-3′	5′-CCAGCGAGATGTCAGAATCC-3′
CytC	5′-TTGCACTTACACCGGTACTTAAGC-3′	5′-ACGTCCCCACTCTCTAAGTCCAA-3
GLS1	5′-TGCTACCTGTCTCCATGGCTT-3′	5′-CTTAGATGGCACCTCCTTTGG-3′

4.8. Western Blotting Analysis

Cells were lysed and separated using electrophoresis as previously described [5]. Cells were washed with ice cold PBS containing 1 mM Na4VO3, and lysed in 100 mL of cell RIPA lysis buffer (Merck Millipore, Vimodrone, Milan, Italy) containing PMSF (Sigma-Aldrich), sodium orthovanadate (Sigma-Aldrich) and protease inhibitor cocktail (Calbiochem). Aliquots of supernatants containing equal amounts of protein (40 mg) in Laemmli buffer were separated on Bolt® Bis-Tris Plus gels 4e12% precast polyacrylamide gels (Life Technologies, Monza, Italy). Fractionated proteins were transferred from the gel to a PVDF (polyvinylidene difluoride) membrane using iBlot 2 system (Life Technologies, Monza, Italy). Blots were stained with Ponceau red to ensure equal loading and complete transfer of proteins, and then they were blocked for 1 h, at room temperature, with Odyssey blocking buffer (Dasit Science, Cornaredo, Milan, Italy). Subsequently, the membrane was probed at 4 °C overnight with primary antibodies diluted in a solution of 1:1 Odyssey blocking buffer/T-PBS buffer. The primary antibodies were: rabbit anti-PKM2 (1:1000, Cell Signaling Technology, Danvers, MA, USA), rabbit anti-MCT-4 (1:1000, Santa Cruz Biotechnology, Santa Cruz, CA, USA), rabbit anti-GLUT1 (1:1000, Cell Signaling Technology). The membrane was washed in T-PBS buffer, incubated for 1 h at room temperature with goat anti-rabbit IgG Alexa Flour 750 antibody or with goat antimouse IgG Alexa Fluor 680 antibody (Invitrogen, Monza, Italy), and then visualized by an Odyssey Infrared Imaging System (LI-COR® Bioscience, Lincoln, NE, USA). Mouse anti-β tubulin monoclonal antibody (1:1000, Cell Signaling Technology) was used to assess equal amount of protein loaded in each lane.

4.9. Statistical Analysis

Densitometric data are expressed as means ± standard errors of the mean (SEM) depicted by vertical bars from representative experiment of at least three independent experiments. Statistical analysis of the data was performed by ANOVA and Tukey's multiple comparisons test, and $p \leq 0.05$ was considered statistically significant.

5. Conclusions

The rapid growth of cancer cells mainly depends on their high glycolytic metabolism. Indeed, tumor cells, compared to normal tissues, prefer to exploit the glycolytic pathway even in the presence of sufficient oxygen to sustain the oxidative phosphorylation. This is likely due to the fact that glycolysis guarantees a rapid availability of metabolic intermediates, assuring not only sufficient energy for their survival, but also an efficient production of nucleotides, amino acids and lipids needed to duplicate cell content before mitosis. In this study we show that the natural product OLEO is able to reduce the glycolytic rate of a wide range of solid and liquid tumor cells, without affecting their basal respiration but rather down-regulating the expression of three key effectors of the glycolytic pathway, i.e., GLUT-1, PKM2 and MCT4, likely resulting in a decreased glucose entrance and biomass production. Thus, the inhibition of glycolysis by OLEO acquires a great significance for the targeting of cancer cell growth and expansion. Our findings, together with previous evidence showing the anti-cancer effects exerted by OLEO, pure Ole and analogs [23,53,54], reinforce the hypothesis to propose the use of these natural compounds in combination with conventional therapy used in the treatment of cancer.

Supplementary Materials: The following are available online at http://www.mdpi.com/2072-6694/12/2/317/s1, Figure S1: Cell growth of human MSC treated with OLEO 200 µM, Figure S2: Detailed information of protein expression analysis by Western blot: (A) Original blot for the Figure 3B, (B) Densitometry and intensity ratio of each band.

Author Contributions: Conceptualization, J.R., C.N. and L.C.; validation, S.P., E.A., F.B.; formal analysis, J.R., S.P., E.A.; investigation, J.R., S.P., F.B., G.C., K.T.; resources, S.U., A.R., C.N., L.C.; writing—original draft preparation, J.R., S.P., L.C.; writing—review and editing, F.B., E.A., C.N., A.R., S.U., G.C.; visualization, J.R., S.P.; supervision, L.C., J.R.; project administration, L.C., C.N.; funding acquisition, L.C., C.N., A.R. All authors have read and agreed to the published version of the manuscript.

Funding: This research was funded by Istituto Toscano Tumori and Ente Cassa di Risparmio di Firenze.

References

1. Barker, J.; Khan, M.A.; Solomos, T. Mechanism of the Pasteur effect. *Nature* **1964**, *201*, 1126–1127. [CrossRef]

2. Warburg, O. On respiratory impairment in cancer cells. *Science* **1956**, *124*, 269–270. [PubMed]

3. Vander Heiden, M.G.; Cantley, L.C.; Thompson, C.B. Understanding the Warburg effect: The metabolic requirements of cell proliferation. *Science* **2009**, *324*, 1029–1033. [CrossRef] [PubMed]

4. Peppicelli, S.; Bianchini, F.; Torre, E.; Calorini, L. Contribution of acidic melanoma cells undergoing epithelial-to-mesenchymal transition to aggressiveness of non-acidic melanoma cells. *Clin. Exp. Metastasis* **2014**, *31*, 423–433. [CrossRef] [PubMed]

5. Ruzzolini, J.; Peppicelli, S.; Andreucci, E.; Bianchini, F.; Margheri, F.; Laurenzana, A.; Fibbi, G.; Pimpinelli, N.; Calorini, L. Everolimus selectively targets vemurafenib resistant BRAFV600E melanoma cells adapted to low pH. *Cancer Lett.* **2017**, *408*, 43–54. [CrossRef] [PubMed]

6. Peppicelli, S.; Ruzzolini, J.; Bianchini, F.; Andreucci, E.; Nediani, C.; Laurenzana, A.; Margheri, F.; Fibbi, G.; Calorini, L. Anoikis Resistance as a Further Trait of Acidic-Adapted Melanoma Cells. *J. Oncol.* **2019**, *2019*, 1–13. [CrossRef] [PubMed]

7. Peppicelli, S.; Bianchini, F.; Calorini, L. Extracellular acidity, a "reappreciated" trait of tumor environment driving malignancy: Perspectives in diagnosis and therapy. *Cancer Metastasis Rev.* **2014**, *33*, 823–832. [CrossRef] [PubMed]

8. Shim, H.; Dolde, C.; Lewis, B.C.; Wu, C.S.; Dang, G.; Jungmann, R.A.; Dalla-Favera, R.; Dang, C.V. c-Myc transactivation of LDH-A: Implications for tumor metabolism and growth. *Proc. Natl. Acad. Sci. USA* **1997**, *94*, 6658–6663. [CrossRef]

9. Dong, G.; Mao, Q.; Xia, W.; Xu, Y.; Wang, J.; Xu, L.; Jiang, F. PKM2 and cancer: The function of PKM2 beyond glycolysis. *Oncol. Lett.* **2016**, *11*, 1980–1986. [CrossRef]

10. Sun, Q.; Chen, X.; Ma, J.; Peng, H.; Wang, F.; Zha, X.; Wang, Y.; Jing, Y.; Yang, H.; Chen, R.; et al. Mammalian target of rapamycin up-regulation of pyruvate kinase isoenzyme type M2 is critical for aerobic glycolysis and tumor growth. *Proc. Natl. Acad. Sci. USA* **2011**, *108*, 4129–4134. [CrossRef]

11. Del Rio, D.; Rodriguez-Mateos, A.; Spencer, J.P.; Tognolini, M.; Borges, G.; Crozier, A. Dietary (poly)phenolics in human health: Structures, bioavailability, and evidence of protective effects against chronic diseases. *Antioxid. Redox Signal.* **2013**, *18*, 1818–1892. [CrossRef] [PubMed]

12. Romani, A.; Ieri, F.; Urciuoli, S.; Noce, A.; Marrone, G.; Nediani, C.; Bernini, R. Health Effects of Phenolic Compounds Found in Extra-Virgin Olive Oil, By-Products, and Leaf of Olea europaea L. *Nutrients* **2019**, *11*, 1776. [CrossRef] [PubMed]

13. Castelli, V.; Grassi, D.; Bocale, R.; d'Angelo, M.; Antonosante, A.; Cimini, A.; Ferri, C.; Desideri, G. Diet and Brain Health: Which Role for Polyphenols? *Curr. Pharm. Des.* **2018**, *24*, 227–238. [CrossRef] [PubMed]

14. Boss, A.; Bishop, K.S.; Marlow, G.; Barnett, M.P.; Ferguson, L.R. Evidence to Support the Anti-Cancer Effect of Olive Leaf Extract and Future Directions. *Nutrients* **2016**, *8*, 513. [CrossRef]

15. Milanizadeh, S.; Reza Bigdeli, M. Pro-Apoptotic and Anti-Angiogenesis Effects of Olive Leaf Extract on Spontaneous Mouse Mammary Tumor Model by Balancing Vascular Endothelial Growth Factor and Endostatin Levels. *Nutr. Cancer* **2019**, *71*, 1374–1381. [CrossRef] [PubMed]

16. Butler, M.S.; Robertson, A.A.B.; Cooper, M.A. Natural product and natural product derived drugs in clinical trials. *Nat. Prod. Rep.* **2014**, *31*, 1612–1661. [CrossRef]

17. Seca, A.; Pinto, D. Plant Secondary Metabolites as Anticancer Agents: Successes in Clinical Trials and Therapeutic Application. *Int. J. Mol. Sci.* **2018**, *19*, 263. [CrossRef]

18. Goldsmith, C.D.; Bond, D.R.; Jankowski, H.; Weidenhofer, J.; Stathopoulos, C.E.; Roach, P.D.; Scarlett, C.J. The Olive Biophenols Oleuropein and Hydroxytyrosol Selectively Reduce Proliferation, Influence the Cell Cycle, and Induce Apoptosis in Pancreatic Cancer Cells. *Int. J. Mol. Sci.* **2018**, *19*, 1937. [CrossRef]

19. Siddique, A.B.; Ayoub, N.M.; Tajmim, A.; Meyer, S.A.; Hill, R.A.; El Sayed, K.A. (−)-Oleocanthal Prevents Breast Cancer Locoregional Recurrence After Primary Tumor Surgical Excision and Neoadjuvant Targeted Therapy in Orthotopic Nude Mouse Models. *Cancers* **2019**, *11*, 637. [CrossRef]

20. Zhou, Q.; Pan, H.; Li, J. Molecular Insights into Potential Contributions of Natural Polyphenols to Lung Cancer Treatment. *Cancers* **2019**, *11*, 1565. [CrossRef]

21. Juli, G.; Oliverio, M.; Bellizzi, D.; Gallo Cantafio, M.E.; Grillone, K.; Passarino, G.; Colica, C.; Nardi, M.; Rossi, M.; Procopio, A.; et al. Anti-tumor Activity and Epigenetic Impact of the Polyphenol Oleacein in Multiple Myeloma. *Cancers* **2019**, *11*, 990. [CrossRef] [PubMed]

22. De Stefanis, D.; Scimè, S.; Accomazzo, S.; Catti, A.; Occhipinti, A.; Bertea, C.M.; Costelli, P. Anti-Proliferative Effects of an Extra-Virgin Olive Oil Extract Enriched in Ligstroside Aglycone and Oleocanthal on Human Liver Cancer Cell Lines. *Cancers* **2019**, *11*, 1640. [CrossRef] [PubMed]

23. Nediani, C.; Ruzzolini, J.; Romani, A.; Calorini, L. Oleuropein, a Bioactive Compound from Olea europaea L., as a Potential Preventive and Therapeutic Agent in Non-Communicable Diseases. *Antioxidants* **2019**, *8*, 578. [CrossRef] [PubMed]

24. Shamshoum, H.; Vlavcheski, F.; Tsiani, E. Anticancer effects of oleuropein. *BioFactors* **2017**, *43*, 517–528. [CrossRef]

25. Margheri, F.; Menicacci, B.; Laurenzana, A.; Del Rosso, M.; Fibbi, G.; Cipolleschi, M.G.; Ruzzolini, J.; Nediani, C.; Mocali, A.; Giovannelli, L. Oleuropein aglycone attenuates the pro-angiogenic phenotype of senescent fibroblasts: A functional study in endothelial cells. *J. Funct. Foods* **2019**, *53*, 219–226. [CrossRef]

26. Ruzzolini, J.; Peppicelli, S.; Andreucci, E.; Bianchini, F.; Scardigli, A.; Romani, A.; la Marca, G.; Nediani, C.; Calorini, L. Oleuropein, the Main Polyphenol of Olea europaea Leaf Extract, Has an Anti-Cancer Effect on Human BRAF Melanoma Cells and Potentiates the Cytotoxicity of Current Chemotherapies. *Nutrients* **2018**, *10*, 1950. [CrossRef] [PubMed]

27. Guerra, A.R.; Duarte, M.F.; Duarte, I.F. Targeting Tumor Metabolism with Plant-Derived Natural Products: Emerging Trends in Cancer Therapy. *J. Agric. Food Chem.* **2018**, *66*, 10663–10685. [CrossRef]

28. Gerhäuser, C. Cancer cell metabolism, epigenetics and the potential influence of dietary components-A perspective. *Biomed. Res.* **2012**, *23*, 69–89.

29. Gao, J.L.; Chen, Y.G. Natural compounds regulate glycolysis in hypoxic tumor microenvironment. *Biomed. Res. Int.* **2015**, *2015*, 354143. [CrossRef]

30. Salehi, B.; Zucca, P.; Sharifi-Rad, M.; Pezzani, R.; Rajabi, S.; Setzer, W.N.; Varoni, E.M.; Iriti, M.; Kobarfard, F.; Sharifi-Rad, J. Phytotherapeutics in cancer invasion and metastasis. *Phytother. Res.* **2018**, *32*, 1425–1449. [CrossRef]

31. Sharma, S.H.; Thulasingam, S.; Chellappan, D.R.; Chinnaswamy, P.; Nagarajan, S. Morin and Esculetin supplementation modulates c-myc induced energy metabolism and attenuates neoplastic changes in rats challenged with the procarcinogen 1,2-dimethylhydrazine. *Eur. J. Pharmacol.* **2017**, *796*, 20–31. [CrossRef] [PubMed]

32. Gómez de Cedrón, M.; Vargas, T.; Madrona, A.; Jiménez, A.; Pérez-Pérez, M.J.; Quintela, J.C.; Reglero, G.; San-Félix, A.; Ramírez de Molina, A. Novel Polyphenols That Inhibit Colon Cancer Cell Growth Affecting Cancer Cell Metabolism. *J. Pharmacol. Exp. Ther.* **2018**, *366*, 377–389. [CrossRef] [PubMed]

33. Simmons, R.A. Cell Glucose Transport and Glucose Handling During Fetal and Neonatal Development. In *Fetal and Neonatal Physiology*, 5th ed.; Elsevier: Amsterdam, The Netherlands, 2017; pp. 428–435.

34. Barbosa, A.M.; Martel, F. Targeting Glucose Transporters for Breast Cancer Therapy: The Effect of Natural and Synthetic Compounds. *Cancers* **2020**, *12*, 154. [CrossRef] [PubMed]

35. Zambrano, A.; Molt, M.; Uribe, E.; Salas, M. Glut 1 in Cancer Cells and the Inhibitory Action of Resveratrol as A Potential Therapeutic Strategy. *Int. J. Mol. Sci.* **2019**. [CrossRef]

36. Wiese, E.K.; Hitosugi, T. Tyrosine Kinase Signaling in Cancer Metabolism: PKM2 Paradox in the Warburg Effect. *Front. Cell Dev. Biol.* **2018**, *6*, 79. [CrossRef]

37. Nájera, L.; Alonso-Juarranz, M.; Garrido, M.; Ballestín, C.; Moya, L.; Martínez-Díaz, M.; Carrillo, R.; Juarranz, A.; Rojo, F.; Cuezva, J.M.; et al. Prognostic implications of markers of the metabolic phenotype in human cutaneous melanoma. *Br. J. Dermatol.* **2019**, *181*, 114–127. [CrossRef]

38. Zhang, H.; Feng, C.; Zhang, M.; Zeng, A.; Si, L.; Yu, N.; Bai, M. miR-625-5p/PKM2 negatively regulates melanoma glycolysis state. *J. Cell Biochem.* **2019**, *120*, 2964–2972. [CrossRef]

39. Iqbal, M.A.; Bamezai, R.N. Resveratrol inhibits cancer cell metabolism by down regulating pyruvate kinase M2 via inhibition of mammalian target of rapamycin. *PLoS ONE* **2012**, *7*, e36764. [CrossRef]

40. Li, W.; Ma, X.; Li, N.; Liu, H.; Dong, Q.; Zhang, J.; Yang, C.; Liu, Y.; Liang, Q.; Zhang, S.; et al. Resveratrol inhibits Hexokinases II mediated glycolysis in non-small cell lung cancer via targeting Akt signaling pathway. *Exp. Cell Res.* **2016**, *349*, 320–327. [CrossRef]

41. Saunier, E.; Antonio, S.; Regazzetti, A.; Auzeil, N.; Laprévote, O.; Shay, J.W.; Coumoul, X.; Barouki, R.; Benelli, C.; Huc, L.; et al. Resveratrol reverses the Warburg effect by targeting the pyruvate dehydrogenase complex in colon cancer cells. *Sci. Rep.* **2017**, *7*, 6945. [CrossRef]

42. Siddiqui, F.A.; Prakasam, G.; Chattopadhyay, S.; Rehman, A.U.; Padder, R.A.; Ansari, M.A.; Irshad, R.; Mangalhara, K.; Bamezai, R.N.K.; Husain, M.; et al. Curcumin decreases Warburg effect in cancer cells by down-regulating pyruvate kinase M2 via mTOR-HIF1α inhibition. *Sci. Rep.* **2018**, *8*, 8323. [CrossRef]

43. Pinheiro, C.; Miranda-Gonçalves, V.; Longatto-Filho, A.; Vicente, A.L.; Berardinelli, G.N.; Scapulatempo-Neto, C.; Costa, R.F.; Viana, C.R.; Reis, R.M.; Baltazar, F.; et al. The metabolic microenvironment of melanomas: Prognostic value of MCT1 and MCT4. *Cell Cycle* **2016**, *15*, 1462–1470. [CrossRef]

44. Romani, A.; Mulas, S.; Heimler, D. Polyphenols and secoiridoids in raw material (Olea europaea L. leaves) and commercial food supplements. *Eur. Food Res. Technol.* **2017**, *243*, 429–435. [CrossRef]

45. Romani, A.; Scardigli, A.; Pinelli, P. An environmentally friendly process for the production of extracts rich in phenolic antioxidants from Olea europaea L. and Cynara scolymus L. matrices. *Eur. Food Res. Technol.* **2017**, *243*, 1229–1238. [CrossRef]

46. Guex, C.G.; Reginato, F.Z.; Figueredo, K.C.; da Silva, A.R.H.D.; Pires, F.B.; Jesus, R.D.S.; Lhamas, C.L.; Lopes, G.H.H.; Bauermann, L.F. Safety assessment of ethanolic extract of Olea europaea L. leaves after acute and subacute administration to Wistar rats. *Regul. Toxicol. Pharmacol.* **2018**, *95*, 395–399. [CrossRef]

47. Sepporta, M.V.; Fuccelli, R.; Rosignoli, P.; Ricci, G.; Servili, M.; Fabiani, R. Oleuropein Prevents Azoxymethane-Induced Colon Crypt Dysplasia and Leukocytes DNA Damage in A/J Mice. *J. Med. Food* **2016**, *19*, 983–989. [CrossRef]

48. Giner, E.; Recio, M.C.; Ríos, J.L.; Cerdá-Nicolás, J.M.; Giner, R.M. Chemopreventive effect of oleuropein in colitis-associated colorectal cancer in c57bl/6 mice. *Mol. Nutr. Food Res.* **2016**, *60*, 242–255. [CrossRef]

49. Park, S.; Choi, Y.; Um, S.J.; Yoon, S.K.; Park, T. Oleuropein attenuates hepatic steatosis induced by high-fat diet in mice. *J. Hepatol.* **2011**, *54*, 984–993. [CrossRef]

50. Peppicelli, S.; Bianchini, F.; Toti, A.; Laurenzana, A.; Fibbi, G.; Calorini, L. Extracellular acidity strengthens mesenchymal stem cells to promote melanoma progression. *Cell Cycle* **2015**, *14*, 3088–3100. [CrossRef]

51. Peppicelli, S.; Ruzzolini, J.; Andreucci, E.; Bianchini, F.; Kontos, F.; Yamada, T.; Ferrone, S.; Calorini, L. Potential Role of HLA Class I Antigens in the Glycolytic Metabolism and Motility of Melanoma Cells. *Cancers* **2019**, *11*, 1249. [CrossRef]
52. Livak, K.J.; Schmittgen, T.D. Analysis of Relative Gene Expression Data Using Real-Time Quantitative PCR and the 2−ΔΔCT Method. *Methods* **2001**, *25*, 402–408. [CrossRef]
53. Acquaviva, R.; Di Giacomo, C.; Sorrenti, V.; Galvano, F.; Santangelo, R.; Cardile, V.; Gangia, S.; D'Orazio, N.; Abraham, N.G.; Vanella, L. Antiproliferative effect of oleuropein in prostate cell lines. *Int. J. Oncol.* **2012**, *41*, 31–38. [CrossRef]
54. Samara, P.; Christoforidou, N.; Lemus, C.; Argyropoulou, A.; Ioannou, K.; Vougogiannopoulou, K.; Aligiannis, N.; Paronis, E.; Gaboriaud-Kolar, N.; Tsitsilonis, O.; et al. New semi-synthetic analogs of oleuropein show improved anticancer activity in vitro and in vivo. *Eur. J. Med. Chem.* **2017**, *137*, 11–29. [CrossRef]

Article

Anti-tumor Activity and Epigenetic Impact of the Polyphenol Oleacein in Multiple Myeloma

Giada Juli [1], Manuela Oliverio [2], Dina Bellizzi [3], Maria Eugenia Gallo Cantafio [1], Katia Grillone [1], Giuseppe Passarino [3], Carmela Colica [4], Monica Nardi [2], Marco Rossi [1], Antonio Procopio [2], Pierosandro Tagliaferri [1], Pierfrancesco Tassone [1,*] and Nicola Amodio [1,*]

[1] Department of Experimental and Clinical Medicine, *Magna Graecia* University of Catanzaro, 88100 Catanzaro, Italy

[2] Department of Health Science, *Magna Graecia* University of Catanzaro, 88100 Catanzaro, Italy

[3] Department of Biology, Ecology and Earth Sciences (DiBEST), University of Calabria, 87036 Arcavacata di Rende, Italy

[4] CNR, IBFM UOS of Germaneto, *Magna Graecia* University of Catanzaro, 88100, Catanzaro Italy

* Correspondence: tassone@unicz.it (P.T.); amodio@unicz.it (N.A.); Tel.: +39-0961-3694159 (P.T. & N.A.); Fax: +39-0961-3697077 (P.T. & N.A.)

Received: 28 June 2019; Accepted: 11 July 2019; Published: 16 July 2019

Abstract: Olive oil contains different biologically active polyphenols, among which oleacein, the most abundant secoiridoid, has recently emerged for its beneficial properties in various disease contexts. By using in vitro models of human multiple myeloma (MM), we here investigated the anti-tumor potential of oleacein and the underlying bio-molecular sequelae. Within a low micromolar range, oleacein reduced the viability of MM primary samples and cell lines even in the presence of bone marrow stromal cells (BMSCs), while sparing healthy peripheral blood mononuclear cells. We also demonstrated that oleacein inhibited MM cell clonogenicity, prompted cell cycle blockade and triggered apoptosis. We evaluated the epigenetic impact of oleacein on MM cells, and observed dose-dependent accumulation of both acetylated histones and α-tubulin, along with down-regulation of several class I/II histone deacetylases (HDACs) both at the mRNA and protein level, providing evidence of the HDAC inhibitory activity of this compound; conversely, no effect on global DNA methylation was found. Mechanistically, HDACs inhibition by oleacein was associated with down-regulation of Sp1, the major transactivator of HDACs promoter, *via* Caspase 8 activation. Of potential translational significance, oleacein synergistically enhanced the in vitro anti-MM activity of the proteasome inhibitor carfilzomib. Altogether, these results indicate that oleacein is endowed with HDAC inhibitory properties, which associate with significant anti-MM activity both as single agent or in combination with carfilzomib. These findings may pave the way to novel potential anti-MM epi-therapeutic approaches based on natural agents.

Keywords: experimental therapeutics; HDAC; multiple myeloma; oleacein

---

1. Introduction

Multiple myeloma (MM) is a clonal B cell malignancy characterized by the accumulation of tumor plasma cells (PCs) in the bone marrow (BM), where different cell types establish a complex microenvironment that supports survival, proliferation and drug-resistance of the malignant clone. The last few years have witnessed a rapid development of drugs for the treatment of this malignancy, leading to increased extent and frequency of response and to the improvement in median overall survival of patients. However, despite such therapeutic advancements, MM eventually evolves into a drug-resistant phase leading to patients' death [1]. This finding has stimulated continuous investigation on new therapeutic options, as single agents or in combination with established anti-MM drugs. In this regard, natural compounds have recently emerged as novel chemopreventive and/or therapeutic tools able to target oncogenic pathways involved in the pathogenesis of human malignancies. Significant anti-inflammatory, anti-oxidant and cytotoxic effects of natural agents have been demonstrated also in MM, either by epidemiologic or animal studies, and even by clinical trials [2]. Several natural compounds from various plants, fungi and marine organisms have been shown to target epigenetic events underpinning tumorigenesis, such as DNA methylation, histone modifications (methylation, acetylation and phosphorylation), and non-coding RNAs [3], known to be deeply dysregulated and representing valuable therapeutic targets in MM [2].

Polyphenols are important constituents of several plants and vegetables, recognized as powerful anti-oxidants endowed with anti-inflammatory, antimicrobial and antitumor activities [4]. Indeed, a major source of polyphenols is represented by extra virgin olive oil (EVOO), whose polyphenolic fraction includes simple phenols (tyrosol and hydroxytyrosol), secoiridoids (oleuropein, oleocanthal and oleacein), and lignans. EVOO-derived secoiridoids, characterized by the presence of elenoic acid or its derivatives in their molecular structure, have been shown to prevent obesity, osteoporosis, and neurodegeneration. Oleocanthal has shown anti-tumor activity in different types of tumors, including MM, hepatocellular carcinoma, breast, prostate, pancreatic cancers and melanoma [5–9]. The most abundant secoiridoid of EVOO is the dialdehydic form of elenolic acid conjugated with 3,4-(dihydroxyphenyl)ethanol (3,4-DHPEA-EDA), also known as oleacein, whose anti-oxidant, anti-inflammatory, and anti-microbial properties have recently emerged [10], while its effects on tumor biology are still poorly defined.

We here aimed to investigate the anti-tumor potential of oleacein against MM. Our results highlight a previously unknown epigenetic impact of oleacein on MM cells, with potential implications for the management of MM and possibly other malignancies.

2. Results

2.1. Inhibitory Effects of Oleacein on MM Cell Viability and Survival

Oleacein, whose chemical structure is reported in Figure 1A, was obtained by a green semi-synthetic modification of oleuropein as previously reported [11]. By using a panel of eight different MM cell lines carrying the major cytogenetic aberrations of MM, we sought to analyze the impact of oleacein on cell viability. MM cells were exposed to increasing doses of oleacein, and cell viability was assessed by Cell Titer Glo (CTG) assay. Noteworthy, a dose-dependent inhibition of cell viability was observed 48 h after oleacein treatment, with IC50s ranging from 5.0 to 20.0 μM (Figure 1B); conversely, oleacein did not affect the viability of PBMCs from healthy donors (Figure 1C), suggesting a favorable therapeutic index. We next evaluated the effects of oleacein on MM cells in the presence of the BM *milieu*, which is known to trigger drug-resistance [1]. Importantly, the inhibitory effect of oleacein was maintained even when MM cell lines, or primary CD138$^+$ cells purified from MM patients, were cultured in the presence of HS-5 stromal cells, thus suggesting that oleacein can overcome BM microenvironment-mediated pro-survival effects (Figure 1D). In addition, oleacein drastically suppressed the clonogenicity of MM cells in methylcellulose cultures (Figure 1E). Collectively, these data unveil an inhibitory activity of oleacein on MM cell viability and survival.

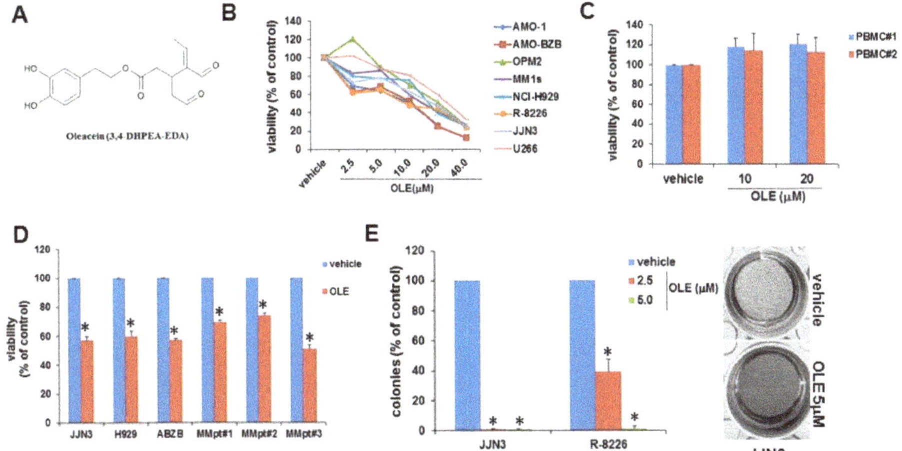

Figure 1. Effects of oleacein on multiple myeloma (MM) cell survival. (**A**) Chemical structure of oleacein. (**B**) Cell viability of MM cell lines as determined by Cell Titer Glo (CTG) assay 48 h after treatment with increasing doses of oleacein or vehicle (DMSO). (**C**) CTG assay performed on peripheral blood mononuclear cells (PBMCs) from three different healthy donors treated with oleacein for 48 h. (**D**) CTG assay in MM cell lines and primary CD138$^+$ cells from three MM patients (MM pt#1, #2 and#3) co-cultured on HS-5 stromal cells and treated for 48 h with 5.0µM oleacein. (**E**) Colony formation assay performed on MM cell lines treated for 14 days with oleacein; representative pictures of JJN3 colonies at day 14 are shown in the right panel (5× magnification). * $p < 0.05$ as compared to vehicle-treated cells.

2.2. Oleacein Triggers Cell Cycle Arrest and Apoptosis

To unravel the biological sequelae of oleacein in MM, we first analyzed by flow cytometry the cell cycle profile of oleacein-treated cells after propidium iodide staining. As shown in Figure 2A, oleacein increased the percentage of hypodiploid cells (sub-G0 phase), and also induced the accumulation of cells in the G0/G1 phase; WB analysis showed a dose-dependent increase of cell cycle inhibitors p27^{KIP1} and p21^{CIP1} protein expression (Figure 2B), strengthening the capability of oleacein to trigger cell cycle blockade.

In order to confirm apoptosis induction, we performed Annexin V/7-AAD staining on MM cell lines after oleacein treatment. We found an increase in late apoptotic events, which ranged from 20 to 30% after treatment with oleacein 5.0 and 10.0 µM, respectively (Figure 2C); the increase in cleaved PARP1, caspase-3 and caspase-8 on oleacein-treated MM cell lines, as shown by Western Blot (WB), further confirmed apoptosis induction (Figure 2D); no activation of caspase-7 and -9 was observed (Supplementary Figure S1), thus indicating that oleacein predominantly activates the extrinsic apoptotic pathway. These results therefore indicate that oleacein may elicit anti-MM activity through modulation of cell cycle and apoptosis.

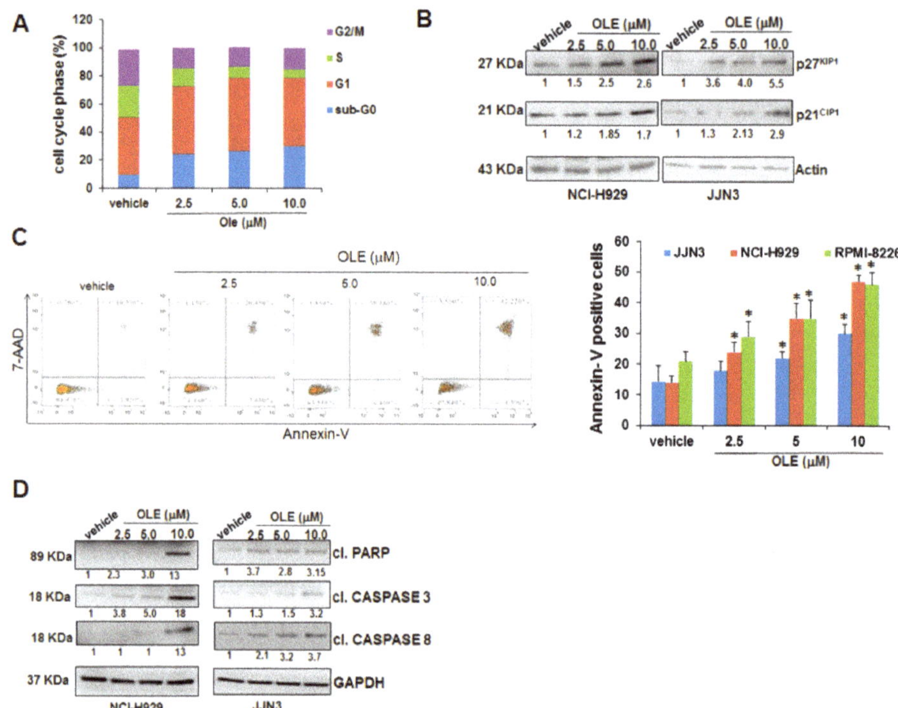

Figure 2. Oleacein triggers cell cycle blockade and apoptosis. (**A**) Cell cycle analysis was performed on NCI-H929 cells by PI staining, 24 h after treatment with oleacein or vehicle (DMSO). (**B**)Western Blot (WB) analysis of p27^{KIP1} and p21^{CIP1} in whole cell lysates from MM cells after treatment with oleacein for 24 h; actin was used as loading control. (**C**) Annexin V/7-AAD staining of MM cells after treatment with oleacein for 48 h; a representative experiment on NCI-H929 cells is shown on the left side. (**D**) WB of PARP1, cleaved caspase-3 and cleaved caspase-8 in NCI-H929 and JJN3 cell lines after 24 h of oleacein treatment; GAPDH was used as loading control. * $p < 0.05$ as compared to vehicle-treated cells.

2.3. HDAC Inhibitory Activity of Oleacein in MM

Aberrant epigenetic patterns are common in MM, where they are frequently associated with disease onset and/or progression to advanced stages [12–14]. A large body of literature has highlighted the capability of several natural compounds to revert the altered epigenome of MM cells by counteracting key oncogenic epigenetic regulators [2]. On this basis, we investigated the epigenetic impact of oleacein on MM cells, by analyzing its effects both on global DNA methylation (GDM) and histone acetylation, the two major epigenetic mechanisms dysregulated in MM. Oleacein did not significantly modify the whole content of methylated cytosines in DNA from NCI-H929 and JJN3 cell lines (Figure 3A); in line with the latter finding, no significant changes in mRNA or protein levels of DNA methyltransferases (DNMT1, DNMT3A and DNMT3B) were observed upon oleacein treatment, as shown by QRT-PCR (Figure 3B) and WB analyses (Figure 3C). Conversely, oleacein induced a significant increase in acetylated histone H3, histone H4 (Figure 3D) and α-tubulin (Figure 3E). Collectively, these findings suggest that oleacein is able to modulate the acetylome of MM cells.

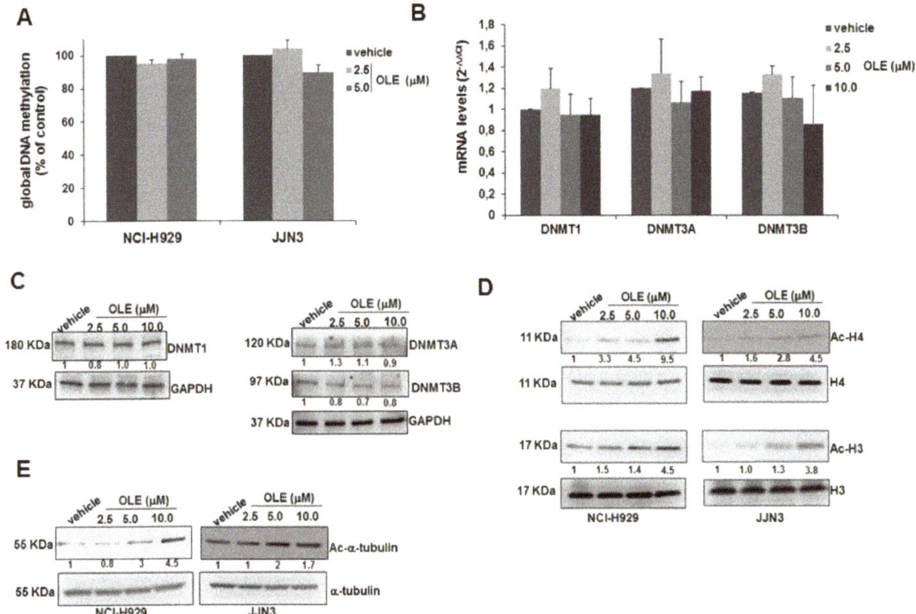

Figure 3. Oleacein affects the acetylome but not the methylome of MM cells. (**A**) Global DNA methylation was measured in MM cells treated for 24 h with oleacein, as reported in materials and methods. Quantitative Real Time PCR (QRT-PCR) (**B**) and WB analysis (**C**) of DNMT1, DNMT3A and DNMT3B in JJN3 cells treated for 24 h with oleacein; GAPDH was used as loading control. WB analysis of acetylated histone H3, histone H3, acetylated histone H4, histone H4 (**D**) and acetylated α-tubulin (**E**) in NCI-H929 and JJN3 cells treated with oleacein for 24 h; GAPDH was used as loading control.

Aberrant expression and/or activity of HDACs drive malignant transformation of tumor cells, thus making HDACs valuable therapeutic targets in MM [13,15]. We analyzed, in oleacein-treated JJN3 cells, the mRNA and protein expression of HDACs with established oncogenic role in MM. Intriguingly, oleacein induced down-regulation of several class I/II HDACs, namely HDAC1/2/3/4/6, both at mRNA (Figure 4A) and protein level (Figure 4B); moreover, biochemical fractionation experiments indicated that oleacein reduced both the nuclear and the cytoplasmic fraction of class II HDAC4 and HDAC6, which are known to shuttle between the nucleus and the cytoplasm (Figure 4C). To understand whether oleacein could act as a canonic HDAC inhibitor, we carried out an in vitro HDAC activity assay using JJN3 nuclear extracts. Incubation with oleacein did not induce any change in the HDAC activity recovered from nuclear extracts, differently from trichostatin A (TSA) or SAHA (Supplementary Figure S2), that were used as positive controls. This finding suggests that the impact of oleacein on the acetylome of MM cells does not occur via enzymatic HDAC inhibition. We therefore explored additional mechanisms accounting for oleacein effects on HDACs, and hypothesized that oleacein could transcriptionally regulate HDACs *via* Sp1. In fact, Sp1, a ubiquitous transcription factor endowed with oncogenic activity in hematologic and solid malignancies [12,13,16,17], was proven to act as a transcriptional activator of HDACs [18]. Interestingly, oleacein treatment induced down-regulation of Sp1 (Figure 4D), and this effect occurred in a caspase 8-dependent fashion, since it was abrogated by Z-IETD-FMK, a selective caspase 8 inhibitor (Figure 4E). These results indicate that oleacein effects on HDACs expression might be mediated by Sp1.

Sp1 is involved in negative feedback loops with miRNAs, like miR-29b [19–21] and miR-22 [22], both acting as tumor suppressors in MM [23,24]. As expected, oleacein-induced Sp1 inhibition was paralleled by the upregulation of miR-29b and miR-22 (Supplementary Figure S3), thus strengthening the role of Sp1 pathway's inhibition in the anti-MM activity of oleacein.

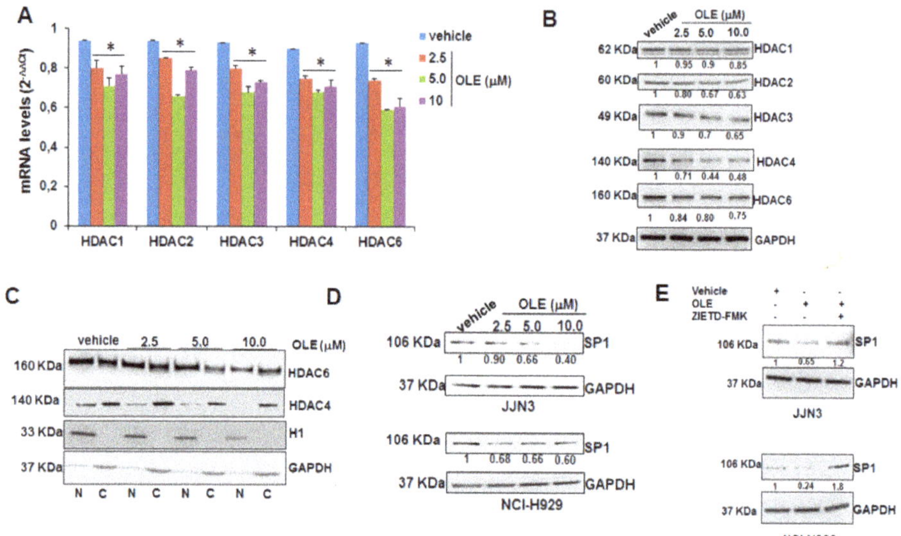

Figure 4. Oleacein targets HDACs. QRT-PCR (**A**) and WB analysis (**B**) of HDAC1, HDAC2, HDAC3, HDAC4, HDAC6 in JJN3 cells treated with oleacein for 24 h; GAPDH was used as loading control. (**C**) WB analysis of HDAC4 and HDAC6 in nuclear (N) and cytoplasmic (C) protein fractions from JJN3 cells treated for 24 h with oleacein; histone H1 and GAPDH were used as nuclear and cytoplasmic marker, respectively. (**D**) WB analysis of Sp1 in JJN3 cells treated with oleacein for 24 h. (**E**) WB analysis of Sp1 in JJN3 cells treated with 5.0 μM oleacein with or without 20.0 μM Z-ITED-FMK; GAPDH was used as loading control. * $p < 0.05$ as compared to vehicle-treated cells.

2.4. Oleacein Enhances the Anti-MM Activity of Carfilzomib

HDAC inhibitors (HDACi) are part of the therapeutic armamentarium against MM, and clinical studies have shown promising therapeutic activity of pan- or selective-HDACi when used within combination regimens [13]. On this basis, we investigated whether, similarly to pan-HDACi, oleacein treatment could trigger synergistic anti-MM activity in combination with clinically-relevant proteasome inhibitors. With this aim, NCI-H929 cells were treated with different concentrations of oleacein with or without bortezomib or carfilzomib, and subsequently cell viability was analyzed by CTG; the occurrence of synergism was assessed by Calcusyn. Interestingly, oleacein synergistically enhanced the effects of carfilzomib (CI < 1.0) on the inhibition of cell viability (Figure 5A), while combination with bortezomib was generally antagonistic (CI > 1.0; Supplementary Figure S4). Annexin V-7AAD staining of NCI-H929-treated cells indicated a higher apoptotic rate when oleacein was combined with carfilzomib, as compared to single agent treatment (Figure 5B). Accordingly, WB analysis showed increased cleavage of caspase 3 and superior Sp1 downregulation upon oleacein *plus* carfilzomib combination, thus confirming enhancement of apoptosis; moreover, oleacein *plus* carfilzomib enhanced downregulation of HDAC2, HDAC3, HDAC4 and HDAC6, with respect to single-agent treatment, which associated with increased histone H4 acetylation (Figure 5C). Collectively, these results indicate that the combination of olecein with carfilzomib results in significant acetylome derangement and apoptosis triggering of MM cells.

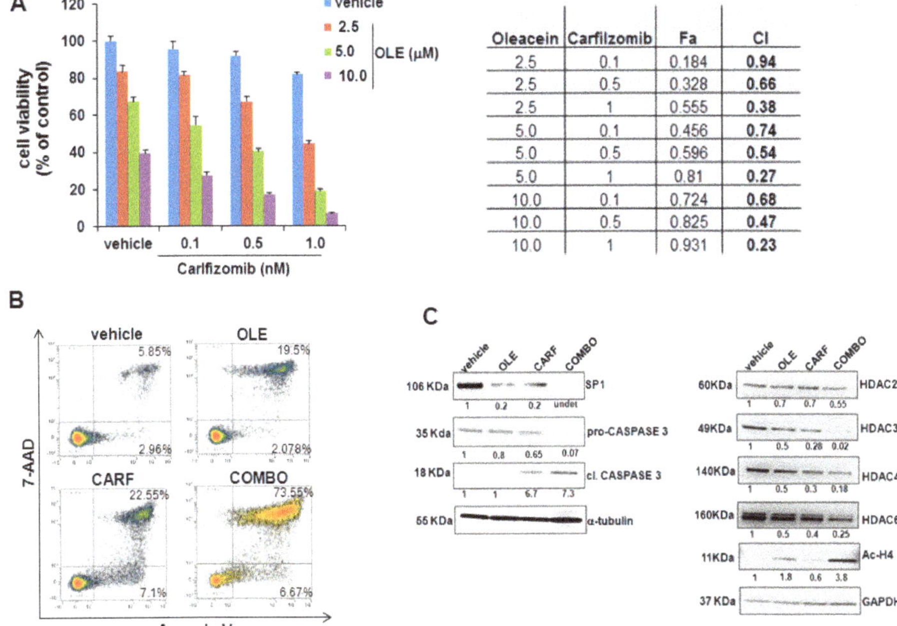

Oleacein	Carfilzomib	Fa	CI
2.5	0.1	0.184	**0.94**
2.5	0.5	0.328	**0.66**
2.5	1	0.555	**0.38**
5.0	0.1	0.456	**0.74**
5.0	0.5	0.596	**0.54**
5.0	1	0.81	**0.27**
10.0	0.1	0.724	**0.68**
10.0	0.5	0.825	**0.47**
10.0	1	0.931	**0.23**

Figure 5. Oleacein enhances the anti-MM activity of carfilzomib. (**A**) CTG assay was performed on NCI-H929 cells treated with oleacein (2.5, 5.0 or 10.0 µM) and carfilzomib (0.1, 0.5 and 1.0 nM). Results are expressed as percentage of the viability of vehicle-treated cells. The right panel reports values of fraction affected (Fa) and combination indexes (CI) in a triplicate experiment, as calculated by the Calcusyn software. (**B**) Annexin V/7-AAD staining of NCI-H929 cells after treatment with vehicle (DMSO), 5.0 µM oleacein and 1.0 nM carfilzomib for 24 h; a representative FACS experiment is reported. (**C**). WB analysis of pro-Caspase 3, cleaved caspase 3, SP1, HDAC2, HDAC3, HDAC4, HDAC6, and acetylated histone H4 in NCI-H929 cells treated with carfilzomib (1.0 nM), oleacein (5.0 µM) or a combination of the two; α-tubulin or GAPDH were used as loading controls.

3. Discussion

Naturally occurring compounds endowed with anti-tumor activity have been found in different sources, such as vegetables, fruits, herbs and fermented products. These agents may act either by preventing the onset of primary cancer, or by antagonizing the evolution of pre-malignant and malignant lesions towards more aggressive stages. Noteworthy, experimental findings on a variety of natural compounds, including curcumin, resveratrol, celastrol and many others, have demonstrated significant advantages for the management of MM [2]. The molecular mechanisms underlying the anti-tumor activity of such compounds are diverse and only partially understood, with inhibition of oncogenic signal transduction pathways and modulation of the cellular epigenome being the most well documented. Regarding the epigenetic-modulating effects, it has been demonstrated that several natural agents, by targeting DNMTs [25], HDACs [26] or non-coding-RNAs [27], may revert aberrant epigenetic patterns implicated in the pathogenesis of human neoplasias, including MM [2].

Polyphenols found in the EVOO, a major component of mediterranean diet, have demonstrated to be protective against several diseases, including those of cardiovascular and metabolic origin [28]. The pro-active ingredient oleuropein and its derivative hydroxytyrosol have been widely studied, demonstrating many beneficial effects, both in vitro and in vivo, in experimental preclinical models [29]. Moreover, many studies have disclosed remarkable anti-tumor activity of oleuropein and hydroxytyrosol against several types of cancers [30]. However, the amount of oleuropein in the EVOO is too low to fully explain the beneficial effects deriving from EVOO assumption with diet. In fact, the endogenous β-glucosidase released during the olive oil extraction process hydrolyzes oleuropein, generating a series of degradation products, all of which are less hydrophilic than the original natural secoiridoids, and therefore more soluble in the oily matrix extracted from the drupes. Thus, EVOO is scarce in oleuropein and much more abundant in its degradation product oleacein [31], thus making it one of the most plausible effectors of the biological activity of EVOO [11,32].

Oleacein can be obtained by a simple and environmentally friendly method, starting from the easily available natural oleuropein [11]. Recent preclinical studies have highlighted anti-microbial [33], anti-inflammatory [34], and protective effects of oleacein against diet-dependent metabolic alterations [35]; conversely, its anti-antitumor activity remains poorly characterized. We here provided the first evidence of the anti-tumor activity of oleacein against MM cells: importantly, oleacein triggered cell cycle arrest and apoptosis and reduced clonogenicity, without exerting any toxic effect on healthy PBMCs, thus suggesting a favorable therapeutic index of this agent. Moreover, oleacein cytotoxic effects were also observed against primary MM cells co-cultured with BM-derived stromal cells, demonstrating the capability of oleacein to overcome BM microenvironment-dependent drug resistance.

We sought to shed light on the molecular mechanisms underlying oleacein anti-tumor activity in MM. To this aim, we focused on epigenetic mechanisms, known to be a major target of several natural agents, and whose dysregulation has been largely implicated in the pathogenesis of MM [14]. Indeed, aberrant expression of effectors of the epigenetic machinery, including DNMTs [36], HDACs [13], polycomb genes [37,38] and non-coding RNAs [39,40], has been reported in MM and has been harnessed in the context of novel anti-tumor strategies [41].

Collectively, our data indicate a strong increase in acetylation of histones and of α-acetyl-tubulin upon oleacein treatment, while no effect on GDM could be observed. This finding highlights a novel HDAC inhibitory activity of oleacein in MM. We attempted to characterize the mechanisms underlying such HDAC inhibitory effects, and found out that oleacein could transcriptionally inhibit HDACs expression likely *via* targeting of Sp1, a known transactivator of HDACs' promoter. Since previous findings indicated that bortezomib-evoked transcriptional repression of HDACs by Sp1 occurs in a caspase 8-dependent fashion [18], we investigated whether oleacein effect on Sp1 could be similarly mediated by caspase 8. In agreement with this hypothesis, oleacein-induced Sp1 down-regulation was abrogated by the caspase 8 inhibitor Z-IETD-FMK.

Dysregulated transcription factors may drive down-regulation of tumor suppressor miRNAs in MM [12,40,42]. By establishing molecular feedback loops, Sp1, a pleiotropic transcription factor endowed with oncogenic activity in human malignancies [17], was shown to negatively affect the expression of miRNAs [40]. We have previously reported that miRNA dysregulation features prominently in the pathobiology of MM, with certain miRNAs, such as miR-125a-5p [43], miR-21 [44], miR-221 [45] and miR-17-92 cluster [46], highly expressed in MM and acting as oncogenes, while others such as miR-29b [20], miR-22 [24] and miR-125b [47], behave as tumor suppressors.

Consistent with inhibition of Sp1, oleacein triggered upregulation of tumor suppressive miRNAs, namely miR-29b [20] and miR-22 [24], which are known to be negatively regulated by this transcription factor [19,22]. These findings underscore the ability of oleacein to trigger a tumor suppressive miRNA network likely contributing to its cytotoxicity against MM cells.

Non-selective HDAC inhibitors, such as romidepsin, vorinostat and panobinostat, have shown a remarkable anti-MM effect in preclinical and clinical studies, with significant efficacy, along with reduced side effects, when given within combination regimens [13]; amongst pan-HDACi, panobinostat has been approved by FDA for MM treatment [48]. Having demonstrated its pan-HDAC inhibitory activity, and taking into account the promising clinical data which emerged from MM patients treated with proteasome inhibitors and pan-HDACi combination therapies, we also explored whether oleacein could enhance the anti-tumor activity of bortezomib or carfilzomib. Notably, when combined with carfilzomib, oleacein synergistically enhanced its in vitro cytotoxicity, with superior Sp1 and HDACs down-regulation and a resultant increase in apoptosis of MM cells. A follow-up investigation is planned to evaluate the anti-tumor effect of oleacein-based treatments in the context of validated in vivo preclinical models of human MM.

4. Materials and Methods

4.1. Chemicals

A green semi-synthetic procedure was carried out to purify oleacein from Coratina cultivar olive leaves of Olea Europaea L. as reported; in detail, oleacein was directly extracted from a water solution of oleuropein in the presence of NaCl under microwave assistance at 180 °C; the crude extract was purified by flash chromatography on silica gel (eluent mixture: $CHCl_3$/MeOH 95:5 v/v) [49]. The purity was determined by RP-HPLC, HRMS-ESI, ^{1}H-and ^{13}C-NMR, as previously reported [11]. Bortezomib and carfilzomib were purchased from Selleckchem (Houston, TX, USA) as DMSO stock-solutions.

4.2. Cell Cultures

MM cell lines NCI-H929, RPMI-8226, U266, MM1s and JJN3 were purchased from DSMZ, which certified authentication performed by short tandem repeat DNA typing; the bone marrow stromal cell line HS-5 was purchased from the American Type Culture Collection (Rockville, MD, USA); AMO-1 and AMO-BZB cells were kindly provided by Dr. C. Driessen (University of Tubingen, Tubingen, Germany). The most relevant characteristics of the MM cell lines used are reported in Supplementary Table S1. All these cell lines were immediately frozen and used from the original stock within 6 months. Human MM cell lines were cultured in RPMI-1640 media containing 10% FBS, 2 μmol/L glutamine, 100 U/mL penicillin, and 100 μg/mL streptomycin (GIBCO; Life Technologies, Carlsbad, CA, USA) and tested for mycoplasma contamination. Peripheral blood mononuclear cells (PBMCs) and $CD138^+$ cells from BM of MM patients were isolated by Ficoll-hypaque (Lonza Group, Basel, Switzerland), followed by anti-CD138 microbeads (Miltenyi Biotec, Bergish Gladbach, Germany) selection, in accordance with the Declaration of Helsinki following informed consent and Institutional Review Board (University of Catanzaro, Catanzaro, Italy) approval, as previously reported (institutional approval: n.120/2015) [50]; the purity of immunoselected cells was assessed by flow cytometry using a phycoerythrin-conjugated CD138 monoclonal antibody (BD Pharmingen, San Jose, CA, USA; clone DL-101) and was higher than 95%. In co culture experiments, primary $CD138^+$ MM cells (2.5×10^5 cells) were plated in 24-well plates, and left separated from HS-5 stromal cells (2.5×10^5 cells) growing adherent to the plate by a transwell insert of 0.4 μm pore size (Corning, New York, NY, USA).

4.3. Cell Viability, Apoptosis and Cell Cycle Assay

Cell viability was evaluated by Cell Titer-Glo (CTG; Promega, Madison, Wisconsin, USA), as previously reported [12]. For colony formation assay, 200 cells were plated in triplicate in 1 mL of mixture composed of 1.1% methylcellulose (MethoCult^{TM} STEMCELL Technologies, Cambridge, UK) in RPMI-1640 + 10% FBS. Crystal violet-stained colonies were scored after 2 weeks under an inverted microscope (Leica DM IL LED, Wetzlar, Germany) at 5× magnification using a grid. Apoptosis was evaluated by flow cytometric (FACS) analysis following Annexin V-7AAD staining (BD Pharmingen, San Jose, CA, USA). Drug interactions were assessed by CalcuSyn 2.0 software (Biosoft, Novosibirsk, Russia), which is based on the Chou-Talalay method. When combination index (CI) = 1, this equation represents the conservation isobologram and indicates additive effects; CI < 1 indicates synergism; CI > 1 indicates antagonism. Cell cycle distribution was evaluated by FACS analysis on MM cells previously treated with oleacein for 24 h, after staining with Propidium Iodide (PI). Cells were collected, washed twice with phosphate-buffered saline (PBS) and fixed in cold 70% ethanol at −20 °C. Before FACS analysis, cells were washed with PBS and stained in 50 μg/mL PI, 100 μg/mL RNase, 0.05% Nonidet P-40 for 1 h at room temperature in the dark. Cell cycle profiles were obtained using Attune NxT Flow Cytometer (Thermo Fisher Scientific, Waltham, MA, USA).

4.4. Western Blot and Antibodies

Whole cell protein extracts were prepared using NP40 lysis buffer containing Halt Protease Inhibitor cocktail (Invitrogen, Thermo Scientific, Carlsbad, CA, USA), separated using 4–12% Novex Bis-Tris SDS-acrylamide gels (Invitrogen), and electrotransferred on nitrocellulose membranes (Bio-Rad, Hercules, CA, USA), as described [39]. Then, nitrocellulose membranes were blocked with milk and probed over-night with primary antibodies at 4 °C; then membranes were washed three times in PBS-Tween and incubated with a secondary antibody conjugated with horseradish peroxidase for 2 h at room temperature. Chemiluminescence was detected using SuperSignal West Pico PLUS Chemiluminescent Substrate (Thermo Scientific). Western blot (WB) was performed using Cell Signaling antibodies: PARP (#9532), -Caspase-8 (#9746), -Caspase-3 (#9665), AcH3-Lysin 8 (K9) (#9649P), SP1 (#9389S), HDAC1 (#5356T), HDAC2 (#5113P), HDAC3 (3949P), HDAC4 (#7628S), HDAC6 (#7558P), histone H4 (#2935), histone H3 (#4499), α-tubulin (#2125). Ac-α tubulin (sc-23950), Ac-H4 Ser1/Lys 5/Lys8/Lys Lys 12 (sc-34263), Actin (sc-1616) and -GAPDH (sc-25778) were from Santa Cruz Biotechnology (Dallas, TX, USA); Dnmt1 (ab 13537), Dnmt3a (ab 13888) and Dnmt3b (ab 2851) were from abcam (Cambridge, UK). Densitometric analysis of blots was performed by LI-COR Image Studio Digits Ver 5.0 (Bad Homburg, Germany), expressed as a relative protein unit after normalization with appropriate housekeeping, and reported under each blot. Whole blots of all experiments presented in this study are reported as Supplementary Figures S5–S9.

4.5. Reverse Transcription and Quantitative Real Time PCR (qRT-PCR)

Total RNA was extracted from cells using TRIzol® reagent (Gibco, Life Technologies, Carlsbad, CA, USA), following the manufacturer's instructions. The RNA quantity and quality were assessed through NanoDrop® ND-1000 Spectrophotometer (Waltham, MA, USA). To evaluate transcript changes, 1000 ng of total RNA was reverse-transcribed to cDNA using the "High Capacity cDNA Reverse Transcription Kit" (Applied Biosystems, Carlsbad, CA, USA). The following single-tube TaqMan assays (Applied Biosystems, Carlsbad, CA, USA) were used to detect and quantify genes using the Viia7 DX real time PCR instrument (Life Technologies, Waltham, MA, USA): DNMT1 (Hs00154749_m1), DNMT3a (Hs01027166_m1), DNMT3b (Hs00171876_m1), HDAC1 (Hs02621185_s1), HDAC2 (Hs00231032_m1), HDAC3 (Hs00187320_m1), HDAC4 (Hs01041638_m1), HDAC6 (Hs00195869_m1), and GAPDH (Hs02786624 g1). miRNA expression levels were determined by TaqMan RT-PCR, using the single-tube TaqMan miRNA assays (hsa-miR-29b, assay ID 000413; hsa-miR-22, assay ID 000398, Applied Biosystems) to quantify mature miRNAs, by the use of the StepOne Thermocycler (Thermo Fisher Scientific, Waltham, MA, USA) and the sequence detection system, as previously reported [51]; miRNAs expression levels were normalized on RNU44 (assay ID 001094). Comparative real-time polymerase chain reaction (RT-PCR) was performed in triplicate.

4.6. HDAC Activity Assay

Nuclear extracts prepared using NE-PER Nuclear and Cytoplasmic Extraction Reagents kit (Thermo Scientific, catalog #78833) were mixed with Oleacein or DMSO, and then HDAC activity was determined according to the manufacturer's instructions (BioVision, Zurich, Switzerland; catalog #K331-100). Trichostatin A (TSA) and SAHA were used as positive controls.

4.7. Quantification of Global 5-Methylcytosine Levels

Global DNA methylation levels were determined by using 5-mC DNA ELISA kit (Zymo Research, Irvine, CA, USA) as described [52]. Briefly, 100ng of genomic DNA, brought to final volume to 100 µL with 5-mC coating buffer, was denatured at 98 °C for 5min, put in ice for 10' and then coated on the surface of the ELISA plate wells. After incubation at 37 °C for 1 h, the wells were washed thrice with 200 µL of 5-mC ELISA buffer and then incubated at 37 °C for 1 h with an antibody mix consisting of anti-5-mC (1:2000) and secondary (1:1000) antibodies. Then, the antibody mix was removed from the wells through three consecutive washes with 200 µL of 5-mC ELISA buffer. One-hundred microliters of HRP developer was added to each well and incubated at room temperature for 1 h. Absorbance at 405 nm was measured using an ELISA plate reader. The percentage of 5-mC was calculated using the second-order regression equation of the standard curve that was constructed by using mixtures of the fully unmethylated and methylated control DNAs, provided by the manufacturer, to generate standards of known 5-mC percentage (0, 5, 10, 25, 50, 75 and 100%).

4.8. Statistical Analysis

Each experiment was performed at least three times, and values were reported as mean ± standard deviation. Data were analyzed using Student's t tests for two group comparisons or a one-way analysis of variance (ANOVA) for multiple comparisons using the Graphpad software (GraphPad Software, La Jolla, CA, USA). p-value < 0.05 was considered significant.

5. Conclusions

Our results indicate that oleacein, the most abundant EVOO secoiridoid, elicits significant anti-tumor activity by promoting cell cycle arrest and apoptosis, either as a single agent or in combination with the proteasome inhibitor carfilzomib. Moreover, our data highlight an epigenetic impact of oleacein in MM, as demonstrated by the impairment of the MM acetylome, likely *via* Sp1-dependent transcriptional inhibition of HDACs. Altogether, these findings provide the molecular rationale for potential epi-therapeutic anti-MM strategies based on natural agents.

Supplementary Materials: The following are available online at http://www.mdpi.com/2072-6694/11/7/990/s1, Figure S1: WB of pro-caspase 7, cleaved caspase 7, pro-caspase 9 and cleaved caspase 9 in NCI-H929 cells after 24 h of oleacein treatment, Figure S2: HDAC activity was determined in JJN3 cells treated with oleacein, as reported in materials and methods; TSA was used as positive control. Results are expressed as percentage of HDAC activity as compared to DMSO-treated cells, Figure S3: miR-29b and miR-22 expression levels were determined by qRT-PCR in JJN3 cells treated for 24 h with oleacein; miRNA expression was normalized on RNU44, Figure S4: CTG assay was performed on NCI-H929 cells treated with oleacein (2.5, 5.0 or 10.0 µM) and bortezomib (1.0, 2.0 and 5.0 nM). Results are expressed as percentage of the viability of vehicle-treated cells. The right panel reports values of fraction affected (Fa) and combination indexes (CI) in a triplicate experiment, as calculated by the Calcusyn software, Figure S5: whole blots for Figure 2, Figure S6: whole blots for Figure 3, Figure S7: whole blots for Figure 4, Figure S8: whole blots for Figure 5, Figure S9: whole blots for Figure S1. Table S1: Characteristics of the MM cell lines used in this study.

Author Contributions: Data curation, K.G. and M.N.; Formal analysis, M.R.; Investigation, G.J., M.O., D.B. and N.A.; Methodology, M.E.G.C., G.P. and A.P.; Resources, P.T.; Software, C.C.; Writing—original draft, P.T. (Pierosandro Tagliaferri), P.T. (Pierfrancesco Tassone) and N.A.; Writing—review & editing, N.A.

Funding: This work has been supported by Italian Association for Cancer Research (AIRC), "Innovative Immunotherapeutic Treatments of Human Cancer" Multi Unit Regional Project No. 16695 (co-financed by AIRC and CARICAL foundation) to Pierfrancesco Tassone.

Acknowledgments: We thank Ivana Criniti for editorial and laboratory assistances.

Conflicts of Interest: The authors declare no conflict of interest. The funders had no role in the design of the study; in the collection, analyses, or interpretation of data; in the writing of the manuscript, or in the decision to publish the results.

References

1. Anderson, K.C. Progress and Paradigms in Multiple Myeloma. *Clin. Cancer Res.* **2016**, *22*, 5419–5427. [CrossRef] [PubMed]
2. Raimondi, L.; De Luca, A.; Giavaresi, G.; Barone, A.; Tagliaferri, P.; Tassone, P.; Amodio, N. Impact of natural dietary agents on multiple myeloma prevention and treatment: Molecular insights and potential for clinical translation. *Curr. Med. Chem.* **2018**. [CrossRef] [PubMed]
3. Aggarwal, R.; Jha, M.; Shrivastava, A.; Jha, A.K. Natural Compounds: Role in Reversal of Epigenetic Changes. *Biochemistry* **2015**, *80*, 972–989. [CrossRef] [PubMed]
4. Pandey, K.B., Rizvi, S.I. Plant polyphenols as dietary antioxidants in human health and disease. *Oxid. Med. Cell Longev.* **2009**, *2*, 270–278. [CrossRef] [PubMed]

5. Akl, M.R.; Elsayed, H.E.; Ebrahim, H.Y.; Haggag, E.G.; Kamal, A.M.; El Sayed, K.A. 3-O-[N-(p-fluorobenzenesulfonyl)-carbamoyl]-oleanolic acid, a semisynthetic analog of oleanolic acid, induces apoptosis in breast cancer cells. *Eur. J. Pharmacol.* **2014**, *740*, 209–217. [CrossRef] [PubMed]

6. Khanfar, M.A.; Bardaweel, S.K.; Akl, M.R.; El Sayed, K.A. Olive Oil-derived Oleocanthal as Potent Inhibitor of Mammalian Target of Rapamycin: Biological Evaluation and Molecular Modeling Studies. *Phytother. Res.* **2015**, *29*, 1776–1782. [CrossRef] [PubMed]

7. LeGendre, O.; Breslin, P.A.; Foster, D.A. (-)-Oleocanthal rapidly and selectively induces cancer cell death via lysosomal membrane permeabilization. *Mol. Cell. Oncol.* **2015**, *2*, e1006077. [CrossRef] [PubMed]

8. Scotece, M.; Gomez, R.; Conde, J.; Lopez, V.; Gomez-Reino, J.J.; Lago, F.; Smith, A.B., 3rd; Gualillo, O. Oleocanthal inhibits proliferation and MIP-1alpha expression in human multiple myeloma cells. *Curr. Med. Chem.* **2013**, *20*, 2467–2475. [CrossRef] [PubMed]

9. Fogli, S.; Arena, C.; Carpi, S.; Polini, B.; Bertini, S.; Digiacomo, M.; Gado, F.; Saba, A.; Saccomanni, G.; Breschi, M.C.; et al. Cytotoxic Activity of Oleocanthal Isolated from Virgin Olive Oil on Human Melanoma Cells. *Nutr. Cancer* **2016**, *68*, 873–877. [CrossRef]

10. Naruszewicz, M.; Czerwinska, M.E.; Kiss, A.K. Oleacein. translation from Mediterranean diet to potential antiatherosclerotic drug. *Curr. Pharm. Des.* **2015**, *21*, 1205–1212. [CrossRef]

11. Costanzo, P.; Bonacci, S.; Cariati, L.; Nardi, M.; Oliverio, M.; Procopio, A. Simple and efficient sustainable semi-synthesis of oleacein [2-(3,4-hydroxyphenyl) ethyl (3S,4E)-4-formyl-3-(2-oxoethyl)hex-4-enoate] as potential additive for edible oils. *Food Chem.* **2018**, *245*, 410–414. [CrossRef] [PubMed]

12. Fulciniti, M.; Amodio, N.; Bandi, R.L.; Cagnetta, A.; Samur, M.K.; Acharya, C.; Prabhala, R.; D'Aquila, P.; Bellizzi, D.; Passarino, G.; et al. miR-23b/SP1/c-myc forms a feed-forward loop supporting multiple myeloma cell growth. *Blood Cancer J.* **2016**, *6*, e380. [CrossRef]

13. Cea, M.; Cagnetta, A.; Gobbi, M.; Patrone, F.; Richardson, P.G.; Hideshima, T.; Anderson, K.C. New insights into the treatment of multiple myeloma with histone deacetylase inhibitors. *Curr. Pharm. Des.* **2013**, *19*, 734–744. [CrossRef] [PubMed]

14. Amodio, N.; D'Aquila, P.; Passarino, G.; Tassone, P.; Bellizzi, D. Epigenetic modifications in multiple myeloma: Recent advances on the role of DNA and histone methylation. *Expert Opin. Ther. Targets* **2017**, *21*, 91–101. [CrossRef] [PubMed]

15. Imai, Y.; Hirano, M.; Kobayashi, M.; Futami, M.; Tojo, A. HDAC Inhibitors Exert Anti-Myeloma Effects through Multiple Modes of Action. *Cancers* **2019**, *11*. [CrossRef]

16. Fulciniti, M.; Amodio, N.; Bandi, R.L.; Munshi, M.; Yang, G.; Xu, L.; Hunter, Z.; Tassone, P.; Anderson, K.C.; Treon, S.P.; et al. MYD88-independent growth and survival effects of Sp1 transactivation in Waldenstrom macroglobulinemia. *Blood* **2014**, *123*, 2673–2681. [CrossRef]

17. Safe, S.; Abbruzzese, J.; Abdelrahim, M.; Hedrick, E. Specificity Protein Transcription Factors and Cancer: Opportunities for Drug Development. *Cancer Prev. Res.* **2018**, *11*, 371–382. [CrossRef]

18. Kikuchi, J.; Wada, T.; Shimizu, R.; Izumi, T.; Akutsu, M.; Mitsunaga, K.; Noborio-Hatano, K.; Nobuyoshi, M.; Ozawa, K.; Kano, Y.; et al. Histone deacetylases are critical targets of bortezomib-induced cytotoxicity in multiple myeloma. *Blood* **2010**, *116*, 406–417. [CrossRef]

19. Amodio, N.; Di Martino, M.T.; Foresta, U.; Leone, E.; Lionetti, M.; Leotta, M.; Gulla, A.M.; Pitari, M.R.; Conforti, F.; Rossi, M.; et al. miR-29b sensitizes multiple myeloma cells to bortezomib-induced apoptosis through the activation of a feedback loop with the transcription factor Sp1. *Cell Death Dis.* **2012**, *3*, e436. [CrossRef]

20. Amodio, N.; Stamato, M.A.; Gulla, A.M.; Morelli, E.; Romeo, E.; Raimondi, L.; Pitari, M.R.; Ferrandino, I.; Misso, G.; Caraglia, M.; et al. Therapeutic Targeting of miR-29b/HDAC4 Epigenetic Loop in Multiple Myeloma. *Mol. Cancer Ther.* **2016**, *15*, 1364–1375. [CrossRef]

21. Amodio, N.; Rossi, M.; Raimondi, L.; Pitari, M.R.; Botta, C.; Tagliaferri, P.; Tassone, P. miR-29s: A family of epi-miRNAs with therapeutic implications in hematologic malignancies. *Oncotarget* **2015**, *6*, 12837–12861. [CrossRef] [PubMed]

22. Xia, S.S.; Zhang, G.J.; Liu, Z.L.; Tian, H.P.; He, Y.; Meng, C.Y.; Li, L.F.; Wang, Z.W.; Zhou, T. MicroRNA-22 suppresses the growth, migration and invasion of colorectal cancer cells through a Sp1 negative feedback loop. *Oncotarget* **2017**, *8*, 36266–36278. [CrossRef] [PubMed]

23. Raimondi, L.; De Luca, A.; Morelli, E.; Giavaresi, G.; Tagliaferri, P.; Tassone, P.; Amodio, N. MicroRNAs: Novel Crossroads between Myeloma Cells and the Bone Marrow Microenvironment. *Biomed. Res. Int.* **2016**, *2016*, 6504593. [CrossRef] [PubMed]

24. Caracciolo, D.; Di Martino, M.T.; Amodio, N.; Morelli, E.; Montesano, M.; Botta, C.; Scionti, F.; Talarico, D.; Altomare, E.; Gallo Cantafio, M.E.; et al. miR-22 suppresses DNA ligase III addiction in multiple myeloma. *Leukemia* **2019**, *33*, 487–498. [CrossRef] [PubMed]

25. Saldivar-Gonzalez, F.I.; Gomez-Garcia, A.; Chavez-Ponce de Leon, D.E.; Sanchez-Cruz, N.; Ruiz-Rios, J.; Pilon-Jimenez, B.A.; Medina-Franco, J.L. Inhibitors of DNA Methyltransferases From Natural Sources: A Computational Perspective. *Front. Pharmacol.* **2018**, *9*, 1144. [CrossRef]

26. Singh, A.K.; Bishayee, A.; Pandey, A.K. Targeting Histone Deacetylases with Natural and Synthetic Agents: An Emerging Anticancer Strategy. *Nutrients* **2018**, *10*. [CrossRef] [PubMed]

27. Huang, D.; Cui, L.; Ahmed, S.; Zainab, F.; Wu, Q.; Wang, X.; Yuan, Z. An overview of epigenetic agents and natural nutrition products targeting DNA methyltransferase, histone deacetylases and microRNAs. *Food Chem. Toxicol.* **2019**, *123*, 574–594. [CrossRef]

28. Bulotta, S.; Celano, M.; Lepore, S.M.; Montalcini, T.; Pujia, A.; Russo, D. Beneficial effects of the olive oil phenolic components oleuropein and hydroxytyrosol: Focus on protection against cardiovascular and metabolic diseases. *J. Transl. Med.* **2014**, *12*, 219. [CrossRef]

29. Barbaro, B.; Toietta, G.; Maggio, R.; Arciello, M.; Tarocchi, M.; Galli, A.; Balsano, C. Effects of the olive-derived polyphenol oleuropein on human health. *Int. J. Mol. Sci.* **2014**, *15*, 18508–18524. [CrossRef]

30. Fabiani, R. Anti-cancer properties of olive oil secoiridoid phenols: A systematic review of in vivo studies. *Food Funct.* **2016**, *7*, 4145–4159. [CrossRef]

31. Flemmig, J.; Rusch, D.; Czerwinska, M.E.; Rauwald, H.W.; Arnhold, J. Components of a standardised olive leaf dry extract (Ph. Eur.) promote hypothiocyanite production by lactoperoxidase. *Arch. Biochem. Biophys.* **2014**, *549*, 17–25. [CrossRef] [PubMed]

32. Nardi, M.; Bonacci, S.; De Luca, G.; Maiuolo, J.; Oliverio, M.; Sindona, G.; Procopio, A. Biomimetic synthesis and antioxidant evaluation of 3,4-DHPEA-EDA [2-(3,4-hydroxyphenyl) ethyl (3S,4E)-4-formyl-3-(2-oxoethyl)hex-4-enoate]. *Food Chem.* **2014**, *162*, 89–93. [CrossRef] [PubMed]

33. Koutsoni, O.S.; Karampetsou, K.; Kyriazis, I.D.; Stathopoulos, P.; Aligiannis, N.; Halabalaki, M.; Skaltsounis, L.A.; Dotsika, E. Evaluation of total phenolic fraction derived from extra virgin olive oil for its antileishmanial activity. *Phytomedicine* **2018**, *47*, 143–150. [CrossRef] [PubMed]

34. Filipek, A.; Czerwinska, M.E.; Kiss, A.K.; Wrzosek, M.; Naruszewicz, M. Oleacein enhances anti-inflammatory activity of human macrophages by increasing CD163 receptor expression. *Phytomedicine* **2015**, *22*, 1255–1261. [CrossRef] [PubMed]

35. Lombardo, G.E.; Lepore, S.M.; Morittu, V.M.; Arcidiacono, B.; Colica, C.; Procopio, A.; Maggisano, V.; Bulotta, S.; Costa, N.; Mignogna, C.; et al. Effects of Oleacein on High-Fat Diet-Dependent Steatosis, Weight Gain, and Insulin Resistance in Mice. *Front. Endocrinol.* **2018**, *9*, 116. [CrossRef] [PubMed]

36. Amodio, N.; Leotta, M.; Bellizzi, D.; Di Martino, M.T.; D'Aquila, P.; Lionetti, M.; Fabiani, F.; Leone, E.; Gulla, A.M.; Passarino, G.; et al. DNA-demethylating and anti-tumor activity of synthetic miR-29b mimics in multiple myeloma. *Oncotarget* **2012**, *3*, 1246–1258. [CrossRef]

37. Rizq, O.; Mimura, N.; Oshima, M.; Saraya, A.; Koide, S.; Kato, Y.; Aoyama, K.; Nakajima-Takagi, Y.; Wang, C.; Chiba, T.; et al. Dual Inhibition of EZH2 and EZH1 Sensitizes PRC2-Dependent Tumors to Proteasome Inhibition. *Clin. Cancer Res.* **2017**, *23*, 4817–4830. [CrossRef]

38. Stamato, M.A.; Juli, G.; Romeo, E.; Ronchetti, D.; Arbitrio, M.; Caracciolo, D.; Neri, A.; Tagliaferri, P.; Tassone, P.; Amodio, N. Inhibition of EZH2 triggers the tumor suppressive miR-29b network in multiple myeloma. *Oncotarget* **2017**, *8*, 106527–106537. [CrossRef]

39. Amodio, N.; Stamato, M.A.; Juli, G.; Morelli, E.; Fulciniti, M.; Manzoni, M.; Taiana, E.; Agnelli, L.; Cantafio, M.E.G.; Romeo, E.; et al. Drugging the lncRNA MALAT1 via LNA gapmeR ASO inhibits gene expression of proteasome subunits and triggers anti-multiple myeloma activity. *Leukemia* **2018**, *32*, 1948–1957. [CrossRef]

40. Amodio, N.; Di Martino, M.T.; Neri, A.; Tagliaferri, P.; Tassone, P. Non-coding RNA: A novel opportunity for the personalized treatment of multiple myeloma. *Expert Opin. Biol. Ther.* **2013**, *13*, S125–137. [CrossRef]

41. Rossi, M.; Amodio, N.; Di Martino, M.T.; Tagliaferri, P.; Tassone, P.; Cho, W.C. MicroRNA and multiple myeloma: From laboratory findings to translational therapeutic approaches. *Curr. Pharm. Biotechnol.* **2014**, *15*, 459–467. [CrossRef] [PubMed]

42. Calura, E.; Bisognin, A.; Manzoni, M.; Todoerti, K.; Taiana, E.; Sales, G.; Morgan, G.J.; Tonon, G.; Amodio, N.; Tassone, P.; et al. Disentangling the microRNA regulatory milieu in multiple myeloma: Integrative genomics analysis outlines mixed miRNA-TF circuits and pathway-derived networks modulated in t(4;14) patients. *Oncotarget* **2016**, *7*, 2367–2378. [CrossRef] [PubMed]

43. Leotta, M.; Biamonte, L.; Raimondi, L.; Ronchetti, D.; Di Martino, M.T.; Botta, C.; Leone, E.; Pitari, M.R.; Neri, A.; Giordano, A.; et al. A p53-dependent tumor suppressor network is induced by selective miR-125a-5p inhibition in multiple myeloma cells. *J. Cell. Physiol.* **2014**, *229*, 2106–2116. [CrossRef]

44. Pitari, M.R.; Rossi, M.; Amodio, N.; Botta, C.; Morelli, E.; Federico, C.; Gulla, A.; Caracciolo, D.; Di Martino, M.T.; Arbitrio, M.; et al. Inhibition of miR-21 restores RANKL/OPG ratio in multiple myeloma-derived bone marrow stromal cells and impairs the resorbing activity of mature osteoclasts. *Oncotarget* **2015**, *6*, 27343–27358. [CrossRef] [PubMed]

45. Gulla, A.; Di Martino, M.T.; Gallo Cantafio, M.E.; Morelli, E.; Amodio, N.; Botta, C.; Pitari, M.R.; Lio, S.G.; Britti, D.; Stamato, M.A.; et al. A 13 mer LNA-i-miR-221 Inhibitor Restores Drug Sensitivity in Melphalan-Refractory Multiple Myeloma Cells. *Clin. Cancer Res.* **2016**, *22*, 1222–1233. [CrossRef] [PubMed]

46. Morelli, E.; Biamonte, L.; Federico, C.; Amodio, N.; Di Martino, M.T.; Gallo Cantafio, M.E.; Manzoni, M.; Scionti, F.; Samur, M.K.; Gulla, A.; et al. Therapeutic vulnerability of multiple myeloma to MIR17PTi, a first-in-class inhibitor of pri-miR-17-92. *Blood* **2018**, *132*, 1050–1063. [CrossRef] [PubMed]

47. Morelli, E.; Leone, E.; Cantafio, M.E.; Di Martino, M.T.; Amodio, N.; Biamonte, L.; Gulla, A.; Foresta, U.; Pitari, M.R.; Botta, C.; et al. Selective targeting of IRF4 by synthetic microRNA-125b-5p mimics induces anti-multiple myeloma activity in vitro and in vivo. *Leukemia* **2015**, *29*, 2173–2183. [CrossRef]

48. Fenichel, M.P. FDA approves new agent for multiple myeloma. *J. Natl. Cancer Inst.* **2015**, *107*, djv165. [CrossRef]

49. Procopio, A.; Alcaro, S.; Nardi, M.; Oliverio, M.; Ortuso, F.; Sacchetta, P.; Pieragostino, D.; Sindona, G. Synthesis, biological evaluation, and molecular modeling of oleuropein and its semisynthetic derivatives as cyclooxygenase inhibitors. *J. Agric. Food Chem.* **2009**, *57*, 11161–11167. [CrossRef]

50. Amodio, N.; Gallo Cantafio, M.E.; Botta, C.; Agosti, V.; Federico, C.; Caracciolo, D.; Ronchetti, D.; Rossi, M.; Driessen, C.; Neri, A.; et al. Replacement of miR-155 Elicits Tumor Suppressive Activity and Antagonizes Bortezomib Resistance in Multiple Myeloma. *Cancers* **2019**, *11*. [CrossRef]

51. Di Sanzo, M.; Chirillo, R.; Aversa, I.; Biamonte, F.; Santamaria, G.; Giovannone, E.D.; Faniello, M.C.; Cuda, G.; Costanzo, F. shRNA targeting of ferritin heavy chain activates H19/miR-675 axis in K562 cells. *Gene* **2018**, *657*, 92–99. [CrossRef] [PubMed]

52. Guarasci, F.; D'Aquila, P.; Mandala, M.; Garasto, S.; Lattanzio, F.; Corsonello, A.; Passarino, G.; Bellizzi, D. Aging and nutrition induce tissue-specific changes on global DNA methylation status in rats. *Mech. Ageing Dev.* **2018**, *174*, 47–54. [CrossRef] [PubMed]

Article

Divergent Effects of Daidzein and Its Metabolites on Estrogen-Induced Survival of Breast Cancer Cells

Emiliano Montalesi, Manuela Cipolletti, Patrizio Cracco, Marco Fiocchetti and Maria Marino *

Department of Science, University Roma Tre, Viale Guglielmo Marconi 446, I-00146 Roma, Italy; emiliano.montalesi@uniroma3.it (E.M.); manuela.cipolletti@uniroma3.it (M.C.); patrizio.cracco95@live.it (P.C.); marco.fiocchetti@uniroma3.it (M.F.)

* Correspondence: maria.marino@uniroma3.it; Tel.: +39-065733-6320

Received: 28 November 2019; Accepted: 23 December 2019; Published: 9 January 2020

Abstract: Although soy consumption is associated with breast cancer prevention, the low bioavailability and the extensive metabolism of soy-active components limit their clinical application. Here, the impact of daidzein (D) and its metabolites on estrogen-dependent anti-apoptotic pathway has been evaluated in breast cancer cells. In estrogen receptor α-positive breast cancer cells treated with D and its metabolites, single or in mixture, ERα activation and Neuroglobin (NGB) levels, an anti-apoptotic estrogen/ERα-inducible protein, were evaluated. Moreover, the apoptotic cascade activation, as well as the cell number after stimulation was assessed in the absence/presence of paclitaxel to determine the compound effects on cell susceptibility to a chemotherapeutic agent. Among the metabolites, only D-4′-sulfate maintains the anti-estrogenic effect of D, reducing the NGB levels and rendering breast cancer cells more prone to the paclitaxel treatment, whereas other metabolites showed estrogen mimetic effects, or even estrogen independent effects. Intriguingly, the co-stimulation of D and gut metabolites strongly reduced D effects. The results highlight the important and complex influence of metabolic transformation on isoflavones physiological effects and demonstrate the need to take biotransformation into account when assessing the potential health benefits of consumption of soy isoflavones in cancer.

Keywords: estrogen; estrogen receptor alpha; polyphenols; daidzein; daidzein metabolites; paclitaxel; apoptosis; breast cancer cells

---

1. Introduction

Plant-derived polyphenols are naturally occurring nonsteroidal compounds that play important roles in ecological functions such as pollinator attraction or protection from herbivores and UV irradiation [1]. Due to their molecular structure and size, some of these molecules, including lignans, flavonoids, and stilbenes, have a chemical structure that resembles that of human estrogens, in particular to 17-β-estradiol (E2) [2]. Among other, isoflavones, a class of flavonoids ranked among the most estrogenic compounds, bind to estrogen receptor subtypes (i.e., ERα and ERβ) [3,4] exerting estrogenic and/or antiestrogenic effects [1]. For isoflavones, the key to their bioactivity in human and animals seems to rely on their (anti)estrogenic activity. Indeed, due their antiestrogenic activities, isoflavone enriched diets are associated with a lower incidence of a variety of estrogen-related cancers, including breast, endometrial, and ovarian cancers [5,6].

The main dietary source of isoflavones in humans are soybean and soybean products, which contain mainly daidzein (7,4′-dihydroxyisoflavone, D) and genistein (7, 4′-dihydroxy-6-methoxyisoflavone), whose potential efficacy against breast cancer is well documented [7–12]. Although these data are promising for the use of these compounds as anticancer therapeutic agents, the therapeutic application of isoflavones is still limited, mainly due to their scarce bioavailability in human beings. In particular, D is almost completely metabolized by the gut microbiota and liver, resulting in water-soluble metabolites (e.g., equol, D-sulfates, o-desmethylangolensin) [13]. Thus, not only D has a low concentration and persistence in the bloodstream, even when consumed in high quantities, but also its metabolites strongly overcome the concentrations of the precursor probably affecting D biological activities [14–18]. Nowadays, the possibility that the metabolites may mimic the anti-carcinogenic effect of D in endocrine-related cancers or may act as synergistic or antagonistic molecules of their precursor is still unknown. Previously, the ability of D, metabolites, to modulate ERβ subtype activities activating a pro-apoptotic cascade in HeLa cancer cells transfected with ERβ expression vector, has been reported [19]. The possibility that a similar scenario could be found also in the modulation of ERα-dependent activities important for breast cancer cell progression is intriguing.

Breast cancer is one of the most common fatal diseases in women. A considerably higher ERα/ERβ ratio is reported in some breast cancer types, when compared to a healthy tissue, namely because of a reduction in the ERβ level [20,21]. The 70% of breast cancers are ERα-positive, where this subtype of receptor mediates E2-induced cancer cell survival and proliferation [22–25]. In particular, we recently demonstrated that E2 stimulation rapidly enhances the ERα activity (Ser118 phosphorylation and PI3K/AKT pathway activation) in breast cancer cells increasing the intracellular levels of an anti-apoptotic globin, neuroglobin (NGB) [22]. E2-induced NGB upregulation in cancer cells represents an inducible defense mechanism of E2-related human breast cancer rendering them insensitive to several injury including chemotherapy [22,26,27]. Indeed, NGB displays a pivotal role in the E2/ERα-induced anti-apoptotic pathway that abrogates the cell death induced by a chemotherapeutic agent (paclitaxel, Pacl) [22]. Intriguingly, the stilbene Resveratrol decreases NGB levels interfering with E2/ERα-induced NGB up-regulation potentiating Pacl pro-apoptotic effects [4]. In this study, we investigated the potential interference of daidzein on this pathway and evaluated if its metabolites produced mainly from gut microbiota (i.e., equol, Eq, and O-desmethylangolensin, O-DMA) and from both liver and gut enzymes (D-4′-sulfate, D4S, D-7-sulfate, D7S, and D-4′,7-disulfate, DDS) mimic D effect or may act as synergistic or antagonistic molecules. The ERα positive breast cancer cells, MCF-7 and T47D, have been used as the experimental models.

2. Results

2.1. Effect of D and Its Metabolites on NGB Levels in Breast Cancer Cells

NGB levels were evaluated in MCF-7 cells pre-treated for 24 h with different concentrations of D (Figure 1a) and its metabolites ranging between 0.1 and 10 µM (Figure 1). E2 (10 nM, 24 h) was used as positive control. Eq and O-DMA (Figure 1b) and D7S, D4S, and DDS (Figure 1c) were selected as prototypes of the D metabolites mainly produced by gut microbiota and the liver, respectively. The results clearly indicate that daidzein (1–10 µM) and D4S (0.1–1 µM) reduced the basal level of NGB levels in MCF-7 cells. On the other hand, Eq, O-DMA, D7S, and DDS, like E2, increased the level of NGB (Figure 1a,b). For the successive experiments, D and its mimetic sulphate metabolite (i.e., D4S) were selected. In addition, Eq, one gut metabolite, was selected as negative control. All compounds were used at 1 µM concentration in successive experiments.

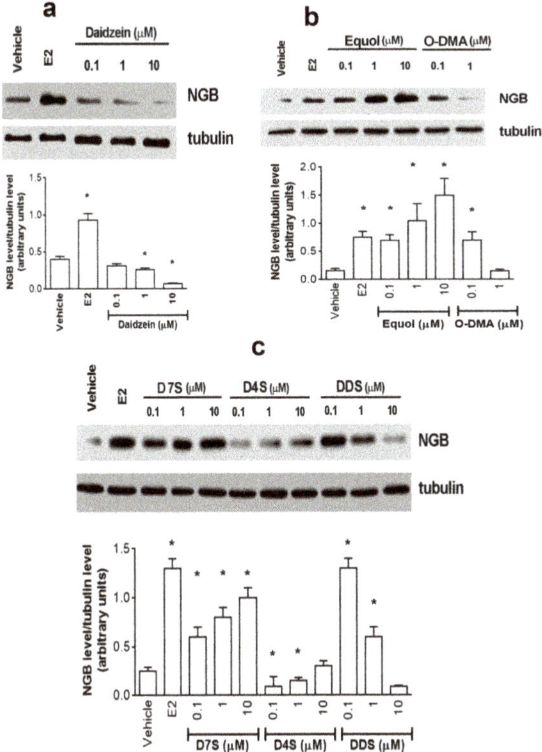

Figure 1. Effects of daidzein (**a**), equol and O-desmethylangolesin (**b**), daidzein-7-sulfate, daidzein-4′-sulfate and daidzein-7,4′-disulfate (**c**) on neuroglobin intracellular levels. (**a**–**c**) Western blot (top) and densitometric analyses (bottom) of NGB protein levels in MCF-7 cells treated for 24 h with the vehicle (DMSO), E2 (10 nM), D and its metabolites (0.1, 1.0, and 10 μM). The amount of proteins was normalized by comparison with tubulin levels. Data are the mean ± SD of four different experiments. $p < 0.001$ was determined with Student t test with respect to the vehicle (*) treated samples. E2: estradiol; NGB: neuroglobin; DMSO: dimethyl sulfoxide; D: daidzein; Eq: equol; O-DMA: O-desmethylangolesin; D7S: daidzein-7-sulfate; D4S: daidzein-4′-sulfate; DDS: daidzein-7,4′-disulfate.

The modulation of NGB levels by D, D4S, and Eq (1 μM, 24 h) was also confirmed in T47D cells (Figure 2). Indeed, also in these ERα-positive cells, D and D4S significantly reduced the basal level of NGB, whereas Eq, like E2, increased the globin level (Figure 2).

2.2. Mechanisms of D-, D4S-, and Eq Induced Modulation of NGB Levels

The involvement of ERα in the effects of D and its metabolites has been confirmed by pre-treating MCF-7 cells with 100 nM of the ERα inhibitor Endoxifen (Endo) before compound stimulation. As shown in Figure 3a, endoxifen pre-treatment completely impairs E2- and Eq-induced NGB up-regulation as well as D- and D4S-induced NGB down-regulation, strongly corroborating the necessity of an active ERα to modulate NGB levels. In particular, E2 rapidly down-regulates ERα levels maintaining high its phosphorylation status (Figure 3b) while neither D nor its metabolites modify the receptor levels but still increase ERα phosphorylation, although at lower level than E2 (Figure 3b). As expected, endoxifen pre-treatment completely prevents the ERα activation by all compounds considered (Figure 3b).

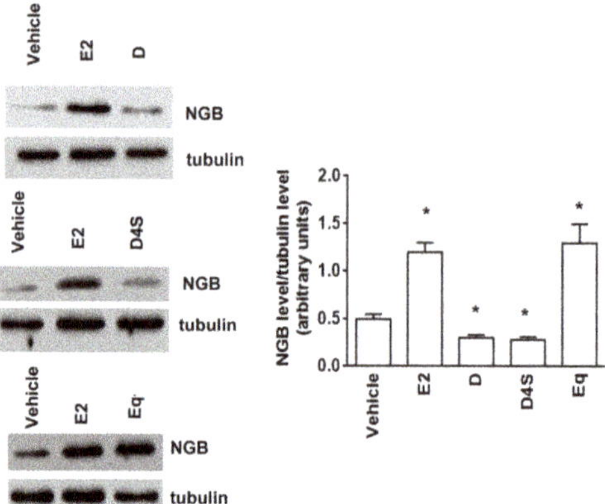

Figure 2. Effects of daidzein, daidzein-4′-sulfate and equol on neuroglobin intracellular levels in T47D cells. Western blot (left) and densitometric analyses (right) of NGB protein levels in T47D cells treated for 24 h with the vehicle (DMSO), E2 (10 nM), D (1 μM), D4S (1 μM), or Eq (1 μM). The amount of proteins was normalized by comparison with tubulin levels. Data are the mean ± SD of three different experiments. $p < 0.001$ was determined with Student t-test with respect to the vehicle (*) treated samples. DMSO: dimethyl sulfoxide; E2: estradiol; NGB: neuroglobin; D: daidzein; D4S: daidzein-4′-sulfate; Eq: equol.

ERα activation is the first step of a signal pathway triggered by E2 to enhance NGB levels. The activation of AKT is necessary to rapidly impair NGB degradation and assure NGB gene transcription via the CREBP transcription factor [28]. On the other hand, the ability of flavonoids (i.e., naringenin) to trigger the ERα-dependent activation of p38 has been demonstrated [29]. These evidences prompted us to evaluate if D and its metabolites trigger the activation of these kinases. Figure 4 shows that, as expected, E2 elicits the rapid and persistent activation of AKT enhancing its phosphorylation status in MCF-7 cells after both 1 h and 24 h of stimulation (Figure 4a,b). On the other hand, the hormone rapidly activates p38 phosphorylation (Figure 4c), but 24 h after stimulation, the phosphorylation status of p38 return similar to the control (Figure 4d). Completely different is the effect of D and its metabolites. Indeed, both D and D4S do not activate AKT phosphorylation, but these compounds, in particular D4S, trigger the rapid and persistent activation of p38 phosphorylation (Figure 4c,d). Endoxifen pretreatment prevents the D and D4S effects as well as that of E2, although 1 h after D4S stimulation the ER inhibitor does not completely impede p38 activation (Figure 4c), suggesting that an ERα-independent mechanism is at the root of the very high p38 phosphorylation induced by this sulphate metabolite. Similarly, Eq stimulation of MCF-7 cells rapidly activates AKT phosphorylation by an ERα-independent pathway, not prevented by endoxifen pre-treatment (Figure 4a). However, the Eq-induced AKT activation is transient, and indeed the level of kinase phosphorylation was similar to the control 24 h after Eq stimulation. Like D and D4S, Eq triggers the rapid and persistent ERα-dependent p38 activation (Figure 4c,d). As a whole, these data strongly sustain that daidzein does not share similar action mechanisms with all of its metabolites, at least 1 h after treatment.

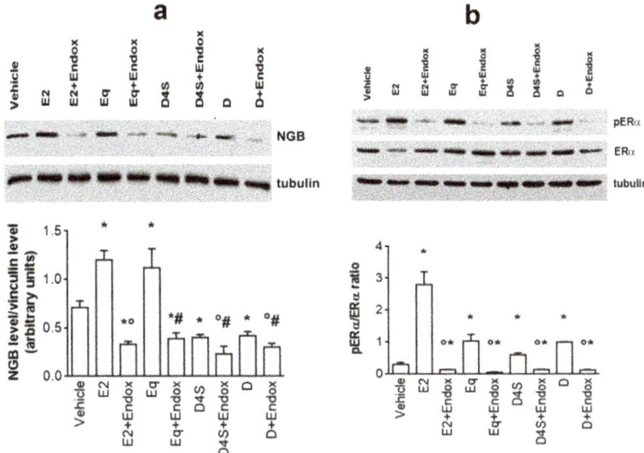

Figure 3. Daidzein, daidzein-4′-sulfate and equol effect on ERα activation status. (**a**) Western blot (top) and densitometric analyses (bottom) of NGB protein levels in MCF-7 cells treated for 24 h with either vehicle (DMSO) or E2 (10 nM) or D, D4S and Eq (1 μM) in presence or absence of the ERα inhibitor Endoxifen (1 μM; 30 min pretreatment). The amount of proteins was normalized by comparison with tubulin levels. Data are the mean ± SD of three different experiments. $p < 0.001$ was determined with Student's t test with respect to the vehicle (*) or E2-treated (°) samples. (**b**) ERα activation by daidzein, daidzein-4′-sulfate and equol. The panel represents the ERαSer118 phosphorylation status calculated as the ratio pERα/ERα). Determined by Western blot analysis in MCF-7 cells exposed for 1h to either vehicle (DMSO) or E2 (10 nM) or D, D4S and Eq (1 μM) in presence or absence of ERα inhibitor Endoxifen (1 μM; 30 min pretreatment). The nitrocellulose was stripped and then probed with anti-ERα antibody. The pERα/ERα ratio was calculated with respect to tubulin obtained by densitometric analyses of three different experiments (mean ± SD). $p < 0.001$ was determined by Student t test with respect to vehicle (*), E2-treated (°) or Endox-untreated samples (#). DMSO: dimethyl sulfoxide; E2: estradiol; Endox: endoxifen; ERα: estrogen receptor α; NGB: neuroglobin; D: daidzein; D4S: daidzein-4′-sulfate; Eq: equol.

2.3. Physiological Outcomes of D- and D4S-Induced E2/ERα/NGB Pathway Avoidance

We previously demonstrated that the ability of Resveratrol to impair NGB accumulation rendered cancer cells more prone to the anticancer effect of the chemotherapeutic agent paclitaxel (Pacl) [4]. This evidence prompted us to evaluate if D and D4S also exhibit this ability. As expected, Pacl treatment (100 nM) reduces NGB levels (Figure 5a–c) with the parallel increase of cleaved PARP-1 (i.e., 86 kDa band), a well-known biomarker of late apoptotic events (Figure 5d–f), and reduction of cell number (Figure 5g). Cell pre-treatment with E2 strongly prevents all Pacl effects in MCF-7 cells still enhancing NGB levels (Figure 5a–c), cell number (Figure 5g), and strongly reducing Pacl-induced PARP-1 cleavage (Figure 5d–f). Although neither D nor D4S pre-treatment affected Pacl effects, both these compounds restored Pacl effects in the presence of E2 (Figure 5). The ability of D and D4S to restore Pacl effects on cell number is also confirmed in T47D cell line (Figure 5h).

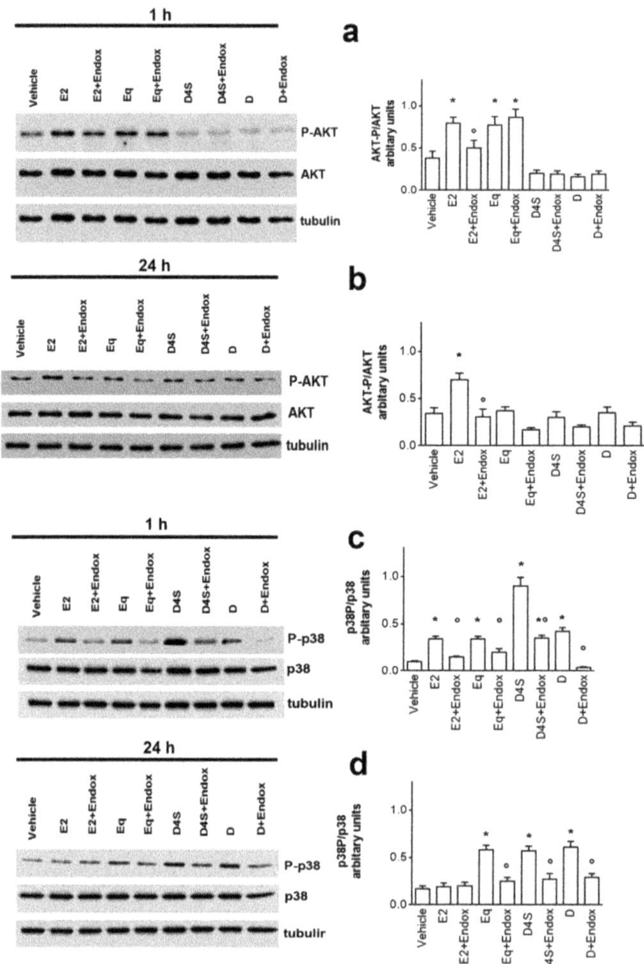

Figure 4. Daidzein, daidzein-4′-sulfate and equol action mechanism. The phosphorylation of the Ser473 residue of AKT (pAKT) (**a**,**b**) and Thr180/Tyr182 residues on P-38 (**c**,**d**) was determined by western blot analysis in MCF-7 cells exposed for 1 h (**a**,**c**) and 24 h (**b**,**d**) to either vehicle (DMSO) or E2 (10 nM) in presence or absence of D, D4S and Eq (1 μM). The nitrocellulose was stripped and then and then probed with anti-AKT or anti-P38 antibodies. In the panels, the pAKT/AKT (**a**,**b**) and pP38/P38 (**c**,**d**) ratios are represented. These ratios are calculated with respect to tubulin obtained by densitometric analyses of three different experiments (mean ± SD). $p < 0.001$ was determined by Student t-test with respect to vehicle (*) or Endox-untreated (°) samples. AKT: protein kinase B; E2: estradiol; Endox: endoxifen; ERα: estrogen receptor α; NGB: neuroglobin; p38: p38 mitogen-activated protein kinase; D: daidzein; D4S: daidzein-4′-sulfate; Eq: equol.

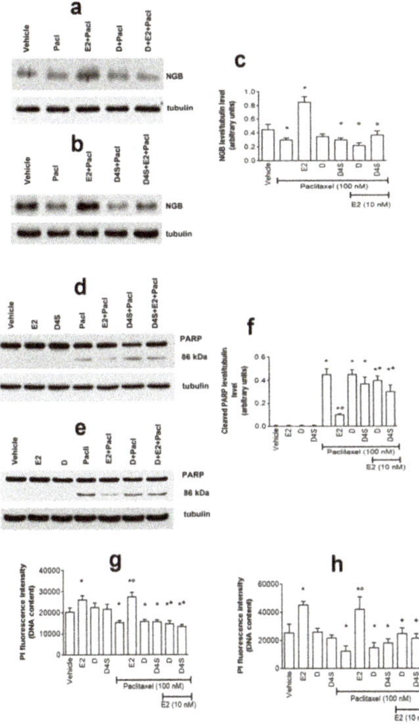

Figure 5. Physiological outcomes of D- and D4S-induced E2/ERα/NGB pathway avoidance. Western blot (left) and densitometric analyses (right) of NGB levels (**a–c**) and PARP-1 cleavage (**d–f**) in MCF-7 cells. Cells were treated with D (**a,e**) or D4S (**b,d**) (1 μM) in presence or absence of Pacl (100 nM) for 24 h, with either vehicle or E2; some Pacl treated cells were co-stimulated with either E2 or D or E2 together with D (**a,e**) or E2 or D4S or E2 together with D4S (**b,d**). (**c,f**) panels are densitometries that summarize the D- and D4S-induced modulation of NGB and PARP-1 cleavage respectively. The amount of protein was normalized by comparison with tubulin levels. Data are the mean ± SD of three different experiments. $p < 0.001$ was determined with Student t test with respect to the vehicle (*), Pacl-treated (°) samples or D- and D4S-treated samples, co-stimulated with Pacl but not with E2 (+). Effects of cellular DNA content obtained from PI assay on MCF-7 (**g**) or T47D (**h**) cells. The cells were treated for 24 h with either vehicle (DMSO) or E2 (10 nM) or Pacl (100 nM; 24 h); some samples were treated with D or D4S (1 μM) in presence or absence of Pacl and in presence of absence of E2. Data are mean ± SD of five different experiments. (*) $p < 0.001$ was calculated with Student t test versus vehicle or Pacl-treated (°) samples or D- and D4S-treated samples, co-stimulated with Pacl but not with E2 (+). E2: estradiol; NGB: neuroglobin; DMSO: dimethyl sulfoxide; D: daidzein; Eq: equol; D4S: daidzein-4′-sulfate; Pacl: paclitaxel.

2.4. Physiological Outcomes of D in Mixture with Its Metabolites

Due to its extensive biotransformation in the human body, D in circulation and in tissue is mainly present as a mixture with its metabolites. In order to evaluate if D maintains its anti-estrogenic actions described before, MCF-7 cells were treated with a mixture of compounds containing microbiota-produced metabolites (i.e., Eq and O-DMA, 1 μM each, gut metabolites) or metabolites produced by the gut and in liver enzymes (i.e., D7S, DDS, and D4S, 1 μM each, S metabolites) in presence or absence of D (1 μM). Figure 6a shows that the concentration of mixtures does not exert cytotoxic effects, indeed, the DNA content, and consequently the cell number, remain constant in MCF-7 cells stimulated with mixture or with the single compounds. D maintains the ability to reduce

NGB levels only in the co-stimulation with the mixture of sulphate metabolites (S metab), whereas this isoflavone effect is completely impaired by co-stimulation with the gut metabolites (Figure 6b). Notably, the mixture of S metabolites reduces NGB levels with respect to the control, while the mixture of gut metabolites increases NGB levels with respect to the control (Figure 6b). More intriguing is the effect of mixtures on Pacl-induced apoptosis (Figure 6c,d). Like the single compounds (Figure 5), neither sulphate nor gut metabolites induce the PARP cleavage at the concentration tested, however, sulphate metabolites preserve the Pacl-induced PARP-1 cleavage even in the presence of D or of E2 (Figure 6c), while gut metabolites reduce the Pacl effects even in the presence of D (Figure 6d). This latter effect is more evident in the presence of E2 (Figure 6d). As per other experiments, identical results have been obtained in T47D (data not showed), confirming that the mixture effects is not dependent on cellular context.

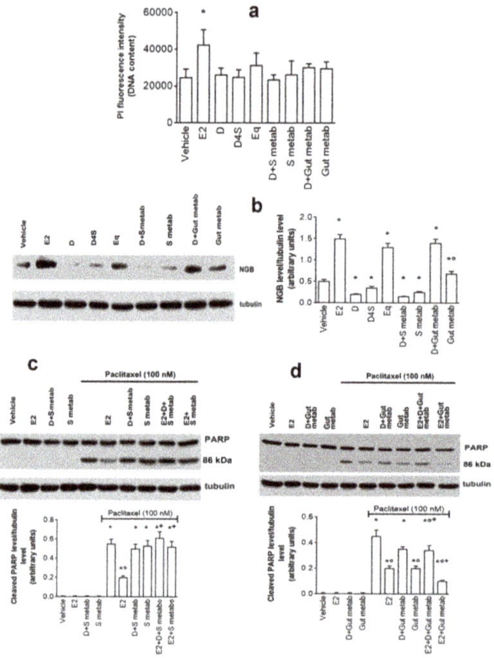

Figure 6. Physiological outcomes of D in mixture with its metabolites. (**a**) Analyses of cellular DNA content obtained from PI assay, Western blot (left) and densitometric analyses (right) of NGB levels (**b**) in MCF-7 cells. (**c**,**d**) Western blot (up) and densitometric analyses (bottom) of PARP-1 cleavage in MCF-7 cells. The MCF-7 cells were treated for 24 h with either vehicle (DMSO) or E2 (10 nM) or D, D4S or Eq (1 µM); some samples were treated with all the sulfate metabolites (D4S, D7S and DDS, 1 µM each) or all the gut metabolites (Eq and O-DMA, 1 µM each) in presence or absence of D (1 µM). Data are mean ± SD of three different experiments. (*) $p < 0.001$ was calculated with Student t test versus vehicle in (**a**). In **b**, data are the mean ± SD of five different experiments: $p < 0.001$ was determined with Student t test with respect to the vehicle (*) or to D-treated samples co-stimulated with metabolites (°). In (**c**,**d**), data are the mean ± SD of three different experiments. $p < 0.001$ was determined with Student *t* test with respect to the vehicle (*) or Pacl-treated (°) or E2 untreated samples co-stimulated with the metabolites (+). E2: estradiol; NGB: neuroglobin; DMSO: dimethyl sulfoxide; D: daidzein; Eq: equol; O-DMA: o-desmethylangolesin; D7S: daidzein-7-sulfate; D4S: daidzein-4′-sulfate; DDS: daidzein-7,4′-disulfate; Pacl: paclitaxel.

3. Discussion

Soy-derived isoflavones consumption has been largely recommended to the Western population for their possible vital role in maintaining human health through the regulation of metabolism and body weight, concurrent to the prevention of chronic and degenerative diseases including cancer and neurodegenerative disorders. Still, their short- and long-term effects have not been fully characterized, although some have cautioned that there may be harmful effects of overconsumption, especially in cases where compounds are isolated rather than consumed in a food matrix [30,31]. To further complicate this picture, isoflavones, once consumed as either aglycone or glycosides, enter a complex pathway of biotransformation that renders almost negligible the presence of the original molecule. The relative concentration of different metabolites in both plasma and tissues is determined by the specific contribution of intestinal microbiota and by de-conjugation/conjugation processes within the human body. Nowadays, available data on the estrogenic activity of D metabolites are restricted largely to Eq, whose production depends on the individual ability to host specific intestinal bacteria [32]. Once absorbed, daidzein is efficiently re-conjugated in the gut with either glucuronic acid or sulfate. Conjugation with sulfonic acid takes place also in the liver by hepatic sulfotransferase enzymes. Therefore, the plasma level of isoflavones in people on a soy-rich diet is very low (about 1–5 μM) [5,14], while they are present in the circulation predominantly in their glucuronide and sulfate forms [33].

Nowadays, the ability of D metabolites to maintain the effects of their precursor is largely unknown. The main aim of this study was to determine whether D metabolites produced by sulfotransferase and by microbiota enzymes maintain their anticancer effects, consisting in affecting ERα activities that are important for E2-induced resistance of breast cancer cells to chemotherapeutic injury. For this purpose, we utilized two human ERα positive breast cancer cells in which the pathway E2/ERα/NGB has been previously identified as pivotal for breast cancer cell susceptibility to the chemotherapeutic agent paclitaxel [22].

Our data indicate that, unlike E2 that induced NGB overexpression, 1–10 μM D reduced NGB levels under the basal level (i.e., vehicle -treated samples) in both MCF-7 and T47D breast cancer cells. This D effect was mimicked only by the D4S metabolite that reduced NGB levels at lower concentrations than D (i.e., 0.1–1 μM); while Eq and D7S showed an E2 like behavior, increasing NGB levels in a concentration dependent manner. Surprisingly, DDS and O-DMA increased NGB levels only at very low concentrations (i.e., 0.1 μM), and were ineffective at high concentrations (10 μM for DDS and 1 μM for O-DMA, respectively).

The differences between D and its metabolites in modulating NGB levels, as well as the different concentrations necessary to obtain the effect, suggest that D and its metabolites may trigger different signal transduction pathways. D4S and Eq were selected as representatives of the two contrasting effects to determine their mechanisms. Like E2, D, D4S, and Eq trigger ERαS118 phosphorylation, even if only E2 reduces the receptor levels, an important mechanism for E2-induced cancer cell proliferation [34]. Moreover, ERα is necessary for D and its metabolites to regulate NGB levels, confirming that D and its metabolites could bind to and activate ERα, as already reported [3]. However, downstream of ERα activation, D and its metabolites trigger divergent signal pathways. Indeed, differently from E2, D and D4S do not trigger AKT activation, which is pivotal for E2-induced NGB accumulation [28]. Instead, these compounds rapidly (1 h) and persistently (24 h) activate p38 kinase, whose activation is commonly shared among flavonoids (e.g., naringenin, quercetin) [29,35]. Upon receptor binding, naringenin modifies ERα conformation driving the receptor far from the plasma membrane (i.e., receptor de-palmitoylation) and decoupling its association with the active sub-unit of PI3K, but not with p38 kinase [24,29]. Consequently, AKT is not activated, whereas the persistent activation of p38 occurs, driving cancer cells to the activation of the apoptotic cascade that culminates with PARP cleavage [29]. A similar signal transduction pathway seems to be activated by D and D4S that at the concentrations used here (1 μM) do not affect the phosphorylation status of AKT, allowing the persistent p38 activation. On the other hand, Eq rapidly and transiently activates AKT phosphorylation that even if does not impair the persistent p38 activation, as E2 does, is sufficient to accumulate NGB into the cells. Note that the D4S-induced p38 and Eq-induced AKT phosphorylation is ERα-independent, at least partially for D4S, sustaining that these metabolites could bind to other cellular receptors, including the Arylic Receptor [36,37], which, in turn, can interfere with the estrogenic signal. As a whole, the chemical structure of D and its metabolites allows different ERα conformations that, in turn, drive cells to physiological outcomes that differ from that triggered by E2 [24,38,39].

Paclitaxel, a first line therapeutic agent for breast cancer, is a prototype of chemotherapeutic agents, which action mechanism is well known [4]. As reported above, E2/ERα-induced NGB accumulation in cancer cells represents an anti-apoptotic pathway, which abrogates the cell death induced by a chemotherapeutic agent (paclitaxel, Pacl) [4]. The ability of NGB accumulation to act as a shield against Pacl has been further confirmed here. In fact, E2 pretreatment impairs Pacl reduction of NGB as well as its ability to reduce cell number and activate PARP cleavage. Although at the selected concentrations (1 μM) they do not affect cell number or PARP cleavage, D and D4S completely prevent E2 effects, allowing the Pacl-induced activation of a pro-apoptotic cascade, even in the presence of the hormone, sustaining that the anti-estrogenic activity is pivotal for rendering cancer cells more vulnerable to the chemotherapeutic drug. The anti-proliferative action of isoflavones and other plant-derived polyphenols in cancer cells has been widely reported and disputed. Nowadays, it is quite accepted that only high concentrations of isoflavones (≤10 μM), very far from the concentration present in the plasma after a meal rich in polyphenols, could activate a mitochondrial-dependent apoptotic cascade, while low concentrations of these compounds result less active or, even, increase cancer cell proliferation [6–9]. Our data confirm that low concentrations of D and D4S are unable to trigger apoptosis or enhance Pacl effects, but acting as anti-estrogenic compounds, they can avoid the E2-induced anti-apoptotic effect on this chemotherapeutic drug.

The current study indicates, for the first time, that just D4S but no other metabolites retain the D anti-estrogenic activity. This effect is maintained when cancer cells were co-treated with D and mixtures of S metabolites, containing also D7S and DDS. On the contrary, the mixture of gut metabolites, containing Eq and O-DMA, completely impairs D effects, resulting in estrogen mimetic action. Identical results have been obtained in both MCF-7 and T47D (not completely reported), suggesting that the effect is not dependent on the cellular context. These results highlight the need to use physiologically relevant metabolites when investigating the putative beneficial properties of polyphenols against cancer.

4. Materials and Methods

4.1. Reagents

The Bradford protein assay and the chemiluminescence reagents for Western blot Clarity Western ECL Substrate were obtained from Bio-Rad Laboratories (Hercules, CA, USA). The anti-phospho-ERα (pERα Ser118) antibody and anti-phospho-AKT (pAKT Ser473) were purchased from Cell Signalling Technology Inc. (Beverly, MA, USA). The anti-α-tubulin and anti-NGB antibodies were purchased from Merck (Darmstadt, D). Specific antibodies against AKT, ERα (HC20), and poly [ADP-ribose] polymerase 1 (PARP-1) were obtained from Santa Cruz Biotechnology (Santa Cruz, CA, USA). The ERα inhibitor endoxifen was purchased from Tocris (Ballwin, MO, USA). All the other products were from Merck. Analytical and reagent grade products were used without further purification.

4.2. Cell Culture and Treatment

Human breast cancer cells MCF-7 and T47D (American Type Culture Collection, LGC Standards S.r.l., Milano, Italy) were grown in air containing 5% CO2 in either modified, phenol red free, Dulbecco's Modified Eagle's Medium (DMEM) medium. Ten percent (*v/v*) of charcoal-stripped fetal calf serum, L-glutamine (2 mM), gentamicin (0.1 mg/mL), and penicillin (100 U/mL) were added to the media before use. Cells were used at passage 4–8, as previously described [28]. Cell line authentication was periodically performed by the amplification of multiple short tandem repeat loci by BMR genomics

S.r.l (Padova, Italy). Cells were treated for 24 h with either vehicle (DMSO/phosphate-buffered saline, 1:10; *v/v*) or E2 (1 or 10 nM) or D and/or its metabolites (0.1, 1.0, and 10 μM) or Pacl (100 nM) or the metabolites mixtures. Two different metabolite mixtures were used for cell treatment, Sulphate metabolites (S metab) and Gut metabolites (Gut metab). The mixture of S metab was constituted by D4S (1 μM), D7S (1 μM) and DDS (1 μM), while that of Gut metab was constituted by Equol (1 μM) and O-DMA (1 μM). When indicated, endoxifen (1 μM) was added 30 min before compound administration, or E2 (10 nM) was added 4 h before Pacl addition (0.1, 1, and 100 nM) for 24 h. The E2 concentrations were selected on the bases of dose-response experiments already performed [40].

4.3. Western Blot Assay

Briefly, after the treatments, cells were lysed and proteins were solubilized in the YY buffer (50 mM HEPES at pH 7.5, 10% glycerol, 150 mM NaCl, 1% Triton X-100, 1 mM EDTA, and 1 mM EGTA) containing 0.70% (*w/v*) sodium dodecyl sulfate (SDS). Total proteins were quantified using the Bradford protein assay. Solubilized proteins (20 μg) were resolved by 7% or 15% SDS-polyacrylamide gel electrophoresis at 100 V for 1 h at 24.0 °C and then transferred to nitrocellulose with the Trans-Blot Turbo Transfer System (Bio-Rad) for 7 min. The nitrocellulose was treated with 5% (*w/v*) bovine serum albumin in 138.0 mM NaCl, 25.0 mM Tris, pH 8.0, at 24.0 °C for 1 h. Nitrocellulose was probed overnight at 4.0 °C with either anti-NGB (final dilution, 1:1000), or anti-pERα (final dilution, 1:1000), or anti-pAKT (final dilution, 1:1000), or anti p-38 (final dilution 1:1000), or anti-PARP-1 (final dilution, 1:1000) antibodies. The nitrocellulose was stripped by Restore Western Blot Stripping Buffer (Pierce Chemical, Rockford, IL) for 10 min at room temperature and then probed with either anti-ERα (final dilution, 1:1000) or anti-AKT (final dilution 1:1000) or anti-p38 (final dilution, 1:1000) or anti-α-tubulin (final dilution, 1:30,000) antibodies to normalize the protein loaded. The antibody reactivity was detected with ECL chemiluminescence Western blotting detection reagent using a ChemiDoc XRS+ Imaging System (Bio-Rad Laboratories, Hercules, CA, USA). Densitometric analyses were performed by the ImageJ software for Microsoft Windows (National Institute of Health, Bethesda, Rockville, MD, USA).

4.4. Cellular DNA Content, Propidium Iodide (PI) Assay

Cells were grown up to 80% confluence in 96-well plate and treated with the selected compounds. The cells were fixed and permeabilized with cold ethanol (EtOH) 70% for 15 min at −20 °C. EtOH solution were removed and the cells were incubated with PI buffer for 30 min in the dark. The solution was removed, and the cells were rinsed with PBS solution. The fluorescence was revealed (excitation wavelength: 537 nm; emission wavelength: 621 nm) with TECAN Spark 20 M multimode microplate reader (Life Science, Milano, Italy).

4.5. Statistical Analysis

The statistical analysis was performed by Student's t test to compare two sets of data by INSTAT software system for Windows. In all cases, $p < 0.05$ was considered significant.

5. Conclusions

Breast cancer, as well as lung, bronchial, and colorectal cancer, are estimated to be the three most commonly diagnosed types of malignancies. In particular, breast cancer alone accounts for 29% of all new cancers among women and the age of its onset is rapidly decreasing [41]. ERα activation status and signaling is considered one of the major drivers stimulating breast cancer proliferation, survival, and invasion [23–25]. The importance of ERα lies within its prognostic value, as it identifies patients that could benefit from the endocrine therapy [25]. Although the use of ERα antagonists has led to a considerable decline in breast cancer mortality, many patients become resistant to this therapy and developed metastatic tumors [24,41]. These observations have sparked the need for alternative approaches, increasing a sustained interest in soy isoflavones as a promising therapeutic option in breast cancer chemoprevention or as an adjuvant for chemotherapeutic agents. These claims lead patients with increased breast cancer risk to take into consideration supplementing their diet with soy or soy derivates, assuming that a high consumption might reduce the cancer risk [30,42]. Unfortunately, the increased economic interests in soy isoflavones has not yet been paralleled by the deciphering of the cellular and molecular mechanisms underlying their potential chemo-preventive role, which remains obscure.

The data reported here strongly sustain the need to examine in deep the effect and the action mechanisms of soy isoflavone. In particular, the activity of gut microbiota should be investigated in patients before isoflavone consumption, due to its key role in the metabolism and bioavailability of isoflavones [43], microflora influencing factors also require consideration. As examples, a high-carbohydrate milieu causes more extensive biotransformation, greatly enhancing Eq formation, and a *Clostridium* sp. strain that converts D principally to O-DMA has been identified from gut microbiota. These bio-transformations could render isoflavones less active [44] and reduce the efficacy of the chemotherapeutic treatment. Moreover, concerns could arise also among healthy individuals who regularly consume soy products. Indeed, Equol and O-DMA producer individuals may be subjected to a long exposure to potent estrogenic compounds. Those who are not, on the other hand, will be mostly exposed to anti-estrogenic compounds.

Author Contributions: Conceptualization, M.M. and E.M.; methodology, E.M., M.C., and P.C.; formal analysis, M.M. and E.M.; investigation, E.M., M.C., and P.C.; resources, M.M and M.F.; data curation, M.M. and E.M.; writing-original draft preparation, M.M.; writing-review and editing, M.M., M.C., M.F., E.M, and P.C.; supervision and funding acquisition, M.M. All authors have read and agreed to the published version of the manuscript.

Funding: This work was supported by a grant from PRIN-MIUR 2017 (Prot. 2017SNRXH3) to M.M. The grant of Excellence Department, MIUR (Legge 232/2016, Articolo 1, Comma 314–337), is gratefully acknowledged.

Conflicts of Interest: The authors declare no conflict of interest.

References

1. Křížová, L.; Dadáková, K.; Kašparovská, J.; Kašparovský, T. Isoflavones. *Molecules* **2019**, *24*, 1076. [CrossRef]
2. Di Gioia, F.; Petropoulos, S.A. Phytoestrogens, phytosteroids and saponins in vegetables: Biosynthesis, functions, health effects and practical applications. *Adv. Food Nutr. Res.* **2019**, *90*, 351–421.
3. Lacroix, S.; Klicic Badoux, J.; Scott-Boyer, M.P.; Parolo, S.; Matone, A.; Priami, C.; Morine, M.J.; Kaput, J.; Moco, S. A computationally driven analysis of the polyphenol-protein interactome. *Sci. Rep.* **2018**, *8*, 2232. [CrossRef] [PubMed]
4. Cipolletti, M.; Montalesi, E.; Nuzzo, M.T.; Fiocchetti, M.; Ascenzi, P.; Marino, M. Potentiation of paclitaxel effect by resveratrol in human breast cancer cells by counteracting the 17β-estradiol/estrogen receptor α/neuroglobin pathway. *J. Cell. Physiol.* **2019**, *234*, 3147–3157. [CrossRef]
5. Sarkar, F.H.; Li, Y. Soy isoflavones and cancer prevention. *Cancer Investig.* **2003**, *21*, 744–757. [CrossRef] [PubMed]
6. Sahin, I.; Bilir, B.; Ali, S.; Sahin, K.; Kucuk, O. Soy Isoflavones in Integrative Oncology: Increased Efficacy and Decreased Toxicity of Cancer Therapy. *Integr. Cancer Ther.* **2019**, *18*. [CrossRef] [PubMed]
7. Guo, J.; Wang, Q.; Zhang, Y.; Sun, W.; Zhang, S.; Li, Y.; Wang, J.; Bao, Y. Functional daidzein enhances the anticancer effect of topotecan and reverses BCRP-mediated drug resistance in breast cancer. *Pharmacol. Res.* **2019**, *147*, 104387. [CrossRef]
8. Choi, E.J.; Kim, G.H. Antiproliferative activity of daidzein and genistein may be related to ERα/c-erbB-2 expression in human breast cancer cells. *Mol. Med. Rep.* **2013**, *7*, 781–784. [CrossRef]
9. Zhu, Y.; Yao, Y.; Shi, Z.; Everaert, N.; Ren, G. Synergistic Effect of Bioactive Anticarcinogens from Soybean on Anti-Proliferative Activity in MDA-MB-231 and MCF-7 Human Breast Cancer Cells In Vitro. *Molecules* **2018**, *23*, 1557. [CrossRef]
10. Bao, C.; Namgung, H.; Lee, J.; Park, H.C.; Ko, J.; Moon, H.; Ko, H.W.; Lee, H.J. Daidzein suppresses tumor necrosis factor-α induced migration and invasion by inhibiting hedgehog/Gli1 signaling in human breast cancer cells. *J. Agric. Food Chem.* **2014**, *62*, 3759–3767. [CrossRef]

11. Liu, X.; Suzuki, N.; Santosh Laxmi, Y.R.; Okamoto, Y.; Shibutani, S. Anti-breast cancer potential of daidzein in rodents. *Life Sci.* **2012**, *91*, 415–419. [CrossRef] [PubMed]

12. Jin, S.; Zhang, Q.Y.; Kang, X.M.; Wang, J.X.; Zhao, W.H. Daidzein induces MCF-7 breast cancer cell apoptosis via the mitochondrial pathway. *Ann. Oncol.* **2010**, *21*, 263–268. [CrossRef] [PubMed]

13. Heinonen, S.; Wähälä, K.; Adlercreutz, H. Identification of isoflavone metabolites Dihydrodaidzein, Dihydrogenistein, 6-OH-O-DMA, and cis-4-OH-equol in human urine by gas chromatography–mass spectroscopy using authentic reference compounds. *Anal. Biochem.* **1999**, *274*, 211–219. [CrossRef] [PubMed]

14. King, R.A.; Bursill, D.B. Plasma and urinary kinetics of the isoflavones daidzein and genistein after a single soy meal in humans. *Am. J. Clin. Nutr.* **1998**, *67*, 867–872. [CrossRef]

15. Gardana, C.; Canzi, E.; Simonetti, P. The role of diet in the metabolism of daidzein by human faecal microbiota sampled from Italian volunteers. *J. Nutr. Biochem.* **2009**, *20*, 940–947. [CrossRef]

16. Chang, Y.C.; Nair, M.G.; Nitiss, J.L. Metabolites of daidzein and genistein and their biological activities. *J. Nat. Prod.* **1995**, *58*, 1901–1905. [CrossRef]

17. Soukup, S.T.; Helppi, J.; Müller, D.R.; Zierau, O.; Watzl, B.; Vollmer, G.; Diel, P.; Bub, A.; Kulling, S.E. Phase II metabolism of the soy isoflavones genistein and daidzein in humans, rats and mice: A cross-species and sex comparison. *Arch. Toxicol.* **2016**, *90*, 1335–1347. [CrossRef]

18. Chang, Y.C.; Nair, M.G. Metabolism of daidzein and genistein by intestinal bacteria. *J. Nat. Prod.* **1995**, *58*, 1892–1896. [CrossRef]

19. Totta, P.; Acconcia, F.; Virgili, F.; Cassidy, A.; Weinberg, P.D.; Rimbach, G.; Marino, M. Daidzein-sulfate metabolites affect transcriptional and antiproliferative activities of estrogen receptor-beta in cultured human cancer cells. *J. Nutr.* **2005**, *135*, 2687–2693. [CrossRef]

20. Lazennec, G.; Bresson, D.; Lucas, A.; Chauveau, C.; Vignon, F. ERβ inhibits proliferation and invasion of breast cancer cells. *Endocrinology* **2001**, *142*, 4120–4130. [CrossRef]

21. Marino, M.; Ascenzi, P. Membrane association of estrogen receptor alpha and beta influences 17beta-estradiol-mediated cancer cell proliferation. *Steroids* **2008**, *73*, 853–858. [CrossRef]

22. Fiocchetti, M.; Cipolletti, M.; Leone, S.; Ascenzi, P.; Marino, M. Neuroglobin overexpression induced by the 17β-Estradiol-Estrogen receptor-α Pathway reduces the sensitivity of MCF-7 Breast cancer cell to paclitaxel. *IUBMB Life* **2016**, *68*, 645–651. [CrossRef] [PubMed]

23. Nilsson, S.; Gustafsson, J.-Å. Estrogen receptors: Therapies targeted to receptor subtypes. *Clin. Pharmacol. Ther.* **2011**, *89*, 44–55. [CrossRef] [PubMed]

24. Ascenzi, P.; Bocedi, A.; Marino, M. Structure-function relationship of estrogen receptor alpha and beta: Impact on human health. *Mol. Asp. Med.* **2006**, *27*, 299–402. [CrossRef] [PubMed]

25. Thomas, C.; Gustafsson, J.-Å. The different roles of ER subtypes in cancer biology and therapy. *Nat. Rev. Cancer* **2011**, *11*, 597–608. [CrossRef] [PubMed]

26. Fiocchetti, M.; Solar Fernandez, V.; Montalesi, E.; Marino, M. Neuroglobin: A novel player in the oxidative stress response of cancer cells. *Oxidative Med. Cell. Longev.* **2019**, *2019*, 6315034. [CrossRef] [PubMed]

27. Fiocchetti, M.; Cipolletti, M.; Leone, S.; Naldini, A.; Carraro, F.; Giordano, D.; Verde, C.; Ascenzi, P.; Marino, M. Neuroglobin in breast cancer cells: Effect of hypoxia and oxidative stress on protein level, localization, and anti-apoptotic function. *PLoS ONE* **2016**, *11*, e0154959. [CrossRef] [PubMed]

28. Fiocchetti, M.; Cipolletti, M.; Ascenzi, P.; Marino, M. Dissecting the 17β-estradiol pathways necessary for neuroglobin anti-apoptotic activity in breast cancer. *J. Cell. Physiol.* **2018**, *233*, 5087–5103. [CrossRef]

29. Galluzzo, P.; Ascenzi, P.; Bulzomi, P.; Marino, M. The nutritional flavanone naringenin triggers antiestrogenic effects by regulating estrogen receptor alpha-palmitoylation. *Endocrinology* **2008**, *149*, 2567–2575. [CrossRef]

30. Crowe, K.M.; Francis, C. Position of the academy of nutrition and dietetics: Functional foods. *J. Acad. Nutr. Diet.* **2013**, *113*, 1096–1103. [CrossRef]

31. Cory, H.; Passarelli, S.; Szeto, J.; Tamez, M.; Mattei, J. The role of polyphenols in human health and food systems: A mini-review. *Front. Nutr.* **2018**, *5*, 87. [CrossRef] [PubMed]

32. Selvaraj, V.; Zakroczymski, M.A.; Naaz, A.; Mukai, M.; Ju, Y.H.; Doerge, D.R.; Katzenellenbogen, J.A.; Helferich, W.G.; Cooke, P.S. Estrogenicity of the isoflavone metabolite equol on reproductive and non-reproductive organs in mice. *Biol. Reprod.* **2004**, *71*, 966–972. [CrossRef]

33. Karakaya, S. Bioavailability of phenolic compounds. *Crit. Rev. Food Sci. Nutr.* **2004**, *44*, 453–464. [CrossRef] [PubMed]

34. Acconcia, F.; Fiocchetti, M.; Marino, M. Xenoestrogen regulation of ERα/ERβ balance in hormone-associated cancers. *Mol. Cell. Endocrinol.* **2017**, *457*, 3–12. [CrossRef] [PubMed]

35. Bulzomi, P.; Galluzzo, P.; Bolli, A.; Leone, S.; Acconcia, F.; Marino, M. The pro-apoptotic effect of quercetin in cancer cell lines requires ERβ-dependent signals. *J. Cell. Physiol.* **2012**, *227*, 1891–1898. [CrossRef]

36. Park, H.; Jin, U.H.; Orr, A.A.; Echegaray, S.P.; Davidson, L.A.; Allred, C.D.; Chapkin, R.S.; Jayaraman, A.; Lee, K.; Tamamis, P.; et al. Isoflavones as Ah Receptor Agonists in Colon-Derived Cell Lines: Structure-Activity Relationships. *Chem. Res. Toxicol.* **2019**, *32*, 2353–2364. [CrossRef]

37. Lee, J.E.; Safe, S. 3′,4′-dimethoxyflavone as an aryl hydrocarbon receptor antagonist in human breast cancer cells. *Toxicol. Sci.* **2000**, *58*, 235–242. [CrossRef]

38. Marino, M.; Pellegrini, M.; La Rosa, P.; Acconcia, F. Susceptibility of estrogen receptor rapid responses to xenoestrogens: Physiological outcomes. *Steroids* **2012**, *77*, 910–917. [CrossRef]

39. La Rosa, P.; Pellegrini, M.; Totta, P.; Acconcia, F.; Marino, M. Xenoestrogens alter estrogen receptor (ER) α intracellular levels. *PLoS ONE* **2014**, *9*, e88961. [CrossRef]

40. Fiocchetti, M.; Nuzzo, M.T.; Totta, P.; Acconcia, F.; Ascenzi, P.; Marino, M. Neuroglobin, a pro-survival player in estrogen receptor α-positive cancer cells. *Cell Death Dis.* **2014**, *5*, e1449. [CrossRef]

41. Uifălean, A.; Schneider, S.; Ionescu, C.; Lalk, M.; Iuga, C.A. Soy Isoflavones and Breast Cancer Cell Lines: Molecular Mechanisms and Future Perspectives. *Molecules* **2015**, *21*, 13. [CrossRef] [PubMed]

42. Fang, C.Y.; Tseng, M.; Daly, M.B. Correlates of soy food consumption in women at increased risk for breast cancer. *J. Am. Diet. Assoc.* **2005**, *105*, 1552–1558. [CrossRef] [PubMed]

43. Rafii, F.; Davis, C.; Park, M.; Heinze, T.M.; Beger, R.D. Variations in metabolism of the soy isoflavonoid daidzein by human intestinal microfloras from different individuals. *Arch. Microbiol.* **2003**, *180*, 11–16. [CrossRef] [PubMed]

44. Hur, H.G.; Beger, R.D.; Heinze, T.M.; Lay, J.O., Jr.; Freeman, J.P.; Dore, J.; Rafii, F. Isolation of an anaerobic intestinal bacterium capable of cleaving the C-ring of the isoflavonoid daidzein. *Arch. Microbiol.* **2002**, *178*, 8–12. [CrossRef] [PubMed]

Article

Gigantol Targets Cancer Stem Cells and Destabilizes Tumors via the Suppression of the PI3K/AKT and JAK/STAT Pathways in Ectopic Lung Cancer Xenografts

Nattanan Losuwannarak [1,2], **Arnatchai Maiuthed** [1], **Nakarin Kitkumthorn** [3], **Asada Leelahavanichkul** [4], **Sittiruk Roytrakul** [5] **and Pithi Chanvorachote** [1,2,*]

[1] Cell-Based Drug and Health Product Development Research Unit, Faculty of Pharmaceutical Sciences, Chulalongkorn University, Bangkok 10330, Thailand; los.nattanan@hotmail.com (N.L.); m.arnatchai@gmail.com (A.M.)

[2] Department of Pharmacology and Physiology, Faculty of Pharmaceutical Sciences, Chulalongkorn University, Bangkok 10330, Thailand

[3] Department of Oral Biology, Faculty of Dentistry, Mahidol University, Bangkok 10400, Thailand; nakarinkit@gmail.com

[4] Department of Microbiology, Faculty of Medicine, Chulalongkorn University, Bangkok 10330, Thailand; a_leelahavanit@yahoo.com

[5] Proteomics Research Laboratory, 113 Thailand Science Park, Phahonyothin Road, Khlong 1, Khlong Luang, Pathum Thani 12120, Thailand; sittiruk@biotec.or.th

[*] Correspondence: pithi_chan@yahoo.com; Tel.: +662-218-8344

Received: 5 November 2019; Accepted: 11 December 2019; Published: 17 December 2019

Abstract: Lung cancer has long been recognized as an important world heath concern due to its high incidence and death rate. The failure of treatment strategies, as well as the regrowth of the disease driven by cancer stem cells (CSCs) residing in the tumor, lead to the urgent need for a novel CSC-targeting therapy. Here, we utilized proteome alteration analysis and ectopic tumor xenografts to gain insight on how gigantol, a bibenzyl compound from orchid species, could attenuate CSCs and reduce tumor integrity. The proteomics revealed that gigantol affected several functional proteins influencing the properties of CSCs, especially cell proliferation and survival. Importantly, the PI3K/AKT/mTOR and JAK/STAT related pathways were found to be suppressed by gigantol, while the JNK signal was enhanced. The in vivo nude mice model confirmed that pretreatment of the cells with gigantol prior to a tumor becoming established could decrease the cell division and tumor maintenance. The results indicated that gigantol decreased the relative tumor weight with dramatically reduced tumor cell proliferation, as indicated by Ki-67 labeling. Although gigantol only slightly altered the epithelial-to-mesenchymal and angiogenesis statuses, the gigantol-treated group showed a dramatic loss of tumor integrity as compared with the well-grown tumor mass of the untreated control. This study reveals the effects of gigantol on tumor initiation, growth, and maintain in the scope that the cells at the first step of tumor initiation have lesser CSC property than the control untreated cells. This study reveals novel insights into the anti-tumor mechanisms of gigantol focused on CSC targeting and destabilizing tumor integrity via suppression of the PI3K/AKT/mTOR and JAK/STAT pathways. This data supports the potential of gigantol to be further developed as a drug for lung cancer.

Keywords: gigantol; AKT; JAK/STAT; cancer stem cell; tumor maintenance; tumor density; lung cancer; proteomics

1. Introduction

A new paradigm shift in the field of cancer cell biology is being driven by the concept of a key cancer cell population controlling the whole tumor, termed "cancer stem cells (CSCs)" [1]. CSCs from various types of cancers share a number of conservation properties, such as self-renewal ability, the generation of multiple types of differentiated cancer cells to drive tumor growth and heterogeneity, and resistance to chemotherapy via an upsurge of the DNA repair system and drug efflux transporter [2]. Therapeutic strategies targeting CSCs, including CSC direct eradication, CSC differentiation into tumor bulk cells, deletion of the supportive signals from a CSC niche, and suppression of CSC pathways, could lead to effective cancer therapy [3].

In lung adenocarcinoma, CSCs from patients were found to be less than 1.5% of the whole tumor cell population [4], but this small subpopulation was still substantial for tumorigenesis and tumor relapse [5]. The key driving pathways of CSCs, such as the PI3K/AKT/mTOR and JAK/STAT3 signals, were found to be significantly increased in cancers with high CSC properties, and hence investigations of many small molecules targeting such pathways are ongoing in clinical trials [6,7]. Protein kinase B (PKB) or AKT, which is, in fact, frequently upregulated in lung cancer plays a key role in cell survival and proliferation [8]. The activation of AKT was shown to be related with cisplatin resistance in lung cancer cells [9]. The roles of AKT on the properties of CSCs and their survival have been demonstrated in several key studies [10,11]. Likewise, signal transducer and activator of transcription 3 (STAT3) activation has been associated with poor prognosis as well as augmented CSCs [12]. A higher level of phosphorylated STAT3 (active STAT3) contributed to epithelial-to-mesenchymal transition (EMT) as well as increased CSC-like phenotypes of non-small cell lung cancer cells (NSCLCs), while the inhibition of STAT3 caused the opposite effects [13]. Instead of bulk non-CSC tumor clearance, the targeting of AKT and STAT3 is believed to be a promising anti-cancer strategy that could lead to the tumor collapse and prevention of the relapse of the disease [14,15].

Recently, natural compounds from plants have garnered increasing attention either as potential drugs or lead compounds in drug discovery research [16,17]. The key benefits of natural compounds are the abundance of plants, compound diversity, and cost effectiveness. In focusing on CSCs and tumor growth inhibition, previous studies have reported the promising activities of the bibenzyl derivative chrysotoxine in the suppression of AKT and Src [18]. In vivo studies further revealed that the bibenzyl derivative moscatilin reduced tumor volumes of lung and esophageal cancer xenografts [19,20]. Gigantol, a bibenzyl compound, is one of the polyphenolic components frequently found in traditional Chinese medicine, and has been shown to have several pharmacological effects, e.g., anti-inflammatory, amelioration of diabetic nephropathy and cataract, and anti-cancer [21–23]. The structure of gigantol consists of a bibenzyl core (Figure 1A). In vitro studies reported that gigantol triggered the apoptotic cell death of lung cancer cell lines but was not toxic to keratinocytes [24].

Our previous studies revealed several effects of noncytotoxic concentrations of gigantol on NSCLCs [25–28]. Pretreatment of 5 to 20 µM of gigantol showed a reduction of the tumor-forming capacity of NSCLCs, represented by significantly suppressing the anchorage-independent growth. In addition, with a single pretreatment of gigantol, the ability of cancer cells to form spheroids, a critical hallmark of CSCs, was abundantly suppressed [25]. These data indicated that the cancer cells had lost their self-renewal capability, which was confirmed by Western blot results showing the downregulation of octamer-binding transcription factor 4 (Oct 4) and Nanog, essential transcription factors for self-renewal and CSC-like phenotype maintenance [25]. Altogether, gigantol has the potential to attenuate tumorigenesis. However, certain information regarding the tumor growth attenuation mechanism and key evidence in animal models are still required. In this study, a subcutaneous xenograft model, as well as proteomic analysis of tumor growth regulatory pathways, were performed to help illustrate a clearer picture of how gigantol could suppress lung cancer.

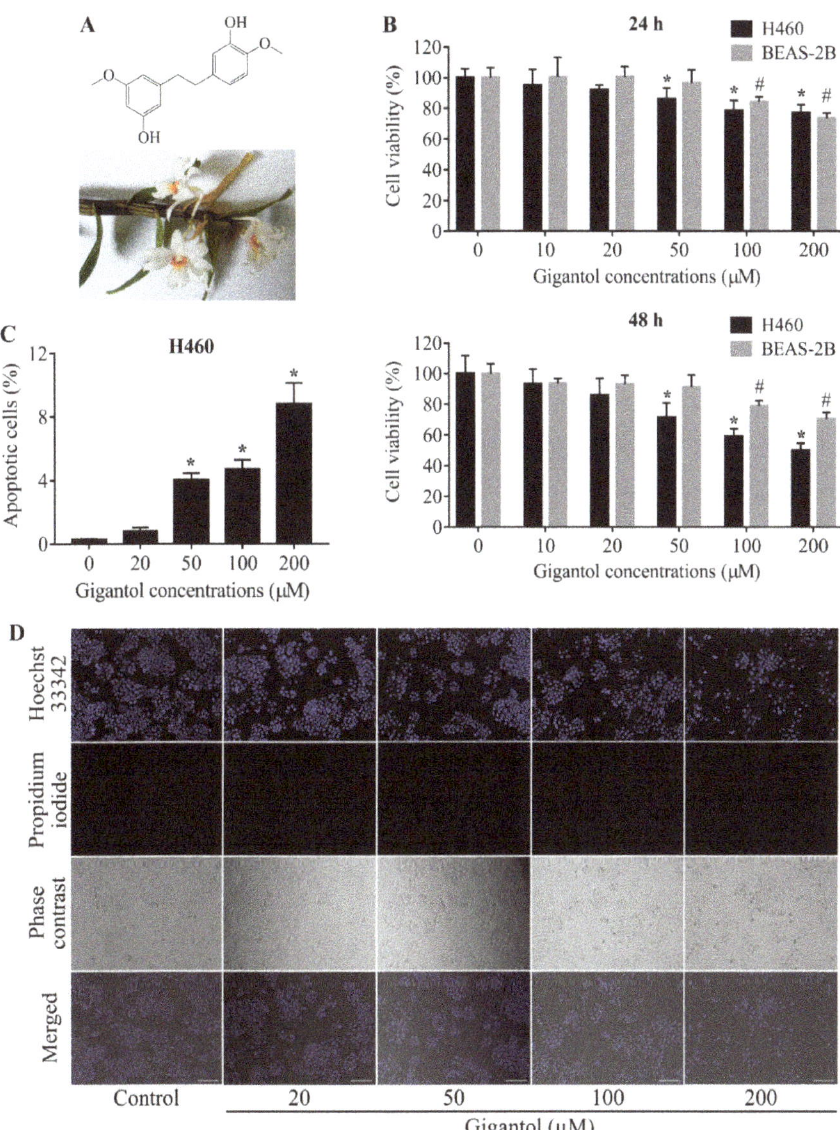

Figure 1. (**A**) Gigantol structure and the plant specimen, Dendrobium draconis Rchb.f. (**B**) Graphs showing the percentages of H460 and BEAS-2B cells viability. Both cell lines were treated with 0 to 200 μM of gigantol or vehicle for 24 to 48 h and then analyzed by MTT assay for cell viability. The percentages of viable cells were compared to their untreated controls. (**C**) Graph showing the percentages of apoptotic cell death after 24 h of gigantol exposure. Necrotic cells could not be detected. (**D**) Photographs of Hoechst 33342, Propidium iodide (PI), and phase contrast fields showing cancer cells morphologies after 24 h of gigantol treatment, the scale bars represent 100 μm and the magnification is 100×. Each experiment was performed in biological triplicates, * indicates $p < 0.05$ as compared with untreated group of H460, # indicates $p < 0.05$ as compared with untreated group of BEAS-2B (one-way ANOVA, Dunnett's test).

2. Results

2.1. Determination of Noncytotoxic Concentrations of Gigantol

Treatment of human NSCLCs H460 with 10 to 20 μM of gigantol for 24 and 48 h had a nonsignificant effect on survival of the cells, while a significant reduction of cell survival could be first detected in response to gigantol at a concentration of 50 μM (Figure 1B). Moreover, cell viability evaluation revealed that gigantol exhibited less toxicity to human lung epithelial cells BEAS-2B as compared with lung cancer cells. Confirmation of cell death, either via apoptosis or necrosis, was detected under a fluorescent microscope after staining with Hoechst 33342 and propidium iodide (PI), as described in the Materials and Methods section. The nuclear staining results revealed that condensed and fragmented nuclei of apoptosis cells could be observed only in the cells treated with gigantol at 200 μM. It is worth indicating that treatment with gigantol at all concentrations (0 to 200 μM) caused no necrosis (Figure 1C,D). Noncytotoxic concentrations of gigantol (0 to 20 μM) were used in subsequent experiments.

2.2. Functional Classification and Enrichment Analysis of the Down- and Upregulated Proteins in Gigantol-Treated Cells

In total, 4351 proteins were identified from the control cells, while 3745 proteins were identified from the gigantol-treated cells. The protein lists from the control and gigantol-treated cells were input to a Venn diagram and 2373 proteins (54.54%) were identified as being only from the control cells, 1767 proteins (47.18%) only from the gigantol-treated cells, and 1978 proteins from both groups (Figure 2A). The protein lists that were uniquely found in the control or gigantol-treated cells were subjected to further bioinformatic analysis (the lists of proteins are in Table S1).

The down- and upregulated protein lists were categorized into three ontologies, molecular function, biological process, and cellular component using Panther software (conducted on 8 October 2019). The most overrepresented molecular functions were binding (38.3% down- and 35.6% upregulated proteins) and catalytic activity (32.0% down- and 35.6% upregulated proteins) (Figure 2B). The most overrepresented biological process categories were cellular process (31.8% down- and 32.6% upregulated proteins), and metabolic process (21.0% down- and 19.8% upregulated proteins) (Figure 2C). The most overrepresented cellular component categories were cell (39.4% down- and 40.2% upregulated proteins) and organelle (33.6% down- and 32.2% upregulated proteins) (Figure 2D).

The two protein lists were input to Enrichr software to identify the enriched biological processes associated with the down- and upregulated proteins after treating with gigantol (conducted on 8 October 2019). Enrichment terms from the Gene Ontology (GO) biological process of downregulated proteins in gigantol-treated cells are involved in macromolecule biosynthesis, DNA-templated transcription, gene expression, protein phosphorylation, cytoskeleton organization, and telomere maintenance. In contrast, enrichment biological processes of upregulated proteins in gigantol-treated cells are involved in intracellular signal transduction, protein phosphorylation, gene expression, and protein biosynthesis processes (Table 1; The full lists of the enriched biological processes of down- and upregulated proteins altered by gigantol are in Tables S2 and S3).

2.3. Protein–Protein Interaction Networks of the Down- and Upregulated Proteins in Gigantol-Treated Cells

Kinases are vital enzymes that regulate intracellular signaling. Several oncogenes and tumor suppressor genes are kinase enzymes or proteins linked to protein kinase activity. Therefore, the proteins that linked to the GO term "protein phosphorylation (GO:0006468)" obtained from Enrichr were subjected to protein-protein interaction network analysis with the Search Tool for Retrieval of Interacting Genes/Proteins (STRING) database in order to determine the significant kinase pathways affected by gigantol.

Table 1. First 10 ranking enrichment terms from the GO biological process of down- and upregulated proteins in gigantol-treated cells.

Term	Overlap	*p*-Value
enriched biological processes associated with the down-regulated proteins		
regulation of cellular macromolecule biosynthetic process (GO:2000112)	129/632	2.65×10^{-10}
regulation of nucleic acid-templated transcription (GO:1903506)	121/608	4.65×10^{-9}
protein phosphorylation (GO:0006468)	97/471	2.82×10^{-8}
regulation of transcription, DNA-templated (GO:0006355)	254/1599	3.09×10^{-7}
regulation of gene expression (GO:0010468)	174/1038	9.47×10^{-7}
phosphorylation (GO:0016310)	76/387	5.61×10^{-6}
protein autophosphorylation (GO:0046777)	41/176	1.44×10^{-5}
cytoskeleton organization (GO:0007010)	29/127	$3.49E \times 10^{-4}$
membrane depolarization during action potential (GO:0086010)	13/39	3.58×10^{-4}
regulation of telomere maintenance (GO:0032204)	10/26	4.67×10^{-4}
enriched biological processes associated with the up-regulated proteins		
regulation of intracellular signal transduction (GO:1902531)	62/423	4.82×10^{-5}
protein phosphorylation (GO:0006468)	67/471	6.24×10^{-5}
ribosomal large subunit biogenesis (GO:0042273)	16/64	1.02×10^{-4}
cyclic purine nucleotide metabolic process (GO:0052652)	10/31	2.17×10^{-4}
regulation of gene expression (GO:0010468)	124/1038	2.75×10^{-4}
nucleotide biosynthetic process (GO:0009165)	9/27	3.40×10^{-4}
DNA replication checkpoint (GO:0000076)	7/17	3.59×10^{-4}
RNA splicing, via transesterification reactions with bulged adenosine as nucleophile (GO:0000377)	37/237	4.55×10^{-4}
regulation of mRNA processing (GO:0050684)	9/29	6.19×10^{-4}
mRNA processing (GO:0006397)	42/284	6.22×10^{-4}

The 97 proteins that were downregulated and 67 proteins that were upregulated in gigantol-treated cells obtained from the GO term "protein phosphorylation (GO: 0006468)" were separately input to the STRING software (conducted on 8 October 2019). The resulting networks were presented (Figure 3A,C), and the significant nodes were determined from both the down- and upregulated protein lists. The top 10% of the downregulated proteins that had the most protein interactions were the following: MTOR, PIK3CA, JAK1, JAK2, PIK3CD, ERBB2, CHEK1, IGF1R, PTK2, ALK, and JAK3. Whereas, the top 10% of the upregulated proteins that had the most protein interactions were the following: AKT1, MAPK1, ABL1, MAPK8, CDK2, and PAK2.

The significant nodes of downregulated proteins were then analyzed for the pathways involved in CSCs using STRING, as shown in Figure 3A. According to the Kyoto Encyclopedia of Genes and Genomes (KEGG) pathway database (https://www.genome.jp/kegg/), the significant proteins that were downregulated by gigantol treatment were involved primarily in the PI3K/AKT and JAK/STAT signaling pathways (Figure 3B). These pathways were indicated as signaling pathways regulating the pluripotency of stem cells. Whereas, the significant nodes of the upregulated proteins were related to the ErbB signaling pathway, which supported the CSCs' properties (Figure 3D). Interestingly, gigantol was previously shown to potentially suppress cancer cells growth, anoikis-resistance, and cancer stemness [25–27]. Regarding the data, the downregulated kinase proteins that were affected by gigantol treatment were linked to the many well-known pathways associated with tumorigenesis and CSC maintenance.

We further confirmed that the gigantol target pathways were crucial for CSC maintenance. The list of proteins involved in CSC regulation was extracted from the KEGG pathway database, using the term "hsa04550: Signaling pathways regulating pluripotency of stem cells—Homo sapiens (human)", and mapped with our H460 proteomic profiles (conducted on 28 October 2019.). In total, 50 proteins were represented in the KEGG pathway, 12 proteins were significantly upregulated, 20 proteins were significantly downregulated, and 18 proteins were not significantly altered by gigantol (Figure 4A). Remarkably, the downregulated proteins affected by gigantol were mostly linked to the PI3K/AKT and JAK/STAT pathways (protein lists of the two pathways were obtained from KEGG pathway database "hsa04151: PI3K-AKT signaling pathway—Homo sapiens (human)" and "hsa04630: JAK-STAT signaling pathway—Homo sapiens (human)"; conducted on 28 October 2019.). Nevertheless, the levels of proteins belonging to the Wnt pathway, another pathway related to CSCs, were not significantly changed.

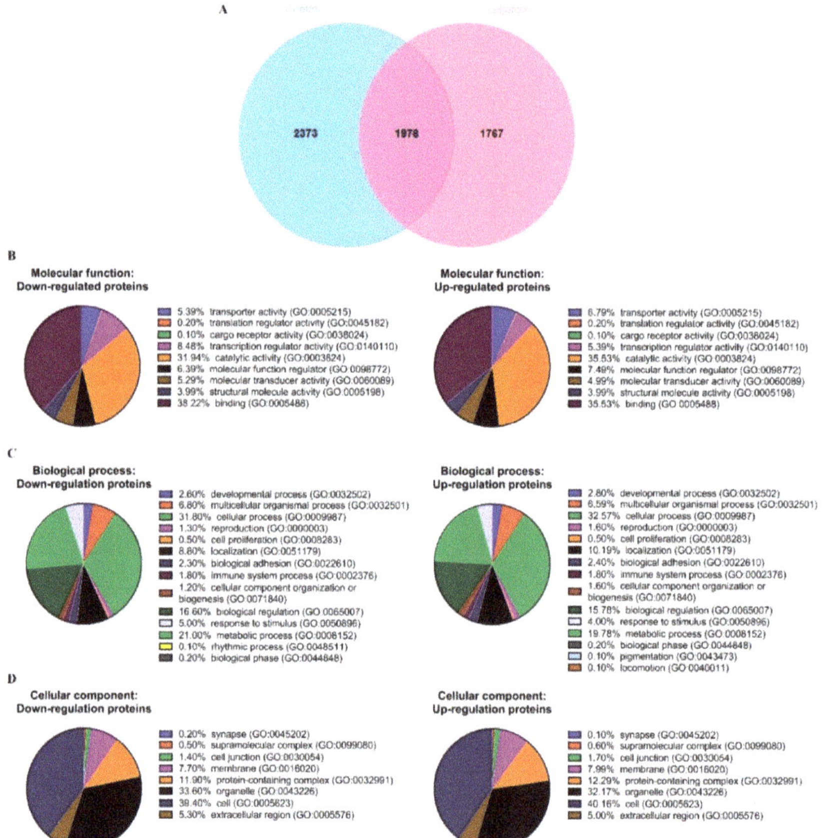

Figure 2. H460 cells were treated with 20 µM of gigantol or its vehicle (0.004% DMSO) for 24 h before the whole-cell lysates were collected. The experiment was performed in biological triplicates. (**A**) Venn diagram showing the difference in proteins expressions between the control and gigantol-treated H460 cells. Three functional classifications of the 2373 down- and 1767 upregulated proteins affected by gigantol treatment using Panther software: (**B**) molecular function, (**C**) biological process, and (**D**) cellular component.

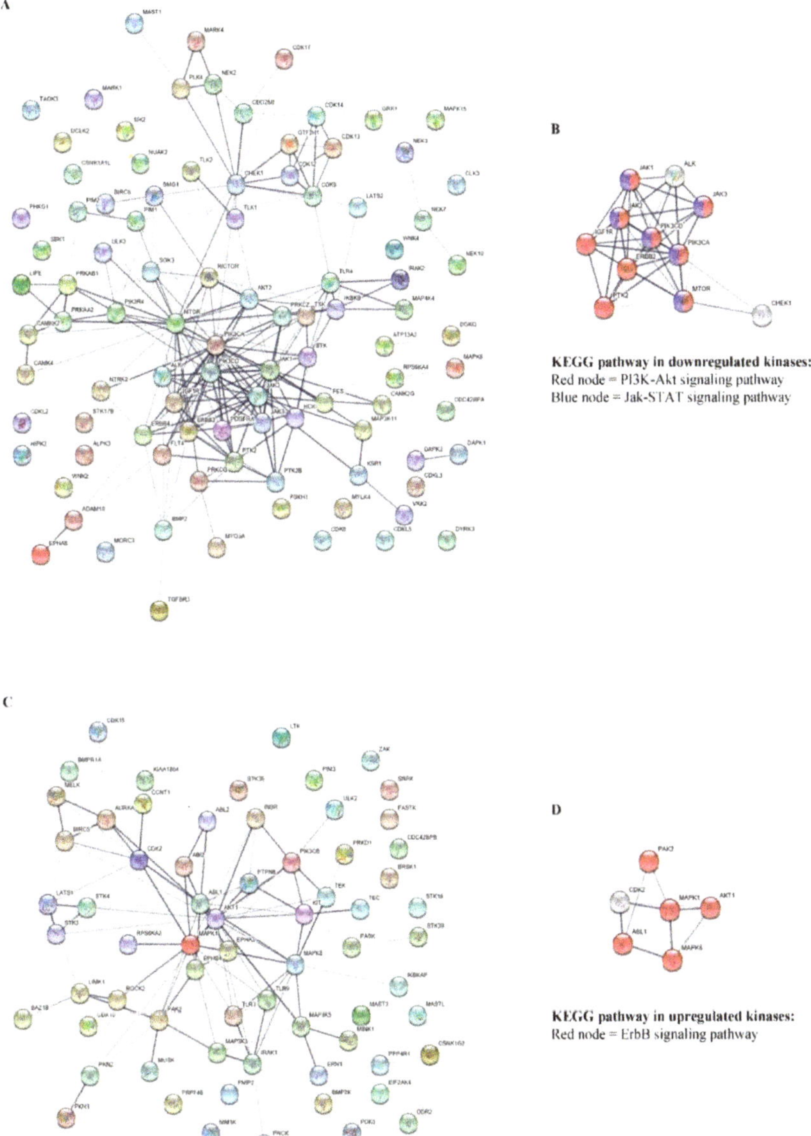

Figure 3. Networks presenting the functional protein-protein interactions of the (**A**) 97 down- and (**C**) 67 upregulated proteins related to the GO term "protein phosphorylation" (GO:0006468). The significant nodes of each network are identified and rebuilt as a network of CSC linked pathways. (**B**) According to the KEGG pathways database, significant nodes of the downregulated proteins were labeled with red for the PI3K-AKT signaling pathway (hsa04151) with FDR 4.08e−13 and blue for the JAK-STAT signaling pathway (hsa04630) with false discovery rate (FDR) 2.80e−09. (**D**) Significant nodes of upregulated proteins were labeled with red for the ErbB signaling pathway (hsa04012) with FDR 1.28×10^{-9}

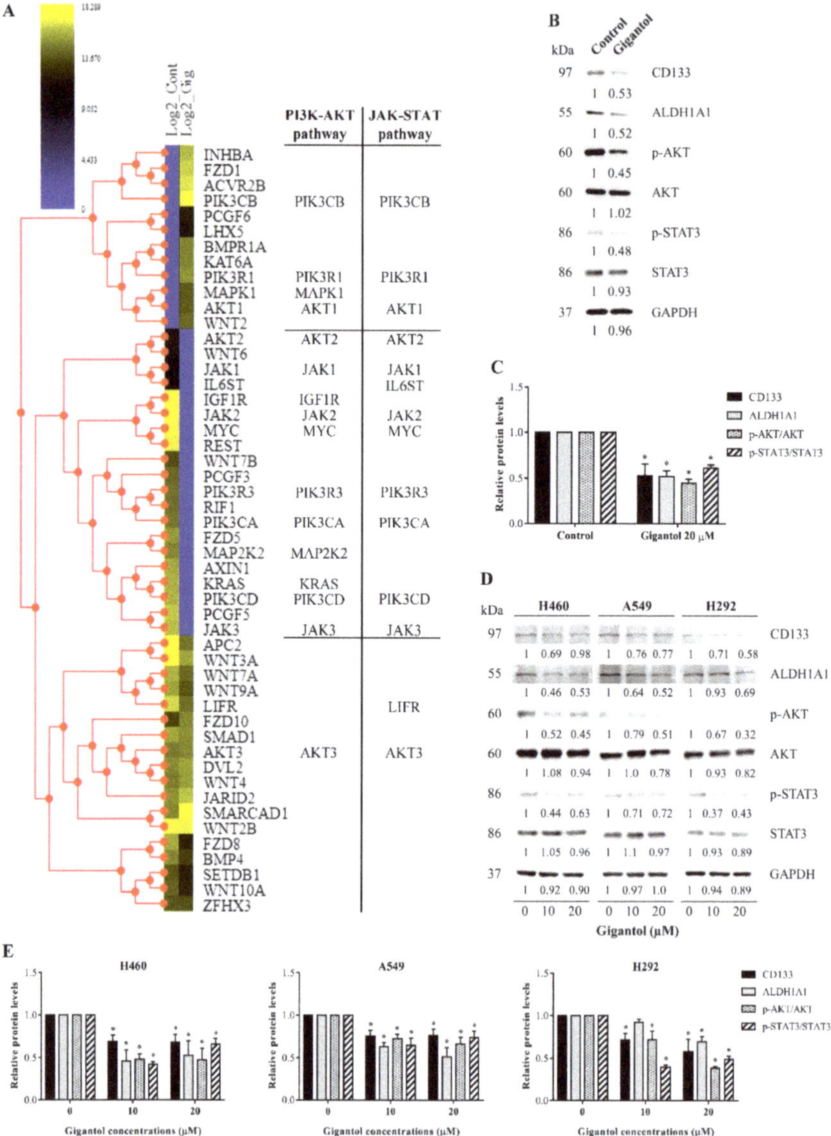

Figure 4. (**A**) Heatmap representing the levels of proteins associated with the signaling pathways regulating the pluripotency of stem cells in the control and gigantol-treated H460 cells (left and right columns of the heatmap, respectively). Proteins belonging to each pathway are listed to the right. (**B**) CSC markers and key kinases of AKT and STAT3 were determined by Western blotting and (**C**) the immunoblot signal intensities were quantified by densitometry. The uncropped protein bands are in Figure S2 (S2A: The protein bands from Figure 4B; S2B: The protein bands from Figure 4D). (**D**) The effects of gigantol on AKT, STATS3, and CSC markers were confirmed in two other NSCLC cell lines, A549 and H292, and (**E**) the relative protein levels were quantified. The mean data from each experiment was normalized to the GAPDH results. The experiment was performed biologically triplicated. Data represent the means ± SD ($n = 3$) * indicates $p < 0.05$ as compared with the control group (student's *t*-test).

To confirm, the key proteins of PI3K/AKT and JAK/STAT pathways including AKT, phosphorylated Akt (S473), STAT3, phosphorylated STAT3 (S727), and CSC markers were determined by Western blot analysis using the same cell population of proteomics and xenograft experiments. The band density of active form of Akt (phosphorylated AKT) and active STAT3 (phosphorylated STAT3) was normalized with their own total forms in order to determine the levels of activation. The results showed that gigantol could inhibit the activation of AKT and STAT3. In addition, the CSC makers (CD133 and ALHD1A1) were found to be significantly reduced in response to gigantol treatment (Figure 4B,C). Moreover, the effect of gigantol on PI3K/AKT, JAK/STAT, and CSC markers was confirmed in other NSCLC cells (A549 and H292 cells) (Figure 4D,E). It was quite clear that the PI3K/AKT and JAK/STAT signaling pathways were the target pathways of gigantol on CSC maintenance in NSCLCs. It was possible that gigantol could show some effects on tumor formation and its integrity in vivo by means of CSC suppression.

2.4. Gigantol Negatively Regulates Tumor Cell Growth in Vivo

The concept of the in vivo xenograft was to compare the ability to form and maintain a tumor between the untreated control and gigantol-treated H460 cells. This experiment revealed the effects of gigantol on the cancer cells whether the CSC or other survival signals were suppressed by the treatment at the time of inoculation. The pretreatment procedure excluded the direct anticancer effect of the compound on the tumor cell after mice implantation.

After injection of lung cancer cells into two flanks of each mouse, most mice generated palpable tumors on day seven and most control tumors had reached their endpoint size on day 13. Figure 5A demonstrates that every mouse had a similar growth rate (indicated by body weight) in a normal range. The results showed that most gigantol-treated tumors were lighter than their paired control tumors (Figure 5B,C). However, the average weights of the tumors were only slightly different (control group mean = 966 ± 154.4 mg, gigantol group mean = 698 ± 154.5 mg, $n = 5$, $p = 0.255$, Student's t-test). Tumor growth rates varied between the mice, but the mean tumor growth rates of the control and gigantol groups were not different (Figure 5D,E). The dissected tumor densities were compared, and the results showed that the gigantol groups had lower tumor densities than their paired untreated controls (Figure 5F).

2.5. Histological Observation Showed Lower Viable Tumor Areas in the Gigantol-Treated Tumors

Having shown that gigantol pretreatment caused tumors with a lesser density as compared with the untreated control, we wish to emphasize this phenomenon as previous studies have indicated that changes in tumor density, as indicated by CT imaging showing a loss of tumor mass, can be a potential assessment for anti-cancer drug action [29,30]. Cross-section slices of the tumors were co-stained by hematoxylin and eosin (H&E), and, then, were photographed. The macro-morphology of the tumor structure was similar among all the tumors (small nodules packed within a tumor lobe, surrounded with fibrous tissues), whereas the percentage of intact and non-viable tumor cells of the two groups were dramatically different. Figure 6 demonstrates that while the control tumors exhibited a dense viable tumor mass, the gigantol tumors showed a substantial loss of tumor mass, as indicated by a hollowing with the magenta staining of cells or pale pink cells without nuclear staining.

2.6. Gigantol Suppresses Tumor Cells Proliferation but not Tumor Vasculature.

Two pairs of tumors were selected for Ki-67 and α-smooth muscle actin (α-SMA) immunohistochemistry (IHC) staining. The hot spots and cold spots of Ki-67 positive cells are shown in Figure 7A. The mean percentage of Ki-67 positive cells of the control group was 62.45 ± 0.3951 and that of the gigantol-treated group was 49.49 ± 0.7348 (p-value = 0.0041, Student's t-test, $n = 2$, Figure 7B).

The α-SMA signals from cancer cells in all tumors were so low that they could not be scored (Figure 7C). This result indicated that both the control and gigantol groups had a low level of mesenchymal-like phenotypes. Figure 7D presents mature vessels covered by pericytes. The number of vessels per area detected by α-SMA staining was similar in all the tumors.

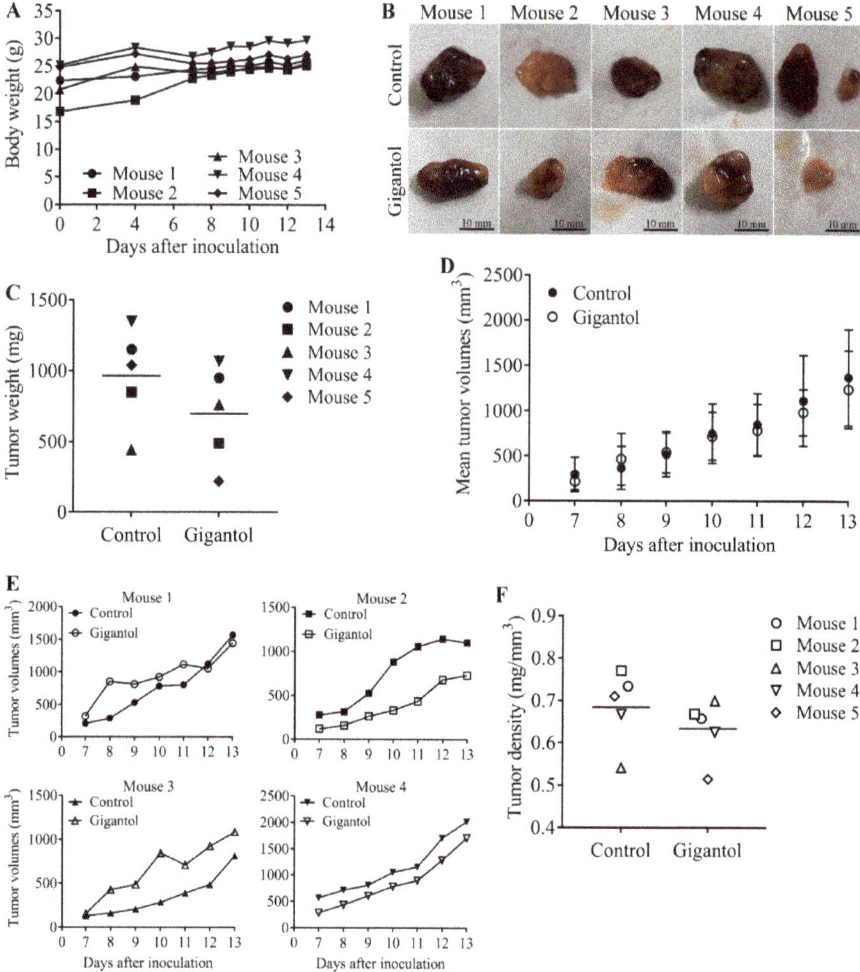

Figure 5. (**A**) Graph showing mice body weights starting at the day of cancer cell inoculation. There was no significant change of the body weights until the day of termination. (**B**) Untreated (upper row) and gigantol-treated (lower row) tumors were dissected and photographed at day 13 after inoculation. Scale bars represent 10 mm in length. (**C**) Graph showing grouped means of the control and gigantol tumor weights. The 5 different markers represent each pair of tumors. The gigantol-treated tumors had lower tumor weights as compared with their own control tumors, except for mouse 3. (**D**) Graph presenting the mean growth rate of the control and gigantol groups. (**E**) Four graphs demonstrating the individual tumor growth rate of each mouse (tumor growth of mouse 5 could not be accomplished because the gigantol-treated tumor was not palpable and measured until the day of termination). (**F**) Tumor density was calculated as weight by volume. The horizon lines represent means of each group.

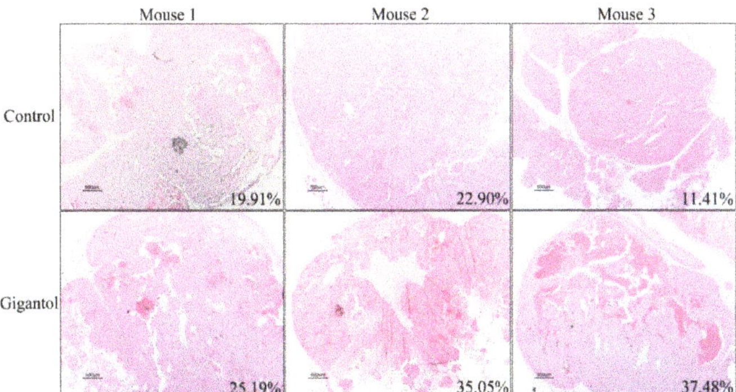

Figure 6. Hematoxylin and eosin staining showing intratumor morphology (20×). Percentages of necrotic areas as compared with their total areas are shown at the lower-right edge of each picture. Scale bars at the lower-left edge of each picture represent 500 µm lengths.

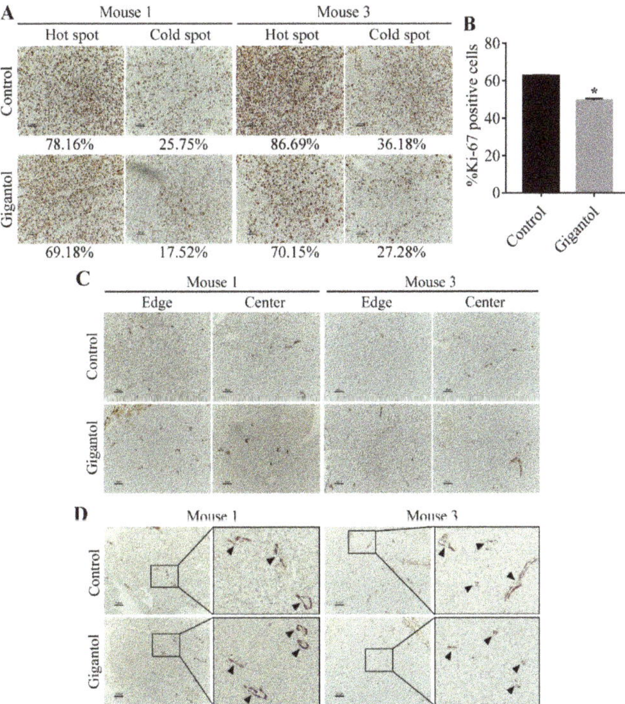

Figure 7. (**A**) Immunohistochemistry (IHC) staining demonstrating 200-fold magnified pictures of hot spots and cold spots from the control and gigantol-treated tumors. The percentages of Ki-67 positive cells as compared with total cells are displayed under their pictures. (**B**) Graph showing the means of %Ki-67 positive cells. The gigantol-treated tumors have lower Ki-67 positive cells than the control tumors. * indicates $p < 0.05$ as compared with the control group (Student's *t*-test). (**C**) α-SMA IHC staining of cancer cells in both edge and center areas of tumors showing no difference of signal levels (200×). The numbers of mature tumor vessels per areas between the control and gigantol groups were not different. (**D**) Pictures showing vessel distribution among the tumor mass (100×). Arrow indicates a vessel (400×).

3. Discussion

According to the increasing trend of cancer incidence and mortality, the development of novel anti-cancer therapies is highly needed. Among malignant tumors, lung cancer has been shown to be the main cause of cancer-related mortality and treatment failure [31], leading to the requirement for effective therapeutic options. Previous studies have reported the potential anti-cancer activities of gigantol, one of the most widely studied bibenzyls. Gigantol has exhibited cytotoxicity against various types of cancer cells, such as breast, liver, and lung cancer cells [24,32,33]. Moreover, gigantol could attenuate certain aggressive phenotypes that bring about tumor progression and metastasis, including proliferation, migration, invasion, anoikis-resistance, and anchorage-independent growth [25–28].

Proteomics analysis (Figure S1).demonstrated that PI3K/AKT/mTOR and JAK/STAT were among the most affected proteins in response to gigantol treatment. The key kinases belonging to the PI3K/AKT/mTOR axis, including phosphoinositide 3-kinases (PI3Ks, α and δ isoforms) and mammalian target of rapamycin (mTOR), were significantly decreased in the gigantol-treated cells (Figure 3A,B). Both isoforms of PI3K can activate phosphatidylinositol (3,4,5)-trisphosphate (PIP3), an upstream activator of AKT [34]. In addition, PI3K can trigger an AKT-independent mechanism, which transduces signals through serine/threonine-protein kinase Sgk3 (SGK3) and mTOR complex 2 (mTORC2) [35]. Consistently, our proteomic results showed the suppression of key proteins of the PI3K-mediated AKT-independent pathway, such as PIK3CA, PIK3CD, SGK3, MTOR, and RICTOR, which were simultaneously downregulated (Table S1).

Janus kinase 1 and 2 (JAK1 and JAK2) are transducers of the heteromeric receptors of interleukin 6 and 10 (IL-6 and IL-10), which activate signal transducer and activator of transcription 3 (STAT3). STAT3 was shown to mediate cancer cell survival, proliferation, angiogenesis, and metastasis, as well as maintaining the CSC phenotypes [7,36]. Although the STAT3 protein could not be detected in our proteomic profiles due to its low abundance, its downstream target genes, including cyclin D1 and c-Myc, were downregulated [12] (Table S1). JAK3 is an upstream regulator of STAT5 and STAT6. An accumulating data exercise revealed that inhibition of the JAK3 signaling could reduce cancer progression [37]. It is possible that the suppression of JAK/STAT signaling by gigantol should attenuate CSC in lung cancer. In addition, the mitogen-activated protein kinase 8 (MAPK8) or c-Jun N-terminal kinase (JNK) protein level was found to be induced by gigantol treatment (Figure 3D). JNK plays a role in controlling cancer cell death. The activation of JNK is necessary for intrinsic and extrinsic apoptosis, and autophagic cell death [38]. JNK signaling has been reported as a vital molecular mechanism of many anti-cancer-agents-induced cancer cell death and inactivation of such a protein led to cancer cell resistance to death stimuli [39,40]. An early upregulation of JNK by gigantol before the cancer cells encountered the stressful conditions in the tumor possibly led to stress induced JNK hyperactivation, which subsequently promoted the expression of proapoptotic proteins [38].

Recent evidence has suggested that CSCs functions as a seed of tumors. Not only do the CSCs use their ability of self-renewal and differentiation for tumor establishing, but also implicate cancer progression, metastasis, and disease relapse [1]. Regarding this matter, our previous work unraveled new information that gigantol could suppress CSC activity and discontinue their role in maintaining tumor [2,41]. This finding is quite in agreement with the previous study indicating that CSCs play a key role in tumor maintenance. This study revealed the effect of gigantol of PI3K/AKT and JAK/STAT3 suppression on the tumor initiation, growth, and maintenance based on the concept that the cells at the first step of tumor initiation had lesser CSC property than the control untreated cells.

In this study, the lung cancer cells were treated with a noncytotoxic concentration of gigantol prior to inoculation into mice subcutaneous skins. The same populations of gigantol-treated cells were subjected to proteomics. This experimental design displays the clear mechanism of gigantol treatment in attenuation of the CSC-supportive PI3K/AKT/mTOR and JAK/STAT3 signals at the time of tumor initiation. Although this experiment used only a single treatment with low dose, gigantol could inhibit the tumor growth rate (Figure 5E). Tumors from the gigantol-pretreated cells had lower weights and densities (Figure 5C,F). Furthermore, the histological tumor integrity was determined.

Previous studies have either demonstrated or proposed that the tumor density can be a promising assessment of anti-cancer drug evaluation as they have given more accuracy on the assessment of an anti-cancer drug response and have contributed to better treatment outcomes [29,30,42]. We observed the cross-sectional histology of the tumors to assess the integrity of intact cell viable areas as compared with the cell death areas as recommended in the guideline [43]. Interestingly, our results indicated that most of the gigantol-treated tumors had a dramatic loss of tumor mass as compared with those of the untreated controls (Figure 6). Consistently, the intratumor structure and tumor phenotype of Ki-67 labeling showed that the gigantol-treated tumors had lower proliferative cancer cells (Figure 7A,B). However, we found that the EMT properties of cancer cells observed by α-SMA staining was not altered by treatment with gigantol (Figure 7C). Also, the angiogenic capability of both groups of tumors was not different (Figure 7D). The phenotypic observation revealed that the pretreatment with gigantol did not have an effect on tumor neoangiogenesis. Further investigation on gigantol-mediated stromal cell-induced angiogenesis is thus suggested.

This study was designed in a manner of a pharmacological study that minimized the confounding factors in the system and focused on the effect of gigantol on the cancer cells. We could assume from the results that gigantol treatment altered the tumor-promoting activities of the cells prior to the process of tumor inoculation and such alteration attenuated the ability of the cancer cell to grow and maintain a tumor, resulting in a reduced tumor mass with viable cancer cell loss. Although our results helped us scope the direct action of the compound on NSCLCs, further investigation is necessary, including the injection of the substance into a tumor or animal after tumor formation to gain more insights.

4. Materials and Methods

4.1. Cell Line Cultures

Human NSCLC H460 and normal bronchus epithelial cell BEAS-2B lines were purchased from the American Type Culture Collection (Manassas, VA, USA) and were cultured in Roswell Park Memorial Institute (RPMI) 1640 medium and Dulbecco's modified Eagle medium (DMEM), respectively, supplemented with 10% fetal bovine serum, 2 mM L-glutamine, and 100 units/mL each of penicillin and streptomycin in a humidified atmosphere with 5% CO_2 at 37 °C.

4.2. Animals

Six-week old male BALB/cAJcl nude mice were purchased from Nomura Siam International (Samut Prakan, Thailand). Five mice were maintained in one cage under strictly hygiene housing with controlled temperature (23 ± 2 °C) and light/dark cycle (12 h light/12 h dark) at the Animal House of Faculty of Medicine, Chulalongkorn University. The study was approved by the Institutional Animal Care and Use Committee of the Faculty of Medicine, Chulalongkorn University, Bangkok, Thailand (ethical reference number CULAC 001/2561). Animal welfare and experimental procedures were strictly carried out in accordance with The Eighth Edition of the Guide for the Care and Use of Laboratory Animals (NRC 2011) [44]. All efforts were made to minimize animals' suffering and to reduce the number of animals used.

4.3. Chemicals and Reagents

Gigantol was extracted from stems of Dendrobium draconis Rchb.f., as previously described [45] and dissolved in dimethylsulfoxide (DMSO) at the indicated working concentrations. 3-(4,5-Dimethylthiazol-2-yl) 2,5-diphenyltetrazolium bromide (MTT), Hoechst 33342, propidium iodide (PI), bovine serum albumin (BSA), dimethyl sulfoxide (DMSO), cocktail protease inhibitor, hematoxylin, and eosin were purchased from Sigma chemical, Inc. (Chemical Express, Bangkok, Thailand). RPMI-1640 medium, DMEM, phosphate buffer saline (PBS), glutamine, penicillin, and streptomycin were purchased from Gibco company (Gibthai, Bangkok, Thailand). Primary antibodies against CD133, ALDH1A1, total AKT, phosphorylated AKT (Ser473), total STAT3, phosphorylated STAT3

(Ser727), and GAPDH, horseradish peroxidase labeled secondary antibodies, and RIPA lysis buffer were purchase from Cell Signaling Technology (Theera Trading, Bangkok, Thailand). Pentobarbital sodium injection was purchased from Ceva Sante Animal (VET AGRITECH, Nonthaburi, Thailand). 3,3′-Diaminobenzidine tetrahydrochloride hydrate was purchased from TCI Co., LTD (Chemical Express, Bangkok, Thailand). Primary antibodies of Ki-67 and α-SMA and matched secondary antibodies were purchased from DAKO (Medicare Supply, Bangkok, Thailand).

4.4. Cell Viability Assay

In order to elucidate the possible tumor suppression activity of gigantol, first, we selected the concentrations of the compound that caused no toxicity to the cancer cells. Cell viability was determined by plating cells at a density of 10,000 cells per well in 96-well plates. The cells were allowed to adhere overnight, medium was removed, and medium with various concentrations of gigantol (0 to 200 μM) or 0.1% DMSO was added. After 24 to 48 h of treatment, the number of viable cells were measured with the use of MTT assay. Medium was aspirated and 0.4 mg/mL of MTT in PBS was added to each well. The plate was then incubated at 37 °C, 5% CO2 for 3 h. Afterwards, the resulting formazan crystal was dissolved in 100 μL of DMSO and subjected to a 570 nm absorbance reading via a microplate reader (ClarioStar, BMG Labtech, Germany). The assay was performed biological triplicate.

4.5. Cell Death Determination Assay

Nuclear co-staining with Hoechst 33342 and propidium iodide (PI) was used to determine apoptotic and necrotic cell death. Cells were treated with gigantol as described in cell viability assay. Then, cells were incubated with 10 μM of Hoechst 33342 and 5 μM PI for 30 min at 37 °C. Cells were visualized and imaged under a fluorescence microscope (Nikon eclipse Ts2 with Nikon DS Fi3 camera). Apoptotic cell could be detected by Hoechst 33342 nuclear staining, showing condensed nucleus and fragmented nuclei of apoptotic bodies. Necrotic cell could be detected by PI staining.

4.6. Sample Preparation

H460 cells were treated with 20 μM gigantol or 0.01% DMSO (vehicle) for 24 h. The cells were lysed with 0.5% SDS. Total protein amount collected from each sample was measured with Lowry assay with bovine serum albumin as a standard [46]. Equal protein amount from 3 independent biological samples were pooled. Fifty micrograms of protein from control or gigantol treated cells were subjected to in-solution digestion. Samples were completely dissolved in 10 mM ammonium bicarbonate (AMBIC), reduced disulfide bonds using 5 mM dithiothreitol (DTT) in 10 mM AMBIC at 60 °C for 1 h and alkylation of sulfhydryl groups by using 15 mM Iodoacetamide (IAA) in 10 mM AMBIC, at room temperature for 45 min in the dark. For digestion, samples were mixed with 50 ng/μL of sequencing grade trypsin (1:20 ratio) (Promega, Walldorf, Germany) and incubated at 37 °C overnight. Prior to LC-MS/MS analysis, the digested samples must be dried and protonated with 0.1% formic acid before injection into LC-MS/MS.

4.7. Liquid Chromatography-Tandem Mass Spectrometry (LC-MS/MS)

The LC-MS/MS was used to determine the quantification of the peptides from the digested samples. The tryptic peptide samples were prepared for injection into an Ultimate3000 Nano/Capillary LC System (Thermo Scientific, Gloucester, UK) coupled to a Hybrid quadrupole Q-Tof impact II™ (Bruker Daltonics, Coventry, UK) equipped with a Nano-captive spray ion source. Briefly, peptides were enriched on a μ-Precolumn 300 μm i.d. × 5 mm C18 Pepmap 100, 5 μm, 100 A (Thermo Scientific, UK), separated on a 75 μm I.D. × 15 cm and packed with Acclaim PepMap RSLC C18, 2 μm, 100Å, nanoViper (Thermo Scientific, UK). Solvent A and B containing 0.1% formic acid in water and 0.1% formic acid in 80% acetonitrile, respectively, were supplied on the analytical column. A gradient of 5% to 55% solvent B was used to elute the peptides at a constant flow rate of 0.30 μL/min for 30 min. Electrospray ionization was carried out at 1.6 kV using the CaptiveSpray. Mass spectra (MS) and

MS/MS spectra were obtained in the positive-ion mode over the range (m/z) 150–2200 (Compass 1.9 software, Bruker Daltonics).

4.8. Bioinformatics and Data Analysis

MaxQuant 1.6.6.0 was used to quantify the proteins in individual samples using the Andromeda search engine to correlate MS/MS spectra to the Uniprot Homo sapiens database [47]. The following parameters were used for data processing: maximum of two miss cleavages, mass tolerance of 20 ppm for main search, trypsin as digesting enzyme, carbamidomethylation of cysteine as fixed modification, and the oxidation of methionine and acetylation of the protein N-terminus as variable modifications. Only peptides with a minimum of 7 amino acids, as well as at least one unique peptide, were required for protein identification. Only proteins with at least two peptides, and at least one unique peptide, were considered as being identified and used for further data analysis.

The gene list enrichment analysis was conducted using Enrichr software (https://amp.pharm.mssm.edu/Enrichr/) [48]. Protein organization and biological action was investigated conforming to protein analysis through evolutionary relationships (Panther software; http://pantherdb.org/) protein classification [49]. A Venn diagram (analyzed by jVenn software; http://jvenn.toulouse.inra.fr/app/index.html) was used to show the differences between protein lists originating from different differential analyses [50]. The Search Tool for Retrieval of Interacting Genes/Proteins (STRING) software version 11 (https://string-db.org/cgi/input.pl) was used to analyze the common and the forecasted functional interaction networks between identified proteins [51]. Cytoscape 3.7.2 (https://cytoscape.org/) was utilized to analyze the significant nodes from protein–protein interaction networks [52]. The significant nodes analysis was modified from Rezaei-Tavirani (2017) [53]. The degree values, which were determined by an amount of interacted proteins with the node, were analyzed and the top 10% of the nodes based on degree value were selected as significant nodes. The heatmap visualization and statistical analyses were conducted using the MultiExperiment Viewer (MeV) in the TM4 suite software (http://mev.tm4.org/#/welcome) [54].

4.9. Western Blot Analysis

Cells were lysed with RIPA lysis buffer containing 20 mM Tris-HCl (pH 7.5), 150 mM NaCl, 1 mM Na$_2$EDTA, 1 mM EGTA, 1% NP-40, 1% sodium deoxycholate, 2.5 mM sodium pyrophosphate, 1 mM beta-glycerophosphate, 1 mM Na$_3$VO$_4$, 1 µg/mL leupeptin, and cocktail protease inhibitor mixture for 30 minutes on ice. The protein contents of the cell lysates were evaluated by Lowry assay. Samples with equal amounts of protein (60 µg) were run in the SDS-PAGE before they were transferred onto 0.45 mm nitrocellulose membranes (Bio-Rad, Hercules, California, United States). Transferred membranes were blocked for 1 h in 5% non-fat dry milk in Tris-buffered saline with Tween 20 (25mM Tris-HCl, pH 7.5, 125 mM NaCl, and 0.05% Tween 20) and incubated overnight with specific primary antibodies against CD133, ALDH1A1, total AKT, phosphorylated AKT (Ser473), total STAT3, phosphorylated STAT3 (Ser727), and GAPDH. Membranes were washed three times with Tris-buffered saline with Tween 20 and incubated with appropriate horseradish peroxidase labeled secondary antibodies for 2 h at room temperature. The immune complexes were detected by Clarity and Clarity Max ECL Western Blotting Substrates (Bio-Rad) and imaged with ImageQuant LAS 4000 biomolecular imager (GE Healthcare, Chicago, Illinois, United States).

4.10. Subcutaneous Tumor Xenograft Procedure

The scheme of experimental design is shown in Figure 8. The human NSCLCs were prepared prior to the tumor establishment. The H460 cells were cultured in medium with 20 µM of gigantol or vehicle for 48 h (5 individual sets of cancer cell cultures). Then, the 70% confluent monolayer lung cancer cells were trypsinized, suspended in Hank's saline buffer solution and counted by TC20 automated cell counter (Bio-Rad). Each cell suspension was adjusted to a concentration of viable

5×106 cells per 100 µL. The cancer cell suspensions were kept on ice and rapidly transferred to an in vivo subcutaneous xenograft operation.

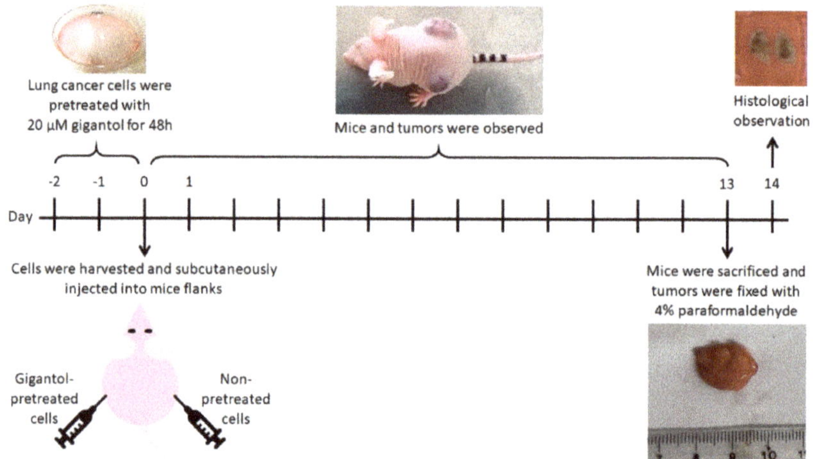

Figure 8. Scheme showing the in vivo experimental procedures.

To minimize variation between animal bodies, one mouse was assigned to bear both control and its paired gigantol-treated tumor. One flank of a mouse was inoculated subcutaneously with viable 5×106 cells of untreated cells and another flank with gigantol pretreated cells. Mice were weighed, and the tumors were observed every 2 days. When a tumor was palpable, the mouse would be observed daily. Vernier Caliper was used to measure the most length and its own orthogonal most width of each tumor. Tumor volumes were calculated by the formula (length × width × width)/2. Tumor growth rates were verified by means of plotting calculated tumor volumes by days. Mice were not exposed to gigantol throughout the experiment. Once control or treatment tumor reached its endpoint size (20 mm in diameter), the tumor-bearing mouse was euthanized by intraperitoneal injection of pentobarbital sodium solution (>150 mg/kg) [55] and then the tumors were dissected, washed with ice-cold PBS, weighed, and photographed with a ruler. The tumors were weighed and immediately fixed with 4% paraformaldehyde for 24 h. Tumors were embedded in paraffin blocks, sliced and stained with hematoxylin and eosin (H&E) for further histologic observation. Necrotic and total areas of tumor slices were determined using ImageJ software [56].

4.11. Immunohistochemistry Staining of Ki-67 and α-SMA

Two pairs of tumor slides were selected for staining with Ki-67 (dilution 1:300) and α-smooth muscle actin (α-SMA, dilution 1:100) antibodies and, then, were visualized by incubation with 3,3′-diaminobenzidine. Slides were observed under a brightfield microscope (Nikon eclipse E600 with Nikon DXM1200F camera).

The level of Ki-67 assessment was modified from Jang (2017) [57]. Areas with the highest (hot spot) and the lowest (cold spot) numbers of positive cells (indicated by dark brown staining in nucleus) were selected and the percentages of the positive cells as compared with total cells were calculated. Averages of %Ki-67 positive cells were calculated from the summed total and Ki-67 positive cells from all hot spots and cold spots of the two mice.

α-SMA, a marker of mesenchymal phenotype, was used to detect cancer cells with EMT-like phenotype, endothelial cells, and vascular pericytes [58]. For angiogenesis determination, edge and center areas of each tumor were selected and the number of mature blood vessels (indicated by circular lining of cells labeled with high signal of α-SMA) were counted.

4.12. Statistical Analysis

One-way analysis of variance (one-way ANOVA) and student's *t*-test were performed to conduct statistical analysis (GraphPad Prism 7.0). Data were expressed as mean ± standard deviation (SD) and values of $p < 0.05$ were indicative of significant differences.

5. Conclusions

Data from this study demonstrated that pretreatment with gigantol can suppress tumor growth, reduce tumor density, and attenuate the tumor maintenance of NSCLCs. This information can benefit and encourage further investigation of this useful compound to be used for anti-cancer approaches.

Supplementary Materials: The following are available online at http://www.mdpi.com/2072-6694/11/12/2032/s1, Figure S1: The workflow for proteomics analysis, Figure S2: The uncropped images of Western blot bands, Table S1: The lists of proteins which were downregulated, upregulated, and not significantly altered by the effects of gigantol, Table S2: GO biological process enrichment analysis results of the significantly downregulated proteins from Enrichr software, Table S3: GO biological process enrichment analysis results of the significantly upregulated proteins from Enrichr software.

Author Contributions: Conceptualization, P.C.; data curation, N.L. and P.C.; formal analysis, N.L., N.K., and S.R.; funding acquisition, P.C.; investigation, N.L.; methodology, P.C., N.L., and A.M.; project administration, P.C.; resources, P.C., S.R., and A.L.; supervision, P.C., S.R., and A.L.; validation, N.L., A.M., N.K., and P.C.; visualization, N.L.; writing—original draft, P.C. and N.L.; writing—review and editing, P.C. and N.L.

Funding: This work was supported by a grant from the Thailand Research Fund (grant number RSA6180036).

Acknowledgments: The authors would like to thank Boonchoo Sritularak for offering orchid extract, Yodying Yingchutrakul for LC-MS/MS technical support, and Panomwat Amornphimoltham for in vivo experiment support. In addition, the authors would like to acknowledge the Pharmaceutical Research Instrument Center, Faculty of Pharmaceutical Sciences, Chulalongkorn University for providing equipment.

Conflicts of Interest: The authors declare no conflict of interest.

References

1. Ayob, A.Z.; Ramasamy, T.S. Cancer stem cells as key drivers of tumour progression. *J. Biomed. Sci.* **2018**, *25*, 20. [CrossRef] [PubMed]
2. Batlle, E.; Clevers, H. Cancer stem cells revisited. *Nat. Med.* **2017**, *23*, 1124–1134. [CrossRef] [PubMed]
3. Garber, K. Cancer stem cell pipeline flounders. *Nat. Rev. Drug Discov.* **2018**, *17*, 771–773. [CrossRef] [PubMed]
4. Carney, D.; Gazdar, A.; Bunn, J.P.; Guccion, J. Demonstration of the stem cell nature of clonogenic tumor cells from lung cancer patients. *Stem Cells* **1982**, *1*, 149–164. [PubMed]
5. Gomez-Casal, R.; Bhattacharya, C.; Ganesh, N.; Bailey, L.; Basse, P.; Gibson, M.; Epperly, M.; Levina, V. Non-small cell lung cancer cells survived ionizing radiation treatment display cancer stem cell and epithelial-mesenchymal transition phenotypes. *Mol. Cancer* **2013**, *12*, 94. [CrossRef] [PubMed]
6. Papadimitrakopoulou, V. Development of PI3K/AKT/mTOR pathway inhibitors and their application in personalized therapy for non-small-cell lung cancer. *J. Thorac. Oncol.* **2012**, *7*, 1315–1326. [CrossRef] [PubMed]
7. Loh, C.Y.; Arya, A.; Naema, A.F.; Wong, W.F.; Sethi, G.; Looi, C.Y. Signal Transducer and Activator of Transcription (STATs) Proteins in Cancer and Inflammation: Functions and Therapeutic Implication. *Front. Oncol.* **2019**, *9*, 48. [CrossRef]
8. Lawlor, M.A.; Alessi, D.R. PKB/AKT. *J. Cell Sci.* **2001**, *114*, 2903.
9. Wang, M.; Liu, Z.M.; Li, X.C.; Yao, Y.T.; Yin, Z.X. Activation of ERK1/2 and AKT is associated with cisplatin resistance in human lung cancer cells. *J. Chemother.* **2013**, *25*, 162–169. [CrossRef]
10. Oo, A.K.K.; Calle, A.S.; Nair, N.; Mahmud, H.; Vaidyanath, A.; Yamauchi, J.; Khayrani, A.C.; Du, J.; Alam, M.J.; Seno, A.; et al. Up-Regulation of PI 3-Kinases and the Activation of PI3K-AKT Signaling Pathway in Cancer Stem-Like Cells Through DNA Hypomethylation Mediated by the Cancer Microenvironment. *Transl. Oncol.* **2018**, *11*, 653–663. [CrossRef]

11. Hambardzumyan, D.; Becher, O.J.; Rosenblum, M.K.; Pandolfi, P.P.; Manova-Todorova, K.; Holland, E.C. PI3K pathway regulates survival of cancer stem cells residing in the perivascular niche following radiation in medulloblastoma in vivo. *Genes Dev.* **2008**, *22*, 436–448. [CrossRef] [PubMed]

12. Galoczova, M.; Coates, P.; Vojtesek, B. STAT3, stem cells, cancer stem cells and p63. *Cell. Mol. Biol. Lett.* **2018**, *23*, 12. [CrossRef] [PubMed]

13. Song, L.; Rawal, B.; Nemeth, J.A.; Haura, E.B. JAK1 activates STAT3 activity in non-small-cell lung cancer cells and IL-6 neutralizing antibodies can suppress JAK1-STAT3 signaling. *Mol. Cancer Ther.* **2011**, *10*, 481–494. [CrossRef] [PubMed]

14. Bahmad, H.F.; Mouhieddine, T.H.; Chalhoub, R.M.; Assi, S.; Araji, T.; Chamaa, F.; Itani, M.M.; Nokkari, A.; Kobeissy, F.; Daoud, G.; et al. The AKT/mTOR pathway in cancer stem/progenitor cells is a potential therapeutic target for glioblastoma and neuroblastoma. *Oncotarget* **2018**, *9*, 33549–33561. [CrossRef] [PubMed]

15. Shibata, M.; Hoque, M.O. Targeting Cancer Stem Cells: A Strategy for Effective Eradication of Cancer. *Cancers* **2019**, *11*, 732. [CrossRef] [PubMed]

16. Chanvorachote, P.; Chamni, S.; Ninsontia, C.; Phiboonchaiyanan, P.P. Potential Anti-metastasis Natural Compounds for Lung Cancer. *Anticancer Res.* **2016**, *36*, 5707–5717. [CrossRef] [PubMed]

17. Shin, S.A.; Moon, S.Y.; Kim, W.Y.; Paek, S.M.; Park, H.H.; Lee, C.S. Structure-Based Classification and Anti-Cancer Effects of Plant Metabolites. *Int. J. Mol. Sci.* **2018**, *19*, 2651. [CrossRef]

18. Bhummaphan, N.; Pongrakhananon, V.; Sritularak, B.; Chanvorachote, P. Cancer Stem Cell-Suppressing Activity of Chrysotoxine, a Bibenzyl from *Dendrobium pulchellum*. *J. Pharmacol. Exp. Ther.* **2018**, *364*, 332–346. [CrossRef]

19. Tsai, A.C.; Pan, S.L.; Liao, C.H.; Guh, J.H.; Wang, S.W.; Sun, H.L.; Liu, Y.N.; Chen, C.C.; Shen, C.C.; Chang, Y.L.; et al. Moscatilin, a bibenzyl derivative from the India orchid *Dendrobrium loddigesii*, suppresses tumor angiogenesis and growth in vitro and in vivo. *Cancer Lett.* **2010**, *292*, 163–170. [CrossRef]

20. Chen, W.K.; Chen, C.A.; Chi, C.W.; Li, L.H.; Lin, C.P.; Shieh, H.R.; Hsu, M.L.; Ko, C.C.; Hwang, J.J.; Chen, Y.J. Moscatilin Inhibits Growth of Human Esophageal Cancer Xenograft and Sensitizes Cancer Cells to Radiotherapy. *J. Clin. Med.* **2019**, *8*, 187. [CrossRef]

21. Won, J.H.; Kim, J.Y.; Yun, K.J.; Lee, J.H.; Back, N.I.; Chung, H.G.; Chung, S.A.; Jeong, T.S.; Choi, M.S.; Lee, K.T. Gigantol isolated from the whole plants of *Cymbidium goeringii* inhibits the LPS-induced iNOS and COX-2 expression via NF-kappaB inactivation in RAW 264.7 macrophages cells. *Planta Med.* **2006**, *72*, 1181–1187. [CrossRef] [PubMed]

22. Wu, J.; Li, X.; Wan, W.; Yang, Q.; Ma, W.; Chen, D.; Hu, J.; Chen, C.O.; Wei, X. Gigantol from *Dendrobium chrysotoxum* Lindl. Binds and inhibits aldose reductase gene to exert its anti-cataract activity: An in vitro mechanistic study. *J. Ethnopharmacol.* **2017**, *198*, 255–261. [CrossRef] [PubMed]

23. Chen, M.F.; Liou, S.S.; Hong, T.Y.; Kao, S.T.; Liu, I.M. Gigantol has Protective Effects against High Glucose-Evoked Nephrotoxicity in Mouse Glomerulus Mesangial Cells by Suppressing ROS/MAPK/NF-kappaB Signaling Pathways. *Molecules* **2018**, *24*, 80. [CrossRef] [PubMed]

24. Charoenrungruang, S.; Chanvorachote, P.; Sritularak, B.; Pongrakananon, V. Gigantol-induced apoptosis in lung cancer cell through mitochondrial-dependent pathway. *Thai J. Pharm. Sci.* **2014**, *38*, 67–73.

25. Bhummaphan, N.; Chanvorachote, P. Gigantol Suppresses Cancer Stem Cell-Like Phenotypes in Lung Cancer Cells. *Evid. Based Complement. Altern. Med.* **2015**, *2015*, 836564. [CrossRef]

26. Charoenrungruang, S.; Chanvorachote, P.; Sritularak, B.; Pongrakhananon, V. Gigantol, a bibenzyl from *Dendrobium draconis*, inhibits the migratory behavior of non-small cell lung cancer cells. *J. Nat. Prod.* **2014**, *77*, 1359–1366. [CrossRef]

27. Unahabhokha, T.; Chanvorachote, P.; Sritularak, B.; Kitsongsermthon, J.; Pongrakhananon, V. Gigantol Inhibits Epithelial to Mesenchymal Process in Human Lung Cancer Cells. *Evid. Based Complement. Altern. Med.* **2016**, *2016*, 4561674. [CrossRef]

28. Unahabhokha, T.; Chanvorachote, P.; Pongrakhananon, V. The attenuation of epithelial to mesenchymal transition and induction of anoikis by gigantol in human lung cancer H460 cells. *Tumor Biol.* **2016**, *37*, 8633–8641. [CrossRef]

29. Faivre, S.; Zappa, M.; Vilgrain, V.; Boucher, E.; Douillard, J.Y.; Lim, H.Y.; Kim, J.S.; Im, S.A.; Kang, Y.K.; Bouattour, M.; et al. Changes in tumor density in patients with advanced hepatocellular carcinoma treated with sunitinib. *Clin. Cancer Res.* **2011**, *17*, 4504–4512. [CrossRef]

30. Hale, M.D.; Nankivell, M.G.; Mueller, W.; West, N.P.; Stenning, S.P.; Wright, A.I.; Treanor, D.; Langley, R.E.; Ward, L.C.; Allum, W.H.; et al. The relationship between tumor cell density in the pretreatment biopsy and survival after chemotherapy in OE02 trial esophageal cancer patients. *J. Clin. Oncol.* **2014**, *32* (Suppl. 3), 49. [CrossRef]

31. Siegel, R.L.; Miller, K.D.; Jemal, A. Cancer statistics, 2018. *CA Cancer J. Clin.* **2018**, *68*, 7–30. [CrossRef] [PubMed]

32. Chen, H.; Huang, Y.; Huang, J.; Lin, L.; Wei, G. Gigantol attenuates the proliferation of human liver cancer HepG2 cells through the PI3K/AKT/NF-kappaB signaling pathway. *Oncol Rep* **2017**, *37*, 865–870. [CrossRef] [PubMed]

33. Yu, S.; Wang, Z.; Su, Z.; Song, J.; Zhou, L.; Sun, Q.; Liu, S.; Li, S.; Li, Y.; Wang, M.; et al. Gigantol inhibits Wnt/beta-catenin signaling and exhibits anticancer activity in breast cancer cells. *BMC Complement. Altern. Med.* **2018**, *18*, 59. [CrossRef] [PubMed]

34. Thorpe, L.M.; Yuzugullu, H.; Zhao, J.J. PI3K in cancer: Divergent roles of isoforms, modes of activation and therapeutic targeting. *Nat. Rev. Cancer* **2015**, *15*, 7–24. [CrossRef] [PubMed]

35. Lien, E.C.; Dibble, C.C.; Toker, A. PI3K signaling in cancer: Beyond AKT. *Curr. Opin. Cell Biol.* **2017**, *45*, 62–71. [CrossRef] [PubMed]

36. Johnson, D.E.; O'Keefe, R.A.; Grandis, J.R. Targeting the IL-6/JAK/STAT3 signalling axis in cancer. *Nat. Rev. Clin. Oncol.* **2018**, *15*, 234–248. [CrossRef] [PubMed]

37. Valle-Mendiola, A.; Weiss-Steider, B.; Rocha-Zavaleta, L.; Soto-Cruz, I. IL-2 enhances cervical cancer cells proliferation and JAK3/STAT5 phosphorylation at low doses, while at high doses IL-2 has opposite effects. *Cancer Investig.* **2014**, *32*, 115–125. [CrossRef]

38. Dhanasekaran, D.N.; Reddy, E.P. JNK-signaling: A multiplexing hub in programmed cell death. *Genes Cancer* **2017**, *8*, 682–694.

39. Hu, L.; Zhang, T.; Liu, D.; Guan, G.; Huang, J.; Proksch, P.; Chen, X.; Lin, W. Notoamide-type alkaloid induced apoptosis and autophagy via a P38/JNK signaling pathway in hepatocellular carcinoma cells. *RSC Adv.* **2019**, *9*, 19855–19868. [CrossRef]

40. Bai, Y.; Liu, X.; Qi, X.; Liu, X.; Peng, F.; Li, H.; Fu, H.; Pei, S.; Chen, L.; Chi, X.; et al. PDIA6 modulates apoptosis and autophagy of non-small cell lung cancer cells via the MAP4K1/JNK signaling pathway. *EBioMedicine* **2019**, *42*, 311–325. [CrossRef]

41. Fukunaga-Kalabis, M.; Roesch, A.; Herlyn, M. From cancer stem cells to tumor maintenance in melanoma. *J. Investig. Dermatol.* **2011**, *131*, 1600–1604. [CrossRef] [PubMed]

42. Yang, D.; Woodard, G.; Zhou, C.; Wang, X.; Liu, Z.; Ye, Z.; Li, K. Significance of different response evaluation criteria in predicting progression-free survival of lung cancer with certain imaging characteristics. *Thorac. Cancer* **2016**, *7*, 535–542. [CrossRef] [PubMed]

43. Lencioni, R.; Llovet, J.M. Modified RECIST (mRECIST) assessment for hepatocellular carcinoma. *Semin. Liver Dis.* **2010**, *30*, 52–60. [CrossRef] [PubMed]

44. *Guide for the Care and Use of Laboratory Animals—Portuguese Edition*; The National Academics Press: Washington, DC, USA, 1996; p. 130.

45. Sritularak, B.; Anuwat, M.; Likhitwitayawuid, K. A new phenanthrenequinone from *Dendrobium draconis*. *J. Asian Nat. Prod. Res.* **2011**, *13*, 251–255. [CrossRef] [PubMed]

46. Lowry, O.H.; Rosebrough, N.J.; Farr, A.L.; Randall, R.J. PROTEIN MEASUREMENT WITH THE FOLIN PHENOL REAGENT. *J. Biol. Chem.* **1951**, *193*, 265–275.

47. Tyanova, S.; Temu, T.; Cox, J. The MaxQuant computational platform for mass spectrometry-based shotgun proteomics. *Nat. Protoc.* **2016**, *11*, 2301–2319. [CrossRef]

48. Kuleshov, M.V.; Jones, M.R.; Rouillard, A.D.; Fernandez, N.F.; Duan, Q.; Wang, Z.; Koplev, S.; Jenkins, S.L.; Jagodnik, K.M.; Lachmann, A.; et al. Enrichr: A comprehensive gene set enrichment analysis web server 2016 update. *Nucleic Acids Res.* **2016**, *44*, W90–W97. [CrossRef] [PubMed]

49. Mi, H.; Muruganujan, A.; Ebert, D.; Huang, X.; Thomas, P.D. PANTHER version 14: More genomes, a new PANTHER GO-slim and improvements in enrichment analysis tools. *Nucleic Acids Res.* **2019**, *47*, D419–D426. [CrossRef]

50. Bardou, P.; Mariette, J.; Escudié, F.; Djemiel, C.; Klopp, C. jvenn: An interactive Venn diagram viewer. *BMC Bioinform.* **2014**, *15*, 293. [CrossRef] [PubMed]

51. Szklarczyk, D.; Gable, A.L.; Lyon, D.; Junge, A.; Wyder, S.; Huerta-Cepas, J.; Simonovic, M.; Doncheva, N.T.; Morris, J.H.; Bork, P.; et al. STRING v11: Protein-protein association networks with increased coverage, supporting functional discovery in genome-wide experimental datasets. *Nucleic Acids Res.* **2019**, *47*, D607–D613. [CrossRef]

52. Shannon, P.; Markiel, A.; Ozier, O.; Baliga, N.S.; Wang, J.T.; Ramage, D.; Amin, N.; Schwikowski, B.; Ideker, T. Cytoscape: A software environment for integrated models of biomolecular interaction networks. *Genome Res.* **2003**, *13*, 2498–2504. [CrossRef] [PubMed]

53. Rezaei-Tavirani, M.; Rezaei-Taviran, S.; Mansouri, M.; Rostami-Nejad, M.; Rezaei-Tavirani, M. Protein-Protein Interaction Network Analysis for a Biomarker Panel Related to Human Esophageal Adenocarcinoma. *Asian Pac. J. Cancer Prev.* **2017**, *18*, 3357–3363. [PubMed]

54. Howe, E.A.; Sinha, R.; Schlauch, D.; Quackenbush, J. RNA-Seq analysis in MeV. *Bioinformatics* **2011**, *27*, 3209–3210. [CrossRef] [PubMed]

55. Leary, S. AVMA Guidelines for the Euthanasia of Animals: 2013 Edition. American Veterinary Medical Association. Journal of the American Veterinary Medical Association 2013. Available online: https://www.avma.org/KB/Policies/Documents/euthanasia.pdf (accessed on 8 October 2019).

56. Schneider, C.A.; Rasband, W.S.; Eliceiri, K.W. NIH Image to ImageJ: 25 years of image analysis. *Nat. Methods* **2012**, *9*, 671–675. [CrossRef] [PubMed]

57. Jang, M.H.; Kim, H.J.; Chung, Y.R.; Lee, Y.; Park, S.Y. A comparison of Ki-67 counting methods in luminal Breast Cancer: The Average Method vs. the Hot Spot Method. *PLoS ONE* **2017**, *12*, e0172031. [CrossRef] [PubMed]

58. Lee, H.W.; Park, Y.M.; Lee, S.J.; Cho, H.J.; Kim, D.H.; Lee, J.I.; Kang, M.S.; Seol, H.J.; Shim, Y.M.; Nam, D.H.; et al. Alpha-smooth muscle actin (ACTA2) is required for metastatic potential of human lung adenocarcinoma. *Clin. Cancer Res.* **2013**, *19*, 5879–5889. [CrossRef] [PubMed]

Article

The Chalcone Lonchocarpin Inhibits Wnt/β-Catenin Signaling and Suppresses Colorectal Cancer Proliferation

Danilo Predes [1],[†], Luiz F. S. Oliveira [1],[†], Laís S. S. Ferreira [1],[†], Lorena A. Maia [1],[†], João M. A. Delou [1], Anderson Faletti [1], Igor Oliveira [1], Nathalia G. Amado [1], Alice H. Reis [1], Carlos A. M. Fraga [1], Ricardo Kuster [2], Fabio A. Mendes [1], Helena L. Borges [1] and Jose G. Abreu [1],[*]

[1]   Program of Cell and Developmental Biology, Institute of Biomedical Sciences,
      Federal University of Rio de Janeiro, Rio de Janeiro 21941-902, Brazil; danilopredes@gmail.com (D.P.);
      oliveira.lfs2@gmail.com (L.F.S.O.); ssflais@gmail.com (L.S.S.F.); agostini.maia@gmail.com (L.A.M.);
      jmdelou@gmail.com (J.M.A.D.); 7andersonfaletti7@gmail.com (A.F.); igor.oliveira93@outlook.com (I.O.);
      ngamado@gmail.com (N.G.A.); alicehreis@gmail.com (A.H.R.); cmfraga@ccsdecania.ufrj.br (C.A.M.F.);
      famendes@gmail.com (F.A.M.); hborges@icb.ufrj.br (H.L.B.)
[2]   Department of Chemistry, Federal University of Espírito Santo, Espírito Santo 29075-910, Brazil;
      kusterrm@gmail.com
[*]   Correspondence: garciajr@icb.ufrj.br; Tel.: +55-21-3938-6486
[†]   These authors contributed equally.

Received: 9 October 2019; Accepted: 13 November 2019; Published: 7 December 2019

Abstract: The deregulation of the Wnt/β-catenin signaling pathway is a central event in colorectal cancer progression, thus a promising target for drug development. Many natural compounds, such as flavonoids, have been described as Wnt/β-catenin inhibitors and consequently modulate important biological processes like inflammation, redox balance, cancer promotion and progress, as well as cancer cell death. In this context, we identified the chalcone lonchocarpin isolated from *Lonchocarpus sericeus* as a Wnt/β-catenin pathway inhibitor, both in vitro and in vivo. Lonchocarpin impairs β-catenin nuclear localization and also inhibits the constitutively active form of TCF4, dnTCF4-VP16. *Xenopus laevis* embryology assays suggest that lonchocarpin acts at the transcriptional level. Additionally, we described lonchocarpin inhibitory effects on cell migration and cell proliferation on HCT116, SW480, and DLD-1 colorectal cancer cell lines, without any detectable effects on the non-tumoral intestinal cell line IEC-6. Moreover, lonchocarpin reduces tumor proliferation on the colorectal cancer AOM/DSS mice model. Taken together, our results support lonchocarpin as a novel Wnt/β-catenin inhibitor compound that impairs colorectal cancer cell growth in vitro and in vivo.

Keywords: anticancer drugs; flavonoids; natural compounds; *Xenopus laevis*; AOM/DSS model

1. Introduction

Colorectal cancer (CRC) is the third most commonly diagnosed cancer and the second most common cause of cancer death. According to World Health Organization (WHO), it is expected that there were 1.8 million cases and 862,000 deaths in 2018 [1]. Sporadic CRC initiation, promotion, and progression is mostly driven by a sequence of known genetic mutations in key signaling pathways, frequently related to DNA damage response and sustained proliferation in the absence of growth factors. In CRC, 93% of the cases have at least one mutation of one Wnt/β-catenin pathway component [2]. The most frequently mutated gene in colorectal cancer is *APC* (adenomatous polyposis coli) that is a β-catenin destruction complex component. *APC* mutation occurs in 81% of non-hypermutated colorectal cancers cases and in 51% of hypermutated colorectal cancer cases, triggering tumorigenesis in intestinal polyps of patients with familial adenomatous polyposis [3]. The Wnt/β-catenin signaling pathway coordinates several cell behavior aspects, such as cell proliferation, differentiation, stemness, polarity, and migration [4,5]. In the absence of Wnt ligands, the destruction complex is active in the cytoplasm, phosphorylating β-catenin, a key component of the canonical Wnt pathway, leading to its degradation by the proteasome [6,7]. Wnt interaction with its receptors Frizzled (Fzd) and LDL receptor-related protein 5/6 (LRP5/6) disrupts the destruction complex assembly leading to β-catenin stabilization, cytoplasmic accumulation, translocation to the nucleus and binding to the T-cell factor/lymphoid enhancer factor (TCF/LEF), allowing Wnt target gene transcription [8]. Despite the crucial role of Wnt signaling on colorectal tumorigenesis, there is no Wnt/β-catenin inhibitor approved for clinical use [9]. Due to the importance of Wnt/β-catenin and its frequent mutations upstream to β-catenin translocation to the nucleus, it is crucial to find anticancer drugs that target the pathway downstream to this phenomenon [2].

Addressing normal and pathological Wnt/β-catenin signaling functioning requires multidisciplinary experiments combining in vitro and in vivo approaches. Among different models for studying Wnt/β-catenin signaling in vivo, *Xenopus laevis* stands out for its liability and efficiency. Wnt/β-catenin signaling plays a key role in two fundamental steps during the Xenopus early development that can be exploited for the screening of new drug candidates: the dorso-ventral and the antero-posterior axis patterning [10–12]. Indeed, the Xenopus model system has been explored to discover Pyrvinium, an FDA approved compound, as a Wnt signaling inhibitor that acts downstream of β-catenin. Pyrvinium impaired Xenopus embryo secondary axis induction in a dose-dependent manner and decreased colon cancer cells viability [13].

In addition, the AOM/DSS mouse model stands as a relevant preclinical inflammation-associated CRC model with histologic and phenotypic features that recapitulates the aberrant crypt foci-adenoma-carcinoma found in the human CRC [14]. Consistent with CRC development, in the AOM/DSS murine model, β-catenin nuclear translocation is observed in both flat and polypoid lesions likely due to β-catenin mutation [15]. In this context, the study of synthetic and natural compounds able to inhibit the Wnt/β-catenin signaling pathway have been explored as possible antitumor prototypes. Among the small natural molecules studied, the flavonoids, polyphenolic compounds found in many plants with a wide range of biological effects, stand out. Many flavonoids have been described as inhibitors of Wnt signaling and potential antitumor compounds, such as apigenin, EGCG, silibin, kaempferol, isorhamnetin, quercetin, isoquercitrin, derricin, and derricidin [16–25]. However, the specific mechanism by which some of these compounds affect Wnt/β-catenin signaling as well as its capacity to impair CRC growth is still not elucidated. Along the flavonoid biosynthesis pathway, the chalcones are well known as precursors of the flavonoids. Lonchocarpin is a chalcone first isolated from *Lonchocarpus sericeus* (as known as *Derris sericeu*) by Baudrenghien and colleagues in 1949 [26], and its correct structure was elucidated by the same researchers in 1953. The cytotoxic effects of the chalcone lonchocarpin have been previously described in neuroblastoma and leukemia cell lines [27], however, its role in CRC and the Wnt/β-catenin pathway is unknown.

In the present work, we describe lonchocarpin ability of inhibiting Wnt/β-catenin both in vitro, in colon cancer cell lines, and in vivo, with Wnt specific *Xenopus laevis* embryonic assays. In addition, acute administration of lonchocarpin in a preclinical CRC mouse model reduced cell proliferation in adenocarcinomas. Altogether, our data show lonchocarpin as a potent Wnt/β-catenin inhibitor that impairs cancer cell proliferation both in vitro and in vivo, and a promising compound for further antitumor clinical investigation and development.

2. Results

2.1. Lonchocarpin Inhibits Wnt/β-Catenin Pathway and Reduces Nuclear β-Catenin Levels

It has been shown that natural compounds, including chalcones, have growth-inhibitory properties in cancer cell lines by modulating Wnt/β-catenin signaling [17,18]. We employed an RKO pBAR/Renilla based screening of natural compounds and found lonchocarpin as a Wnt signaling modulator hit (data not shown). In this context, we evaluated whether lonchocarpin, a chalcone isolated from *Lonchocarpus sericeus*, is able to inhibit the Wnt signaling pathway in human colorectal cancer cell lines RKO pBAR/Renilla stimulated with Wnt3a CM (conditioned medium) for 24 h and non-stimulated SW480 pBAR/Renilla (Figure 1A). SW480 does not need Wnt stimulation, since it harbors an APC mutation that overactivates the canonical Wnt signaling. Cells were treated with lonchocarpin overnight, with or without the conditioned medium. Lonchocarpin decreased luciferase reporter activity in a concentration-dependent manner, starting at 3 μM in RKO pBAR/Renilla and 5 μM in SW480 pBAR/Renilla (Figure 1B,C), and presented the half maximal inhibitory concentration (IC50) of 4 μM in SW480 pBAR/Renilla (Figure 1D).

In order to validate the Wnt/β-catenin reporter gene assay inhibition, we performed immunocytochemistry on SW480 cells. In order to activate the pathway, we treated RKO cells with Wnt3a CM for 24 h and performed immunocytochemistry to assess nuclear β-catenin cell count. These cells were cotreated with DMSO (vehicle), 10 or 20 μM lonchocarpin. L-cell CM treated cells displayed 34% β-catenin positive nuclei (Figure 1E–E"), while Wnt3a CM treatment increased the mean to 86% (Figure 1F–F"). RKO cells cotreated with 10 μM (Figure 1G–G") or 20 μM Figure 1H–H") lonchocarpin showed 54% and 30% nuclear β-catenin positive cells, respectively (Figure 1J). This data show that lonchocarpin treatment decreased β-catenin nuclear translocation in a concentration-dependent manner (Figure 1E–H"). To assess whether lonchocarpin affects total β-catenin protein level, we performed immunoblotting assay of RKO cells stimulated overnight with Wnt3a CM or 1 μM BIO, cotreated with DMSO or 20 μM lonchocarpin (Figure 1K). Immunoblot assay showed that lonchocarpin did not expressively reduced β-catenin total levels (Figure 1K), suggesting that β-catenin degradation, or β-catenin stabilization impairment might not be a lonchocarpin mechanism of action. In order to further validate our immunocytochemistry analysis, we performed immunoblotting of RKO cytosolic and nuclear fractions (Figure 1L). These cells were treated accordingly to previous assay. Immunoblotting showed that lonchocarpin reduces β-catenin nuclear level, suggesting an inhibition of nuclear translocation, or, possibly, an inhibition of β-catenin nuclear interaction with other proteins or DNA. Thus, this data suggests that lonchocarpin inhibits Wnt/β-catenin signaling pathway and impairs β-catenin nuclear localization.

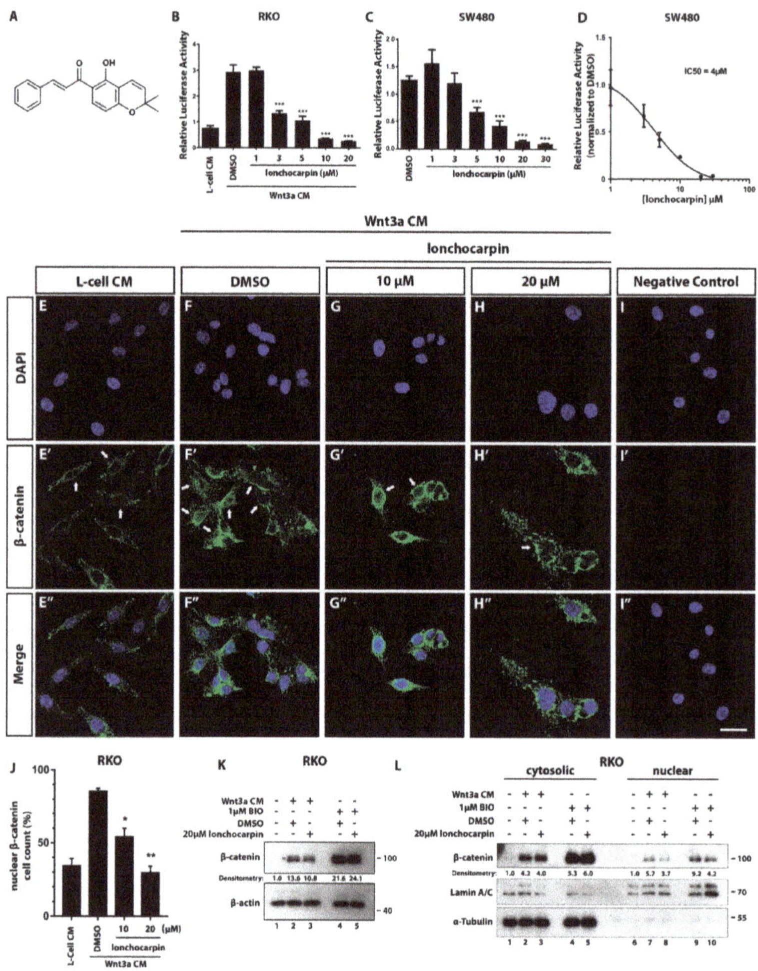

Figure 1. Lonchocarpin inhibits the Wnt/β-catenin pathway. Treatment with lonchocarpin inhibits Wnt reporter activity. (**A**) Lonchocarpin chemical structure, (**B**) RKO-pBAR/Renilla, and (**C**) SW480-pBAR/Renila gene reporter luciferase assay. (**D**) Lonchocarpin half maximal inhibitory concentration is 4 μM in the SW480-pBAR/Renilla cell lineage. Graph bars represent mean and SD. (**E–I**) SW480 β-catenin and DAPI immunocytochemistry staining showed that lonchocarpin decreases β-catenin translocation after lonchocarpin treatment. RKO cells were treated with (**E–E″**) L-cell control conditioned medium or Wnt3a conditioned medium with (**F–F″**) DMSO, (**G–G″**) 10 μM or (**H–H″**) 20 μM lonchocarpin. (**I–I″**) Immunocytochemistry negative control. (**J**) Quantification of nuclear β-catenin positive cell count ratio. Graph bars represent mean and SEM. (**K**) Immunoblotting depicting β-catenin and β-actin total levels of RKO cells treated with Wnt3a CM (conditioned medium) for 7 h. Densitometry is shown as the ratio of β-catenin/β-actin. (**K**) Immunoblot of cell lysate of RKO cells treated with 1-L-cell CM, 2-Wnt3a CM + DMSO, 3-Wnt3a CM + lonchocarpin 20 μM, 4-BIO 1 μM + DMSO, 5-BIO 1 μM + lonchocarpin 20 μM. (**L**) Immunoblot of cytosolic and nuclear fractions of RKO cells treated with 1,6-L-cell CM, 2,7-Wnt3a CM + DMSO, 3,8-Wnt3a CM + lonchocarpin 20 μM, 4,9-BIO 1 μM + DMSO, 5,10-BIO 1 μM + lonchocarpin 20 μM. Cytosolic densitometry was calculated considering α-Tubulin as the loading control, while nuclear densitometry considered Lamin A/C as the loading control. * $p < 0.05$, ** $p < 0.01$, *** $p < 0.001$. Scale bar represents 20 μm.

2.2. Lonchocarpin Inhibits the Canonical Wnt Pathway Downstream of the Destruction Complex

SW480 harbors a mutation in APC that deletes its carboxyl-terminus domain, preventing the destruction complex assembly [28,29]. Considering that lonchocarpin inhibited the Wnt reporter gene in SW480 pBAR/Renilla (Figure 1C,D), and also prevented β-catenin nuclear localization (Figure 1L), we speculated that the flavonoid could act downstream of the destruction complex. To test this hypothesis, we performed epistasis assay using Wnt/β-catenin-specific reporter (Super TOPFLASH) transfected cells induced by Wnt3a CM (Figure 2A), or co-transfected with wild type β-catenin (Figure 2B), constitutively active β-catenin S33A (Figure 2C), or dnTCF4 VP16 (Figure 2D), a constitutive active TCF4 form that does not rely on β-catenin for inducing the pathway.

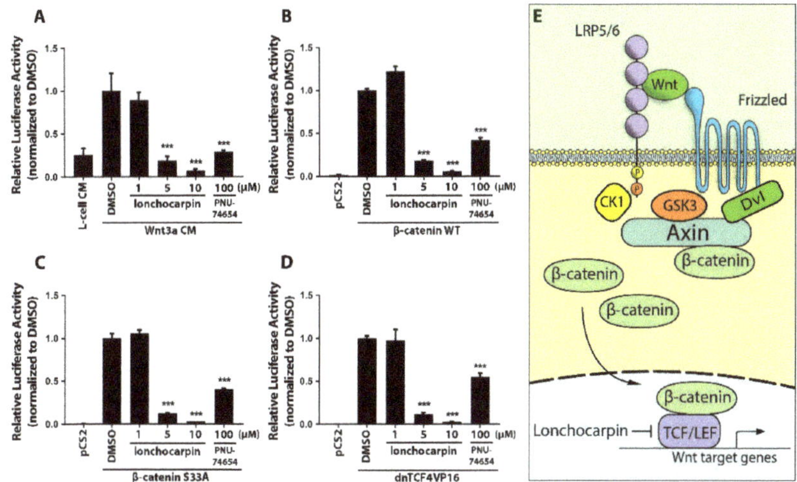

Figure 2. Lonchocarpin inhibits Wnt/β-catenin pathway downstream to TCF4. Lonchocarpin suppresses Wnt/β-catenin pathway induced by Wnt3a CM treatment (**A**) or by transfection with (**B**) β-catenin, (**C**) β-catenin S33A, or (**D**) dnTCF4VP16 in HEK293T cells. (**E**) Proposed lonchocarpin mechanism of action. Graph bars represent mean and SD. *** $p < 0.001$.

Our results showed that lonchocarpin inhibits Wnt/β-catenin signaling activation when pathway activation is triggered by wild-type β-catenin, β-catenin S33A, or dnTCF4 VP16 overexpression (Figure 2A–D). Interestingly, lonchocarpin showed a more potent and efficient inhibitory effect than the previously published compound PNU-74654 (Figure 2A–D) [30]. In this scenario, we propose that lonchocarpin acts downstream of the destruction complex and impairs TCF4 mediated transcription.

Together, these data suggest that lonchocarpin inhibits Wnt/β-catenin signaling by impairing β-catenin nuclear localization while also hindering TCF activity.

2.3. Lonchocarpin Treatment Disturbs Xenopus laevis Embryos Axial Patterning and Rescues Wnt Overactivation Phenotypes

Our in vitro data suggest that lonchocarpin inhibits Wnt/β-catenin by impairing β-catenin nuclear localization (Figure 1E–L) while also suppressing TCF mediated transcription (Figure 2D). Next, we investigated if this inhibitory effect is also observed in vivo. Manipulation of *Xenopus laevis* embryonic development has been successfully used to validate compounds targeting Wnt/β-catenin signaling through the interpretation of embryonic phenotypes and Wnt signaling overactivation [11]. In this context, injection of 1.6 pmol lonchocarpin in the embryo animal dorsal blastomeres (Figure 3A) induced head defects, characterized by reduction of anterior structures in 23% of injected embryos, such as the cement gland and diminished eyes (Figure 3D) while uninjected or DMSO-injected embryos developed normally (Figure 3B,C). Considering that Wnt/β-catenin is active in the dorsal side of the Xenopus embryo, which will organize the anterior-posterior axis, this result suggests an inhibition of the pathway by lonchocarpin. In order to confirm this result, we injected 10 pmol of lonchocarpin into the blastocoele space of stage 9 embryos (Figure 3E). Lonchocarpin injection induced anterior structures enlargement, such as the head and cement gland in 48% of the embryos (Figure 3H), while uninjected and DMSO injected embryos developed normally (Figure 3F,G). This lonchocarpin effect is consistent with Wnt/β-catenin inhibition in the embryo anterior region, since the signaling at this embryonic stage induces posterior structures.

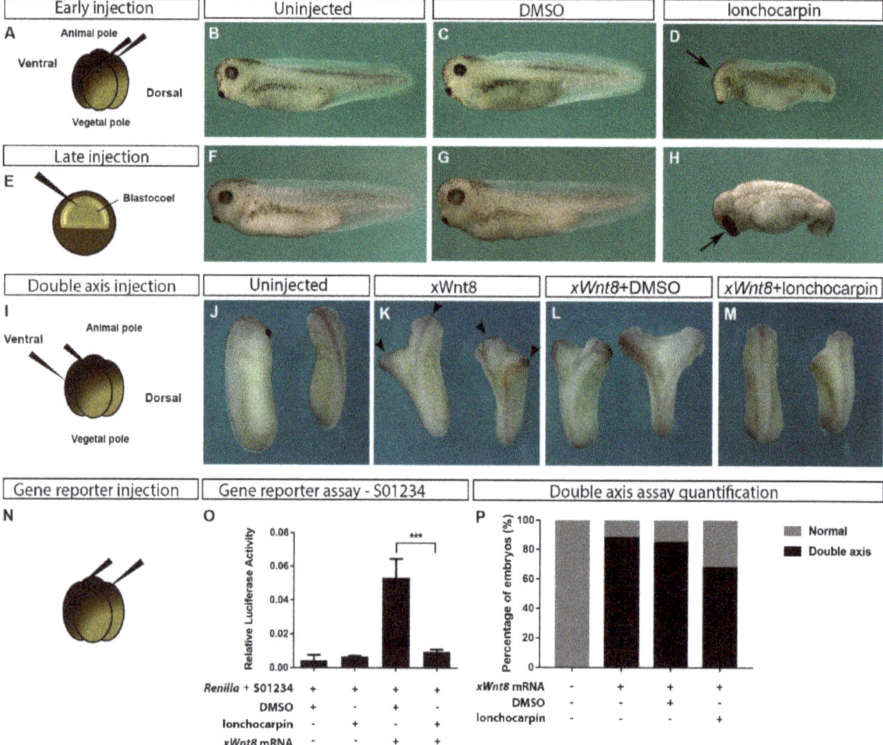

Figure 3. Lonchocarpin alters axial patterning in *Xenopus laevis* embryos and inhibits Wnt-8 induced axis and S01234-luciferase reporter. The 4-cell stage injected embryos display a smaller head compared uninjected and DMSO-injected embryos (**A–D**). Stage 9 blastulae injected embryos display a larger head (arrow) and cement gland (arrow) compared to uninjected or DMSO-injected embryos (**E–H**). Injection of xWnt8 mRNA into ventral blastomere induced ectopic axis (arrowhead) compared to uninjected embryos (**I–K,P**). Lonchocarpin inhibited Wnt8-induced secondary axis (**M,P**), but not DMSO (**L,P**). Lonchocarpin injection inhibits Wnt8-induced S01234-luciferase specific reporter activation in *Xenopus laevis* embryo (**N,O**) *** $p < 0.001$. Graph bars represent mean and SD (see also Figure S1).

It is well established that Wnt/β-catenin signaling overactivation in the Xenopus embryo 4-cell stage ventral side induces ectopic axis containing head and dorsal structures [10,12,31]. We injected 10 pg of xWnt8 mRNA combined or not with 10 pmol lonchocarpin or DMSO into two ventral blastomeres at 4-cell stage embryo (Figure 3I). xWnt8 or xWnt8 + DMSO-injected embryos developed a secondary axis in 90% of the embryos, while only 60% of the lonchocarpin-injected embryos developed a secondary axis (Figure 3K–M,P).

Thus, these results strongly suggest lonchocarpin induces phenotypes consistent with Wnt/β-catenin inhibition in Xenopus embryos (see also Figure S1 for quantification). To confirm whether lonchocarpin inhibits Wnt signaling in Xenopus, we coinjected a Wnt/β-catenin specific gene reporter S01234-luciferase with xWnt8 mRNA and 2.4 pmol lonchocarpin or DMSO into 4-cell *stage Xenopus laevis* embryos (Figure 3N). Lonchocarpin suppressed 82% of Wnt/β-catenin signaling gene reporter activation (Figure 3O).

These data support that lonchocarpin inhibits Wnt/β-catenin signaling pathway activation both in vivo and in vitro.

2.4. Lonchocarpin Reduces HCT116, SW480, and DLD-1 Cell Proliferation

Since lonchocarpin inhibited Wnt/β-catenin downstream to the destruction complex both in vitro and in vivo, we asked whether lonchocarpin has antitumor effects on colorectal cancer cell lines where Wnt/β-catenin signaling has a critical role on tumorigenesis. Canonical Wnt signaling activation leads to proliferation in many cell types, including colorectal cancer cells, thus we asked whether lonchocarpin inhibits colon cancer cell proliferation. We treated HCT116, SW480, DLD-1, and IEC-6 cell lines with 5, 10, or 20 μM lonchocarpin for 24 h and performed the Click-it EdU proliferation assay.

Lonchocarpin inhibited 33% of HCT116 EdU positive cell count at 5 and 10 μM and reduced 75% of EdU positive cell count at 30 μM (Figure 4B–E). Lonchocarpin inhibited 50% and 85% of SW480 EdU positive cell count at 10 and 20 μM, respectively (Figure 4H–J). Lonchocarpin inhibited 40% and 75% of DLD-1 proliferative cells ratio at 10 and 20 μM, respectively (Figure 4M–O). However, lonchocarpin did not affect the non-tumoral intestinal cell line IEC-6 proliferating cells ratio (Figure 4Q–T). These data show that lonchocarpin suppresses colorectal cancer cell lines proliferation while not affecting the non-tumoral cell line IEC-6 proliferation. Next, we asked whether lonchocarpin suppresses proliferation through cell toxicity. In order to assess cellular viability, we performed MTT assay after treatment with 10, 20, 30, 40, and 50 μM lonchocarpin for 24, 48, or 72 h (Figure 4). We noticed that 20 to 50 μM lonchocarpin at all treatment intervals reduced relative 570 nm absorbance of HCT116, SW480, and DLD-1 colorectal cancer cell lines but 10 μM had no effect (Figure 4U–W). Curiously, 20 μM lonchocarpin did not decrease relative 570 nm absorbance, but maintained the same reading throughout the experiment, suggesting a proliferation inhibition.

However, in the non-tumoral intestinal cell line IEC-6, lonchocarpin decreased relative 570 nm absorbance only at 50 μM. Exclusively at 72 h of treatment there was noticeable absorbance reduction in 40 μM lonchocarpin (Figure 4X). These data show that lonchocarpin inhibits cell proliferation and viability preferentially in the colorectal cancer cell lines.

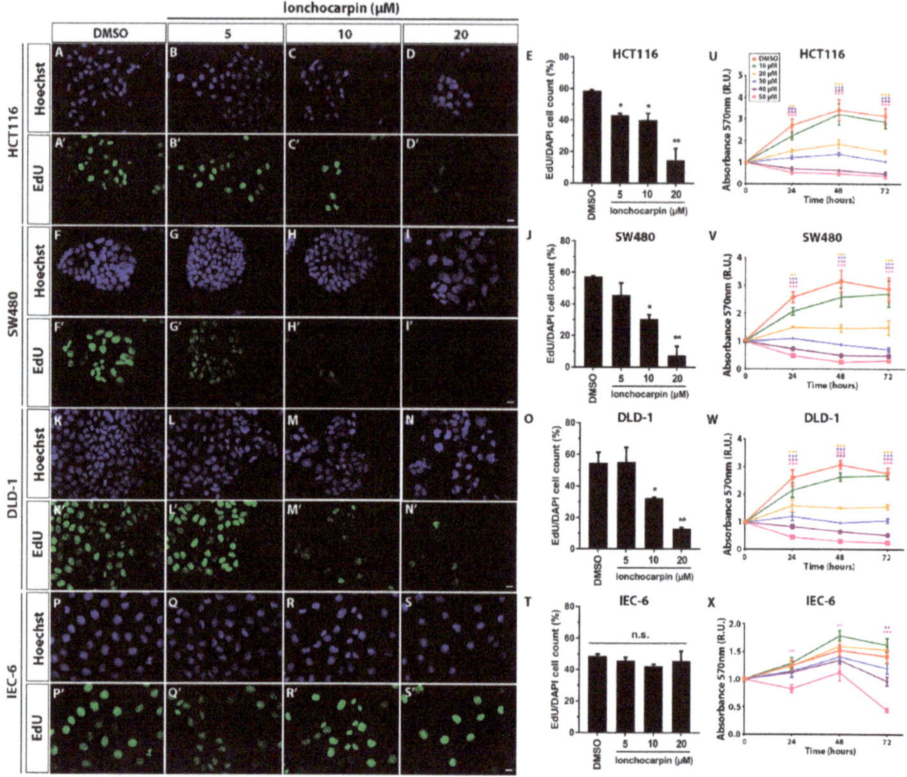

Figure 4. Lonchocarpin inhibits HCT116, SW480, and DLD-1 colorectal cancer cell lines proliferation. Proliferation assay shows that lonchocarpin suppresses proliferation of HCT116, SW480, and DLD-1 colorectal cancer cell lines, while not affecting IEC-6 non-tumor small intestine cell line proliferation. DAPI and EdU stained cells micrographs acquired 24 h post 5, 10, and 20 µM lonchocarpin treatment. (**A–E**) HCT116, (**F–J**) SW480, (**K–O**) DLD-1, (**P–T**) IEC-6. Graphs show the percentage of EdU positive cells. Scale bar represents 10 µm. Graph bars represent mean and SD. MTT assay shows that lower lonchocarpin concentrations are less cytotoxic to the non-tumor cell line IEC-6 compared to the tumor cell lines. MTT assay performed after treatment with 10, 20, 30, 40, and 50 µM lonchocarpin of (**U**) HCT116, (**V**) SW480, (**W**) DLD-1, and (**X**) IEC-6 cells during 24, 48, and 72 h. R.U. (Relative Units). * $p < 0.1$, ** $p < 0.01$. Graphs show mean and SEM.

2.5. Lonchocarpin Reduces Cell Migration in HCT116, SW480, and DLD-1 Colorectal Cancer Cell Lines

Canonical Wnt signaling key protein β-catenin interacts with adhesion proteins in the membrane that may affect cell adhesion and migration. We evaluated whether lonchocarpin affects HCT116, SW480 and DLD-1 colorectal cancer cell lines and IEC-6 non-tumoral intestinal cell line migration by performing scratch assay. HCT116 lonchocarpin treatment reduced 55% of scratch closure at 20 µM (Figure 5A–E). Lonchocarpin treatment of SW480 reduced 40% and 55% of scratch closure at 10 and 20 µM, respectively (Figure 5F–J). Likewise, DLD-1 lonchocarpin treatment reduced 45% of scratch closure at 20 µM (Figure 5K–O). However, lonchocarpin treatment did not affect cell migration of the non-tumoral cell line IEC-6 (Figure 5P–T). These data show that lonchocarpin treatment impairs cell migration of colorectal cell lines, while not affecting the migration of the non-tumoral cell line.

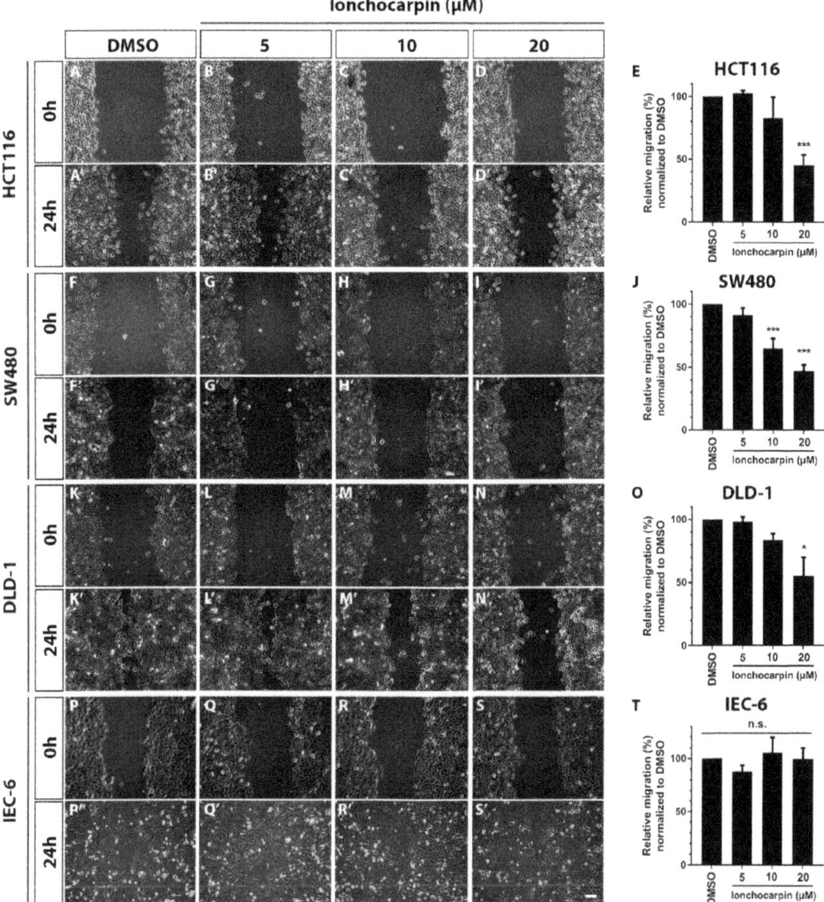

Figure 5. Lonchocarpin inhibits cell migration in HCT116, SW480, and DLD-1 colorectal cancer cell lines. Scratch assay shows that lonchocarpin impairs migration of HCT116, SW480, and DLD-1 colorectal cancer cell lines while not interfering with the IEC-6 non-tumor intestine cell line migration. Images show cell migration through the scratch healing 24 h post treatment with 5, 10, and 20 μM lonchocarpin (**A–E**) HCT116, (**F–J**) SW480, (**K–O**) DLD-1, (**P–T**) IEC-6. Graph shows relative wound area closure relative to time 0 h. All conditions were normalized to DMSO. * $p < 0.1$, *** $p < 0.001$. Scale bar represents 100 μm. Graph bars represent mean and SD.

2.6. Lonchocarpin Decreases Cell Proliferation in Azoxymethane (AOM)/Dextran Sulfate Sodium (DSS) Induced Adenocarcinomas

Lonchocarpin inhibits Wnt/β-catenin in vitro and in vivo, while also presenting antitumor effects in vitro. We asked whether lonchocarpin also has antitumor effects in vivo. We assessed the efficacy of lonchocarpin therapeutic administration in an azoxymethane (AOM)/dextran sulfate sodium (DSS)-induced model of colon cancer. After three cycles of DSS, when colon tumors were expected in most animals, lonchocarpin (50 or 100 mg·kg^{-1}·day^{-1}) was injected intraperitoneally for four days, and the mice were assessed 3 h after the last injection (Figure 6A). Colon tumors were macroscopically observed in almost all mice submitted to the protocol (77%), and histopathological analyses revealed several changes in the intestinal mucosa (Figure S2). The most frequent alterations included no presence of mononuclear and polymorphonuclear leukocyte infiltrates in the lamina propria and submucosa, hyperplastic epithelium, in addition several adenomas and adenocarcinomas (Figure 6B). Lonchocarpin did not show any toxicity to the treated animals, but significantly reduced tumor proliferation (Figure 6C–F″). Lonchocarpin at 100 mg·kg^{-1}·day^{-1} significantly decreased 31% and 38% of proliferative Ki-67 and BrdU positive cells in adenocarcinomas of the treated mice compared either with vehicle groups, respectively (Figure 6G,H, Tables S1 and S2). However, lonchocarpin showed a more efficient antiproliferative effect when administered at the highest dosage (100 mg·kg^{-1}·day^{-1}) in comparison to the lower dosage (50 mg·kg^{-1}·day^{-1}), in which no statistical relevance was found (Figure 6G,H, Tables S1 and S2).

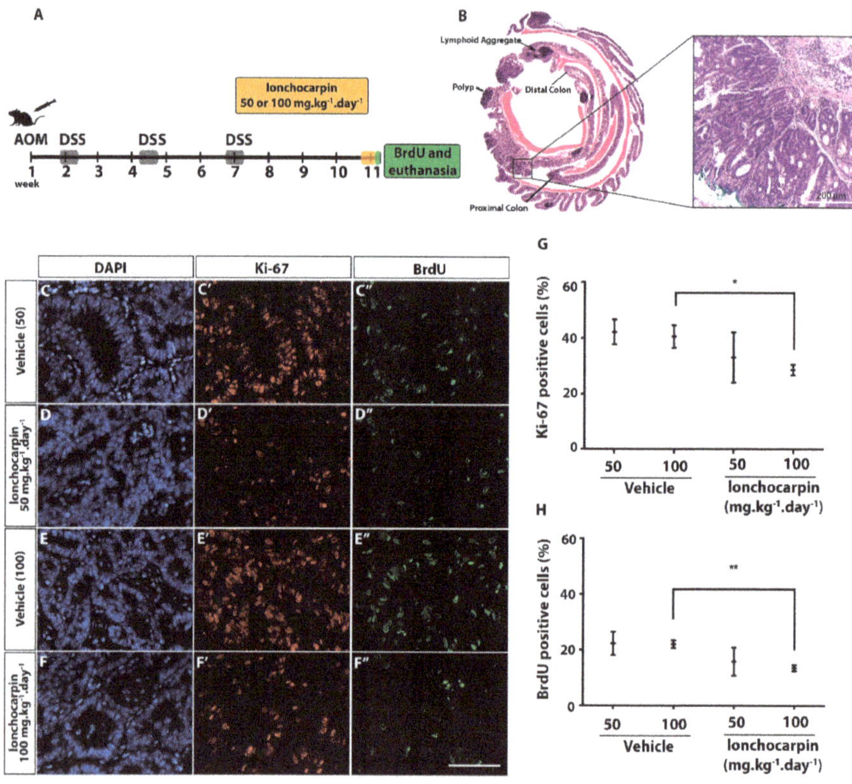

Figure 6. Lonchocarpin decreases cell proliferation in adenocarcinomas. (**A**) Timeline of inflammation-driven colon cancer tumorigenesis model (azoxymethane (AOM)/dextran sulfate sodium (DSS)) protocol in adult mice. Groups were divided as follows, vehicle 50 or 100: i.p. vehicle corresponding to volume for the respective dose in the treated group; treated groups: i.p. injections of 2.5 mg/mL lonchocarpin 50 or 100 mg·kg^{-1}·day^{-1} for four days. Vehicle: PEG400 30% in sterile NaCl 0.9%. BrdU i.p. injection (100 µg/kg) 1 h before euthanasia. (**B**) Representative H&E swiss-roll image of a colon section from a mouse subjected to AOM/DSS protocol. The zoom picture shows the corresponding area of an adenocarcinoma. Scale bar represents 200 µm. (**C**) Representative immunofluorescence photomicrographs of adenocarcinoma areas from colon sections stained for proliferation markers (**C′–F′**) Ki-67 (red, Cy3) and (**C″–F″**) BrdU (green, Alexa 488). (**C–F**) Nuclei stained with DAPI (blue). (**D,E**) Quantification of the percentage of (**G**) Ki-67 (**H**) or BrdU positive cells per indicated group. Scale bar represents 100 µm. Graphs represent mean and SEM. * $p < 0.05$ ** $p < 0.01$ Student *t*-test of lonchocarpin treatment condition in comparison to vehicle.

Taken together, our data demonstrate that lonchocarpin suppress the colorectal cancer cell growth in vitro and in vivo.

3. Discussion

The Wnt/β-catenin pathway plays a key role in colorectal tumorigenesis Integrated analysis of 195 colorectal tumors revealed that Wnt signaling pathway components were mutated in 94% of all tumors, and these mutations occurred mostly downstream to APC[10]. Hence, describing the novel canonical Wnt signaling pathway inhibiting small molecules that act downstream to APC is a recurrent strategy to improve colorectal cancer treatability.

Several studies show the antitumor effect of natural compounds that act as inhibitors of multiple components of the Wnt/β-catenin pathway [17]. Quercetin has been described to disrupt TCF/β-catenin interaction [23,32]. Epigallocatechin-3-gallate (EGCG) has been shown as a Wnt/β-catenin inhibitor by promoting β-catenin degradation [33]. Isoquercitrin has been described as an inhibitor of the Wnt signaling both in vitro and in vivo, and impairing tumor growth in vitro [34]. The chalcones derricin and derricidin have also been reported as canonical Wnt signaling inhibitors impairing tumor growth in vitro [18].

The present work is the first to identify lonchocarpin as a negative modulator of the Wnt/β-catenin pathway in RKO and SW480 colon tumor cell lines and in the HEK293T embryonic kidney cell line. We further elucidate that lonchocarpin acts downstream to β-catenin stabilization, probably at the TCF level, since it inhibits the overactivation of the Wnt/β-catenin pathway induced by the transfection of wild-type β-catenin, constitutively active β-catenin S33A or the constitutively active dnTCF4 VP16 in the HEK293T epistasis assay, while also inhibiting the Wnt/β-catenin pathway reporter in SW480 pBAR/Renilla cells, which harbors an APC truncation. We also show that lonchocarpin inhibits proliferation, migration, and cell viability in most of the three colorectal cancer cell lines, HCT116, SW480, and DLD-1, while not altering any of these aspects of the non-tumoral intestinal rat cell line IEC-6.

Recent work has identified that 24 h treatment with 50 μM lonchocarpin of SK-N-SH neuroblastoma line induces AMPK phosphorylation, which results in increased glucose uptake and inhibits protein synthesis [27]. Although our data shows effects at lower concentrations, this work corroborates our findings that lonchocarpin decreases cell viability. Thus, the antitumor effects that we described may be also a consequence of modulation of other signaling pathways besides Wnt/β-catenin.

Previous work has also measured cell viability following lonchocarpin treatment through MTT assay, indicating that the IC_{50} for cell growth in the CEM leukemia cell line is 10.4 μg/mL, the same as 3.4 μM of lonchocarpin [35]. Although not explored by the authors, leukemia cell lines are known to have Wnt signaling activating mutations, and lonchocarpin growth inhibiting effect could be due to Wnt signaling inhibition [36]. In this same study, authors show that derricin also inhibits leukemia cell growth [35], a chalcone also described as a canonical Wnt signaling inhibitor [18]. Comparatively, our cell viability data in HCT116, SW480, and DLD-1 colorectal cancer cell lines show that lonchocarpin reduces cell viability starting at 20 μM. This disparity may be due to the different origin of the cell lines. Lonchocarpin has also been shown as inhibiting lung cancer cells H292 growth in vitro and murine sarcoma S180 graft growth in vivo by inducing Caspase-3 mediated cell death [37]. Curiously, increased cleaved Caspase-3 levels were found at 48 h [37], but not at 24 h, suggesting that canonical Wnt signaling is inhibited prior apoptosis induction. Thus, previous lonchocarpin biological effect descriptions confirm that the antitumor effect of this chalcone is not exclusive to colorectal cancer.

Additionally, we show that lonchocarpin inhibits Wnt/β-catenin signaling in vivo in the *Xenopus laevis* embryo model. Xenopus embryo is a robust and reliable system to approach Wnt/β-catenin signaling in vivo since it is critical for axis patterning [10]. In that sense, Wnt signaling plays two major roles in early Xenopus development: a prior role that is composed by maternal Wnt components, and a latter role that is regulated by genes transcribed by the embryo itself. We show that injection of lonchocarpin at 4-cell stage, a stage where the embryo has no functioning transcription machinery, induced microcephaly and impaired Wnt8-induced axis, although a not very strong effect, most likely due to low lonchocarpin concentration into the embryo blastomere (Figure 3). Consistently, lonchocarpin injection into the blastocoele as well as co-injection with Wnt8-specific reporter (S01234), both addressing a moment where the embryo transcribes genes by itself, induced head enlargement and suppressed 82% of Wnt/β-catenin signaling gene reporter activation (Figure 3). These results are consistent with previous epistasis assay showing that lonchocarpin inhibits Wnt pathway downstream to β-catenin stabilization level, by impairing TCF mediated transcription (Figure 2).

Considering the relevance of the lonchocarpin Wnt/β-catenin inhibition and its functional effects on CRC cell lines as well as in Xenopus embryo, we tested lonchocarpin therapeutic administration in an inflammation-associated CRC mouse model. The AOM/DSS model has been widely used for CRC studies since it is highly reproducible, and recapitulates human cancer histological features and the major driven mutations. Indeed, it has become a powerful platform for chemopreventive and anticancer drug discovery studies. AOM is a procarcinogen metabolized into alkylating agent methylazocymethanol (MAM) in the liver. After excretion into the bile, it induces mutagenesis of the colonic epithelium. Colonic tumors are accelerated by a heparin-like polysaccharide DSS, which causes colonic epithelial damage, mirroring some of the features of inflammatory bowel disease [38]. Acute lonchocarpin i.p. administration in mice containing fully developed carcinomas reduced Ki-67 positive and BrdU positive cell count (Figure 6). Therefore, indicating lonchocarpin as an acute anti-proliferative agent in AOM/DSS induced adenocarcinoma. Although very promising, only the lonchocarpin highest dose produced anti-proliferative effects suggesting that its biological activity should be further enhanced through the synthesis of novel optimized analogues, or by using alternative administration approaches. In the middle of the 1950s, it was first isolated from the stem wood of *Camptotheca acuminata* the precursor of one of the most used chemotherapeutic agents for colorectal cancer treatment, the alkaloid camptothecin [39]. First described as showing antileukemic and antitumor activities, for the following three decades many camptothecin analogues have been described in order to enhance its antitumor properties. One of these analogues was CPT-11, currently known as Irinotecan, that is widely used to treat colorectal cancer [40]. Together with Irinotecan, 5-FU is also widely used clinically. Intriguingly, Wnt signaling inhibition has been shown to decrease resistance of colorectal cells to these chemotherapy drugs' treatment [41]. Interestingly, MEK signaling inhibitors have been shown to increase canonical Wnt signaling in CRC [42], and the co-treatment of MEK inhibitors and Wnt signaling inhibitors resulted in a reduction of tumor growth [42]. Thus, MEK inhibitors should also be addressed as a cotreatment with lonchocarpin.

There are currently 55 clinical trials aiming to inhibit Wnt signaling pathway in cancers, in which 21 are CRC studies (clinicaltrials.gov). Among the 21 CRC clinical trials, only two tested compounds inhibit canonical Wnt signaling at the transcriptional level, PRI-724 and resveratrol. In this context, in comparison with these compounds, the in vitro IC50 of lonchocarpin Wnt signaling inhibition is noticeable. PRI-724 inhibited the Wnt signaling pathway at 25 μM in vitro [43] (authors used cyclin D1 Western blot to check Wnt signaling modulation), resveratrol inhibited at 20 μM [44] (authors used a Wnt signaling specific gene reporter assay), whereas 4 uM lonchocarpin reached the IC50. The PRI-724 clinical trial has been withdrawn due to supply issues (NCT02413853) and resveratrol results have not been published yet (NCT00256334). We believe that lonchocarpin anticancer effects should be further addressed in preclinical studies, so it can be a possible clinical trial candidate.

The use of natural compounds as drug candidates has been improved by the use of alternative delivery approaches such as controlled delivery systems. Indeed, nanostructuration has been used to circumvent solubility issues [45–47]. These strategies can solve instability and poor water solubility issues [48,49] and improve a drug candidate pharmacokinetic profile. These systems could be used to further enhance lonchocarpin anticancer properties in vivo.

At last, the similarity of lonchocarpin and derricin chemical structure deserves to be noticed. These two natural compounds inhibit Wnt signaling through similar mechanisms, while derricidin has a different one [18]. This comparison paves a new way for structure-function studies, and the quest for new Wnt signaling inhibitor pharmacophores.

4. Materials and Methods

4.1. Cell Lines and Chemical Compounds

HEK293T, SW480, HCT116, DLD-1, and IEC-6 cell lines were purchased from ATCC and RKO-pBAR/Renilla and SW480-pBAR/Renilla were a gift from Professor Xi He (Harvard Medical School). All cell lines were maintained in DMEM-F-12 (Gibco, Life Technologies Limited, Paisley, UK) enriched with 10% fetal bovine serum (Gibco). The chalcone lonchocarpin was purified by Nascimento and Mors, 1972 [50] and kindly donated for this study by professor Ricardo Kuster (Federal University of Espirito Santo). PNU-74654 was synthesized at the Laboratory of Evaluation and Synthesis of Bioactives substances (Biomedical Sciences Institute, UFRJ). Both compounds were diluted in DMSO (Sigma-Aldrich, Saint Louis, MO, USA) at the concentration of 10 mM. PNU-74654 has been previously described as an inhibitor of Wnt/β-catenin pathway by blocking the interaction between β-catenin and TCF [30]. L-cell conditioned medium (CM) and Wnt3a CM were obtained according to the ATCC protocol. L-cells and L-Wnt3a cells were plated into 75 cm^2 flasks at 50% confluence with 10 mL DMEM medium containing 10% FBS. After 4 days, the first batch of medium was obtained and kept at 4 °C. Three days later, the last batch of medium was obtained and combined with the first one. Finally, L-cell CM and Wnt3a CM were passed through a 0.22 μm filter, and kept at 20 °C.

4.2. Wnt/β-Catenin Luciferase Reporter Assay

First, 1.2×10^4 cells/well RKO-pBAR/Renilla and SW480-pBAR/Renilla cells were cultured on 96-well plates in DMEM/F-12 containing 10% fetal bovine serum (Gibco). Then, 24 h later, cells were treated with lonchocarpin at the concentrations of 1, 3, 5, 10, 20, and 30 μM in the presence of Wnt3a conditioned medium for an additional 24 h. L-cell CM was used as negative control, and 0.3% DMSO was used as vehicle control. After 24 h of treatment, Firefly and Renilla luciferase activities were detected according to the manufacturer's protocol (Dual Luciferase Reporter Assay System, Promega, Madison, WI, USA).

The 1.2×10^4 cells/well HEK293T cells were cultured on 96-well plates in DMEM/F-12 containing 10% fetal bovine serum (Gibco). At 70% confluence, each well was transfected with 100 ng TOP-Flash plasmid, 10 ng Renilla-luciferase plasmid, and 100 ng wild type β-catenin or 100 ng β-catenin S33A using Lipofectamine 3000 (Invitrogen, Life Technologies Corporation, Carlsbad, CA, USA). Then, 18 h after transfection, cells were treated with 1, 5, and 10 μM of lonchocarpin. After 24 h, Firefly and Renilla luciferase activities were detected according to the manufacturer's protocol (Dual Luciferase Reporter Assay System, Promega).

4.3. Immunocytochemistry

SW480 cells were cultured in 24-well plates with 4.0×10^4 cells/well in DMEM-F12 media (Gibco) containing 10% fetal bovine serum (Gibco). Cells were fixed in 4% paraformaldehyde, washed with phosphate buffered saline, and permeabilized with 0.3% Triton X-100. Then, each well was blocked for 1 h with 5% bovine serum albumin. The rabbit anti-β-catenin primary antibody (Sigma Aldrich) was diluted in PBS containing 1% bovine serum albumin (1:200) and incubated overnight. The secondary antibody anti-rabbit Alexa Fluor 546 (Sigma Aldrich) was diluted (1:500) with 1% bovine serum albumin (1:500) and incubated for 2 h at room temperature. 4,6-diamidino-2-phenylindole staining (Cell Signaling) was performed for 15 min, and then slides were mounted with Fluoromount (Sigma Aldrich). Images were captured using the confocal microscope Leica TCS SP5.

4.4. Immunoblotting and Cell Fractionation

First, 2×10^5 cells/well RKO-pBAR/Renilla cells were cultured on 12-well plates and treated with 20 μM lonchocarpin for 24 h. Then, the cells were harvested in ice-cold PBS 1X, followed by cytosolic lysis in digitonin buffer with protease inhibitors (150 mM NaCl, 50 mM Tris, 25 μg/mL digitonin, pH 7.4) for 5 min. Then, the lysate was centrifuged for 10 min at 2000 G at 4 °C and the supernatant was collected and considered as the cytosolic fraction. Following digitonin extraction, the remaining cell pellets were washed in ice-cold PBS 1X and lysed in NP-40 buffer (150 mM NaCl, 50 mM Tris, 1%NP-40, pH 7.4) with protease inhibitors for 15 min and centrifuged for 10 min at 7000 G at 4 °C. The supernatant was collected and considered as the membrane fraction. After that, the cell pellets were washed again in ice-cold PBS 1X and lysed in RIPA buffer with protease inhibitors for 20 min and centrifugated after at 16,000 G at 4 °C. The supernatant was collected and considered as the nuclear fraction. Whole cell lysates were prepared using Triton buffer (150 mM NaCl, 50 mM Tris, 1% Triton X-100, 1 mM EDTA, 10% Glycerol, pH 7.5). Finally, cell lysates were denatured with Laemmli buffer at 95 for 5 min, and the protein samples were subjected to SDS-PAGE and transferred into PVDF membranes (Millipore, Merck KGaA, Darmstadt, Germany). The membrane was then blocked with 2% PVP (Sigma Aldrich) in TBS-Tween-20, incubated with primary monoclonal antibodies β-catenin (BD, 1:500), β-actin (SCBT, 1:2000); lamin A/C (CST, 1:500) and α-tubulin (Sigma, 1:2000) overnight at 4 °C. After three washes with TBS-T, the membranes were incubated for 1 h with HRP-conjugated secondary antibodies (CST). The immunoblots were visualized by chemiluminescence using SuperSignal West Pico and West Femto (ThermoFisher, Life Technologies Corporation, Carlsbad, CA, USA).

4.5. Xenopus laevis Embryo Manipulations

Frog experiments were carried out according to the guidelines granted by the Animal Care and Use Ethic Committee from the Federal University of Rio de Janeiro and were approved by this committee under the permission number 152/13. Female adult frogs (Nasco Inc., Fort Atkinson, WI, USA) were stimulated with 1000 IU human chorionic gonadotropin (Ferring Pharmaceuticals, Kiel, Germany). *Xenopus laevis* embryos were obtained through in vitro fertilization and staged according to Nieuwkoop and Farber [51]. All experiments were performed at 22 °C. For the Wnt/β-catenin signaling specific reporter assay, two transverse blastomeres of 4-cell stage embryos were injected with 4 nL containing 280 pg of S01234-luciferase plasmid, 50 pg of TK-Renilla plasmid, 1 pg of xWnt8 mRNA, and 1.2 pmol of lonchocarpin or DMSO each, for a total of 2.4 pmol of lonchocarpin per embryo. For synthetic xWnt8 mRNA, the plasmid was linearized with NotI and transcribed with SP6 RNA polymerase using the mMessage mMachine kit (Applied Biosystems, Austin, TX, USA). After microinjections, embryos were cultivated in 0.1× Barth (8.89 mM NaCl; 0.1 mM KCl; 0.24 mM NaHCO$_3$; 0.08 mM MgSO$_4$·7H$_2$O; 1 mM Hepes; 0.03 mM Ca(NO$_3$)$_2$·4H$_2$O; 0.04 mM CaCl$_2$·2H$_2$O; pH 7.7) until sibling control embryos reached stage 10. Triplicates of four embryos were lysed using 50 μL of 1× Passive Lysis Buffer (Promega). Then, 10 μL of embryo lysate was used for gene reporter activity detection. Firefly and Renilla luciferase activities were detected according to the manufacturer's protocol (Dual Luciferase Reporter Assay System, Promega).

In order to modulate the maternal wave, 4-cell stage embryos dorsal blastomeres were injected in the animal pole with 4 nL containing 200 μM of lonchocarpin (0.8 pmol/embryo) or DMSO each, for a total of 1.6 pmol of lonchocarpin per embryo. In order to modulate the zygotic wave, stage 9 embryos were injected with 50 nL containing 200 μM of lonchocarpin (10 pmol/embryo) or DMSO into the blastocoel. After injection, embryos were maintained in 0.1X Barth, until stage 35, when the phenotypes were analyzed.

4.6. MTT Assay

3-(4,5-Dimethylthiazol-2-yl)-2,5-diphenyltetrazolium bromide (MTT) was used to assay mitochondrial activity in viable cells. Cells were plated at a concentration of 1.2×10^4 cells/well in

96-well tissue culture plates in DMEM/F-12 containing 10% fetal bovine serum (Gibco) and cultured for 24 h. Cells were treated for 24 h with 10, 20, 30, 40, or 50 µM of lonchocarpin and 0.5% DMSO was used as the vehicle control. MTT was added to each well at a final concentration of 0.25 mg/mL for 1 h. The formazan reaction product was dissolved with 100% DMSO and quantified spectrophotometrically at 570 nm (Modulus™ II Microplate Multimode Reader, Promega).

4.7. Scratch Assay

HCT116, SW480, DLD-1, and IEC-6 were cultured on 12-well plates in DMEM/F-12 containing 10% fetal bovine serum (Gibco). Confluent wells were scratched in the center of each well with a pipette tip and treated with the proliferation inhibitor Ara-C at 10 µM. After scratch, cells were treated with 5, 10, or 20 µM lonchocarpin for 24 h. Images were taken at 0 h and the wound areas were measured at 0 and 24 h after treatment. Each experiment was carried out in triplicate, and three fields were counted per well.

4.8. Cell Proliferation Assay

HCT116, SW480, DLD-1, and IEC-6 were cultured on 24-well plates with 4.0×10^4 cells/well in DMEM/F-12 containing 10% fetal bovine serum (Gibco). Then, 24 h later, cells were treated with 20 or 30 µM lonchocarpin. DMSO 0.3% was used as vehicle control. Then, 18 h later, we added EdU to the cells, and 6 h later cells were fixed with paraformaldehyde 4% and the experiment was conducted according to Click-iT EdU (Life Technologies Corporation, Carlsbad, CA, USA) manufacturer protocol.

4.9. AOM/DSS Protocol

Animal procedures were approved and carried out according to the guidelines by the Animal Care and Use Ethic Committee from the Federal University of Rio de Janeiro under register 85/15. Male and female 129SvJxC57BL6 mixed mice (8–12 weeks) were housed in microisolator cages and maintained at 23 °C with a 12/12-h light/dark cycle and free access to food and water. A total of 26 experimental mice were pre-treated with vermifuge (vetmax plus 0.04% and ivermectin 0.4%) in quarantine for 5 days. All mice were intraperitoneally injected with AOM (12.5 mg/kg; Sigma-Aldrich) once. Five days later, all animals received the first cycle of DSS treatment composed by 2% DSS (MP Biochemicals, Solon, OH, USA) in drinking water for five days, followed by a 2-week rest period without DSS. The 2% DSS treatment cycle was repeated once and followed by a last cycle of 1.5% DSS for 4 days. Mice were monitored every day. Any mouse that lost greater than 20% body weight, demonstrate hunched posture, or moved in a limited fashion was euthanized. Four weeks after the last DSS cycle the animals were randomized into five groups as follows: no treatment group (only AOM/DSS); i.p. injections of 50 or 100 mg·kg^{-1}·day^{-1}, of vehicle (30% polyethylene glycol 400 (PEG400) with 0.9% saline); and i.p. injections of 2.5 mg/mL lonchocarpin 50 or 100 mg·kg^{-1}·day^{-1} for four days. Due to limitation of the lonchocarpin solubility, to reach 100 mg·kg^{-1}·day^{-1} i.p was performed every 12 h with 50 mg/kg. Euthanasia was performed after the treatment and the animals received an i.p. BrdU injection (100 mg/kg) 1 h before euthanasia.

4.10. Tissue Processing, Histopathology, H&E, and Immunofluorescence

After euthanasia, the colons were removed, longitudinally opened, cleaned with phosphate-buffered saline (PBS) and fixed in 4% buffered paraformaldehyde for 24 h at 4 °C. The swiss-rolls were processed in sequential ethanol and xilol for paraplast inclusion and the tissues were sectioned in Leica RM2125 RTS microtome and stained with hematoxylin and eosin (H&E) for microscopic identification of lesions, adenomas, and adenocarcinomas. Indirect immunofluorescence was performed after serial deparaffinization in xilol and ethanol. Heat induced epitope retrieval was performed in sodium citrate buffer (10 mM Sodium citrate, 0.05% Tween 20, pH 6.0) in a steamer and nonspecific binding sites were blocked with bovine serum albumin 3% in PBS. The sections were incubated with monoclonal antibodies rat anti-Ki-67 (Invitrogen #14569882; 1:100) and mouse anti-BrdU (GE Healthcare #RPN202;

1:3) overnight at 4 °C. Anti-rat biotinylated and anti-mouse Alexa 488-conjugated secondary antibodies (Invitrogen) were used to visualize Ki-67 and BrdU, respectively and the nuclei were stained with DAPI. Images were captured using the Olympus Light Microscope BX53 with a LUCPLFLN 20XPH objective and a SC50 color camera (Olympus Life Science Solutions America, Waltham, MA, USA). Only the adenocarcinoma areas confirmed in H&E staining were considered in immunofluorescence analysis. All images were manually and independently counted by at least two authors (LFSO, JMAD and AF). The proportion of positive-stained nuclei in the epithelial crypt cells found in the adenocarcinoma area were calculated and compared between the groups.

4.11. Statistical Analysis

For MTT assays statistical analysis we used a two-way ANOVA following a Bonferroni post-test (GraphPad Prism version 6.0), error bars represent standard error. For all other results, we used a one-way ANOVA (GraphPad Prism version 6.0). Figures show the mean of three replicates performed three times; standard deviation and statistical significance was set at $* p < 0.05 ** p < 0.01 *** p < 0.001$.

5. Conclusions

In summary, our data describes lonchocarpin, a flavonoid from the chalcone class, as a potent inhibitor of the Wnt/β-catenin pathway that acts downstream to β-catenin stabilization level and impairs TCF mediated transcription. In addition, lonchocarpin presents anti-tumor growth properties in vitro and inhibits adenocarcinoma proliferation in an in vivo CRC model. Further studies should be conducted in order to improve its activity and perhaps propose lonchocarpin as an alternative in CRC treatment.

Supplementary Materials: The following are available online at http://www.mdpi.com/2072-6694/11/12/1968/s1, Figure S1: *Xenopus laevis* embryo assays quantification. Figure S2: AOM/DSS tumor induction protocol. Table S1: Ki67 positive cell means of each treatment group. Table S2: BrdU positive cell means of each treatment group.

Author Contributions: Conceptualization: D.P., L.F.S.O., L.S.S.F., L.A.M., J.M.A.D., N.G.A., A.H.R., F.A.M., H.L.B. and J.G.A. Methodology: D.P., L.F.S.O., L.A.M., J.M.A.D., N.G.A., A.H.R., F.A.M, H.L.B. and J.G.A. Investigation: D.P., L.F.S.O., L.S.S.F., L.A.M., J.M.A.D., A.F., I.O. Resources: C.A.M.F, R.K., F.A.M., H.L.B. and J.G.A. Writing: D.P., L.F.S.O., L.S.S.F., L.A.M., J.M.A.D., F.A.M., H.L.B. and J.G.A. Review and editing: D.P., L.F.S.O., L.S.S.F., L.A.M., J.M.A.D., N.G.A., A.H.R., C.A.M.F., R.K., F.A.M., H.L.B. and J.G.A. Visualization: D.P., L.F.S.O., L.S.S.F., L.A.M., F.A.M. and J.G.A. Supervision: J.G.A. Project administration: J.G.A. Funding acquisition: F.A.M., H.L.B., and J.G.A.

Funding: This research was funded by Conselho Nacional de Desenvolvimento Científico e Tecnológico (CNPq), by the Fundação Carlos Chagas Filho de Amparo à Pesquisa do Estado do Rio de Janeiro (FAPERJ), grant numbers 433522/2018-6-18/02/2019, E-26/203.009/2018 (239212) and by the Fundação do Câncer/Oncobiology Program.

Acknowledgments: We thank Walter B. Mors, in memoriam, one of the founders of the Research Institute of Natural Products at Federal University of Rio de Janeiro, for donation of lonchocapin purified compound. We thank Simone Rodrigues and Fabio Jorge da Silva for animal care. We thank Carolina Batista for providing technical assistance. This work was supported by the Conselho Nacional de Desenvolvimento Científico e Tecnológico (CNPq) and the Fundação Carlos Chagas Filho de Amparo à Pesquisa do Estado do Rio de Janeiro (FAPERJ).

Conflicts of Interest: The authors declare no conflict of interest.

References

1. Siegel, R.L.; Miller, K.D.; Jemal, A. Cancer statistics, 2018. *CA A Cancer J. Clin.* **2018**, *68*, 7–30. [CrossRef] [PubMed]
2. Network, C.G.A. Comprehensive molecular characterization of human colon and rectal cancer. *Nature* **2012**, *487*, 330–337. [CrossRef] [PubMed]
3. Kinzler, K.; Nilbert, M.; Vogelstein, B.; Bryan, T.; Levy, D.; Smith, K.; Preisinger, A.; Hedge, P.; Markham, A. Identification of a gene located at chromosome 5q21 that is mutated in colorectal cancers. *Science* **1991**, *251*, 1366–1370. [CrossRef] [PubMed]
4. Miki, T.; Yasuda, S.Y.; Kahn, M. Wnt/β-catenin Signaling in Embryonic Stem Cell Self-renewal and Somatic Cell Reprogramming. *Stem Cell Rev. Rep.* **2011**, *7*, 836–846. [CrossRef]

5. Van Amerongen, R.; Nusse, R. Towards an integrated view of Wnt signaling in development. *Development* **2009**, *136*, 3205–3214. [CrossRef]

6. Kim, S.-E.; Huang, H.; Zhao, M.; Zhang, X.; Zhang, A.; Semonov, M.V.; MacDonald, B.T.; Zhang, X.; Abreu, J.G.; Peng, L.; et al. Wnt stabilization of β-catenin reveals principles for morphogen receptor-scaffold assemblies. *Science* **2013**, *340*, 867–870. [CrossRef]

7. MacDonald, B.T.; Tamai, K.; He, X. Wnt/β-Catenin Signaling: Components, Mechanisms, and Diseases. *Dev. Cell* **2009**, *17*, 9–26. [CrossRef]

8. Tortelote, G.G.; Reis, R.R.; de Almeida Mendes, F.; Abreu, J.G. Complexity of the Wnt/β-catenin pathway: Searching for an activation model. *Cell. Signal.* **2017**, *40*, 30–43. [CrossRef]

9. Kahn, M. Can we safely target the WNT pathway? *Nat. Rev. Drug Discov.* **2014**, *13*, 513–532. [CrossRef]

10. Larabell, C.A.; Torres, M.; Rowning, B.A.; Yost, C.; Miller, J.R.; Wu, M.; Kimelman, D.; Moon, R.T. Establishment of the Dorso-ventral Axis in Xenopus Embryos Is Presaged by Early Asymmetries in β-Catenin That Are Modulated by the Wnt Signaling Pathway. *J. Cell Biol.* **1997**, *136*, 1123–1136. [CrossRef] [PubMed]

11. Maia, L.A.; Velloso, I.; Abreu, J.G. Advances in the use of Xenopus for successful drug screening. *Expert Opin. Drug Discov.* **2017**, *12*, 1153–1159. [CrossRef] [PubMed]

12. McMahon, A.P.; Moon, R.T. Ectopic expression of the proto-oncogene int-1 in Xenopus embryos leads to duplication of the embryonic axis. *Cell* **1989**, *58*, 1075–1084. [CrossRef]

13. Thorne, C.A.; Hanson, A.J.; Schneider, J.; Tahinci, E.; Orton, D.; Cselenyi, C.S.; Jernigan, K.K.; Meyers, K.C.; Hang, B.I.; Waterson, A.G.; et al. Small-molecule inhibition of Wnt signaling through activation of casein kinase 1α. *Nat. Chem. Biol.* **2010**, *6*, 829. [CrossRef] [PubMed]

14. Tanaka, T.; Kohno, H.; Suzuki, R.; Yamada, Y.; Sugie, S.; Mori, H. A novel inflammation-related mouse colon carcinogenesis model induced by azoxymethane and dextran sodium sulfate. *Cancer Sci.* **2003**, *94*, 965–973. [CrossRef]

15. Tanaka, T.; Suzuki, R.; Kohno, H.; Sugie, S.; Takahashi, M.; Wakabayashi, K. Colonic adenocarcinomas rapidly induced by the combined treatment with 2-amino-1-methyl-6-phenylimidazo[4,5-b]pyridine and dextran sodium sulfate in male ICR mice possess β-catenin gene mutations and increases immunoreactivity for β-catenin, cyclooxygenase-2 and inducible nitric oxide synthase. *Carcinogenesis* **2005**, *26*, 229–238.

16. Amado, N.G.; Cerqueira, D.M.; Menezes, F.S.; Da Silva, J.F.M.; Neto, V.M.; Abreu, J.G. Isoquercitrin isolated from Hyptis fasciculata reduces glioblastoma cell proliferation and changes β-catenin cellular localization. *Anti-Cancer Drugs* **2009**, *20*, 543–552. [CrossRef]

17. Amado, N.G.; Predes, D.; Moreno, M.M.; Carvalho, I.O.; Mendes, F.A.; Abreu, J.G. Flavonoids and Wnt/β-Catenin Signaling: Potential Role in Colorectal Cancer Therapies. *Int. J. Mol. Sci.* **2014**, *15*, 12094–12106. [CrossRef]

18. Fonseca, B.F.; Predes, D.; Cerqueira, D.M.; Reis, A.H.; Amado, N.G.; Cayres, M.C.L.; Kuster, R.M.; Oliveira, F.L.; Mendes, F.A.; Abreu, J.G. Derricin and Derricidin Inhibit Wnt/β-Catenin Signaling and Suppress Colon Cancer Cell Growth In Vitro. *PLoS ONE* **2015**, *10*, e0120919. [CrossRef]

19. Gao, Z.; Xu, Z.; Hung, M.S.; Lin, Y.C.; Wang, T.; Gong, M.; Zhi, X.; Jablon, D.M.; You, L. Promoter demethylation of WIF-1 by epigallocatechin-3-gallate in lung cancer cells. *Anticancer Res.* **2009**, *29*, 2025–2030.

20. Kaur, M.; Velmurugan, B.; Tyagi, A.; Agarwal, C.; Singh, R.P.; Agarwal, R. Silibinin Suppresses Growth of Human Colorectal Carcinoma SW480 Cells in Culture and Xenograft through Down-regulation of β-Catenin-Dependent Signaling. *Neoplasia* **2010**, *12*, 415–424. [CrossRef]

21. Landesman-Bollag, E.; Song, D.H.; Mourez, R.R.; Sussman, D.J.; Cardiff, R.D.; Sonenshein, G.E.; Seldin, D.C. Protein kinase CK2: Signaling and tumorigenesis in the mammary gland. *Mol. Cell. Biochem.* **2001**, *227*, 153–165. [CrossRef] [PubMed]

22. Li, Y.; Wang, Z.; Kong, D.; Li, R.; Sarkar, S.H.; Sarkar, F.H. Regulation of Akt/FOXO3a/GSK-3β/AR signaling network by isoflavone in prostate cancer cells. *J. Biol. Chem.* **2008**, *283*, 27707–27716. [CrossRef] [PubMed]

23. Park, C.H.; Chang, J.Y.; Hahm, E.R.; Park, S.; Kim, H.K.; Yang, C.H. Quercetin, a potent inhibitor against β-catenin/Tcf signaling in SW480 colon cancer cells. *Biochem. Biophys. Res. Commun.* **2005**, *328*, 227–234. [CrossRef] [PubMed]

24. Lee, J.; Lee, J.; Jung, E.; Hwang, W.; Kim, Y.S.; Park, D. Isorhamnetin-induced anti-adipogenesis is mediated by stabilization of β-catenin protein. *Life Sci.* **2010**, *86*, 416–423. [CrossRef]

25. Song, D.H.; Sussman, D.J.; Seldin, D.C. Endogenous Protein Kinase CK2 Participates in Wnt Signaling in Mammary Epithelial Cells. *J. Biol. Chem.* **2000**, *275*, 23790–23797. [CrossRef]

26. Baudrenghien, J.; Jadot, J.; Huis, R. Etablissement de la formule de structure de la lonchocarpine. *Bull. Soc. Roy. Sci. Liège* **1949**, *18*, 52.

27. Cursino, L.M.C.; Lima, N.M.; Murillo, R.; Nunez, C.V.; Merfort, I.; Humar, M. Isolation of Flavonoids from Deguelia duckeana and Their Effect on Cellular Viability, AMPK, eEF2, eIF2 and eIF4E. *Molecules* **2016**, *21*, 192. [CrossRef]

28. Korinek, V.; Barker, N.; Morin, P.J.; Van Wichen, D.; De Weger, R.; Kinzler, K.W.; Vogelstein, B.; Clevers, H. Constitutive transcriptional activation by a β-catenin-Tcf complex in APC−/− colon carcinoma. *Science* **1997**, *275*, 1784–1787. [CrossRef]

29. Munemitsu, S.; Albert, I.; Rubinfeld, B.; Polakis, P. Deletion of an amino-terminal sequence beta-catenin in vivo and promotes hyperphosporylation of the adenomatous polyposis coli tumor suppressor protein. *Mol. Cell. Biol.* **1996**, *16*, 4088–4094. [CrossRef]

30. Trosset, J.Y.; Dalvit, C.; Knapp, S.; Fasolini, M.; Veronesi, M.; Mantegani, S.; Gianellini, L.M.; Catana, C.; Sundström, M.; Stouten, P.F. Inhibition of protein–protein interactions: The discovery of druglike β-catenin inhibitors by combining virtual and biophysical screening. *Proteins Struct. Funct. Bioinform.* **2006**, *64*, 60–67. [CrossRef]

31. Zhang, X.; Abreu, J.G.; Yokota, C.; MacDonald, B.T.; Singh, S.; Coburn, K.L.A.; Cheong, S.-M.; Zhang, M.M.; Ye, Q.-Z.; Hang, H.C.; et al. Tiki1 is required for head formation via Wnt cleavage-oxidation and inactivation. *Cell* **2012**, *149*, 1565–1577. [CrossRef] [PubMed]

32. Amado, N.G.; Fonseca, B.F.; Cerqueira, D.M.; Reis, A.H.; Simas, A.B.C.; Kuster, R.M.; Mendes, F.A.; Abreu, J.G. Effects of natural compounds on Xenopus embryogenesis: A potential read out for functional drug discovery targeting Wnt/β-catenin signaling. *Curr. Top. Med. Chem.* **2012**, *12*, 2103–2113. [CrossRef] [PubMed]

33. Oh, S.; Gwak, J.; Park, S.; Yang, C.S. Green tea polyphenol EGCG suppresses Wnt/β-catenin signaling by promoting GSK-3β- and PP2A-independent β-catenin phosphorylation/degradation. *BioFactors* **2014**, *40*, 586–595. [CrossRef]

34. Amado, N.G.; Predes, D.; Fonseca, B.F.; Cerqueira, D.M.; Reis, A.H.; Dudenhoeffer, A.C.; Borges, H.L.; Mendes, F.A.; Abreu, J.G. Isoquercitrin suppresses colon cancer cell growth in vitro by targeting the Wnt/β-catenin signaling pathway. *J. Biol. Chem.* **2014**, *289*, 35456–35467. [CrossRef] [PubMed]

35. Cunha, G.M.D.; Fontenele, J.B.; Nobre Júnior, H.V.; de Sousa, F.C.; Silveira, E.R.; Nogueira, N.A.; de Moraes, M.O.; Viana, G.S.; Costa-Lotufo, L.V. Cytotoxic activity of chalcones isolated from *Lonchocarpus sericeus* (Pocr.) Kunth. *Phytother. Res.* **2003**, *17*, 155–159. [CrossRef] [PubMed]

36. Staal, F.; Famili, F.; Garcia Perez, L.; Pike-Overzet, K. Aberrant Wnt signaling in leukemia. *Cancers* **2016**, *8*, 78. [CrossRef]

37. Chen, G.; Zhou, D.; Li, X.Z.; Jiang, Z.; Tan, C.; Wei, X.Y.; Ling, J.; Jing, J.; Liu, F.; Li, N. A natural chalcone induces apoptosis in lung cancer cells: 3D-QSAR, docking and an in vivo/vitro assay. *Sci. Rep.* **2017**, *7*, 10729. [CrossRef]

38. De Robertis, M.; Massi, E.; Poeta, M.L.; Carotti, S.; Morini, S.; Cecchetelli, L.; Signori, E.; Fazio, V.M. The AOM/DSS murine model for the study of colon carcinogenesis: From pathways to diagnosis and therapy studies. *J. Carcinog.* **2011**, *10*. [CrossRef]

39. Wall, M.E.; Wani, M.C.; Cook, C.; Palmer, K.H.; McPhail, A.A.; Sim, G. Plant antitumor agents. I. The isolation and structure of camptothecin, a novel alkaloidal leukemia and tumor inhibitor from camptotheca acuminata. *J. Am. Chem. Soc.* **1966**, *88*, 3888–3890. [CrossRef]

40. Wall, M.E.; Wani, M.C. Discovery to clinic. *Ann. N. Y. Acad. Sci.* **1996**, *803*, 1–12. [CrossRef]

41. Chikazawa, N.; Tanaka, H.; Tasaka, T.; Nakamura, M.; Tanaka, M.; Onishi, H.; Katano, M. Inhibition of Wnt signaling pathway decreases chemotherapy-resistant side-population colon cancer cells. *Anticancer Res.* **2010**, *30*, 2041–2048.

42. Zhan, T.; Ambrosi, G.; Wandmacher, A.; Rauscher, B.; Betge, J.; Rindtorff, N.; Häussler, R.; Hinsenkamp, I.; Bamberg, L.; Hessling, B.; et al. MEK inhibitors activate Wnt signalling and induce stem cell plasticity in colorectal cancer. *Nat. Commun.* **2019**, *10*, 2197. [CrossRef] [PubMed]

43. Fang, F.; VanCleave, A.; Helmuth, R.; Torres, H.; Rickel, K.; Wollenzien, H.; Sun, H.; Zeng, E.; Zhao, J.; Tao, J. Targeting the Wnt/β-catenin pathway in human osteosarcoma cells. *Oncotarget* **2018**, *9*, 36780–36792. [CrossRef] [PubMed]

44. Chen, H.J.; Hsu, L.S.; Shia, Y.T.; Lin, M.W.; Lin, C.M. The β-catenin/TCF complex as a novel target of resveratrol in the Wnt/β-catenin signaling pathway. *Biochem. Pharmacol.* **2012**, *84*, 1143–1153. [CrossRef] [PubMed]

45. Arbain, N.H.; Salim, N.; Wui, W.T.; Basri, M.; Rahman, M.B.A. Optimization of Quercetin loaded Palm Oil Ester Based Nanoemulsion Formulation for Pulmonary Delivery. *J. Oleo Sci.* **2018**, *67*, 933–940. [CrossRef] [PubMed]

46. Kakran, M.; Shegokar, R.; Sahoo, N.G.; Al Shaal, L.; Li, L.; Müller, R.H. Fabrication of quercetin nanocrystals: Comparison of different methods. *Eur. J. Pharm. Biopharm.* **2012**, *80*, 113–121. [CrossRef]

47. Kumari, A.; Yadav, S.K.; Pakade, Y.B.; Singh, B.; Yadav, S.C. Development of biodegradable nanoparticles for delivery of quercetin. *Colloids Surf. B Biointerfaces* **2010**, *80*, 184–192. [CrossRef]

48. Ader, P. Bioavailability and metabolism of the flavonol quercetin in the pig. *Free Radic. Biol. Med.* **2000**, *28*, 1056–1067. [CrossRef]

49. Pool, H.; Quintanar, D.; Figueroa, J.D.D.; Mano, C.M.; Bechara, J.E.H.; Godínez, L.A.; Mendoza, S.; Godí Nez, L.A. Antioxidant Effects of Quercetin and Catechin Encapsulated into PLGA Nanoparticles. *J. Nanomater.* **2012**, *2012*, 86. [CrossRef]

50. Nascimento, M.C.D.; Mors, W. Chalcones of the root bark of Derris sericea. *Phytochemistry* **1972**, *11*, 3023–3028. [CrossRef]

51. Nieuwkoop, P.; Faber, J. *Normal Table of Xenopus Laevis*; North-Holland Publishing, Co.: Amsterdam, The Netherlands, 1967.

Article

Isorhamnetin Induces Cell Cycle Arrest and Apoptosis Via Reactive Oxygen Species-Mediated AMP-Activated Protein Kinase Signaling Pathway Activation in Human Bladder Cancer Cells

Cheol Park [1,†], Hee-Jae Cha [2,†], Eun Ok Choi [3,4], Hyesook Lee [3,4], Hyun Hwang-Bo [3,5], Seon Yeong Ji [3,5], Min Yeong Kim [3,4], So Young Kim [3,4], Su Hyun Hong [3,4], JaeHun Cheong [5], Gi-Young Kim [6], Seok Joong Yun [7], Hye Jin Hwang [8], Wun-Jae Kim [7,*] and Yung Hyun Choi [3,4,*]

[1] Department of Molecular Biology, College of Natural Sciences, Dong-eui University, Busan 47340, Korea; parkch@deu.ac.kr

[2] Department of Parasitology and Genetics, Kosin University College of Medicine, Busan 49267, Korea; hcha@kosin.ac.kr

[3] Anti-Aging Research Center, Dong-eui University, Busan 47227, Korea; nakajo39@naver.com (E.O.C.); 14769@deu.ac.kr (H.L.); hbhyun2003@naver.com (H.H.-B.); 14602@deu.ac.kr (S.Y.J.); ilytoo365@deu.ac.kr (M.Y.K.); 14731@deu.ac.kr (S.Y.K.); hongsh@deu.ac.kr (S.H.H.)

[4] Department of Biochemistry, Dong-eui University College of Korean Medicine, Busan 47227, Korea

[5] Department of Molecular Biology, Pusan National University, Busan 46241, Korea; molecule85@pusan.ac.kr

[6] Department of Marine Life Sciences, School of Marine Biomedical Sciences, Jeju National University, Jeju 63243, Korea; immunkim@jejunu.ac.kr

[7] Department of Urology, College of Medicine, Chungbuk National University, Chungbuk 8644, Korea; sjyun@chungbuk.ac.kr

[8] Department of Food and Nutrition, College of Nursing, Healthcare Sciences & Human Ecology, Dong-Eui University, Busan 47340, Korea; hhj2001@deu.ac.kr

[*] Correspondence: wjkim@chungbuk.ac.kr (W.-J.K.); choiyh@deu.ac.kr (Y.H.C.); Tel.: +82-43-269-6136 (W.-J.K.); +82-51-850-7413 (Y.H.C.)

[†] Equally contributed.

Received: 9 July 2019; Accepted: 5 September 2019; Published: 4 October 2019

Abstract: Isorhamnetin is an O-methylated flavonol that is predominantly found in the fruits and leaves of various plants, which have been used for traditional herbal remedies. Although several previous studies have reported that this flavonol has diverse health-promoting effects, evidence is still lacking for the underlying molecular mechanism of its anti-cancer efficacy. In this study, we examined the anti-proliferative effect of isorhamnetin on human bladder cancer cells and found that isorhamnetin triggered the gap 2/ mitosis (G2/M) phase cell arrest and apoptosis. Our data showed that isorhamnetin decreased the expression of Wee1 and cyclin B1, but increased the expression of cyclin-dependent kinase (Cdk) inhibitor p21$^{WAF1/CIP1}$, and increased p21 was bound to Cdk1. In addition, isorhamnetin-induced apoptosis was associated with the increased expression of the Fas/Fas ligand, reduced ratio of B-cell lymphoma 2 (Bcl-2)/Bcl-2 associated X protein (Bax) expression, cytosolic release of cytochrome *c*, and activation of caspases. Moreover, isorhamnetin inactivated the adenosine 5′-monophosphate-activated protein kinase (AMPK) signaling pathway by diminishing the adenosine triphosphate (ATP) production due to impaired mitochondrial function. Furthermore, isorhamnetin stimulated production of intracellular reactive oxygen species (ROS); however, the interruption of ROS generation using a ROS scavenger led to an escape from isorhamnetin-mediated G2/M arrest and apoptosis. Collectively, this is the first report to show that isorhamnetin inhibited the proliferation of human bladder cancer cells by ROS-dependent arrest of the cell cycle at the G2/M phase and induction of apoptosis. Therefore, our results provide an important basis for the interpretation of the anti-cancer mechanism of isorhamnetin in bladder cancer cells and support the rationale for the need to evaluate more precise molecular mechanisms and in vivo anti-cancer properties.

Keywords: isorhamnetin; G2/M arrest; apoptosis; ROS; AMPK

---

1. Introduction

Although new therapies for treating cancer patients are being developed, chemotherapy is still the main approach for the treatment of cancer. However, some limitations, such as adverse side effects, drug resistance, and limited efficacy, remain to be solved [1,2]. Therefore, urgent new therapeutic strategies that minimize these limitations and have high therapeutic efficacy are required. In this respect, there is an increasing interest in the importance of compounds derived from natural resources that have been traditionally used for the prevention and treatment of various diseases [3–5]. In particular, numerous naturally occurring agents have been reported to cause cell cycle arrest and induce apoptosis, which are important strategies for the control of proliferation in cancer cells, without inducing toxicity in normal cells [6,7]. These agents have also emerged as an alternative to chemopreventive and chemotherapeutic agents because they can specifically regulate various cellular signaling pathways in cancer cells [8,9].

Isorhamnetin (3'-methoxy-3,4',5,7-tetrahydroxyflavone) is a flavonol aglycone found in some medicinal plants, such as *Hippophae rhamnoides* L., *Oenanthe javanica*, and *Ginkgo biloba* L., which are used as traditional medicines for the treatment of rheumatism, hemorrhage, cardiovascular disease, and cancer [10,11]. As one of the metabolites of quercetin, isorhamnetin is structurally similar to kaempferol, and is also called 3-O-methyl quercetin [12–14]. Isorhamnetin displays a number of biological properties due to its antioxidant, anti-inflammatory, and metabolic properties [15–19], and is also considered to have potential as an anti-cancer agent based on the results of various cancer cell models. For example, isorhamnetin has been reported to inhibit human leukemia, breast, colon, and cervical cancer cell proliferation through the gap 2/ mitosis (G2/M) phase arrest [20–23], and to induce mitotic block in non-small cell lung carcinoma cells, thus enhancing cisplatin- and carboplatin-induced G2/M arrest [24]. However, isorhamnetin induced S-phase arrest in some cancer cells [25,26], indicating that cell cycle arrest by isorhamnetin is dependent on the type of cancer cell line.

In addition, the anti-cancer effects of isorhamnetin in various cancer cell lines have been shown to involve the death receptor (DR)-dependent extrinsic and/or mitochondria-dependent intrinsic pathways [19,24,27–31], which are representative apoptosis inducing pathways. It was also found that the anti-cancer effect of isorhamnetin was accompanied by the disturbance of various cellular signaling pathways [20,25,32]. Furthermore, isorhamnetin showed a strong cytotoxic effect through a reactive oxygen species (ROS)-dependent apoptosis pathway in breast cancer cells [26]. In particular, isorhamnetin was able to induce high cytotoxicity at low doses compared to quercetin in cancer cells, including hepatocellular carcinoma and leukemia cells [33,34]. Although the possibility of the growth inhibitory activity of isorhamnetin in bladder cancer cells has recently been proposed [35], no molecular mechanism has been reported to support its effect. Therefore, in this study, we investigated the anti-cancer efficacy of isorhamnetin in human bladder cancer cells, focusing on the mechanisms associated with the induction of cell cycle arrest and apoptosis.

2. Results

2.1. Isorhamnetin Inhibited Cell Viability in Bladder Cancer Cells

To examine the cytotoxic effect of isorhamnetin, four bladder cancer T24 cell lines (T24, 5637, and 2531J) were treated with various concentrations of isorhamnetin, and then the 3-(4,5-dimethyl-2-thiazolyl)-2,5-diphenyltetra-zolium bromide (MTT) assay was conducted. Although there are some differences depending on the cell line, the cell viability was significantly decreased in a concentration-dependent manner in isorhamnetin-treated cells (Figure 1A), without affecting normal cultured human keratinocyte HaCaT cells and Chang liver cells under the same conditions. In addition, the 50% inhibitory concentration (IC_{50}) values of isorhamnetin on T24 and

5637 cells were 127.86 µM and 145.75 µM, respectively. The microscopic examination demonstrated that the phenotypic characteristics of isorhamnetin-treated T24 and 5637 cells showed irregular cell outlines, a decrease of cell density, shrinkage, and an increase of detached cells (Figure 1B, upper panel). In addition, 2531J cells showed similar results from the isorhamnetin treatment.

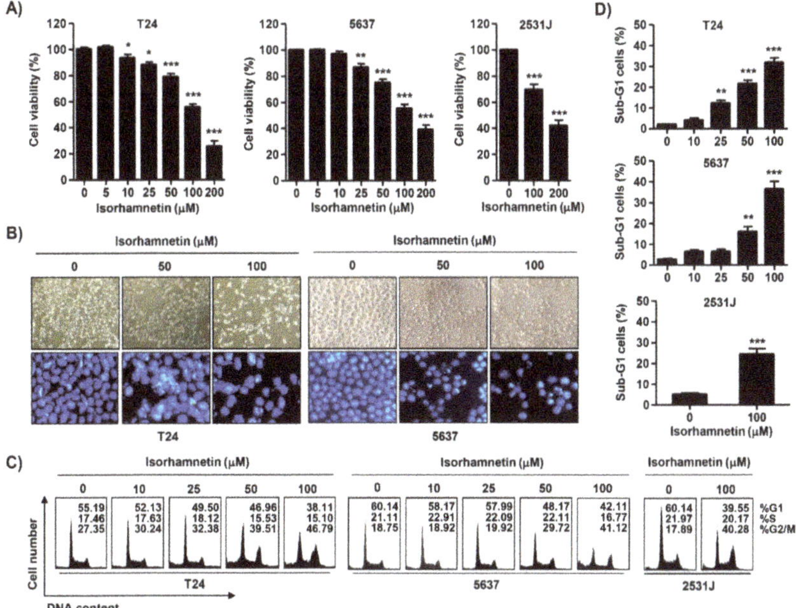

Figure 1. The inhibition of cell viability and induction of cell cycle arrest at gap 2/ mitosis (G2/M) phase using isorhamnetin in bladder cancer cells. T24, 5637, and 2531J cells were treated with the indicated concentrations of isorhamnetin for 48 h. (**A**) The cell viability was assessed using 3-(4,5-dimethyl-2-thiazolyl)-2,5-diphenyltetra-zolium bromide (MTT) assay. Each bar represents the mean ± standard deviation (SD) of three independent experiments (* $p < 0.05$ and *** $p < 0.0001$ compared to the control). (**B**, Upper panel) Morphological changes of T24 and 5637 cells were observed using phase-contrast microscopy. (B, Lower panel) The 4′,6-diamidino-2-phenylindole (DAPI)-stained nuclei were pictured under a fluorescence microscope. Representative photographs of the morphological changes are presented. (**C,D**) The cells were stained with propidium iodide (PI) solution for flow cytometry analysis. Quantification of the cell population (in percent) in different cell cycle phases of viable cells is shown. (D) Sub-G 1% was calculated as the percentage of the number of cells in the sub-G1 population relative to the number of total cells. Data were expressed as the mean ± SD of three independent experiments (* $p < 0.05$, ** $p < 0.001$, and *** $p < 0.0001$ compared to the control).

2.2. Isorhamnetin Induced G2/M Phase Arrest and Apoptosis in Bladder Cancer Cells

To examine the mechanism responsible for the isorhamnetin-induced anti-proliferative effect, the cell cycle distribution profile was examined. Flow cytometry data demonstrated that the percentage of cells arrested at G2/M phase was increased with increasing isorhamnetin treatment concentration, coupled with a decrease in the proportion of cells in the G1 and S phases (Figure 1C). Meanwhile, a significant increase of the cells in the sub-G1 phase, which was used as an index of apoptotic cells, was observed in isorhamnetin-treated cells (Figure 1D). Therefore, 4′,6-diamidino-2-phenylindole (DAPI) staining was performed to investigate whether apoptosis was involved in cell the growth inhibition induced by isorhamnetin. Figure 1B (lower panel) indicates that morphological changes of the nuclei, which were observed in apoptosis-inducing cells, such as nuclear fragmentation and chromatin condensation, were dominantly found in isorhamnetin-treated T24 and 5637 cells. Since 2531J cells also had the same results, the following experiments were performed on T24 and 5637 cells. To quantify the apoptosis triggered by isorhamnetin, an annexin V-fluorescein isothiocyanate (FITC)/propidium iodide (PI) double staining assay was conducted. As indicated in Figure 2A,B, after treatment with isorhamnetin, the populations of annexin V-staining positive cells were significantly increased, as compared to the control. On the other hand, T24 cells showed slightly increase in necrotic death upon 100 μM of isorhamnetin, but not 5637 cells (Figure 2B). Consistent with this, the results from agarose gel electrophoresis showed that as the isorhamnetin concentration increased and more fragmented DNA was observed (Figure 2C), indicating that an isorhamnetin-induced G2/M phase arrest was associated with the induction of apoptosis.

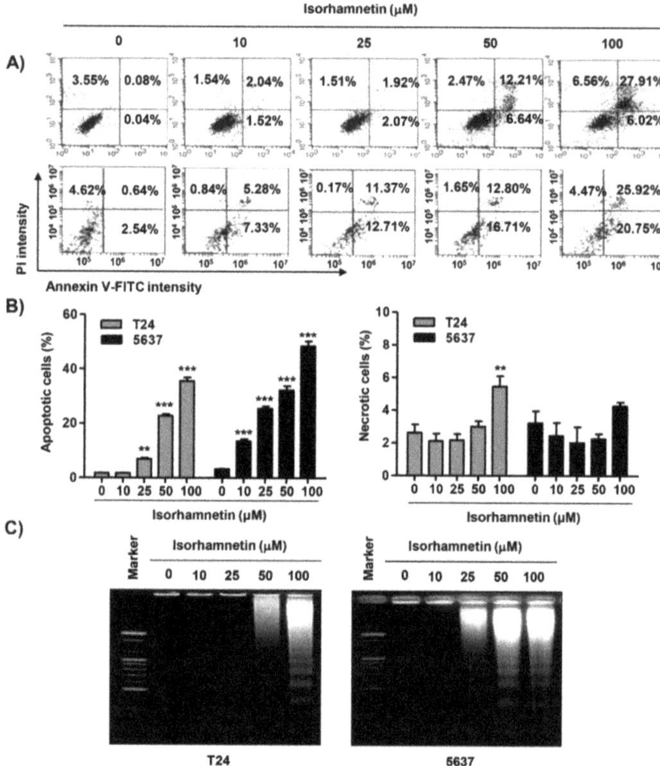

Figure 2. Induction of apoptosis using isorhamnetin in bladder cancer cells. (**A,B**) After treatment with different concentrations of isorhamnetin for 48 h, the cells were fixed and stained with annexin V-fluorescein isothiocyanate (FITC) and PI for flow cytometry analysis. (A) Representative profiles. The results show early apoptosis, defined as annexin V$^+$ and PI$^-$ cells (lower right quadrant), and late apoptosis, defined as annexin V$^+$ and PI$^+$ (upper right quadrant) cells. (**B**) The percentages of apoptotic cells (left) and necrotic cells (right) were determined by expressing the numbers of Annexin V$^+$ cells as percentages of all the present cells. The results are presented as the mean ± SD of three independent experiments (** $p < 0.001$ and *** $p < 0.0001$ compared to the control). (**C**) DNA fragmentation in the cells cultured under the same conditions was analyzed via the extraction of genomic DNA, electrophoresis in agarose gel, and then visualization using ethidium bromide (EtBr) staining.

2.3. Isorhamnetin Regulated the Expression of G2/M Phase-Associated Proteins in Bladder Cancer Cells

To explore the biochemical event of the isorhamnetin-elicited cell cycle arrest, the levels of G2/M phase-associated proteins were analyzed. The immunoblotting results revealed that following isorhamnetin treatment, the levels of Wee1 and cyclin B1 were reduced, and the effect was concentration-dependent, while the expression of cyclin-dependent kinase (Cdk) 1 (also called cell division cycle 2, Cdc2) was maintained at the level of the control group (Figure 3A,B). However, the expression of Cdk inhibitor p21$^{WAF1/CIP1}$ was markedly increased in response to isorhamnetin exposure. Next, we performed co-immunoprecipitation to investigate the role of isorhamnetin-induced p21, and found that this increased p21 via treatment with isorhamnetin complexed with Cdk1 (Figure 3C). These results suggest that increased p21 protein in isorhamnetin-treated cells contributed to G2/M phase arrest by inhibiting its activity through binding to Cdk1.

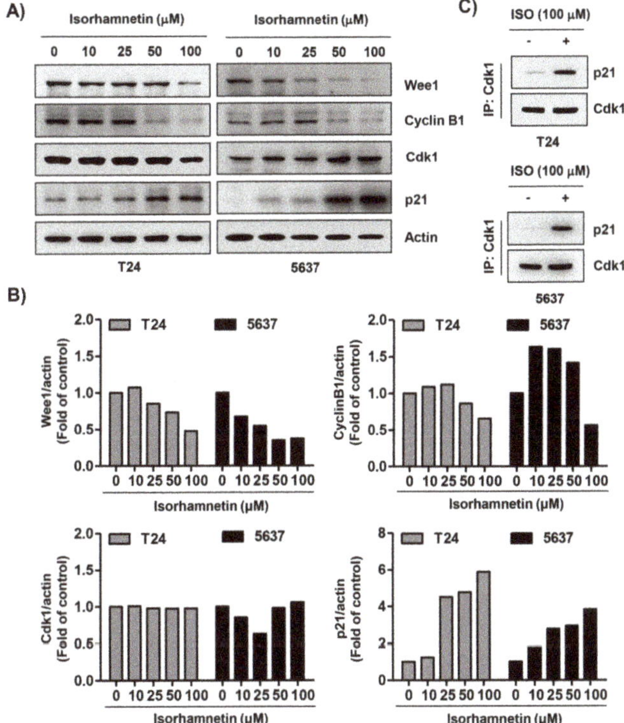

Figure 3. Effects of isorhamnetin on the levels of cell cycle regulatory proteins in bladder cancer cells. (**A**) T24 and 5637 cells were treated with the indicated concentrations of isorhamnetin for 48 h, and then total cell lysates were prepared. Western blotting was then performed using the indicated antibodies and an enhanced chemiluminescence (ECL) detection system. Actin was used as an internal control. (**B**) The expression of each protein was indicated as a fold change relative to the control. Quantitative analysis of mean pixel density was performed using the ImageJ® software. (**C**) Cells were incubated without or with 100 µM isorhamnetin for 48 h, and then equal amounts of proteins were immunoprecipitated with the anti-cyclin-dependent kinase (Cdk) 1 antibody. Western blotting using immunocomplexes was performed using anti-p21 or anti-Cdk1 antibodies and an ECL detection system (IP, immunoprecipitation).

2.4. Isorhamnetin Modulated the Expression of Apoptosis-Regulatory Proteins in Bladder Cancer Cells

To investigate the pathway of isorhamnetin-induced apoptosis, caspases activities were determined. Figure 4A,B shows that the protein levels of pro-caspase-8, -9, and -3 were concentration-dependently decreased, which was associated with the degradation of poly(ADP-ribose) polymerase (PARP). Therefore, we quantitatively assessed each caspase activity in the presence of isorhamnetin using fluorogenic substrates to determine whether these immunoblotting results were directly related to activation of the corresponding caspases and found that treatment with isorhamnetin significantly stimulated the activation of these caspases in a concentration-dependent manner in comparison with untreated control cells (Figure 4C). In addition, the effects of isorhamnetin on the expression of the Fas/Fas ligand (FasL) and B-cell lymphoma 2 (Bcl-2) family members were determined. Figure 4A,B shows that both Fas and FasL protein levels were up-regulated, and the level of Bcl-2-associated X protein (Bax), a pro-apoptotic protein, was also increased, while the level of Bcl-2, an anti-apoptotic protein, was reduced in isorhamnetin-treated cells. Furthermore, isorhamnetin promoted the release of cytochrome *c* from mitochondria into cytosol (Figure 4D).

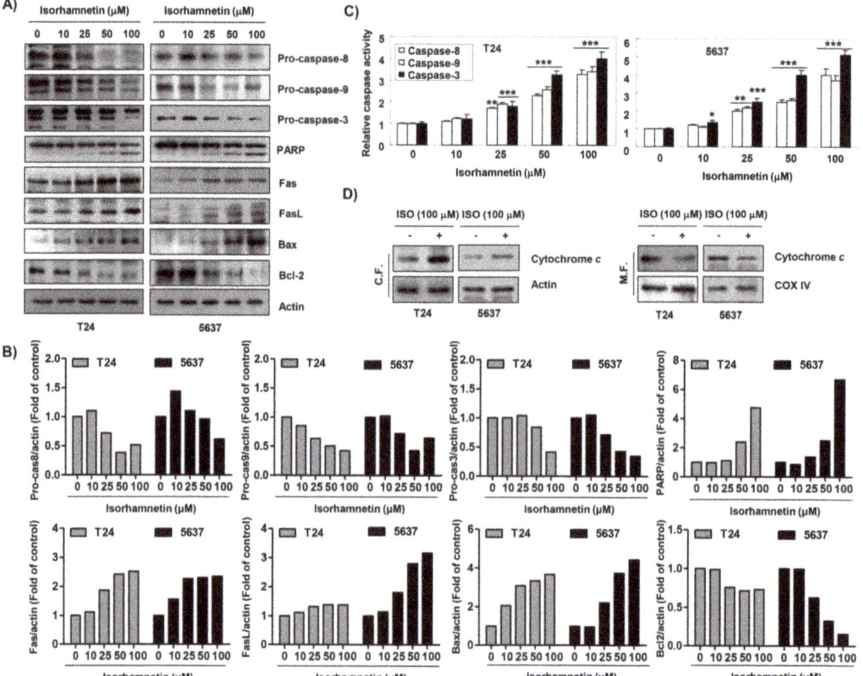

Figure 4. Modulation of apoptosis-regulatory factors using isorhamnetin in bladder cancer cells. (**A**) After treatment with isorhamnetin for 48 h, Western blotting was performed using the indicated antibodies and an ECL detection system. Actin was used as an internal control. (**B**) The expression of each protein was indicated as a fold change relative to the control. Quantitative analysis of mean pixel density was performed using the ImageJ® software. (**C**) The activities of caspases were evaluated using caspase colorimetric assay kits. The data were expressed as the mean ± SD of three independent experiments (* $p < 0.05$, ** $p < 0.001$, and *** $p < 0.0001$ compared to the control). (**D**) After treatment without or with 100 µM isorhamnetin for 48 h, cytosolic and mitochondrial proteins were prepared and analyzed for cytochrome *c* expression using Western blot analysis. Equal protein loading was confirmed via the analysis of actin and cytochrome oxidase subunit VI (COX VI) in each protein extract (C.F.—cytosolic fraction; M.F.—mitochondrial fraction).

2.5. Isorhamnetin Increased ROS Generation but Decreased ATP Content in Cancer Cells

To investigate the involvement of ROS on the cytotoxic effect of isorhamnetin, we performed flow cytometry analysis using a fluorescent probe, 5,6-carboxy-2′,7′-dichlorodihydrofluorescein diacetate (DCF-DA). Our data indicated that the production of ROS showed a significant increase within 1 h of the isorhamnetin treatment, and then gradually decreased, while the antioxidant N-acetyl-L-cysteine (NAC) suppressed it to the control level (Figure 5A–C). In addition, an ATP colorimetric assay kit was used to measure the content of mitochondrial ATP in the cells. Figure 5D shows that the concentration of ATP in the isorhamnetin-treated cells decreased in a concentration-dependent manner. However, under the condition that NAC existed, it was markedly weakened, indicating that the decrease in ATP levels was associated with ROS production.

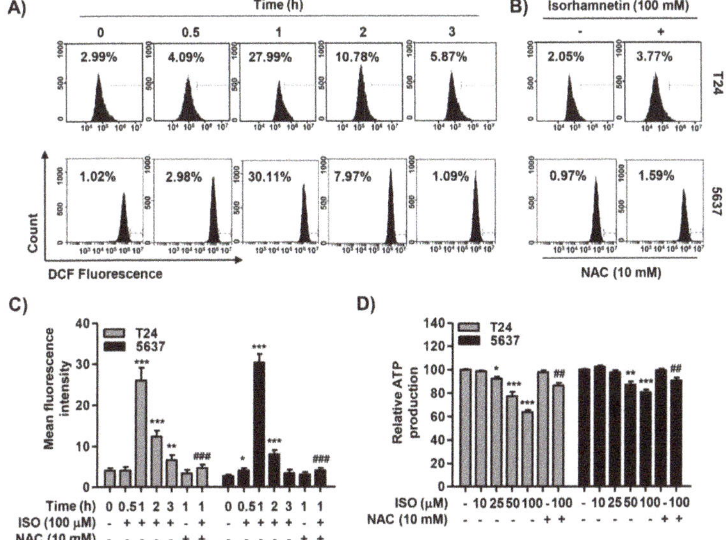

Figure 5. Accumulation of reactive oxygen species (ROS) and decrease of ATP content using isorhamnetin in bladder cancer cells. (**A**) Cells were treated with 100 μM isorhamnetin for the indicated times. (**B**) The cells were pre-treated with or without 10 mM N-acetyl-L-cysteine (NAC) for 1 h before isorhamnetin treatment for 1 h. (A,B) The medium was discarded and the cells were incubated for 20 min with medium containing 5,6-carboxy-2′,7′-dichlorodihydrofluorescein diacetate (DCF-DA). ROS generation was measured using flow cytometry. (**C**) Each bar represents the mean ± SD of three independent experiments. (**D**) After treatment with the indicated concentrations of isorhamnetin in the presence or absence of NAC, the content of intracellular ATP was measured. Each point represents the mean ± SD of three independent experiments (* $p < 0.05$, ** $p < 0.001$, and *** $p < 0.0001$ compared to control; ## $p < 0.001$ and ### $p < 0.0001$ compared to isorhamnetin-treated cells). ISO—isorhamnetin.

2.6. Isorhamnetin Reduced Mitochondrial Membrane Potential (MMP, ΔΨm) and Activated Adenosine 5′-Monophosphate-Activated Protein Kinase (AMPK) Signaling in Bladder Cancer Cells

We assessed the level of MMP to investigate whether the inhibition of ROS-dependent ATP production by isorhamnetin was associated with impaired mitochondrial function. According to the results of flow cytometry using 5,5′,6,6′-tetrachloro-1,1′,3,3′-tetraethylimidacarbocyanine iodide (JC-1) dyes, the formation of JC-1 aggregates in mitochondria was maintained at a relatively high rate in cells not treated with isorhamnetin, while the ratio of JC-1 monomers increased with increasing isorhamnetin treatment concentration, indicating a remarkable depletion of MMP after isorhamnetin treatment (Figure 6A,B). Furthermore, isorhamnetin increased the phosphorylated level of AMPK, as well as its downstream factor acetyl-CoA carboxylase (ACC) (Figure 6C,F), indicating that the AMPK signaling pathway was activated as a result of the loss of ATP. Additionally, we evaluated the effects of isorhamnetin on the phosphorylation of the mechanistic target of rapamycin (mTOR), p70S6K, and Unc-51-like kinase (ULK1), which are AMPK downstream molecules that regulate cell proliferation, apoptosis, and autophagy [36]. Exposure of T24 and 5637 cells with isorhamnetin led to down-regulation in the phosphorylation of mTOR and p76S6K in a dose-dependent manner (Figure 6F,G). Interestingly, we found the isorhamnetin inhibited autophagy via down-regulation of the expression and phosphorylation of ULK1. In addition, our supplementary result showed that the expression of autophagy-related markers was down-regulated using the isorhamnetin treatment, similar to ULK1 (Supplementary Figure S1). However, the presence of NAC or compound C, an antagonist of AMPK, significantly prevented the isorhamnetin-induced loss of MMP (Figure 6A,B), and NAC or compound C also markedly abolished enhanced activation of the AMPK signaling by isorhamnetin (Figure 6E). These data indicate that isorhamnetin-promoted mitochondrial dysfunction associated with the disturbance of ATP production was mediated through an ROS-dependent pathway.

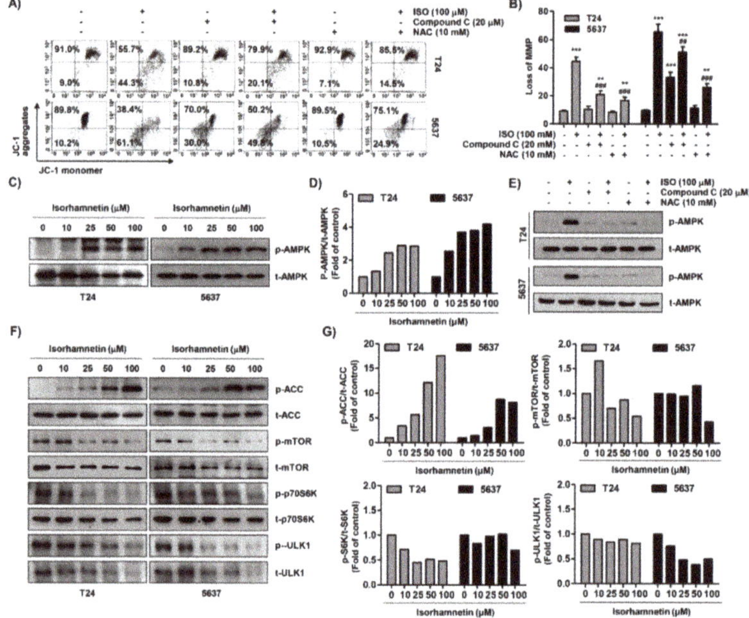

Figure 6. Mitochondrial dysfunction and activation of adenosine 5′-monophosphate-activated protein kinase (AMPK) signaling pathway using isorhamnetin in bladder cancer cells. (**A,B**) Cells were treated with 100 µM isorhamnetin for 48 h, or pre-treated with 20 µM compound C or 10 mM NAC for 1 h before isorhamnetin treatment for 48 h. (A) The cells were stained with 5,5′,6,6′-tetrachloro-1,1′,3,3′-tetraethylimidacarbocyanine iodide (JC-1) dye, and were then analyzed using flow cytometry in order to evaluate the changes in mitochondrial membrane potential (MMP). (**B**) Each bar represents the percentage of cells with JC-1 monomers (mean ± SD of triplicate determinations, ** $p < 0.001$ and *** $p < 0.0001$ compared to the control; ## $p < 0.001$ and ### $p < 0.0001$ compared to the isorhamnetin-treated cells). (**C,F**) After treatment with the indicated concentrations of isorhamnetin for 48 h, total cell lysates were prepared and Western blotting was then performed using the indicated antibodies and an ECL detection system. (**D,G**) The expression of each protein was indicated as a fold change relative to the control. Quantitative analysis of mean pixel density was performed using the ImageJ® software. (**E**) The cells cultured under the same conditions as A and B were collected, and Western blotting was then performed.

2.7. ROS Acted as an Upstream Regulator of Isorhamnetin-Mediated Apoptosis and Cell Cycle Blockade in Bladder Cancer Cells

The effect of ROS on isorhamnetin-mediated apoptosis and G2/M phase arrest was further investigated to determine the role of ROS in the anti-cancer activity of isorhamnetin. As depicted in the results of the DAPI staining and flow cytometry analysis, artificially blocking the production of ROS using NAC drastically attenuated isorhamnetin-induced apoptosis (Figure 7A–C). In parallel, pretreatment with NAC protected isorhamnetin-mediated G2/M arrest, which was related to a decrease in the number of sub-G1 phase cells (Figure 7E). Consistent with these results, inhibiting ROS production greatly restored reduced cell viability using isorhamnetin (Figure 7F), demonstrating that ROS generation was shown to be necessary for the contribution of apoptosis and G2/M arrest using isorhamnetin.

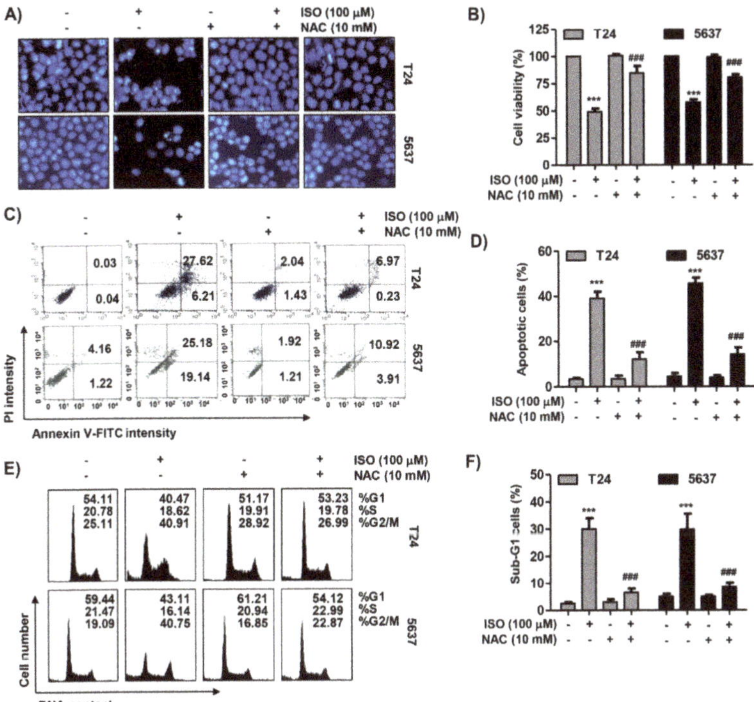

Figure 7. Roles of ROS in isorhamnetin-induced apoptosis and cell cycle arrest in bladder cancer cells. Cells were either treated with 100 µg/mL isorhamnetin for 48 h or pre-treated with 10 mM NAC for 1 h before isorhamnetin treatment, and were then collected. (**A**) The DAPI-stained nuclei were pictured under a fluorescence microscope. (**B,C**) The cells were stained with annexin V-FITC and PI for flow cytometry analysis. (B) Representative profiles. The results show early apoptosis, defined as annexin V^+ and PI^- cells (lower right quadrant), and late apoptosis, defined as annexin V^+ and PI^+ (upper right quadrant) cells. (C) The percentages of apoptotic cells were determined by expressing the numbers of Annexin V^+ cells as percentages of all the present cells. (**D**) The cells were stained with PI solution for flow cytometry analysis. Quantification of the cell population (in percent) in different cell cycle phases of viable cells is shown. (**E**) The percentages of apoptotic sub-G1 were calculated as the percentage of the number of cells in the sub-G1 population relative to the number of total cells. (**F**) The cell viability was assessed using an MTT assay. Each bar represents the mean ± SD of three independent experiments (*** $p < 0.0001$ compared to the control; ### $p < 0.0001$ compared to the isorhamnetin-treated cells).

3. Discussion

In many previous studies, it is clear that the induction of apoptosis by many anti-cancer agents is associated with cell cycle arrest at specific checkpoints [6,37]. In particular, the deregulation of cell cycle control is clearly implicated in the development and progression of most tumors, and the interruption of this progression is considered to be an important strategy to inhibit the proliferation of cancer cells [6,37]. Therefore, we first investigated whether the suppression of bladder cancer cell proliferation by isorhamnetin was associated with cell cycle arrest. The results of flow cytometry analysis showed that isorhamnetin caused G2/M phase arrest, similar to the results of previous studies in several human cancer cell lines [20–23], suggesting that G2/M phase arrest is one of the mechanisms of the growth inhibitory effects of isorhamnetin in human bladder cancer cells. The progression of the cell cycle in eukaryotic cells is tightly controlled by the interaction of cyclins and Cdks with their inhibitory factors [38,39]. In this process, the transition from G2 to M phase is achieved through the increased activity of Cdk1 by cyclin B1 complexing with Cdk1. In addition, Wee1 is a tyrosine kinase that induces phosphorylation of Cdk1, resulting in inhibition of cyclin B-Cdk1 activity and preventing cell mitotic entry [40,41]. In the current study, exposure of bladder cancer cells to isorhamnetin markedly reduced the expression of cyclin B1 and Wee1, without significant changes in the expression of Cdk1.

p21, a typical Cdk inhibitor belonging to the kinase inhibitory protein/ CDK interacting protein (KIP/CIP) family, has a broad-spectrum of specificity in the cell cycle proteins [38,39]. p21 was first reported to be a major inducer of tumor suppressor p53-dependent cell cycle arrest induced by DNA damage, but it could act as a mediator of p53-independent cell arrest in various types of cancer cells [42,43]. As a Cdk inhibitor, when p21 expression increases, it forms complexes with Cdks, reducing their kinase activity and inhibiting cell cycle progression [42,44]. According to our data, isorhamnetin dramatically increased p21 levels with increasing treatment concentration, and increased p21 complexed with Cdk1, which might have contributed to the inhibition of Cdk1 kinase activity. In addition, since T24 and 5637 cells are mutant p53 gene-bearing cell lines [45], increased p21 expression using isorhamnetin was thought to contribute to G2/M arrest, regardless of p53 gene status. Collectively, our data suggest that isorhamnetin-triggered G2/M arrest was due to the decreased expression of Wee1 and cyclin B1, and inactivation of p53-independent p21-mediated Cdk1 kinase.

Because the induction of apoptosis in cancer cells along with cell cycle arrest is a promising approach to cancer therapy, we assessed whether G2/M arrest using isorhamnetin was associated with apoptosis induction. Based on the results of morphological changes, DNA fragmentation, and flow cytometry analysis, we found that the cytotoxic effect of isorhamnetin was achieved through the induction of apoptosis associated with G2/M arrest. As is well known, apoptosis can be largely categorized into extrinsic and intrinsic pathways in mammalian cells [37,46]. The extrinsic pathway is characterized by the activation of caspase-8 by the formation of the death-inducing signal complex through the binding of death ligands to the cell surface DRs. For example, when FasL, one of the typical death ligands, binds to the corresponding DR, Fas, caspase-8 is sequentially activated [46,47]. On the other hand, the intrinsic pathway begins via the activation of caspase-9 through the release of mitochondrial pro-apoptotic proteins, such as cytochrome *c*, from mitochondria to cytoplasm due to increased mitochondrial permeability. This pathway is tightly regulated by the Bcl-2 protein family that includes pro- and anti-apoptotic proteins, which guard mitochondrial integrity and control the release of cytochrome *c* through the mitochondrial transition pore [48,49]. Caspases-8 and -9, which correspond to the initiator caspases of each pathway, ultimately activate apoptosis through the cleavage of various cellular substrates, such as PARP, by activating downstream executioner caspases, including caspase-3 and -7 [37,50]. In addition, these pathways are strictly controlled by a variety of cellular signaling pathways and regulatory molecules [50,51]. Our results show that isorhamnetin increased the expression of Fas and FasL; activated caspase-8, -9, and -3; and induced the cleavage of PARP. In addition, consistent with previous studies in non-small cell lung cancer cells and Lewis lung cancer cells [31,52], mitochondrial dysfunction was induced, as confirmed by the loss of MMP in isorhamnetin-treated cells. Moreover, the loss of MMP was accompanied by a down-regulation in the Bcl-2/Bax ratio and the promotion of cytosolic release of cytochrome *c*. Therefore, based on those observations, we speculated that the pro-apoptotic effect of isorhamnetin in bladder cancer cells could occur by simultaneously activating extrinsic and intrinsic pathways.

Growing evidence demonstrates that many anti-cancer agents induce apoptosis through pro-oxidant properties, such as increasing ROS accumulation or destroying cellular antioxidant systems [8,53]. In particular, mitochondria are the major subcellular organelles responsible for the production of ROS in the cells and are also a major target of ROS [54,55]. Therefore, elevating intracellular levels of ROS production is considered to be one of the ideal mechanisms for killing cancer cells through the activation of intrinsic pathways. Intriguingly, in various cell types, ROS are involved in activating the signaling system of AMPK, a key sensor that regulates energy balance and cell fate [56–58]. Mitochondrial dysfunction, due to excessive production of ROS, leads to a loss of function of the respiratory chain in the mitochondrial inner membrane, which can lower intracellular ATP levels and activate AMPK [56,57]. Choi et al. first reported that ROS induces concentration-dependent activation of AMPK [59]. More recently, it has been described that AMPK can be activated by ROS, thereby leading to an increase of glycolysis [60,61]. Furthermore, Corton et al. reported that hypoxic activation of AMPK was dependent on the levels of the mitochondrial ROS [62], and Tavazzi et al. demonstrated that AMPK activation was caused by ROS-mediated intracellular ATP depletion [63]. On the contrary, it has been reported that treatment of the potent ROS scavengers, including NAC and dimethyl sulfoxide (DMSO), significantly abolished oxidative stress-induced AMPK activation and ATP depletion [60,61,64]. Consistent with a previous study in breast cancer cells [22], our results show that isorhamnetin treatment markedly increased the levels of ROS production; however, the ROS scavenger, NAC, greatly weakened the accumulation of ROS by isorhamnetin. The quenching of ROS generation also significantly diminished the isorhamnetin-induced disruption of MMP to the control level, followed by significant ATP restoration, indicating that ROS act as upstream signaling molecules to enhance isorhamnetin-mediated mitochondrial dysfunction. Our results also demonstrate that the activation of the AMPK signaling pathway was increased in cells exposed to isorhamnetin, probably due to decreased ATP content. Furthermore, the presence of NAC markedly attenuated isorhamnetin-induced phosphorylation of AMPK, while their total protein levels were kept at an equivalent level, suggesting that the isorhamnetin-induced activation of AMPK signaling pathway is dependent on ROS production. Subsequently, NAC pretreatment also significantly reversed the enhanced apoptosis, G2/M phase arrest, and viability reduction induced by isorhamnetin, confirming that increasing ROS may serve as a key contributor to the anti-cancer effects of isorhamnetin. The AMPK acts as a metabolic mater switch that controls cell fate, such as cell survival, apoptosis, and autophagy [65]. Indeed, fatty acid synthesis is a critical energy-consuming process for the differentiation of tumor cells, and it has been demonstrated that AMPK inhibits lipid synthesis by the phosphorylation and inactivation of acetyl-CoA carboxylase 1 (ACC1) [65]. Furthermore, AMPK directly inhibits mTOR complex I, which regulates p70S6K, an enhancer of protein synthesis. In this sense, AMPK plays a critical role as a cell growth suppressor by inhibiting protein, rRNA, and lipid synthesis [65,66]. In the present study, we conjecture that the AMPK-mediated interruption of the mTOR/p70S6K/ACC1 signaling pathway may contribute to isorhamnetin-induced cell cycle arrest and apoptosis. On the other hand, there are conflicting opinions on the relationship between AMPK and autophagy. Although increasing evidence described that AMPK activation can induce the autophagy through the inhibition of mTOR and phosphorylation of ULK1 [67,68], a few studies reported that ROS attenuated autophagy by the down-regulation of ULK-1 [69,70]. Interestingly, based on our results, we found the isorhamnetin inhibited autophagy by down-regulation of the expression and phosphorylation of ULK1. Therefore, our data suggested that isorhamnetin-induced ROS activates AMPK, and subsequently down-regulates the mTOR/ACC1/ULK1 signaling pathway, which results in promoting cell apoptosis and inhibits autophagy at the same time.

The current results lead us to suggest that the production of ROS by isorhamnetin plays a critical role in the induction of G2/M arrest and apoptosis through simultaneous initiation of both extrinsic and intrinsic pathways in human bladder cancer cells. In addition, ROS act as an upstream signal related to the effect of isorhamnetin on the activation of the AMPK signaling pathway. However, further studies are warranted to identify the molecular mechanisms of isorhamnetin-mediated activation of AMPK signaling on autophagy and mitochondrial energy metabolism in bladder cancer cells. In addition, further studies are required to identify and understand the role of intracellular organelles involved in ROS generation by isorhamnetin, including in vivo animal experiments.

4. Materials and Methods

4.1. Cell Culture and Isorhamnetin Treatment

The human bladder cancer cell lines (T24, 5637, 2531J, and EJ) were purchased from the American Type Culture Collection (Manassas, MD, USA). The cells were cultured in Dulbecco's modified Eagle's medium supplemented with 10% fetal bovine serum and 1% penicillin-streptomycin (100 U/mL penicillin and 100 μg/mL streptomycin, all from WelGENE Inc., Daegu, Republic of Korea) at 37 °C under a humidified 5% CO_2. The cells were sub-cultured every 3–4 days to maintain logarithmic growth, and were allowed to grow for 24 h before treatments were applied. Isorhamnetin was obtained from Sigma-Aldrich Chemical Co. (St. Louis, MO, USA), and was dissolved in dimethyl sulfoxide (DMSO, Sigma-Aldrich Chemical Co.) to a final concentration of 100 M. Prior to use, the stock solution was diluted with culture medium to the desired concentration.

4.2. Cell Viability Assay

Cell viability was determined using an MTT assay, as previously described [71]. Briefly, cells (1×10^4 cells/well) were seeded onto 96-well plates in 100 μL medium. After overnight incubation, the cells were exposed to a series of concentrations of isorhamnetin for 48 h. Thereafter, the MTT reagent (Sigma-Aldrich Chemical Co.) at 50 μg/mL final concentration was added to each well and cells were incubated continuously at 37 °C for 2 h. The medium was then removed and 100 μL DMSO was added to each well to dissolve the formed blue formazan crystals, followed by measurement at 540 nm in a microplate reader (Molecular Device Co., Sunnyvale, CA, USA). All results were performed in three independent experiments and the cell survival rate was expressed as a percentage of the control. The morphological changes of cells were directly observed and photographed using phase-contrast microscopy (Carl Zeiss, Oberkochen, Germany).

4.3. Determination of Cell Cycle Distribution Using Flow Cytometric Analysis

PI staining was applied to analyze the DNA content and cell cycle distribution. In brief, cells were exposed to different concentrations of isorhamnetin for 48 h, and then the cells were harvested and fixed gently in 70% ice-cold ethanol (in phosphate-buffered saline, PBS, WelGENE Inc., Daegu, Korea) at 4 °C for 30 min. The cells were re-suspended in PBS containing 40 μg/mL PI, 100 μg/mL RNase A, and 0.1% triton X-100 (all from Sigma-Aldrich Chemical Co.) in a dark room at 37 °C for 30 min, and subjected to flow cytometry (BD Biosciences, San Jose, CA, USA), to determine the cell cycle distribution and apoptotic cells (sub-G1 phase).

4.4. Determination of Apoptotic Cell Death by Flow Cytometric Analysis

The Annexin V-FITC staining kit from BD Biosciences (San Jose, CA, USA) was used to determine and quantify the apoptotic cells using flow cytometry, according to the manufacturer's instruction. In brief, the collected cells were suspended in the supplied binding buffer, and then stained with FITC-conjugated annexin V and PI at room temperature (RT) for 20 min in the dark. The fluorescent intensities of the cells were detected using flow cytometry, and the annexin V^+/PI^- and annexin V^+/PI^+ cell populations were considered indicators of apoptotic cells.

4.5. Nuclear Staining and Deoxyribonucleic Acid (DAN) Fragmentation Assay

The changes of nuclear morphology for assessing apoptosis were assessed using DAPI staining. Briefly, cells were cultured with or without different concentrations of isorhamnetin for 48 h, and were then fixed with 4% paraformaldehyde (Sigma-Aldrich Chemical Co.) for 10 min at RT. The cells were rinsed with PBS, and incubated with 1 µg/mL DAPI solution (Sigma-Aldrich Chemical Co.) at 37 °C for 10 min. Stained cells were visualized and photographed using fluorescence microscopy (Carl Zeiss, Oberkochen, Germany). For DNA fragmentation assay, the collected cells were lysed in a buffer containing 10 mM Tris-HCl (pH 7.4), 150 mM NaCl, 5 mM ethylenediaminetetraacetic acid, and 0.5% Triton X-100 for 30 min. The fragmented DNA in the supernatant was extracted using an equal volume of neutral phenol:chloroform:isoamyl alcohol (25:24:1, Sigma-Aldrich Chemical Co.), analyzed electrophoretically on 1% agarose gel containing EtBr (Sigma-Aldrich Chemical Co.), and photographed under a Fusion FX Image system (Vilber Lourmat, Torcy, France).

4.6. Protein Extraction, Co-Immunoprecipitation, and Western Blot Analysis

After treatment, both adherent and floating cells were harvested, and the whole cellular proteins were prepared using the Bradford protein assay kit (Bio-Rad Laboratories, Hercules, CA, USA), according to the manufacturer's protocol. For the preparation of mitochondrial and cytosolic proteins from the cells, NE-PER nuclear and cytoplasmic extraction reagents (Thermo Fisher Scientific Inc., Waltham, UT, USA) were applied. Protein concentration was measured using the Bio-Rad protein assay kit (Bio-Rad Laboratories, Hercules, CA, USA), according to the manufacturer's instructions. For the co-immunoprecipitation assay, the 500 µg of cell lysates from each sample was precleaned with normal rabbit immunoglobulin G (IgG) and a protein-A-sepharose bead slurry (Amersham, Arlington Heights, IL, USA), and immunoprecipitation was conducted using 1 µg of anti-Cdk1 antibody (Santa Cruz Biotechnology, Inc., Santa Cruz, CA, USA) and protein-A-sepharose (Sigma-Aldrich Chemical Co.). The protein complex was then prepared according to the previously described method [62]. For Western blot analysis, equal amounts of protein samples or immunoprecipitated proteins were separated using sodium dodecyl sulphate-polyacrylamide gel electrophoresis and transferred to polyvinylidene difluoride membranes (Millipore, Bedford, MA, USA) (whole blot figures can be found at the Supplementary). The membranes were blocked with Tris-buffered saline (10 mM Tris-Cl, pH 7.4) containing 0.5% Tween-20 and 5% nonfat dry milk for 1 h at RT, and then probed with the indicated primary antibodies (Santa Cruz Biotechnology, Inc., and Cell Signaling Technology, Danvers, MA, USA), to react with the blotted membranes at 4 °C overnight. Afterwards, the membranes were incubated with the corresponding horseradish peroxidase-conjugated secondary antibodies (Santa Cruz Biotechnology, Inc.), developed using an ECL detection kit (GE Healthcare Life Sciences, Little Chalfont, U.K.), and then visualized using a Fusion FX Image system. Densitometric analysis of the data was performed using the ImageJ® software (v1.48, NIH, Bethesda, MD, USA).

4.7. Caspase Activity Assay

The activity of caspases was measured according to the manufacturer's instructions for the Caspase colorimetric assay kits (R&D Systems, Minneapolis, MN, USA). Briefly, cells were harvested and lysed in the lysis buffer provided in the kit on ice for 10 min, and then centrifuged at $10,000 \times g$ for 1 min. The supernatants containing equal proteins were incubated with the supplied reaction mixtures, including the fluorogenic peptide substrate (Asp-Glu-Val-Asp specific for caspase-3, Ile-Glu-Thr-Asp for caspase-8, and Leu-Glu-His-Asp specific for caspase-9) labeled with p-nitroaniline (pNA) for 1 h at 37 °C in the dark. The amounts of released pNA was measured using a microplate reader using excitement at 405 nm and emitting at 510 nm.

4.8. Measurement of ROS Production and MMP

The production of ROS was measured using DCF-DA, as described previously [72]. At the end of the treatment with isorhamnetin for defined periods in the presence or absence of NAC (Sigma-Aldrich Chemical Co.), cells were washed with PBS and incubated with 10 μM DCF-DA (Invitrogen, Carlsbad, CA, USA) in the dark at 37 °C for 20 min. Subsequently, cells were analyzed for DCF fluorescence using flow cytometry at 480 nm/520 nm. To measure MMP, JC-1 staining was performed according to the manufacturer's instructions. After treatment with isorhamnetin for 48 h in the presence or absence of NAC or compound C, cells were exposed to 10 μM JC-1 (Invitrogen) for 30 min at 37 °C, and then analyzed using flow cytometry at 488 nm/575 nm, as previously described [73].

4.9. Detection of ATP Levels

The firefly luciferase-based ATP Bioluminescence assay kit (Roche Applied Science, Indianapolis, IN, USA) was used for the detection of intracellular ATP levels, according to the manufacturer's instructions. Briefly, cells treated with isorhamnetin for 48 h with or without NAC were lysed with the lysis buffer provided in the kit, and the supernatants were collected via centrifugation at 12,000× *g* for 10 min at 4 °C. Subsequently, an equal amount of supernatants and ATP detection reagent, which catalyzed the light production from ATP and luciferin, were mixed. Firefly luciferase activity was immediately measured using a luminometer and the ATP level was calculated according to the ATP standard curve. Intracellular ATP levels were calculated as a percentage of the untreated control.

4.10. Statistical Analysis

All experiments were performed at least three times. Data were analyzed using GraphPad Prism software (version 5.03; GraphPad Software, Inc., La Jolla, CA, USA), and expressed as the mean ± standard deviation (SD). Differences between groups were assessed using analysis of variance, followed by ANOVA-Tukey's post hoc test, and $p < 0.05$ was considered to indicate a statistically significant difference.

5. Conclusions

Our findings demonstrate that isorhamnetin exerted an anti-proliferative effect on human bladder cancer cells through the induction of cell cycle arrest during the G2/M phase and apoptosis. Isorhamnetin-induced G2/M arrest was attributed to the decrease in Wee1 and cyclin B1 expression and the upregulation of p21. Isorhamnetin also induced apoptosis by activating caspase-8 and -9, which belong to the initiator caspases of the extrinsic and intrinsic pathways, respectively, followed by the activation of effector caspase-3, leading to the degradation of PARP. In addition, isorhamnetin enhanced the mitochondrial dysfunction, which was associated with an increase in Bax/Bcl-2 expression ratio and cytochrome *c* release into the cytoplasm. Moreover, the induction of G2/M arrest and apoptosis by isorhamnetin was accompanied by activation of the AMPK signaling pathway, and excessive production of ROS. However, artificial interception of the AMPK signaling pathway attenuated isorhamnetin-induced apoptosis, and the interruption of ROS generation led cells to escape from G2/M arrest and apoptosis. Based on these finding, we suggest that isorhamnetin has chemopreventive potential by inducing G2/M arrest and apoptosis through ROS-dependent activation of the AMPK signaling pathway in bladder cancer cells.

Supplementary Materials: The following are available online at http://www.mdpi.com/2072-6694/11/10/1494/s1, Whole blot figures.

Author Contributions: Y.H.C., C.P., and W.-J.K. conceived and designed the experiments; H.-J.C., E.O.C., H.L., H.H.-B., S.Y.J., M.Y.K., S.Y.K., and S.H.H. performed the experiments; G.-Y.K., S.J.Y., J.H.C., and H.J.H. analyzed the data; C.P. wrote the paper; and Y.H.C. edited the paper.

Funding: This research was funded by the Basic Science Research Program through a National Research Foundation of Korea (NRF) grant funded by the Korea government (2017R1D1A1B03032689 and 2018R1A2B2005705) and a grant (0820050) from the National R&D Program for Cancer Control, Ministry of Health and Welfare, Republic of Korea.

Conflicts of Interest: The authors declare no conflict of interest.

References

1. Kumar, A.; Jaitak, V. Natural products as multidrug resistance modulators in cancer. *Eur. J. Med. Chem.* **2019**, *176*, 268–291. [CrossRef] [PubMed]
2. Fu, B.; Wang, N.; Tan, H.Y.; Li, S.; Cheung, F.; Feng, Y. Multi-component herbal products in the prevention and treatment of chemotherapy-associated toxicity and side effects: A review on experimental and clinical evidences. *Front. Pharmacol.* **2018**, *9*, 1394. [CrossRef] [PubMed]
3. Butler, M.S.; Robertson, A.A.; Cooper, M.A. Natural product and natural product derived drugs in clinical trials. *Nat. Prod. Rep.* **2014**, *31*, 1612–1661. [CrossRef] [PubMed]
4. Nobili, S.; Lippi, D.; Witort, E.; Donnini, M.; Bausi, L.; Mini, E.; Capaccioli, S. Natural compounds for cancer treatment and prevention. *Pharmacol. Res.* **2009**, *59*, 365–378. [CrossRef]
5. Efferth, T.; Fu, Y.J.; Zu, Y.G.; Schwarz, G.; Konkimalla, V.S.; Wink, M. Molecular target-guided tumor therapy with natural products derived from traditional Chinese medicine. *Curr. Med. Chem.* **2007**, *14*, 2024–2032. [CrossRef]
6. Medema, R.H.; Macůrek, L. Checkpoint control and cancer. *Oncogene* **2012**, *31*, 2601–2613. [CrossRef]
7. Hanahan, D.; Weinberg, R.A. Hallmarks of cancer: The next generation. *Cell* **2011**, *144*, 646–674. [CrossRef]
8. Bolhassani, A. Cancer chemoprevention by natural carotenoids as an efficient strategy. *Anticancer Agents Med. Chem.* **2015**, *15*, 1026–1231. [CrossRef]
9. Schnekenburger, M.; Dicato, M.; Diederich, M. Plant-derived epigenetic modulators for cancer treatment and prevention. *Biotechnol. Adv.* **2014**, *32*, 1123–1132. [CrossRef]
10. Gu, Q.; Duan, G.; Yu, X. Bioconversion of flavonoid glycosides from *Hippophae rhamnoides* leaves into flavonoid aglycones by *Eurotium amstelodami*. *Microorganisms* **2019**, *7*, 122. [CrossRef]
11. Zhang, Q.; Cui, H. Simultaneous determination of quercetin, kaempferol, and isorhamnetin in phytopharmaceuticals of *Hippophae rhamnoides* L. by high-performance liquid chromatography with chemiluminescence detection. *J. Sep. Sci.* **2005**, *28*, 1171–1178. [CrossRef] [PubMed]
12. Chen, C.; Xu, X.M.; Chen, Y.; Yu, M.Y.; Wen, F.Y.; Zhang, H. Identification, quantification and antioxidant activity of acylated flavonol glycosides from sea buckthorn (*Hippophae rhamnoides* ssp. sinensis). *Food Chem.* **2013**, *141*, 1573–1579. [CrossRef] [PubMed]
13. Castrillo, J.L.; Vanden Berghe, D.; Carrasco, L. 3-Methylquercetin is a potent and selective inhibitor of poliovirus RNA synthesis. *Virology* **1986**, *152*, 219–227. [CrossRef]
14. Rösch, D.; Krumbein, A.; Mügge, C.; Kroh, L.W. Structural investigations of flavonol glycosides from sea buckthorn (*Hippophaë rhamnoides*) pomace by NMR spectroscopy and HPLC-ESI-MS(n). *J. Agric. Food Chem.* **2004**, *52*, 4039–4046. [CrossRef] [PubMed]
15. Qi, F.; Sun, J.H.; Yan, J.Q.; Li, C.M.; Lv, X.C. Anti-inflammatory effects of isorhamnetin on LPS-stimulated human gingival fibroblasts by activating Nrf2 signaling pathway. *Microb. Pathog.* **2018**, *120*, 37–41. [CrossRef] [PubMed]
16. Abdallah, H.M.; Esmat, A. Antioxidant and anti-inflammatory activities of the major phenolics from *Zygophyllum simplex* L. *J. Ethnopharmacol.* **2017**, *205*, 51–56. [CrossRef]
17. Seo, S.; Seo, K.; Ki, S.H.; Shin, S.M. Isorhamnetin inhibits reactive oxygen species-dependent hypoxia inducible factor (HIF)-1α accumulation. *Biol. Pharm. Bull.* **2016**, *39*, 1830–1838. [CrossRef]
18. Yang, J.H.; Kim, S.C.; Kim, K.M.; Jang, C.H.; Cho, S.S.; Kim, S.J.; Ku, S.K.; Cho, I.J.; Ki, S.H. Isorhamnetin attenuates liver fibrosis by inhibiting TGF-β/Smad signaling and relieving oxidative stress. *Eur. J. Pharmacol.* **2016**, *783*, 92–102. [CrossRef]
19. Luo, Y.; Sun, G.; Dong, X.; Wang, M.; Qin, M.; Yu, Y.; Sun, X. Isorhamnetin attenuates atherosclerosis by inhibiting macrophage apoptosis via PI3K/AKT activation and HO-1 induction. *PLoS ONE* **2015**, *10*, e0120259. [CrossRef]

20. Zhang, H.W.; Hu, J.J.; Fu, R.Q.; Liu, X.; Zhang, Y.H.; Li, J.; Liu, L.; Li, Y.N.; Deng, Q.; Luo, Q.S.; et al. Flavonoids inhibit cell proliferation and induce apoptosis and autophagy through downregulation of III9K$_\gamma$ mediated PI3K/AKT/mTOR/p70S6K/ULK signaling pathway in human breast cancer cells. *Sci. Rep.* **2018**, *8*, 11255. [CrossRef]

21. Wei, J.; Su, H.; Bi, Y.; Li, J.; Feng, L.; Sheng, W. Anti-proliferative effect of isorhamnetin on HeLa cells through inducing G2/M cell cycle arrest. *Exp. Ther. Med.* **2018**, *15*, 3917–3923. [CrossRef] [PubMed]

22. Wu, X.; Chen, X.; Dan, J.; Cao, Y.; Gao, S.; Guo, Z.; Zerbe, P.; Chai, Y.; Diao, Y.; Zhang, L. Characterization of anti-leukemia components from *Indigo naturalis* using comprehensive two-dimensional K562/cell membrane chromatography and in silico target identification. *Sci. Rep.* **2016**, *6*, 25491. [CrossRef] [PubMed]

23. Li, C.; Yang, X.; Chen, C.; Cai, S.; Hu, J. Isorhamnetin suppresses colon cancer cell growth through the PI3K-Akt-mTOR pathway. *Mol. Med. Rep.* **2014**, *9*, 935–940. [CrossRef] [PubMed]

24. Zhang, B.Y.; Wang, Y.M.; Gong, H.; Zhao, H.; Lv, X.Y.; Yuan, G.H.; Han, S.R. Isorhamnetin flavonoid synergistically enhances the anticancer activity and apoptosis induction by cisplatin and carboplatin in non-small cell lung carcinoma (NSCLC). *Int. J. Clin. Exp. Pathol.* **2015**, *8*, 25–37. [PubMed]

25. Wang, J.L.; Quan, Q.; Ji, R.; Guo, X.Y.; Zhang, J.M.; Li, X.; Liu, Y.G. Isorhamnetin suppresses PANC-1 pancreatic cancer cell proliferation through S phase arrest. *Biomed. Pharmacother.* **2018**, *108*, 925–933. [CrossRef] [PubMed]

26. Wu, Q.; Kroon, P.A.; Shao, H.; Needs, P.W.; Yang, X. Differential effects of quercetin and two of its derivatives, isorhamnetin and isorhamnetin-3-glucuronide, in inhibiting the proliferation of human breast-cancer MCF-7 cells. *J. Agric. Food Chem.* **2018**, *66*, 7181–7189. [CrossRef]

27. Sak, K.; Lust, H.; Kase, M.; Jaal, J. Cytotoxic action of methylquercetins in human lung adenocarcinoma cells. *Oncol. Lett.* **2018**, *15*, 1973–1978. [CrossRef]

28. Huang, S.P.; Ho, T.M.; Yang, C.W.; Chang, Y.J.; Chen, J.F.; Shaw, N.S.; Horng, J.C.; Hsu, S.L.; Liao, M.Y.; Wu, L.C.; et al. Chemopreventive potential of ethanolic extracts of luobuma leaves (*Apocynum venetum* L.) in androgen insensitive prostate cancer. *Nutrients* **2017**, *9*, 948. [CrossRef]

29. Li, Q.; Ren, F.Q.; Yang, C.L.; Zhou, L.M.; Liu, Y.Y.; Xiao, J.; Zhu, L.; Wang, Z.G. Anti-proliferation effects of isorhamnetin on lung cancer cells in vitro and in vivo. *Asian Pac. J. Cancer. Prev.* **2015**, *16*, 3035–3042. [CrossRef]

30. Antunes-Ricardo, M.; Moreno-García, B.E.; Gutiérrez-Uribe, J.A.; Aráiz-Hernández, D.; Alvarez, M.M.; Serna-Saldivar, S.O. Induction of apoptosis in colon cancer cells treated with isorhamnetin glycosides from *Opuntia ficusindica* pads. *Plant Foods Hum. Nutr.* **2014**, *69*, 331–336. [CrossRef]

31. Lee, H.J.; Lee, H.J.; Lee, E.O.; Ko, S.G.; Bae, H.S.; Kim, C.H.; Ahn, K.S.; Lu, J.; Kim, S.H. Mitochondria-cytochrome *c*-caspase-9 cascade mediates isorhamnetin-induced apoptosis. *Cancer Lett.* **2008**, *270*, 342–353. [CrossRef]

32. Hu, S.; Huang, L.; Meng, L.; Sun, H.; Zhang, W.; Xu, Y. Isorhamnetin inhibits cell proliferation and induces apoptosis in breast cancer via Akt and mitogen-activated protein kinase kinase signaling pathways. *Mol. Med. Rep.* **2015**, *12*, 6745–6751. [CrossRef] [PubMed]

33. Choi, K.C.; Chung, W.T.; Kwon, J.K.; Yu, J.Y.; Jang, Y.S.; Park, S.M.; Lee, S.Y.; Lee, J.C. Inhibitory effects of quercetin on aflatoxin B1-induced hepatic damage in mice. *Food Chem. Toxicol.* **2010**, *48*, 2747–2753. [CrossRef] [PubMed]

34. Hibasami, H.; Mitani, A.; Katsuzaki, H.; Imai, K.; Yoshioka, K.; Komiya, T. Isolation of five types of flavonol from seabuckthorn (*Hippophae rhamnoides*) and induction of apoptosis by some of the flavonols in human promyelotic leukemia HL-60 cells. *Int. J. Mol. Med.* **2005**, *15*, 805–809. [CrossRef] [PubMed]

35. Prasain, J.K.; Rajbhandari, R.; Keeton, A.B.; Piazza, G.A.; Barnes, S. Metabolism and growth inhibitory activity of cranberry derived flavonoids in bladder cancer cells. *Food Funct.* **2016**, *7*, 4012–4019. [CrossRef]

36. Ebrahimi, S.; Hosseini, M.; Shahidsales, S.; Maftouh, M.; Ferns, G.A.; Ghayour-Mobarhan, M.; Hassanian, S.M.; Avan, A. Targeting the Akt/PI3K Signaling Pathway as a Potential Therapeutic Strategy for the Treatment of Pancreatic Cancer. *Curr. Med. Chem.* **2017**, *24*, 1321–1331. [CrossRef] [PubMed]

37. Hassan, M.; Watari, H.; AbuAlmaaty, A.; Ohba, Y.; Sakuragi, N. Apoptosis and molecular targeting therapy in cancer. *Biomed. Res. Int.* **2014**, *2014*, 150845. [CrossRef]

38. Bai, J.; Li, Y.; Zhang, G. Cell cycle regulation and anticancer drug discovery. *Cancer Biol. Med.* **2017**, *14*, 348–362.

39. Sánchez-Martínez, C.; Gelbert, L.M.; Lallena, M.J.; De Dios, A. Cyclin dependent kinase (CDK) inhibitors as anticancer drugs. *Bioorg. Med. Chem. Lett.* **2015**, *25*, 3420–3435. [CrossRef]

40. Matheson, C.J.; Backos, D.S.; Reigan, P. Targeting WEE1 kinase in cancer. *Trends Pharmacol. Sci.* **2016**, *37*, 872–881. [CrossRef]

41. Bulavin, D.V.; Demidenko, Z.N.; Phillips, C.; Moody, S.A.; Fornace, A.J., Jr. Phosphorylation of Xenopus Cdc25C at Ser285 interferes with ability to activate a DNA damage replication checkpoint in pre-midblastula embryos. *Cell Cycle* **2003**, *2*, 263–266. [CrossRef]

42. Karimian, A.; Ahmadi, Y.; Yousefi, B. Multiple functions of p21 in cell cycle, apoptosis and transcriptional regulation after DNA damage. *DNA Repair* **2016**, *42*, 63–71. [CrossRef] [PubMed]

43. Reinhardt, H.C.; Schumacher, B. The p53 network: Cellular and systemic DNA damage responses in aging and cancer. *Trends Genet.* **2012**, *28*, 128–136. [CrossRef] [PubMed]

44. Hydbring, P.; Malumbres, M.; Sicinski, P. Non-canonical functions of cell cycle cyclins and cyclin-dependent kinases. *Nat. Rev. Mol. Cell Biol.* **2016**, *17*, 280–292. [CrossRef] [PubMed]

45. Cooper, M.J.; Haluschak, J.J.; Johnson, D.; Schwartz, S.; Morrison, L.J.; Lippa, M.; Hatzivassiliou, G.; Tan, J. p53 mutations in bladder carcinoma cell lines. *Oncol. Res.* **1994**, *6*, 569–579. [PubMed]

46. Pfeffer, C.M.; Singh, A.T.K. Apoptosis: A target for anticancer therapy. *Int. J. Mol. Sci.* **2018**, *19*, 448. [CrossRef]

47. Kantari, C.; Walczak, H. Caspase-8 and bid: Caught in the act between death receptors and mitochondria. *Biochim. Biophys. Acta* **2011**, *1813*, 558–563. [CrossRef] [PubMed]

48. Edlich, F. BCL-2 proteins and apoptosis: Recent insights and unknowns. *Biochem. Biophys. Res. Commun.* **2018**, *500*, 26–34. [CrossRef]

49. Birkinshaw, R.W.; Czabotar, P.E. The BCL-2 family of proteins and mitochondrial outer membrane permeabilisation. *Semin. Cell Dev. Biol.* **2017**, *72*, 152–162. [CrossRef]

50. Kiraz, Y.; Adan, A.; Kartal Yandim, M.; Baran, Y. Major apoptotic mechanisms and genes involved in apoptosis. *Tumour Biol.* **2016**, *37*, 8471–8486. [CrossRef]

51. Schultz, D.R.; Harrington, W.J., Jr. Apoptosis: Programmed cell death at a molecular level. *Semin. Arthritis Rheum.* **2003**, *32*, 345–369. [CrossRef] [PubMed]

52. Ruan, Y.; Hu, K.; Chen, H. Autophagy inhibition enhances isorhamnetin-induced mitochondria-dependent apoptosis in non-small cell lung cancer cells. *Mol. Med. Rep.* **2015**, *12*, 5796–5806. [CrossRef] [PubMed]

53. Badrinath, N.; Yoo, S.Y. Mitochondria in cancer: In the aspects of tumorigenesis and targeted therapy. *Carcinogenesis* **2018**, *39*, 1419–1430. [CrossRef] [PubMed]

54. Moloney, J.N.; Cotter, T.G. ROS signalling in the biology of cancer. *Semin. Cell Dev. Biol.* **2018**, *80*, 50–64. [CrossRef] [PubMed]

55. Galadari, S.; Rahman, A.; Pallichankandy, S.; Thayyullathil, F. Reactive oxygen species and cancer paradox: To promote or to suppress? *Free Radic. Biol. Med.* **2017**, *104*, 144–164. [CrossRef] [PubMed]

56. Thirupathi, A.; De Souza, C.T. Multi-regulatory network of ROS: The interconnection of ROS, PGC-1 alpha, and AMPK-SIRT1 during exercise. *J. Physiol. Biochem.* **2017**, *73*, 487–494. [CrossRef]

57. Wu, S.B.; Wu, Y.T.; Wu, T.P.; Wei, Y.H. Role of AMPK-mediated adaptive responses in human cells with mitochondrial dysfunction to oxidative stress. *Biochim. Biophys. Acta* **2014**, *1840*, 1331–1344. [CrossRef] [PubMed]

58. Zou, M.H.; Kirkpatrick, S.S.; Davis, B.J.; Nelson, J.S.; Wiles, W.G., 4th; Schlattner, U.; Neumann, D.; Brownlee, M.; Freeman, M.B.; Goldman, M.H. Activation of the AMP-activated protein kinase by the anti-diabetic drug metformin in vivo. Role of mitochondrial reactive nitrogen species. *J. Biol. Chem.* **2004**, *279*, 43940–43951. [CrossRef]

59. Choi, S.L.; Kim, S.J.; Lee, K.T.; Kim, J.; Mu, J.; Birnbaum, M.J.; Soo Kim, S.; Ha, J. The regulation of AMP activated protein kinase by H_2O_2. *Biochem. Biophys. Res. Commun.* **2001**, *287*, 92–97. [CrossRef]

60. Wu, S.B.; Wei, Y.H. AMPK-mediated Increase of glycolysis as an adaptive response to oxidative stress in human cells: Implication of the cell survival in mitochondrial diseases. *Biochim. Biophys. Acta* **2012**, *1822*, 233–247. [CrossRef]

61. Cao, C.; Lu, S.; Jiang, Q.; Wang, W.J.; Song, X.; Kivlin, R.; Wallin, B.; Bagdasarian, A.; Tamakloe, T.; Chu, W.M.; et al. EGFR activation confers protections against UV-induced apoptosis in cultured mouse skin dendritic cells. *Cell Signal* **2018**, *20*, 1830–1838. [CrossRef]

62. Corton, J.M.; Gillespie, J.G.; Hardi, D.G. Role of the AMP-activated protein kinase in the cellular stress response. *Curr. Biol.* **1994**, *4*, 315–324. [CrossRef]
63. Tavazzi, B.; Di Pierro, D.; Amorini, A.M.; Fazzina, G.; Tuttobene, M.; Giardina, B.; Lazzarino, G. Energy metabolism and lipid peroxidation of human erythrocytes as a function of increased oxidative stress. *Eur. J. Biochem.* **2000**, *267*, 684–689. [CrossRef] [PubMed]
64. Chen, L.; Xu, B.; Liu, L.; Luo, Y.; Yin, J.; Zhou, H.; Chen, W.; Shen, T.; Han, X.; Huang, S. Hydrogen peroxide inhibits mTOR signaling by activation of AMPK alpha leading to apoptosis of neuronal cells. *Lab. Investig.* **2010**, *90*, 762–773. [CrossRef] [PubMed]
65. Vara-Ciruelos, D.; Russell, F.M.; Hardie, D.G. The strange case of AMPK and cancer: Dr Jekyll or Mr Hyde? *Open Biol.* **2019**, *9*, 190099. [CrossRef]
66. Vander Heiden, M.G.; Cantley, L.C.; Thompson, C.B. Understanding the Warburg effect: The metabolic requirements of cell proliferation. *Science* **2009**, *324*, 1029–1033. [CrossRef] [PubMed]
67. Kim, J.; Kundu, M.; Viollet, B.; Guan, K.L. AMPK and mTOR regulate autophagy through direct phosphorylation of Ulk1. *Nat. Cell Biol.* **2011**, *13*, 132–141. [CrossRef] [PubMed]
68. Egan, D.F.; Shackelford, D.B.; Mihaylova, M.M.; Gelino, S.; Kohnz, R.A.; Mair, W.; Vasquez, D.S.; Joshi, A.; Gwinn, D.M.; Taylor, R.; et al. Phosphorylation of ULK1 (hATG1) by AMP-activated protein kinase connects energy sensing to mitophagy. *Science* **2011**, *331*, 456–461. [CrossRef]
69. Wong, C.H.; Iskandar, K.B.; Yadav, S.K.; Hirpara, J.L.; Loh, T.; Pervaiz, S. Simultaneous induction of non-canonical autophagy and apoptosis in cancer cells by ROS-dependent ERK and JNK activation. *PLoS ONE* **2010**, *5*, e9996. [CrossRef] [PubMed]
70. Ci, Y.; Shi, K.; An, J.; Yang, Y.; Hui, K.; Wu, P.; Shi, L.; Xu, C. ROS inhibit autophagy by downregulating ULK1 mediated by the phosphorylation of p53 in selenite-treated NB4 cells. *Cell Death Dis.* **2014**, *5*, e1542. [CrossRef]
71. Park, C.; Jeong, N.Y.; Kim, G.Y.; Han, M.H.; Chung, I.M.; Kim, W.J.; Yoo, Y.H.; Choi, Y.H. Momilactone B induces apoptosis and G1 arrest of the cell cycle in human monocytic leukemia U937 cells through downregulation of pRB phosphorylation and induction of the cyclin-dependent kinase inhibitor p21Waf1/Cip1. *Oncol. Rep.* **2014**, *31*, 1653–1660. [CrossRef] [PubMed]
72. Koh, E.M.; Lee, E.K.; Song, C.H.; Song, J.; Chung, H.Y.; Chae, C.H.; Jung, K.J. Ferulate, an active component of wheat germ, ameliorates oxidative stress-induced PTK/PTP imbalance and PP2A inactivation. *Toxicol. Res.* **2018**, *34*, 333–341. [CrossRef] [PubMed]
73. Kim, D.Y.; Kim, J.H.; Lee, J.C.; Won, M.H.; Yang, S.R.; Kim, H.C.; Wie, M.B. Zinc oxide nanoparticles exhibit both cyclooxygenase- and lipoxygenase-mediated apoptosis in human bone marrow-derived mesenchymal stem cells. *Toxicol. Res.* **2019**, *35*, 83–91. [CrossRef] [PubMed]

Article

XIAP as a Target of New Small Organic Natural Molecules Inducing Human Cancer Cell Death

Diego Muñoz [1,2], Martina Brucoli [3], Silvia Zecchini [4], Adrian Sandoval-Hernandez [1], Gonzalo Arboleda [5], Fabian Lopez-Vallejo [1], Wilman Delgado [1], Matteo Giovarelli [4], Marco Coazzoli [4], Elisabetta Catalani [6], Clara De Palma [7], Cristiana Perrotta [4], Luis Cuca [1], Emilio Clementi [4,7,8] and Davide Cervia [6,*]

[1] Departamento de Química, Facultad de Ciencias, Universidad Nacional de Colombia, Bogotá 111321, Colombia; drmunozc@unal.edu.co (D.M.); agsandovalh@unal.edu.co (A.S.-H.); fhlopezv@unal.edu.co (F.L.-V.); wadelgadoa@unal.edu.co (W.D.); lecucas@unal.edu.co (L.C.)

[2] Facultad de Ciencias, Universidad de Ciencias Aplicadas y Ambientales, Bogotá 111166, Colombia

[3] Tumour Cell Death Laboratory, Cancer Research UK Beatson Institute, Glasgow G61 1BD, UK; 2310786B@student.gla.ac.uk

[4] Department of Biomedical and Clinical Sciences "Luigi Sacco" (DIBIC), Università degli Studi di Milano, 20157 Milano, Italy; silvia.zecchini@unimi.it (S.Z.); matteo.giovarelli@unimi.it (M.G.); marco.coazzoli@unimi.it (M.C.); cristiana.perrotta@unimi.it (C.P.); emilio.clementi@unimi.it (E.C.)

[5] Grupo de Muerte Celular, Instituto de Genética, Universidad Nacional de Colombia, Bogotá 111321, Colombia; gharboledab@unal.edu.co

[6] Department for Innovation in Biological, Agro-food and Forest systems (DIBAF), Università degli Studi della Tuscia, 01100 Viterbo, Italy; ecatalani@unitus.it

[7] Unit of Clinical Pharmacology, University Hospital "Luigi Sacco"-ASST Fatebenefratelli Sacco, 20157 Milano, Italy; clara.depalma@asst-fbf-sacco.it

[8] Scientific Institute IRCCS "Eugenio Medea", 23842 Bosisio Parini, Italy

* Correspondence: d.cervia@unitus.it; Tel.: +39-0761-357040

Received: 27 August 2019; Accepted: 4 September 2019; Published: 9 September 2019

Abstract: X-linked inhibitor of apoptosis protein (XIAP) is an emerging crucial therapeutic target in cancer. We report on the discovery and characterisation of small organic molecules from *Piper* genus plants exhibiting XIAP antagonism, namely erioquinol, a quinol substituted in the 4-position with an alkenyl group and the alkenylphenols eriopodols A–C. Another isolated compound was originally identified as gibbilimbol B. Erioquinol was the most potent inhibitor of human cancer cell viability when compared with gibbilimbol B and eriopodol A was listed as intermediate. Gibbilimbol B and eriopodol A induced apoptosis through mitochondrial permeabilisation and caspase activation while erioquinol acted on cell fate via caspase-independent/non-apoptotic mechanisms, likely involving mitochondrial dysfunctions and aberrant generation of reactive oxygen species. In silico modelling and molecular approaches suggested that all molecules inhibit XIAP by binding to XIAP-baculoviral IAP repeat domain. This demonstrates a novel aspect of XIAP as a key determinant of tumour control, at the molecular crossroad of caspase-dependent/independent cell death pathway and indicates molecular aspects to develop tumour-effective XIAP antagonists.

Keywords: phytochemicals; small organic agents; *Piper eriopodon*, alkenylphenols; human cancer cells; cell death; apoptosis; caspase-independent cell death; XIAP antagonists; XIAP-BIR3 domain

---

1. Introduction

The characterisation of small molecules (whose molecular weight does not exceed 900 Daltons) with well-defined chemical structures is a good approach to develop new therapeutic agents in proliferative, infectious, or neurodegenerative disorders [1–5]. Natural products possess enormous structural and chemical diversity that cannot be matched by any synthetic libraries of small molecules and continue to show a great translational potential [6–10]. In some cases, the complex chemical composition of some natural products has made difficult their isolation, structure elucidation and characterisation, thus prompting the search of new efficient synthetic pathways. In recent years the interest in the fundamental understanding of natural products and their engineered variants has been strongly renewed [6].

The simple active chemical structures of phenolic compounds from plants make them optimal lead candidates because of their broad biological activity, especially the protective, anti-oxidant and anti-tumour effects [11–14]. Plants of the genus *Piper* (Piperaceae family), are a very common food resource in neotropical forests and are widely used to obtain culinary spices. *Piper* genus constitutes one major class of medicinal plants and contains a valuable resource of phenolic bioactive compounds [15–21]. Among them, piplartine, hydroxychavicol, 4-nerodlidylcatechol and gibbilimbols A–D displayed potent cytotoxic/anti-tumoural effects in a variety of human cancer cells in vitro and in vivo [19,22–29].

Apoptosis, a closely regulated programmed cell death mechanism, is an essential process to maintain tissue homeostasis and its escape it is one of the hallmarks of cancer [30]. Substantial advances have been made on apoptosis-based anti-cancer therapeutics [31]. The most potent human IAP currently identified is the X-linked inhibitor of apoptosis protein (XIAP), a 57 kDa protein with three zinc-binding baculovirus IAP repeat (BIR) domains (BIR 1–3) which may also have actions additional to regulation of apoptosis [32]. The anti-apoptotic function of XIAP is antagonised by the second mitochondria-derived activator of caspases or direct IAP binding protein with low pI (Smac/DIABLO), a mitochondria protein released during apoptosis. The key role of XIAP and its potential clinical relevance is well established in tumours and several XIAP inhibitors have been developed or discovered as cytotoxic agents [32–43]. Despite different small molecules that inhibit XIAP have been identified and are moving through the pipeline of clinical development, the need of new ones to refine further therapeutic approaches based on XIAP antagonism is undeniable in translational research [41].

Herein we wish to report the discovery and chemical/biological characterisation of novel natural small compounds from *Piper* genus. Furthermore, a deeper insight into their cell death mechanism in human cells provides a proof-of-concept study of their pharmaceutical potential as antagonists of XIAP that may open important insights on XIAP as a suitable turning point for multiple cellular pathways.

2. Results and Discussion

2.1. Structural Identification of New Piper Genus-Derived Compounds

The chemical structures of compounds isolated from leaves of *P. eriopodon* (Figure 1A) were identified by interpretation of their corresponding high resolution electrospray ionisation mass spectrometry (HRESIMS), ^{1}H- and ^{13}C-NMR (nuclear magnetic resonance) spectral data, including attached proton test (APT), correlated spectroscopy (COSY), heteronuclear multiple quantum coherence (HMQC) and heteronuclear multiple bond correlation (HMBC) experiments, as well as by comparison of the spectral data with those reported in the literature.

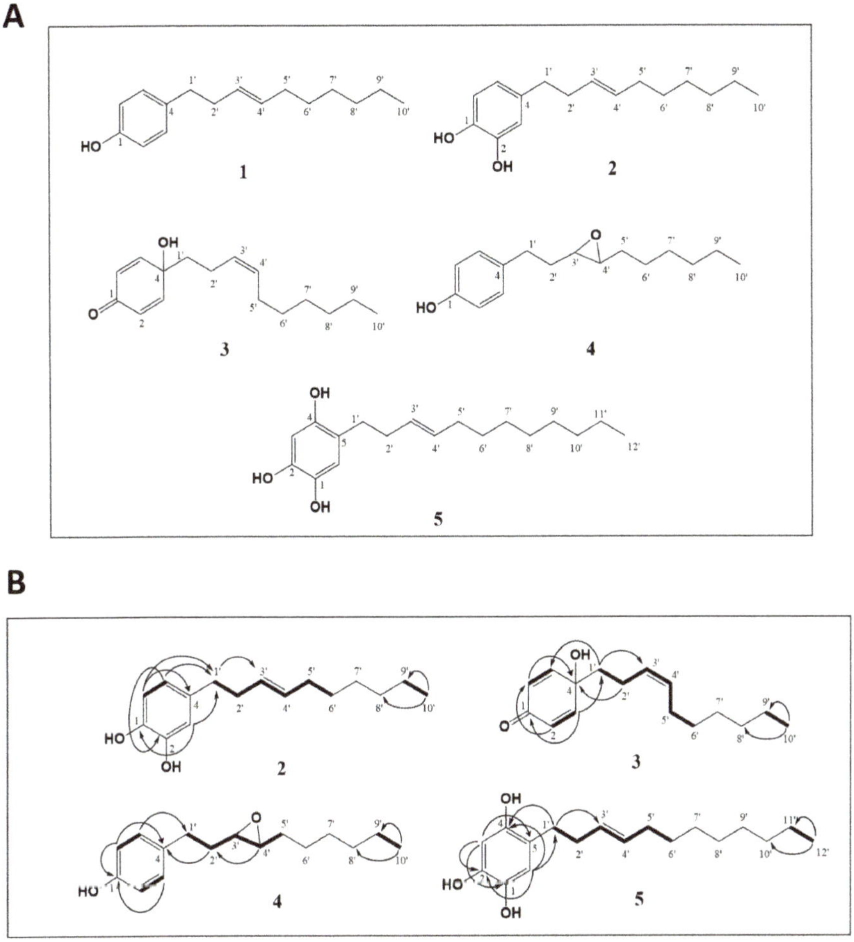

Figure 1. Identification of new *Piper* genus-derived compounds. (**A**) Structures of compounds 1–5. (**B**) Key correlated spectroscopy (COSY) (bold) and heteronuclear multiple bond correlation (HMBC) (H→C) for compounds **2–5**.

Compound 1 (Figure S1, Tables S1 and S2) was obtained as colorless oil and identified unequivocally as gibbilimbol B ((*E*)-4-(dec-3′-enyl)phenol) [19].

Compound 2 (Figure S2, Tables S1 and S2) was obtained as pale yellow oil. The molecular formula for compound **2** was established as $C_{16}H_{24}O_2$ based on the HRESIMS peak at m/z 247.1706 [M-H]⁻ (calcd. 247.1703). The ¹H- NMR spectrum showed clear signals for a 1,2,4-trisubstituted aromatic ring δH 6.77 (1H, d, *J* = 7.6 Hz, H-6), 6.71 (1H, s, H-3), 6.60 (1H, d, *J* = 7.5 Hz, H-5) and an alkenyl fragment. The ¹³C-NMR spectrum showed ten signals, practically the same as the alkenyl chain of gibbilimbol B, including the double bond position in C-3′, which was confirmed by correlations observed in both COSY and HMBC experiments (Figure 1B). Based on the ¹³C-NMR chemical shifts of the allylic carbons δ_C 34.6 (C-2′) and δ_C 32.6 (C-5′), the configuration of the double bond for compound 2 was assigned as *E* [18], by comparison with the ¹³C-NMR chemical shift of the allylic carbons in the *E* analogue gibbilimbol B (δ_C 34.6 (C-2′) and δ_C 32.6 (C-5′)), which differed significantly from the chemical shift values reported for the *Z* analogue climacostol [δ_C 33.2 (C-1′) and δ_C 27.3 (C-4′)] [44].

Thus, the chemical structure of compound **2** was elucidated as (*E*)-4-(dec-3'-enyl)benzene-1,2-diol and it was given the common name of eriopodol A.

Compound **3** (Figure S3, Tables S1 and S2) was obtained as clear oil and its molecular formula was deduced as $C_{16}H_{24}O_2$ from the HRESIMS spectrum, which exhibited a molecular ion peak at *m/z* 247.1706 [M-H]$^-$ (calcd. 247.1703). The ^{1}H-NMR spectrum for compound 3 showed signals for an alkenyl chain and two signals in δH 6.11 (2H, d, *J* = 9.94 Hz) and 6.81(2H, d, *J* = 9.96 Hz). The ^{13}C-NMR spectrum for compound **3** showed signals for an α-β unsaturated carbonyl in $δ_C$ 185.9, an oxygenated quaternary carbon in $δ_C$ 69.6 and ten signals for the typical side chain of the alkenyl fragment. Based on the correlations observed in COSY and HMBC experiments (Figure 1B), the structure of 3 was determined as a quinol derivative, substituted in the 4-position with an alkenyl group. The position and geometry of the double bond of compound 3 was assigned by comparing the chemical shift values of the allylic carbons $δ_C$ 32.4 (C-2') and $δ_C$ 26.6 (C-5') as explained above for eriopodol A. The geometry of compound 3 was determined as *Z* and its chemical structure was elucidated as (*Z*)-4-(dec-3'-enyl)-4-hydroxycyclohexa-2,5-dien-1-one. The common name of erioquinol was then assigned.

Compound **4** (Figure S4, Tables S1 and S2) was obtained as pale yellow oil. The molecular formula for compound 4 was confirmed to be $C_{16}H_{24}O_2$ based on the HRESIMS peak at *m/z* 247.1715 [M-H]$^-$ (calcd. 247.1703). The ^{1}H-NMR and ^{13}C-NMR spectra of compound 4 showed almost the same chemical shifts as the alkenylphenol gibbilimbol B, but without the unsaturated signal in the ^{1}H-NMR spectrum. Therefore, the carbons C-3' and C-4' showed chemical shifts in $δ_C$ 59.6 (C-3') and $δ_C$ 58.6 (C-4'), corresponding two oxygenated methines from an epoxide group, which was confirmed with COSY and HMBC experiments (Figure 1B). The structure of compound **4** was elucidated as 4-(3',4'-epoxydecenyl)phenol and the common name of eriopodol B was assigned.

Compound **5** (Figure S5, Tables S1 and S2) was obtained as pale yellow amorphous solid (m.p. 138.5 °C). The molecular formula for compound **5** was established as $C_{18}H_{28}O_3$ based on the HRESIMS peak at *m/z* 291.1973 [M-H]$^-$ (calcd. 291.1966). The NMR data for compound **5** were very close to those of eriopodol A, although it contains one additional hydroxyl group in the benzene ring and two additional carbons at the end of the alkenyl chain (Figure 1B). The position and geometry of the double bound for compound **5** was assigned as explained above. The structure of compound **5** was elucidated as (*E*)-5-(dodec-3'-enyl)benzene-1,2,4-triol and the common name of eriopodol C was assigned.

Taken together, phytochemical investigation of leaves from P. eriopodon yielded four new alkenyl derivatives and one known compound. In particular, erioquinol is a new quinol substituted in the 4-position with an alkenyl group and eriopodols A-C correspond to new alkenylphenols. The known isolated compound was originally identified as gibbilimbol B, from the medicinal plant P. gibbilimbum [19] and, more recently, from P. malacophyllum [21] and P. eriopodon [29]. The simple chemical structure of alkenylphenols are characterised by hydroxylated benzenes, substituted by side alkyl chains of different lengths with at least one double bond, generally with E geometry. Alkenylphenols with different reported biological properties, such as antibacterial, anti-parasitic, anti-inflammatory and cytotoxic activities, are widely found in the Piper genus [17,19–21,45]. Quinols are 4-hydroxycyclohexa-2,5-dien-1-ones which rarely occur as derivatives of some natural products [46–48]. An important feature of quinols substituted in the 4-position with aryl groups, is that they represent a class of potent anti-tumour molecules with activities against colon, renal, and breast cancer cells [49–51].

2.2. Piper Genus-Derived Compounds Exhibit Cytotoxic Effects

Several recent studies in glioblastoma and breast cancer cells have reported that extracts or active compounds isolated from *Piper* genus possess anti-tumoural/pro-apoptotic properties [52–61]. In order to assess whether the compounds we isolated could be developed further for therapeutic applications, we tested their cytotoxic action in the human cancer cells, U373 (glioblastoma astrocytoma) and MCF7 (breast adenocarcinoma) cell lines, since they are widely used as suitable in vitro models of cancer research. We first examined the effects of gibbilimbol B, eriopodols A–C, and erioquinol on cell viability. Gibbilimbol B was used as a reference compound of *Piper* genus derivatives, since its cytotoxic action has been previously tested in various tumour cells, including MCF7 [19,29]. In our experiments, cell viability was analysed by 3-(4,5-dimethylthiazol-2-yl)-2,5-diphenyltetrazolium bromide (MTT) assay after treatment with previously mentioned compounds at increasing concentrations for 24 h. As shown in Figure 2, a concentration-dependent inhibition of MTT absorbance was observed for all compounds with an IC_{50} (the concentration producing half the maximum inhibition) ranging from 1.78 to 31.91 µg/mL; the rank order of potencies was: erioquinol > eriopodol A > eriopodol C > gibbilimbol B > eriopodol B and erioquinol > eriopodol A > eriopodol C/gibbilimbol B > eriopodol B for U373 and MCF7 cells, respectively (Table 1). Their effects were maximal (E_{max}—concentration producing the maximum effect—nearly 100% inhibition) between 10–100 µg/mL.

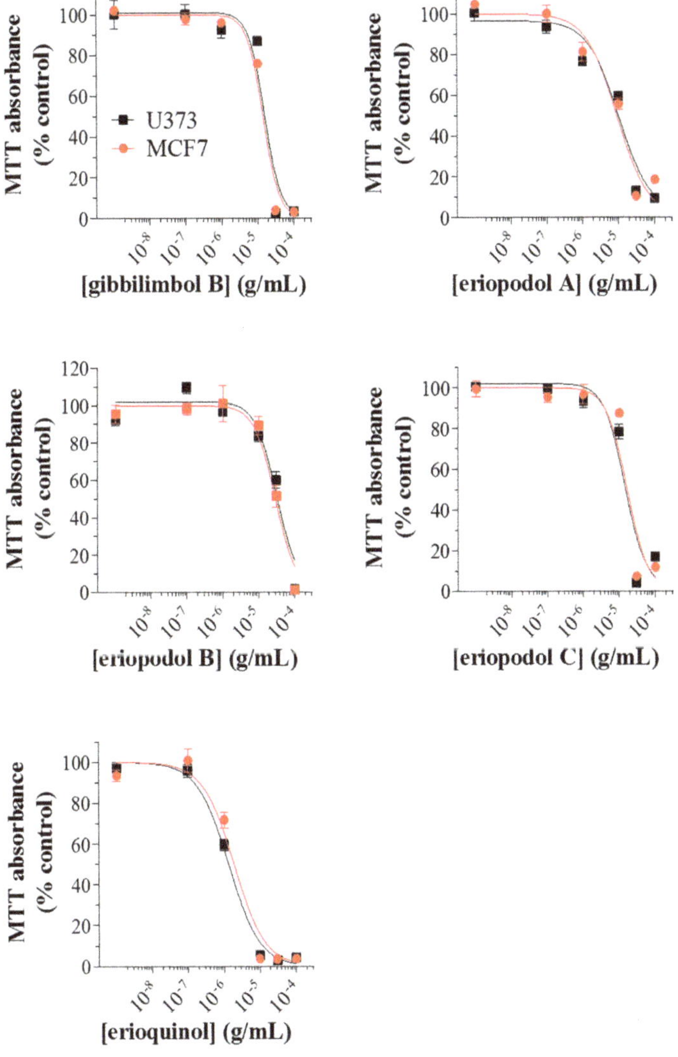

Figure 2. *Piper* genus-derived compounds exhibit cytotoxic effects in human cancer cells. U373 and MCF7 cells were treated with increasing concentrations of gibbilimbol B, eriopodol A, eriopodol B, eriopodol C,

and erioquinol for 24 h before 3-(4,5-dimethylthiazol-2-yl)-2,5-diphenyltetrazolium bromide (MTT) assay. Results are expressed by setting the absorbance of the reduced MTT in the respective control (vehicle-treated) samples, i.e., absence of compounds, as 100%. The data points are representative of four independent experiments.

Table 1. Inhibitory effects of *Piper* genus-derived compounds on human cancer cell viability.

Compound	IC$_{50}$ (µg/mL)	
	U373 Cells	**MCF7 Cells**
Gibbilimbol B	16.79	16.44
Eriopodol A	11.12	10.12
Eriopodol B	31.91	29.36
Eriopodol C	14.30	16.30
Erioquinol	1.78	2.63

3-(4,5-dimethylthiazol-2-yl)-2,5-diphenyltetrazolium bromide (MTT) assay was performed treating cells for 24 h in the absence (vehicle) or in the presence of increasing concentrations of *Piper* genus-derived compounds. The results have been obtained in four independent experiments.

Eriopodol A and erioquinol were selected for further investigation, as they displayed the most potent inhibitory effects on cell viability. Gibbilimbol B (available in high quantity) was also included. When compared with gibbilimbol B [29], the higher cytotoxic effect of eriopodol A and erioquinol (24 h), was also shown by MTT assays using additional cell lines, like human A549 lung (IC$_{50}$ of eriopodol A and erioquinol: 6.12 and 2.65 µg/mL, respectively) and PC-3 prostate (IC$_{50}$ of eriopodol A and erioquinol: 1.84 and 2.21 µg/mL, respectively) cancer cells, further confirming enhanced pharmacological activity of these new *Piper* genus derivatives (Figure 3A).

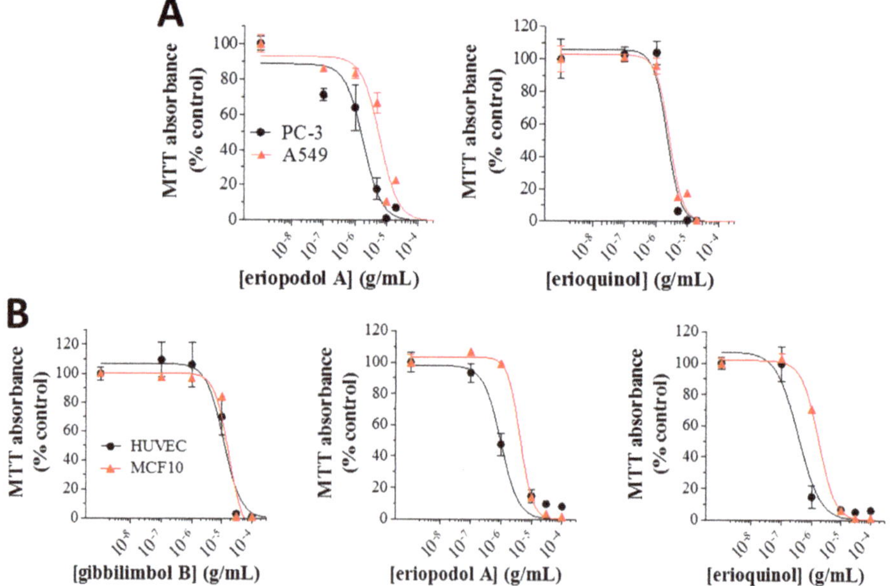

Figure 3. *Piper* genus-derived compounds exhibit cytotoxic effects in cancer and non-transformed human cells. (**A**) PC-3/A549 cells were treated with increasing concentrations of eriopodol A and erioquinol while

(**B**) human umbilical vein endothelial cells (HUVEC)/MCF10 cells were treated with increasing concentrations of gibbilimbol B, eriopodol A, and erioquinol, for 24 h before 3-(4,5-dimethylthiazol-2-yl)-2,5-diphenyltetrazolium bromide (MTT) assay. Results are expressed by setting the absorbance of the reduced MTT in the respective control (vehicle-treated) samples, i.e., absence of compounds, as 100%. The data points are representative of four independent experiments.

Similar results were obtained in human umbilical vein endothelial cells (HUVEC) (IC$_{50}$ of 24 h gibbilimbol B, eriopodol A, and erioquinol: 11.49, 0.99, and 0.36 µg/mL, respectively) and the non-tumourigenic human breast MCF10 cells (IC$_{50}$ of 24 h gibbilimbol B, eriopodol A, and erioquinol: 17.11, 4.27, and 1.70 µg/mL, respectively) (Figure 3B). The fact that the potency of the compounds was even slightly higher in these non-transformed/high proliferating cells suggests that their effects are not necessarily correlated to the cancerous origin of cells, in agreement with other small molecules we have recently characterised [62]. On the other hand, many cytotoxic compounds, including chemotherapy agents, are specifically designed to primarily affect rapidly proliferating cells, and many "normal" cells are also highly proliferative, such as cells in the bone marrow. The possibility that *Piper* genus-derived compounds preferentially affect high proliferating vs. low proliferating cells remains to be elucidated.

We then measured the concentration-dependent inhibition of MTT absorbance at increasing times of exposure in MCF7 cells, used as reference cell line. Our results indicated that the potency of gibbilimbol B did not substantially change (IC$_{50}$-6 h: 20.31 µg/mL; 12 h: 27.36 µg/mL; 24 h: 16.44 µg/mL) while the potency of eriopodol A increased at 24 h (IC$_{50}$ - 6 h: 31.19 µg/mL; 12 h: 32.75 µg/mL; 24 h: 11.13 µg/mL) (Figure 4). Of interest, the potency of erioquinol was greater than gibbilimbol B and eriopodol A at each time-point, even increasing over time (IC$_{50}$-6 h: 14.72 µg/mL; 12 h: 4.25 µg/mL; 24 h: 1.93 µg/mL). These comparative data indicate that erioquinol is the most potent compound with faster kinetics when compared with gibbilimbol B; eriopodol A has a somewhat intermediate behavior.

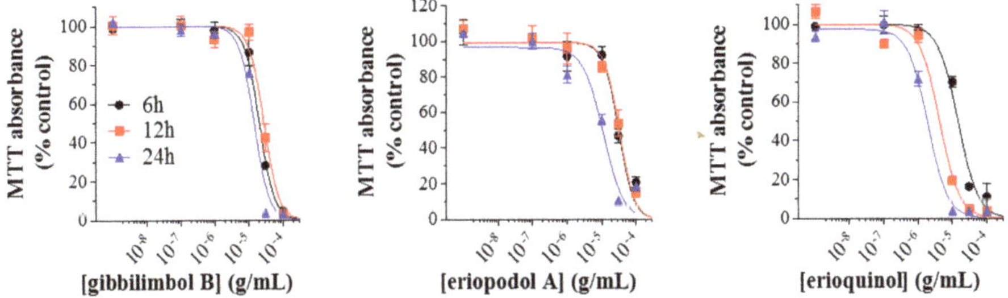

Figure 4. Time-response of *Piper* genus-derived compounds on cell viability. MCF7 cells were treated with increasing concentrations of gibbilimbol B, eriopodol A, and erioquinol for 6, 12, and 24 h before 3-(4,5-dimethylthiazol-2-yl)-2,5-diphenyltetrazolium bromide (MTT) assay. Results are expressed by setting the absorbance of the reduced MTT in the respective control (vehicle-treated) samples, i.e., absence of compounds, as 100%. The data points are representative of four independent experiments.

2.3. *Piper Genus-Derived Compounds Induce Cell Death*

MCF7 cells treated for 12 h with gibbilimbol B and eriopodol A (30 µg/mL) showed an inter-nucleosomal degradation of genomic DNA typical of late apoptotic cells, as determined by a terminal deoxynucleotidyl transferase dUTP nick end labeling (TUNEL) assay (Figure 5A), while DNA fragmented cells were few following erioquinol (10 µg/mL) treatment. Bright field microscopy demonstrated that cells exposed to increasing concentrations of gibbilimbol B and eriopodol A at 6 h (a temporal window sufficient to determine their cytotoxic effects) had morphological hallmarks of apoptosis, such as progressive roundness, shrunken cytoplasm and the formation of condensed nuclei (Figure 5B). In contrast, cells treated with erioquinol displayed a translucent cytoplasm and no overall nuclei condensation. Of interest, 4′,6-diamidine-2′-phenylindole dihydrochloride (DAPI) staining clearly revealed the nuclei of cells undergoing apoptosis in the presence of gibbilimbol B and eriopodol A (30 µg/mL) for 6 h, while erioquinol (10 µg/mL) treatment was associated with the appearence of multinucleated cells (Figure 5C). Accordingly, when analysed by flow cytometry using Annexin V and propidium iodide (PI) staining, MCF7 cells treated with erioquinol showed a progressive and marked increase of membrane disruption, as shown by early positivity to both Annexin V and PI staining, while the typical early apoptotic pattern, evidenced as Annexin V$^+$/PI$^-$ was almost undetectable over time (Figure 5D).

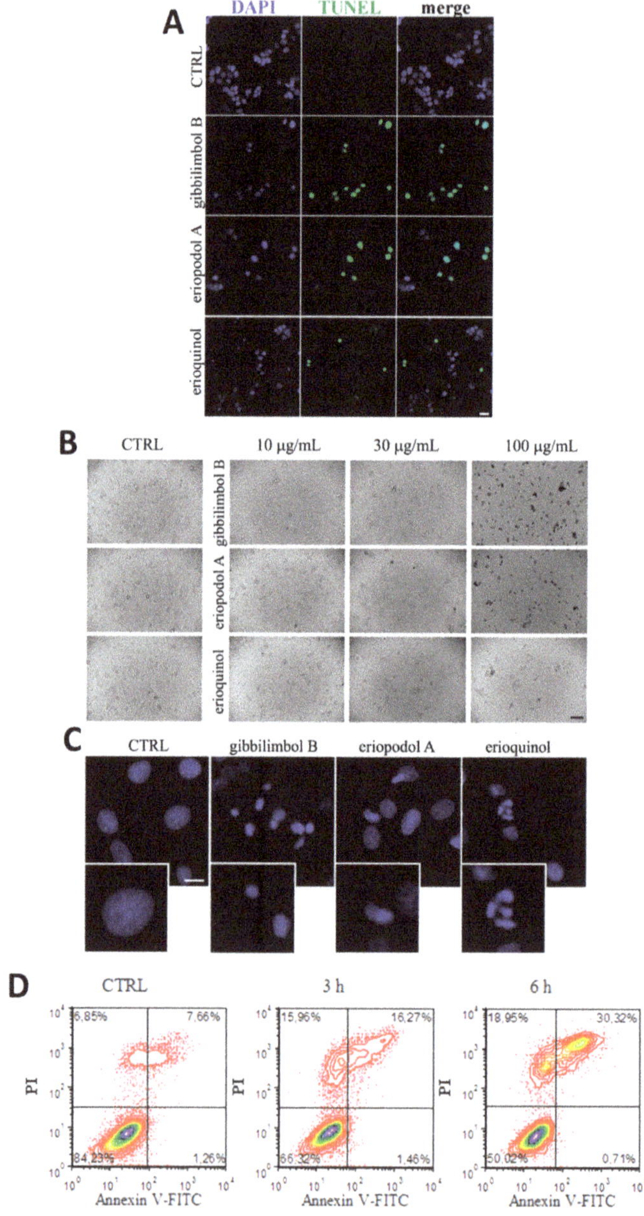

Figure 5. *Piper* genus-derived compounds induce cell death. (**A**) terminal deoxynucleotidyl transferase

dUTP nick end labeling (TUNEL) staining of MCF7 cells treated for 12 h in the absence (CTRL, control) and in the presence of gibbilimbol B/eriopodol A (30 μg/mL) or erioquinol (10 μg/mL). 4′,6-diamidine-2′-phenylindole dihydrochloride (DAPI) was used for nuclei detection. Scale bar = 50 μm. (**B**) Bright field microscopy of MCF7 cells treated for 6 h in the absence (CTRL) and in the presence of gibbilimbol B, eriopodol A, or erioquinol at increasing concentrations. Scale bar = 100 μm. (**C**) 4′,6-diamidine-2′-phenylindole dihydrochloride (DAPI) staining of MCF7 cells treated for 6 h in the absence (CTRL) and in the presence of gibbilimbol B/eriopodol A (30 μg/mL) or erioquinol (10 μg/mL). Scale bar = 10 μm. Lower panels represent enlarged image details. (**D**) Evaluation by flow cytometry of Annexin V-fluorescein isothiocyanate(FITC)/propidium iodide (PI) staining in MCF7 cells treated in the absence (CTRL) and in the presence of 10 μg/mL erioquinol, for 3 and 6 h. Quadrants are drawn, and relative proportion of labelled cells is indicated. The events shown in the lower left-hand quadrant are unlabeled cells. Images and data are representative of four independent experiments.

In addition, cells treated with 30 μg/mL gibbilimbol B and eriopodol A displayed activation of caspase 9 and 7 at 3 h and 6 h, as showed by western blot analysis (Figure 6A,B). On the other hand, erioquinol (10 μg/mL) treated cells did not display any sign of caspase 7 activity even at later time point (Figure 6C). These results were confirmed by immunofluorescence experiments. Indeed, a time-dependent and intensive cleaved-caspase 7 staining was detected in the cytoplasm of MCF7 cells in the presence of gibbilimbol B and eriopodol A while positive cells were absent following the administration of erioquinol (Figure 6D). The activation of caspase 7 by 6 h gibbilimbol B and eriopodol A (30 μg/mL) but not erioquinol (10 μg/mL) was achieved also in U373 cells (Figure 7A). The fact that these cells displayed apoptotic and non-apoptotic features in the presence of gibbilimbol B/eriopodol A and erioquinol, respectively (Figure 7B), similarly to what obtained in MCF7 cells, indicate that cell death mechanisms of the compounds are comparable among cell lines. Accordingly, the activation of caspase 7 by gibbilimbol B and eriopodol A but not erioquinol was observed also in MCF10 cells (Figure S6A).

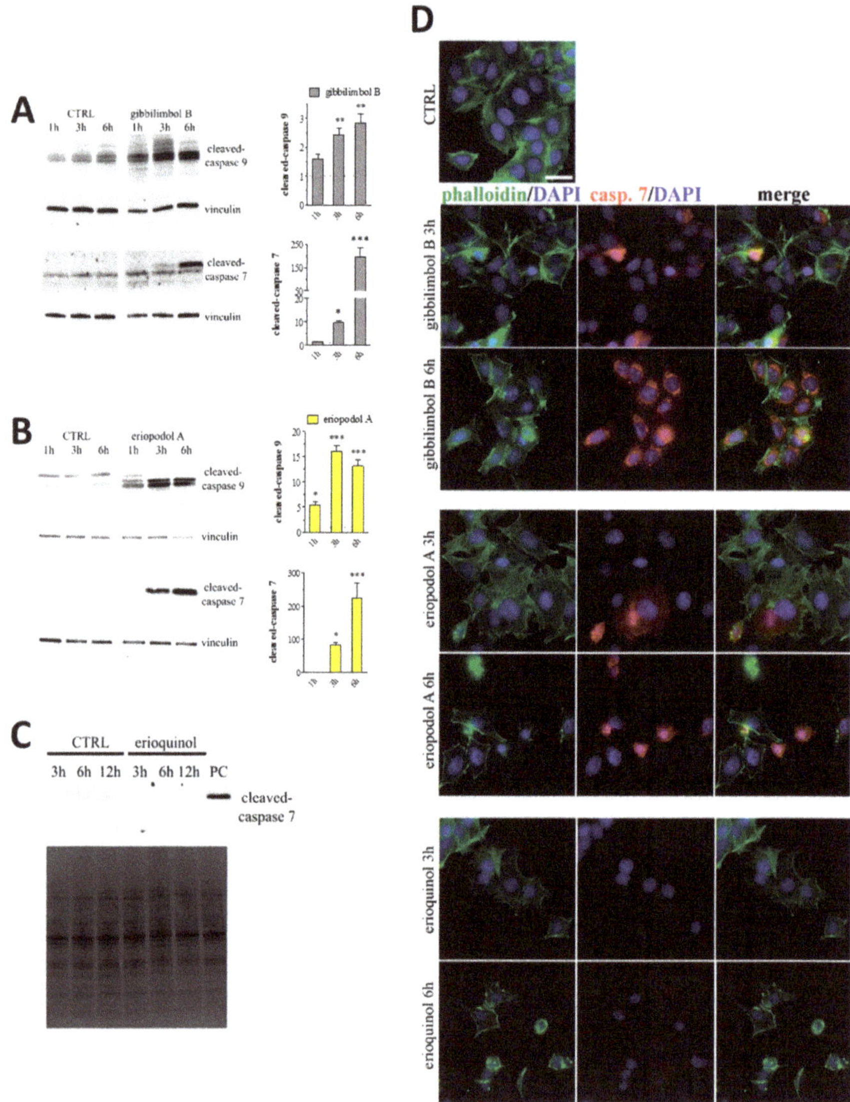

Figure 6. *Piper* genus-derived compounds induce cell death. Western blot analysis of cleaved-caspase 9 and 7 in MCF7 cells treated for increasing times in the absence (CTRL, control) and in the presence of (**A**) gibbilimbol B or (**B**) eriopodol A. Vinculin was used as internal standard. Right panels: densitometric analysis expressed as fold change of CTRL. Images and data are representative of three-five independent experiments. * $p < 0.01$, ** $p < 0.001$, and *** $p < 0.0001$ relative to CTRL. (**C**) Western blot analysis of cleaved-caspase 7 in MCF7 cells treated for increasing times in the absence (CTRL) and in the presence of 10 µg/mL erioquinol. The stain-free gel was used as loading control. Images are representative of three independent experiments. PC: positive control. (**D**) Immunofluorescence imaging of cleaved-caspase 7 (punctate red pattern) in MCF7 cells treated for 3 and 6 h in the absence (CTRL) and in the presence of gibbilimbol B/eriopodol A (30 µg/mL) or erioquinol (10 µg/mL). 4′,6-diamidine-2′-phenylindole dihydrochloride (DAPI) (blue) and phalloidin (green) were used for nuclei and cytoskeleton detection, respectively. Images are representative of four independent experiments. Scale bar = 25 µm.

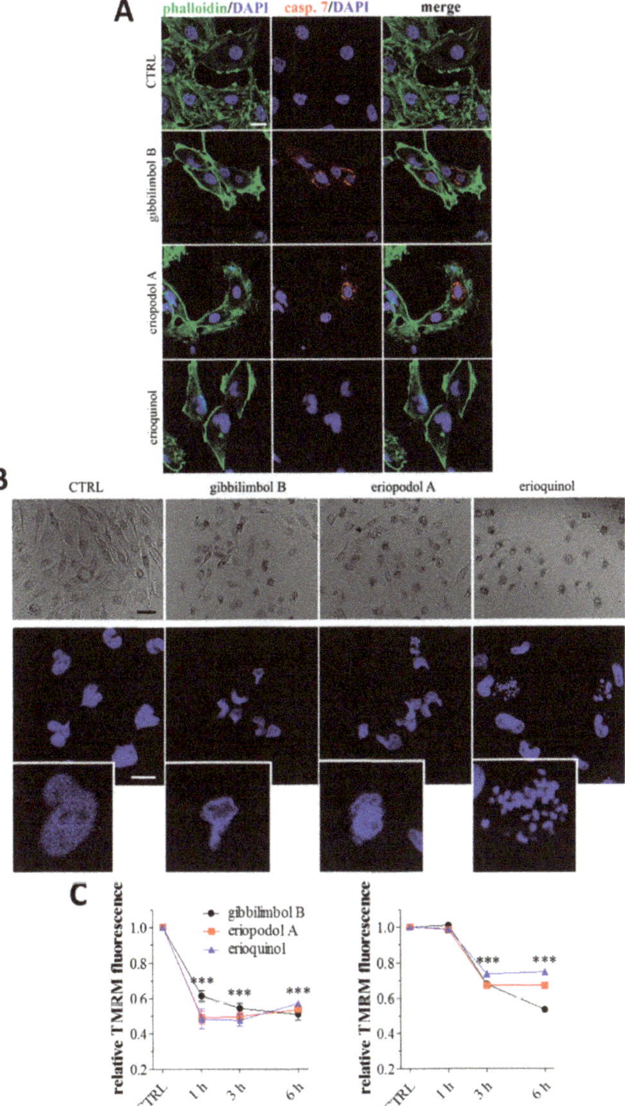

Figure 7. *Piper* genus-derived compounds induce cell death and mitochondrial dysfunction. (**A**) Immunofluorescence (confocal) imaging of cleaved-caspase 7 (punctate red pattern) in U373 cells treated for 6 h in the absence (CTRL, control) and in the presence of gibbilimbol B/eriopodol A (30 µg/mL) or erioquinol (10 µg/mL). 4′,6-diamidine-2′-phenylindole dihydrochloride (DAPI) (blue) and phalloidin (green) were used for nuclei and cytoskeleton detection, respectively. Scale bar = 25 µm. (**B**) Bright field microscopy (upper panels) and DAPI staining (lower panels) of U373 cells treated for 6 h in the absence (CTRL) and in the presence of of gibbilimbol B/eriopodol A (30 µg/mL) or erioquinol (10 µg/mL). Scale bars = 50 µm (bright field) and 10 µm (DAPI). Lower panels represent enlarged image details. (**C**) Quantitative analysis of tetramethylrhodamine methyl ester (TMRM) fluorescence changes over time in MCF7 (left panel) and U373 (righ panel) cells in the absence (CTRL) and in the presence of gibbilimbol B/eriopodol A (30 µg/mL) or erioquinol (10 µg/mL). Results are expressed by setting TMRM fluorescence in the respective control (vehicle-treated) samples, i.e., absence of compounds, as 1. *** $p < 0.0001$ relative to CTRL. Images and data are representative of four independent experiments.

In order to better describe the mechanism behind the activity of the compounds, we investigated mitochondria functionality with tetramethylrhodamine methyl ester (TMRM), a red fluorescent dye that is sequestered by active mitochondria. Of note, MCF7 and U373 cells treated for increasing time with 30 µg/mL gibbilimbol B/eriopodol A or 10 µg/mL erioquinol presented a comparable decrease in TMRM fluorescence vs. control, with MCF7 cells full responding within 1 h (Figure 7C). This indicates low mitochondria membrane potential likely associated to the destabilisation of the mitochondrial membrane systems.

The fact that the three compounds similarly induce mitochondria membrane permeabilisation both in MCF7 and U373 cells, was further confirmed by the subcellular location of cytochrome c. As shown in Figure 8A,B, 3 h administration of gibbilimbol B/eriopodol/erioquinol induced an alteration in the cytochrome c staining pattern from mitochondrial (co-localisation with COX IV, a marker for mitochondria), to a more cytosolic distribution (presence of many clusters which did not overlap with COX IV), indicating a release of cytochrome c from the dysfunctional mitochondria.

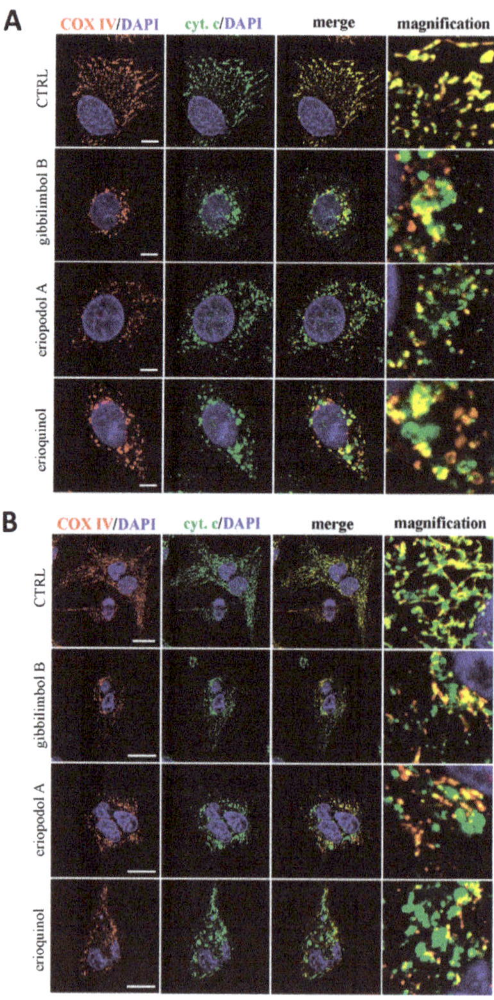

Figure 8. Confocal microscopy for co-localisation of cytochrome c with mitochondria. (**A**) MCF7 and

(**B**) U373 cells were treated for 3 h in the absence (CTRL, control) and in the presence of gibbilimbol B/eriopodol A (30 μg/mL) or erioquinol (10 μg/mL). Cells were then stained for cytochrome c (green) and the mitochondrial marker COX IV). 4′,6-diamidine-2′-phenylindole dihydrochloride (DAPI) (blue) was used for nuclei detection. The images are representative of three independent experiments. Scale bars: 10 μm (MCF7) and 25 μm (U373). Panels on the right represent enlarged image details.

Damaged mitochondria are considered as the main source of reactive oxygen species (ROS) which play major roles in the fate of cancer cells [63]. Noteworthily, MCF7 and U373 cells staining with 2′-7′dichlorofluorescin diacetate (DCFH-DA), a permeant fluorogenic dye cell reagent that measures hydroxyl, peroxyl and other ROS activity, revealed that erioquinol effect (10 μg/mL, 6 h) is characterised by marked accumulation of ROS, which are absent in cells treated with gibbilimbol B and eriopodol A (30 μg/mL, 6 h) (Figure 9A,B). Together with lack of caspase activation, aberrant ROS production is another divergence between gibbilimbol B/eriopodol A and erioquinol-induced cell death. In this respect, erioquinol is likely inducing a robust mitochondrial stress which results in ROS production and release into the cytoplasm.

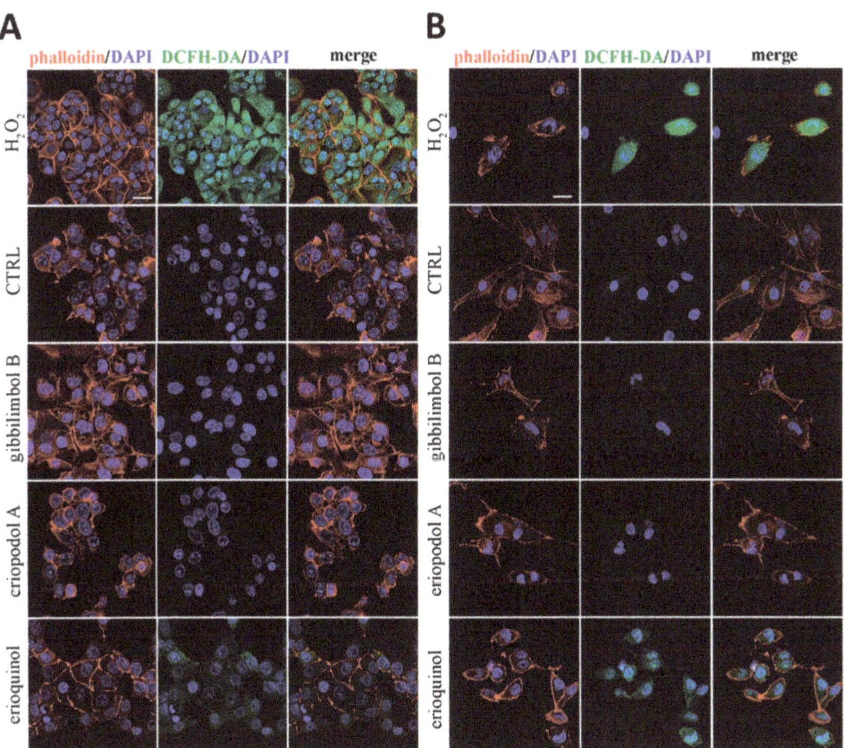

Figure 9. Confocal microscopy for reactive oxygen species (ROS) detection. (**A**) MCF7 and (**B**) U373 cells were treated for 6 h in the absence (CTRL, control) and in the presence of gibbilimbol B/eriopodol A (30 μg/mL) or erioquinol (10 μg/mL). Cells were then stained for ROS (2′-7′dichlorofluorescin diacetate - DCFH-DA, green). 4′,6-diamidine-2′-phenylindole dihydrochloride (DAPI) (blue) and phalloidin (red) were used for nuclei and cytoskeleton detection, respectively. The images are representative of three independent experiments. Scale bar: 25 μm.

Finally we confirmed as apoptotic the effect of gibbilimbol B and eriopodol A by inhibiting their cytotoxic activity with the pan-caspase inhibitor Z-VAD-(OMe)-FMK. As displayed by MTT assays (Figure 10A), the loss of cell viability in MCF7 cells treated with 30 μg/mL gibbilimbol B and eriopodol A was significantly inhibited when 50 µM Z-VAD-(OMe)-FMK was simultaneously added to the 6 h treatment protocol, demonstrating the dependency on caspases of the two compounds. However, the simultaneous addition of Z-VAD-(OMe)-FMK did not affect the activity of 10 µg/mL erioquinol. Taken together our data data demonstrate that gibbilimbol B and eriopodol A effectively induced intrinsic apoptosis triggered by mitochondrial membrane permeabilisation, release of cytochrome c, an early induction of initiator caspase 9, and a consecutive activation of effector caspase 7. Erioquinol, although it affects comparably mitochondrial functions, appears to act in a different manner, i.e., involving mitochondrial ROS release and non-apoptotic/caspase-independent mechanisms. Caspase-independent cell death was first described to affect mitochondria potential, and eventually mitochondrial outer membrane permeabilisation [64], although not followed by caspase activation. Those features resemble the outcome of erioquinol treatment.

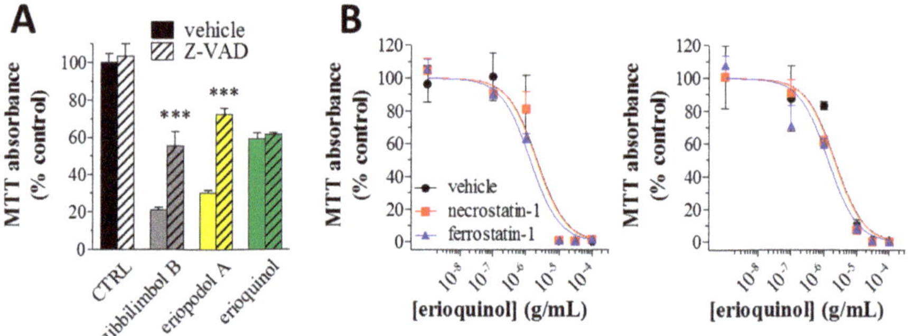

Figure 10. *Piper* genus-derived compounds induce caspase-dependent and independent loss of cell viability. (**A**) MCF7 cells were cultured in the absence (CTRL, control) and in the presence of 30 µg/mL gibbilimbol B/eriopodol A or 10 µg/mL erioquinol for 6 h, before 3-(4,5-dimethylthiazol-2-yl)-2,5-diphenyltetrazolium bromide (MTT) assay. The pan-caspase inhibitor Z-VAD-(OMe)-FMK (50 µM) or its vehicle were used as well. Results are expressed by setting the absorbance of the reduced MTT in the CTRL, as 100%. Data are representative of four-twelve independent experiments. *** $p < 0.0001$ relative to the respective compound alone, i.e., + Z-VAD vehicle. (**B**) MCF7 (left panel) and U373 (right panel) cells were treated with increasing concentrations of erioquinol for 24 h before MTT assay. Erioquinol was administered both in the absence (vehicle) or in the presence of 50 µM necrostatin-1 and 10 µM ferrostatin-1 (2 h pre-treatment), a necroptosis and ferroptosis inhibitor, respectively. Results are expressed by setting the absorbance of the reduced MTT in the control samples (absence of erioquinol) as 100%. The data points are representative of four independent experiments.

Several forms of regulated cell death manifest with a morphology different from apoptosis [65,66], and many compounds from nature source can induce non-apoptotic programmed cell death in cancer cells [67]. Among them, necroptosis can be partially rescued by the receptor-interacting serine-threonine kinase 1 inhibitor necrostatin-1 and ferroptosis by ferrostatin-1, an inhibitor of lipid peroxidation. We thus treated MCF7 and U373 cells with increasing concentrations of erioquinol (24 h) with or without 50 µM necrostatin-1 and 10 µM ferrostatin-1 (2 h pre-treatment). As shown in Figure 10B, the concentration-dependent inhibition of MTT absorbance did not change, suggesting erioquinol-induced death was independent from necroptosis and ferroptosis, two cell death pathways known to be caspase-independent [65,66]. ROS were recently linked to a caspase-independent form of cell death, which cannot be rescue by necrostatin-1 or ferrostatin-1 treatment, and therefore not imputable to either necroptosis or ferroptosis [68]. Treatment with erioquinol might lead to a similar cascade of events, although additional work is required to fully characterise the role of ROS and the cell death process induced by this *Piper* genus-derived compound.

2.4. XIAP as a Molecular Target of Piper Genus-Derived Compounds

XIAP-mediated inhibition of apoptosis goes through its reversible binding to active caspase-9, via its BIR3 domain, and caspase-3/7 when stabilised to XIAP-BIR2 domain [69–71]. It has been also demonstrated that XIAP controls different pathways functionally uncoupled to caspases, leading to the possibility that XIAP system might control cell death/survival through multiple mechanisms [32,34,72–82].

Embelin, a natural benzoquinone with potential therapeutic interest, has been isolated from the fruit of the *Embelia* ribes and discovered through molecular docking analysis of over 8200 molecules as a potent small molecule XIAP inhibitor that binds to the XIAP-BIR3 domain [83–86]. It should be noted that embelin displays chemical features similar to those of erioquinol, eriopodol A, and gibbilimbol B [83]. We assessed if erioquinol, eriopodol A, and gibbilimbol B are able to bind to the XIAP-BIR3 domain in a similar way of embelin. Using molecular docking analysis and molecular dynamics simulations for embelin and isolated new compounds, it was found the structural basis of the predicted interactions with the BIR3 domain of XIAP. Figure 11A provides a general view of the docked conformations obtained for gibbilimbol B, eriopodol A, erioquinol, and embelin. Interestingly, the binding site for gibbilimbol B, eriopodol A, and erioquinol is the same binding site of embelin and with similar energy and binding mode. All docked compounds fits in to the P1, P2 and P3 of the P1–P4 pockets reported for the binding site of the XIAP-BIR3 domain in complex with Smac, the endogenous antagonist ligand of IAPs [35,87,88].

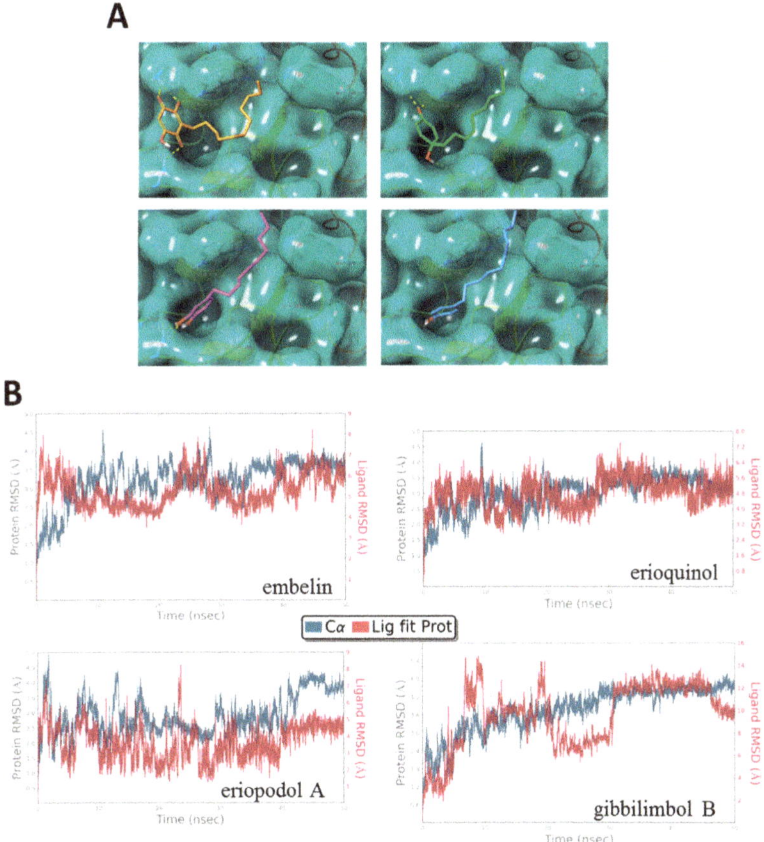

Figure 11. *Cont.*

C

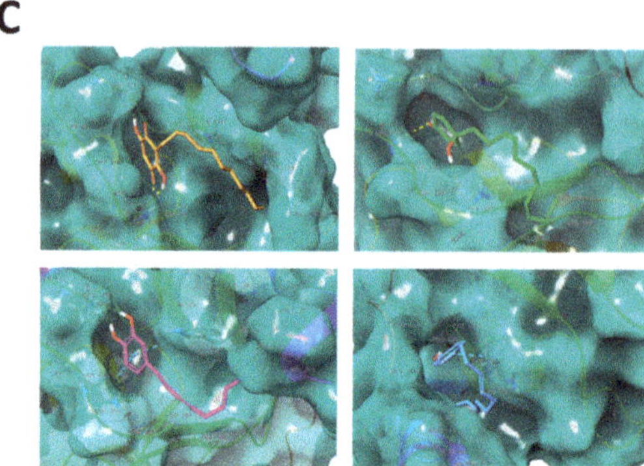

Figure 11. X-linked inhibitor of apoptosis protein (XIAP) as a molecular target of *Piper* genus-derived compounds. (**A**) Molecular docking and (**C**) dynamics analysis of embelin (orange), erioquinol (green), eriopodol A (purple) and gibbilimbol B (blue) in complex with the baculovirus IAP repeat (BIR)-3 domain of XIAP (PDB code 5C83). Interacting residues are displayed in wireframe, hydrogen bonds are displayed in yellow dot lines and π-π stacking interactions are displayed in blue dot lines. (**B**) Protein-ligand root mean square deviation (RMSD) trajectory of the atomic positions for ligands (red, Lig fit Prot) and the receptor (blu, Cα positions) BIR-3 domain of XIAP, for the dynamics trajectory of 50 ns.

Experimental structures of the XIAP-BIR3 domain in different complexes with embelin, Smac or Smac mimetics and non-peptidomimetics small molecules, revealed that residues GLY306, THR308, GLU314, TRP323 and TYR324 are crucial residues involved in the interaction with the BIR3 domain of XIAP [35,41,86,89]. The results of the docking experiments show a possible binding mode for gibbilimbol B, eriopodol A, and erioquinol. Accordingly, the phenolic ring of gibbilimbol B and eriopodol A forms hydrogen bonds with LYS311 and GLU314 (Figure 11A), the quinol ring of erioquinol forms three hydrogen bonds with THR308, LYS322, and TRP323, while residues GLY306, LEU307, TRP323, and TYR324 of the XIAP-BIR3 domain forms hydrophobic interactions with the tail of the alkenyl derivatives.

In addition, molecular dynamics simulations for 50 ns were carried out to assess the stability of the protein-ligand complexes between the docked compounds and the BIR3 domain of XIAP. The stability of the modelled complex of alkenyl derivatives and embelin was confirmed during the period of simulation by little variations in the root mean square deviation (RMSD) trajectory (Figure 11B). Although some changes were observed in the interacting residues of XIAP BIR-3 domain after molecular dynamics simulations (Figure S7), the preferred location of the binding mode for all evaluated ligands were maintained in the pockets P1-P4 of BIR-3 domain of XIAP during the period of simulation (Figure 11C). Also, the binding mode obtained in the docking and dynamics simulations for embelin are according to the interactions pattern determined experimentally by NMR studies in the XIAP-embelin complex, which revealed that TRP323 of the BIR3 domain of XIAP are crucial in the binding of embelin [86]. These findings strongly suggest the highly stable complex formation between the BIR-3 domain of XIAP and the alkenyl derivatives.

XIAP is highly expressed in different human tumour cells and cancer specimens from patients and plays an important role in conferring chemoresistance [33,90]. Because XIAP blocks apoptosis at the downstream effector phase, where multiple signalling events may converge, it represents an attractive molecular target for the design of new anti-cancer drugs [32–41,43]. Two broad approaches have been taken to develop clinical inhibitors of XIAP—antisense oligonucleotides, targeting the entire protein, and small molecule inhibitors, binding a single domain. Small molecule inhibitors offer the potential of more rapid inhibition of their target in vivo and more predictable duration of action [34,41]. Among the small molecule phytochemicals, the XIAP inhibitor embelin exhibited cytotoxic activity in various human tumoural cells, including breast cancer [83–86,91]. In addition, the withaferin-A induced cytotoxicity in human breast cancer cells was associated with suppression of XIAP protein [92] and berberine was shown to induce apoptosis in tumours, likely through the inhibition of XIAP [93]. The just mentioned molecular modelling of our new molecules binding to XIAP-BIR3 domain drove us to examine if they shared a similar activity with already described XIAP inhibitors. With the aim of understanding the role of XIAP in the cell death phenotype, we first determined if our cellular model is anyhow affected by XIAP depletion. Using the Lipofectamine reagent, MCF7 cells were transiently transfected with a XIAP-specific or a scrambled targeting siRNA. When treated with 50 nM of siRNA for 24 h, the protein levels of XIAP markedly decreased to ca. 45% compared to control siRNA transfected samples (Figure 12A) indicating a partial depletion of XIAP. In agreement with previous indications [94–96], the outcome in viability of XIAP knockdown in MCF7 cells, which showed a significant reduction (ca. 40%) in MTT absorbance upon depletion of XIAP (Figure 12B), led us to the conclusion that MCF7 cells depend on XIAP for survival since death mechanisms are neutralised by physiological levels of XIAP. We then tried to add clues on the involvement of XIAP in the cytotoxic effect of *Piper* genus-derived compounds. As shown in Figure 12B, XIAP downregulation in MCF7 cells significantly enhanced the toxicity, as measured by MTT absorbance, of 6 h administration of gibbilimbol B (30 µg/mL), eriopodol A (30 µg/mL) and erioquinol (10 µg/mL) indicating their combined action with XIAP siRNA in inhibiting cell viability. Since 100% knockdown was never achieved with siRNA technique (the absence of detectable XIAP after siRNA transfection, i.e., by XIAP siRNA at 100 nM for 24 h, paralleled the increase of cleaved-caspase 7 levels and the complete loss of MCF7 cell viability (Figure S6D and data not shown)), it is reasonable to assume that that the effects of gibbilimbol B/eriopodol A/erioquinol on the residual XIAP protein in the siRNA-treated cells further induced MCF7 cell death. On the other hand, similar results (additive effect) would be achieved if the compounds target cytotoxic pathways other than XIAP. However, although this is not a formal biological evidence, the simplest explanation of the combined action is a XIAP-mediated mechanism accounting for, at least in part, the cytotoxicity of our new compounds. Accordingly, the positive effects of gibbilimbol B and eriopodol A on caspase 7 activity robustly increased after XIAP silencing (Figure 12C). Downregulation of XIAP by siRNA is known to sensitise human breast cancer cells to death mediated by different chemical agents [94,97]. Finally, using real-time PCR and western blot assays to measure XIAP expression, we found that cell exposure to gibbilimbol B, eriopodol A, and erioquinol at increasing times did not significantly modify mRNA (Figure 12D) and protein levels (Figure 12E) of XIAP. Overall, our data exclude a role of gibbilimbol B/eriopodol A/erioquinol on the regulation of XIAP expression but rather are consistent with the antagonism of XIAP activity through their binding to XIAP-BIR3 domain.

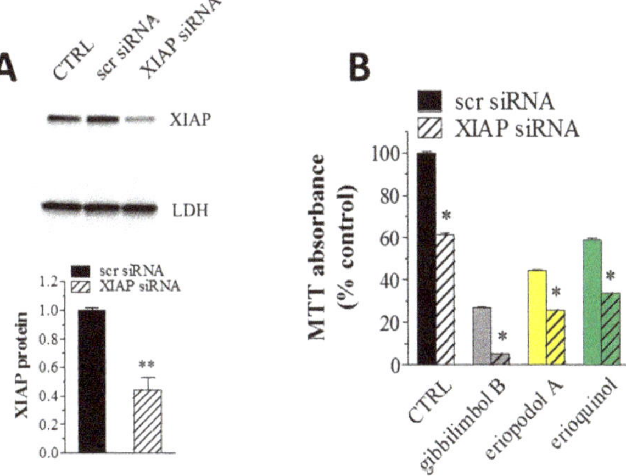

Figure 12. *Cont.*

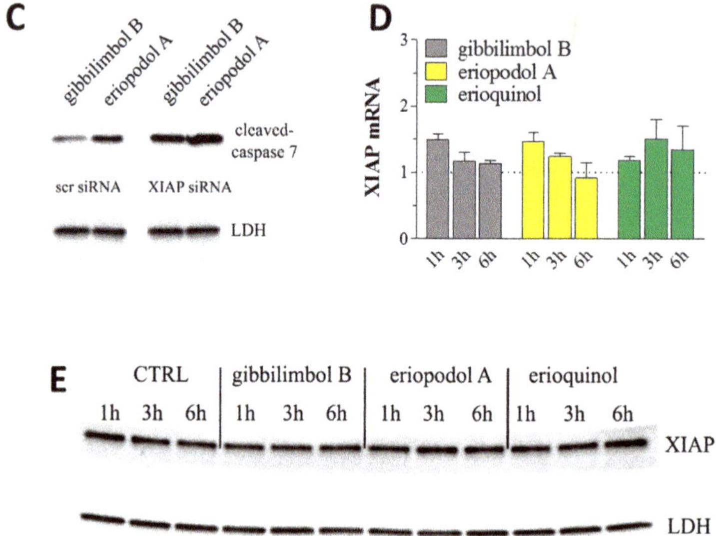

Figure 12. X-linked inhibitor of apoptosis protein (XIAP) as a molecular target of *Piper* genus-derived compounds. (**A**) Western blot analysis of XIAP in MCF7 cells both untransfected (CTRL, control) or transfected for 24 h with a XIAP-specific and scrambled targeting (scr) siRNA (50 nM). Lactate dehydrogenase (LDH) was used as internal standard. Low panel: densitometric analysis expressed as fold change of scr siRNA. Images and data are representative of three independent experiments. **P < 0.001 relative to scr siRNA. (**B**) MCF7 cells were transfected for 24 h with a XIAP-specific or scr siRNA (50 nM) and then cultured in the absence (CTRL) and in the presence of 30 μg/mL gibbilimbol B/eriopodol A or 10 μg/mL erioquinol for 6 h, before 3-(4,5-dimethylthiazol-2-yl)-2,5-diphenyltetrazolium bromide (MTT) assay. Results are expressed by setting the absorbance of the reduced MTT in the scr siRNA CTRL, as 100%. Data are representative of three independent experiments. * $p < 0.0001$ relative to the respective scr siRNA. (**C**) Western blot analysis of cleaved-caspase 7 in MCF7 cells transfected for 24 h with a XIAP-specific or scr siRNA (50 nM) and then cultured in the presence of 30 μg/mL gibbilimbol B and eriopodol A for 6 h. LDH was used as internal standard. Images are representative of three independent experiments. (**D**) Real-time PCR and (**E**) Western blot analysis of XIAP mRNA and protein expression, respectively, in MCF7 cells treated for increasing times in the absence (CTRL) and in the presence of 30 μg/mL gibbilimbol B/eriopodol A or 10 μg/mL erioquinol. β-actin (PCR) and LDH (Western blot) were used as internal standards. PCR results are expressed as fold change of respective CTRL, set as 1. Images and data are representative of three independent experiments.

Since escape from apoptosis is one of the preeminent features of cancer, pharmacological interest in targeting endogenous apoptosis inhibitors, such as B-cell lymphoma (BCL)-2 and IAPs family members, has been constant [32–43,82,98,99]. The efforts, including clinical trials, directed towards identifying small molecules inhibitors of the BCL-2 family of proteins and promote apoptosis with the so-called BH3 mimetics, that mimic the action of certain BH3-only proteins [98], proved the releasing of "apoptosis brakes" as a winning strategy to induce primary cell death in cancer or to sensitise tumour to chemotherapy. Differently to BCL-2 family members, IAPs, and in particular XIAP, have a late role in the apoptotic timeline, and they target already active caspases to prevent cell death. The structural data surrounding the interaction between the BIR3 domain of XIAP and caspases suggest that small molecules that bind the BIR3 pocket of XIAP could mimic the action of Smac and inhibit the interaction between XIAP and caspase [34,41]. Interestingly its multi-domain structure makes XIAP a component of multiple cellular pathways, not only the ones leading to apoptosis. XIAP versatility has been highlighted in inflammation and inflammatory cell death, such as necroptosis [32,79–82].

Even though these aspects are yet to be completely elucidated, we suggest here - in addition to the widely described activity of XIAP inhibitors in apoptosis induction (gibbilimbol B and eriopodol A) —an interesting example of how the pharmacological targeting of XIAP-BIR3 domain can go beyond the simple induction of apoptosis—and extends its influence in modulating cell death signalling events other than caspase-activation (erioquinol). The relevance of non-apoptotic cell death in cancer treatment has recently gained interest as a means to simultaneously targetting tumours and enhancing the inflammatory response [100]; XIAP, in this context, is an interesting crossroad of pathways involved in both cell death and inflammation.

3. Materials and Methods

3.1. Extraction and Isolation of Natural Compounds

P. eriopodon was collected in Fusagasuga, in the Department of Cundinamarca (Colombia). The plant material was identified by Dr. Adolfo Jara Muñoz at Herbario Nacional Colombiano and a voucher specimen (COL516757) was deposited at the Instituto de Ciencias Naturales, Universidad Nacional de Colombia.

Dried and powdered leaves of *P. eriopodon* (1.14 Kg) were extracted exhaustively with ethanol 96% (3 × 5L) at room temperature. After filtration, the solvent was evaporated under reduced pressure, to yield 103.6 g of crude extract. The crude extract (100.0 g) was subjected to silica gel flash chromatography and eluted with a step gradient of toluene/ethyl acetate (0:100, 20:80, 40:60, 60:40, 80:20 and 0:100 (V/V)) to afford eight fractions. Fraction 1 (34.2 g) was chromatographed over silica gel, eluting with a mixture of a three-phase *n*-hexane/dichloromethane/ethyl acetate (25:70:5) solvent system to afford ten fractions (A to J). Fraction E (10.0 g) was chromatographed over Sephadex LH-20 (4.5 × 45 cm, *n*-hexane/chloroform/methanol, 2:2:1) to give six fractions (E1 to E6). In agreement with a previous report [29], compound **1** (7.93 g) was obtained from fraction E3, after column chromatography on Sephadex LH-20 (4.5 × 30 cm, *n*-hexane/acetone/methanol, 2:2:1). Fraction E4 (974.6 mg) was submitted to column chromatography on Sephadex LH-20 (4.0 × 20 cm, *n*-hexane/acetone/methanol, 2:2:1) to yield six fractions (E4.1 to E4.6). Compound **3** (33.2 mg) was obtained from fraction E4.4 (378.8 mg) through Sephadex LH-20 (2.0 × 25 cm, *n*-hexane/acetone/methanol, 2:2:1) and silica gel column chromatography eluted with *n*-hexane/acetone 8:2. Fraction E5 (2.19 g) was subjected to column chromatography on silica gel using a mixture of toluene/ethyl acetate (9:1) to afford ten fractions (E5.1 to E5.10). Fraction E5.3 (153.2 mg) was purified by flash chromatography to yield compound 4 (20.0 mg).

Fraction 2 (8.0 g) was submitted to silica gel column chromatography eluted with *n*-hexane/ethyl acetate 8:2, yielding seven fractions (K–Q). Fraction Q was subjected to flash chromatography eluted with dichloromethane/acetone (7:3) to yield six fractions (Q1–Q6). Fraction Q3 (1.44 g) was subjected to column chromatography over Sephadex LH-20 (4.0 x 20 cm, hexane/acetone/methanol, 2:2:1) to afford six fractions (Q3.1 to Q3.6). Fraction Q3.4 (239.4 mg) was chromatographed over Sephadex LH-20 (4.0 × 20 cm, hexane/chloroform/methanol, 2:2:1) and then purified by silica gel column chromatography eluted with *n*-hexane/acetone (7:3) to yield compound **5** (4.0 mg). Fraction Q6 (570.5 mg) was subjected to flash chromatography eluted with *n*-hexane/acetone 7:3 to afford seven fractions (Q6.1–Q6.7). Compound **2** (166.0 mg) was obtained from fraction Q6.1.

3.2. General Chemical Methods

Flash chromatography was carried out with silica gel (230–400 mesh; Merck, Darmastadt, Germany), column chromatography was performed using silica gel (70–230 mesh; Merck) and Shepadex® LH20 (Sigma-Aldrich, St. Louis, MO, USA), analytical thin layer chromatography was performed using precoated silica gel plates 60 F_{254} (0.25 mm, Merck). ^{1}H and ^{13}C NMR 1D and 2D (COSY, HMQC and HMBC) spectra, were recorded on an Avance 400 spectrometer (Bruker, Millerica, MA, USA) at 400 MHz for ^{1}H and 100 MHz for ^{13}C using the solvent peaks as internal references, the spectra were recorded in $CDCl_3$ and MeOD (Merck). High-resolution mass data were collected on an Accurate-Mass quadrupole Time-of-Flight (q-TOF) (Agilent Technologies, Santa Clara, CA, USA) mass spectrometer, ESI negative mode, Nebuliser 50 (psi), Gas Flow 10 L/min, Gas Temp 350 °C. Fragmentor 175 V, Skimmer 75 V, Vpp 750 V.

3.3. Cell Culture and Chemicals

Human U373 glioma, MCF7 breast cancer, A549 lung cancer and PC-3 prostate cancer cells were grown in Dulbecco's Modified Eagle Medium (DMEM), supplemented with 10% foetal bovine serum, 2 mM glutamine, 100 U/mL penicillin/streptomycin, at 37 °C in a humidified atmosphere containing 5% CO_2 (logarithmic growth phase, routine passages every 3 days). The human breast epithelial cell line MCF10 was cultured in DMEM/F12 Ham's Mixture supplemented with 5% horse serum, epithelial growth factor 20 ng/mL, insulin 10 μg/mL, hydrocortisone 0.5 mg/mL, cholera toxin 100 ng/mL, and 100 U/mL penicillin/streptomycin. HUVEC were grown in EGM-2 Endothelial Cell Growth Medium-2 BulletKit (Lonza, Basel, Switzerland), according to the manufacturer's protocol.

Foetal bovine serum, horse serum, glutamine and penicillin/streptomycin were obtained from Euroclone (Milano, Italy). TMRM was purchased from ThermoFisher Scientific (Waltham, MA, USA) while necrostatin-1 and ferrostatin-1 were obtained from Santa Cruz Biotechnology (Dallas, TX, USA). Where not indicated, the reagents were purchased from Sigma-Aldrich.

3.4. MTT Assay

U373, MCF7, A549, PC-3, HUVEC, and MCF10 cell viability was determined by MTT assay using published protocols [101–105]. MTT absorbance was quantified spectrophotometrically using a Glomax Multi Detection System microplate reader (Promega, Milano, Italy).

3.5. TUNEL Assay

Using published protocols [106,107], MCF7 or U373 cells cultured in 120-mm coverslips were fixed in 4% paraformaldehyde in 0.1 M phosphate buffer (PB), pH 7.4, for 10 min. The TUNEL method (DeadEnd Fluorometric TUNEL System, Promega) was used to assay apoptosis, according to the manufacturer's protocol. DAPI (nuclei detection) staining was also performed.

3.6. Immunofluorescence Microscopy Analysis

Using published protocols [106,108], MCF7 or U373 cells cultured in 120-mm coverslips were fixed in 4% paraformaldehyde in 0.1 M PB, pH 7.4, for 10 min. Cells were pre-incubated for 1 h min with 5% of normal goat serum (Life Technologies, Monza, Italy) in 0.1 M PB (pH 7.4) containing 0.1% Triton X-100, before overnight incubation with the rabbit monoclonal anti-cleaved caspase 7 (Cell Signaling Technology, Danvers, MA, USA). In double-label immunofluorescence experiments, the mouse monoclonal anti-cytochrome c primary antibody (Cell Signaling Technology) was used in conjunction with the rabbit monoclonal primary antibody directed to COX IV (Cell Signaling Technology). For fluorescence detection, coverslips were stained with the appropriate Alexa Fluor secondary antibodies (Life Technologies) and mounted on glass slides in a ProLong Gold Antifade Mountant (Life Technologies). DAPI and/or fluorescein phalloidin (cytoskeleton detection) staining was also used. Cells were analysed with a DMI4000 B automated inverted microscope equipped with a DCF310 digital camera (Leica Microsystems, Wetzlar, Germany). When indicated, confocal imaging was performed with a TCS SP8 System (Leica Microsystems). Image acquisitions were controlled by the Leica Application Suite X.

3.7. Annexin V Staining

MCF7 cells were incubated with 5 μg/mL Annexin V-fluorescein isothiocyanate (FITC) to assess the phosphatidylserine exposure on the outer leaflet of the plasma membrane, and 5 μg/mL PI

(DNA-binding probe) to exclude necrotic cells in binding buffer (10 mM HEPES, 140 mM NaCl, 2.5 mM CaCl$_2$) [109]. Cell staining was analysed by Gallios Flow Cytometer (Beckman-Coulter, Brea, CA, USA) and the software FCS Express 4 (De Novo System, Portland, OR, USA).

3.8. Western Blotting

Using published protocols [107,110,111], MCF7 and MCF10 cells were homogenised in RIPA lysis buffer, supplemented with a cocktail of protease inhibitors (cOmplete; Roche Diagnostics, Milano, Italy). Equal amounts of proteins were separated by 4–20% SDS-polyacrylamide gel electrophoresis (Criterion TGX Stain-free precast gels and Criterion Cell system; Bio-Rad, Hercules, CA, USA) and transferred onto nitrocellulose membrane using a Bio-Rad Trans-Blot Turbo System. When indicated, the membranes were probed using the rabbit monoclonal anti-cleaved caspase 7 and anti-XIAP (Cell Signaling Technology) primary antibodies. After the incubation with the appropriate horseradish-peroxidase-conjugated secondary antibody (Cell Signaling Technology), bands were visualised using the Clarity Western ECL substrate with a ChemiDoc MP imaging system (Bio-Rad). To monitor for potential artefacts in loading and transfer among samples in different lanes, the blots were routinely treated with the Restore Western Blot Stripping Buffer (ThermoFisher Scientific) and re-probed with the goat anti-Lactate dehydrogenase (LDH)-A (Santa Cruz Biotechnology) and the mouse anti-vinculin primary antibodies. The stain-free gel was used as loading control as well. When appropriated, bands were quantified for densitometry using the Bio-Rad Image Lab software.

3.9. Mitochondrial Membrane Potential Analysis

Using published protocols [112], mitochondria of MCF7 and U373 cells were labeled using TMRM, a voltage-sensitive cationic lipophilic dye, partitioning and accumulating in the mitochondrial matrix based upon the Nernst equation. After treatments, cells were trypsinised, counted and incubated with 100 nM TMRM for 30 min at 37 °C. Fluorescence was measured by using a Glomax Multi Detection System microplate reader (Promega), excitation wavelength: 525 nm; emission wavelength: 580–640 nm). After background subtraction, the data were normalised on cell number.

3.10. Measurement of ROS

MCF7 or U373 cells cultured in 120-mm coverslips were exposed to 30 µM DCFH-DA (0.1 M PB, pH 7.4) and fixed in 4% paraformaldehyde for 20 min. For fluorescence detection, coverslips were mounted on glass slides and observed with a laser-scanning confocal microscope (TCS SP8 System and Application suite X, Leica Microsystems). DAPI and fluorescein phalloidin (nuclei and cytoskeleton detection, respectively) staining was also used.

3.11. Molecular Modeling

AutoDock4 was used to carry out the molecular docking. The Protein Data Bank crystallographic structure PDB 5C83 was considered as receptor model [87]. The preparation of the macromolecule was made with PyMOL (version 2.0, PyMol Molecular Graphics, Schrodinger, New York, NY, USA) System) and XIAP-BIR3 domain was selected as receptor [113]. Energy maps was established with Autogrid4 involving all atom types. After 25 million of energy evaluations in the binding pocket and using a grid of 50 × 50 × 50 points, all conformations of the ligand were clustered according to the energy and conformations. The docking results were visualised using the computational program Maestro 11.6. The molecular dynamics simulations were carried out with Desmond simulation package of Maestro (Desmond Molecular Dynamics System; D. E. Shaw Research, New York, NY, USA, 2016) using the OPLS 2005 force field parameters. A solvated system (TIP3P) and a predefined model for electrically neutral system (physiological concentrations of monovalent ions, NaCl 0.15 M) were used in an orthorhombic box and maintained at constant temperature of 300 K for all simulations. The dynamics simulations were analysed using the Simulation Interaction Diagram tool of Desmond package, monitoring the behaviour and stability of simulations by RMSD of the ligand and protein atom positions in time.

3.12. RNA Interference

Gene silencing of XIAP in MCF7 cells was performed as previously published [106]. Briefly, according to the manufacturer's protocol, iBONI siRNA Pool (Riboxx, Radebeul, Germany) targeting human XIAP were mixed to Lipofectamine RNAiMax transfection reagent (Life Technologies). iBONI siRNA Pool negative control (Riboxx) (scrambled targeting siRNAs) was also used. The mix was added to cultured MCF7 cells at a siRNA concentration of 50 nM for 24 h.

3.13. Real-Time PCR

The analysis of mRNA expression was performed as previously described [106,114,115]. Briefly, total RNA from MCF7 cells was extracted with the High Pure RNA Isolation Kit (Roche Applied Science, Mannheim, Germany), according to the manufacturer's protocol. First-strand cDNA was generated from 1 μg of total RNA using iScript Reverse Transcription Supermix (Bio-Rad). Primer pairs (Eurofins Genomics, Milano, Italy) for XIAP (NM_001167; forward ACCGTGCGGTGCTTTAGTT, reverse TGCGTGGCACTATTTTCAAGATA) and β-actin (NM_001101; forward ATAGCACAGCCTGGATAGCAACGTAC, reverse CACCTTCTACAAT GAGCTGCGTGTG) were designed to hybridise to unique regions of the appropriate gene sequence. PCR was performed using SsoAdvanced Universal SYBR Green Supermix and the CFX96 Touch Real-Time PCR Detection System (Bio-Rad). The fold change was determined relative to the selected control sample after normalising to β-actin (internal standard) by the formula $2^{-\Delta\Delta CT}$.

3.14. Statistics

Statistical significance of raw data between the groups in each experiment was evaluated using unpaired Student's *t*-test (single comparisons) or one-way ANOVA followed by the Newman-Keuls post-test (multiple comparisons). The IC_{50} and E_{max} concentration were determined by non-linear regression curve analysis of the concentration-effect responses. Potency values among concentration-response curves were compared with the F-test. Data belonging from different experiments were represented and averaged in the same graph. The GraphPad Prism software package (GraphPad Software, San Diego, CA, USA) was used. The results were expressed as means ± standard error of mean (SEM) of the indicated n values.

4. Conclusions

This study adds to the renewed biological interest in natural derived compounds, by presenting a chemical and biological characterisation of new small organic molecules derived from *Piper* genus plants. Following a recent preliminary report of gibbilimbol B as cytotoxic in breast cancer cell lines, we explored this observation by comparing it to similarly structured new molecules. Erioquinol that appeared to be the most potent compound versus gibbilimbol B and eriopodol A was listed as an intermediate. A more detailed investigation of the biological mechanism behind these molecules' activity in shaping cell viability revealed induction of caspase-dependent apoptosis following exposure of tumour cells to gibbilimbol B and eriopodol A and, interestingly, display of caspase-independent/non-apoptotic features in cell treated with erioquinol. In silico modelling and molecular approaches gave us a first preliminary insight into the molecular target of *Piper* genus compounds, the anti-apoptotic protein XIAP (Figure 13). Of note, an already identified XIAP inhibitor shared structural and binding similarities with them. The appeal of XIAP as a therapeutic target in cancer is not restricted to inhibition of apoptosis, but comprehends the regulation of other cellular physiological aspects, such as control of caspase-independent cell death. The molecular signature behind our observation opens important implications to further dissect the role of XIAP and for the development of novel XIAP antagonists for cancer treatment.

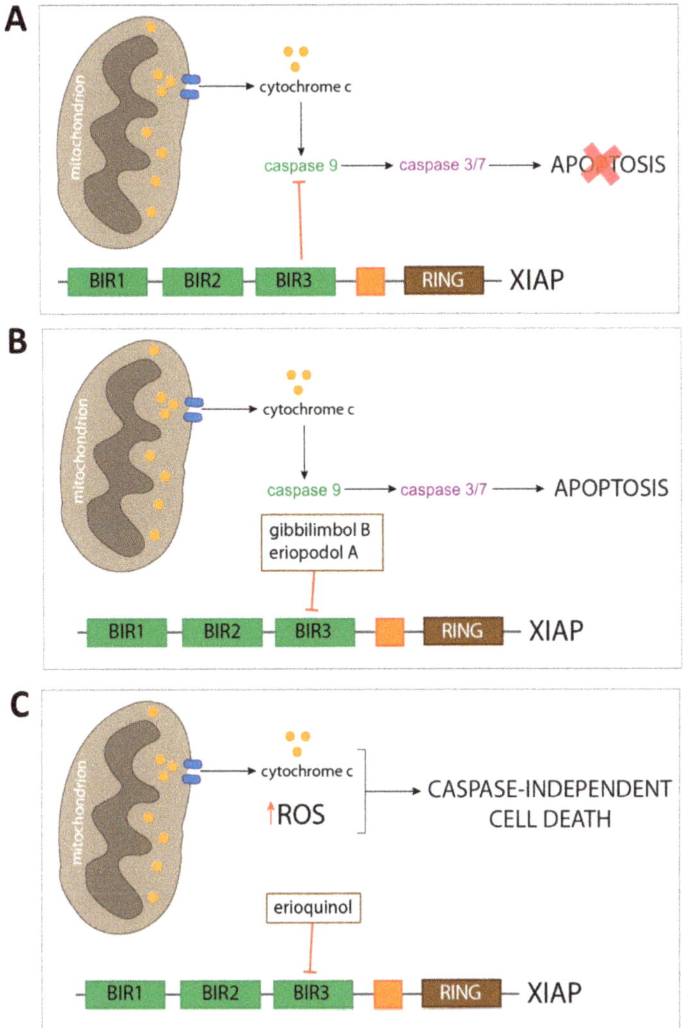

Figure 13. Schematic picture depicting cell death mechanisms of *Piper* genus-derived compounds. Escape of both intrinsic and extrinsic apoptosis is a common feature of cancer cells. (**A**) This hallmark is often carried out by overexpressing anti-apoptotic proteins, such as X-linked inhibitor of apoptosis protein (XIAP), which prevents the execution of apoptosis by binding of its baculovirus IAP repeat (BIR)

3 domain to already active initiator caspase 9. In order to counteract this resistance to cell death, several cancer pharmacological therapies have the aim of removing the 'molecular brakes' to apoptosis sensitising cancer cell to undergo loss of viability. The approach we described includes the use of three compounds from *Piper* genus plants which were predicted to bind XIAP-BIR3 domain. (**B**) Two of the compounds (gibbilimbol B and eriopodol A) were shown to induce a classical pro-apoptotic response, including mitochondrial outer membrane polarisation, release of cytochrome c, and subsequent activation of both initiator and effector caspases. (**C**) Despites triggering a similar response at the mitochondria level, erioquinol does not act through the apoptotic machinery, and results in a caspase-independent cell death characterised by cytoplasmic reactive oxygen species (ROS) accumulation.

Supplementary Materials: The following are available online at http://www.mdpi.com/2072-6694/11/9/1336/s1, Figure S1: NMR spectroscopy (400 MHz, CDCl$_3$) and HRESIMS of compound 1, Figure S2: NMR spectroscopy (400 MHz, CDCl$_3$) and HRESIMS of compound 2, Figure S3: NMR spectroscopy (400 MHz, CDCl$_3$) and HRESIMS of compound 3, Figure S4: NMR spectroscopy (400 MHz, CDCl$_3$) and HRESIMS of compound 4, Figure S5: NMR spectroscopy (400 MHz, CDCl$_3$) and HRESIMS of compound 5, Figure S6: Western blot, Figure S7: Protein-ligand interactions fraction for evaluated ligands and X-linked inhibitor of apoptosis protein (XIAP) baculovirus IAP repeat (BIR)-3 domain during the molecular dynamics trajectory of 50 ns, Table S1: ^{1}H NMR (400 MHz) data for compounds 1–4 in CDCl$_3$ and compound 5 in MeOD, Table S2: ^{13}C NMR (100 MHz) data for compounds 1–4 in CDCl$_3$ and compound 5 in MeOD.

Author Contributions: D.M.: design and conception of the experimental plan and analysis work, isolation and characterisation of the compounds, acquisition analysis and interpretation of human cell data, in silico analysis, contribution to article writing. M.B., S.Z.: design, acquisition, analysis and interpretation of human cell data, contribution to the experimental plan and article writing. F.L.-V.: in silico analysis, contribution to the supervision of the experiments, interpretation and processing of the results. A.S-H., M.G., M.C., E.C. (Elisabetta Catalani): acquisition and processing of the results, contribution to the experimental plan. C.D.P., C.P., E.C. (Emilio Clementi): contribution to the design, conception and interpretation of human cell results, contribution to article writing. G.A., W.D., L.C.: supervision of the experiments and interpretation of the results, contribution to article writing. D.C.: conception and coordination of the experimental and analysis work, supervision of human cell experiments, analysis/interpretation and processing of data, article writing. All authors provided critical feedback, edited and approved the final manuscript.

Funding: The research has been supported by grants from the Italian Ministry of Education, University and Research: "PRIN2015" to E.Clem./D.C. and "Departments of Excellence-2018" Program (Dipartimenti di Eccellenza) to DIBAF (University of Tuscia, Viterbo, Italy) (Project "Landscape 4.0 - food, wellbeing and environment"). Authors also acknowledge the financial support provided by the Universidad Nacional de Colombia (Bogotá) and the Administrative Department of Science, Technology and Innovation from Colombia "COLCIENCIAS" (Grant No. 528-2011) to D.M.

Acknowledgments: We are grateful to Francesca Proietti Serafini (University of Tuscia, Viterbo, Italy) for the help with viability cell assays and to Debora Parolin (University of Milan, Milano, Italy) for HUVEC supply.

Conflicts of Interest: The authors declare no competing financial interests.

References

1. Gurevich, E.V.; Gurevich, V.V. Therapeutic potential of small molecules and engineered proteins. *Handb. Exp. Pharmacol.* **2014**, *219*, 1–12.

2. Cheng, B.; Yuan, W.E.; Su, J.; Liu, Y.; Chen, J. Recent advances in small molecule based cancer immunotherapy. *Eur. J. Med. Chem.* **2018**, *157*, 582–598. [CrossRef]

3. Huck, B.R.; Kotzner, L.; Urbahns, K. Small Molecules Drive Big Improvements in Immuno-Oncology Therapies. *Angew. Chem. Int. Ed. Engl.* **2018**, *57*, 4412–4428. [CrossRef]

4. Schiavone, S.; Trabace, L. Small Molecules: Therapeutic Application in Neuropsychiatric and Neurodegenerative Disorders. *Molecules* **2018**, *23*, 411. [CrossRef]

5. Dhanak, D.; Edwards, J.P.; Nguyen, A.; Tummino, P.J. Small-Molecule Targets in Immuno-Oncology. *Cell Chem. Biol.* **2017**, *24*, 1148–1160. [CrossRef]

6. Shen, B. A New Golden Age of Natural Products Drug Discovery. *Cell* **2015**, *163*, 1297–1300. [CrossRef]

7. Nobili, S.; Lippi, D.; Witort, E.; Donnini, M.; Bausi, L.; Mini, E.; Capaccioli, S. Natural compounds for cancer treatment and prevention. *Pharmacol. Res.* **2009**, *59*, 365–378. [CrossRef]

8. Harvey, A.L.; Edrada-Ebel, R.; Quinn, R.J. The re-emergence of natural products for drug discovery in the genomics era. *Nat. Rev. Drug Discov.* **2015**, *14*, 111–129. [CrossRef]

9. Newman, D.J.; Cragg, G.M. Natural Products as Sources of New Drugs from 1981 to 2014. *J. Nat. Prod.* **2016**, *79*, 629–661. [CrossRef]

10. Catalani, E.; Proietti Serafini, F.; Zecchini, S.; Picchietti, S.; Fausto, A.M.; Marcantoni, E.; Buonanno, F.; Ortenzi, C.; Perrotta, C.; Cervia, D. Natural products from aquatic eukaryotic microorganisms for cancer therapy: Perspectives on anti-tumour properties of ciliate bioactive molecules. *Pharmacol. Res.* **2016**, *113*, 409–420. [CrossRef]

11. Carocho, M.; Ferreira, I.C. The role of phenolic compounds in the fight against cancer—A review. *Anticancer Agents Med. Chem.* **2013**, *13*, 1236–1258. [CrossRef]

12. Guerra, A.R.; Duarte, M.F.; Duarte, I.F. Targeting Tumor Metabolism with Plant-Derived Natural Products: Emerging Trends in Cancer Therapy. *J. Agric. Food Chem.* **2018**, *66*, 10663–10685. [CrossRef]

13. Jafari, S.; Saeidnia, S.; Abdollahi, M. Role of natural phenolic compounds in cancer chemoprevention via regulation of the cell cycle. *Curr. Pharm. Biotechnol.* **2014**, *15*, 409–421. [CrossRef]

14. Tungmunnithum, D.; Thongboonyou, A.; Pholboon, A.; Yangsabai, A. Flavonoids and Other Phenolic Compounds from Medicinal Plants for Pharmaceutical and Medical Aspects: An Overview. *Medicines* **2018**, *5*, 93. [CrossRef]

15. Parmar, V.S.; Jain, S.C.; Gupta, S.; Talwar, S.; Rajwanshi, V.K.; Kumar, R.; Azim, A.; Malhotra, S.; Kumar, N.; Jain, R.; et al. Polyphenols and alkaloids from Piper species. *Phytochemistry* **1998**, *49*, 1069–1078. [CrossRef]

16. Xiang, C.P.; Shi, Y.N.; Liu, F.F.; Li, H.Z.; Zhang, Y.J.; Yang, C.R.; Xu, M. A Survey of the Chemical Compounds of Piper spp. (Piperaceae) and Their Biological Activities. *Nat. Prod. Commun.* **2016**, *11*, 1403–1408. [CrossRef]

17. Valdivia, C.; Marquez, N.; Eriksson, J.; Vilaseca, A.; Munoz, E.; Sterner, O. Bioactive alkenylphenols from Piper obliquum. *Bioorg. Med. Chem.* **2008**, *16*, 4120–4126. [CrossRef]

18. Yang, S.X.; Sun, Q.Y.; Yang, F.M.; Hu, G.W.; Luo, J.F.; Wang, Y.H.; Long, C.L. Sarmentosumols A to F, new mono- and dimeric alkenylphenols from Piper sarmentosum. *Planta Med.* **2013**, *79*, 693–696. [CrossRef]

19. Orjala, J.; Mian, P.; Rali, T.; Sticher, O. Gibbilimbols A-D, cytotoxic and antibacterial alkenylphenols from Piper gibbilimbum. *J. Nat. Prod.* **1998**, *61*, 939–941. [CrossRef]

20. Yoshida, N.C.; Benedetti, A.M.; dos Santos, R.A.; Ramos, C.S.; Batista, R.; Yamaguchi, L.F.; Kato, M.J. Alkenylphenols from Piper dilatatum and P. diospyrifolium. *Phytochem. Lett.* **2018**, *25*, 136–140. [CrossRef]

21. De Oliveira, A.; Mesquita, J.T.; Tempone, A.G.; Lago, J.H.G.; Guimaraes, E.F.; Kato, M.J. Leishmanicidal activity of an alkenylphenol from Piper malacophyllum is related to plasma membrane disruption. *Exp. Parasitol.* **2012**, *132*, 383–387. [CrossRef] [PubMed]

22. Bezerra, D.P.; Pessoa, C.; de Moraes, M.O.; Saker-Neto, N.; Silveira, E.R.; Costa-Lotufo, L.V. Overview of the therapeutic potential of piplartine (piperlongumine). *Eur. J. Pharm. Sci.* **2013**, *48*, 453–463. [CrossRef] [PubMed]

23. D'Sousa Costa, C.O.; Araujo Neto, J.H.; Baliza, I.R.S.; Dias, R.B.; Valverde, L.F.; Vidal, M.T.A.; Sales, C.B.S.; Rocha, C.A.G.; Moreira, D.R.M.; Soares, M.B.P.; et al. Novel piplartine-containing ruthenium complexes: Synthesis, cell growth inhibition, apoptosis induction and ROS production on HCT116 cells. *Oncotarget* **2017**, *8*, 104367–104392. [CrossRef] [PubMed]

24. Piska, K.; Gunia-Krzyzak, A.; Koczurkiewicz, P.; Wojcik-Pszczola, K.; Pekala, E. Piperlongumine (piplartine) as a lead compound for anticancer agents—Synthesis and properties of analogues: A mini-review. *Eur. J. Med. Chem.* **2018**, *156*, 13–20. [CrossRef] [PubMed]

25. Benfica, P.L.; Avila, R.I.; Rodrigues, B.D.S.; Cortez, A.P.; Batista, A.C.; Gaeti, M.P.N.; Lima, E.M.; Rezende, K.R.; Valadares, M.C. 4-Nerolidylcatechol: Apoptosis by mitochondrial mechanisms with reduction in cyclin D1 at G0/G1 stage of the chronic myelogenous K562 cell line. *Pharm. Biol.* **2017**, *55*, 1899–1908. [CrossRef] [PubMed]

26. Cortez, A.P.; de Avila, R.I.; da Cunha, C.R.; Santos, A.P.; Menegatti, R.; Rezende, K.R.; Valadares, M.C. 4-Nerolidylcatechol analogues as promising anticancer agents. *Eur. J. Pharmacol.* **2015**, *765*, 517–524. [CrossRef] [PubMed]

27. Gundala, S.R.; Yang, C.; Mukkavilli, R.; Paranjpe, R.; Brahmbhatt, M.; Pannu, V.; Cheng, A.; Reid, M.D.; Aneja, R. Hydroxychavicol, a betel leaf component, inhibits prostate cancer through ROS-driven DNA damage and apoptosis. *Toxicol. Appl. Pharmacol.* **2014**, *280*, 86–96. [CrossRef] [PubMed]

28. Hemamalini, V.; Velayutham, D.P.M.; Lakshmanan, L.; Muthusamy, K.; Sivaramakrishnan, S.; Premkumar, K. Inhibitory potential of Hydroxychavicol on Ehrlich ascites carcinoma model and in silico interaction on cancer targets. *Nat. Prod. Res.* **2018**, 1–6. [CrossRef]

29. Munoz, D.R.; Sandoval-Hernandez, A.G.; Delgado, W.A.; Arboleda, G.H.; Cuca, L.E. In vitro anticancer screening of Colombian plants from Piper genus (Piperaceae). *J. Pharmacogn. Phytother.* **2018**, *10*, 174–181.

30. Hanahan, D.; Weinberg, R.A. Hallmarks of cancer: The next generation. *Cell* **2011**, *144*, 646–674. [CrossRef]

31. Pfeffer, C.M.; Singh, A.T.K. Apoptosis: A Target for Anticancer Therapy. *Int. J. Mol. Sci.* **2018**, *19*, 448. [CrossRef] [PubMed]

32. Lalaoui, N.; Vaux, D.L. Recent advances in understanding inhibitor of apoptosis proteins. *F1000Research* **2018**, *7*. [CrossRef] [PubMed]

33. Rathore, R.; McCallum, J.E.; Varghese, E.; Florea, A.M.; Busselberg, D. Overcoming chemotherapy drug resistance by targeting inhibitors of apoptosis proteins (IAPs). *Apoptosis* **2017**, *22*, 898–919. [CrossRef] [PubMed]

34. Schimmer, A.D.; Dalili, S.; Batey, R.A.; Riedl, S.J. Targeting XIAP for the treatment of malignancy. *Cell Death Differ.* **2006**, *13*, 179–188. [CrossRef]

35. Tamanini, E.; Buck, I.M.; Chessari, G.; Chiarparin, E.; Day, J.E.H.; Frederickson, M.; Griffiths-Jones, C.M.; Hearn, K.; Heightman, T.D.; Iqbal, A.; et al. Discovery of a Potent Nonpeptidomimetic, Small-Molecule Antagonist of Cellular Inhibitor of Apoptosis Protein 1 (cIAP1) and X-Linked Inhibitor of Apoptosis Protein (XIAP). *J. Med. Chem.* **2017**, *60*, 4611–4625. [CrossRef]

36. Sun, H.; Stuckey, J.A.; Nikolovska-Coleska, Z.; Qin, D.; Meagher, J.L.; Qiu, S.; Lu, J.; Yang, C.Y.; Saito, N.G.; Wang, S. Structure-based design, synthesis, evaluation, and crystallographic studies of conformationally constrained Smac mimetics as inhibitors of the X-linked inhibitor of apoptosis protein (XIAP). *J. Med. Chem.* **2008**, *51*, 7169–7180. [CrossRef]

37. Fakler, M.; Loeder, S.; Vogler, M.; Schneider, K.; Jeremias, I.; Debatin, K.M.; Fulda, S. Small molecule XIAP inhibitors cooperate with TRAIL to induce apoptosis in childhood acute leukemia cells and overcome Bcl-2-mediated resistance. *Blood* **2009**, *113*, 1710–1722. [CrossRef] [PubMed]

38. Vogler, M.; Walczak, H.; Stadel, D.; Haas, T.L.; Genze, F.; Jovanovic, M.; Bhanot, U.; Hasel, C.; Moller, P.; Gschwend, J.E.; et al. Small molecule XIAP inhibitors enhance TRAIL-induced apoptosis and antitumor activity in preclinical models of pancreatic carcinoma. *Cancer Res.* **2009**, *69*, 2425–2434. [CrossRef]

39. Dean, E.J.; Ward, T.; Pinilla, C.; Houghten, R.; Welsh, K.; Makin, G.; Ranson, M.; Dive, C. A small molecule inhibitor of XIAP induces apoptosis and synergises with vinorelbine and cisplatin in NSCLC. *Br. J. Cancer* **2010**, *102*, 97–103. [CrossRef]

40. Obexer, P.; Ausserlechner, M.J. X-linked inhibitor of apoptosis protein—A critical death resistance regulator and therapeutic target for personalized cancer therapy. *Front. Oncol.* **2014**, *4*, 197. [CrossRef]

41. Cong, H.; Xu, L.; Wu, Y.; Qu, Z.; Bian, T.; Zhang, W.; Xing, C.; Zhuang, C. Inhibitor of Apoptosis Protein (IAP) Antagonists in Anticancer Agent Discovery: Current Status and Perspectives. *J. Med. Chem.* **2019**. [CrossRef]

42. Fulda, S. Promises and Challenges of Smac Mimetics as Cancer Therapeutics. *Clin. Cancer Res.* **2015**, *21*, 5030–5036. [CrossRef] [PubMed]

43. Schimmer, A.D.; Welsh, K.; Pinilla, C.; Wang, Z.; Krajewska, M.; Bonneau, M.J.; Pedersen, I.M.; Kitada, S.; Scott, F.L.; Bailly-Maitre, B.; et al. Small-molecule antagonists of apoptosis suppressor XIAP exhibit broad antitumor activity. *Cancer Cell* **2004**, *5*, 25–35. [CrossRef]

44. Masaki, M.E.; Harumoto, T.; Terazima, M.N.; Miyake, A.; Usuki, Y.; Iio, H. Climacostol, a defense toxin of the heterotrich ciliate Climacostomum virens against predators. *Tetrahedron Lett.* **1999**, *40*, 8227–8229. [CrossRef]

45. Varela, M.T.; Dias, R.Z.; Martins, L.F.; Ferreira, D.D.; Tempone, A.G.; Ueno, A.K.; Lago, J.H.G.; Fernandes, J.P.S. Gibbilimbol analogues as antiparasitic agents-Synthesis and biological activity against Trypanosoma cruzi and Leishmania (L.) infantum. *Bioorg. Med. Chem. Lett.* **2016**, *26*, 1180–1183. [CrossRef] [PubMed]

46. Carreno, M.C.; Gonzalez-Lopez, M.; Urbano, A. Oxidative de-aromatization of para-alkyl phenols into para-peroxyquinols and para-quinols mediated by oxone as a source of singlet oxygen. *Angew. Chem. Int. Ed.* **2006**, *45*, 2737–2741. [CrossRef]

47. Freitas, G.C.; Batista, J.M.; Franchi, G.C.; Nowill, A.E.; Yamaguchi, L.F.; Vilcachagua, J.D.; Favaro, D.C.; Furlan, M.; Guimaraes, E.F.; Jeffrey, C.S.; et al. Cytotoxic non-aromatic B-ring flavanones from Piper carniconnectivum C. DC. *Phytochemistry* **2014**, *97*, 81–87. [CrossRef] [PubMed]

48. Nishino, C.; Kobayashi, K.; Fukushima, M. Halleridone, a cytotoxic constituent from Cornus controversa. *J. Nat. Prod.* **1988**, *51*, 1281–1282. [CrossRef]

49. Bradshaw, T.D.; Matthews, C.S.; Cookson, J.; Chew, E.H.; Shah, M.; Bailey, K.; Monks, A.; Harris, E.; Westwell, A.D.; Wells, G.; et al. Elucidation of thioredoxin as a molecular target for antitumor quinols. *Cancer Res.* **2005**, *65*, 3911–3919. [CrossRef]

50. Berry, J.M.; Bradshaw, T.D.; Fichtner, I.; Ren, R.; Schwalbe, C.H.; Wells, G.; Chew, E.H.; Stevens, M.F.; Westwell, A.D. Quinols as novel therapeutic agents. 2.(1) 4-(1-Arylsulfonylindol-2-yl)-4-hydroxycyclohexa-2,5-dien-1-ones and related agents as potent and selective antitumor agents. *J. Med. Chem.* **2005**, *48*, 639–644. [CrossRef]

51. McCarroll, A.J.; Bradshaw, T.D.; Westwell, A.D.; Matthews, C.S.; Stevens, M.F. Quinols as novel therapeutic agents. 7.1 Synthesis of antitumor 4-[1-(arylsulfonyl-1H-indol-2-yl)]-4-hydroxycyclohexa-2,5-dien-1-ones by Sonogashira reactions. *J. Med. Chem.* **2007**, *50*, 1707–1710. [CrossRef] [PubMed]

52. Abu Bakar, A.; Akhtar, M.N.; Mohd Ali, N.; Yeap, S.K.; Quah, C.K.; Loh, W.S.; Alitheen, N.B.; Zareen, S.; Ul-Haq, Z.; Shah, S.A.A. Design, Synthesis and Docking Studies of Flavokawain B Type Chalcones and Their Cytotoxic Effects on MCF-7 and MDA-MB-231 Cell Lines. *Molecules* **2018**, *23*, 616. [CrossRef] [PubMed]

53. Sriwiriyajan, S.; Sukpondma, Y.; Srisawat, T.; Madla, S.; Graidist, P. (-)-Kusunokinin and piperloguminine from Piper nigrum: An alternative option to treat breast cancer. *Biomed. Pharm.* **2017**, *92*, 732–743. [CrossRef] [PubMed]

54. Sriwiriyajan, S.; Ninpesh, T.; Sukpondma, Y.; Nasomyon, T.; Graidist, P. Cytotoxicity Screening of Plants of Genus Piper in Breast Cancer Cell Lines. *Trop. J. Pharm. Res.* **2014**, *13*, 921–928. [CrossRef]

55. Fan, L.; Cao, X.; Yan, H.; Wang, Q.; Tian, X.; Zhang, L.; He, X.; Borjihan, G.; Morigen. The synthetic antihyperlipidemic drug potassium piperate selectively kills breast cancer cells through inhibiting G1-S-phase transition and inducing apoptosis. *Oncotarget* **2017**, *8*, 47250–47268. [CrossRef] [PubMed]

56. Park, M.J.; Lee, D.E.; Shim, M.K.; Jang, E.H.; Lee, J.K.; Jeong, S.Y.; Kim, J.H. Piperlongumine inhibits TGF-beta-induced epithelial-to-mesenchymal transition by modulating the expression of E-cadherin, Snail1, and Twist1. *Eur. J. Pharmacol.* **2017**, *812*, 243–249. [CrossRef] [PubMed]

57. De Souza Grinevicius, V.M.; Kviecinski, M.R.; Santos Mota, N.S.; Ourique, F.; Porfirio Will Castro, L.S.; Andreguetti, R.R.; Gomes Correia, J.F.; Filho, D.W.; Pich, C.T.; Pedrosa, R.C. Piper nigrum ethanolic extract rich in piperamides causes ROS overproduction, oxidative damage in DNA leading to cell cycle arrest and apoptosis in cancer cells. *J. Ethnopharmacol.* **2016**, *189*, 139–147. [CrossRef]

58. Deng, Y.; Sriwiriyajan, S.; Tedasen, A.; Hiransai, P.; Graidist, P. Anti-cancer effects of Piper nigrum via inducing multiple molecular signaling in vivo and in vitro. *J. Ethnopharmacol.* **2016**, *188*, 87–95. [CrossRef]

59. Da Nobrega, F.R.; Ozdemir, O.; Nascimento Sousa, S.C.S.; Barboza, J.N.; Turkez, H.; de Sousa, D.P. Piplartine Analogues and Cytotoxic Evaluation against Glioblastoma. *Molecules* **2018**, *23*, 1382. [CrossRef]

60. Abdul Rahman, A.; Jamal, A.R.; Harun, R.; Mohd Mokhtar, N.; Wan Ngah, W.Z. Gamma-tocotrienol and hydroxy-chavicol synergistically inhibits growth and induces apoptosis of human glioma cells. *BMC Complement. Altern. Med.* **2014**, *14*, 213. [CrossRef]

61. Subramanian, U.; Poongavanam, S.; Vanisree, A.J. Studies on the neuroprotective role of Piper longum in C6 glioma induced rats. *Investing. New Drugs* **2010**, *28*, 615–623. [CrossRef] [PubMed]

62. Buonanno, F.; Catalani, E.; Cervia, D.; Proietti Serafini, F.; Picchietti, S.; Fausto, A.M.; Giorgi, S.; Lupidi, G.; Rossi, F.V.; Marcantoni, E.; et al. Bioactivity and Structural Properties of Novel Synthetic Analogues of the Protozoan Toxin Climacostol. *Toxins* **2019**, *11*, 42. [CrossRef] [PubMed]

63. Moloney, J.N.; Cotter, T.G. ROS signalling in the biology of cancer. *Semin. cell Dev. Biol.* **2018**, *80*, 50–64. [CrossRef] [PubMed]

64. Tait, S.W.; Green, D.R. Caspase-independent cell death: Leaving the set without the final cut. *Oncogene* **2008**, *27*, 6452–6461. [CrossRef]

65. Galluzzi, L.; Kepp, O.; Chan, F.K.; Kroemer, G. Necroptosis: Mechanisms and Relevance to Disease. *Annu. Rev. Pathol.* **2017**, *12*, 103–130. [CrossRef] [PubMed]

66. Green, D.R. The Coming Decade of Cell Death Research: Five Riddles. *Cell* **2019**, *177*, 1094–1107. [CrossRef] [PubMed]

67. Ye, J.; Zhang, R.; Wu, F.; Zhai, L.; Wang, K.; Xiao, M.; Xie, T.; Sui, X. Non-apoptotic cell death in malignant tumor cells and natural compounds. *Cancer Lett.* **2018**, *420*, 210–227. [CrossRef]

68. Holze, C.; Michaudel, C.; Mackowiak, C.; Haas, D.A.; Benda, C.; Hubel, P.; Pennemann, F.L.; Schnepf, D.; Wettmarshausen, J.; Braun, M.; et al. Oxeiptosis, a ROS-induced caspase-independent apoptosis-like cell-death pathway. *Nat. Immunol.* **2018**, *19*, 130–140. [CrossRef] [PubMed]

69. Chai, J.; Shiozaki, E.; Srinivasula, S.M.; Wu, Q.; Datta, P.; Alnemri, E.S.; Shi, Y. Structural basis of caspase-7 inhibition by XIAP. *Cell* **2001**, *104*, 769–780. [CrossRef]

70. Shiozaki, E.N.; Chai, J.; Rigotti, D.J.; Riedl, S.J.; Li, P.; Srinivasula, S.M.; Alnemri, E.S.; Fairman, R.; Shi, Y. Mechanism of XIAP-mediated inhibition of caspase-9. *Mol. Cell* **2003**, *11*, 519–527. [CrossRef]

71. Suzuki, Y.; Nakabayashi, Y.; Nakata, K.; Reed, J.C.; Takahashi, R. X-linked inhibitor of apoptosis protein (XIAP) inhibits caspase-3 and -7 in distinct modes. *J. Biol. Chem.* **2001**, *276*, 27058–27063. [CrossRef] [PubMed]

72. Jung, S.; Li, C.; Duan, J.; Lee, S.; Kim, K.; Park, Y.; Yang, Y.; Kim, K.I.; Lim, J.S.; Cheon, C.I.; et al. TRIP-Br1 oncoprotein inhibits autophagy, apoptosis, and necroptosis under nutrient/serum-deprived condition. *Oncotarget* **2015**, *6*, 29060–29075. [CrossRef] [PubMed]

73. Lewis, J.; Burstein, E.; Reffey, S.B.; Bratton, S.B.; Roberts, A.B.; Duckett, C.S. Uncoupling of the signaling and caspase-inhibitory properties of X-linked inhibitor of apoptosis. *J. Biol. Chem.* **2004**, *279*, 9023–9029. [CrossRef] [PubMed]

74. Burstein, E.; Ganesh, L.; Dick, R.D.; van De Sluis, B.; Wilkinson, J.C.; Klomp, L.W.; Wijmenga, C.; Brewer, G.J.; Nabel, G.J.; Duckett, C.S. A novel role for XIAP in copper homeostasis through regulation of MURR1. *EMBO J.* **2004**, *23*, 244–254. [CrossRef] [PubMed]

75. Sanna, M.G.; da Silva Correia, J.; Ducrey, O.; Lee, J.; Nomoto, K.; Schrantz, N.; Deveraux, Q.L.; Ulevitch, R.J. IAP suppression of apoptosis involves distinct mechanisms: The TAK1/JNK1 signaling cascade and caspase inhibition. *Mol. Cell Biol.* **2002**, *22*, 1754–1766. [CrossRef] [PubMed]

76. Lewis, E.M.; Wilkinson, A.S.; Davis, N.Y.; Horita, D.A.; Wilkinson, J.C. Nondegradative ubiquitination of apoptosis inducing factor (AIF) by X-linked inhibitor of apoptosis at a residue critical for AIF-mediated chromatin degradation. *Biochemistry* **2011**, *50*, 11084–11096. [CrossRef] [PubMed]

77. Levkau, B.; Garton, K.J.; Ferri, N.; Kloke, K.; Nofer, J.R.; Baba, H.A.; Raines, E.W.; Breithardt, G. xIAP induces cell-cycle arrest and activates nuclear factor-kappaB: New survival pathways disabled by caspase-mediated cleavage during apoptosis of human endothelial cells. *Circ. Res.* **2001**, *88*, 282–290. [CrossRef] [PubMed]

78. Mufti, A.R.; Burstein, E.; Csomos, R.A.; Graf, P.C.; Wilkinson, J.C.; Dick, R.D.; Challa, M.; Son, J.K.; Bratton, S.B.; Su, G.L.; et al. XIAP Is a copper binding protein deregulated in Wilson's disease and other copper toxicosis disorders. *Mol. Cell* **2006**, *21*, 775–785. [CrossRef] [PubMed]

79. Wicki, S.; Gurzeler, U.; Wei-Lynn Wong, W.; Jost, P.J.; Bachmann, D.; Kaufmann, T. Loss of XIAP facilitates switch to TNFalpha-induced necroptosis in mouse neutrophils. *Cell Death Dis.* **2016**, *7*, e2422. [CrossRef] [PubMed]

80. Yabal, M.; Jost, P.J. XIAP as a regulator of inflammatory cell death: The TNF and RIP3 angle. *Mol. Cell. Oncol.* **2015**, *2*, e964622. [CrossRef] [PubMed]

81. Yabal, M.; Muller, N.; Adler, H.; Knies, N.; Gross, C.J.; Damgaard, R.B.; Kanegane, H.; Ringelhan, M.; Kaufmann, T.; Heikenwalder, M.; et al. XIAP restricts TNF- and RIP3-dependent cell death and inflammasome activation. *Cell Rep.* **2014**, *7*, 1796–1808. [CrossRef] [PubMed]

82. Lawlor, K.E.; Feltham, R.; Yabal, M.; Conos, S.A.; Chen, K.W.; Ziehe, S.; Grass, C.; Zhan, Y.; Nguyen, T.A.; Hall, C.; et al. XIAP Loss Triggers RIPK3- and Caspase-8-Driven IL-1beta Activation and Cell Death as a Consequence of TLR-MyD88-Induced cIAP1-TRAF2 Degradation. *Cell Rep.* **2017**, *20*, 668–682. [CrossRef] [PubMed]

83. Poojari, R. Embelin—A drug of antiquity: Shifting the paradigm towards modern medicine. *Expert Opin. Investig. Drugs* **2014**, *23*, 427–444. [CrossRef] [PubMed]

84. Ko, J.H.; Lee, S.G.; Yang, W.M.; Um, J.Y.; Sethi, G.; Mishra, S.; Shanmugam, M.K.; Ahn, K.S. The Application of Embelin for Cancer Prevention and Therapy. *Molecules* **2018**, *23*, 621. [CrossRef] [PubMed]

85. Prabhu, K.S.; Achkar, I.W.; Kuttikrishnan, S.; Akhtar, S.; Khan, A.Q.; Siveen, K.S.; Uddin, S. Embelin: A benzoquinone possesses therapeutic potential for the treatment of human cancer. *Future Med. Chem.* **2018**, *10*, 961–976. [CrossRef] [PubMed]

86. Nikolovska-Coleska, Z.; Xu, L.; Hu, Z.; Tomita, Y.; Li, P.; Roller, P.P.; Wang, R.; Fang, X.; Guo, R.; Zhang, M.; et al. Discovery of embelin as a cell-permeable, small-molecular weight inhibitor of XIAP through structure-based computational screening of a traditional herbal medicine three-dimensional structure database. *J. Med. Chem.* **2004**, *47*, 2430–2440. [CrossRef]

87. Chessari, G.; Buck, I.M.; Day, J.E.; Day, P.J.; Iqbal, A.; Johnson, C.N.; Lewis, E.J.; Martins, V.; Miller, D.; Reader, M.; et al. Fragment-Based Drug Discovery Targeting Inhibitor of Apoptosis Proteins: Discovery of a Non-Alanine Lead Series with Dual Activity Against cIAP1 and XIAP. *J. Med. Chem.* **2015**, *58*, 6574–6588. [CrossRef]

88. Jin, X.; Lee, K.; Kim, N.H.; Kim, H.S.; Yook, J.I.; Choi, J.; No, K.T. Natural products used as a chemical library for protein-protein interaction targeted drug discovery. *J. Mol. Graph. Model.* **2018**, *79*, 46–58. [CrossRef]

89. Johnson, C.N.; Ahn, J.S.; Buck, I.M.; Chiarparin, E.; Day, J.E.H.; Hopkins, A.; Howard, S.; Lewis, E.J.; Martins, V.; Millemaggi, A.; et al. A Fragment-Derived Clinical Candidate for Antagonism of X-Linked and Cellular Inhibitor of Apoptosis Proteins: 1-(6-[(4-Fluorophenyl)methyl]-5-(hydroxymethyl)-3,3-dimethyl-1H, 2H,3H-pyrrolo[3,2-b]pyridin-1-yl)-2-[(2R,5R)-5-methyl-2-([(3R)-3-methylmorpholin-4-yl]methyl)piperazin-1 -yl]ethan-1-one (ASTX660). *J. Med. Chem.* **2018**, *61*, 7314–7329.

90. Kashkar, H. X-linked inhibitor of apoptosis: A chemoresistance factor or a hollow promise. *Clin. Cancer Res.* **2010**, *16*, 4496–4502. [CrossRef]

91. Shah, P.; Djisam, R.; Damulira, H.; Aganze, A.; Danquah, M. Embelin inhibits proliferation, induces apoptosis and alters gene expression profiles in breast cancer cells. *Pharmacol. Rep.* **2016**, *68*, 638–644. [CrossRef] [PubMed]

92. Hahm, E.R.; Singh, S.V. Withaferin A-induced apoptosis in human breast cancer cells is associated with suppression of inhibitor of apoptosis family protein expression. *Cancer Lett.* **2013**, *334*, 101–108. [CrossRef] [PubMed]

93. Liu, J.; Zhang, X.; Liu, A.; Liu, S.; Zhang, L.; Wu, B.; Hu, Q. Berberine induces apoptosis in p53-null leukemia cells by down-regulating XIAP at the post-transcriptional level. *Cell. Physiol. Biochem.* **2013**, *32*, 1213–1224. [CrossRef] [PubMed]

94. Lima, R.T.; Martins, L.M.; Guimaraes, J.E.; Sambade, C.; Vasconcelos, M.H. Specific downregulation of bcl-2 and xIAP by RNAi enhances the effects of chemotherapeutic agents in MCF-7 human breast cancer cells. *Cancer Gene Ther.* **2004**, *11*, 309–316. [CrossRef] [PubMed]

95. Sensintaffar, J.; Scott, F.L.; Peach, R.; Hager, J.H. XIAP is not required for human tumor cell survival in the absence of an exogenous death signal. *BMC Cancer* **2010**, *10*, 11. [CrossRef] [PubMed]

96. Zhang, Y.; Wang, Y.; Gao, W.; Zhang, R.; Han, X.; Jia, M.; Guan, W. Transfer of siRNA against XIAP induces apoptosis and reduces tumor cells growth potential in human breast cancer in vitro and in vivo. *Breast Cancer Res. Treat* **2006**, *96*, 267–277. [CrossRef] [PubMed]

97. Foster, F.M.; Owens, T.W.; Tanianis-Hughes, J.; Clarke, R.B.; Brennan, K.; Bundred, N.J.; Streuli, C.H. Targeting inhibitor of apoptosis proteins in combination with ErbB antagonists in breast cancer. *Breast Cancer Res. BCR* **2009**, *11*, R41. [CrossRef] [PubMed]

98. Montero, J.; Letai, A. Why do BCL-2 inhibitors work and where should we use them in the clinic? *Cell Death Differ.* **2018**, *25*, 56–64. [CrossRef] [PubMed]

99. Delbridge, A.R.; Strasser, A. The BCL-2 protein family, BH3-mimetics and cancer therapy. *Cell Death Differ.* **2015**, *22*, 1071–1080. [CrossRef]

100. Giampazolias, E.; Zunino, B.; Dhayade, S.; Bock, F.; Cloix, C.; Cao, K.; Roca, A.; Lopez, J.; Ichim, G.; Proics, E.; et al. Mitochondrial permeabilization engages NF-kappaB-dependent anti-tumour activity under caspase deficiency. *Nat. Cell Biol.* **2017**, *19*, 1116–1129. [CrossRef] [PubMed]

101. Armani, C.; Catalani, E.; Balbarini, A.; Bagnoli, P.; Cervia, D. Expression, pharmacology, and functional role of somatostatin receptor subtypes 1 and 2 in human macrophages. *J. Leukoc. Biol.* **2007**, *81*, 845–855. [CrossRef] [PubMed]

102. Cervia, D.; Martini, D.; Garcia-Gil, M.; Di Giuseppe, G.; Guella, G.; Dini, F.; Bagnoli, P. Cytotoxic effects and apoptotic signalling mechanisms of the sesquiterpenoid euplotin C, a secondary metabolite of the marine ciliate *Euplotes crassus*, in tumour cells. *Apoptosis* **2006**, *11*, 829–843. [CrossRef] [PubMed]

103. Cervia, D.; Garcia-Gil, M.; Simonetti, E.; Di Giuseppe, G.; Guella, G.; Bagnoli, P.; Dini, F. Molecular mechanisms of euplotin C-induced apoptosis: Involvement of mitochondrial dysfunction, oxidative stress and proteases. *Apoptosis* **2007**, *12*, 1349–1363. [CrossRef] [PubMed]

104. Perrotta, C.; Buldorini, M.; Assi, E.; Cazzato, D.; De Palma, C.; Clementi, E.; Cervia, D. The thyroid hormone triiodothyronine controls macrophage maturation and functions: Protective role during inflammation. *Am. J. Pathol.* **2014**, *184*, 230–247. [CrossRef]

105. Di Giuseppe, G.; Cervia, D.; Vallesi, A. Divergences in the Response to Ultraviolet Radiation Between Polar and Non-Polar Ciliated Protozoa: UV Radiation Effects in Euplotes. *Microb. Ecol.* **2011**, *63*, 334–338. [CrossRef] [PubMed]

106. Perrotta, C.; Buonanno, F.; Zecchini, S.; Giavazzi, A.; Proietti Serafini, F.; Catalani, E.; Guerra, L.; Belardinelli, M.C.; Picchietti, S.; Fausto, A.M.; et al. Climacostol reduces tumour progression in a mouse model of melanoma via the p53-dependent intrinsic apoptotic programme. *Sci. Rep.* **2016**, *6*, 27281. [CrossRef] [PubMed]

107. Bizzozero, L.; Cazzato, D.; Cervia, D.; Assi, E.; Simbari, F.; Pagni, F.; De Palma, C.; Monno, A.; Verdelli, C.; Querini, P.R.; et al. Acid sphingomyelinase determines melanoma progression and metastatic behaviour via the microphtalmia-associated transcription factor signalling pathway. *Cell Death Differ.* **2014**, *21*, 507–520. [CrossRef]

108. Zecchini, S.; Proietti Serafini, F.; Catalani, E.; Giovarelli, M.; Coazzoli, M.; Di Renzo, I.; De Palma, C.; Perrotta, C.; Clementi, E.; Buonanno, F.; et al. Dysfunctional autophagy induced by the pro-apoptotic natural compound climacostol in tumour cells. *Cell Death Dis.* **2019**, *10*, 10. [CrossRef]

109. Assi, E.; Cervia, D.; Bizzozero, L.; Capobianco, A.; Pambianco, S.; Morisi, F.; De Palma, C.; Moscheni, C.; Pellegrino, P.; Clementi, E.; et al. Modulation of Acid Sphingomyelinase in Melanoma Reprogrammes the Tumour Immune Microenvironment. *Mediat. Inflamm* **2015**, *2015*, 370482. [CrossRef]

110. Cervia, D.; Assi, E.; De Palma, C.; Giovarelli, M.; Bizzozero, L.; Pambianco, S.; Di Renzo, I.; Zecchini, S.; Moscheni, C.; Vantaggiato, C.; et al. Essential role for acid sphingomyelinase-inhibited autophagy in melanoma response to cisplatin. *Oncotarget* **2016**, *7*, 24995–25009. [CrossRef]

111. Perrotta, C.; De Palma, C.; Clementi, E.; Cervia, D. Hormones and immunity in cancer: Are thyroid hormones endocrine players in the microglia/glioma cross-talk? *Front. Cell. Neurosci.* **2015**, *9*, 236. [CrossRef] [PubMed]

112. Vantaggiato, C.; Castelli, M.; Giovarelli, M.; Orso, G.; Bassi, M.T.; Clementi, E.; De Palma, C. The Fine Tuning of Drp1-Dependent Mitochondrial Remodeling and Autophagy Controls Neuronal Differentiation. *Front. Cell. Neurosci.* **2019**, *13*, 120. [CrossRef] [PubMed]

113. Pfisterer, P.H.; Wolber, G.; Efferth, T.; Rollinger, J.M.; Stuppner, H. Natural products in structure-assisted design of molecular cancer therapeutics. *Curr. Pharm. Des.* **2010**, *16*, 1718–1741. [CrossRef] [PubMed]

114. Cazzato, D.; Assi, E.; Moscheni, C.; Brunelli, S.; De Palma, C.; Cervia, D.; Perrotta, C.; Clementi, E. Nitric oxide drives embryonic myogenesis in chicken through the upregulation of myogenic differentiation factors. *Exp. Cell Res.* **2014**, *320*, 269–280. [CrossRef] [PubMed]

115. Cervia, D.; Catalani, E.; Belardinelli, M.C.; Perrotta, C.; Picchietti, S.; Alimenti, C.; Casini, G.; Fausto, A.M.; Vallesi, A. The protein pheromone Er-1 of the ciliate Euplotes raikovi stimulates human T-cell activity: Involvement of interleukin-2 system. *Exp. Cell Res.* **2013**, *319*, 56–67. [CrossRef] [PubMed]

Article

Molecular Mechanisms Underlying Yatein-Induced Cell-Cycle Arrest and Microtubule Destabilization in Human Lung Adenocarcinoma Cells

Shang-Tse Ho [1,2], Chi-Chen Lin [3], Yu-Tang Tung [4,5,6,*] and Jyh-Horng Wu [1,*]

[1]   Department of Forestry, National Chung Hsing University, Taichung 402, Taiwan; s741129@hotmail.com
[2]   Agricultural Biotechnology Research Center, Academia Sinica, Taipei 115, Taiwan
[3]   Institute of Biomedical Science, National Chung Hsing University, Taichung 402, Taiwan; lincc@nchu.edu.tw
[4]   Graduate Institute of Metabolism and Obesity Sciences, Taipei Medical University, Taipei 110, Taiwan
[5]   Nutrition Research Center, Taipei Medical University Hospital, Taipei 110, Taiwan
[6]   Cell Physiology and Molecular Image Research Center, Wan Fang Hospital, Taipei Medical University, Taipei 116, Taiwan
*   Correspondence: f91625059@tmu.edu.tw (Y.-T.T.); eric@nchu.edu.tw (J.-H.W.);
    Tel.: +886-2-27361661 (ext. 7352) (Y.-T.T.); +886-4-22840345 (ext. 136) (J.-H.W.)

Received: 30 July 2019; Accepted: 12 September 2019; Published: 17 September 2019

Abstract: Yatein is an antitumor agent isolated from *Calocedrus formosana* Florin leaves extract. In our previous study, we found that yatein inhibited the growth of human lung adenocarcinoma A549 and CL1-5 cells by inducing intrinsic and extrinsic apoptotic pathways. To further uncover the effects and mechanisms of yatein-induced inhibition on A549 and CL1-5 cell growth, we evaluated yatein-mediated antitumor activity in vivo and the regulatory effects of yatein on cell-cycle progression and microtubule dynamics. Flow cytometry and western blotting revealed that yatein induces G_2/M arrest in A549 and CL1-5 cells. Yatein also destabilized microtubules and interfered with microtubule dynamics in the two cell lines. Furthermore, we evaluated the antitumor activity of yatein in vivo using a xenograft mouse model and found that yatein treatment altered cyclin B/Cdc2 complex expression and significantly inhibited tumor growth. Taken together, our results suggested that yatein effectively inhibited the growth of A549 and CL1-5 cells possibly by disrupting cell-cycle progression and microtubule dynamics.

Keywords: *Calocedrus formosana*; lung cancer; yatein; cell-cycle arrest; xenograft

---

1. Introduction

Natural products have been used for treating disease for thousands of years. More recently, natural products are being continuously developed for pharmaceutical applications, particularly for anticancer, antibacterial, and antiviral applications [1]. To date, explorative and mechanistic studies on new bioactive compounds are still ongoing.

Cells routinely come in contact with endogenous and exogenous stress stimuli, such as toxic chemicals, UV radiation, and reactive oxygen species. These stress stimuli damage chromosomes, affecting DNA replication and chromosome segregation [2]. Each phase of the cell cycle, including G_0, G_1, S, G_2, and M phase, functions to maintain genetic material duplication and cell division. Abnormal expression of cell-cycle proteins disrupts the cell cycle and is thus closely associated with tumorigenesis [3]. Many natural products exhibit anticancer properties by interacting with cell-cycle proteins [4]. Thus, the identification of a new cell-cycle–targeting natural substance may provide new alternatives in cancer chemotherapy.

Calocedrus formosana is a valuable softwood species in Taiwan, which not only has high industrial economic value but also exhibits multiple bioactivities [5–14]. We previously found that the *C. formosana*

extract and its active phytocompound, yatein, inhibited the growth of human lung adenocarcinoma A549 and CL1-5 cells by inducing caspase-related apoptosis [15]. However, whether yatein regulates the cell cycle in human lung adenocarcinoma remains unclear. To uncover the mechanisms of yatein-mediated human lung adenocarcinoma growth inhibition, we examined the effects of yatein on cell-cycle progression, tubulin dynamics, and in vivo tumor growth.

2. Results

2.1. Yatein Induces Cell-Cycle Arrest at G_2/M Phase and Enhances G_2/M Phase-Related Protein Expression in Human A549 and CL1-5 Cells

To elucidate the mechanism underlying the anti-lung adenocarcinoma effects of yatein, cell-cycle distribution was analyzed in the yatein-treated A549 and CL1-5 cells. We found that 5 µM yatein treatment induced cell-cycle arrest at G_2/M phase in both cell lines (Figure 1). We further analyzed the kinetics of the effects of yatein on A549 and CL1-5 cells through flow cytometry (Figure 2). Compared with untreated cells, we found that more cells entered the G_2/M phase at 6 and 12 h after yatein treatment in both cell types. Next, we evaluated the effects of yatein on G_2/M arrest-related protein expression using western blot analysis (Figure 3). To this end, A549 and CL1-5 cells were treated with 5 µM yatein for 6 and 12 h, and the expression of Cdc2, Cdc25c, and cyclin B1 was analyzed (Figure 3). Cdc2, Cdc25c, and cyclin B1 are key regulators of the cell cycle (particularly in the G_2/M phase). Our results revealed that 6 and 12 h of yatein treatment upregulated cyclin B1, but not Cdc2 and Cdc25c, expression in A549 and CL1-5 cells. However, yatein treatment showed an increasing trend of Cdc2 phosphorylation in both cell types. Notably, yatein-induced Cdc2 phosphorylation was higher at 6 h than at 12 h in both the cell types, indicating that Cdc2 was involved in G_2/M phase regulation in A549 and CL1-5 cells at an early stage.

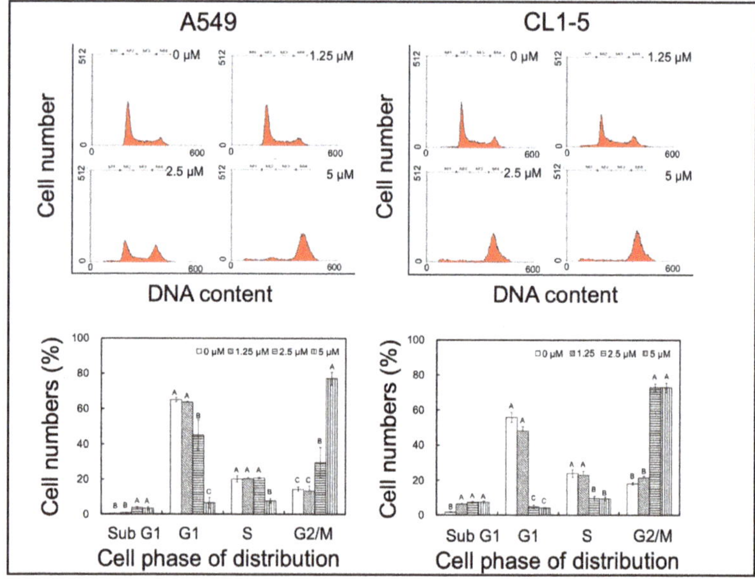

Figure 1. Effects of yatein treatment for 24 h with different concentrations on cell-cycle progression in A549 and CL1-5 cells. The results represent the mean ± SD (n = 3). Different letters indicate significant differences among each group in A549 and CL1-5 cells (p < 0.05).

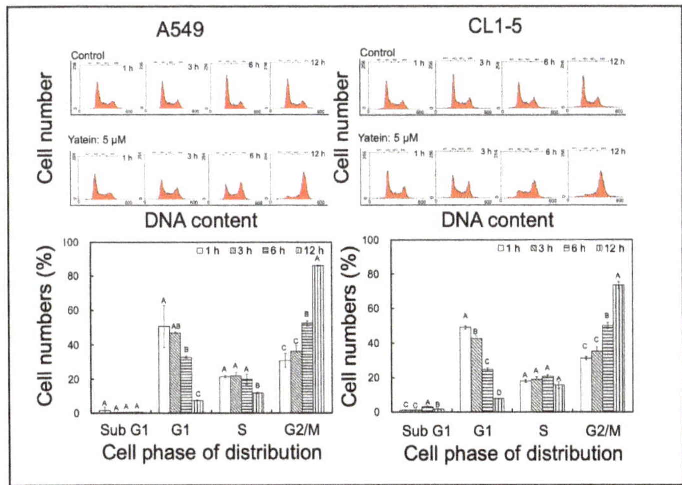

Figure 2. Effect kinetics of 5 μM yatein treatment on cell-cycle progression in A549 and CL1-5 cells. The results represent the mean ± SD (*n* = 3). Different letters indicate significant differences among each group in A549 and CL1-5 cells (*p* < 0.05).

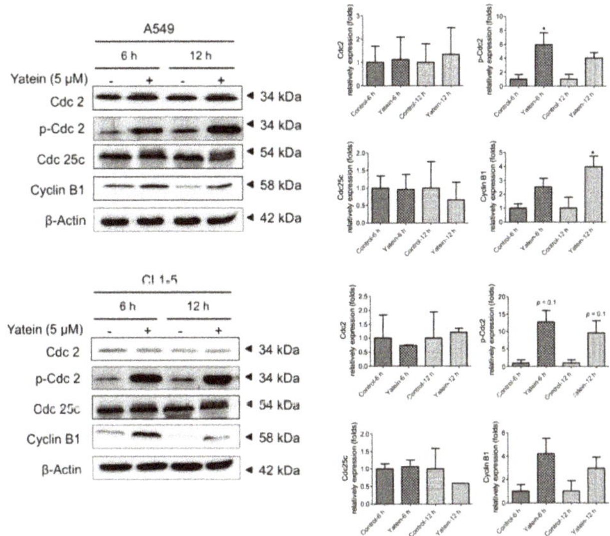

Figure 3. Expression of cell-cycle regulatory proteins in A549 and CL1-5 cells after yatein treatment (5 μM) for 6 h and 12 h. The bands were analyzed using the ImageJ software and normalized to β-actin expression. All data presented are representative of three independent experiments. The quantifications represent the mean ± SEM (*n* = 2-3). * indicates a significant difference compared with the control group (*p* < 0.05).

2.2. Yatein Induces DNA Damage through Activation of the ATM/ATR Pathway in Human A549 and CL1-5 Cells

DNA damage induces cell-cycle arrest and apoptosis in cancer cells [16]. The ATM/ATR pathway is related to DNA damage process. To address whether yatein induced DNA damage in cells, we examined the effects of yatein treatment on the ATM/ATR pathway. We found that yatein treatment showed an increasing trend of ATM and ATR phosphorylation level in A549 and CL1-5 cells for 6 h and 12 h treatments. However, ATM and ATR expression were not affected (Figure 4A). We next evaluated the expression and phosphorylation of Chk1 and Chk2 in A549 and CL1-5 cells after yatein treatment for 6 and 12 h. Our results revealed that yatein treatment showed an increasing trend of Chk1 and Chk2 phosphorylation level in A549 and CL1-5 cells. These results suggested that yatein induced DNA damage and altered cell-cycle distribution by activating the ATM/Chk2 and ATR/Chk1 DNA repair pathways in A549 and CL1-5 cells. In addition to the activation of the ATM/ATR pathway, we found that yatein treatment affected p53, Wee1, and 14-3-3 σ expression in the A549 cells (Figure 4B). Conversely, yatein did not upregulate the p53 and 14-3-3 σ expression in CL1-5 cells. However, Wee1 expression was increased after yatein treatment for 6 h in CL1-5 cells. These results indicated that the p53 pathway might play an additional role in yatein-induced growth inhibition in A549 cells but not in CL1-5 cells.

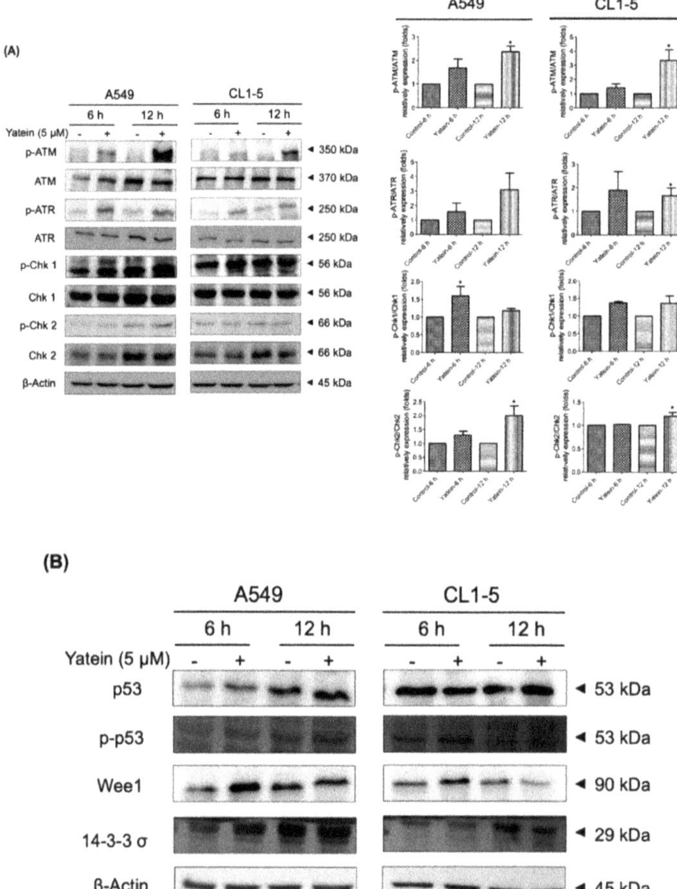

Figure 4. Expression of ATM, ATR, Chk1, and Chk2 (**A**), and p53 related proteins (**B**) in A549 and CL1-5 cells after yatein treatment (5 μM) for 6 and 12 h. All data presented are representative of two to three independent experiments. The quantifications represent the mean ± SEM (n = 2–3). * indicates significant differences compared with the control group in A549 and CL1-5 cells ($p < 0.05$).

2.3. Yatein Influences Microtubule Dynamics in Human A549 and CL1-5 Cells

Apart from DNA damage, inhibition of microtubule dynamics is another mechanism that induces G_2/M cell-cycle arrest. For instance, taxol, a well-known anticancer drug, arrests cells in the G_2/M phase by disrupting microtubule polymerization [17]. We assessed microtubule dynamics in yatein-treated A549 and CL1-5 cells using confocal microscopy and western blot analysis. Our confocal microscopy results showed that 5 µM yatein treatment caused a diffusion of green fluorescence (tubulin) in the yatein-treated A549 and CL1-5 cells compared with control cells within 6 h (Figure 5A). In addition, as shown in Figure 5B, yatein decreased tubulin polymerization in a dose-dependent manner in both A549 and CL1-5 cells after 24 h treatment. These results revealed that the effects of yatein on microtubule dynamics were similar to those of colchicine, a microtubule-depolymerizing agent (Figure 5C). Consistently, yatein-treated cells exhibited a similar pattern of tubulin distribution to that of colchicine-treated cells. By contrast, taxol-treated cells demonstrated a higher level of tubulin polymerization compared with the untreated cells. Taken together, our results indicated that yatein and colchicine affected microtubule dynamics by inhibiting tubulin polymerization.

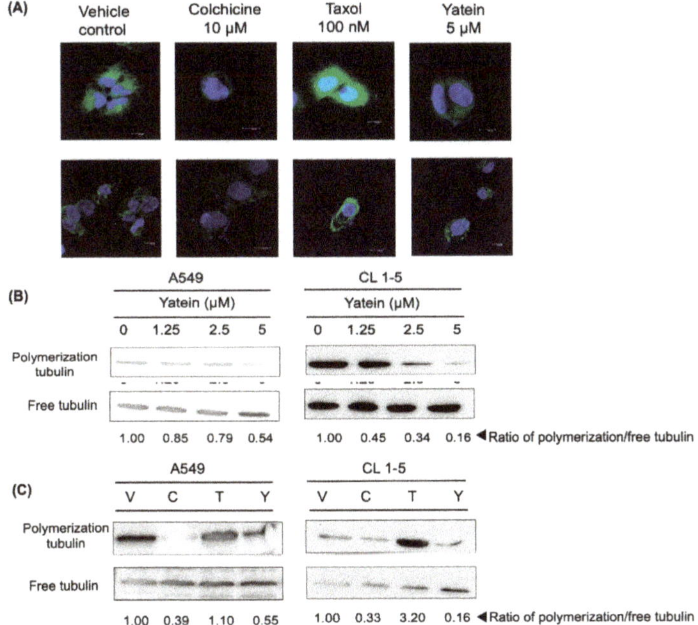

Figure 5. (**A**) Confocal images of tubulin expression of A549 (upper) and CL1-5 (lower) cells after yatein, colchicine, and taxol treatment for 6 h. Scale bar = 100 µm. (**B**) Western blot analysis of tubulin expression of A549 and CL1-5 cells after treatment with various concentrations of yatein for 24 h. (**C**) Western blot analysis of tubulin expression of A549 and CL1-5 cells after yatein and positive control treatment for 24 h. V: Vehicle control, C: 10 µM colchicine, T: 100 nM taxol, Y: 5 µM yatein. The quantitative values are the ratio of polymerization tubulin/free tubulin in different treatment groups.

2.4. Yatein Exhibits In Vivo Antitumor Effects in a Xenograft Mouse Model

To validate the growth inhibitory effects of yatein in an in vivo context, we applied a xenograft mouse model using A549 cells. To this end, NOD/SCID mice were inoculated with A549-luc cells for 10 days and were then given 20 mg/kg yatein (i.p.) five times per week for 42 days. Animal body weight, food intake, and tumor growth were monitored and quantified during the experiment. Our results showed that the body weight and food intake of the control and yatein-treated mice did not differ

(Figure 6A,B), suggesting that yatein was well tolerated in the mice. We found that the tumor volumes of the control mice (41.3 mm^3) and yatein-treated mice (38.5 mm^3) were similar in the first 14 days (Figure 6C). However, within the 21-42 days (end of experiment) time window, tumor growth in the yatein-treated mice was significantly slower compared with the control mice ($p < 0.05$). Consistently, In Vivo Imaging System (IVIS) analysis showed that the tumor tissue luminescence in the yatein-treated group (16.5 × 10^5 photons/s) was lower than that in the vehicle control group (24.0 × 10^5 photons/s). To confirm that yatein exhibited the same antitumor mechanisms in vivo as observed in vitro, we evaluated cyclin B1 expression and Cdc2 phosphorylation in the tumor tissue in the vehicle control and yatein-treated groups (Figure 6D). We found that both cyclin B1 expression and Cdc2 phosphorylation moderately increased in the yatein-treated group compared with the control group (Figure 6D). Taken together, these results suggested that the induction of cyclin B/Cdc2 complex expression and activation were associated with the antitumor effects of yatein in vivo.

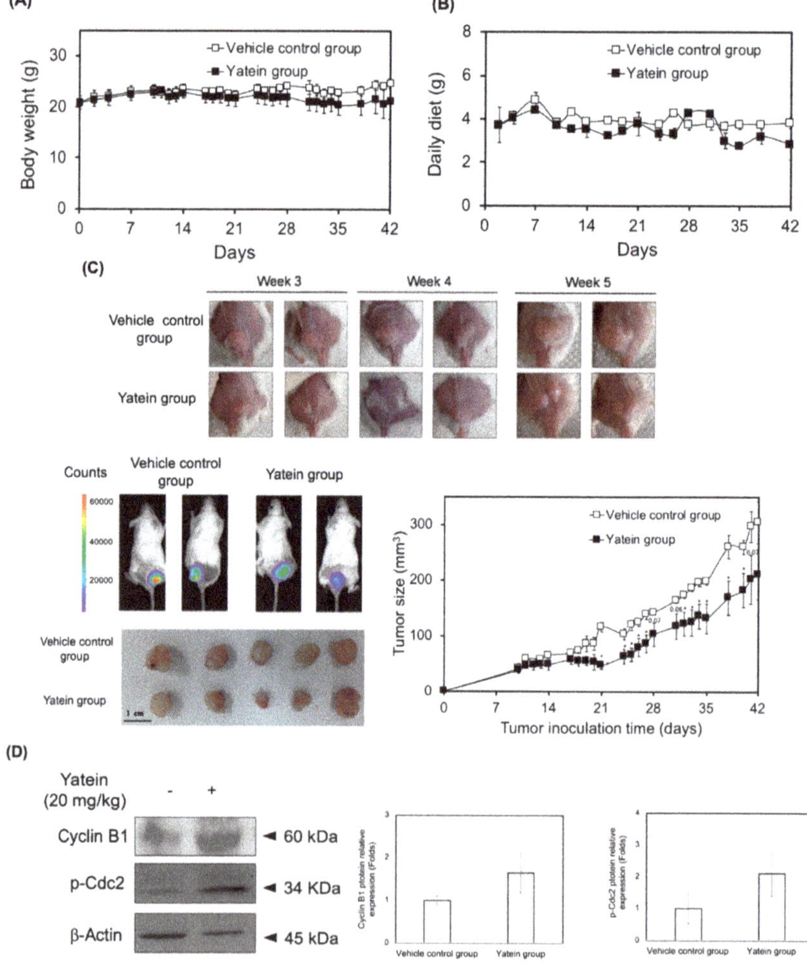

Figure 6. In vivo antitumor effects of yatein. (**A**) Body weight, (**B**) daily diet intake, and (**C**) tumor size and IVIS images of A549-luc cell xenograft mice treated with vehicle or 20 mg/kg of yatein during the experimental period. The results represent the mean ± SEM (*n* = 5). (**D**) Western blot analysis of tumor tissue from the A549-luc xenograft mice (values were mean ± SEM, *n* = 3).

3. Discussion

According to our previous study [15], the IC_{50} values for the 24 h yatein treatment in the A549 and CL1-5 cells were 10.0 and 2.1 µM, respectively. In addition, yatein exhibited excellent growth inhibitory effect in the A549 (IC_{50} values = 3.5 µM) and CL1-5 cells (IC_{50} values = 1.9 µM) after 72 h treatment. Interestingly, yatein was unable to inhibit human nasopharyngeal carcinoma (HONE-1) and human gastric cancer (NUGC) cells growth at a concentration of 50 µg/mL ($\cong$ 125 µM) [10]. In the present study, we found that yatein induced G_2/M cell-cycle arrest in A549 and CL1-5 cells at the first 24 h treatment. Additionally, treatment with yatein for 48 h induced a dose-dependent increase in both early and late stage apoptosis in the A549 and CL1-5 cells [15]. As previously reported, some lignan compounds can inhibit cancer cell growth by inducing cell-cycle arrest. For example, vitexin compound-1 (VB1), a lignan isolated from a plant used in traditional Chinese medicine, namely *Vitex negundo*, inhibits the growth of MDA-MB-435 and SMMC-7721 cells by inducing G_2/M phase cell-cycle arrest after 24 h treatment at a concentration of 10 µM [18]. Similarly, benozofuran lignan shows a dose- and time-dependent induction of G_2/M cell-cycle arrest in Jurkat T-cells [19]. Additionally, the mechanism underlying the effects of yatein on cell-cycle regulation was examined in the present study. Cyclin-dependent kinases (CDKs) collaborate with specific cyclins to tightly regulate cell cycle and cell division. Thus, the CDK/cyclin complex plays a key role in cell-cycle progression [20]. Our results indicated that yatein treatment showed an increasing trend of cyclin B1 expression and Cdc2 phosphorylation level. Aberrant activation of cyclin B1/p-Cdc2 activity is closely associated with mitotic catastrophe. Mitotic catastrophe is a type of cell death that results from mitotic regulation dysfunctions and can be induced by chemical and physical stresses [21]. Liu et al. [22] reported that expression of cyclin B1 and level of Cdc2 phosphorylation were increased in malignant glioma cells treated with a synthetic quinazolinone analog (MJ-66), suggesting that MJ-66 induced G_2/M arrest and mitotic catastrophe in malignant glioma cells. Subramaniam et al. [23] reported that curcumin-induced mitotic catastrophe coupled with increased expression of cyclin B1 and Cdc-2 in MiaPaCa-2 cells. Therefore, we can speculate the cytotoxic effects of yatein on A549 and CL1-5 cells might be partially due to the induction of mitotic catastrophe. The correlation between yatein-induced cell death and mitotic catastrophe still need to be investigated in a future study.

DNA damage is vital in the regulation of cell cycle and apoptosis. In this study, we found that yatein induced cell-cycle arrest by inducing DNA damage in A549 and CL1-5 cells. In general, ATM and ATR kinases are the initiating kinases of the DNA damage response (DDR) signaling pathway that is activated by DNA damage or DNA replication stress [24]. ATM is activated by DNA double-strand breaks, while ATR is activated by various factors related to DNA damage [25]. According to previous studies, several natural products, including berberine, curcumin, and sinularin, induced DNA damage and G_2/M arrest, but not S phase arrest, in various cancer cells [26–28]. The results revealed that these natural products induced DNA damage and arrested cells at G_2/M phase. At the same time, the ATM pathway was up-regulated [26–28]. After ATM activation, several proteins involved in the regulation of ATM on DNA repair, cell-cycle arrest, apoptosis, and other downstream processes, such as Brca1, Chk2, and p53, are phosphorylated [29]. Our results revealed that yatein induced the expression of ATR/Chk1 and ATM/Chk2 in A549 cells. Notably, only ATR/Chk1 expression was upregulated after yatein treatment in CL1-5 cells. We found that the expression and phosphorylation of p53, Wee1, and 14-3-3 σ expression were induced after yatein treatment in A549 cells, suggesting that yatein activated p53-mediated signaling pathway to inhibit growth in A549 cells. Yatein did not affect the expression of p53 and its related proteins in CL1-5 cells; this was because the CL1-5 cell line carries a p53 mutation, which has been implicated in more than 50% of all cancer cases [30]. The p53 mutation causes a defective G_1 checkpoint in cancer cells that results in increased DNA damage at the G_2 checkpoint compared with noncancer cells. Wee1 is a key protein that is closely associated with G_2 checkpoint abrogation and mitotic catastrophe [30,31]. Accordingly, we found that yatein treatment increased Wee1 expression in CL1-5 cells, indicating that Wee1 played a crucial role in the regulation of G_2/M cell-cycle distribution in CL1-5 cells in the context of yatein treatment. On the other hand, we also found yatein is able to induce ROS production in A549 and CL1-5 cells in our previous study [15], suggesting that yatein may induce the oxidative DNA damage. Moreover, our previous results also revealed that the CL1-5 cells were more sensitive for the ROS production than the A549 cells after yatein treatment [15].

Additionally, we showed that yatein affected the microtubule assembly in both A549 and CL1-5 cells after 6 h treatment (Figure 5A) and 24 h treatment (Figure 5B and 5C). The induction of mitotic arrest is associated with dysfunctional microtubule dynamics [32]. Microtubules play vital roles in cell proliferation, trafficking, signaling, and migration [33]. Considering the importance of microtubule dynamics, several small microtubule-targeting molecules have been designed and used as anticancer drugs [30]. Tubulin has three major binding sites: taxane, vinca, and colchicine domains. In this study, we found that yatein exhibited similar effects to that of colchicine, implying that yatein inhibited tubulin polymerization. These results suggested that tubulin polymerization inhibition partially contributed to the ability of yatein to arrest A549 and CL1-5 cells at the G_2/M phase.

The in vivo growth inhibitory effects of yatein on A549 cells were elucidated in this study. Our results revealed that yatein treatment significantly suppressed tumor growth in mice without affecting their body weight and food intake. Additionally, yatein treatment increased cyclin B1 expression and Cdc2 phosphorylation in vivo, indicating that the induction of mitotic catastrophe was involved in the anticancer mechanism of yatein in vivo. Taken together, we found that yatein affected cell-cycle progression and microtubule dynamics, induced DDR, and exhibited anticancer properties in vivo (Figure 7).

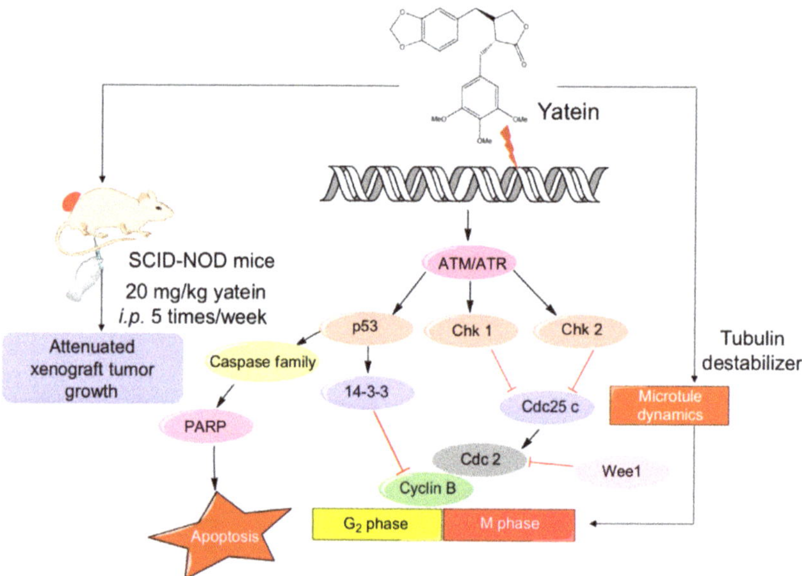

Figure 7. Proposed mechanism of the inhibitory effects of yatein on lung adenocarcinoma cells.

4. Materials and Methods

4.1. Preparation of Yatein

The phytocompound, yatein, was isolated from *C. formosana* leaves extracts. In brief, *C. formosana* leaves were extracted using methanol at room temperature (RT) for one week (twice) to obtain a methanolic extract. The dried samples were further divided to *n*-hexane, ethyl acetate (EtOAc), *n*-butanol, and H_2O fractions using liquid–liquid partition. The *n*-hexane fraction was further fractionated into ten subfractions using normal phase column chromatography (Geduran Si-60, Merck, Darmstadt, Germany). Yatein was isolated and purified from the subfraction 4 by semipreparative high-performance liquid chromatography using a PU-2080 pump (Jasco, Tokyo, Japan) equipped with an RI-2031 detector (Jasco) and a 5 µm Luna silica column (250 mm × 10.0 mm internal diameter; Phenomenex, Torrance, CA, USA). The mobile phase consisted of 30% of EtOAc and 70% of *n*-hexane (*v/v*), and the flow rate was 4 mL/min. The retention time of yatein in HPLC analysis was 18.0 min. The purity and the structure elucidation of yatein was conducted by ^{1}H and ^{13}C NMR, and all spectrum data were consistent with literature [34].

4.2. Cell Culture

Human A549 cell line was purchased from Bioresource Collection and Research Center (BCRC 60124) and cultured in RPMI-1640 (Gibco, Gran Island, NY, USA) supplemented with 10% (*v/v*) fetal bovine serum (Gibco) and 1% (*v/v*) antibiotic–antimycotic agent (Gibco). CL1-5 cell line was provided by Dr. Jeremy J.-W. Chen (National Chung Hsing University, Taichung, Taiwan) and cultured in Dulbecco's modified Eagle's medium (Gibco) supplemented with 10% (*v/v*) fetal bovine serum (Gibco) and 1% (*v/v*) antibiotic–antimycotic agent (Gibco). The cells were incubated in a 37 °C humidified incubator containing 5% CO_2.

4.3. Cell-Cycle Distribution Analysis

The A549 and CL1-5 cells were seeded onto a 6-well plate at a density of 1×10^5 cells/well and incubated overnight. The cells were then treated with various concentrations of the test samples for the indicated durations. After treatment, the cells were collected and mixed with ice-cold 75% ethanol and then fixed overnight at −20 °C. Then, the cells were washed twice using phosphate-buffered saline (PBS), and then incubated with 200 µL of propidium iodide (PI, BD Bioscience, Franklin Lakes, NJ, USA) solution (containing 1 mg/mL PI, 2 mg/mL RNase A, and 0.5% Triton X-100) for 30 min at RT in the dark. After 30 min, the PI-stained cells were immediately analyzed using a flow cytometer (Accuri 5, Accuri Cytometers, Inc., Ann Arbor, MI, USA). The data were analyzed using the C6 Accuri system software (Accuri Cytometers, Inc.).

4.4. Isolation of Microtubule Proteins

The A549 and CL1-5 cells (1×10^6 cells) were seeded onto culture dishes (10 cm) overnight and treated with 5 µM yatein. The cells were harvested through trypsinization and washed with PBS. After centrifugation, the supernatant was removed and the cell pellets were mixed with 200 µL of microtubule stabilizing buffer (1 mM $MgCl_2$, 2 mM Tris-HCl, 2 mM EGTA, and 0.5% Triton-100 in ddI$_2$O) and incubated at RT for 20 min. Subsequently, the mixture was centrifuged at 12,000× *g* for 10 min (4 °C) to obtain the supernatant (monomer tubulin fraction). The remaining cell pellets were washed using the microtubule stabilizing buffer and lysed in a radioimmunoprecipitation assay buffer containing 10% proteinase inhibitor and 10% phosphatase inhibitor (Sigma-Aldrich, St. Louis, MO, USA) at 4 °C for 30 min to obtain the polymer tubulin fraction. The monomer tubulin and polymer tubulin fractions were transferred into microtubes (1.5 mL) and stored at −20 °C until further analysis.

4.5. Western Blot Analysis

The expression of proteins in cells were determined using western blot analysis as previously reported [15]. In this study, the primary antibodies were anti-14-3-3 σ (Santa Cruz, Dallas, TX, USA), anti-β-actin (Santa Cruz), anti-ATM (Santa Cruz), anti-ATR (Santa Cruz), anti-Cdc2 (Abcam, Cambridge, MA, USA), anti-Cdc25c (Santa Cruz), anti-Chk1 (Santa Cruz), anti-Chk2 (Santa Cruz), anti-cyclin B1 (Abcam), anti-p53 (Cell Signaling Technology Inc, Danvers, MA, USA), anti-phospho-p53 Ser 15 (Cell Signaling Technology Inc), anti-phospho-ATM (Santa Cruz), anti-phospho-ATR (Santa Cruz), anti-phospho-Cdc2 (Santa Cruz), anti-phospho-Chk1 (Santa Cruz), anti-phospho-Chk2 (Santa Cruz), anti-β-tubulin (Abcam), and anti-Wee1 (Santa Cruz) antibodies. An enhanced chemiluminescence

(ECL, Sigma-Aldrich) system was used for developing signals of the blots, which were analyzed using a LAS3000 system (Fujifilm, Tokyo, Japan).

4.6. Immunofluorescence

The A549 and CL1-5 cells (1×10^5 cells) were seeded on a chamber slide for 24 h and treated with colchicine (10 µM), taxol (100 nM), and yatein (5 µM) for 6 h. The cells were washed with PBS and fixed with 1% paraformaldehyde (Sigma-Aldrich) at RT for 30 min. The paraformaldehyde was removed, and the cells were washed thrice with PBS. Next, the cells were blocked using 1% BSA in PBS at 4 °C overnight. After they were washed with PBS, the cells were incubated with the anti-β-tubulin antibody (1:200 in 1% BSA solution) at 4 °C overnight. After washing with PBS, an anti-rabbit-FITC antibody was added to the cells and then incubated overnight. Then, the antibody was rinsed off and the cells were washed thrice using PBS. 4,6-Diamidino-2-phenylindole was added and the cells were incubated for 15 min at RT in the dark. A Leica TCS SP2 confocal spectral microscope (Buffalo, NY, USA) was used to observe immunofluorescence staining of microtubule dynamics.

4.7. In Vivo Antitumor Activity

The procedures involving animals were performed according to the guidelines of the Institutional Animal Care and Use Committee of National Chung Hsing University (IACUC no. 107-127). The A549-luc cells were mixed with Matrigel (Sigma-Aldrich) at a 1:1 ratio. The cells were injected subcutaneously into the back of nonobese diabetic and severe combined immunodeficiency (NOD/SCID) mice (male, 6–8 weeks old) at a density of 3.5×10^6 cells/mouse. Tumors were allowed to grow for 10 days and were then treated with an intraperitoneal (i.p.) injection of either 0.5% DMSO in ddH$_2$O to the mice in the vehicle control group ($n = 5$) or 20 mg/kg of yatein (dissolved in 0.5% DMSO in ddH$_2$O) to the mice in the yatein group ($n = 5$). The tumor-bearing mice were sacrificed after 42 days. Tumor volume was measured five times/week and calculated using the following formula: Length × width × thickness × 0.5 (mm^3). An IVIS (Caliper lifescience IVIS Spectrum CT) was used to analyze the luminescence of tumor tissue.

4.8. Statistical Analysis

Data are expressed as mean ± standard deviation (SD) or mean ± standard error of the mean (SEM). Statistical analysis was performed using the shuffle test or non-parametric Kruskal-Wallis test with Dunn's post hoc tests. A p of < 0.05 was considered statistically significant.

5. Conclusions

The present study demonstrated that yatein suppressed lung adenocarcinoma cancer cells growth by inducing cell-cycle arrest, mitotic catastrophe, and microtubule depolymerization. Yatein induced G$_2$/M arrest by upregulating the expression of cyclin B1 and Cdc2 phosphorylation in lung adenocarcinoma cancer cells. In addition, mitotic catastrophe and microtubule depolymerization were involved in yatein-mediated lung adenocarcinoma cancer cells growth inhibition. Furthermore, we confirmed the in vivo antitumor effects of yatein using a xenograft mouse model. These findings provide novel insights into the in vitro and in vivo antitumor efficacy of yatein and demonstrate the potential of this phytocompound as an anticancer lead compound for lung adenocarcinoma cancer treatment.

Author Contributions: Conceptualization, S.-T.H., Y.-T.T., and J.-H.W.; methodology, S.-T.H., Y.-T.T. and J.-H.W.; formal analysis, S.-T.H. and C.-C.L.; resources, Y.-T.T. and J.-H.W.; writing—original draft, S.-T.H.; writing—review and editing, C.-C.L., Y.-T.T. and J.-H.W.; supervision, Y.-T.T. and J.-H.W.

Funding: This research work was supported in part by the Ministry of Science and Technology of Taiwan and Taipei Medical University under Grant Nos. MOST 107-2628-H-038-001-MY3 and 106-6820-001-112, respectively.

Conflicts of Interest: The authors declare that there is no conflict of interest regarding the publication of this paper.

References

1. Newman, D.J.; Cragg, G.M. Natural products as sources of new drugs from 1981 to 2014. *J. Nat. Prod.* **2016**, *79*, 629–661. [CrossRef] [PubMed]
2. Visconti, R.; Della Monica, R.; Grieco, D. Cell cycle checkpoint in cancer: a therapeutically targetable double-edged sword. *J. Exp. Clin. Cancer Res.* **2016**, *35*, 153. [CrossRef] [PubMed]
3. Otto, T.; Sicinski, P. Cell cycle proteins as promising targets in cancer therapy. *Nat. Rev. Cancer* **2017**, *17*, 93–115. [CrossRef] [PubMed]
4. Bailon-Moscoso, N.; Cevallos-Solorzano, G.; Romero-Benavides, J.C.; Orellana, M.I. Natural compounds as modulators of cell cycle arrest: application for anticancer chemotherapies. *Curr. Genom.* **2017**, *18*, 106–131. [CrossRef] [PubMed]
5. Chang, H.T.; Cheng, Y.H.; Wu, C.L.; Chang, S.T.; Chang, T.T.; Su, Y.C. Antifungal activity of essential oil and its constituents from *Calocedrus macrolepis* var. formosana Florin leaf against plant pathogenic fungi. *Bioresour. Technol.* **2008**, *99*, 6266–6270.
6. Yen, T.B.; Chang, H.T.; Hsieh, C.C.; Chang, S.T. Antifungal properties of ethanolic extract and its active compounds from *Calocedrus macrolepis* var. formosana (Florin) heartwood. *Bioresour. Technol.* **2008**, *99*, 4871–4877. [CrossRef] [PubMed]
7. Ho, C.L.; Tseng, Y.H.; Wang, E.I.; Liao, P.C.; Chou, J.C.; Lin, C.N.; Su, Y.C. Composition, antioxidant and antimicrobial activities of the seed essential oil of *Calocedrus formosana* from Taiwan. *Nat. Prod. Commun.* **2011**, *6*, 133–136. [CrossRef]
8. Wang, S.Y.; Wu, J.H.; Cheng, S.S.; Lo, C.P.; Chang, H.N.; Shyur, L.F.; Chang, S.T. Antioxidant activity of extracts from *Calocedrus formosana* leaf, bark, and heartwood. *J. Wood Sci.* **2004**, *50*, 422–426. [CrossRef]
9. Yen, P.L.; Wu, C.L.; Chang, S.T.; Huang, S.L.; Chang, H.T. Antioxidative lignans from phytochemical extract of *Calocedrus formosana* Florin. *BioResources* **2012**, *7*, 4122–4131.
10. Chiang, Y.M.; Liu, H.K.; Lo, J.M.; Chien, S.C.; Chan, Y.F.; Lee, T.H.; Su, J.K.; Kuo, Y.H. Cytotoxic constituents of the leaves of *Calocedrus formosana*. *J. Chin. Chem. Soc.* **2013**, *50*, 161–166. [CrossRef]
11. Yuan, S.Y.; Lin, C.C.; Hsu, S.L.; Cheng, Y.W.; Wu, J.H.; Cheng, C.L.; Yang, C.R. Leaf extracts of *Calocedrus formosana* (Florin) induce G2/M cell cycle arrest and apoptosis in human bladder cancer cells. *Evid.-Based Complement. Altern. Med.* **2011**, *2011*, 380923. [CrossRef]
12. Tsai, C.C.; Chen, C.J.; Tseng, H.W.; Hua, K.F.; Tsai, R.Y.; Lai, M.H.; Chao, L.K.; Chen, S.T. Cytomic screening of immuno-modulating activity compounds from *Calocedrus formosana*. *Comb. Chem. High Throughput Screen.* **2008**, *11*, 834–842. [CrossRef]
13. Chao, K.P.; Hua, K.F.; Hsu, H.Y.; Su, Y.C.; Chang, S.T. Anti-Inflammatory activity of sugiol, a diterpene isolated from *Calocedrus formosana* bark. *Planta Med.* **2005**, *71*, 300–305. [CrossRef]
14. Jayakumar, T.; Liu, C.H.; Wu, G.Y.; Lee, T.Y.; Manubolu, M.; Hsieh, C.Y.; Yang, C.H.; Sheu, J.R. Hinokitiol inhibits migration of A549 lung cancer cells via suppression of MMPs and induction of antioxidant enzymes and apoptosis. *Int. J. Mol. Sci.* **2018**, *19*, 939. [CrossRef]
15. Ho, S.T.; Lin, C.C.; Wu, T.L.; Tung, Y.T.; Wu, J.H. Antitumor agent yatein from *Calocedrus formosana* Florin leaf induces apoptosis in non-small-cell lung cancer cells. *J. Wood Sci.*. under review.
16. Sancar, A.; Lindsey-Boltz, L.A.; Ünsal-Kaçmaz, K.; Linn, S. Molecular mechanisms of mammalian DNA repair and the DNA damage checkpoints. *Annu. Rev. Biochem.* **2004**, *73*, 39–85. [CrossRef]
17. Bacus, S.S.; Gudkov, A.V.; Lowe, M.; Lyass, L.; Yung, Y.; Komarov, A.P.; Keyomarsi, K.; Yarden, Y.; Seger, R. Taxol-Induced apoptosis depends on MAP kinase pathways (ERK and p38) and is independent of p53. *Oncogene* **2001**, *20*, 147. [CrossRef]
18. Xin, H.; Kong, Y.; Wang, Y.; Zhou, Y.; Zhu, Y.; Li, D.; Tan, W. Lignans extracted from *Vitex negundo* possess cytotoxic activity by G2/M phase cell cycle arrest and apoptosis induction. *Phytomedicine* **2013**, *20*, 640–647. [CrossRef]
19. Manna, S.K.; Bose, J.S.; Gangan, V.; Raviprakash, N.; Navaneetha, T.; Raghavendra, P.B.; Babajan, B.; Kumar, C.S.; Jain, S.K. Novel derivative of benzofuran induces cell death mostly by G2/M cell cycle arrest through p53-dependent pathway but partially by inhibition of NF-Kappab. *J. Biol. Chem.* **2010**, *285*, 22318–22327. [CrossRef]
20. Satyanarayana, A.; Kaldis, P. Mammalian cell-cycle regulation: several Cdks, numerous cyclins and diverse compensatory mechanisms. *Oncogene* **2009**, *28*, 2925. [CrossRef]

21. Vakifahmetoglu, H.; Olsson, M.; Zhivotovsky, B. Death through a tragedy: mitotic catastrophe. *Cell Death Differ.* **2008**, *15*, 1153. [CrossRef] [PubMed]

22. Liu, W.T.; Chen, C.; Lu, I.C.; Kuo, S.C.; Lee, K.H.; Chen, T.L.; Song, T.S.; Lu, Y.L.; Gean, P.W.; Hour, M.J. MJ-66 induces malignant glioma cells G2/M phase arrest and mitotic catastrophe through regulation of cyclin B1/Cdk1 complex. *Neuropharmacology* **2014**, *86*, 219–227. [CrossRef] [PubMed]

23. Subramaniam, D.; Ramalingam, S.; Linehan, D.C.; Dieckgraefe, B.K.; Postier, R.G.; Houchen, C.W.; Jensen, R.A.; Anant, S. RNA binding protein CUGBP2/CELF2 mediates curcumin-induced mitotic catastrophe of pancreatic cancer cells. *PLoS ONE* **2011**, *6*. [CrossRef] [PubMed]

24. Maréchal, A.; Zou, L. DNA damage sensing by the ATM and ATR kinases. *Cold Spring Harb. Perspect. Biol.* **2013**, *5*, 012716. [CrossRef] [PubMed]

25. Smith, J.; Tho, L.M.; Xu, N.; Gillespie, D.A. The ATM-Chk2 and ATR-Chk1 pathways in DNA damage signaling and cancer. *Adv. Cancer Res.* **2010**, *108*, 73–112.

26. Sahu, R.P.; Batra, S.; Srivastava, S.K. Activation of ATM/Chk1 by curcumin causes cell cycle arrest and apoptosis in human pancreatic cancer cells. *Br. J. Cancer* **2009**, *100*, 1425–1433. [CrossRef]

27. Wang, Y.; Liu, Q.; Liu, Z.; Li, B.; Sun, Z.; Zhou, H.; Zhang, X.; Gong, Y.; Shao, C. Berberine, a genotoxic alkaloid, induces ATM-Chk1 mediated G2 arrest in prostate cancer cells. *Mutat. Res.* **2012**, *734*, 20–29. [CrossRef]

28. Chung, T.W.; Lin, S.C.; Su, J.H.; Chen, Y.K.; Lin, C.C.; Chan, H.L. Sinularin induces DNA damage, G2/M phase arrest, and apoptosis in human hepatocellular carcinoma cells. *BMC Complement Altern. Med.* **2017**, *17*, 62. [CrossRef]

29. Shiloh, Y. ATM and related protein kinases: Safeguarding genome integrity. *Nat. Rev. Cancer* **2003**, *3*, 155–168. [CrossRef]

30. Schmidt, M.; Rohe, A.; Platzer, C.; Najjar, A.; Erdmann, F.; Sippl, W. Regulation of G2/M transition by inhibition of WEE1 and PKMYT1 kinases. *Molecules* **2017**, *22*, 2045. [CrossRef]

31. Ashwell, S. Chapter 10—Checkpoint Kinase and Wee1 inhibitors as anticancer therapeutics. In *DNA Repair in Cancer Therapy*; Kelley, M.R., Ed.; Academic Press: San Diego, CA, USA, 2012; pp. 211–234.

32. Cotugno, R.; Fortunato, R.; Santoro, A.; Gallotta, D.; Braca, A.; De Tommasi, N.; Belisario, M.A. Effect of sesquiterpene lactone coronopilin on leukaemia cell population growth, cell type-specific induction of apoptosis and mitotic catastrophe. *Cell Prolif.* **2011**, *45*, 53–65. [CrossRef] [PubMed]

33. Dumontet, C.; Jordan, M.A. Microtubule-Binding agents: a dynamic field of cancer therapeutics. *Nat. Rev. Drug Discov.* **2010**, *9*, 790–803. [CrossRef] [PubMed]

34. Ikeda, R.; Nagao, T.; Okabe, H.; Nakano, Y.; Matsunaga, H.; Katano, M.; Mori, M. Antiproliferative constituents in umbelliferae plants. III. Constituents in the root and the ground part of *Anthriscus sylvestris* Hoffm. *Chem. Pharm. Bull.* **1998**, *46*, 871–887. [PubMed]

Review

Bioactive Phenolic Compounds in the Modulation of Central and Peripheral Nervous System Cancers: Facts and Misdeeds

Lorena Perrone [1,2], Simone Sampaolo [1] and Mariarosa Anna Beatrice Melone [1,3,*]

[1] Department of Advanced Medical and Surgical Sciences, 2nd Division of Neurology, Center for Rare Diseases and InterUniversity Center for Research in Neurosciences, University of Campania "Luigi Vanvitelli", Via Sergio Pansini, 5 80131 Naples, Italy; perronelorena1@gmail.com (L.P.); simone.sampaolo@unicampania.it (S.S.)

[2] Department of Chemistry and Biology, University Grenoble Alpes, 38400 Saint-Martin-d'Hères, France

[3] Sbarro Institute for Cancer Research and Molecular Medicine, Department of Biology, Temple University, BioLife Building (015-00)1900 North 12th Street, Philadelphia, PA 19122-6078, USA

* Correspondence: marina.melone@unicampania.it

Received: 16 January 2020; Accepted: 12 February 2020; Published: 15 February 2020

Abstract: Efficacious therapies are not available for the cure of both gliomas and glioneuronal tumors, which represent the most numerous and heterogeneous primary cancers of the central nervous system (CNS), and for neoplasms of the peripheral nervous system (PNS), which can be divided into benign tumors, mainly represented by schwannomas and neurofibromas, and malignant tumors of the peripheral nerve sheath (MPNST). Increased cellular oxidative stress and other metabolic aspects have been reported as potential etiologies in the nervous system tumors. Thus polyphenols have been tested as effective natural compounds likely useful for the prevention and therapy of this group of neoplasms, because of their antioxidant and anti-inflammatory activity. However, polyphenols show poor intestinal absorption due to individual intestinal microbiota content, poor bioavailability, and difficulty in passing the blood–brain barrier (BBB). Recently, polymeric nanoparticle-based polyphenol delivery improved their gastrointestinal absorption, their bioavailability, and entry into defined target organs. Herein, we summarize recent findings about the primary polyphenols employed for nervous system tumor prevention and treatment. We describe the limitations of their application in clinical practice and the new strategies aimed at enhancing their bioavailability and targeted delivery.

Keywords: brain cancer; gliomas; schwannomas; malignant tumors of the peripheral nerve sheath (MPNST); neurofibromas; polyphenols; bioavailability; nanoparticle-based delivery systems

1. Introduction

Primary tumors of the central nervous system (CNS) are a complex heterogeneous group of benign and malignant cancers, each with their unique biology, prognosis, and sensitivity to the proposed therapies. As a general rule, brain tumors are named according to the non-neoplastic cell types that they most closely resemble and/or to their location where they are located in the brain, as classified by the World Health Organization (WHO) [1].

Gliomas and glioneuronal tumors are the most frequent *heterogeneous group of* primary tumors of the CNS. Gliomas, representing about 30% of the whole CNS cancers and 80% of malignant brain tumors [2], develop from glial cells, so called because the ancients thought that these cells served as "glue" between neurons (glia = glue in Greek). It is actually a group of tumors, including astrocytomas, oligodendrogliomas, ependymomas, and mixed gliomas. *Gliomas* can be aggressive (high degree of malignancy) or have a more indolent behavior (low degree of malignancy). The highest-grade astrocytomas are known as glioblastoma. Non-glial tumors constitute the bulk of neoplasms encountered in the CNS. They include a wide variety of tumor types and a spectrum of behavior ranging from indolent *benign* to highly invasive.

Tumors of the peripheral nervous system (PNS) *can be divided into* benign *tumors, mainly represented by* schwannomas and neurofibromas, and malignant tumors of the peripheral nerve sheath (MPNST), which are a type of sarcoma with very low-frequency. Arising from the soft tissue that surrounds nerves, they develop sporadically or in a particular genetic context. Indeed, neurofibromas are part of the diagnostic criteria inclusion for neurofibromatosis type 1 (NF1), also named von Recklinghausen disease (Table 1) [3,4].

Table 1. Criteria for the clinical diagnosis of NF1 *(At least two are required)*.

Criterion	Features
Six or more café-au-lait macules	>5 mm before puberty >15 mm after puberty
Freckling	Axillary, inguinal
Neurofibromas	Two or more neurofibromas or one plexiform neurofibroma
Skeletal dysplasia	Sphenoid or tibial lesion
Lisch nodules	Two or more iris hamartomas
Optic glioma	Detected by neuroimaging (usually MRI)
First degree relative with NF1	Sibling or parent with NF1
Skeletal dysplasia	Sphenoid or tibial lesion

Similarly, the MPNST develop in 50% of cases in a context of NF1. On the other hand, schwannomas, especially tumors of acoustics, are a major diagnostic criterion for neurofibromatosis type 2 (NF2) (Table 2) [5].

Table 2. Diagnostic criteria for Neurofibromatosis type 2 (these include the National Institutes of Health (NIH) criteria with additional criteria).

Main Criteria	Additional Criteria
Bilateral vestibular schwannomas (VS) or family history of NF2 *plus* (1) Unilateral VS *or* (2) Any two of: meningioma, glioma, neurofibroma, schwannoma, posterior subcapsular lenticular opacities NF2—Neurofibromatosis type 2; VS—vestibular schwannomas	Unilateral VS *plus* any two of: meningioma, glioma, neurofibroma, schwannoma, and posterior subcapsular opacities or Multiple meningioma (two or more) plus unilateral VS or any two of: glioma, neurofibroma, schwannoma, and cataract

A recently published survey of the global incidence of brain cancer estimated that between 1990 and 2016 its worldwide incidence increased most in populations grouped in the low quintile of SDI, a socio demographic index indicator of income per capita, educational level, and total fertility rate. Moreover, it is proposed that significant differences in the incidence of CNS cancer between various geographical regions can be due not only to variations in diagnoses and reporting practices but also to genetic and environmental risk factors not yet identified [6].

Unfortunately, the sole accurate targeting of genetic lesions has been shown to be an incomplete strategy, unable to extend the survival of brain tumor patient [7,8]. In the past time, cancer has been considered as a set of diseases that are caused by the accumulation of genetic mutations, and that aberrant regulation of epigenetic mechanisms may lead to human diseases, including cancer. Contrary to genetic mutations, epigenetic modifications are reversible. For this reason, epigenetic alterations are considered more effective therapeutic targets. Indeed, recent studies confirm the relevance of diet and bioactive dietary compounds for the prevention of epigenetic alterations in cancer. In fact, while former epidemiological studies did not support any or very little association between consumption of vegetable and fruit and reduced risk of cancer, clinical studies based on case-control analysis as well as data produced by large clinical cohort studies indicate an inverse correlation between the incidence of certain types of cancer and the regular consumption of fruits and vegetables [9]. For this reason, it has been hypothesized that only certain types of fruits and vegetables, likely those containing polyphenols, can exert a protective effect against cancer [9]. Correspondingly, several studies investigating the effect of diet on cancer risk and cancer progression provided evidence that polyphenols derived from tea, red wine, cocoa, certain fruits, and olive oil have an impact on carcinogenesis and tumor progression [10]. Several polyphenols, in fact, exhibit potent anti-tumor activity through their capability to reverse epigenetic alterations leading to oncogene activation and down-regulation of tumor suppressor genes [11] by interacting with oxidative reactive intermediates [12] and mutagens [13], modulating the signaling molecules involved in cell-cycle regulation [14], regulating the expression of cancer-related genes [15], or inducing apoptosis [16,17].

Furthermore, the transition from normal to cancer cell is characterized by altered cellular energy metabolism [18,19]. Indeed, the energy metabolism drives the cascade of events which lead a cell to proliferate or to die.

It is known, for example, that tumors are characterized by enhanced glucose uptake and an increased glycolysis rate [9]. Indeed, polyphenols act as inhibitors of glucose absorption and metabolism in cancer [20,21].

Oxidative stress is a common alteration present in adult-onset brain tumors as well as in hereditary cancer of the nervous system, such as Neurofibromatosis type 1 (NF1) and Tuberous Sclerosis (TSC) [22,23]. Since polyphenols show a powerful antioxidant activity, several studies have analyzed their therapeutic potential to counteract tumor progression [23].

Taking into account recent studies, the present review aims to provide up-to-date data on the effect of polyphenols in preventing the progression of central and peripheral nervous system tumors, which helps to explore their therapeutic values for future clinical settings.

2. Polyphenols

Phenols are "molecules possessing at minimum one aromatic ring with one or more hydroxyl groups attached" [23]. They are subdivided into flavonoids and non-flavonoids [24] (Figure 1).

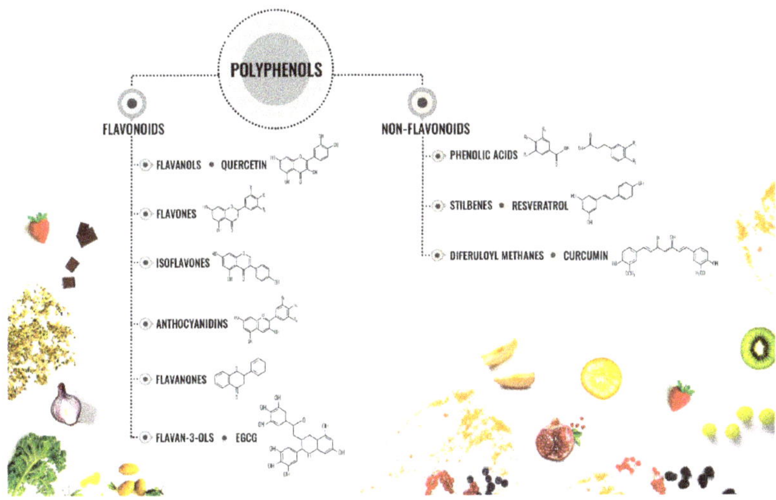

Figure 1. Chemical structures of different subtypes of polyphenols.

Flavonoids. They contain 15 carbons and two aromatic rings linked by a three-carbon bridge. This class includes: flavanols, flavones, isoflavones, flavanones, anthocyanidins, and flavan-3-ols [24].

a. Flavanols are represented in the whole plant kingdom excluding the algae. Flavonols include kaempferol, quercetin, isorhamnetin, and myricetin [24].
b. Flavones are contained in some herbs, such as Celery (*Apium graveolens*) and parsley (*Petrosilinum hortense*). They include apigenin, luteolin, wogonin, and baicalein [24].
c. Isoflavones are contained in leguminous plants. Soybeans (*Glycine max*) present a large amount of daidzein and genistein [24]. Chickpeas are enriched in biochanin A, while peanuts contain high levels of genistein. Since isoflavones show a structure similar to estrogens, they are also defined as phytoestrogens [24].
d. Flavanones are contained at elevated levels in the flavedo of citrus fruits. Hesperetin-7-O-rutinoside is also called hesperidin and is the most common member of this subclass and it is present also in apricots, plums, and bilberries [24].
e. Anthocyanidins are plant pigments, which react with organic acids and sugars, leading to the formation of molecules of different colors. The major components of this subclass are peonidin, pelargonidin, petunidin, cyanidin, delphinidin, and malvidin [24].
f. Flavan-3-ols. They can range from monomers (epicatechin and catechin) to oligomeric and polymeric proanthocyanidins (also called tannins). High concentrations of epicatechin-3-gallate (ECG) and epigallocatechin-3-gallate (EGCG) are present in green tea (*Camelia sinensis*) [24].

Non-flavonoids. They include the C_6-C_1 phenolic acids, which have a dietary relevance. The most common is the gallic acid, which is contained in numerous fruits and plants. Ellagitannins are contained in strawberries (*Fragaria ananassa*), raspberries (*Rubus idaeus*), blackberries (*Rubus spp*), persimmon (*Diospyros kaki*), pomegranates (*Punica granatum*), hazelnuts (*Corylus avellana*), and walnuts (*Juglans regia*) [24]. Belonging to this class are also: (i) secoiridoids and ligstroside, which are present in olive oil, (ii) stilbenes, whose major representing is resveratrol and is found in wine, and (iii) diferuloylmethanes that include curcumin, a curcuminoid of turmeric (Curcuma longa) [23].

3. Mechanisms of Cancer Modulation by Polyphenols

3.1. Regulation of Glucose Homeostasis

High proliferating tumor cells need to produce extra ATP to maintain their energy status, produce an increased biosynthesis of macromolecules, and maintain the cellular redox status [25]. For this reason, tumor cells need to reprogram the own metabolic flux with the aim of obtaining a surplus of ATP that is needed for an increased rate of proliferation [25]. Indeed, cancer cells are characterized by metabolic reprogramming involving an altered energy metabolism, called the Warburg effect, that shows elevated uptake and consumption of glucose and also enhanced creation of lactate buildup despite the presence of oxygen [26]. This metabolic change in tumor cells promotes cell proliferation and tumor progression by generating an increased level of glycolytic intermediates needed for the synthesis of new molecules. In addition, it augments the metabolism of glucose, which counteracts the excess metabolic formation of reactive oxygen species (ROS) in cancer cells [27]. Because of the necessity of extra ATP, glucose deprivation displays a higher cytotoxic effect on several cancer cells compared to normal cells. For this reason, inhibitors of glucose uptake and of oxidative metabolism (glycolysis inhibitors) are considered as therapeutic strategies against cancer [26]. Importantly, the Warburg effect, also defined as "aerobic lactatogenesis" [28], is considered an excellent target against cancer because therapeutic strategies directed against this metabolic shift will induce less negative side effects, leading to the reduction of treatment-associated morbidities. Indeed, several drugs targeting glucose metabolism are in trial or already approved compounds to treat cancer [29].

Several polyphenols exert an inhibitory effect on various steps of energy pathways in cancer cells, including inhibition of the uptake of glucose and block of enzymes involved in glucose metabolism.

Polyphenols interfere with the glucose transporters in various cell types, affecting glucose uptake. Gossypol decreases glucose uptake in various cell types by a competitive mechanism [30]. Naringenin inhibits basal as well insulin-induced uptake of glucose in tumoral cells by blocking PIP3/Akt and MAPK activity [31]. This inhibition of glucose uptake results in an anti-proliferative effect of naringenin treatment in cancer cells [31]. Genistein reduces glucose uptake in cancer cell lines [32]. Resveratrol inhibits glycolysis in cancer cells, where it also blocks glucose uptake by lowering the expression of the glucose transporter GLUT1 [33]. Resveratrol reduces cell viability in cancer cells. Such resveratrol-dependent reduction of GLUT1 depends on a reduction of ROS and subsequent down-regulation of the transcription factor HIF-1 -alpha [33]. Moreover, resveratrol reduces cell viability by directly inhibiting the 6-Phosphofructo-1-kinase-1 (PKF), an essential enzyme of the glycolytic pathway, leading to a decreased glucose utilization and reduced ATP levels in cancer cells [34]. The flavone hesperetin reduces basal glucose uptake in cancer cells by down-regulating GLUT1 [35]. It also impairs insulin-dependent glucose uptake in cancer cells by impairing GLUT4 translocation to the plasma membrane [35]. The flavonoids epigallocatechin-3-gallate (EGCG) and quercetin reduce glucose uptake and lactate production in tumoral cells [36]. The reduction of glucose uptake is independent from signaling pathways modulated by PKC, PKG, PKA, and calcium-calmodulin [36]. The reduction of glucose uptake and lactate production due to these flavonoids results in a cytotoxic and anti-proliferative effect [36]. Luteolin reduces the glycolytic flux in cancer cells but does not affect the glucose uptake [37]. 1,2,3,4,6-penta-O-galloyl-beta-d-glucose (PGG) down-regulates the genes encoding enzymes of the pyruvate metabolism, such as acylphosphatase, pyruvate carboxylase, and aldehyde dehydrogenase (ALDH3B1) in cancer cells [38].

3.2. Modulation of Advanced Glycation Endproducts (AGEs) and their Receptor (RAGE)

Aging, altered metabolism, and diet induce the formation of advanced glycation endproducts (AGEs), which participate in the progression of several diseases, including cancer [39–43]. Additionally, members of the AGEs family, such as N-carboxymethyllysine (CML) and argpyrimidine, are elevated in several tumors and are implicated in cancer progression [43]. Several polyphenols inhibit AGEs formation both in vitro and in vivo. Indeed, Tomato paste blocks the formation of AGEs [44], tea polyphenols interfere with AGEs formation in physiological conditions [45], and EGCG inhibits AGEs formation [46]. The receptor for AGEs (RAGE) is also involved in cancer cell invasion and metastasis [47]. Polyphenols interfere with RAGE activation and signaling. EGCG inhibits RAGE activation [48]. Quercetin reduces RAGE expression in cancer cells, inducing cell cycle arrest, autophagy, apoptosis, and chemo-sensitivity [49].

3.3. Modulation of Oxidative Stress and Related Signaling Pathways

Reactive oxygen species (ROS) induce tumor formation by producing genetic mutation, inducing oncogenes, and promoting oxidative stress. The last has an effect on cell proliferation, apoptosis, and survival. Tumor cells are characterized by enhanced ROS production, leading to redox unbalance. ROS induce the oxidation of protein cysteine residues, which may induce cellular proliferation [50]. ROS play a key function in tumor initiation, promotion, and progression [51]. ROS activate oncogenes, such as Ras and c-Myc, and promote p53-dependent DNA repair [52]. In addition, ROS promote cancer progression also by inducing the activation of several signaling pathways, such as the phosphoinositide-3-kinase (PI3K)/protein kinase B (AKT), mitogen-activated protein kinase (MAPK)/extracellular-signal-regulated kinase (ERK), the inhibitor of kappa B (IκB), kinase (IKK)/nuclear factor κB (NFκB), the protein kinase D (PKD), JNK, and PI3K, which in turn modulate the activity of several transcription factors that participate to cancer initiation/progression [52].

Polyphenols possess both anti-oxidant and pro-oxidant activity, which modulate cell proliferation and apoptotic pathways [53]. Quercetin scavenges superoxide anions, leading to the formation of H_2O_2, semiquinone, and quercetin radicals. The latter deplete the intracellular anti-oxidant systems [52]. Indeed, quercetin leads to ROS-dependent apoptosis, necrosis, and autophagy in cancer cells [52].

In addition, quercetin induces cell cycle arrest by modulating p21^{WAF}, cyclin B, and p27^{KIP1} in cancer cells [52]. Notably, curcumin exerts opposite effects in cancer cells compared to its effects in normal cells, suggesting that curcumin can be beneficial in preventing cancer without affecting the homeostasis of normal cells. In normal cells, curcumin acts as an anti-oxidant by scavenging hydroxyl radicals, superoxide, nitric oxide, H2O2, and peroxynitrite [54]. In addition, curcumin regulates the expression of HO-1, GPX, and SOD in normal cells [54]. On the other hand, in cancer cells curcumin has a pro-oxidative effect that induces apoptosis [55]. Indeed, curcumin induces ROS formation in cancer cells [56,57]. Curcumin's pro-oxidative activity occurs in the mitochondrial membranes of cancer cells, leading to depolarization of mitochondrial and down-regulation of ATP synthesis, ultimately inducing apoptosis [58]. Curcumin-dependent ROS production in cancer cells induces Erk1/2 and p38 MAPK pathways that lead to autophagy [59]. Curcumin blocks the survival and proliferation of cancer stem cell through a ROS-mediated inhibition of NFκB and STAT3 in glioblastoma [60]. Capsaicin has been shown to exert beneficial effects against glioma. Indeed, in glioma, capsaicin (trans-8-methyl-N-vanillyl-6-nonenamide) affects the mitochondrial membrane potential, leading to ROS elevation and subsequent caspase 3 activation [61]. Capsaicin not only induces apoptosis but also cell cycle arrest [62]. EGCG possesses both anti-oxidant and pro-oxidant activity. It produces ROS by auto-oxidation [63]. EGCG induces apoptosis by inhibiting the PI3k/AKT pathway. In addition, it decreases the mitochondrial membrane potential, leading to apoptosis [52]. Phenethyl isothiocyanate (PEITC) decreases intracellular GSH, leading to enhanced ROS accumulation and mitochondrial dysfunction exclusively in cancer cells [64]. Benzyl isothiocyanate (BITC) induces oxidative stress in glioma cells by exhausting SOD and GSH levels, leading to caspase-mediated apoptosis [65]. Piperine blocks tumor growth by inducing oxidative stress, mitochondria dysfunction, and subsequent apoptosis [66]. Resveratrol inhibits tum initiation and progression by inducing apoptosis in neuroblastoma cells [67]. It promotes apoptosis by inducing the death receptors for TRAIL and FasL, ROS-dependent caspase activation, and p53 [68]. P-Coumaric acid (p-CoA) promotes apoptosis in cancer cells by enhancing ROS formation and inducing mitochondrial depolarization [69]. Naringenin promotes apoptosis in cancer cells through induction of ROS formation, which in turn leads to p38 MAPK-dependent caspase activation [70]. Gallic acid inhibits cancer cell growth by inducing ROS production [71]. Thus, we may conclude that several polyphenols counteract tumor progression by inducing cancer-specific ROS production.

3.4. Other Mechanisms

Polyphenols modify the metabolism of pro-carcinogens through mechanisms that alter the levels of cytochromes P450 (CYPs), which plays an essential role in cancer promotion [72]. Polyphenols produce epigenetic changes that have a preventive effect against cancer [73]. Polyphenols also show a preventive effect by inhibiting inflammation [74]. They also modulate the autophagy flux [75].

Polyphenols prevent metastasis formation by affecting the activity of urokinase and matrix metalloproteinases and by inhibiting angiogenesis through modulation of the vascular endothelial growth factor expression and receptor phosphorylation [76].

In addition, dietary polyphenols are employed together with conventional pharmacological therapy or cytotoxic agents used to treat drug-resistant cancer cells [74].

Several studies indicated herein tested the efficacy of polyphenols at concentrations equivalent to oral sub ministration of juices/extracts of natural substances containing polyphenols (range 100–800 mg/day), while other studies analyzed higher concentrations (range 10–75 μM). Notably, concentrations of polyphenols equivalent to diet intake show also beneficial effects.

4. Polyphenol Effects on Central Nervous System Cancers

Several risk factors are involved in the onset and progression of adult brain tumors: aging, diet, environmental exposure, head trauma, infections, and cigarette smoking [77]. Primary brain tumors are named gliomas and classified according to their putative original cell type. Glioblastoma multiform

(GBM) shows the highest aggressive phenotype (grade IV) representing the 60% of age-related brain tumors [78]. At present there is not definite cure for GBM; thus, researchers are looking at innovative therapeutic strategies [23].

Recent studies in vitro and in vivo have underlined the therapeutic efficacy against cancer of several polyphenols, such as quercetin, epigallocatechin-3-gallate (EGCG), resveratrol, and curcumin [79]. Moreover, it has been reported that the polyphenols' therapeutic potential is further enhanced when they are used in combination or added to pharmaceutical compounds [74,79].

4.1. Curcumin Effects on Central Nervous System Cancers

The effective anti-tumor activity of curcumin treatment in GBM has been shown by several studies [23,80]. Curcumin exerts a pro-differentiative effect in glioma-stem cells because of its activation of the autophagy flux [81]. In human glioblastoma T98G cells, this polyphenol induces the activation of both receptor-mediated and mitochondria-mediated proteolytic pathways, which in turn promote apoptosis [55]. Curcumin down-regulates the expression of cancer signaling pathways (i.e., Notch1 and pAKT), leading to blockade of cell growth, apoptosis, and inhibition of migration and invasion [82]. In human glioma cells, it lowers the protein levels of neuronal precursor cell-expressed developmentally down-regulated 4–1 (NEDD4) [82]. Delivery of curcumin into the brain of GBM mice produces the remission of tumor in 50% of the animals and modifies the phenotype of the microglia surrounding the tumor [83]. Curcumin exerts an efficient induction of autophagy [75], which can lead to apoptosis in cancer cells. Moreover, curcumin treatment in A172 human glioblastoma cells leads to cell death by inducing autophagy flux [84]. Another study confirmed that curcumin promotes autophagy in glioblastoma cells while it inhibits mitophagy [85]. Curcumin also blocks invasion and migration potential of glioblastoma U87 cells by decreasing the expression of fascin, an actin-binding protein involved in migration and invasion [86]. Noteworthy, curcumin exerts a radiosensitizing effect on GBM [87]. Moreover, curcumin acts as a photosensitizer in sNB-19 glioblastoma cells, showing that it can be used to improve the photodynamic therapy for GBM treatment [88].

4.2. Resveratrol Effects on Central Nervous System Cancers

Several studies demonstrate the efficacy of resveratrol in lowering tumorigenesis and preventing metastasis [23,89,90]. Resveratrol has a powerful capability of down-regulating the self-renewal and tumor-initiating capability of glioma stem cells obtained from GBM patients by inducing the p53/ p21 pathway [91]. Resveratrol possesses a potent effect in inhibiting the invasion and migration capability of glioblastoma cells by activating the RhoA/ROCK pathway [92]. Resveratrol decreases cell growth and motility, enhances cell death, and interferes with the epithelial-mesenchymal transition modulating the Wnt signaling pathway [93]. Resveratrol lowers tumorigenic potential and improves the effects of radiotherapy in vitro and in vivo against GBM-derived tumor stem cells to by inhibiting the signal transducer and activator of transcription 3 (STAT3) [94]. Resveratrol blocks the growth of U-87MG glioblastoma cells and lowers the expression of human telomerase reverse transcriptase (hTERT) as well as the catalytic subunit of the telomerase and a biomarker of cell immortalization, confirming that resveratrol can be used as a therapeutic agent for GBM [95]. The postoperative administration of resveratrol results in a significant prognosis amelioration of rat-advanced orthotopic glioblastoma by reducing growth, inducting apoptosis, and suppressing STAT3 signaling [96]. In addition, resveratrol blocks epithelial-mesenchymal transition in GBM by modulating Smad signaling [97].

In combination with Paclitaxel, resveratrol enhances the oxidant and apoptotic effect of the pharmacological compound by activating the TRPM2 channel in glioblastoma cells [98].

4.3. EGCG Effects on Central Nervous System Cancers

Several studies show the beneficial effect of EGCG as a therapeutic agent for brain tumors [23]. EGCG exerts an inhibitory effect in three glioma cell lines by modulating the epidermal growth factor-1 (EGF-1) [99]. EGCG potentiates the effects of ionizing radiation (IR) in GBM by modulating the activity of Ras homolog gene family member A (RhoA) and survivin, with the last being involved in the regulation of apoptosis. Treatment with EGCG combined with radiotherapy ameliorates the efficacy of IR treatments [100]. EGCG also enhances the anti-cancer activity of cytotoxic agents [101]. Indeed, in a mouse model of glioblastoma, EGCG enhances the anti-cancer potential of temozolomide, which promotes DNA damage [101]. Treatment with EGCG alone or in combination with temozolomide affects glioma stem cell survival and migration capability, as well as inhibits neurosphere formation [102]. In addition, such treatments induce apoptosis by down-regulating p-Akt and Bcl-2 [102]. In human glioblastoma U251 cells, EGCG promotes apoptosis and blocks cell-growth because of inhibiting the JAK2/STAT3 signaling pathway [103]. EGCG suppresses the invasion properties of human glioblastoma T-98G cells by down-regulating MMP-2 and MMP-9 expression [104]. Interestingly, EGCG inhibits the effects of the glucose-regulated protein 78 (GRP78), which is up-regulated in GBM by direct protein–protein interaction that results in a conformational change in GRP78, probably leading to its inactivation [105]. At the low concentration of 100 nM, EGCG activates endogenous repair pathways while at higher concentrations, EGCG induces ROS production and autophagy [106]. These data suggest that drinking green tea containing low concentrations of EGCG may exert a chemo-preventive effect against GBM, while higher concentrations (500 μM) show a therapeutic effect [106]. Importantly, EGCG inhibits the expression of O^6-Methylguanine DNA-Methyltransferase (MGMT) in GBM-derived cells only, which is an essential regulator of the resistance to temozolomide (TMZ) in glioblastomas. EGCG treatment in two GBM cell lines (GBM-XD and T98G) results in suppression of MGMT expression, abolishes TMZ resistance, and prevents β catenin translocation into the nucleus [107]. On the contrary, the addition of EGCG to non-tumor glial cell culture (GliaX) enhances MGMT expression by inhibiting the methylation of the MGMT promoter [107]. Recently, it has been shown that EGCG induces telomere shortening in U251 glioblastoma, leading to senescence [108]. In addition, it also promotes telomere-independent genotoxicity [108].

4.4. Quercetin Effects on Central Nervous System Cancers

The therapeutic potential of quercetin for the cure of GBM has been extensively analyzed [23]. The glycoside form of quercetin called Rutin exerts an anti-proliferative effect on human GBM cells [109]. Rutin reduces the survival and proliferation of GL-15 cell lines, resulting in a decrement of phosphorylated extracellular signal-regulated protein kinases 1 and 2 (ERK1/2), which exert an essential role in cell proliferation and apoptosis modulation [109]. In GBM cell cultures, rutin induces astroglial differentiation and apoptosis [109]. In GBM cultures, treatment with quercetin alone or together with temozolomide, induces apoptosis, whereas it does not affect autophagy [110]. Treatment with quercetin or its addition during with irradiation promotes apoptosis. This anti-cancer effect is due to activation of caspase-3 and poly [ADP-ribose] polymerase 1 (PARP-1), which are concomitant to the inhibition of the Akt pathway [111]. In GBM cells, co-treatment with rutin and temozolomide results in enhanced cytotoxicity because of inhibition of the autophagy flux [112]. Studies in subcutaneous and orthotopic xenograft using concomitant treatment with temozolomide and rutin show a decreased tumor volumes, while treatment with temozolomide or rutin alone is less effective [112]. In U251 glioblastoma human cells, quercetin inhibits cell proliferation and viability as well as invasion and migration properties [113,114]. Quercetin shows a pro-apoptotic effect also because it regulates the expression of apoptotic genes and because it induces the cell cycle arrest [113]. Using U87MG, C6, and U138 glioblastoma cultures, we demonstrated that the water extract of *Ruta graveolens* L. promotes cell death. We also found that rue activates ERK1/2 and AKT, resulting in an inhibition of cell growth. We also show that rutin, the major component of the *Ruta graveolens* water extract, is unable to induce cell death [115]. Quercetin in combination with sodium butyrate promotes apoptosis in rat C6 and human T98G GBM cells by inhibiting autophagy [116].

5. Polyphenol Effects on Tumors of the Peripheral Nervous System

Neurofibromatosis type 1 (NF1) is an autosomal dominant disorder showing complex phenotypes and it is caused by inherited mutations in the NF1 gene, which is a tumor suppressor. Almost all NF1 patients develop pigmentary lesions (café-au-lait macules, skinfold freckling, and Lisch nodules) and dermal neurofibromas (Table 1). In some patients are also present brain tumors (glioblastoma and optic pathway gliomas), peripheral nerve tumors (plexiform neurofibromas, spinal neurofibromas, and malignant peripheral nerve sheath tumors), skeletal abnormalities (tibial pseudarthrosis, orbital dysplasia, and scoliosis), attention deficits, learning disabilities, and social and behavioral problems, which impair the quality of life [117].

Neurofibromatosis type 2 (NF2) is a genetic disorder characterized by the presence of multiple benign tumors of the peripheral and central nervous system (including meningiomas, schwannomas, and ependymomas) (Table 2).

NF2 patients are almost always diagnosed late in life, around the second or third decade of life [118,119]. NF2 is characterized by the presence of benign tumors. However, such tumors can induce mortality that is associated to the location of the tumors as well as can be promoted by the treatments. Currently, the only therapy available consists in a local treatment of the tumors and is not effective. Thus, there is a need to develop systemic therapies aimed to improve the outcome of NF2 [23,119].

It is well documented that a healthy diet including a high consumption of fruit and vegetables has a preventive effect against cancer and results in a lower incidence of tumor development and tumor-induced mortality [120].

Recently, researchers started to investigate the chemopreventive and/or chemotherapeutic potential of polyphenolic compounds [120]. Since polyphenols possess anti-oxidant proprieties, their consumption has a beneficial effect against the high levels of oxidative stress produced by cancer cells [51].

5.1. Curcumin Effects on Tumors of the Peripheral Nervous System

Curcumin reduces proliferation and enhances the apoptosis rate in HEI-193 human schwannoma cells [118]. These results indicate that administration of curcumin to patients with NF2 schwannomas may exert a beneficial effect. We describe the first experience with curcumin supplementation in NF1 patients. We show that a therapeutic strategy involving a high adherence to the Mediterranean diet together with the administration of 1200 mg/day of curcumin results in a significant reduction of the number and volume of cutaneous neurofibromas in NF1 patients [121]. Notably, we demonstrate by Magnetic Resonance Imaging that in one patient this therapeutic strategy results in a sensible reduction in volume (28%) of a large cranial plexiform neurofibroma. On the contrary, administration of curcumin in association with a Western diet has not effect on NF1 tumors, suggesting that some components in the Mediterranean diet may improve curcumin bioavailability and activity [121]. A recent study revealed that calebin-A, derived from turmeric *Curcuma longa*, (a) inhibits the cell growth in the malignant peripheral nerve sheath tumor (MPNST) transformed from NF1-related plexiform neurofibroma, and (b) blocks cell growth in primary neurofibroma cells [122]. Calebin-A induces the cell cycle arrest and decreases hTERT, phosphorylated- ERK1/2, -AKT, and surviving [122].

5.2. EGCG Effects on Tumors of the Peripheral Nervous System

Only one study reported that EGCG reduces the proliferation of an MPNST transformed from NF1-related plexiform neurofibroma [122].

6. Bioavailability of Dietary Polyphenols

Several studies have indicated that dietary polyphenols exert neuroprotective functions. However, their clinical application is still limited. In fact, polyphenols exert poor effect in vivo when compared to their activity in vitro [123,124]. The difference between in vitro and in vivo effects of polyphenols is mainly associated to their poor absorption, rapid metabolism, and massive system elimination, which represent a limitation regarding their therapeutic action and clinical application [125]. Several studies underline that the chemical structure of polyphenols plays a key role in modulating the rate and extent of their absorption upon ingestion [74]. Noteworthy, the individual variability in drug absorption and metabolism has a key role in modulating the effects of polyphenols in vivo. Indeed, the absorption is regulated by the local microflora, by the metabolic activity and by the hepatic function [126]. In addition, the CNS is protected by the blood brain barrier (BBB), which regulates the transport of molecules into the CNS. Thus, the transport across the BBB further limits the therapeutic potential of dietary polyphenols. For this reason, several ongoing research projects are studying how to improve the bioavailability of polyphenols [125]. These studies aim at ameliorating the biochemical stability and transport across the BBB of the polyphenols as well as decreasing their degradation [125]. Another study investigated the transport across the BBB of bioavailable phenolic sulfates derived from the colonic metabolism of berries [127]. They found that these compounds show a differential transport across the BBB, which was related to their chemical structure. In addition, they discovered that these compounds were further metabolized by the endothelial cells, leading to the production of novel molecules with potential bioactivity [127]. This study also demonstrated that pre-treatment with these compounds (a) ameliorated the response to oxidative stress and toxicity and (b) reduced the inflammatory response by modulating NF-kB activity [127]. Thus, this study demonstrated that these polyphenols cross the BBB and exert a neuroprotective and anti-inflammatory function. Furthermore, it has been shown that the gut microbiota metabolize the dietary polyphenols, promoting the production of bioactive molecules that cross the BBB and modulate the neuronal function by acting as neurotransmitters [128]. Interestingly, dietary polyphenols modulate the bacterial composition of the gut microbiota, acting on the microbiota-gut-brain axis, which is considered as a neuroendocrine system [128]. These studies support the hypothesis that dietary polyphenols exert a beneficial effect by modulating the gut microbiota, leading to a neuroprotective effect via the gut-brain axis. Thus, they may have a therapeutic role in the prevention of diseases affecting the nervous system [128].

However, several challenges remain. These include (a) the exploration of the therapeutic interplay between polyphenols or other natural substances contained in the Mediterranean diet [121,129], (b) their molecular characterization, and (c) the definition of optimal absorption levels and bioavailability improvement. This is necessary to ensure therapeutic efficacy, that these substances cross the intestinal and blood–brain barriers, and that matrices can be developed for the release of product formulations. Indeed, nanotechnology can provide new materials for the delivery of polyphenols, improving their absorption and efficacy [130]. These technologies can provide food-based nanodelivery vehicles with different surface properties. To date, several nanovehicles, such as nanoemulsions, protein-polysaccharide coacervas, liposomes, and small cochlear structures, are produced only on a laboratory scale. In the future these systems will have applications in the development of functional foods at an industry scale [131].

7. Clinical Trials

To date, the US National Institute of Health database shows only two completed clinical trials using curcumin and polyphenols for the treatment of GBM (http://www.clinicaltrial.gov/; searching for: "Glioblastoma multiforme" and "Curcumin" and "polyphenols"; and http://www.clinicaltrial.gov/; searching for: "Glioblastoma multiforme" and "Curcumin"). There are no any curcumin-based clinical trials for NF2 or NF1 treatments (http://www.clinicaltrial.gov/; searching for: "Neurofibromatosis type 2" and "Curcumin").

8. Conclusions

The incidence of brain tumors has been increasing recently. Despite considerable efforts to find an effective therapy, the treatment of some cancers of the nervous system still remains a challenge in a war in which, thus far, few battles have been won. The numerous metabolic aspects underlying the tumors of the CNS and PNS have opened the way to new therapeutic approaches that see an interesting therapeutic strategy in diet and, in particular, foods with anti-oxidant activity. In particular, several studies have underlined the beneficial effect of dietary polyphenols for the prevention of tumors of the CNS and PNS.

Furthermore, recent studies have revealed the positive effect of polyphenols on the microbioma-intestine-brain axis, demonstrating the therapeutic potential of dietary polyphenols in the prevention of diseases affecting the nervous system.

However, the low bioavailability of dietary polyphenols is still a limitation for their introduction into clinical practice. A promising solution lies in polymeric nanoparticle-based polyphenol delivery systems that prevent the degradation of bioactive compounds and enhance their absorption and bioavailability.

Author Contributions: L.P., S.S., and M.A.B.M. wrote the manuscript. All the authors reviewed the final version and agree to publication. All authors have read and agreed to the published version of the manuscript.

Funding: M.A.B.M. and S.S. thank the Regione Campania (RIS 3—POR FESR 2007/2013—Obiettivo 2.1, DIP. 54-DG 91 n. 403, 15/10/2015), Inter-University Center for Research in Neurosciences and University of Campania "Luigi Vanvitelli," (project V:ALERE 2019 Id343) Naples, Italy, for financial support. L.P. and M.A.B.M. are financed by Italian Ministry of Economic Development (MiSE)—Fund for Sustainable Development—Call "HORIZON2020" PON I&C 2014-2020, FOR.TUNA project, code No. F/050347/01_03/X32.

Acknowledgments: We are grateful to Antonia Auletta for preparing the figures.

Conflicts of Interest: The authors declare no conflict of interest.

References

1. Louis, D.N.; Perry, A.; Reifenberger, G.; von Deimling, A.; Figarella-Branger, D.; Cavenee, W.K.; Ohgaki, H.; Wiestler, O.D.; Kleihues, P.; Ellison, D.W. The 2016 World Health Organization Classification of Tumors of the Central Nervous System: A summary. *Acta Neuropathol.* **2016**, *131*, 803–820. [CrossRef] [PubMed]

2. Goodenberger, M.L.; Jenkins, R.B. Genetics of adult glioma. *Cancer Genet.* **2012**, *205*, 613–621. [CrossRef] [PubMed]

3. National Institutes of Health. Consensus Development Conference Statement: Neurofibromatosis. Bethesda, Md., USA, July 13–15, 1987. *Neurofibromatosis* **1988**, *1*, 172–178.

4. Giugliano, T.; Santoro, C.; Torella, A.; Del Vecchio Blanco, F.; Grandone, A.; Onore, M.E.; Melone, M.A.B.; Straccia, G.; Melis, D.; Piccolo, V.; et al. Clinical and Genetic Findings in Children with Neurofibromatosis Type 1, Legius Syndrome, and Other Related Neurocutaneous Disorders. *Genes* **2019**, *10*, 580. [CrossRef]

5. Evans, D.G. Neurofibromatosis type 2 (NF2): A clinical and molecular review. *Orphanet. J. Rare Dis.* **2009**, *4*, 16. [CrossRef]

6. GBD 2016 Brain and Other CNS Cancer Collaborators. Global, regional, and national burden of brain and other CNS cancer, 1990–2016: A systematic analysis for the Global Burden of Disease Study 2016. *Lancet Neurol.* **2019**, *18*, 376–393. [CrossRef]

7. Dhez, A.C.; Benedetti, F.; Antonosante, A.; Panella, G.; Ranieri, B.; Florio, T.M.; Cristiano, L.; Angelucci, F.; Giansanti, F.; Di Leandro, L.; et al. Targeted therapy of human glioblastoma via delivery of a toxin through a peptide directed to cell surface nucleolin. *J. Cell Physiol.* **2018**, *233*, 4091–4105. [CrossRef]

8. Mack, S.C.; Hubert, C.G.; Miller, T.E.; Taylor, M.D.; Rich, J.N. An epigenetic gateway to brain tumor cell identity. *Nat. Neurosci.* **2016**, *19*, 10–19. [CrossRef]

9. Vauzour, D.; Rodriguez-Mateos, A.; Corona, G.; Oruna-Concha, M.J.; Spencer, J.P. Polyphenols and human health: Prevention of disease and mechanisms of action. *Nutrients* **2010**, *2*, 1106–1131. [CrossRef]

10. Middleton, E., Jr.; Kandaswami, C.; Theoharides, T.C. The effects of plant flavonoids on mammalian cells: Implications for inflammation, heart disease, and cancer. *Pharm. Res.* **2000**, *52*, 673–751.

11. Carlos-Reyes, Á.; López-González, J.S.; Meneses-Flores, M.; Gallardo-Rincón, D.; Ruíz-García, E.; Marchat, L.A.; Astudillo-de la Vega, H.; Hernández de la Cruz, O.N.; López-Camarillo, C. Dietary Compounds as Epigenetic Modulating Agents in Cancer. *Front. Genet.* **2019**, *10*, 79. [CrossRef] [PubMed]

12. Duthie, S.J.; Dobson, V.L. Dietary flavonoids protect human colonocyte DNA from oxidative attack in vitro. *Eur. J. Nutr.* **1999**, *38*, 28–34. [CrossRef] [PubMed]

13. Calomme, M.; Pieters, L.; Vlietinck, A.; Vanden Berghe, D. Inhibition of bacterial mutagenesis by Citrus flavonoids. *Planta Med.* **1996**, *62*, 222–226. [CrossRef] [PubMed]

14. Plaumann, B.; Fritsche, M.; Rimpler, H.; Brandner, G.; Hess, R.D. Flavonoids activate wild-type p53. *Oncogene* **1996**, *13*, 1605–1614.

15. van Erk, M.J.; Roepman, P.; van der Lende, T.R.; Stierum, R.H.; Aarts, J.M.; van Bladeren, P.J.; van Ommen, B. Integrated assessment by multiple gene expression analysis of quercetin bioactivity on anticancer-related mechanisms in colon cancer cells in vitro. *Eur. J. Nutr.* **2005**, *44*, 143–156. [CrossRef]

16. Fabiani, R.; De Bartolomeo, A.; Rosignoli, P.; Servili, M.; Montedoro, G.F.; Morozzi, G. Cancer chemoprevention by hydroxytyrosol isolated from virgin olive oil through G1 cell cycle arrest and apoptosis. *Eur. J. Cancer Prev.* **2002**, *11*, 351–358. [CrossRef]

17. Mantena, S.K.; Baliga, M.S.; Katiyar, S.K. Grape seed proanthocyanidins induce apoptosis and inhibit metastasis of highly metastatic breast carcinoma cells. *Carcinogenesis* **2006**, *27*, 1682–1691. [CrossRef]

18. Cirillo, A.; Di Salle, A.; Petillo, O.; Melone, M.A.; Grimaldi, G.; Bellotti, A.; Torelli, G.; De'Santi, M.S.; Cantatore, G.; Marinelli, A.; et al. High grade glioblastoma is associated with aberrant expression of ZFP57, a protein involved in gene imprinting, and of CPT1A and CPT1C that regulate fatty acid metabolism. *Cancer Biol. Ther.* **2014**, *15*, 735–741. [CrossRef]

19. Melone, M.A.B.; Valentino, A.; Margarucci, S.; Galderisi, U.; Giordano, A.; Peluso, G. The carnitine system and cancer metabolic plasticity. *Cell Death Dis.* **2018**, *9*, 228. [CrossRef]

20. León, D.; Uribe, E.; Zambrano, A.; Salas, M. Implications of Resveratrol on Glucose Uptake and Metabolism. *Molecules* **2017**, *22*, 398. [CrossRef]

21. Siddiqui, F.A.; Prakasam, G.; Chattopadhyay, S.; Rehman, A.U.; Padder, R.A.; Ansari, M.A.; Irshad, R.; Mangalhara, K.; Bamezai, R.; Husain, M.; et al. Curcumin decreases Warburg effect in cancer cells by down-regulating pyruvate kinase M2 via mTOR-HIF1α inhibition. *Sci. Rep.* **2018**, *8*, 8323. [CrossRef] [PubMed]

22. Incecik, F.; Avcıoğlu, G.; Erel, Ö.; Neşelioğlu, S.; Besen, S.; Altunbaşak, S. Dynamic thiol/disulphide homeostasis in children with neurofibromatosis type 1 and tuberous sclerosis. *Acta Neurol. Belg.* **2019**, *119*, 419–422. [CrossRef] [PubMed]

23. Squillaro, T.; Schettino, C.; Sampaolo, S.; Galderis, U.; Di Iorio, G.; Giordano, A.; Melone, M.A.B. Adult-onset brain tumors and neurodegeneration: Are polyphenols protective? *J. Cell Physiol.* **2017**, *233*, 3955–3967. [CrossRef] [PubMed]

24. Del Rio, D.; Rodriguez-Mateos, A.; Spencer, J.P.; Tognolini, M.; Borges, G.; Crozier, A. Dietary (poly)phenolics in human health: Structures, bioavailability, and evidence of protective effects against chronic diseases. *Antioxid. Redox Signal.* **2013**, *18*, 1818–1892. [CrossRef]

25. Cairns, R.A.; Harris, I.S.; Mak, T.W. Regulation of cancer cell metabolism. *Nat. Rev. Cancer* **2011**, *11*, 85–95. [CrossRef]

26. Hanahan, D.; Weinberg, R.A. Hallmarks of cancer: The next generation. *Cell* **2011**, *144*, 646–674. [CrossRef]

27. Aykin-Burns, N.; Ahmad, I.M.; Zhu, Y.; Oberley, L.W.; Spitz, D.R. Increased levels of superoxide and H2O2 mediate the differential susceptibility of cancer cells versus normal cells to glucose deprivation. *Biochem. J.* **2009**, *418*, 29–37. [CrossRef]

28. Martel, F.; Guedes, M.; Keating, E. Effect of polyphenols on glucose and lactate transport by breast cancer cells. *Breast Cancer Res. Treat.* **2016**, *157*, 1–11. [CrossRef]

29. Luengo, A.; Gui, D.Y.; Vander-Heiden, M.G. Targeting metabolism for cancer therapy. *Cell Chem. Biol.* **2017**, *24*, 1161–1180. [CrossRef]

30. Pérez, A.; Ojeda, P.; Valenzuela, X.; Ortega, M.; Sánchez, C.; Ojeda, L.; Castro, M.; Cárcamo, J.G.; Rauch, M.C.; Concha, I.I.; et al. Endofacial competitive inhibition of the glucose transporter 1 activity by gossypol. *Am. J. Physiol. Cell Physiol.* **2009**, *297*, C86–C93. [CrossRef]

31. Harmon, A.W.; Patel, Y.M. Naringenin inhibits glucose uptake in MCF-7 breast cancer cells: A mechanism for impaired cellular proliferation. *Breast Cancer Res. Treat.* **2004**, *85*, 103–110. [CrossRef] [PubMed]

32. Lim, H.A.; Kim, J.H.; Kim, J.H.; Sung, M.K.; Kim, M.K.; Park, J.H.; Kim, J.S. Genistein induces glucose-regulated protein 78 in mammary tumor cells. *J. Med. Food* **2006**, *9*, 28–32. [CrossRef]

33. Jung, K.H.; Lee, J.H.; Thien-Quach, C.H.; Paik, J.Y.; Oh, H.; Park, J.W.; Lee, E.J.; Moon, S.H.; Lee, K.H. Resveratrol suppresses cancer cell glucose uptake by targeting reactive oxygen species-mediated hypoxia-inducible factor-1α activation. *J. Nucl. Med.* **2013**, *54*, 2161–2167. [CrossRef] [PubMed]

34. Gomez, L.S.; Zancan, P.; Marcondes, M.C.; Ramos-Santos, L.; Meyer-Fernandes, J.R.; Sola-Penna, M.; Da Silva, D. Resveratrol decreases breast cancer cell viability and glucose metabolism by inhibiting 6-phosphofructo-1-kinase. *Biochimie* **2013**, *98*, 1336–1343. [CrossRef] [PubMed]

35. Yang, Y.; Wolfram, J.; Boom, K.; Fang, X.; Shen, H.; Ferrari, M. Hesperetin impairs glucose uptake and inhibits proliferation of breast cancer cells. *Cell Biochem. Funct.* **2013**, *31*, 374–379. [CrossRef]

36. Moreira, L.; Araújo, I.; Costa, T.; Correia-Branco, A.; Faria, A.; Martel, F.; Keating, E. Quercetin and epigallocatechin gallate inhibit glucose uptake and metabolism by breast cancer cells by an estrogen receptor-independent mechanism. *Exp. Cell Res.* **2013**, *319*, 1784–1795. [CrossRef]

37. Du, G.J.; Song, Z.H.; Lin, H.H.; Han, X.F.; Zhang, S.; Yang, Y.M. Luteolin as a gly-colysis inhibitor offers superior efficacy and lesser toxicity of doxorubicin in breast cancer cells. *Biochem. Biophys. Res. Commun.* **2008**, *372*, 497–502. [CrossRef]

38. Yu, W.S.; Jeong, S.J.; Kim, J.H.; Lee, H.J.; Song, H.S.; Kim, M.S.; Ko, E.; Lee, H.J.; Khil, J.H.; Jang, H.J.; et al. The genome-wide expression profile of 1,2,3,4,6-penta-O-galloyl-β-D-glucose-treated MDA-MB-231 breast cancer cells: Molecular target on cancer metabolism. *Mol. Cells* **2011**, *32*, 123–132. [CrossRef]

39. Anzilotti, S.; Giampà, C.; Laurenti, D.; Perrone, L.; Bernardi, G.; Melone, M.A.; Fusco, F.R. Immunohistochemical localization of receptor for advanced glycation end (RAGE) products in the R6/2 mouse model of Huntington's disease. *Brain Res. Bull.* **2012**, *87*, 350–358. [CrossRef]

40. Aragno, M.; Mastrocola, R. Dietary Sugars and Endogenous Formation of Advanced Glycation Endproducts: Emerging Mechanisms of Disease. *Nutrients* **2017**, *9*, 385. [CrossRef]

41. Perrone, L.; Grant, W.B. Observational and ecological studies of dietary advanced glycation end products in national diets and Alzheimer's disease incidence and prevalence. *J. Alzheimers Dis.* **2015**, *45*, 965–979. [CrossRef] [PubMed]

42. Perrone, L.; Sbai, O.; Nawroth, P.P.; Bierhaus, A. The Complexity of Sporadic Alzheimer's Disease Pathogenesis: The Role of RAGE as Therapeutic Target to Promote Neuroprotection by Inhibiting Neurovascular Dysfunction. *Int. J. Alzheimers Dis.* **2012**, *2012*, 734956. [CrossRef] [PubMed]

43. van Heist, J.; Niessen, H.; Hoekman, K.; Schalkwij, C. Advanced glycation end products in human cancer tissues: Detection of Nepsilon-(carboxymethyl)lysine and argpryrimidine. *Ann. N. Y. Acad. Sci.* **2005**, *1043*, 725–733. [CrossRef] [PubMed]

44. Kiho, T.; Usui, S.; Hirano, K.; Aizawa, K.; Inakuma, T. Tomato paste fraction inhibiting the formation of advance glycation end-products. *Biosci. Biotechnol. Biochem.* **2004**, *1*, 200–205. [CrossRef]

45. Lo, C.-Y.; Li, S.; Tan, D.; Pan, M.-H.; Sang, S.; Ho, C.-T. Trapping reactions of reactive carbonyl species with tea polyphenols in simulated physiological conditions. *Mol. Nutr. Food Res.* **2006**, *50*, 1118–1128. [CrossRef]

46. Sang, S.; Shao, X.; Bai, N.; Lo, C.-Y.; Yang, C.; Ho, C.-T. Tea polyphenol (−)-Epigallocatechin-3-Gallate: A new trapping agent of reactive dicarbonyl species. *Chem. Res. Toxicol.* **2007**, *20*, 1862–1870. [CrossRef]

47. Palanissami, G.; Paul, S.F.D. RAGE and Its Ligands: Molecular Interplay Between Glycation, Inflammation, and Hallmarks of Cancer-a Review. *Horm. Cancer* **2018**, *9*, 295–325. [CrossRef]

48. Takada, M.; Ku, Y.; Toyama, H.; Suzuki, Y.; Kuroda, Y. Suppressive effects of tea polyphenol and conformational changes with receptor for advanced glycation en products (RAGE) expression in human hepatoma cells. *Hepatogastroenterology* **2002**, *49*, 928–931.

49. Lan, C.Y.; Chen, S.Y.; Kuo, C.W.; Lu, C.C.; Yen, G.C. Quercetin facilitates cell death and chemosensitivity through RAGE/PI3K/AKT/mTOR axis in human pancreatic cancer cells. *J. Food Drug Anal.* **2019**, *27*, 887–896. [CrossRef]

50. Schieber, M.; Chandel, N.S. ROS function in redox signaling and oxidative stress. *Curr. Biol.* **2014**, *24*, R453–R462. [CrossRef]

51. Liou, G.Y.; Storz, P. Reactive oxygen species in cancer. *Free Radic. Res.* **2010**, *44*, 479–496. [CrossRef] [PubMed]

52. NavaneethaKrishnan, S.; Rosales, J.L.; Lee, K.Y. ROS-Mediated Cancer Cell Killing through Dietary Phytochemicals. *Oxid. Med. Cell Longev.* **2019**, *2019*, 905154. [CrossRef] [PubMed]

53. Gibellini, L.; Pinti, M.; Nasi, M.; De Biasi, S.; Roat, E.; Bertoncelli, L.; Cossarizza, A. Interfering with ROS Metabolism in Cancer Cells: The Potential Role of Quercetin. *Cancers* **2010**, *2*, 1288–1311. [CrossRef] [PubMed]

54. Gibellini, L.; Bianchini, E.; De Biasi, S.; Nasi, M.; Cossarizza, A.; Pinti, M. Natural compounds modulating mitochon-drial functions. *Evid. Based Complement. Altern. Med.* **2015**, *2015*, 527209. [CrossRef]

55. Karmakar, S.; Banik, N.L.; Patel, S.J.; Ray, S.K. Curcumin activated both receptor-mediated and mitochondria-mediated proteolytic pathways for apoptosis in human glioblastoma T98G cells. *Neurosci. Lett.* **2006**, *407*, 53–58. [CrossRef]

56. Bhaumik, S.; Anjum, R.; Rangaraj, N.; Pardhasaradhi, B.V.V.; Khar, A. Curcumin mediated apoptosis in AK-5 tumor cells involves the production of reactive oxygen intermediates. *FEBS Lett.* **1999**, *456*, 311–314. [CrossRef]

57. Thayyullathil, F.; Chathoth, S.; Hago, A.; Patel, M.; Galadari, S. Rapid reactive oxygen species (ROS) generation induced by curcumin leads to caspase-dependent and -independent apoptosis in L929 cells. *Free Radic. Biol. Med.* **2008**, *45*, 1403–1412. [CrossRef]

58. Morin, D.; Barthelemy, S.; Zini, R.; Labidalle, S.; Tillement, J.P. Curcumin induces the mitochondrial permeability transition pore mediated by membrane protein thiol oxidation. *FEBS Lett.* **2001**, *495*, 131–136. [CrossRef]

59. Lee, Y.J.; Kim, N.Y.; Suh, Y.A.; Lee, C. Involvement of ROS in curcumin-induced autophagic cell death. *Korean J. Physiol. Pharmacol.* **2011**, *15*, 1–7. [CrossRef]

60. Gersey, Z.C.; Rodriguez, G.A.; Barbarite, E.; Sanchez, A.; Walters, W.M.; Ohaeto, K.C.; Komotar, R.J.; Graham, R.M. Curcumin decreases malignant characteristics of glioblastoma stem cells via induction of reactive oxygen species. *BMC Cancer* **2017**, *17*, 99. [CrossRef]

61. Xie, L.; Xiang, G.H.; Tang, T.; Tang, Y.; Zhao, L.Y.; Liu, D.; Zhang, Y.R.; Tang, J.T.; Zhou, S.; Wu, D.H. Capsaicin and dihydrocapsaicin induce apoptosis in human glioma cells via ROS and Ca2+-mediated mitochondrial pathway. *Mol. Med. Rep.* **2016**, *14*, 4198–4208. [CrossRef]

62. Lin, C.H.; Lu, W.C.; Wang, C.W.; Chan, Y.C.; Chen, M.K. Capsaicin induces cell cycle arrest and apoptosis in human KB cancer cells. *BMC Complement. Altern. Med.* **2013**, *13*, 46. [CrossRef]

63. Min, K.J.; Kwon, T.K. Anticancer effects and molecular mechanisms of epigallocatechin-3-gallate. *Integr. Med. Res.* **2014**, *3*, 16–24. [CrossRef]

64. Trachootham, D.; Zhou, Y.; Zhang, H.; Demizu, Y.; Chen, Z.; Pelicano, H.; Chiao, P.J.; Achanta, G.; Arlinghaus, R.B.; Liu, J.; et al. Selective killing of oncogenically transformed cells through a ROS-mediated mechanism by beta-phenylethyl isothiocyanate. *Cancer Cell* **2006**, *10*, 241–252. [CrossRef]

65. Zhu, Y.; Zhuang, J.X.; Wang, Q.; Zhang, H.Y.; Yang, P. Inhibitory effect of benzyl isothiocyanate on proliferation in vitro of human glioma cells. *Asian Pac. J. Cancer Prev.* **2013**, *14*, 2607–2610. [CrossRef]

66. Rather, R.A.; Bhagat, M. Cancer chemoprevention and piperine: Molecular mechanisms and therapeutic opportunities. *Front. Cell Dev. Biol.* **2018**, *6*, 10. [CrossRef]

67. Chen, Y.; Tseng, S.H.; Lai, H.S.; Chen, W.J. Resveratrol-induced cellular apoptosis and cell cycle arrest in neuroblas-toma cells and antitumor effects on neuroblastoma in mice. *Surgery* **2004**, *136*, 57–66. [CrossRef]

68. Shankar, S.; Chen, Q.; Siddiqui, I.; Sarva, K.; Srivastava, R.K. Sensitization of TRAIL-resistant LNCaP cells by resveratrol (3, 4′, 5 tri-hydroxystilbene): Molecular mechanisms and therapeutic potential. *J. Mol. Signal.* **2007**, *2*, 7. [CrossRef]

69. Rosa, L.D.S.; Jordão, N.A.; Soares, N.D.; DeMesquita, J.F.; Monteiro, M.; Teodoro, A.J. Pharmacokinetic, Antipro-liferative and Apoptotic Effects of Phenolic Acids in Human Colon Adenocarcinoma Cells Using In Vitro and In Silico Approaches. *Molecules* **2018**, *23*, 2569. [CrossRef]

70. Totta, P.; Acconcia, F.; Leone, S.; Cardillo, I.; Marino, M. Mechanisms of Naringenin-induced Apoptotic Cascade in Cancer Cells: Involvement of Estrogen Receptor α and β Signalling. *IUBMB Life* **2004**, *56*, 491–499. [CrossRef]

71. You, B.R.; Kim, S.Z.; Kim, S.H.; Park, W.H. Gallic acid-induced lung cancer cell death is accompanied by ROS increase and glutathione depletion. *Mol. Cell. Biochem.* **2011**, *357*, 295–303. [CrossRef]

72. Scalbert, A.; Manach, C.; Morand, C.; Remesy, C.; Jimenez, L. Dietary polyphenols and the prevention of diseases. *Crit. Rev. Food Sci. Nutr.* **2005**, *45*, 287–306. [CrossRef]

73. Thakur, V.S.; Gupta, K.; Gupta, S. The chemopreventive and chemotherapeutic potentials of tea polyphenols. *Curr. Pharm. Biotechnol.* **2012**, *13*, 191–199. [CrossRef]

74. Pandey, K.B.; Rizvi, S.I. Plant polyphenols as dietary antioxidants in human health and disease. *Oxidative Med. Cell. Longev.* **2009**, *2*, 270–278. [CrossRef]

75. Perrone, L.; Squillaro, T.; Napolitano, F.; Terracciano, C.; Sampaolo, S.; Melone, M.A.B. The Autophagy Signaling Pathway: A Potential Multifunctional Therapeutic Target of Curcumin in Neurological and Neuromuscular Diseases. *Nutrients* **2019**, *11*, 1881. [CrossRef]

76. Beltz, L.A.; Bayer, D.K.; Moss, A.L.; Simet, I.M. Mechanisms of cancer prevention by green and black tea polyphenols. *Anti-Cancer Agents Med. Chem.* **2006**, *6*, 389–406. [CrossRef]

77. Anand, P.; Kunnumakkara, A.B.; Sundaram, C.; Harikumar, K.B.; Tharakan, S.T.; Lai, O.S.; Sung, B.; Aggarwal, B.B. Cancer is a preventable disease that requires major lifestyle changes. *Pharm. Res.* **2008**, *25*, 2200. [CrossRef]

78. Stoll, E.A.; Horner, P.J.; Rostomily, R.C. The impact of age on oncogenic potential: Tumor-initiating cells and the brain microenvironment. *Aging Cell* **2013**, *12*, 733–741. [CrossRef]
79. Niedzwiecki, A.; Roomi, M.W.; Kalinovsky, T.; Rath, M. Anticancer efficacy of polyphenols and their combinations. *Nutrients* **2016**, *8*, 552. [CrossRef]
80. Shahcheraghi, S.H.; Zangui, M.; Lotfi, M.; Ghayour-Mobarhan, M.; Ghorbani, A.; Jaliani, H.Z.; Sadeghnia, H.R.; Sahebkar, A. Therapeutic Potential of Curcumin in the Treatment of Glioblastoma Multiforme. *Curr. Pharm. Des.* **2019**, *25*, 333–342. [CrossRef]
81. Zhuang, W.; Long, L.; Zheng, B.; Ji, W.; Yang, N.; Zhang, Q.; Liang, Z. Curcumin promotes differentiation of glioma-initiating cells by inducing autophagy. *Cancer Sci.* **2012**, *103*, 684–690. [CrossRef]
82. Wang, X.; Deng, J.; Yuan, J.; Tang, X.; Wang, Y.; Chen, H.; Liu, Y.; Zhou, L. Curcumin exerts its tumor suppressive function via inhibition of NEDD4 oncoprotein in glioma cancer cells. *Int. J. Oncol.* **2017**, *51*, 467–477. [CrossRef]
83. Mukherjee, S.; Baidoo, J.; Fried, A.; Atwi, D.; Dolai, S.; Boockvar, J.; Symons, M.; Ruggieri, R.; Raja, K.; Banerjee, P. Curcumin changes the polarity of tumor-associated microglia and eliminates glioblastoma. *Int. J. Cancer* **2016**, *139*, 2838–2849. [CrossRef]
84. Lee, J.E.; Yoon, S.S.; Moon, E.Y. Curcumin-Induced Autophagy Augments Its Antitumor Effect against A172 Human Glioblastoma Cells. *Biomol. Ther.* **2019**, *27*, 484–491. [CrossRef]
85. Maiti, P.; Scott, J.; Sengupta, D.; Al-Gharaibeh, A.; Dunbar, G.L. Curcumin and Solid Lipid Curcumin Particles Induce Autophagy, but Inhibit Mitophagy and the PI3K-Akt/mTOR Pathway in Cultured Glioblastoma Cells. *Int. J. Mol. Sci.* **2019**, *20*, 399. [CrossRef]
86. Park, K.S.; Yoon, S.Y.; Park, S.H.; Hwang, J.H. Anti-Migration and Anti-Invasion Effects of Curcumin via Suppression of Fascin Expression in Glioblastoma Cells. *Brain Tumor. Res. Treat.* **2019**, *7*, 16–24. [CrossRef]
87. Sak, K. Radiosensitizing Potential of Curcumin in Different Cancer Models. *Nutr. Cancer* **2019**, *24*, 1–14. [CrossRef]
88. Kielbik, A.; Wawryka, P.; Przystupski, D.; Rossowska, J.; Szewczyk, A.; Saczko, J.; Kulbacka, J.; Chwiłkowska, A. Effects of Photosensitization of Curcumin in Human Glioblastoma Multiforme Cells. *In Vivo* **2019**, *33*, 1857–1864. [CrossRef]
89. Pawlowska, E.; Szczepanska, J.; Szatkowska, M.; Blasiak, J. An Interplay between Senescence, Apoptosis and Autophagy in Glioblastoma Multiforme-Role in Pathogenesis and Therapeutic Perspective. *Int. J. Mol. Sci.* **2018**, *19*, 889. [CrossRef]
90. Pistollato, F.; Bremer-Hoffmann, S.; Basso, G.; Cano, S.S.; Elio, I.; Vergara, M.M.; Giampieri, F.; Battino, M. Targeting Glioblastoma with the Use of Phytocompounds and Nanoparticles. *Target Oncol.* **2016**, *11*, 1–16. [CrossRef]
91. Sato, A.; Okada, M.; Shibuya, K.; Watanabe, E.; Seino, S.; Suzuki, K.; Narita, Y.; Shibui, S.; Kayama, T.; Kitanaka, C. Resveratrol promotes proteasome-dependent degradation of Nanog via p53 activation and induces differentiation of glioma stem cells. *Stem. Cell Res.* **2013**, *11*, 601–610. [CrossRef]
92. Xiong, W.; Yin, A.; Mao, X.; Zhang, W.; Huang, H.; Zhang, X. Resveratrol suppresses human glioblastoma cell migration and invasion via activation of RhoA/ROCK signaling pathway. *Oncol. Lett.* **2016**, *11*, 484–490. [CrossRef]
93. Cilibrasi, C.; Riva, G.; Romano, G.; Cadamuro, M.; Bazzoni, R.; Butta, V.; Paoletta, L.; Dalprà, L.; Strazzabosco, M.; Lavitrano, M.; et al. Resveratrol Impairs Glioma Stem Cells Proliferation and Motility by Modulating the Wnt Signaling Pathway. *PLoS ONE* **2017**, *12*, e0169854. [CrossRef]
94. Yang, Y.P.; Chang, Y.L.; Huang, P.I.; Chiou, G.Y.; Tseng, L.M.; Chiou, S.H.; Chen, M.H.; Chen, M.T.; Shih, Y.H.; Chang, C.; et al. Resveratrol suppresses tumorigenicity and enhances radiosensitivity in primary glioblastoma tumor initiating cells by inhibiting the STAT3 axis. *J. Cell Physiol.* **2012**, *227*, 976–993. [CrossRef]
95. Mirzazadeh, A.; Kheirollahi, M.; Farashahi, E.; Sadeghian-Nodoushan, F.; Sheikhha, M.H.; Aflatoonian, B. Assessment Effects of Resveratrol on Human Telomerase Reverse Transcriptase Messenger Ribonucleic Acid Transcript in Human Glioblastoma. *Adv. Biomed. Res.* **2017**, *6*, 73.
96. Song, X.; Shu, X.H.; Wu, M.L.; Zheng, X.; Jia, B.; Kong, Q.Y.; Liu, J.; Li, H. Postoperative resveratrol administration improves prognosis of rat orthotopic glioblastomas. *BMC Cancer* **2018**, *18*, 871. [CrossRef]
97. Song, Y.; Chen, Y.; Li, Y.; Lyu, X.; Cui, J.; Cheng, Y.; Zheng, T.; Zhao, L.; Zhao, G. Resveratrol Suppresses Epithelial-Mesenchymal Transition in GBM by Regulating Smad-Dependent Signaling. *Biomed. Res. Int.* **2019**, *2019*, 1321973. [CrossRef]
98. Öztürk, Y.; Günaydın, C.; Yalçın, F.; Nazıroğlu, M.; Braidy, N. Resveratrol Enhances Apoptotic and Oxidant Effects of Paclitaxel through TRPM2 Channel Activation in DBTRG Glioblastoma Cells. *Oxid. Med. Cell Longev.* **2019**, *2019*, 4619865. [CrossRef]

99. Yokoyama, S.; Hirano, H.; Wakimaru, N.; Sarker, K.P.; Kuratsu, J. Inhibitory effect of epigallocatechin-gallate on brain tumor cell lines in vitro. *Neuro-Oncology* **2001**, *3*, 22–28. [CrossRef]

100. McLaughlin, N.; Annabi, B.; Bouzeghrane, M.; Temme, A.; Bahary, J.P.; Moumdjian, R.; Beliveau, R. The Survivin-mediated radio-resistant phenotype of glioblastomas is regulated by RhoA and inhibited by the green tea polyphenol (−)-epigallocatechin-3-gallate. *Brain Res.* **2006**, *1071*, 1–9. [CrossRef]

101. Chen, T.C.; Wang, W.; Golden, E.B.; Thomas, S.; Sivakumar, W.; Hofman, F.M.; Louie, S.G.; Schönthal, A.H. Green tea epigallocatechin gallate enhances therapeutic efficacy of temozolomide in orthotopic mouse glioblastoma models. *Cancer Lett.* **2011**, *302*, 100–108. [CrossRef]

102. Zhang, Y.; Wang, S.X.; Ma, J.W.; Li, H.Y.; Ye, J.C.; Xie, S.M.; Du, B.; Zhong, X.Y. EGCG inhibits properties of glioma stem-like cells and synergizes with temozolomide through downregulation of P-glycoprotein inhibition. *J. Neurooncol.* **2015**, *121*, 41–52. [CrossRef]

103. Sui, X.M.; Wang, J.X.; Zhu, Q.W.; Zhang, Q.F. Epigallocatechin-3-gallate induces apoptosis and proliferation inhibition of glioma cell through suppressing JAK2/STAT3 signaling pathway. *Int. J. Clin. Exp. Med.* **2016**, *9*, 10995–11001.

104. Roomi, M.W.; Kalinovsky, T.; Rath, M.; Niedzwiecki, A. Modulation of MMP-2 and MMP-9 secretion by cytokines, inducers and inhibitors in human glioblastoma T-98G cells. *Oncol. Rep.* **2017**, *37*, 1907–1913. [CrossRef]

105. Gurusinghe, K.R.D.S.N.S.; Mishra, A.; Mishra, S. Glucose-regulated protein 78 substrate-binding domain alters its conformation upon EGCG inhibitor binding to nucleotide-binding domain: Molecular dynamics studies. *Sci. Rep.* **2018**, *8*, 5487. [CrossRef]

106. Grube, S.; Ewald, C.; Kögler, C.; Lawson-McLean, A.; Kalff, R.; Walter, J. Achievable Central Nervous System Concentrations of the Green Tea Catechin EGCG Induce Stress in Glioblastoma Cells in Vitro. *Nutr. Cancer* **2018**, *70*, 1145–1158. [CrossRef]

107. Xie, C.R.; You, C.G.; Zhang, N.; Sheng, H.S.; Zheng, X.S. Epigallocatechin Gallate Preferentially Inhibits O6-Methylguanine DNA-Methyltransferase Expression in Glioblastoma Cells Rather than in Nontumor Glial Cells. *Nutr. Cancer* **2018**, *70*, 1339–1347. [CrossRef]

108. Udroiu, I.; Marinaccio, J.; Sgura, A. Epigallocatechin-3-gallate induces telomere shortening and clastogenic damage in glioblastoma cells. *Environ. Mol. Mutagen.* **2019**, *60*, 683–692. [CrossRef]

109. Santos, B.L.; Silva, A.R.; Pitanga, B.P.; Sousa, C.S.; Grangeiro, M.S.; Fragomeni, B.O.; Coelho, P.L.; Oliveira, M.N.; Menezes-Filho, N.J.; Costa, M.F.; et al. Antiproliferative, proapoptotic and morphogenic effects of the flavonoid rutin on human glioblastoma cells. *Food Chem.* **2011**, *127*, 404–411. [CrossRef]

110. Jakubowicz-Gil, J.; Langner, E.; Badziul, D.; Wertel, I.; Rzeski, W. Apoptosis induction in human glioblastoma multiforme T98G cells upon temozolomide and quercetin treatment. *Tumour Biol. J. Int. Soc. Oncodev. Biol. Med.* **2013**, *34*, 2367–2378. [CrossRef]

111. Pozsgai, E.; Bellyei, S.; Cseh, A.; Boronkai, A.; Racz, B.; Szabo, A.; Sumegi, B.; Hocsak, E. Quercetin increases the efficacy of glioblastoma treatment compared to standard chemoradiotherapy by the suppression of PI-3-kinase-Akt pathway. *Nutr. Cancer* **2013**, *65*, 1059–1066. [CrossRef]

112. Zhang, P.; Sun, S.; Li, N.; Ho, A.S.W.; Kiang, K.M.Y.; Zhang, X.; Cheng, Y.S.; Poon, M.W.; Lee, D.; Pu, J.K.S.; et al. Rutin increases the cytotoxicity of temozolomide in glioblastoma via autophagy inhibition. *J. Neurooncol.* **2017**, *132*, 393–400. [CrossRef]

113. Liu, Y.; Tang, Z.G.; Lin, Y.; Qu, X.G.; Lv, W.; Wang, G.B.; Li, C.L. Effects of quercetin on proliferation and migration of human glioblastoma U251 cells. *Biomed. Pharm.* **2017**, *92*, 33–38. [CrossRef]

114. Liu, Y.; Tang, Z.G.; Yang, J.Q.; Zhou, Y.; Meng, L.H.; Wang, H.; Li, C.L. Low concentration of quercetin antagonizes the invasion and angiogenesis of human glioblastoma U251 cells. *Onco Targets Ther.* **2017**, *10*, 4023–4028. [CrossRef]

115. Gentile, M.T.; Ciniglia, C.; Reccia, M.G.; Volpicelli, F.; Gatti, M.; Thellung, S.; Florio, T.; Melone, M.A.; Colucci-D'Amato, L. Ruta graveolens L. induces death of glioblastoma cells and neural progenitors, but not of neurons, via ERK 1/2 and AKT activation. *PLoS ONE* **2015**, *10*, e0118864. [CrossRef]

116. Taylor, M.A.; Khathayer, F.; Ray, S.K. Quercetin and Sodium Butyrate Synergistically Increase Apoptosis in Rat C6 and Human T98G Glioblastoma Cells Through Inhibition of Autophagy. *Neurochem. Res.* **2019**, *44*, 1715–1725. [CrossRef]

117. Tora, M.S.; Xenos, D.; Texakalidis, P.; Boulis, N.M. Treatment of neurofibromatosis 1-associated malignant peripheral nerve sheath tumors: A systematic review. *Neurosurg. Rev.* **2019**, 1–8. [CrossRef]

118. Angelo, L.S.; Wu, J.Y.; Meng, F.; Sun, M.; Kopetz, S.; McCutcheon, I.E.; Slopis, J.M.; Kurzrock, R. Combining curcumin (diferuloylmethane) and heat shock protein inhibition for neurofibromatosis 2 treatment: Analysis of response and resistance pathways. *Mol. Cancer* **2011**, *10*, 2094–2103. [CrossRef]

119. Blakeley, J. Development of drug treatments for neurofibromatosis type 2-associated vestibular schwannoma. *Curr. Opin. Otolaryngol. Head Neck Surg.* **2012**, *20*, 372–379. [CrossRef]

120. Turrini, E.; Ferruzzi, L.; Fimognari, C. Potential effects of pomegranate polyphenols in cancer prevention and therapy. *Oxidative Med. Cell. Longev.* **2015**, *2015*, 938475. [CrossRef]

121. Esposito, T.; Schettino, C.; Polverino, P.; Allocca, S.; Adelfi, L.; D'Amico, A.; Capaldo, G.; Varriale, B.; Di Salle, A.; Peluso, G.; et al. Synergistic Interplay between Curcumin and Polyphenol-Rich Foods in the Mediterranean Diet: Therapeutic Prospects for Neurofibromatosis 1 Patients. *Nutrients* **2017**, *9*, 783. [CrossRef] [PubMed]

122. Lee, M.J.; Tsai, Y.J.; Lin, M.Y.; You, H.L.; Kalyanam, N.; Ho, C.T.; Pan, M.H. Calebin-A induced death of malignant peripheral nerve sheath tumor cells by activation of histone acetyltransferase. *Phytomedicine* **2019**, *57*, 377–384. [CrossRef] [PubMed]

123. Hu, M.; Wu, B.; Liu, Z. Bioavailability of polyphenols and flavonoids in the era of precision medicine. *Mol Pharm* **2017**, *14*, 2861–2863. [CrossRef] [PubMed]

124. Teng, H.; Chen, L. Polyphenols and bioavailability: An update. *Crit. Rev. Food Sci. Nutr.* **2019**, *59*, 2040–2051. [CrossRef]

125. Pandareesh, M.D.; Mythri, R.B.; Srinivas Bharath, M.M. Bioavailability of dietary polyphenols: Factors contributing to their clinical application in CNS diseases. *Neurochem. Int.* **2015**, *89*, 198–208. [CrossRef]

126. Squillaro, T.; Peluso, G.; Melone, M.A.B. Nanotechnology-based polyphenol delivery: A novel therapeutic strategy for the treatment of age-related neurodegenerative disorder. *Austin Aging Res.* **2017**, *1*, 1004–1009.

127. Figueira, I.; Garcia, G.; Pimpão, R.C.; Terrasso, A.P.; Costa, I.; Almeida, A.F.; Tavares, L.; Pais, T.F.; Pinto, P.; Ventura, M.R.; et al. Polyphenols journey through blood-brain barrier towards neuronal protection. *Sci. Rep.* **2017**, *7*, 11456. [CrossRef]

128. Filosa, S.; Di Meo, F.; Crispi, S. Polyphenols-gut microbiota interplay and brain neuromodulation. *Neural. Regen. Res.* **2018**, *13*, 2055–2059.

129. Finicelli, M.; Squillaro, T.; Di Cristo, F.; Di Salle, A.; Melone, M.A.B.; Galderisi, U.; Peluso, G. Metabolic syndrome, Mediterranean diet, and polyphenols: Evidence and perspectives. *J. Cell Physiol.* **2019**, *234*, 5807–5826. [CrossRef]

130. Squillaro, T.; Cimini, A.; Peluso, G.; Giordano, A.; Melone, M.A.B. Nano-delivery systems for encapsulation of dietary polyphenols: An experimental approach for neurodegenerative diseases and brain tumors. *Biochem. Pharm.* **2018**, *154*, 303–317. [CrossRef]

131. Singh, H. Nanotechnology Applications in Functional Foods; Opportunities and Challenges. *Prev. Nutr. Food Sci.* **2016**, *21*, 1–8. [CrossRef]

Review

Targeting Glucose Transporters for Breast Cancer Therapy: The Effect of Natural and Synthetic Compounds

Ana M. Barbosa [1] and Fátima Martel [2,3,]*

[1] Instituto de Ciências Biomédicas Abel Salazar, University of Porto, 4169-007 Porto, Portugal; anamargabarbosa@gmail.com

[2] Unit of Biochemistry, Department of Biomedicine, Faculty of Medicine, University of Porto, 4200-319 Porto, Portugal

[3] Instituto de Investigação e Inovação em Saúde, University of Porto, 4200-135 Porto, Portugal

* Correspondence: fmartel@med.up.pt; Tel.: +351-22-042-6654

Received: 6 December 2019; Accepted: 7 January 2020; Published: 8 January 2020

Abstract: Reprogramming of cellular energy metabolism is widely accepted to be a cancer hallmark. The deviant energetic metabolism of cancer cells-known as the Warburg effect-consists in much higher rates of glucose uptake and glycolytic oxidation coupled with the production of lactic acid, even in the presence of oxygen. Consequently, cancer cells have higher glucose needs and thus display a higher sensitivity to glucose deprivation-induced death than normal cells. So, inhibitors of glucose uptake are potential therapeutic targets in cancer. Breast cancer is the most commonly diagnosed cancer and a leading cause of cancer death in women worldwide. Overexpression of facilitative glucose transporters (GLUT), mainly GLUT1, in breast cancer cells is firmly established, and the consequences of GLUT inhibition and/or knockout are under investigation. Herein we review the compounds, both of natural and synthetic origin, found to interfere with uptake of glucose by breast cancer cells, and the consequences of interference with that mechanism on breast cancer cell biology. We will also present data where the interaction with GLUT is exploited in order to increase the efficiency or selectivity of anticancer agents, in breast cancer cells.

Keywords: breast cancer; glucose transport; drugs; natural compounds

1. Introduction

According to the last Global Cancer Statistics (GLOBOCAN 2018), breast cancer represented 12% of all cancers, being the second most frequent cancer worldwide, after lung cancer, and caused about 7% of the total cancer deaths in 2018 [1]. In women, breast cancer is the leading type of cancer and the leading cause of cancer death worldwide [1].

Screening programs and adjuvant chemotherapy have had a significant impact on the prognosis of breast cancer patients, having significantly improved their overall survival, disease-free survival, and death rates related to breast-cancer since the early 1990s [2,3]. Nevertheless, efforts must continue in order to reduce not only the incidence but also the mortality and treatment-associated morbidities associated with this disease. In this context, discovery of new molecular targets and the refinement of lead compounds constitute a priority in breast cancer research.

2. Metabolic Reprogramming in Cancer Cells

Metabolic reprogramming and altered energetics is firmly established as a hallmark of cancer and constitutes an active area of basic, translational, and clinical cancer research in recent years [4].

One of the cancer metabolic hallmarks is a deviant energetic metabolism-known as the Warburg effect-characterized by a very high rate of glycolysis and production of lactate, even in the presence of oxygen [5]. Cancer cells have a high dependence on the glycolytic pathway to supply their need of high amounts of adenosine triphosphate (ATP) and also of metabolic intermediates that contribute to several biosynthetic pathways, crucial for cancer progression [4] and to compensate for excess metabolic production of reactive oxygen species (ROS) [6]. So, they shift their main ATP-producing process from oxidative phosphorylation to glucose fermentation, even in aerobic conditions [4]. This altered metabolism may be not only a consequence of genetic mutations, but also a contributing factor or cause of tumorigenesis [7].

More recently, a 'two-compartment' model, also named 'the reverse Warburg effect' or "metabolic coupling", has been proposed to reconsider metabolism in tumors, because it was realized that another type of metabolism occurs in certain types of cancers, having high mitochondrial respiration and low glycolysis rate [8]. According to this model, tumor cells and adjacent stromal fibroblasts form a two-compartment model of cancer metabolism, in which fibroblasts perform aerobic glycolysis (because of the acidic microenvironment induced by cancer cells), and the generated metabolites (such as pyruvate, ketone bodies, fatty acids, and lactate), are transferred to tumor cells, to fuel the Krebs cycle and maintain ATP generation [9]. This metabolic coupling is found in some forms of breast cancer [10], and may contribute drug resistance and therapeutic failure in some types of cancers [11], as observed with tamoxifen-resistance in breast cancer MCF7 cells [12].

3. Upregulation of Glucose Transport in Breast Cancer Cells

Since the energetic metabolic shift in cancer cells produces less ATP per glucose molecule, the demand for glucose in these cells is higher than in normal cells. Therefore, cancer cells rely on higher rates of glucose uptake in order to support their increased energy, biosynthesis and redox needs. This increased rates of cellular uptake of glucose is met by overexpression of glucose transporters, which is observed in most cancer cells [13].

Two families of glucose transporters mediate glucose uptake in mammalian cells: the Na^+-dependent glucose co-transporters (SGLTs) and the facilitative glucose transporters (GLUTs).

The SGLT family (gene symbol *SLC5A*) are secondary active transporters that transfer glucose against its concentration gradient coupled with Na^+ transport down its concentration gradient, which is maintained by the Na^+/K^+ pump. For every glucose molecule that is transported, two Na^+ are also transported. SGLT transporters have 14 transmembrane domains and a high affinity for glucose. At physiological extracellular Na^+ concentration and membrane potential, an apparent Km of 0.5 mM of SGLT1 for glucose was described, but glucose is transported with a lower affinity when the plasma membrane is depolarized and/or the extracellular Na^+ concentration is low [14,15]. SGLT1 and SGLT2 overexpression is present in some types of cancer, such as pancreas, prostate, lung, liver, and ovarian cancer, but these transporters have not been described in breast cancer [16].

The GLUT family (gene symbol *SLC2A*) are facilitative transporters that mediate the transport of glucose down its concentration gradient. This family of transporters is composed of 14 members: GLUT1-GLUT12, GLUT14, and the H^+/myo-inositol transporter. All GLUTs are predicted to have 12 transmembrane domains connected by hydrophilic loops. Each of the GLUT transport protein possesses different affinities for glucose and other hexoses such as fructose. GLUT1, GLUT3, and GLUT4 have a high affinity for glucose (e.g., the Km of GLUT1 for glucose is 1–3 mM), allowing transport of glucose at a high rate under normal physiological conditions [17].

Increased cellular uptake of glucose in tumor cells is associated with increased and deregulated expression of GLUT transporters [13]. Among GLUT family members, overexpression of GLUT1 has been consistently observed in many different cancers, including breast, lung, renal, colorectal, and pancreatic cancers [13,18,19]. Consistent with its overexpression, GLUT1 is crucial for uptake of glucose by breast cancer cells [20–22] and is also the main glucose transporter in breast cancer cell lines (e.g., MCF-7 and MDA-MB-231) [21,23]. GLUT1 is a transporter ubiquitously expressed in most mammalian tissues (abundantly in brain and erythrocytes), being responsible for basal glucose cellular uptake in the majority of tissues [16,17]; it is also the predominant isoform present in human and bovine mammary glands [24,25]. Glucose uptake mediated by GLUT1 appears to be especially critical in the early stages of breast cancer development, affecting cell transformation and tumor formation [26,27]. Indeed, GLUT1 overexpression, which occurs early during the transformation process, induces a change in breast epithelial cell metabolism that precedes morphological changes in breast cancer, and thus may be a fundamental part of the neoplastic process [18]. Interestingly, the loss of even a single GLUT1 allele is sufficient to impose a strong break in breast tumor development in a mouse model [26]. A strong correlation between *GLUT1* gene expression and breast cancers of higher grade and proliferative index and lower degree of differentiation [28] and higher malignant potential, invasiveness, and consequently poorer prognosis [29] exists. GLUT1 is thus considered an oncogene [18–20,30].

One of the factors responsible for the upregulation of GLUT1 in breast tumor cells is hypoxia. The promoters of GLUT1 contain hypoxia-response elements, which bind the hypoxia-inducible factor (HIF-1) to facilitate transcription. Since an increase in the levels of HIF-1α protein is a phenomenon seen in most cancers, it provides a molecular mechanism for cancer-associated overexpression of GLUT1 [18,31]. Additionally, hypoxia appears to increase GLUT1 transport activity in the MCF-7 breast cancer cell line, independently of changes in transporter expression [32]. Besides HIF-1, the ovarian hormone estrogen is also known to induce GLUT1 expression in breast cancer [18,33]. Moreover, the histone deacetylase SIRT6, the cellular oncogene product c-MYC (V-Myc Avian Myelocytomatosis Viral Oncogene Homolog), the pro-survival protein kinase Akt (Protein Kinase B) and mutant p53, all of which induce the expression of GLUT1 [31,34], can also be involved in GLUT1 overexpression in breast cancer.

In addition to GLUT1, which is consistently found to be expressed in breast tumors and cell lines, other GLUT family members can also contribute to glucose uptake by breast cancer cells. More specifically, GLUT2 [19,23] and GLUT3 [18] are also expressed in several breast cancer cell lines. Additionally, GLUT4 expression [30,35–37] and insulin-stimulated glucose uptake were also described in some cancer cell lines [38–40]. Moreover, the involvement of GLUT4 in basal glucose uptake was described in two breast cancer cell lines [41]. Finally, a second insulin-stimulated transporter, GLUT12, was also described in MCF-7 cells [18,42]. Similar to GLUT1, the expression of GLUT3 and GLUT12 correlate with poor prognosis [18,19]. Importantly, increased expression of GLUT1 and GLUT3 was also associated with resistance of cancer cells to radio or chemotherapy [43–45], but the underlying mechanisms linking GLUT and chemo- or radio-resistance remain largely unknown.

Increased glucose uptake by cancer cells has been exploited clinically in diagnosis and follows up of cancer via the use of ^{18}fluoro-2-deoxy-D-glucose (FDG), a radiolabeled glucose analogue, in Positron Emission Tomography (PET) [46]. This radiotracer enters cells via GLUTs, being then phosphorylated by hexokinases into FDG-6-phosphate that cannot be further metabolized and thus accumulates in the cytoplasm. Importantly, the sensitivity of this technique varies depending on the type of cancer, and this heterogeneity has been particularly associated with GLUT1 or GLUT3 tumor expression [23,47].

4. Glucose Transporters as Therapeutic Targets in Breast Cancer

Since cancer cells depend on increased utilization of glucose as compared to normal healthy cells, glucose deprivation is considered an effective anticancer therapy and as a potential strategy for cancer prevention, and many compounds targeting cancer cell energy metabolism are currently on trial or approved as therapeutic agents against cancer [48,49]. These include specific inhibitors of monocarboxylate transporter 1, hexokinase II, glyceraldehyde-3-phosphate dehydrogenase (GAPDH), pyruvate dehydrogenase, pyruvate dehydrogenase kinase 1, cancer-specific mutant isocitrate dehydrogenase, lactate dehydrogenase A, phosphoglycerate mutase 1, phosphofructokinase, or pyruvate kinase M2 [48,50]. In support of glucose deprivation as a molecular target in cancer, high-fat and low-carbohydrate diet appear to provide therapeutic benefits for increased survival by reducing angiogenesis, peri-tumoral edema, cancer migration, and invasion [51]. According to some authors, inhibition of glucose metabolism will not only deplete cancer cells of ATP, but also will lead to enhanced oxidative stress-related cytotoxicity [6].

Additionally, because tumor cells have an increased dependence in relation to extracellular glucose, GLUTs constitute also an anticancer target [18,52–54]. A direct approach to this therapeutic target is to block GLUT-mediated glucose uptake, which would abolish entry of glucose into the cancer cell. Alternatively, new approaches consist in the design and development of "GLUT-transportable anticancer agents", or the use of GLUT antibodies to selectively deliver an anticancer agent to cancer cells.

In this review, we will list compounds, both of natural and synthetic origin, found to interfere with glucose uptake by breast cancer cells, and present the consequences of GLUT inhibition and/or knockout on breast cancer cell biology. We will also present data where the interaction of defined molecules with GLUT is exploited in order to increase its efficiency or selectivity, in breast cancer cells.

5. Effect of Synthetic and Natural Compounds on Glucose Uptake by Breast Cancer Cells

5.1. Effect of Synthetic Compounds

5.1.1. GLUT Inhibitors

WZB117 and STF-31

The effect of two recently described GLUT1 inhibitors, WZB117 and STF-31, on breast cancer cells was studied by some authors. WZB117 is a representative of a group of novel small compounds that were recently reported to inhibit basal glucose transport and cell growth in vitro and in vivo [55,56]. STF-31 is a small molecule that was firstly reported to selectively target von Hippel-Lindau (VHL)-deficient renal cell carcinoma (RCC) cells [55].

These two GLUT1 inhibitors were able to inhibit cell proliferation and induce apoptosis in several breast cancer cell lines (MCF-7, MDA-MB-231, HBL100, and BT549), and these effects were accompanied by interference with cellular glucose handling, increasing the levels of extracellular glucose, and decreasing the levels of extracellular lactate, suggesting an inhibitory effect upon glucose uptake and/or glycolysis. Of interest, STF31 (30 µM) potentiated the antiproliferative effect of metformin (3 mM) in MDA-MB-231 cells [57]. Although the effect on glucose uptake was not studied, GLUT1 inhibition (with WZB117) blocked transformation of MCF10A-ERBB2 cells (a breast epithelial cell line used as a model to study the early events leading to transformation) induced by activated ERBB2 through reduced cell proliferation [26] (Table 1).

In addition of testing these GLUT1 inhibitors alone as a targeted therapy, GLUT1 inhibition in combination with other cancer therapeutics has also been evaluated (Table 1). In one study, WZB117 was found to reduce GLUT1 mRNA and protein levels and glucose uptake and lactate production in two breast cancer cell lines (MCF-7 and MDA-MB-231). The interaction of this agent with radiation was investigated. Glucose metabolism and GLUT1 expression were found to be significantly stimulated by radiotherapy. Interestingly, radioresistant breast cancer cells exhibited upregulated GLUT1 expression and glucose metabolism but combination of WZB117 and radiation re-sensitized the radioresistant cancer cells to radiation [58]. A synergic antitumoral effect was also found between WZB117 and the anticancer drugs cisplatin and paclitaxel, in MCF-7 cells [59]. Finally, the possibility that a combined treatment with a GLUT1 inhibitor could overcome resistance to another breast cancer therapeutic agent (adriamycin) was also investigated. Resistance to adriamycin is a common obstacle occurring during therapy of breast cancer patients. WZB117 was found to resensitize MCF-7/ADR cells (adriamycin-resistant) to adriamycin [60]. Therefore, GLUT1 inhibition could overcome resistance to adriamycin and radiation.

WZB27 and WZB115

Two other GLUT1 inhibitors, WZB27 and WZB115, were synthesized and tested against several cell types, including a breast cancer cell line (MCF-7). These compounds reduced basal glucose uptake and cell proliferation, induced apoptosis, and led to cell cycle arrest in G1/S phase, without affecting much the normal cell line MCF12A. Importantly, their inhibitory effect on cancer cell growth was ameliorated when additional glucose was present, suggesting that the inhibition was due, at least in part, to inhibition of basal glucose uptake. Moreover, when used in combination, the test compounds demonstrated synergistic effects with the anticancer drugs cisplatin and paclitaxel (Table 1) [56].

Bay876

BAY-876 is a highly selective GLUT1 inhibitor under preclinical study for oncolytic treatment [61]. In a recent report, the interaction between GLUT1 and bromodomains (BRDs) was investigated. BRDs are conserved protein interaction modules, which recognize acetyl-lysine modifications, and BRD-containing proteins are components of the transcription factor and chromatin-modifying complexes and determinants of epigenetic memory [62]. BAY876 decreased glucose uptake by a triple-negative breast cancer cell line, and a vulnerability of these breast cancer cells to inhibition of BRPF2/3 BRDs, under conditions of glucose deprivation or GLUT1 inhibition, was reported (Table 1) [63].

2-deoxy-D-glucose

2-deoxy-D-glucose (2-DG) is a synthetic non-metabolizable glucose analogue. 2-DG inhibits the glycolytic pathway, because the product of its phosphorylation by hexokinase cannot be further metabolized and, additionally, is a non-competitive inhibitor of hexokinase, thus causing ATP depletion. Additionally, 2-DG competes with glucose for GLUT [64]. In a triple-negative breast cancer cell line (MDA-MB-231), but not in an estrogen receptor (ER)-positive cell line (MCF-7), 2-DG was able to reduce glucose uptake (Table 1) [65].

GLUT1 shRNA

Another strategy that is being tested to target GLUT1 is by RNA interference (RNAi) using short hairpin RNA (shRNA). Silencing of GLUT1 expression with an shRNA led to a significant decrease in glucose uptake in vitro in both a triple-negative (MDA-MB-468) and a HER2-positive cell line (SK-BR3), together with a decrease of the growth of xenograft tumors (MDA-MB-468 cells) [66]. Similarly, shRNA targeting GLUT1 decreased glucose transport and consumption, reduced lactate secretion, and inhibited growth of the mouse mammary tumor cell line 78617GL, both in vitro and in vivo (Table 1) [27].

A similar negative effect of GLUT1 shRNA on glucose uptake was found in two other triple-negative breast cancer cell lines (MDA-MB-231 and Hs578T), together with a decrease in cell proliferation, migration, and invasion, which was concluded to result from GLUT1-mediated modulation of Epidermal Growth Factor Receptor (EGFR)/Mitogen-Activated Protein Kinase (MAPK), and integrin β1/Src/FAK signaling pathways [67]. However, the same group verified that, contrary to the expected, ablation of GLUT1 attenuated apoptosis and increased drug resistance in triple-negative breast cancer cells (MDA-MB-231 cells), via upregulation of p-Akt/p-GSK-3β (Ser9)/β-catenin/surviving (Table 1) [52]. Not only is the prognostic of triple-negative breast cancer (TNBC) usually poor due to aggressive tumor phenotypes, but also because conventional chemotherapy cannot be used. Therefore, and because TNBC have higher levels of GLUT1, this transporter is seen as a potential therapeutic target, sensitizing cells to chemotherapy. The results of this later study, however, indicate that the potential of GLUT1 as a therapeutic target in TNBC should be carefully re-evaluated [68].

Anti-GLUT1-antibody

An anti-GLUT1 monoclonal antibody was able to decrease glucose uptake in breast cancer cells (MDA-MB-231), and to reduce cell proliferation and stimulate apoptosis (MCF-1 and T47D). Importantly, when associated with chemotherapeutic agents (5 μM cisplatin, 5 μM paclitaxel, or 10 μM gefitinib), it potentiated the anti-proliferative and pro-apoptotic effects of these agents in MCF-7 cells (Table 1). The authors concluded that the use of antibodies to GLUT1 may be a viable but an as yet unexplored therapeutic strategy in tumors that overexpress GLUT1 [69].

GLUT4 shRNA

By stably silencing GLUT4 expression by lentiviral expression of a GLUT4 shRNA, GLUT4 was concluded to have a prominent role in basal glucose uptake in MCF7 and MDA-MB-231 breast cancer cells (Table 1). Moreover, GLUT4 specific downregulation in these two different breast cancer cell lines, with different degrees of malignancy and differentiation, promoted metabolic reprogramming and affected cell proliferation and viability. According to these authors, their study provides proof-of-principle for the feasibility of using pharmacological approaches to inhibit GLUT4 in order to induce metabolic reprogramming in vivo in breast cancer models [41].

5.1.2. Antidiabetics

Biguanides, including metformin and phenformin, are inhibitors of mitochondrial respiratory chain complex I, and have been shown to reduce cancer incidence and cancer-related death [68].

Metformin

Metformin is the most prescribed oral antidiabetic drug used for the treatment of diabetes mellitus. Metformin was associated with reduced risk of developing cancer in diabetic patients in 2005 [70]. Since then, a large amount of studies confirmed this observation, and the role of metformin in breast cancer has been evaluated [71,72]. In this context, the effect of metformin on glucose uptake and metabolism by breast cancer cells has been evaluated in a few studies (Table 1).

In a first study, metformin was found to decrease glucose utilization both in vitro (MDA-MB-231) and in vivo (using MDA-MB-231 cells orthotopically implanted in a mammary fat pad). However, this effect was concluded to be related to a direct inhibitory effect on the glycolytic enzyme hexokinase and drug effects on transmembrane glucose transport were excluded, because glucose uptake and glucose transporters expression levels were not affected by metformin [73].

In the study by Amaral et al. [65], short-term exposure to metformin inhibited glucose uptake, probably by direct inhibition of GLUT1, and, in contrast, long-term exposure to metformin led to a significant increase in glucose uptake, which was not associated with changes in GLUT1 mRNA levels. It was suggested that the increase in glucose uptake induced by long-term metformin, is a compensatory mechanism in response to cellular ATP depletion resulting from its inhibitory effect on oxidative phosphorylation and that this metformin-induced dependence on glycolytic pathway, associated with an anticarcinogenic effect of the drug, provides a biochemical basis for the design of new therapeutic strategies. The increase in glucose uptake after a long-term exposure to metformin, to compensate for the reduced mitochondrial ATP generation, was corroborated in another study, using two triple-negative breast cancer cell lines (MDA-MB-231 and MDA-MB-436) [74]. Lastly, the interaction between metformin and PPARδ (peroxisome-proliferator-activated receptor δ), known to have a role in inflammation, metabolism, and cancer, was recently evaluated [75]. Metformin was able to block the increase in GLUT1 and SGLT1 mRNA and protein levels, glucose uptake, glucose consumption, and lactate production caused by the PPARδ agonist GW501516 in MCF-7 cells. The effect of metformin in reducing the expression of GLUT1 and SGLT1 was not present with metformin alone; rather, it results from metformin-mediated inhibition of PPARδ activity. Therefore, metformin can block the effect of GW501516, but has no effect of its own in reduction of glucose transporters levels, which is in concordance with the previous studies.

Phenformin

A recent study showed that glucose uptake and utilization affects cancer cell sensitivity to phenformin treatment. More specifically, a correlation between low expression of glucose transporters, including GLUT1, and both a defective glucose uptake/utilization and an increased sensitivity to phenformin treatment was found in several cancer cell lines. Moreover, restoration of GLUT1 expression attenuated the phenformin-sensitivity in the corresponding cancer cells [76]. Additionally, Liu and Gan [77], by using the MDA-MB-231 cell line, demonstrated that phenformin upregulates GLUT1 levels, causing increased glucose uptake and production of lactate. Importantly, they verified that this effect of phenformin is dependent on NBR2 (neighbor of *BRCA1* gene 2), a glucose starvation-induced long non-coding RNA that interacts with AMP-Activated Protein Kinase (AMPK) and regulates AMPK activity. They thus concluded that the NBR2-GLUT1 axis may serve as an adaptive response in breast cancer cells to survive in response to phenformin treatment (Table 1) [77].

Troglitazone

Another antidiabetic drug also associated with an anticarcinogenic effect [78], troglitazone, belongs to the class of thiazolidinediones, which activate peroxisome proliferator-activated receptor-γ (PPARγ), although it has been withdrawn from the market due to its hepatotoxicity. Unlike the antidiabetic effects of this drug, many other actions of troglitazone are thought to occur in a PPARγ-independent manner. Since troglitazone is known to cause mitochondrial dysfunction, its effect on glucose metabolism was investigated [79]. Troglitazone enhanced uptake of glucose in several breast cancer cell lines, but changes in GLUT levels do not seem to play a role in this effect, that rather appears to involve MAPK, AMPK, and EGFR. Interestingly, troglitazone reduced T-47D cell content, and this effect was potentiated by restriction of glucose availability. So, it was concluded that troglitazone stimulates uptake of glucose by cancer cells and shifts its metabolism toward glycolysis, likely as an adaptive response to impaired mitochondrial oxidative respiration (Table 1) [80].

5.1.3. Chemotherapeutic Agents

Cisplatin

Cisplatin (cis-diamminedichloroplatinum II) is a very common used chemotherapeutic agent. It is a platinum-derived agent that interferes with DNA replication, and has also been associated with mitochondrial damage. Wang et al [36] used the MDA-MB-231 cell line in order to study cisplatin's metabolic effects. The compound decreased glucose uptake and lactate production and the expression levels of GLUT1 and GLUT4 (Table 1). Cisplatin downregulation of integrin β5 (ITGB5)/FAK signaling pathway was concluded to be responsible for its effect on the expression of GLUT1 and GLUT4 [36].

Sorafenib

The bisarylurea sorafenib is a multi-kinase inhibitor with anti-proliferative and anti-angiogenic activity, currently under evaluation in a variety of solid tumors. Evidence has shown that sorafenib can inhibit oxidative phosphorylation in some types of cancer cell lines [80,81], and the question if it also affects glucose metabolism was then addressed [82]. In this work, the effect of sorafenib on glucose uptake, utilization, lactate production, and GLUT1 expression was investigated in several breast cancer cell lines. Sorafenib produced distinct early and long-term effects on glucose uptake, metabolism, and GLUT1 expression in MCF-7 (ERα-positive), MDA-MB-231 (triple negative), and SKBR3 (ERα-negative/HER2-positive) cell lines. Fasentin (a GLUT1 inhibitor) inhibited the initial GLUT1 overexpression caused by sorafenib and, importantly, its cytotoxic effect (Table 1). It was concluded that the early-term effects were dependent on AMPK and thought to compensate for the loss of mitochondrial ATP, but that persistent activation of AMPK by sorafenib finally led to the impairment of glucose metabolism in all the cell lines, resulting in cell death [82].

Trastuzumab

Trastuzumab, effective in about 15% of women with breast cancer, targets Human Epidermal Growth Factor Receptor 2 (HER2) and downregulates signaling through Akt/phosphoinositide 3-kinase (PI3K) and MAPK pathways. These pathways modulate glucose metabolism and so it was evaluated if trastuzumab decreased glucose uptake in breast cancer cells. For this, xenografts derived from HER2-overexpressing MDA-MB-453 human breast tumor cells were grown in severe combined immunodeficient mice. Xenografts were significantly smaller and [^{18}F] FDG uptake was also reduced in trastuzumab-treated mice. This observation was accompanied by lower GLUT1 protein levels (Table 1) [83].

Doxorubicin (DOX) and 5-fluorouracil (5FU)

The effect of these two chemotherapeutic agents on the expression and activity of GLUT1 and hexokinase and on glucose uptake by the MCF-7 breast cancer cell line was evaluated [84]. Both agents induced a decrease in glucose uptake together with an increase in GLUT1 mRNA levels. The effect on GLUT1 protein levels were not as marked, which suggest posttranslational alterations in GLUT1. It was concluded that after DOX or 5FU therapy, the relationship between glucose and viable cell number can become disjointed, with transient declines in glucose uptake in excess of the decline in cell number despite increased GLUT1 mRNA levels [84]. In another work, DOX and selenium, either free or in PLGA (poly (d, l-lactide-co-glycolide)) nanoparticles were described to reduce the cellular uptake of glucose by MCF-7 and MDA-MB-231 cells, based on measurements of medium glucose levels [85]. So, no direct measurement of glucose uptake was made (Table 1).

Palbociclib

Dysregulation of the cell cycle is a hallmark of cancer that leads to aberrant cellular proliferation and inhibition of cell cycle regulators such as Cyclin-Dependent Kinase 4 (CDK4) and 6 (CDK6) has become a new therapeutic target for the treatment of breast cancer. Palbociclib, an orally-available inhibitor of CDK4 and CDK6, represents the most widely studied compound among cell cycle inhibitors [86]. Interestingly, palbociclib also seems to be able to inhibit GLUT1 mediated glucose uptake and metabolism in TNBC cells [87,88]. Moreover, combination of palbociclib with a chemotherapeutic agent currently used for the treatment of TNBC patients (paclitaxel) inhibited cell proliferation and increased cell death more efficiently than single treatments, associated with a more marked effect on glucose uptake and consumption and on GLUT1 protein levels [87]. Additionally, combination of palbociclib with a PI3K/mTOR inhibitor (BYL719) enhanced the antitumoral effect of these agents and the negative effect of each of these drugs on glucose uptake and consumption and on GLUT1 protein levels (Table 1) [88].

5.1.4. Other Drugs

Propranolol

Another type of drug that has been getting attention for its recently found anticarcinogenic effect are beta-blockers, more specifically propranolol (PROP). Clinical evidence has strongly indicated that PROP can inhibit cancer growth, metastasis development, and tumor recurrence in breast cancer patients [89]. Treatment with PROP decreased hexokinase-2 expression in vitro and 18F-FDG uptake in vivo, but GLUT1 levels were not affected. This indicates that GLUT1 is not involved in the anticarcinogenic effect of PROP (Table 1) [90].

Saracatinib

Prevention of estrogen receptor negative (ER-) and tamoxifen-resistant (TamR) breast cancer remains an important demand due to gaps in pathobiological understanding of this type of cancer. Transforming (sarcoma-inducing) Gene of Rous Sarcoma Virus (Src) activation appears to be a key signaling event driving ER- and TamR breast cancer progression and thus, targeting Src may prevent ER-breast cancer [91]. Accordingly, Src-targeting agents such as the tyrosine kinase inhibitor saracatinib, have been extensively tested in the clinic for treatment of metastatic breast cancer [92]. In the report by Jain et al. [93], activation of Src kinase was investigated as an early signaling alteration in premalignant breast lesions of women who did not respond to tamoxifen, a widely used ER antagonist for hormonal therapy of breast cancer. They verified that Src plays an essential role in regulating glucose uptake, because knocking down Src significantly reduced glucose uptake. Moreover, they showed that saracatinib inhibited glucose uptake in premalignant breast cell lines (MCF-10A and MCF12A) with or without HER-2 overexpression (Table 1).

P53 Modulators

The tumor protein p53, a well-recognized tumor suppressor, is a key regulator of energy metabolism, playing an important role in preventing the cell from reprogramming its energetic metabolic pathway [94]. The p53-reactivating compound RITA (Reactivating p53 and Inducing Tumor Apoptosis) activates p53 in cells expressing oncogenes, whereas its effect in non-transformed cells is almost negligible [95]. This agent decreased GLUT1 mRNA expression in MCF-7 cells. Further, another p53 activator, nutlin3a [96], also caused repression of GLUT1 mRNA expression. Interestingly, the p53 inhibitor pifithrin-α p53 [95] induced the expression of GLUT1 mRNA and abolished the effect of RITA upon GLUT1 mRNA levels (Table 1) [94]. This study shows that reinstatement of p53 function targets the dependence of cancer cells on glycolysis, which can contribute to the selective killing of cancer cells by pharmacologically activated p53.

Akt Inhibitors

The protein kinase Akt is involved in various cellular processes, including cell proliferation, growth and metabolism, and hyperactivation of Akt is commonly observed in human tumors [97]. Three non-ATP-competitive allosteric Akt inhibitors (Akt1i, Akt2i, and Akt1/2i) reduced glucose transport into T-47D breast cancer cells, by interfering with a process distinct from the Akt signaling pathway (involved in movement of GLUT4 to the plasma membrane, e.g., in adipocytes). Among other evidences, the PI3K inhibitor wortmannin was devoid of effect on glucose uptake. It was concluded that these drugs, at least in part, inhibit tumorigenesis through inhibition of glucose transport in tumor cells (Table 1) [98].

PGC1β and HKDC1 shRNA

The peroxisome proliferator-activated receptor-γ (PPARγ) co-activator-1b (PGC1b) promotes tumorigenesis by modulation of mitochondrial function and glycolysis metabolism [99]. On the other hand, hexokinase domain component 1 (HKDC1), recently discovered as a putative hexokinase [100], may be a novel potential therapeutic target for cancer [101]. A recent study demonstrated that knockdown of either PGC1β or HKDC1 resulted in a decrease in glucose uptake in MCF-7 cells (Table 1) [102].

miRNA-34a Inhibitor

miRNA-34a is a tumor suppressor that is expressed in a variety of different types of cancer, including breast cancer. A recent report showed that miRNA-34a inhibition promoted cancer cell proliferation, accelerated glucose uptake and upregulated GLUT1 expression in two triple-negative breast cancer cell lines used, but interestingly, was devoid of effect in the normal human breast epithelial cell line (Table 1) [103].

miRNA-186-3p

Recently, it was verified that systemic delivery of cholesterol-modified agomiR-186-3p to mice bearing tamoxifen-resistant breast tumors effectively attenuates both tumor growth and ^{18}F-FDG uptake (Table 1) [104].

Table 1. Effect of synthetic compounds on glucose uptake by breast cancer cells.

Compound	Concentration (Time)	Cell Line/Model	Effect	Mechanism of Action	Ref
GLUT inhibitors					
WZB117	3–30 µM (24 h)	MCF-7, HBL100	↑ extracellular glucose levels ↔ (HBL100) or ↓ (MCF-7) extracellular lactate levels	-	[57]
WZB117	0.6 µM (16 h)	MCF-7, MDA-MB-231	↓ glucose uptake, lactate production and extracellular levels	↓ GLUT1 mRNA and protein levels	[58]
STF31	0.01–1 µM (24 h)	MCF-7, HBL100	↓ glucose and ↔ lactate in the extracellular medium (HBL100) ↑ glucose and ↓ lactate in the extracellular medium (MCF-7)	-	[57]
WZB27 and WZB115	30 µM and 10 µM, respectively (15 min)	MCF-7	↓ glucose uptake	-	[56]
2-deoxy-D-glucose	2 mM (24 h)	MCF-7 MDA-MB-231	↔ glucose uptake ↓ glucose uptake	- -	[65]
GLUT4 shRNA	transfection	MCF-7	↓ glucose uptake	-	[41]
GLUT1 shRNA	transfection	SK-3R3, MDA-MB-468	↓ glucose uptake and lactate production	-	[66]
GLUT1 shRNA	transfection	78617GL	↓ glucose uptake, consumption and lactate production	-	[27]
GLUT1 shRNA	transfection	MDA-MB-231, Hs578T	↓ glucose uptake	-	[67]
BAY 876	3 µM (5 days)	MDA-MB-436	↓ glucose uptake	-	[63]
Anti-GLUT1 antibody	0.1 mg/mlL (18 h)	MDA-MB-231	↓ glucose uptake	-	[69]
Anti-diabetics					
Metformin	1–10 mM (24–48 h)	MDA-MB-231 Orthotopically implanted MDA-MB-231 cells in mice	↔ glucose uptake	↔ GLUT1, GLUT2, GLUT3, and GLUT4 mRNA levels	[73]
Metformin	0.05–5 mM (26 min) 0.5–1 mM (24 h)	MCF-7, MDA-MB-231	↓ glucose uptake ↑ glucose uptake and lactate production	- ↔ GLUT1 mRNA levels	[65]
Metformin	10 mM (12 h)	MCF-7	↔ GLUT1 and SGLT1 protein levels	↓ PPARδ agonist-induced ↑ glucose uptake, consumption, lactate production and GLUT1 and SGLT1 mRNA and protein levels	[75]
Metformin	2 mM (8 weeks)	MDA-MB-231, MDA-MB-436	↑ glucose uptake	-	[74]
Phenformin	2 mM (12 h)	MDA-MB-231	↑glucose uptake, lactate production	↑ GLUT1 mRNA and protein levels NBR2-dependent	[77]
Troglitazone	20 µM (1 h)	MCF-7, MDA-MB-231, MDA-MB-468, T47D	↑ glucose uptake and lactate production	↔ GLUT1 protein levels PPARγ-independent, MAPK-, AMPK-, and EGFR-dependent	[79]

Table 1. *Cont.*

Compound	Concentration (Time)	Cell Line/Model	Effect	Mechanism of Action	Ref
Chemotherapeutic agents					
Cisplatin	20 μM (48 h)	MDA-MB-231	↓ glucose uptake, lactate production and GLUT1 and GLUT4 mRNA levels	↓ integrin β5/FAK signaling pathway ↓ GLUT1 and GLUT4 mRNA levels	[36]
Sorafenib	7.5 μM (6–24–48 h)	MCF-7 MDA-MB-231 SKBR3	6 h: ↑ glucose uptake, utilization, lactate production (MCF-7 and SKBR3), no effect (MDA-MB-231) 24 h: ↓ glucose uptake and utilization (MCF-7 and SKBR3), ↑ glucose uptake and utilization (MDA-MB-231)	6 h: ↑ GLUT1 protein levels (MCF-7 and SKBR3), no effect (MDA-MB-231) 24 h: ↓ GLUT1 protein levels (MCF-7 and SKBR3), ↑ GLUT1 protein levels (MDA-MB-231) 48: ↓ GLUT1 protein levels (all cell lines) AMPK-dependent inhibition of mTORC1 pathway	[82]
Trastuzumab	initial dose of 4 mg/kg and 2 mg/kg on the 8th and 15th day	xenografts derived from HER2-overexpressing MDA-MB-453 human breast tumour grown in SCID mice	↓ glucose uptake	↓ GLUT1 protein levels	[83]
Doxorrubicin	1 μM (24 h); analysis for 3 d	MCF-7	↓ glucose uptake	↑ GLUT1 mRNA and protein levels	[84]
5-Fluorouracil	200 μM (24 h); analysis for 3 d	MCF-7	↓ glucose uptake	↑ GLUT1 mRNA and protein levels	[84]
Selenium and doxorubicin, free and nanoparticles	Se, nano-Se: 10 μM (24 h) DOX, nano-DOX: 50 μM (24 h)	MCF-7, MDA-MB-231	↑ extracelular glucose levels	-	[85]
Palbociclib + BYL719	0.5 μM palbociclib and 5 μM BYL719 alone or in combination; normoxic or hypoxic conditions (24 h)	MDA-MB-231	Alone: ↓ glucose uptake and consumption When combined: enhancement of effect, in both normoxic and hypoxic conditions	Alone: ↓ GLUT1 protein levels When combined: enhancement of effect, in both normoxic and hypoxic conditions The greater efficacy of the combination ascribed to inhibition of both PI3K/mTOR signaling and c-myc expression	[88]
Palbociclib + paclitaxel	0.5 μM palbociclib and 10 nM paclitaxel alone (48 h) or palbociclib (24 h) + paclitaxel (24 h); normoxic or hypoxic conditions	MDA-MB-231	Alone: ↓ glucose uptake and consumption When combined: enhancement of effect, in both normoxic and hypoxic conditions	Alone: ↓ GLUT1 protein levels When combined: enhancement of effect, in both normoxic and hypoxic conditions The greater efficacy of the combination ascribed to enhancement of inhibition of Rb/E2F/c-myc signaling	[87]
Others					
Tamoxifen	2 μM (72 h)	MCF-7	-	↓ of the ↑ in GLUT1 protein levels induced by E2	[33]
Saracatinib	0.5–1 μM	MCF-10A, MCF12A, with or without HER-2 overexpression	↓ glucose uptake	↓ of ERK1/2-MNK1-eIF4E-mediated cap-dependent translation of c-Myc and transcription of the glucose transporter GLUT1	[93]

Table 1. *Cont.*

Compound	Concentration (Time)	Cell Line/Model	Effect	Mechanism of Action	Ref
RITA, nutlin 3a (P53 activating compounds)	1 μM (8 h)	MCF-7	-	↓ GLUT1 mRNA levels Not related to induction of apoptosis	[94]
Pifithrin-α (PFTα; P53 inhibitor)	Not mentioned	MCF-7	-	↑ GLUT1 mRNA levels, abolishes the effect of RITA on GLUT1 mRNA levels Not related to apoptosis induction	[94]
Wortmannin (PI3K inhibitor)	100 nM (30 min)	T-47D	↔ glucose uptake	-	[98]
Akt1i, Akt2i, Akt1/2i (Akt inhibitors)	10 μM (30 min)	T-47D	↓ glucose uptake	Akt signaling pathway-independent	[98]
PGC1b or HKDC1 shRNA	transfection	MCF-7	↓ glucose uptake	-	[102]
miRNA-34a inhibitor	transfection	ET-20, MDA-MB-231	↓ glucose uptake and GLUT1 protein levels	Inhibition of miRNA34a	[103]
agomiR-186-3p	Systemic delivery of cholesterol-modified agomiR-186-3p	Mice bearing tamoxifen-resistant breast tumors	↓ tumor growth and ^{18}F-FDG uptake	EREG (agonist of EGFR)-mediated	[104]

Legend: ↑, increase; ↓, decrease; ↔ no effect; - not studied.

5.2. Effect of Endogenous Compounds

5.2.1. Hormones

Melatonin

Melatonin is produced and secreted by the pineal gland, and modulates several biological pathways in cancer [105]. Interestingly, GLUT1 appears to be involved in the uptake of melatonin into cancer cells and melatonin appears to bind the glucose binding site of the transporter [106]. Despite the lack of reports elucidating the effects of melatonin targeting the Warburg effect in breast cancer cells, an important study using a xenograft breast cancer model found that glucose uptake and lactate production were inversely correlated with melatonin levels during the 12:12 light:dark cycle [107]. In this context, a recent report evaluated the effect of low pH (6.7) on human breast cancer cell lines (MCF-7 and MDA-MB-231), and the effectiveness of melatonin in the acid tumor microenvironment. Melatonin was able to decrease GLUT1 protein expression levels in both cell lines, both at normal (7.2) and acidic pH (6.7). It was concluded that melatonin treatment increases apoptosis and decreases proliferation and GLUT1 protein expression under acute acidosis conditions in breast cancer cell lines [108] (Table 2).

17β-oestradiol

17β-oestradiol or E2 is a steroid hormone, being the main female sexual hormone. Besides its physiological effects, such as maintenance of reproductive cycle and secondary female characteristics, it plays a major role in the carcinogenesis of breast cancer. A few studies evaluated the effect of E2 on glucose uptake be breast cancer cells (Table 2). In a study using MCF-7 cells, E2 was concluded to have no effect on glucose cellular uptake. In this study, E2 increased culture growth, proliferation rates, cellular viability, and lactate production, but did not affect the uptake of glucose nor GLUT1 mRNA levels. So, it was concluded that the pro-proliferative and cytoprotective effects of E2 are not dependent of stimulation of glucose cellular uptake [109]. In contrast, previous studies found E2 to increase the rate of glucose utilization (although glucose uptake levels were not really measured) [110], to increase glucose uptake and expression/translocation of GLUT4 into the plasma membrane (although E2 showed no effect on the expression/translocation of GLUT1) [30,35], and to increase the expression of GLUT1 [33], or to have no effect on GLUT1 expression levels, although higher rates of glucose uptake were found [111]. So, the effect of E2 on glucose uptake and transporter expression needs to be further clarified. Of note, in the work of Rivenzon-Segal et al. [33], tamoxifen had an opposite effect on GLUT1 and treatment of the cells with both E2 and tamoxifen resulted in a partial (±50%) abolishment of the effect of E2 on GLUT1, demonstrating thus the antiestrogenic activity of tamoxifen with regard to GLUT1 expression.

Progesterone

In the work by Medina et al. [30] the effect of progesterone, which also increases the risk of breast cancer, on GLUT expression levels and glucose uptake by ZR-75-1 cells was analyzed (Table 2). This hormone was found to increase glucose uptake and the expression levels of GLUT1, GLUT3 and GLUT4 [30].

Glucocorticoids

Stress has a vast variety of effects in the human body, one of them being the stimulation of the production of adrenocorticotropic hormone by the anterior pituitary gland. This increases secretion of glucocorticoids, steroid hormones that modulate inflammation and the immune system, cell differentiation, and metabolism [112]. Glucocorticoids are often prescribed in chemotherapy treatments in order to avoid hypersensitivity reactions. Therefore, it is important to investigate if it affects treatment. Additionally, glucocorticoids such as dexamethasone may be involved in resistance processes (chemotherapy desensitization) in various types of solid neoplasms, including breast cancer [113]. Dexamethasone (10.7–10.8 µM; 3 days) showed antiproliferative properties on MCF-7 cells [114]. The antiproliferative effect of dexamethasone in MCF-7 cells was confirmed in a later study; this effect was associated with a slight increase in glucose uptake, a strong increase in GLUT4 expression levels and with the formation of adipocyte-like vesicles. In contrast, dexamethasone did not affect MDA-MB-231 cells proliferation, although it slightly increased glucose uptake and strongly increased GLUT4 expression levels. The authors concluded that dexamethasone treatment induces inhibition of cell growth of dexamethasone-sensitive cancer cells by stimulation of differentiation into adipocyte-like cells [112] (Table 2).

KL1

Klotho is a transmembrane protein that can be shed and act as a circulating hormone in three forms: soluble klotho, KL1, and KL2 [115]. Klotho was proposed to be implicated in aging through inhibition of the Insulin-like Growth Factor 1 (IGF-1) pathway, but it also functions as a tumor suppressor in several types of cancer, including breast cancer [116]. This hormone was recently found to decrease glucose uptake and glycolytic flux in MCF-7 cells, but the mechanism of action was not reported [117] (Table 2).

Insulin

Insulin a peptide hormone secreted by the β cells of the pancreatic islets of Langerhans, with an important role in the maintenance of glucose blood levels. Additionally, insulin exhibits potent anabolic properties and has been implicated in many malignancies, including breast cancer [118]. Insulin is also known to be a modifier of cancer cell metabolism. Indeed, it regulates carbohydrate and lipid metabolism, stimulates DNA synthesis, modulates transcription [118], and stimulates the cellular uptake of various nutrients, including glucose, by facilitated diffusion [119]. Agrawal et al. studied the effect of insulin on the sensitivity of a breast cancer cell line (MCF-7) to 5-fluorouracil (5FU) and cyclophosphamide (CPA) [120]. The chemotherapeutic agents 5FU and CPA are widely used in the clinic and incorporated in the treatment of several cancer, including breast cancer, being associated with increased levels of chemoresistance. Insulin was found to increase the cytotoxic effects of 5FU and CPA in vitro up to two-fold. This effect of insulin was linked to enhancement of apoptosis, activation of apoptotic and autophagic pathways, and to overexpression of GLUT1 and GLUT3 as well as to inhibition of cell proliferation and motility (Table 2). Therefore, it was concluded that insulin sensitization before chemotherapy treatment could overcome chemoresistance [120]. The effect of insulin upon GLUT1 and GLUT3 protein expression levels were hypothesized to be mediated by the PI3K-Akt pathway, but the hypothesis was not tested.

5.2.2. Other Endogenous Compounds

Lactic Acid

Besides being a metabolic fuel, lactate is an important signaling molecule in cancer. This compound induces angiogenesis [121], induces HIF1α, which is associated with cancer cell growth and poor prognosis [122], and stimulates folate uptake by breast cancer cells [38]. Lactic acid interferes also with glucose uptake by breast cancer cells, but either a stimulatory [38] or an inhibitory effect [123] were described (Table 2). In the interesting paper of Turkcan et al, single cells from the core of 4T1 and MDA-MB-231 mice large tumors (>8 mm diameter) grafts were found to take up less glucose than those from the periphery. The authors were able to show that this difference was attributed to an inhibitory effect of lactic acid on glucose uptake [123].

Interleukin-4

Cytokines and chemokines in the tumor microenvironment promote breast cancer progression and metastasis. The interleukin-4 (IL4)/IL4Rα immune signaling axis is a direct promoter of survival and proliferation in breast cancer cells [124]. Venmar et al. [124] investigated whether IL4R-mediated metabolic reprogramming could support tumor growth. They verified that promotion of tumor cell survival and proliferation by IL4 involved an increase in glucose uptake and lactate production by murine 4T1 breast cancer cells, associated with an increase in GLUT1 expression, both in vivo and in vitro (Table 2). Moreover, that concluded that, in addition to IL4, there may also be a role for the second IL4Rα-binding cytokine, IL13, in promoting GLUT1 expression through IL4Rα. Importantly, this effect of IL4 on glucose uptake and transporter expression in murine breast cancer cells was not observed in the human MDA-MB-231 cells [124].

Epidermal Growth Factor

Breast cancer that expresses epidermal growth factor receptors (EGFR) is associated with poor patient prognosis, both in TNBC and in non-TNBC subtypes [125]. In breast cancer patients, EGFR expression is strongly correlated with tumor uptake of the glucose analogue, ^{18}F-FDG [126]. An in vitro study with three breast cancer cell lines showed that Epidermal Growth Factor (EGF) stimulated glucose uptake in EGFR-positive T-47D and MDA-MB-468 cells, but not in the weakly EGFR-positive MCF-7 cells. In T-47D cells, the effect was accompanied by upregulated GLUT1 expression and increased lactate production. EGFR stimulation also increased T47D cell proliferation [127] (Table 2).

Table 2. Effect of endogenous compounds on glucose uptake by breast cancer cells.

Compound	Concentration (Time)	Cell Line/Model	Effect	Mechanism of Action	Ref
Hormones					
Melatonin	1 mM (24 h)	MCF-7, MDA-MB-231	-	↓ GLUT1 protein levels (normal and acidic conditions)	[108]
17β-estradiol	3×10^{-8} M (7 days)	T47D-clcne 11	↑ glucose utilization	-	[110]
17β-estradiol	10 nM (24 h)	ZR-75-1	↑ glucose (0.1 mM) uptake ↔ glucose (15 mM) uptake	↔ GLUT1, GLUT2, GLUT3 and GLUT4 mRNA levels ↔ GLUT1, GLUT2, GLUT3 protein levels, ↑ GLUT4 protein levels	[30]
17β-estradiol	3×10^{-8} M (72 h)	MCF-7, T47D, ZR-75-1	-	↑ GLUT1 protein levels	[33]
17β-estradiol	10 nM (24 h)	T-47D MDA-MB-231 and MDA-MB-468	↑ glucose uptake and lactate production ↔ glucose uptake	↔ GLUT1 protein levels ER- and PI3K–Akt-dependent Non-genomic, membrane-initiated action	[111]
17β-estradiol	10 nM (25–45 min)	MCF-7	↑ glucose uptake	↑ GLUT4 in plasma membrane (but not total protein levels) ↔ GLUT1 (total and plasma membrane protein levels) ERα- and PI3K-dependent Non-genomic, membrane-initiated action	[35]
17β-estradiol	100 nM (48 h)	MCF-7	↔ glucose uptake	↔ GLUT1 mRNA levels	[109]
Progesterone	10 nM (24 h)	ZR-75-1	↑ glucose (0.1 mM) uptake ↔ glucose (15 mM) uptake	↑ GLUT1 and GLUT3 mRNA levels, ↔ GLUT2 and GLUT4 mRNA levels ↑ GLUT1, GLUT3, and GLUT4 protein levels, ↔ GLUT2 protein levels	[30]
Dexamethasone	1 µg/mL (2 weeks)	MCF-7, MDA-MB-231	Small ↑ in glucose uptake	↑ GLUT4 mRNA levels	[112]
KL-1	Not mentioned	MCF-7	↓ glucose uptake and lactate production	-	[117]
Insulin	40 µg/mL (8 h)	MCF-7	-	↑ GLUT1 and GLUT3 protein levels	[120]
Others					
Lactic acid	10–25 mM (48 h)	Single cells isolated from 4T1 and MDA-MB-231 tumors grafts in mice 4T1 and MDA-MB-231	Single cells from the core of tumors grafts took up less glucose than those from the periphery Lactic acid levels were higher in the core of the tumor ↓ glucose uptake	-	[123]
Lactic acid	10 mM (26 min)	T-47D	↑ glucose uptake	-	[38]
Ep. dermal growth factor	100 ng/mL (24 h)	T-47D, MDA-MB-468, MCF-7	↑ glucose uptake (T-47D, MDA-MB-468), ↔ glucose uptake (MCF-7) ↑ lactate production (T47D)	↑ GLUT1 protein levels (T47D) PI3 kinase (PI3K) activation	[127]

Legend: ↑, increase; ↓, decrease; ↔ no effect; - not studied.

5.3. Effect of Exogenous Natural Compounds

5.3.1. Polyphenols

A large class of GLUT inhibitors is represented by polyphenols, a heterogeneous and large family of natural compounds widely distributed in plants and in the human diet (e.g., in fruits, vegetables and beverages such as tea and wine) [128]. Many of these compounds show an appreciable activity on several distinct membrane transporters, including GLUTs [129,130]. Polyphenols possess anticancer effects in relation to several cancer types, including breast cancer. Several distinct mechanisms are involved in their anticarcinogenic effect in breast cancer: interference with redox balance, pro-apoptotic effect, cell cycle arrest, activation of autophagy, inhibition of angiogenesis, anti-inflammatory effect, anti-estrogenic effect, changes in ER expression, aromatase modulation, interference with HER2 signaling, and effect on microbiota [131,132]. Additionally, some polyphenols interfere with glucose cellular uptake by breast cancer cells [133,134], as next described.

Gossypol

This polyphenolic bisnaphthalene aldehyde obtained from the cotton plant markedly increased both glucose consumption and lactate production in MCF-7 cells, but the increase in glucose consumption may not related to an increase in glucose uptake, and rather be the consequence of increased glycolytic rates or increased rates of glycose oxidation not related to glycolysis (e.g., pentose phosphate pathway) [135] (Table 3).

Naringenin

This grapefruit flavanone inhibited both basal and insulin-stimulated glucose uptake in two breast cancer cell lines (MCF-7 and T-47D). The reduction in insulin-stimulated glucose uptake was not associated with changes in GLUT4 protein levels but rather with inhibition of insulin-stimulated PIP3/Akt and p44/p42 MAPK activity [39]. The antiproliferative effect of naringenin was mimicked by low glucose conditions and so it was concluded that it was dependent on impairment of glucose uptake [39] (Table 3).

Genistein

The flavonoid genistein, found in soybean, reduced glucose uptake in both estrogen receptor-positive MCF-7 and -negative (MDA-MB-231) breast cancer cell lines [136]. The inhibitory effect of genistein upon glucose uptake by MCF-7 cells was later confirmed in two studies. In the first, the effect of genistein, daidzein, and a soy seed extract on two distinct breast cancer cell lines were investigated. In MCF-7 cells, these compounds presented an inhibitory effect on cell proliferation that correlated with a decrease in glucose cellular uptake [137]. In the second, exposure to several polyphenols, including genistein (myricetin, genistein, resveratrol, and kaempferol), was shown to reduce glucose uptake by MCF-7 cells, and genistein inhibited glucose uptake with a 50% Inhibitory Concentration (IC_{50}) of 39 μM [138] (Table 3).

Kaempferol

In the work of Azevedo et al. [138], kaempferol was found to be the most potent inhibitor of glucose uptake, with an IC_{50} of 4 μM. Kaempferol (30 μM) decreased glucose uptake and the GLUT1 transcription level. Moreover, low extracellular glucose mimicked, and high extracellular glucose conditions prevented, the antiproliferative and cytotoxic properties of kaempferol. So, it was suggested that inhibition of GLUT1-mediated glucose cellular uptake mediates the anticancer effect of kaempferol in MCF-7 cells [138] (Table 3).

Resveratrol

An inhibitory effect of resveratrol (IC_{50} = 67 µM), found in fruits such as grapes and berries, upon glucose uptake by breast cancer cells was also described in the work by Azevedo et al. [138]. Moreover, an inhibitory effect of this stilbene was previously described in another breast cancer cell line. Resveratrol (150 µM), suppressed uptake of glucose and glycolysis in T-47D breast cancer cells, associated with a reduction in GLUT1 expression and dependent on a reduction in intracellular ROS levels, which decreases HIF-1α accumulation [139] (Table 3).

Hesperitin

This flavanone, found in citrus fruits, reduced both basal and insulin-stimulated glucose uptake in MDA-MB-231 cells. Of note, the negative effect of hesperitin on basal glucose uptake was associated with GLUT1 downregulation, whereas the negative effect on insulin-induced glucose uptake was associated with impaired GLUT4 translocation to the cell membrane [37] (Table 3).

Quercetin and epigallocatechin-3-gallate (EGCG)

The flavonoids quercetin and EGCG (26 min) concentration-dependently inhibited glucose uptake by MCF-7 (IC_{50} = 11–23 µM) and MDA-MB-231 (IC_{50} = 44–16 µM) cells, respectively, associated with a decrease in lactate production. The effects of quercetin and EGCG were independent of estrogen signaling and did not involve Protein Kinase A (PKA), C (PKC), G (PKG) and calcium-calmodulin. A 4 h exposure to quercetin or EGCG induced also a decrease in glucose uptake, which was associated with an increase in GLUT1 transcription rates. Moreover, an antiproliferative and cytotoxic effect of both compounds was described in MCF-7 cells, which was more potent when extracellular glucose was present. So, inhibition of basal glucose uptake and consequently lactate production were concluded to be determinants of the cytotoxic and antiproliferative effects of quercetin and EGCG in breast cancer cells [40]. The inhibitory effect of quercetin on glucose uptake by breast cancer cells was confirmed in later studies, as shown next. Xintaropoulou et al. [57] verified that inhibition of growth of the HBL100 breast cancer cell line by quercetin (50–150 µM) is associated with an increase in the amount of extracellular glucose and a reduction in lactate production, suggesting inhibition of glucose uptake [57]. Quercetin was also found to decrease the mobility of MCF-7 and MDA-MB-231 cells, associated with a decrease glucose uptake, lactate production and GLUT1 protein levels [140], and to decrease glucose uptake and GLUT1 protein levels in MDA-MB-231 cells [141] (Table 3).

In relation to EGCG, a recent study using rodent 4T1 breast carcinoma cancer cells showed that EGCG inhibits breast cancer growth, both in vitro and in vivo, associated with a reduction in glucose and lactic acid levels and GLUT1 mRNA levels in these cells [142] (Table 3).

Phloretin and Phloridzin

The dihydrochalcone phloretin is found primarily in apples and pears and can also be produced when its glycoside phlorizin is consumed and subsequently nearly entirely converted into phloretin by hydrolytic enzymes in the small intestine. Several studies present evidence for an inhibitory effect of phloretin (and also its glycone phloridzin) in relation to glucose uptake by breast cancer cells. Phloretin and phloridzin were found to decrease glucose uptake by a rat breast adenocarcinoma cell line, both in vivo and in vitro [143]. Phloretin was also suggested to decrease glucose uptake (as assessed by the increase in the amount of extracellular glucose and the decrease in the amount of lactate produced) associated with an antiproliferative effect in HBL100, but not MCF-7 cell line [57]. Finally, inhibition of GLUT2 by phloretin was concluded to potentially suppress MDA-MB-231 cell growth and metastasis, although phloretin was found to increase GLUT2 protein levels. The authors concluded that phloretin treatment inhibited uptake of glucose, and, as a consequence, increased GLUT2 protein expression was required for cancer cell survival [63]. In relation to phloridzin, it reduced glucose uptake in several breast cancer cell lines, either alone [38,40] or associated with cytochalasin B [138] (Table 3).

Glabridin

This flavonoid decreased glucose uptake and lactate production, possibly mediated by a decrease in GLUT1 protein levels, in MDA-MB-231 cells [141] (Table 3).

(+)-Catechin

The anticancer efficacy of polyphenols can be enhanced by combining them with compounds such as amino acids and vitamins [144]. In this context, a catechin:Lys complex (Cat:Lys 1:2) was recently tested in MCF-7 and MDA-MB-231 breast cancer cell lines and in the non-tumorigenic breast (MCF12A) cell line. Cat:Lys (24 h) decreased glucose uptake and lactate production in MCF-7 cells but increased glucose uptake and lactate production in MDA-MB-231 and MCF12A cells. Cat:lys (24 h) was also found to increase GLUT1 mRNA expression levels in MDA-MB-231 cells. In contrast, a shorter-term exposure (26 min) of these cell lines to Cat:Lys caused an increase in glucose uptake in MDA-MB-231 and MCF12A cells but no effect on MCF-7 cells [145]. In contrast, (+)-catechin was found to increase glucose uptake by MCF-7 cells [138]. Moreover, in the work of Silva et al. [145], by using a GLUT inhibitor, it was concluded that: (a) there is a contribution of a GLUT-mediated mechanism in glucose uptake in the three breast cell lines, (b) Cat:Lys stimulates GLUT-mediated glucose uptake in MDA-MB-231 and MCF12A cell lines, and (c) Cat:Lys inhibits non-GLUT-mediated glucose uptake in MCF-7 cells, a conclusion that was supported by the results of GLUT1 mRNA expression. So, Cat:Lys shows no consistent effects on glucose uptake by the breast cell lines (Table 3). Thus, apparently, its antitumoral effect is not related to an effect on glucose uptake, because Cat:Lys showed a similar antiproliferative, cytotoxic, antimigratory, and proapoptotic effect on both cancer cell lines and a much less evident effect in the non-tumorigenic cell line [145].

Polyphenolic Esters

Zhang et al. [146] reported inhibition of basal glucose transport in MCF-7 cells and other cell lines (including H1299 lung cancer cell line) by synthesized polyphenolic esters. Although not tested in breast cancer cells, these basal glucose transport inhibitors also inhibited H1299 cell growth, and these two activities appear to be correlated (Table 3).

Curcumin

Very recently, the effect of curcumin (diferuloylmethane), a well-known phytopolyphenolic compound isolated from rhizome of the plant *Curcuma longa* was evaluated and this compound was found to reduce glucose uptake and lactate production in a variety of cancer cell lines, including in MCF-7 cells [147]. Curcumin was concluded to inhibit aerobic glycolysis by downregulating pyruvate kinase M2 expression, which drives the Warburg effect and thus is essential for survival of cancer cells. Nevertheless, a direct effect of curcumin on glucose transporters was not investigated [147] (Table 3).

Cardamonin

This chalcone, isolated from *Alpiniae katsumadai*, reduced glucose uptake as well as lactate production and efflux in the breast cancer MDA-MB-231 cell line [148] (Table 3).

Plant Extracts

Some studies have investigated the effect of plant extracts rather than the effect of individual compounds (Table 3). In one study, *Baeckea frutescens* leaves extracts were found to decrease glucose uptake in two breast cancer cell lines (MCF-7 and MDA-MB-231), associated with a cytotoxic and proapoptotic effect. Importantly, the extracts were devoid of cytotoxic effect in the non-tumoral breast cell line MCF10A and were also devoid of effect on glucose uptake [149]. In another study, an extract of *Petiveria alliacea* leaves and stems reduced glucose uptake and lactate production in the 4T1 breast cell line. However, glucose levels in supernatant, rather than direct measurement of glucose uptake, were measured [150]. Finally, an extract of Kudingcha leaves, one of the *Ligustrum robustum* species, was described to concentration-dependently reduce GLUT1 and GLUT3 protein expression levels and lactate production in two triple-negative breast cancer cell lines [151].

5.3.2. Other Exogenous Natural Compounds

Cytochalasin B

The macrocyclic mycotoxin cytochalasin B is a known GLUT inhibitor used extensively in the literature of GLUT investigation. Cytochalasin is a GLUT1, GLUT2, and GLUT4 inhibitor [152,153]. This compound has been described to interfere also with glucose uptake in several breast cancer cell lines: T-47D [38,98] and MCF-7 and MDA-MB-231 [40] (Table 3).

Genipin

Genipin, an aglycone derived from an iridoid glycoside called geniposide extracted from gardenia fruits, caused a decrease in glucose uptake by the breast cancer cells T-47D and MDA-MB-435, although no effect on MCF-7 and MDA-MB-231 cells was found (Table 3). The effect of genipin was most pronounced in the T-47D cell line; in this cell line, an $IC_{50} = 61$ µM was calculated and a decrease in lactate production was also found. In this report, genipin was concluded to decrease cancer cell glucose uptake by reducing both glycolytic flux and mitochondrial oxidative phosphorylation, an effect that was related to inhibition of Uncoupling Protein 2 (UCP2)-mediated dissipation of energy and restriction of ROS production through proton leakage [154]. However, an effect of genipin on glucose transporters cannot be excluded.

Cantharidin

This sesquiterpenoid bioactive compound is secreted by beetles of the family of Meloidae [155]. Although it has been available for almost a century, its use has not been approved due to its high toxicity to the gastrointestinal tract. However, new anticarcinogenic properties have been found. In a recent study, cantharidin was found to inhibit aerobic glycolysis, associated with a decrease in GLUT1 protein expression levels. It was concluded that cantharidin inhibits nuclear translocation of pyruvate kinase isoform M2 (PKM2), which promotes the transcription of GLUT1. So, cantharidin interferes with the glycolytic metabolic loop between GLUT1 and PKM2 [155] (Table 3).

Betulinic Acid (BA)

BA is a natural pentacyclic terpene reported to be capable of inhibiting various malignancies. BA was recently reported to decrease the viability of breast cancer cell lines MCF-7 and MDA-MB-231, being ineffective against the non-malignant mammary epithelial cell line MCF-10A, indicating that BA might be a highly selective killing agent toward malignant cells. BA was shown to decrease glucose uptake and lactate production in the two cancer cell lines, but it was concluded that suppression of the glycolytic activity mainly occurred at the intracellular level, and no further investigation of its effect upon glucose uptake was done [156] (Table 3).

Benzyl Isothiocyanate (BITC)

Data from numerous preclinical studies advocate BITC, an aromatic isothiocyanate, which occurs naturally in edible cruciferous vegetables, promising for breast cancer chemoprevention [157]. This compound was described to increase glucose uptake by breast cancer cells, both in vivo and in vitro. This effect is probably a compensatory mechanism in response to inhibition of complex III of the mitochondrial respiratory chain and of oxidative phosphorylation caused by this compound [158]. Moreover the effect of BITC upon glucose uptake was found to be dependent on Akt activation (Table 3). So, these results indicate that BITC increased glucose uptake/metabolism in breast cancer cells and suggest that breast cancer chemoprevention by BITC may be augmented by pharmacological inhibition of Akt [159].

Docosahexaenoic Acid (DHA)

n-3 polyunsaturated fatty acids (PUFAs) have been proposed to have anticancer properties, and the effects on cancer cell metabolism constitutes one possible mechanism contributing to their anticancer effect [160]. The effect of DHA on breast cancer and non-cancer cell lines was evaluated (Table 3). It was concluded that DHA contributes to impaired cancer cell growth and survival by altering cancer cell metabolism, including by causing a decrease in glucose uptake, while not affecting non-transformed cells [161].

Vitamin D3 (VD3)

VD3, the bioactive form of Vitamin D, is known to be an important modulator of bone metabolism and diabetes, amongst many other effects. Low levels of VD3 are linked with an increased risk of cancer, whilst high levels of vitamin D3 usually promise better prognosis [162]. Although its positive effects have been shown, not much is known as to the mechanisms involved. Santos et al [163] tested the effect of VD3 on MCF-7 and MDA-MB-231 cells. VD3 significantly reduced GLUT1 mRNA and protein levels and glucose uptake in both cell types. Moreover, lactate production in the highly metastatic MDA-MB-231 cells was significantly reduced (Table 3). This study proved that VD3 decreased breast cancer cell viability along with reduced expression of GLUT1 and key glycolytic enzymes (hexokinase II and lactate dehydrogenase A), causing a decrease in glucose uptake.

γ-Tocotrienol

This member of the vitamin E family of compounds displays potent anticancer effects at doses having little or no effect on normal cell viability [164]. To test the role of glycolysis in the effect caused by γ-tocotrienol, Parajuli et al. [165] used the MCF-7 breast cancer cell line. γ-tocotrienol was found to decrease MCF-7 cell growth, coincident with a concentration-dependent decrease in glucose consumption and lactate production. This effect was correlated with a decrease in the expression of glycolytic enzymes (HK-II, PFK, PKM2, and LDHA) and MCT1, but the compound did not affect GLUT1 expression levels. So, GLUT1 reduction was not considered the mediator of the glycolysis inhibition caused by γ-tocotrienol [165] (Table 3).

d-Allose

This compound, a C-3 epimer of d-glucose with 80% of the sweetness of sucrose, exists in small quantities in nature. This compound is known to possess anticarcinogenic properties via upregulation of thioredoxin interacting protein (TXNIP) [166]. A study involving breast adenocarcinoma, hepatocellular carcinoma, and neuroblastoma cell lines concluded that d-allose downregulates GLUT1 expression and consequently glucose uptake, thus suppressing cancer cell growth, as a result of overexpression of TXNIP [166] (Table 3).

Table 3. Effect of exogenous compounds on glucose uptake by breast cancer cells.

Compound	Concentration (Time)	Cell Line/Model	Effect	Mechanism of Action	Ref
Polyphenols					
Gossypol	10 µM (25 h)	MCF-7	↑ glucose consumption and lactate production ↔ ratio lactate produced/glucose consumed	Quasi-competitive inhibition	[135]
Naringenin	10 µM (15 min)	MCF-7, T-47D	↓ basal and insulin-stimulated glucose uptake	Inhibition of PI3K, Akt, and MAPK pathways	[39]
Genistein	10–100 µM (10 min)	MCF-7, MDA-MB-231	↓ glucose uptake	-	[136]
Genistein, daidzein and a soy seed extract	22 µM, 52 µM and 166 µg/mL, respectively (72 h)	MCF-7, MDA-MB-231	↓ glucose uptake	-	[137]
Myricetin, resveratrol genistein, kaempferol	10–100 µM (26 min)	MCF-7	↓ glucose uptake	Mixed-type inhibition for kaempferol	[138]
Catechin	100 µM (26 min)	MCF-7	↑ glucose uptake	-	[138]
Resveratrol	150 µM (24 h)	T-47D	↓ glucose uptake	↓ GLUT 1 protein levels ↓ intracellular ROS causing ↓ of HIF-1α accumulation	[139]
Hesperetin	50–100 µM (24 h)	MDA-MB-231	↓ basal glucose uptake ↓ insulin-stimulated glucose uptake	↓ GLUT 1 mRNA and protein levels ↓ GLUT4 cell membrane translocation	[37]
Quercetin and EGCG	1–500 µM (26 min) 1–100 µM (4 h)	MCF-7, MDA-MB-231	↓ glucose uptake	Competitive, independent of PKA, PKC, PKG, and calcium-calmodulin intracellular pathways	[40]
Quercetin, Phloretin	50–150 µM (24 h)	HBL100	↓ glucose uptake	-	[57]
Quercetin	30 µM (6.5 h)	MCF-7, MDA-MB-231	↓ glucose uptake, ↓ lactate production	↓ GLUT1 protein levels	[140]
Quercetin	Not mentioned	MDA-MB-231	↓ glucose uptake, ↔ lactate production	↓ GLUT1 protein levels	[141]
Glabridin	Not mentioned	MDA-MB-231	↓ glucose uptake, ↓ lactate production	↓ GLUT1 protein levels	[141]
Phloretin and phloridzin	Not mentioned	Rat mammary adenocarcinoma	↓ glucose uptake in vitro and in vivo	-	[143]
Phloretin	10–150 µM (16 h)	MDA-MB-231	-	↑ GLUT2 protein levels	[167]
Phloridzin	1 mM (26 min)	T-47D	↓ glucose uptake	-	[38]
Phloridzin	0.5–1 mM (26 min)	MCF-7, MDA-MB-231	↓ glucose uptake	-	[40]

Table 3. *Cont.*

Compound	Concentration (Time)	Cell Line/Model	Effect	Mechanism of Action	Ref
Phloridzin + Cytochalasin B	0.5 mM and 50 μM, respectively (26 min)	MCF-7	↓ glucose uptake	-	[138]
EGCG	20–240 μM (24 h) 20–80 μM (24 h) 20 mg/kg (mice)	4T1	↓ glucose and lactic acid levels ↓ glucose and lactic acid levels in tumour	↓ GLUT1 mRNA levels	[142]
Cat:Lys (1:2) complex	0.01–5 mM (26 min)	MCF-7, MDA-MB-231 and MCF12A	↑ glucose uptake (MDA-MB-231 and MCF12A) ↔ glucose uptake (MCF-7)	-	[145]
Cat:Lys (1:2) complex	0.01–1 mM (24 h)	MCF-7, MDA-MB-231 and MCF12A	↑ glucose uptake and lactate production (MDA-MB-231 and MCF12A) ↓ glucose uptake and lactate production (MCF-7)	↑ GLUT1 mRNA levels (MDA-MB-231) ↔ GLUT1 mRNA levels (MCF-7 and MCF12A)	[145]
Compound 1a and 2a (polyphenolic esters)	30 μM (15 min)	MCF-7	↓ glucose uptake	-	[146]
Curcumin	20 μM (24 h)	MCF-7	↓ glucose uptake and lactate production	-	[147]
Cardamonin	20–80 μM (6 h)	MDA-MB-231	↓ glucose uptake and lactate production	-	[148]
Baeckea frutescens extracts	IC50 = 10–127 μg/mL (MCF-7 cells; 72 h)	MCF-7, MDA-MB-231	↓ glucose uptake	-	[149]
Petiveria alliacea extract	3 μg/mL (48 h)	4T1	↓ glucose uptake	-	[150]
Kudingcha extract	100–200 μg/mL (48 h)	MDA-MB-231, HCC1806	- Decrease in lactate levels	↓ GLUT1 and GLUT3 protein levels	[151]
Others					
Cytochalasin B	25 μM (30 min)	T-47D	↓ glucose uptake	-	[98]
Cytochalasin B	50–100 μM (26 min)	T-47D	↓ glucose uptake	-	[38]
Cytochalasin B	10–50 μM (26 min)	MCF-7, MDA-MB-231	↓ glucose uptake	-	[40]
Cantharidin	0.5–2 μM (24 h)	MCF-7, MDA-MB-231	↓ glucose consumption and lactate production	↓ GLUT1 protein levels	[155]
Betulinic acid	5–40 μM (3 h)	MCF-7, MDA-MB-231	↓ glucose uptake and lactate production	-	[156]
Genipin	50–100 μM (24 h)	T-47D, MDA-MB-435 MCF-7, MDA-MB-231	↓ glucose uptake ↔ glucose uptake	-	[154]
Benzyl isothiocyanate	3 mmol/kg diet 2.5 μM (24 h)	MMTV-neu mice mammary tumours MDA-MB-231, SUM159	↑ glucose uptake ↑ glucose uptake	↑ GLUT1 total and membrane-located protein levels Dependent on Akt activation	[159]

Table 3. *Cont.*

Compound	Concentration (Time)	Cell Line/Model	Effect	Mechanism of Action	Ref
Docosahexaenoic acid	100 µM (48 h)	MDA-MB-231, ET-474, MCF-10A	↓ glucose uptake and metabolism (MDA-MB-231, BT-474) ↔ glucose uptake and metabolism (MCF10A)	-	[161]
Vitamin D$_3$	0.5–1 µM (24 h)	MCF-7, MDA-MB-231	↓ glucose uptake (both cell lines) ↓ lactate production (MDA-MB-231 cells)	↓ GLUT1 mRNA and protein levels (both cell lines) AMPK-dependent inhibition of mTOR (?)	[163]
γ-Tocotrienol	6–8 µM (96 h)	MCF-7	↓ glucose consumption and lactate production	↔ GLUT1 mRNA levels	[165]
d-allose	50 mM (7 days)	MDA-MB-231	-	↑ GLUT1 mRNA and protein levels Mediated by overexpression of TXNIP	[166]

Legend: ↑ increase; ↓, decrease; ↔ no effect; - not studied.

6. Effect of Stimulation of the Interaction of Anticancer Agents with GLUT

Conjugation of anticancer agents with glucose or other sugars is a widely exploited technique to design therapeutic agents, in order to improve their uptake into highly glycolytic cancer cells overexpressing GLUTs, thus increasing efficacy while reducing side effects. One possibility is to develop sugar-conjugated agents that can be transported into cancer cells through GLUT without inhibiting GLUTs themselves [168]. Another possibility is to promote interaction of anticancer agents with GLUT by their conjugation with an anti-GLUT antibody. Some of these agents have been tested in breast cancer cell lines, as shown next.

6.1. Adriamycin

Adriamycin (doxorubicin) is effective against many types of solid tumors in clinical applications. However, its use is limited because of systemic toxicity and multidrug resistance. Adriamycin conjugated with a glucose analogue (2-amino-2-deoxy-d-glucose) and succinic acid (2DG-SUC-ADM) was designed to target tumor cells through GLUT1, thus enhancing the selectivity of doxorubicin against cancer cells while reducing its toxicity to healthy cells [169]. In a work using several cancer cell lines, including MCF-7 and MDA-MB-231 cell lines, the complex showed better inhibition to tumor cells and lower toxicity to normal cells, and, most importantly, displayed a potential to reverse multidrug resistance. In vivo experiments also showed that this new complex could significantly decrease organ toxicity and enhance the antitumor efficacy compared with free ADM, indicating 2DG-SUC-ADM as a promising drug for targeted cancer therapy [169]. The GLUT1-mediated transport into the cells explained the specificity of 2DG-SUC-ADM, because uptake of free doxorubicin mainly occurred through diffusion, whereas the uptake of 2DG-SUC-ADM was mostly GLUT1-mediated [169].

Sztandera et al. [170] developed a glucose-modified PAMAM dendrimer for the delivery of doxorubicin (dox) to breast cancer cells, designed to specifically enter tumor cell with enhanced glucose uptake. They verified that PAMAM-dox-glucose conjugate exhibited pH-dependent drug release and an increased cytotoxic activity compared to free drug in MCF-7 cells, in the absence of glucose. They also verified that GLUT1 inhibition eliminated the toxic effect of the conjugate. So, they concluded that the cytotoxic effect of PAMAM-dox-glucose depends on presence of a functional GLUT1, suggesting specific, transporter-dependent internalization as a main route of cellular uptake of glucose-conjugated PAMAM dendrimers [170].

6.2. Paclitaxel

This drug is widely used for the treatment of breast, ovarian, and lung carcinomas, but its low water solubility severely reduces its clinical application. In this context, a new prodrug was designed to enhance its solubility and its selective delivery to cancer by a preferential uptake via GLUTs. More specifically, the glycoconjugation of paclitaxel led to a derivative in which the drug was linked to 1-methyl glucose via a short succinic acid linker. The resulting compound, whose transport was mediated at least in part by GLUT1, showed a comparable cytotoxicity against several cancer cells without toxicity on normal cells. Of note, paclitaxel linked to succinic acid resulted in a lower toxicity against MCF-7 cells than the parent compound, suggesting that the presence of glucose improved its cytotoxicity [171].

6.3. Oxiplatin

The platinum antitumor drug oxaliplatin is a commonly used chemotherapeutic agent; however, its multiple side effects severely limit its benefits. The conjugation with sugar portions was introduced as a strategy to improve the tumor-targeting ability of the drug and also to enhance its water solubility, allowing renal excretion and lower systemic toxicity. A glycosylated (trans-R,R-cyclohexane-1,2-diamine)-malonatoplatinum(II) derivative showed increased cytotoxicity compared to oxaliplatin in all the tested human carcinoma cell lines. Its potency was prevented when human colon cancer (HT29) and breast cancer (MCF-7) cells, which overexpress GLUTs, were treated with the GLUT inhibitor phlorizin, thus confirming that the uptake and the antiproliferative activity of this compound are GLUT-mediated [172].

In summary, a great potential of GLUT-mediated transport of therapeutics into cancer cells opens new roads for targeted delivery of anticancer drugs.

7. Conclusions and Future Perspectives

Despite the high-survival rate in breast cancer patients and the availability of well-designed and effective therapeutic strategies, especially for hormone receptor or HER2-positive breast cancer, more drug research is still needed, particularly regarding triple-negative breast cancer, because of its unresponsiveness to hormone or anti-HER2 therapy. In this context, metabolic targeting of tumors, and more specifically targeting GLUT1-mediated glucose transport, constitutes an interesting approach. In addition to GLUT1, there are a several other potential cancer therapies that target the cellular energetic metabolism pathway in tumors. Indeed, many compounds targeting energy metabolism are currently in trial or approved as therapeutic agents for cancer [48,49]. Preclinical data from these inhibitors are encouraging; therefore, they represent additional options for targeting the enhanced aerobic glycolysis in cancer.

In this review, we show that a wide range of compounds, ranging from endogenous to dietary compounds and synthetic compounds, are able to interfere with glucose uptake by breast cancer cells. Moreover, for some of the presented compounds, their antitumoral effect is concluded to result from the effect on glucose uptake. Since cancer cells are highly dependent on glucose, even if these compounds possess other anticancer-inducing mechanisms, their effect on glucose uptake certainly contributes to their antitumoral effect.

The mechanisms underlying the modulatory effect of the compounds upon glucose uptake are very diverse, ranging from a direct effect upon the transporter, inhibition of transporter gene expression or protein synthesis, impairment of membrane insertion of the transporter, and redox balance modulation. Additionally, the effect of compounds on glucose uptake may be secondary to a decrease in the activity of glycolytic enzymes or signaling pathways. However, it should be pointed out that, for most of the presented compounds, the mechanisms underlying their modulatory effect upon glucose uptake by breast cancer cells have not been investigated. So, more research is needed in this area. A better knowledge of the mechanisms able to interfere with glucose transporter function in cancer cells may open new windows for therapeutic targets in breast cancer.

Most studies on GLUTs in cancer are focused on their role and regulation in the tumor cells. However, solid tumors are composed of several cell types, forming a dynamic and complex network. In this context, a role for GLUT1 in glycolytic reprograming enabling survival, growth, and expansion of effector T lymphocytes has been demonstrated [173]. T lymphocytes in tumors constitute a primary cellular target for immunotherapies including adoptive T cell therapy and immune checkpoint blockade. Moreover, GLUT1 induction was observed in human fibroblasts placed in contact with prostate cancer cells [174]. Cancer-associated fibroblasts (CAFs) are known to promote tumor growth, invasion, chemoresistance, and angiogenesis. So, the requirements for glucose entry and usage by cancer-supporting or cancer-antagonizing cells add on to the complexity of metabolic rewiring in cancer.

Metabolic targeting of tumors using GLUT inhibitors has attracted more and more attention in the past years, which can be demonstrated by the growing number and more recent publications on this subject. Most therapeutic strategies that are being developed to target GLUTs in cancer are in the preclinical phase of drug development. These preclinical data suggest that inactivation of GLUT1, leading to glucose starvation that ultimately leads to cell death, is a viable drug target for cancer therapy [52,175].

Regrettably, therapies designed to target this pathway have not been fully translated to the clinic yet, and clinical trials in cancer patients using GLUT inhibitors to ensure their safety and/or efficacy are still largely lacking. One of the major obstacles to the success of GLUT1-based therapies is the potential systemic toxicity, because GLUT1 is ubiquitously expressed in mammalian tissues. Although it is expected that targeting GLUT1-mediated glucose uptake will have a much more marked effect on cancer cells than in non-cancer cells, because cancer cells are much more sensitive to glucose deprivation than normal cells, a certain degree of side effects may be expected, especially those occurring in organs characterized by high glucose-consumption rates such as the brain, immune system and stem cells. One example is the observation that several glucose transport inhibitors, tested in phase I clinical trials for hepatocellular and prostate cancer, were associated with significant side effects [176]. So, selective blockade of GLUTs in tumor cells still remains a key challenge and research on this subject should be fostered in the near future.

In this context, important points should be considered in future research:

(1). Development of cancer-specific and potent GLUT inhibitors that minimize side effects;

(2). Development of selective targeted delivery of GLUT inhibitors (using recent imaging technology) directly into the tumor or intra-arterially near the tumor, or micro-encapsulation of the inhibitor [52];

(3). More studies on GLUT regulation and function in vivo should be conducted;

(4). Studies on multi-targeted inhibition, allowing lower doses of GLUT inhibitors to be used. For instance, the preliminary results from clinical trials of 2-DG as a monotherapy are inconclusive and ambiguous, with toxic and side effects being reported [49,177]. Currently, 2-DG was reintroduced for use in combination approach, as reported in more recent preclinical and clinical studies, using 2-DG at lower doses to produce synergistic anticancer effects with other chemotherapeutic agents or irradiation. Thus, combination treatments using 2-DG may have encouraging outcomes providing a new opportunity for cancer combination therapy [52,175];

(5). An accurate verification and analysis of the transporter expression profiles, because cancers are extremely heterogeneous diseases and they have unique metabolic features. Moreover, abnormal glycolysis due to defects of mitochondrial oxidative phosphorylation is not absolutely common in spontaneous tumors [8,178];

(6). Many natural compounds interfere with glucose uptake at higher than dietary/physiological concentrations (e.g., polyphenols). For some dietary compounds, these concentrations are not attainable even after the consumption of food supplements due to reduced bioavailability and/or metabolism of these compounds. For these compounds (a) improvement of their bioavailability and delivery, as it would result in the improvement of their biological effects in vivo, (b) the use of natural or synthetic analogs that have better bioavailability or more potency, or (c) combination with other glucose transport inhibitors or with conventional therapy, resulting in a synergistic effect or in improvements of its bioavailability [144,179] are possible strategies.

(7). Resistance resulting from the treatment with GLUT inhibitors, involving development of different routes for energy supply using other energy substrates, is another issue that should not be underestimated. Indeed, cancer cells display metabolic plasticity and can overcome the inhibition of a specific metabolic pathway via the expression or up-regulation of alternative pathways. Moreover, adjacent cells such as fibroblasts can offer metabolic intermediates for the needs of cancer cells. To overcome this problem, combination of other targeted therapy drugs with GLUT inhibitors may be an effective way [180]. Alternatively, combination of two or more antimetabolic agents inhibiting different metabolic pathways simultaneously would decrease resistance and prevent relapse [52].

(8). Research focused on the design of GLUT-transportable chemotherapeutics, which may provide therapeutic selectivity [180];

(9). Characterization of glucose transporters in tumors treated with immunotherapies, to determine if their expression in tumor cells or cells of the tumor environment changes upon treatment, and if their activity is linked to the outcome of therapy [181];

(10). Research focused on overcoming side effects that are expected, especially those occurring in organs such as the brain. In this context, it is known that, in starvation, ketone bodies can replace glucose as fuel for the brain. Therefore, a combined administration of GLUT-interfering agents with either a ketogenic diet or dietary supplements such as triheptanoin (which is currently being tested for the treatment of GLUT1 deficiency [182]), should improve the safety profile of these compounds [53].

In conclusion, in this review we show that several chemically distinct compounds interfere with glucose uptake by breast cancer cells, and these GLUT inhibitors should be used as starting point in future research, which should focus in developing new compounds/combinations/delivery methods to solve specific problems already identified.

Funding: This work received no external funding.

Conflicts of Interest: The authors declare no conflict of interest.

References

1. Bray, F.F.J.; Soerjomataram, I.; Siegel, R.L.; Torre, L.A.; Jemal, A. GLOBOCAN estimates of incidence and mortality worldwide for 36 cancers in 185 countries. *CA Cancer J. Clin.* **2018**, *68*, 394–424. [CrossRef] [PubMed]

2. Rossi, L.; Stevens, D.; Pierga, J.Y.; Lerebours, F.; Reyal, F.; Robain, M.; Asselain, B.; Rouzier, R. Impact of adjuvant chemotherapy on breast cancer survival: A real-world population. *PLoS ONE* **2015**, *10*, e0132853. [CrossRef] [PubMed]

3. Coleman, M.P.; Quaresma, M.; Berrino, F.; Lutz, J.M.; de Angelis, R.; Capocaccia, R.; Baili, P.; Rachet, B.; Gatta, G.; Hakulinen, T.; et al. Cancer survival in five continents: A worldwide population-based study (CONCORD). *Lancet Oncol.* **2008**, *9*, 730–756. [CrossRef]

4. Hanahan, D.; Weinberg, R.A. Hallmarks of cancer: The next generation. *Cell* **2011**, *144*, 646–674. [CrossRef]

5. Warburg, O. On the origin of cancer cells. *Science* **1956**, *123*, 309–314. [CrossRef]

6. Aykin-Burns, N.; Ahmad, I.M.; Zhu, Y.; Oberley, L.W.; Spitz, D.R. Increased levels of superoxide and H_2O_2 mediate the differential susceptibility of cancer cells versus normal cells to glucose deprivation. *Biochem. J.* **2009**, *418*, 29–37. [CrossRef]

7. Yun, J.; Rago, C.; Cheong, I.; Pagliarini, R.; Angenendt, P.; Rajagopalan, H.; Schmidt, K.; Willson, J.K.; Markowitz, S.; Zhou, S.; et al. Glucose deprivation contributes to the development of KRAS pathway mutations in tumor cells. *Science* **2009**, *325*, 1555–1559. [CrossRef]

8. Pavlides, S.; Whitaker-Menezes, D.; Castello-Cros, R.; Flomenberg, N.; Witkiewicz, A.K.; Frank, P.G.; Casimiro, M.C.; Wang, C.; Fortina, P.; Addya, S.; et al. The reverse Warburg effect: Aerobic glycolysis in cancer associated fibroblasts and the tumor stroma. *Cell Cycle* **2009**, *8*, 3984–4001. [CrossRef]

9. Sotgia, F.; Martinez-Outschoorn, U.E.; Pavlides, S.; Howell, A.; Pestell, R.G.; Lisanti, M.P. Understanding the Warburg effect and the prognostic value of stromal caveolin-1 as a marker of a lethal tumor microenvironment. *Breast Cancer Res.* **2011**, *13*, e213. [CrossRef]

10. Sun, K.; Tang, S.; Hou, Y.; Xi, L.; Chen, Y.; Yin, J.; Peng, M.; Zhao, M.; Cui, X.; Liu, M. Oxidized ATM-mediated glycolysis enhancement in breast cancer-associated fibroblasts contributes to tumor invasion through lactate as metabolic coupling. *EBioMedicine* **2019**, *41*, 370–383. [CrossRef]

11. Gonzalez, C.D.; Alvarez, S.; Ropolo, A.; Rosenzvit, C.; Bagnes, M.F.; Vaccaro, M.I. Autophagy, Warburg, and Warburg reverse effects in human cancer. *Biomed Res. Int.* **2014**, *2014*, e926729. [CrossRef] [PubMed]

12. Martinez-Outschoorn, U.E.; Lin, Z.; Ko, Y.H.; Goldberg, A.F.; Flomenberg, N.; Wang, C.; Pavlides, S.; Pestell, R.G.; Howell, A.; Sotgia, F.; et al. Understanding the metabolic basis of drug resistance: Therapeutic induction of the Warburg effect kills cancer cells. *Cell Cycle* **2011**, *10*, 2521–2528. [CrossRef] [PubMed]

13. Szablewski, L. Expression of glucose transporters in cancers. *Biochim. Biophys. Acta* **2013**, *1835*, 164–169. [CrossRef] [PubMed]

14. Wright, E.M.; Loo, D.D.; Hirayama, B.A. Biology of human sodium glucose transporters. *Physiol. Rev.* **2011**, *91*, 733–794. [CrossRef]

15. Wright, E.M.; Turk, E. The sodium/glucose cotransport family SLC5. *Pflugers Arch.* **2004**, *447*, 510–518. [CrossRef]

16. Koepsell, H. The Na(+)-D-glucose cotransporters SGLT1 and SGLT2 are targets for the treatment of diabetes and cancer. *Pharmacol. Ther.* **2017**, *170*, 148–165. [CrossRef]

17. Mueckler, M.; Thorens, B. The SLC2 (GLUT) family of membrane transporters. *Mol. Aspects Med.* **2013**, *34*, 121–138. [CrossRef]

18. Macheda, M.L.; Rogers, S.; Best, J.D. Molecular and cellular regulation of glucose transporter (GLUT) proteins in cancer. *J. Cell Physiol.* **2005**, *202*, 654–662. [CrossRef]

19. Medina, R.A.; Owen, G.I. Glucose transporters: Expression, regulation and cancer. *Biol. Res.* **2002**, *35*, 9–26. [CrossRef]

20. Furuta, E.; Okuda, H.; Kobayashi, A.; Watabe, K. Metabolic genes in cancer: Their roles in tumor progression and clinical implications. *Biochim. Biophys. Acta* **2010**, *1805*, 141–152. [CrossRef]

21. Grover-McKay, M.; Walsh, S.A.; Seftor, E.A.; Thomas, P.A.; Hendrix, M.J. Role for glucose transporter 1 protein in human breast cancer. *Pathol. Oncol. Res.* **1998**, *4*, 115–120. [CrossRef] [PubMed]

22. Lopez-Lazaro, M. The Warburg effect: Why and how do cancer cells activate glycolysis in the presence of oxygen? *Anticancer Agents Med. Chem.* **2008**, *8*, 305–312. [CrossRef] [PubMed]

23. Wuest, M.; Hamann, I.; Bouvet, V.; Glubrecht, D.; Marshall, A.; Trayner, B.; Soueidan, O.M.; Krys, D.; Wagner, M.; Cheeseman, C.; et al. Molecular Imaging of GLUT1 and GLUT5 in breast cancer: A multitracer positron emission tomography imaging study in mice. *Mol. Pharmacol.* **2018**, *93*, 79–89. [CrossRef]

24. Zhao, F.Q. Biology of glucose transport in the mammary gland. *J. Mammary Gland Biol. Neoplasia* **2014**, *19*, 3–17. [CrossRef] [PubMed]

25. Zhao, F.Q.; Keating, A.F. Expression and regulation of glucose transporters in the bovine mammary gland. *J. Dairy Sci.* **2007**, *90* (Suppl. 1), 76–86. [CrossRef]

26. Wellberg, E.A.; Johnson, S.; Finlay-Schultz, J.; Lewis, A.S.; Terrell, K.L.; Sartorius, C.A.; Abel, E.D.; Muller, W.J.; Anderson, S.M. The glucose transporter GLUT1 is required for ErbB2-induced mammary tumorigenesis. *Breast Cancer Res.* **2016**, *18*, e131. [CrossRef]

27. Young, C.D.; Lewis, A.S.; Rudolph, M.C.; Ruehle, M.D.; Jackman, M.R.; Yun, U.J.; Ilkun, O.; Pereira, R.; Abel, E.D.; Anderson, S.M. Modulation of glucose transporter 1 (GLUT1) expression levels alters mouse mammary tumor cell growth in vitro and in vivo. *PLoS ONE* **2011**, *6*, e23205. [CrossRef]

28. Pinheiro, C.; Sousa, B.; Albergaria, A.; Paredes, J.; Dufloth, R.; Vieira, D.; Schmitt, F.; Baltazar, F. GLUT1 and CAIX expression profiles in breast cancer correlate with adverse prognostic factors and MCT1 overexpression. *Histol. Histopathol.* **2011**, *26*, 1279–1286.

29. Krzeslak, A.; Wojcik-Krowiranda, K.; Forma, E.; Jozwiak, P.; Romanowicz, H.; Bienkiewicz, A.; Brys, M. Expression of GLUT1 and GLUT3 glucose transporters in endometrial and breast cancers. *Pathol. Oncol. Res.* **2012**, *18*, 721–728. [CrossRef]

30. Medina, R.A.; Meneses, A.M.; Vera, J.C.; Guzman, C.; Nualart, F.; Astuya, A.; Garcia, M.A.; Kato, S.; Carvajal, A.; Pinto, M.; et al. Estrogen and progesterone up-regulate glucose transporter expression in ZR-75-1 human breast cancer cells. *Endocrinology* **2003**, *144*, 4527–4535. [CrossRef]

31. Ganapathy, V.; Thangaraju, M.; Prasad, P.D. Nutrient transporters in cancer: Relevance to Warburg hypothesis and beyond. *Pharmacol. Ther.* **2009**, *121*, 29–40. [CrossRef] [PubMed]

32. Burgman, P.; Odonoghue, J.A.; Humm, J.L.; Ling, C.C. Hypoxia-Induced increase in FDG uptake in MCF7 cells. *J. Nucl. Med.* **2001**, *42*, 170–175. [PubMed]

33. Rivenzon-Segal, D.; Boldin-Adamsky, S.; Seger, D.; Seger, R.; Degani, H. Glycolysis and glucose transporter 1 as markers of response to hormonal therapy in breast cancer. *Int. J. Cancer* **2003**, *107*, 177–182. [CrossRef] [PubMed]

34. Kato, Y.; Maeda, T.; Suzuki, A.; Baba, Y. Cancer metabolism: New insights into classic characteristics. *Jpn. Dent. Sci. Rev.* **2018**, *54*, 8–21. [CrossRef] [PubMed]

35. Garrido, P.; Moran, J.; Alonso, A.; Gonzalez, S.; Gonzalez, C. 17beta-estradiol activates glucose uptake via GLUT4 translocation and PI3K/Akt signaling pathway in MCF-7 cells. *Endocrinology* **2013**, *154*, 1979–1989. [CrossRef] [PubMed]

36. Wang, S.; Xie, J.; Li, J.; Liu, F.; Wu, X.; Wang, Z. Cisplatin suppresses the growth and proliferation of breast and cervical cancer cell lines by inhibiting integrin beta5-mediated glycolysis. *Am. J. Cancer Res.* **2016**, *6*, 1108–1117. [PubMed]

37. Yang, Y.; Wolfram, J.; Boom, K.; Fang, X.; Shen, H.; Ferrari, M. Hesperetin impairs glucose uptake and inhibits proliferation of breast cancer cells. *Cell Biochem. Funct.* **2013**, *31*, 374–379. [CrossRef]

38. Guedes, M.; Araujo, J.R.; Correia-Branco, A.; Gregorio, I.; Martel, F.; Keating, E. Modulation of the uptake of critical nutrients by breast cancer cells by lactate: Impact on cell survival, proliferation and migration. *Exp. Cell Res.* **2016**, *341*, 111–122. [CrossRef]

39. Harmon, A.W.; Patel, Y.M. Naringenin inhibits glucose uptake in MCF-7 breast cancer cells: A mechanism for impaired cellular proliferation. *Breast Cancer Res. Treat.* **2004**, *85*, 103–110. [CrossRef]

40. Moreira, L.; Araujo, I.; Costa, T.; Correia-Branco, A.; Faria, A.; Martel, F.; Keating, E. Quercetin and epigallocatechin gallate inhibit glucose uptake and metabolism by breast cancer cells by an estrogen receptor-independent mechanism. *Exp. Cell Res.* **2013**, *319*, 1784–1795. [CrossRef]

41. Garrido, P.; Osorio, F.G.; Moran, J.; Cabello, E.; Alonso, A.; Freije, J.M.; Gonzalez, C. Loss of GLUT4 induces metabolic reprogramming and impairs viability of breast cancer cells. *J. Cell Physiol.* **2015**, *230*, 191–198. [CrossRef] [PubMed]

42. Rogers, S.; Macheda, M.L.; Docherty, S.E.; Carty, M.D.; Henderson, M.A.; Soeller, W.C.; Gibbs, E.M.; James, D.E.; Best, J.D. Identification of a novel glucose transporter-like protein-GLUT-12. *Am. J. Physiol. Endocrinol. Metab.* **2002**, *282*, 733–738. [CrossRef] [PubMed]

43. Evans, A.; Bates, V.; Troy, H.; Hewitt, S.; Holbeck, S.; Chung, Y.L.; Phillips, R.; Stubbs, M.; Griffiths, J.; Airley, R. Glut-1 as a therapeutic target: Increased chemoresistance and HIF-1-independent link with cell turnover is revealed through COMPARE analysis and metabolomic studies. *Cancer Chemother. Pharmacol.* **2008**, *61*, 377–393. [CrossRef] [PubMed]

44. Housman, G.; Byler, S.; Heerboth, S.; Lapinska, K.; Longacre, M.; Snyder, N.; Sarkar, S. Drug resistance in cancer: An overview. *Cancers* **2014**, *6*, 1769–1792. [CrossRef] [PubMed]

45. Kuang, R.; Jahangiri, A.; Mascharak, S.; Nguyen, A.; Chandra, A.; Flanigan, P.M.; Yagnik, G.; Wagner, J.R.; de Lay, M.; Carrera, D.; et al. GLUT3 upregulation promotes metabolic reprogramming associated with antiangiogenic therapy resistance. *JCI Insight* **2017**, *2*, e88815. [CrossRef] [PubMed]

46. Bomanji, J.B.; Costa, D.C.; Ell, P.J. Clinical role of positron emission tomography in oncology. *Lancet Oncol.* **2001**, *2*, 157–164. [CrossRef]

47. Zhu, A.; Lee, D.; Shim, H. Metabolic positron emission tomography imaging in cancer detection and therapy response. *Semin. Oncol.* **2011**, *38*, 55–69. [CrossRef]

48. Luengo, A.; Gui, D.Y.; Heiden, M.G.V. Targeting metabolism for cancer therapy. *Cell Chem. Biol.* **2017**, *24*, 1161–1180. [CrossRef]

49. Qian, Y.; Wang, X.; Chen, X. Inhibitors of glucose transport and glycolysis as novel anticancer therapeutics. *World J. Transl. Med.* **2014**, *3*, 37–57. [CrossRef]

50. Michelakis, E.D.; Webster, L.; Mackey, J.R. Dichloroacetate (DCA) as a potential metabolic-targeting therapy for cancer. *Br. J. Cancer* **2008**, *99*, 989–994. [CrossRef]

51. Woolf, E.C.; Syed, N.; Scheck, A.C. Tumor metabolism, the ketogenic diet and beta-hydroxybutyrate: Novel approaches to adjuvant brain tumor therapy. *Front. Mol. Neurosci.* **2016**, *9*, e122. [CrossRef] [PubMed]

52. Abdel-Wahab, A.F.; Mahmoud, W.; Al-Harizy, R.M. Targeting glucose metabolism to suppress cancer progression: Prospective of anti-glycolytic cancer therapy. *Pharmacol. Res.* **2019**, *150*, e104511. [CrossRef] [PubMed]

53. Granchi, C.; Fortunato, S.; Minutolo, F. Anticancer agents interacting with membrane glucose transporters. *Medchemcomm* **2016**, *7*, 1716–1729. [CrossRef] [PubMed]

54. Rodriguez-Enriquez, S.; Marin-Hernandez, A.; Gallardo-Perez, J.C.; Carreno-Fuentes, L.; Moreno-Sanchez, R. Targeting of cancer energy metabolism. *Mol. Nutr. Food Res.* **2009**, *53*, 29–48. [CrossRef] [PubMed]

55. Chan, D.A.; Sutphin, P.D.; Nguyen, P.; Turcotte, S.; Lai, E.W.; Banh, A.; Reynolds, G.E.; Chi, J.T.; Wu, J.; Solow-Cordero, D.E.; et al. Targeting GLUT1 and the Warburg effect in renal cell carcinoma by chemical synthetic lethality. *Sci. Transl. Med.* **2011**, *3*, 94ra70. [CrossRef]

56. Liu, Y.; Zhang, W.; Cao, Y.; Liu, Y.; Bergmeier, S.; Chen, X. Small compound inhibitors of basal glucose transport inhibit cell proliferation and induce apoptosis in cancer cells via glucose-deprivation-like mechanisms. *Cancer Lett.* **2010**, *298*, 176–185. [CrossRef]

57. Xintaropoulou, C.; Ward, C.; Wise, A.; Marston, H.; Turnbull, A.; Langdon, S.P. A comparative analysis of inhibitors of the glycolysis pathway in breast and ovarian cancer cell line models. *Oncotarget* **2015**, *6*, 25677–25695. [CrossRef]

58. Zhao, F.; Ming, J.; Zhou, Y.; Fan, L. Inhibition of Glut1 by WZB117 sensitizes radioresistant breast cancer cells to irradiation. *Cancer Chemother. Pharmacol.* **2016**, *77*, 963–972. [CrossRef]

59. Liu, Y.; Cao, Y.; Zhang, W.; Bergmeier, S.; Qian, Y.; Akbar, H.; Colvin, R.; Ding, J.; Tong, L.; Wu, S.; et al. A small-molecule inhibitor of glucose transporter 1 downregulates glycolysis, induces cell-cycle arrest, and inhibits cancer cell growth in vitro and in vivo. *Mol. Cancer Ther.* **2012**, *11*, 1672–1682. [CrossRef]

60. Chen, Q.; Meng, Y.Q.; Xu, X.F.; Gu, J. Blockade of GLUT1 by WZB117 resensitizes breast cancer cells to adriamycin. *Anticancer Drugs* **2017**, *28*, 880–887. [CrossRef]

61. Siebeneicher, H.; Cleve, A.; Rehwinkel, H.; Neuhaus, R.; Heisler, I.; Muller, T.; Bauser, M.; Buchmann, B. Identification and Optimization of the First Highly Selective GLUT1 Inhibitor BAY-876. *ChemMedChem* **2016**, *11*, 2261–2271. [CrossRef] [PubMed]

62. Zeng, L.; Zhou, M.M. Bromodomain: An acetyl-lysine binding domain. *FEBS Lett.* **2002**, *513*, 124–128. [CrossRef]

63. Wu, Q.; Heidenreich, D.; Zhou, S.; Ackloo, S.; Kramer, A.; Nakka, K.; Lima-Fernandes, E.; Deblois, G.; Duan, S.; Vellanki, R.N.; et al. A chemical toolbox for the study of bromodomains and epigenetic signaling. *Nat. Commun.* **2019**, *10*, e1915. [CrossRef] [PubMed]

64. Kuntz, S.; Mazerbourg, S.; Boisbrun, M.; Cerella, C.; Diederich, M.; Grillier-Vuissoz, I.; Flament, S. Energy restriction mimetic agents to target cancer cells: Comparison between 2-deoxyglucose and thiazolidinediones. *Biochem. Pharmacol.* **2014**, *92*, 102–111. [CrossRef]

65. Amaral, I.; Silva, C.; Correia-Branco, A.; Martel, F. Effect of metformin on estrogen and progesterone receptor-positive (MCF-7) and triple-negative (MDA-MB-231) breast cancer cells. *Biomed. Pharmacother.* **2018**, *102*, 94–101. [CrossRef]

66. Zhang, C.; Liu, J.; Liang, Y.; Wu, R.; Zhao, Y.; Hong, X.; Lin, M.; Yu, H.; Liu, L.; Levine, A.J.; et al. Tumour-associated mutant p53 drives the Warburg effect. *Nat. Commun.* **2013**, *4*, e2935. [CrossRef]

67. Oh, S.; Kim, H.; Nam, K.; Shin, I. Glut1 promotes cell proliferation, migration and invasion by regulating epidermal growth factor receptor and integrin signaling in triple-negative breast cancer cells. *BMB Rep.* **2017**, *50*, 132–137. [CrossRef]

68. Oh, S.; Kim, H.; Nam, K.; Shin, I. Silencing of Glut1 induces chemoresistance via modulation of Akt/GSK-3beta/beta-catenin/survivin signaling pathway in breast cancer cells. *Arch. Biochem. Biophys.* **2017**, *636*, 110–122. [CrossRef]

69. Rastogi, S.; Banerjee, S.; Chellappan, S.; Simon, G.R. Glut-1 antibodies induce growth arrest and apoptosis in human cancer cell lines. *Cancer Lett.* **2007**, *257*, 244–251. [CrossRef]

70. Evans, J.M.; Donnelly, L.A.; Emslie-Smith, A.M.; Alessi, D.R.; Morris, A.D. Metformin and reduced risk of cancer in diabetic patients. *BMJ* **2005**, *330*, 1304–1305. [CrossRef]

71. Grossmann, M.E.; Yang, D.Q.; Guo, Z.; Potter, D.A.; Cleary, M.P. Metformin treatment for the prevention and/or treatment of breast/mammary tumorigenesis. *Curr. Pharmacol. Rep.* **2015**, *1*, 312–323. [CrossRef] [PubMed]

72. Wysocki, P.J.; Wierusz-Wysocka, B. Obesity, hyperinsulinemia and breast cancer: Novel targets and a novel role for metformin. *Expert Rev. Mol. Diagn.* **2010**, *10*, 509–519. [CrossRef] [PubMed]

73. Marini, C.; Salani, B.; Massollo, M.; Amaro, A.; Esposito, A.I.; Orengo, A.M.; Capitanio, S.; Emionite, L.; Riondato, M.; Bottoni, G.; et al. Direct inhibition of hexokinase activity by metformin at least partially impairs glucose metabolism and tumor growth in experimental breast cancer. *Cell Cycle* **2013**, *12*, 3490–3499. [CrossRef] [PubMed]

74. Banerjee, A.; Birts, C.N.; Darley, M.; Parker, R.; Mirnezami, A.H.; West, J.; Cutress, R.I.; Beers, S.A.; Rose-Zerilli, M.J.J.; Blaydes, J.P. Stem cell-like breast cancer cells with acquired resistance to metformin are sensitive to inhibitors of NADH-dependent CtBP dimerization. *Carcinogenesis* **2019**, *40*, 871–882. [CrossRef] [PubMed]

75. Ding, J.; Gou, Q.; Jin, J.; Shi, J.; Liu, Q.; Hou, Y. Metformin inhibits PPARdelta agonist-mediated tumor growth by reducing Glut1 and SLC1A5 expressions of cancer cells. *Eur. J. Pharmacol.* **2019**, *857*, e172425. [CrossRef] [PubMed]

76. Birsoy, K.; Possemato, R.; Lorbeer, F.K.; Bayraktar, E.C.; Thiru, P.; Yucel, B.; Wang, T.; Chen, W.W.; Clish, C.B.; Sabatini, D.M. Metabolic determinants of cancer cell sensitivity to glucose limitation and biguanides. *Nature* **2014**, *508*, 108–112. [CrossRef] [PubMed]

77. Liu, X.; Gan, B. lncRNA NBR2 modulates cancer cell sensitivity to phenformin through GLUT1. *Cell Cycle* **2016**, *15*, 3471–3481. [CrossRef]

78. Elstner, E.; Muller, C.; Koshizuka, K.; Williamson, E.A.; Park, D.; Asou, H.; Shintaku, P.; Said, J.W.; Heber, D.; Koeffler, H.P. Ligands for peroxisome proliferator-activated receptorgamma and retinoic acid receptor inhibit growth and induce apoptosis of human breast cancer cells in vitro and in BNX mice. *Proc. Natl. Acad. Sci. USA* **1998**, *95*, 8806–8811. [CrossRef]

79. Moon, S.H.; Lee, S.J.; Jung, K.H.; Quach, C.H.; Park, J.W.; Lee, J.H.; Cho, Y.S.; Lee, K.H. Troglitazone stimulates cancer cell uptake of ^{18}F-FDG by suppressing mitochondrial respiration and augments sensitivity to glucose restriction. *J. Nucl. Med.* **2016**, *57*, 129–135. [CrossRef]

80. Fiume, L.; Manerba, M.; Vettraino, M.; di Stefano, G. Effect of sorafenib on the energy metabolism of hepatocellular carcinoma cells. *Eur. J. Pharmacol.* **2011**, *670*, 39–43. [CrossRef]

81. Bull, V.H.; Rajalingam, K.; Thiede, B. Sorafenib-induced mitochondrial complex I inactivation and cell death in human neuroblastoma cells. *J. Proteome Res.* **2012**, *11*, 1609–1620. [CrossRef] [PubMed]

82. Fumarola, C.; Caffarra, C.; la Monica, S.; Galetti, M.; Alfieri, R.R.; Cavazzoni, A.; Galvani, E.; Generali, D.; Petronini, P.G.; Bonelli, M.A. Effects of sorafenib on energy metabolism in breast cancer cells: Role of AMPK-mTORC1 signaling. *Breast Cancer Res. Treat.* **2013**, *141*, 67–78. [CrossRef] [PubMed]

83. Smith, T.A.; Appleyard, M.V.; Sharp, S.; Fleming, I.N.; Murray, K.; Thompson, A.M. Response to trastuzumab by HER2 expressing breast tumour xenografts is accompanied by decreased Hexokinase II, glut1 and [18F]-FDG incorporation and changes in ^{31}P-NMR-detectable phosphomonoesters. *Cancer Chemother. Pharmacol.* **2013**, *71*, 473–480. [CrossRef] [PubMed]

84. Engles, J.M.; Quarless, S.A.; Mambo, E.; Ishimori, T.; Cho, S.Y.; Wahl, R.L. Stunning and its effect on ^{3}H-FDG uptake and key gene expression in breast cancer cells undergoing chemotherapy. *J. Nucl. Med.* **2006**, *47*, 603–608.

85. Abd-Rabou, A.A.; Ahmed, H.H.; Shalby, A.B. Selenium overcomes doxorubicin resistance in their nano-platforms against breast and colon cancers. *Biol. Trace Elem. Res.* **2019**. [CrossRef]

86. DeMichele, A.; Clark, A.S.; Tan, K.S.; Heitjan, D.F.; Gramlich, K.; Gallagher, M.; Lal, P.; Feldman, M.; Zhang, P.; Colameco, C.; et al. CDK 4/6 inhibitor palbociclib (PD0332991) in Rb+ advanced breast cancer: Phase II activity, safety, and predictive biomarker assessment. *Clin. Cancer Res.* **2015**, *21*, 995–1001. [CrossRef]

87. Cretella, D.; Fumarola, C.; Bonelli, M.; Alfieri, R.; la Monica, S.; Digiacomo, G.; Cavazzoni, A.; Galetti, M.; Generali, D.; Petronini, P.G. Pre-treatment with the CDK4/6 inhibitor palbociclib improves the efficacy of paclitaxel in TNBC cells. *Sci. Rep.* **2019**, *9*, e13014. [CrossRef]

88. Cretella, D.; Ravelli, A.; Fumarola, C.; la Monica, S.; Digiacomo, G.; Cavazzoni, A.; Alfieri, R.; Biondi, A.; Generali, D.; Bonelli, M.; et al. The anti-tumor efficacy of CDK4/6 inhibition is enhanced by the combination with PI3K/AKT/mTOR inhibitors through impairment of glucose metabolism in TNBC cells. *J. Exp. Clin. Cancer. Res.* **2018**, *37*, e72. [CrossRef]

89. Powe, D.G.; Voss, M.J.; Zanker, K.S.; Habashy, H.O.; Green, A.R.; Ellis, I.O.; Entschladen, F. Beta-blocker drug therapy reduces secondary cancer formation in breast cancer and improves cancer specific survival. *Oncotarget* **2010**, *1*, 628–638. [CrossRef]

90. Kang, F.; Ma, W.; Ma, X.; Shao, Y.; Yang, W.; Chen, X.; Li, L.; Wang, J. Propranolol inhibits glucose metabolism and 18F-FDG uptake of breast cancer through posttranscriptional downregulation of hexokinase-2. *J. Nucl. Med.* **2014**, *55*, 439–445. [CrossRef]

91. Zheng, X.; Resnick, R.J.; Shalloway, D. Apoptosis of estrogen-receptor negative breast cancer and colon cancer cell lines by PTP alpha and src RNAi. *Int. J. Cancer* **2008**, *122*, 1999–2007. [CrossRef] [PubMed]

92. Puls, L.N.; Eadens, M.; Messersmith, W. Current status of SRC inhibitors in solid tumor malignancies. *Oncologist* **2011**, *16*, 566–578. [CrossRef] [PubMed]

93. Jain, S.; Wang, X.; Chang, C.C.; Ibarra-Drendall, C.; Wang, H.; Zhang, Q.; Brady, S.W.; Li, P.; Zhao, H.; Dobbs, J.; et al. Src inhibition blocks c-Myc translation and glucose metabolism to prevent the development of breast cancer. *Cancer Res.* **2015**, *75*, 4863–4875. [CrossRef] [PubMed]

94. Zawacka-Pankau, J.; Grinkevich, V.V.; Hunten, S.; Nikulenkov, F.; Gluch, A.; Li, H.; Enge, M.; Kel, A.; Selivanova, G. Inhibition of glycolytic enzymes mediated by pharmacologically activated p53: Targeting Warburg effect to fight cancer. *J. Biol. Chem.* **2011**, *286*, 41600–41615. [CrossRef] [PubMed]

95. Grinkevich, V.V.; Nikulenkov, F.; Shi, Y.; Enge, M.; Bao, W.; Maljukova, A.; Gluch, A.; Kel, A.; Sangfelt, O.; Selivanova, G. Ablation of key oncogenic pathways by RITA-reactivated p53 is required for efficient apoptosis. *Cancer Cell* **2009**, *15*, 441–453. [CrossRef]

96. Vassilev, L.T.; Vu, B.T.; Graves, B.; Carvajal, D.; Podlaski, F.; Filipovic, Z.; Kong, N.; Kammlott, U.; Lukacs, C.; Klein, C.; et al. In vivo activation of the p53 pathway by small-molecule antagonists of MDM2. *Science* **2004**, *303*, 844–848. [CrossRef]

97. Vivanco, I.; Sawyers, C.L. The phosphatidylinositol 3-Kinase AKT pathway in human cancer. *Nat. Rev. Cancer* **2002**, *2*, 489–501. [CrossRef]

98. Tan, S.X.; Ng, Y.; James, D.E. Akt inhibitors reduce glucose uptake independently of their effects on Akt. *Biochem. J.* **2010**, *432*, 191–197. [CrossRef]

99. Deblois, G.; St-Pierre, J.; Giguere, V. The PGC-1/ERR signaling axis in cancer. *Oncogene* **2013**, *32*, 3483–3490. [CrossRef]

100. Ludvik, A.E.; Pusec, C.M.; Priyadarshini, M.; Angueira, A.R.; Guo, C.; Lo, A.; Hershenhouse, K.S.; Yang, G.Y.; Ding, X.; Reddy, T.E.; et al. HKDC1 is a novel hexokinase involved in whole-body glucose use. *Endocrinology* **2016**, *157*, 3452–3461. [CrossRef]

101. Li, G.H.; Huang, J.F. Inferring therapeutic targets from heterogeneous data: HKDC1 is a novel potential therapeutic target for cancer. *Bioinformatics* **2014**, *30*, 748–752. [CrossRef] [PubMed]

102. Chen, X.; Lv, Y.; Sun, Y.; Zhang, H.; Xie, W.; Zhong, L.; Chen, Q.; Li, M.; Li, L.; Feng, J.; et al. PGC1beta regulates breast tumor growth and metastasis by SREBP1-mediated HKDC1 expression. *Front. Oncol.* **2019**, *9*, e290. [CrossRef] [PubMed]

103. Li, Z.; Gong, X.; Zhang, W.; Zhang, J.; Ding, L.; Li, H.; Tu, D.; Tang, J. Inhibition of miRNA-34a promotes triple negative cancer cell proliferation by promoting glucose uptake. *Exp. Ther. Med.* **2019**, *18*, 3936–3942. [CrossRef] [PubMed]

104. He, M.; Jin, Q.; Chen, C.; Liu, Y.; Ye, X.; Jiang, Y.; Ji, F.; Qian, H.; Gan, D.; Yue, S.; et al. The miR-186-3p/EREG axis orchestrates tamoxifen resistance and aerobic glycolysis in breast cancer cells. *Oncogene* **2019**, *38*, 5551–5565. [CrossRef]

105. de Almeida Chuffa, L.G.; Seiva, F.R.F.; Cucielo, M.S.; Silveira, H.S.; Reiter, R.J.; Lupi, L.A. Mitochondrial functions and melatonin: A tour of the reproductive cancers. *Cell Mol. Life Sci.* **2019**, *76*, 837–863. [CrossRef]

106. Hevia, D.; Gonzalez-Menendez, P.; Quiros-Gonzalez, I.; Miar, A.; Rodriguez-Garcia, A.; Tan, D.X.; Reiter, R.J.; Mayo, J.C.; Sainz, R.M. Melatonin uptake through glucose transporters: A new target for melatonin inhibition of cancer. *J. Pineal Res.* **2015**, *58*, 234–250. [CrossRef]

107. Blask, D.E.; Dauchy, R.T.; Dauchy, E.M.; Mao, L.; Hill, S.M.; Greene, M.W.; Belancio, V.P.; Sauer, L.A.; Davidson, L. Light exposure at night disrupts host/cancer circadian regulatory dynamics: Impact on the Warburg effect, lipid signaling and tumor growth prevention. *PLoS ONE* **2014**, *9*, e102776. [CrossRef]

108. Sonehara, N.M.; Lacerda, J.Z.; Jardim-Perassi, B.V.; de Paula, R., Jr.; Moschetta-Pinheiro, M.G.; Souza, Y.S.T.; de Andrade, J.C.J.; De Campos Zuccari, D.A.P. Melatonin regulates tumor aggressiveness under acidosis condition in breast cancer cell lines. *Oncol. Lett.* **2019**, *17*, 1635–1645. [CrossRef]

109. Nunes, C.; Silva, C.; Correia-Branco, A.; Martel, F. Lack of effect of the procarcinogenic 17beta-estradiol on nutrient uptake by the MCF-7 breast cancer cell line. *Biomed. Pharmacother.* **2017**, *90*, 287–294. [CrossRef]

110. Neeman, M.; Degani, H. Metabolic studies of estrogen- and tamoxifen-treated human breast cancer cells by nuclear magnetic resonance spectroscopy. *Cancer Res.* **1989**, *49*, 589–594.

111. Ko, B.H.; Paik, J.Y.; Jung, K.H.; Lee, K.H. 17beta-estradiol augments ^{18}F-FDG uptake and glycolysis of T47D breast cancer cells via membrane-initiated rapid PI3K-Akt activation. *J. Nucl. Med.* **2010**, *51*, 1740–1747. [CrossRef]

112. Kim, H.I.; Moon, S.H.; Lee, W.C.; Lee, H.J.; Shivakumar, S.B.; Lee, S.H.; Park, B.W.; Rho, G.J.; Jeon, B.G. Inhibition of cell growth by cellular differentiation into adipocyte-like cells in dexamethasone sensitive cancer cell lines. *Anim. Cells Syst.* **2018**, *22*, 178–188. [CrossRef] [PubMed]

113. Zhang, C.; Beckermann, B.; Kallifatidis, G.; Liu, Z.; Rittgen, W.; Edler, L.; Buchler, P.; Debatin, K.M.; Buchler, M.W.; Friess, H.; et al. Corticosteroids induce chemotherapy resistance in the majority of tumour cells from bone, brain, breast, cervix, melanoma and neuroblastoma. *Int. J. Oncol.* **2006**, *29*, 1295–1301. [CrossRef] [PubMed]

114. Buxant, F.; Kindt, N.; Laurent, G.; Noel, J.C.; Saussez, S. Antiproliferative effect of dexamethasone in the MCF-7 breast cancer cell line. *Mol. Med. Rep.* **2015**, *12*, 4051–4054. [CrossRef]

115. Dalton, G.D.; Xie, J.; An, S.W.; Huang, C.L. New insights into the mechanism of action of soluble Klotho. *Front. Endocrinol.* **2017**, *8*, e323. [CrossRef]

116. Wolf, I.; Levanon-Cohen, S.; Bose, S.; Ligumsky, H.; Sredni, B.; Kanety, H.; Kuro-o, M.; Karlan, B.; Kaufman, B.; Koeffler, H.P.; et al. Klotho: A tumor suppressor and a modulator of the IGF-1 and FGF pathways in human breast cancer. *Oncogene* **2008**, *27*, 7094–7105. [CrossRef]

117. Davidov, B.; Shmulevich, R.; Shabtay, A.; Rubinek, T.; Wolf, I. The hormone KL1: A regulator of breast cancer cell metabolism. *Isr. Med. Assoc. J.* **2019**, *21*, e504.

118. Pollak, M. Insulin and insulin-like growth factor signalling in neoplasia. *Nat. Rev. Cancer* **2008**, *8*, 915–928. [CrossRef]

119. Acevedo, C.G.; Marquez, J.L.; Rojas, S.; Bravo, I. Insulin and nitric oxide stimulates glucose transport in human placenta. *Life Sci.* **2005**, *76*, 2643–2653. [CrossRef]

120. Agrawal, S.; Luc, M.; Ziolkowski, P.; Agrawal, A.K.; Pielka, E.; Walaszek, K.; Zduniak, K.; Wozniak, M. Insulin-induced enhancement of MCF-7 breast cancer cell response to 5-fluorouracil and cyclophosphamide. *Tumour Biol.* **2017**, *39*. [CrossRef]

121. Beckert, S.; Farrahi, F.; Aslam, R.S.; Scheuenstuhl, H.; Konigsrainer, A.; Hussain, M.Z.; Hunt, T.K. Lactate stimulates endothelial cell migration. *Wound Repair Regen.* **2006**, *14*, 321–324. [CrossRef] [PubMed]

122. Lu, H.; Forbes, R.A.; Verma, A. Hypoxia-inducible factor 1 activation by aerobic glycolysis implicates the Warburg effect in carcinogenesis. *J. Biol. Chem.* **2002**, *277*, 23111–23115. [CrossRef] [PubMed]

123. Turkcan, S.; Kiru, L.; Naczynski, D.J.; Sasportas, L.S.; Pratx, G. Lactic acid accumulation in the tumor microenvironment suppresses (18)F-FDG uptake. *Cancer Res.* **2019**, *79*, 410–419. [CrossRef] [PubMed]

124. Venmar, K.T.; Kimmel, D.W.; Cliffel, D.E.; Fingleton, B. IL4 receptor alpha mediates enhanced glucose and glutamine metabolism to support breast cancer growth. *Biochim. Biophys. Acta* **2015**, *1853*, 1219–1228. [CrossRef]

125. Masuda, H.; Zhang, D.; Bartholomeusz, C.; Doihara, H.; Hortobagyi, G.N.; Ueno, N.T. Role of epidermal growth factor receptor in breast cancer. *Breast Cancer Res. Treat.* **2012**, *136*, 331–345. [CrossRef]

126. Lee, J.; Lee, E.J.; Moon, S.H.; Kim, S.; Hyun, S.H.; Cho, Y.S.; Choi, J.Y.; Kim, B.T.; Lee, K.H. Strong association of epidermal growth factor receptor status with breast cancer FDG uptake. *Eur. J. Nucl. Med. Mol. Imaging* **2017**, *44*, 1438–1447. [CrossRef]

127. Jung, K.H.; Lee, E.J.; Park, J.W.; Lee, J.H.; Moon, S.H.; Cho, Y.S.; Lee, K.H. EGF receptor stimulation shifts breast cancer cell glucose metabolism toward glycolytic flux through PI3 kinase signaling. *PLoS ONE* **2019**, *14*, e0221294. [CrossRef]

128. Williamson, G. The role of polyphenols in modern nutrition. *Nutr. Bull.* **2017**, *42*, 226–235. [CrossRef]

129. Hussain, S.A.; Sulaiman, A.A.; Alhaddad, H.; Alhadidi, Q. Natural polyphenols: Influence on membrane transporters. *J. Intercult. Ethnopharmacol.* **2016**, *5*, 97–104. [CrossRef]

130. Williamson, G. Possible effects of dietary polyphenols on sugar absorption and digestion. *Mol. Nutr. Food Res.* **2013**, *57*, 48–57. [CrossRef]

131. Losada-Echeberria, M.; Herranz-Lopez, M.; Micol, V.; Barrajon-Catalan, E. Polyphenols as promising drugs against main breast cancer signatures. *Antioxidants* **2017**, *6*, 88. [CrossRef] [PubMed]

132. Mocanu, M.M.; Nagy, P.; Szollosi, J. Chemoprevention of breast cancer by dietary polyphenols. *Molecules* **2015**, *20*, 22578–22620. [CrossRef] [PubMed]

133. Keating, E.; Martel, F. Antimetabolic effects of polyphenols in breast cancer cells: Focus on glucose uptake and metabolism. *Front. Nutr.* **2018**, *5*, e25. [CrossRef]

134. Martel, F.; Guedes, M.; Keating, E. Effect of polyphenols on glucose and lactate transport by breast cancer cells. *Breast Cancer Res. Treat.* **2016**, *157*, 1–11. [CrossRef]

135. Jaroszewski, J.W.; Kaplan, O.; Cohen, J.S. Action of gossypol and rhodamine 123 on wild type and multidrug-resistant MCF-7 human breast cancer cells: 31P nuclear magnetic resonance and toxicity studies. *Cancer Res.* **1990**, *50*, 6936–6943.

136. Lim, H.A.; Kim, J.H.; Kim, J.H.; Sung, M.K.; Kim, M.K.; Park, J.H.; Kim, J.S. Genistein induces glucose-regulated protein 78 in mammary tumor cells. *J. Med. Food* **2006**, *9*, 28–32. [CrossRef]

137. Uifalean, A.; Schneider, S.; Gierok, P.; Ionescu, C.; Iuga, C.A.; Lalk, M. The impact of soy isoflavones on MCF-7 and MDA-MB 231 breast cancer cells using a global metabolomic approach. *Int. J. Mol. Sci.* **2016**, *17*, 1443. [CrossRef]

138. Azevedo, C.; Correia-Branco, A.; Araujo, J.R.; Guimaraes, J.T.; Keating, E.; Martel, F. The chemopreventive effect of the dietary compound kaempferol on the MCF-7 human breast cancer cell line is dependent on inhibition of glucose cellular uptake. *Nutr. Cancer* **2015**, *67*, 504–513. [CrossRef]

139. Jung, K.H.; Lee, J.H.; Quach, C.H.T.; Paik, J.Y.; Oh, H.; Park, J.W.; Lee, E.J.; Moon, S.H.; Lee, K.H. Resveratrol suppresses cancer cell glucose uptake by targeting reactive oxygen species-mediated hypoxia-inducible factor-1alpha activation. *J. Nucl. Med.* **2013**, *54*, 2161–2167. [CrossRef]

140. Jia, L.; Huang, S.; Yin, X.; Zan, Y.; Guo, Y.; Han, L. Quercetin suppresses the mobility of breast cancer by suppressing glycolysis through Akt-mTOR pathway mediated autophagy induction. *Life Sci.* **2018**, *208*, 123–130. [CrossRef]

141. Li, L.J.; Li, G.W.; Xie, Y. Regulatory effects of glabridin and quercetin on energy metabolism of breast cancer cells. *Zhongguo Zhong Yao Za Zhi* **2019**, *44*, 3786–3791. [PubMed]

142. Wei, R.; Mao, L.; Xu, P.; Zheng, X.; Hackman, R.M.; Mackenzie, G.G.; Wang, Y. Suppressing glucose metabolism with epigallocatechin-3-gallate (EGCG) reduces breast cancer cell growth in preclinical models. *Food Funct.* **2018**, *9*, 5682–5696. [CrossRef] [PubMed]

143. Nelson, J.A.; Falk, R.E. Phloridzin and phloretin inhibition of 2-deoxy-D-glucose uptake by tumor cells in vitro and in vivo. *Anticancer Res.* **1993**, *13*, 2293–2299. [PubMed]

144. Niedzwiecki, A.; Roomi, M.W.; Kalinovsky, T.; Rath, M. Anticancer efficacy of polyphenols and their combinations. *Nutrients* **2016**, *8*, 552. [CrossRef] [PubMed]

145. Silva, C.; Correia-Branco, A.; Andrade, N.; Ferreira, A.C.; Soares, M.L.; Sonveaux, P.; Stephenne, J.; Martel, F. Selective pro-apoptotic and antimigratory effects of polyphenol complex catechin:lysine 1:2 in breast, pancreatic and colorectal cancer cell lines. *Eur. J. Pharmacol.* **2019**, *859*, 172533. [CrossRef] [PubMed]

146. Zhang, W.; Liu, Y.; Chen, X.; Bergmeier, S.C. Novel inhibitors of basal glucose transport as potential anticancer agents. *Bioorg. Med. Chem. Lett.* **2010**, *20*, 2191–2194. [CrossRef]

147. Siddiqui, F.A.; Prakasam, G.; Chattopadhyay, S.; Rehman, A.U.; Padder, R.A.; Ansari, M.A.; Irshad, R.; Mangalhara, K.; Bamezai, R.N.K.; Husain, M.; et al. Curcumin decreases Warburg effect in cancer cells by down-regulating pyruvate kinase M2 via mTOR-HIF1alpha inhibition. *Sci. Rep.* **2018**, *8*, e8323. [CrossRef]

148. Jin, J.; Qiu, S.; Wang, P.; Liang, X.; Huang, F.; Wu, H.; Zhang, B.; Zhang, W.; Tian, X.; Xu, R.; et al. Cardamonin inhibits breast cancer growth by repressing HIF-1alpha-dependent metabolic reprogramming. *J. Exp. Clin. Cancer Res.* **2019**, *38*, e377. [CrossRef]

149. Shahruzaman, S.H.; Mustafa, M.F.; Ramli, S.; Maniam, S.; Fakurazi, S.; Maniam, S. The cytotoxic properties of Baeckea frutescens branches extracts in eliminating breast cancer cells. *Evid. Based Complement. Alternat. Med.* **2019**, *2019*, e9607590. [CrossRef]

150. Hernandez, J.F.; Uruena, C.P.; Cifuentes, M.C.; Sandoval, T.A.; Pombo, L.M.; Castaneda, D.; Asea, A.; Fiorentino, S. A Petiveria alliacea standardized fraction induces breast adenocarcinoma cell death by modulating glycolytic metabolism. *J. Ethnopharmacol.* **2014**, *153*, 641–649. [CrossRef]

151. Zhu, S.; Wei, L.; Lin, G.; Tong, Y.; Zhang, J.; Jiang, X.; He, Q.; Lu, X.; Zhu, D.D.; Chen, Y.Q. Metabolic alterations induced by Kudingcha lead to cancer cell apoptosis and metastasis inhibition. *Nutr. Cancer* **2019**, 1–12. [CrossRef] [PubMed]

152. Jung, C.Y.; Rampal, A.L. Cytochalasin B binding sites and glucose transport carrier in human erythrocyte ghosts. *J. Biol. Chem.* **1977**, *252*, 5456–5463. [PubMed]

153. Keller, K.; Mueckler, M. Different mammalian facilitative glucose transporters expressed in Xenopus oocytes. *Biomed. Biochim. Acta* **1990**, *49*, 1201–1203. [PubMed]

154. Cho, Y.S.; Lee, J.H.; Jung, K.H.; Park, J.W.; Moon, S.H.; Choe, Y.S.; Lee, K.H. Molecular mechanism of (18)F-FDG uptake reduction induced by genipin in T47D cancer cell and role of uncoupling protein-2 in cancer cell glucose metabolism. *Nucl. Med. Biol.* **2016**, *43*, 587–592. [CrossRef]

155. Pan, Y.; Zheng, Q.; Ni, W.; Wei, Z.; Yu, S.; Jia, Q.; Wang, M.; Wang, A.; Chen, W.; Lu, Y. Breaking glucose transporter 1/pyruvate kinase M2 glycolytic loop is required for cantharidin inhibition of metastasis in highly metastatic breast cancer. *Front. Pharmacol.* **2019**, *10*, e590. [CrossRef]

156. Jiao, L.; Wang, S.; Zheng, Y.; Wang, N.; Yang, B.; Wang, D.; Yang, D.; Mei, W.; Zhao, Z.; Wang, Z. Betulinic acid suppresses breast cancer aerobic glycolysis via caveolin-1/NF-kappaB/c-Myc pathway. *Biochem. Pharmacol.* **2019**, *161*, 149–162. [CrossRef]

157. Singh, S.V.; Singh, K. Cancer chemoprevention with dietary isothiocyanates mature for clinical translational research. *Carcinogenesis* **2012**, *33*, 1833–1842. [CrossRef]

158. Xiao, D.; Powolny, A.A.; Singh, S.V. Benzyl isothiocyanate targets mitochondrial respiratory chain to trigger reactive oxygen species-dependent apoptosis in human breast cancer cells. *J. Biol. Chem.* **2008**, *283*, 30151–30163. [CrossRef]

159. Roy, R.; Hahm, E.R.; White, A.G.; Anderson, C.J.; Singh, S.V. AKT-dependent sugar addiction by benzyl isothiocyanate in breast cancer cells. *Mol. Carcinog.* **2019**, *58*, 996–1007. [CrossRef]

160. Calder, P.C. Mechanisms of action of (n-3) fatty acids. *J. Nutr.* **2012**, *142*, 592–599. [CrossRef]

161. Mouradian, M.; Kikawa, K.D.; Dranka, B.P.; Komas, S.M.; Kalyanaraman, B.; Pardini, R.S. Docosahexaenoic acid attenuates breast cancer cell metabolism and the Warburg phenotype by targeting bioenergetic function. *Mol. Carcinog.* **2015**, *54*, 810–820. [CrossRef] [PubMed]

162. Krishnan, A.V.; Feldman, D. Mechanisms of the anti-cancer and anti-inflammatory actions of vitamin D. *Annu. Rev. Pharmacol. Toxicol.* **2011**, *51*, 311–336. [CrossRef] [PubMed]

163. Santos, J.M.; Khan, Z.S.; Munir, M.T.; Tarafdar, K.; Rahman, S.M.; Hussain, F. Vitamin D3 decreases glycolysis and invasiveness, and increases cellular stiffness in breast cancer cells. *J. Nutr. Biochem.* **2018**, *53*, 111–120. [CrossRef]

164. Sylvester, P.W.; Akl, M.R.; Malaviya, A.; Parajuli, P.; Ananthula, S.; Tiwari, R.V.; Ayoub, N.M. Potential role of tocotrienols in the treatment and prevention of breast cancer. *Biofactors* **2014**, *40*, 49–58. [CrossRef]

165. Parajuli, P.; Tiwari, R.V.; Sylvester, P.W. Anticancer effects of gamma-tocotrienol are associated with a suppression in aerobic glycolysis. *Biol. Pharm. Bull.* **2015**, *38*, 1352–1360. [CrossRef]

166. Noguchi, C.; Kamitori, K.; Hossain, A.; Hoshikawa, H.; Katagi, A.; Dong, Y.; Sui, L.; Tokuda, M.; Yamaguchi, F. D-Allose Inhibits Cancer Cell Growth by Reducing GLUT1 Expression. *Tohoku J. Exp. Med.* **2016**, *238*, 131–141. [CrossRef]

167. Wu, K.H.; Ho, C.T.; Chen, Z.F.; Chen, L.C.; Whang-Peng, J.; Lin, T.N.; Ho, Y.S. The apple polyphenol phloretin inhibits breast cancer cell migration and proliferation via inhibition of signals by type 2 glucose transporter. *J. Food Drug. Anal.* **2018**, *26*, 221–231. [CrossRef]

168. Calvaresi, E.C.; Hergenrother, P.J. Glucose conjugation for the specific targeting and treatment of cancer. *Chem. Sci.* **2013**, *4*, 2319–2333. [CrossRef]

169. Cao, J.; Cui, S.; Li, S.; Du, C.; Tian, J.; Wan, S.; Qian, Z.; Gu, Y.; Chen, W.R.; Wang, G. Targeted cancer therapy with a 2-deoxyglucose-based adriamycin complex. *Cancer Res.* **2013**, *73*, 1362–1373. [CrossRef]

170. Sztandera, K.; Dzialak, P.; Marcinkowska, M.; Stanczyk, M.; Gorzkiewicz, M.; Janaszewska, A.; Klajnert-Maculewicz, B. Sugar modification enhances cytotoxic activity of PAMAM-Doxorubicin conjugate in glucose-deprived MCF-7 cells—Possible role of GLUT1 transporter. *Pharm. Res.* **2019**, *36*, e140. [CrossRef]

171. Lin, Y.S.; Tungpradit, R.; Sinchaikul, S.; An, F.M.; Liu, D.Z.; Phutrakul, S.; Chen, S.T. Targeting the delivery of glycan-based paclitaxel prodrugs to cancer cells via glucose transporters. *J. Med. Chem.* **2008**, *51*, 7428–7441. [CrossRef] [PubMed]

172. Liu, P.; Lu, Y.; Gao, X.; Liu, R.; Zhang-Negrerie, D.; Shi, Y.; Wang, Y.; Wang, S.; Gao, Q. Highly water-soluble platinum(II) complexes as GLUT substrates for targeted therapy: Improved anticancer efficacy and transporter-mediated cytotoxic properties. *Chem. Commun. Camb.* **2013**, *49*, 2421–2423. [CrossRef] [PubMed]

173. Macintyre, A.N.; Gerriets, V.A.; Nichols, A.G.; Michalek, R.D.; Rudolph, M.C.; Deoliveira, D.; Anderson, S.M.; Abel, E.D.; Chen, B.J.; Hale, L.P.; et al. The glucose transporter Glut1 is selectively essential for CD4 T cell activation and effector function. *Cell Metab.* **2014**, *20*, 61–72. [CrossRef] [PubMed]

174. Fiaschi, T.; Marini, A.; Giannoni, E.; Taddei, M.L.; Gandellini, P.; de Donatis, A.; Lanciotti, M.; Serni, S.; Cirri, P.; Chiarugi, P. Reciprocal metabolic reprogramming through lactate shuttle coordinately influences tumor-stroma interplay. *Cancer Res.* **2012**, *72*, 5130–5140. [CrossRef] [PubMed]

175. Sheng, H.; Tang, W. Glycolysis inhibitors for anticancer therapy: A review of recent patents. *Recent Pat. Anticancer Drug Discov.* **2016**, *11*, 297–308. [CrossRef] [PubMed]

176. Flaig, T.W.; Gustafson, D.L.; Su, L.J.; Zirrolli, J.A.; Crighton, F.; Harrison, G.S.; Pierson, A.S.; Agarwal, R.; Glode, L.M. A phase I and pharmacokinetic study of silybin-phytosome in prostate cancer patients. *Investig. New Drugs* **2007**, *25*, 139–146. [CrossRef]

177. Raez, L.E.; Papadopoulos, K.; Ricart, A.D.; Chiorean, E.G.; Dipaola, R.S.; Stein, M.N.; Lima, C.M.R.; Schlesselman, J.J.; Tolba, K.; Langmuir, V.K.; et al. A phase I dose-escalation trial of 2-deoxy-D-glucose alone or combined with docetaxel in patients with advanced solid tumors. *Cancer Chemother. Pharmacol.* **2013**, *71*, 523–530. [CrossRef]

178. Jose, C.; Bellance, N.; Rossignol, R. Choosing between glycolysis and oxidative phosphorylation: A tumor's dilemma? *Biochim. Biophys. Acta* **2011**, *1807*, 552–561. [CrossRef]

179. Lee, Y.J.; Lee, G.J.; Yi, S.S.; Heo, S.H.; Park, C.R.; Nam, H.S.; Cho, M.K.; Lee, S.H. Cisplatin and resveratrol induce apoptosis and autophagy following oxidative stress in malignant mesothelioma cells. *Food Chem. Toxicol.* **2016**, *97*, 96–107. [CrossRef]

180. Shi, Y.; Liu, S.; Ahmad, S.; Gao, Q. Targeting key transporters in tumor glycolysis as a novel anticancer strategy. *Curr. Top. Med. Chem.* **2018**, *18*, 454–466. [CrossRef]

181. Garcia-Aranda, M.; Redondo, M. Immunotherapy: A challenge of breast cancer treatment. *Cancers* **2019**, *11*, 1822. [CrossRef] [PubMed]

182. Mochel, F.; Hainque, E.; Gras, D.; Adanyeguh, I.M.; Caillet, S.; Heron, B.; Roubertie, A.; Kaphan, E.; Valabregue, R.; Rinaldi, D.; et al. Triheptanoin dramatically reduces paroxysmal motor disorder in patients with GLUT1 deficiency. *J. Neurol. Neurosurg. Psychiatry* **2016**, *87*, 550–553. [CrossRef] [PubMed]

Review

Honokiol: A Review of Its Anticancer Potential and Mechanisms

Chon Phin Ong, Wai Leong Lee, Yin Quan Tang * and Wei Hsum Yap *

School of Biosciences, Faculty of Health and Medical Sciences, Taylor's University Lakeside Campus, No. 1, Jalan Taylor's, Subang Jaya 47500, Malaysia; chonphin96@gmail.com (C.P.O.); leewl0108@gmail.com (W.L.L.)
* Correspondence: yinquan.tang@taylors.edu.my (Y.Q.T.); weihsum.yap@taylors.edu.my (W.H.Y.)

Received: 30 November 2019; Accepted: 19 December 2019; Published: 22 December 2019

Abstract: Cancer is characterised by uncontrolled cell division and abnormal cell growth, which is largely caused by a variety of gene mutations. There are continuous efforts being made to develop effective cancer treatments as resistance to current anticancer drugs has been on the rise. Natural products represent a promising source in the search for anticancer treatments as they possess unique chemical structures and combinations of compounds that may be effective against cancer with a minimal toxicity profile or few side effects compared to standard anticancer therapy. Extensive research on natural products has shown that bioactive natural compounds target multiple cellular processes and pathways involved in cancer progression. In this review, we discuss honokiol, a plant bioactive compound that originates mainly from the *Magnolia* species. Various studies have proven that honokiol exerts broad-range anticancer activity in vitro and in vivo by regulating numerous signalling pathways. These include induction of G0/G1 and G2/M cell cycle arrest (via the regulation of cyclin-dependent kinase (CDK) and cyclin proteins), epithelial–mesenchymal transition inhibition via the downregulation of mesenchymal markers and upregulation of epithelial markers. Additionally, honokiol possesses the capability to supress cell migration and invasion via the downregulation of several matrix-metalloproteinases (activation of 5′ AMP-activated protein kinase (AMPK) and KISS1/KISS1R signalling), inhibiting cell migration, invasion, and metastasis, as well as inducing anti-angiogenesis activity (via the down-regulation of vascular endothelial growth factor (VEGFR) and vascular endothelial growth factor (VEGF)). Combining these studies provides significant insights for the potential of honokiol to be a promising candidate natural compound for chemoprevention and treatment.

Keywords: honokiol; anticancer; mechanism; signalling pathway

---

1. Introduction

Cancer is the outcome of rampant cell division which is associated with cell cycle disorganisation [1], leading to uncontrolled cell proliferation. In addition, it also involves the dysregulation of apoptosis, immune evasion, inflammatory responses, and ultimately, metastatic spread [2]. Over the last few decades, our progressive understanding of the aetiology of cancer together with advancement of cancer treatment, detection, and prevention, have contributed towards receding cancer mortality around the world [3]. However, more than half of cancer cases were diagnosed at a later stage of cancer progression [4]. According to a study by Bray et al. [5], the worldwide estimated number of new cancer cases for the year 2018 was 18.1 million in both sexes and across all ages. Amongst all the cancer types, lung, breast, and colorectum have topped the charts with approximately 2.1 million, 2.1 million, and 1.8 million cases, respectively. On the other hand, the estimated number of deaths was approximately 9.6 million. Asia accounted for more than half of the cancer deaths (57.3%), followed by Europe (20.3%), and America (14.4%). Lung cancer has caused the highest number of deaths due to substandard prognoses. Attempts to develop the effective prevention of cancer may diminish the incidence rate for some cancers, for instance lung cancer in North America and Northern Europe. These western countries have implemented tobacco control in order to avert involuntary exposure to tobacco and minimise active smoking within the community. Unfortunately, a majority of the population are still facing an upsurge of cancer diagnosis, demanding treatment and care [5].

The common treatment regimens for cancer patients include surgery, chemotherapy, and radiotherapy [6]. Although some of these regimens represent the first-in-line options for cancer treatment, the lack of selectivity towards neoplastic cells and the development of drug toxicity has caused these therapeutic effects to recede slowly, rendering it ineffective over the years [7]. Additionally, multidrug resistance tumours pose a severe threat and have been responsible for numerous cancer-related deaths [8]. A modern approach to target multiple cell regulating pathways is mandatory in order to provide highly efficient and targeted cancer therapy. For instance, combination therapy that targets different pathways exhibit significantly lower toxicity towards normal cells compared to mono-therapy [9]. Currently, the development of anticancer drugs possessing the capability to overcome common mechanisms of chemoresistance with minimal toxicity effects would be considered a breakthrough in cancer research [2].

Approximately 70–95% of the world population continues to use traditional medicinal herbs, plants, and fruits which contain valuable bioactive compounds with therapeutic effects to maintain health, as well as to prevent or treat physical and mental illnesses [10]. These biologically active compounds provide extensive opportunities in uncovering competent anticancer agents [2,11]. A majority of the anticancer drugs that are currently in use originate from plants, marine organisms, and microorganisms, such as the well-known plant-derived anti-cancer drugs Paclitaxel (Taxol®) and Camptothecin (CPT) [12].

The *Magnolia* genus is widely distributed throughout the world, especially in East and South-East Asia [13]. Among the *Magnolia* species, *Magnolia officinalis* and *Magnolia obovata* are commonly used in traditional Chinese (known as "Houpu") and Japanese herbal medicine [13,14]. The traditional prescriptions named Hange-koboku-to and Sai-boku-to, which contain the *Magnolia* bark, are still used in modern clinical practice in Japan [15]. There are several potent bioactive compounds in the *Magnolia* species have been identified including honokiol, magnolol, obovatol, 4-*O*-methylhonokiol, and several other neolignan compounds [13,15,16]. This paper highlights the potential anticancer effect of a simple biphenyl neolignan found in this *Magnolia* family, namely honokiol.

Honokiol was traditionally used for anxiety and stroke treatment, as well as the alleviation of flu symptoms [14]. In recent studies, this natural product displayed diverse biological activities, including anti-arrhythmic, anti-inflammatory, anti-oxidative, anti-depressant, anti-thrombocytic, and anxiolytic activities [13,14,16]. Furthermore, it was also shown to exert potent broad-spectrum anti-fungal, antimicrobial, and anti-human immunodeficiency virus (HIV) activities [13]. Due to its ability to cross the blood–brain barrier, honokiol has been deemed beneficial towards neuronal protection through various mechanism, such as the preservation of Na^+/K^+ ATPase, phosphorylation of pro-survival factors, preservation of mitochondria, prevention of glucose, reactive oxgen species (ROS), and inflammatory mediated damage [17]. Hence, honokiol was described as a promiscuous rather than selective agent due to its known pharmacologic effects. Recent studies have been focused on the anti-cancer properties of honokiol, emphasising its tremendous potential as an anticancer agent. In this review, we summarise the anti-cancer properties of honokiol, together with its mechanism of action, based on in vitro and in vivo experimental evidence. In addition, we also summarize the current data on its pharmacological relevance and potential delivery routes for future applications in cancer prevention and treatment.

2. Research Methodology

A systematic search was performed to identify all relevant research papers published on the use of honokiol as a potent anticancer treatment using PubMed (1994–present) and Web of

Sciences (1994–present). The search strategy was performed using several keywords to track down the relevant research articles including 'honokiol', 'cancer', 'cancer statistics', 'structural', 'metabolites', 'mechanism', 'cell death', 'apoptosis', 'anti-inflammatory', 'anti-tumour', 'antioxidant', 'cell proliferation', 'cytotoxicity', 'cell cycle arrest', 'metastasis', 'tumour', 'angiogenesis', 'absorption', 'metabolism', 'toxicity', 'distribution', 'elimination', 'solubility', 'nanoparticles', and 'delivery'.

3. Structure Activity Relationship and Its Derivatives

Honokiol bioactive compounds are easily found in the root and stem bark of the *Magnolia* species, although some studies have also found them in seed cones [13,18]. Due to the structural resemblance of both honokiol and magnolol in the *Magnolia* bark, the extraction of pure honokiol and magnolol cannot be achieved using conventional column chromatography nor thin-layer chromatography. Eventually, their purification process requires a costly alternative like electromigration [16]. The only difference between honokiol and magnolol in terms of structure is only in the position of the hydroxyl group, as shown in Figure 1. In 2007, Chen et al. developed a rapid separation technique using high-capacity high-speed counter-current chromatography (HSCCC) to isolate and purify honokiol and magnolol from crude extracts of *Magnolia* plants. Within 20 min, the resulting fraction has a purity of 98.6% honokiol, indicating that this method exhibited substantial efficiency in honokiol extraction [19]. Two years later, another team of researchers formulated a time-effective synthetic method while providing higher yielding honokiol using Suzuki-Miyaura coupling and Claisen rearrangement as key steps of the synthetic pathway of honokiol. The five steps of the honokiol synthesis pathway includes bromination, Suzuki coupling, allylation, one-pot Claisen's rearrangement, and demethylation, eventually resulting in a 32% overall yield [20]. The emergence of the synthetic method for honokiol has alleviated the risk of extinction of the *Magnolia* species.

Natural bioactive compounds often serve as lead templates and are subjected to structural modification to improve pharmacological activity, physiochemical properties, along with pharmacokinetics, to generate clinically useful structures [21]. According to Anand et al. [22], a comprehensive study of the natural and synthetic analogues of a drug molecule is crucial to determining its fundamental pharmacophores. As seen in Figure 1, honokiol contains two phenyl rings substituted with hydroxyl and allyl groups. In a study conducted by Bohmdorfer et al. [23], it was found that the predominant metabolic pathways of honokiol in the human liver was through sulfation and glucuronidation (Phase II metabolism) of the free hydroxyl groups, inducing rapid excretion and shortening its half-life [23]. Moreover, Lin et al. [24] have hypothesised that the hydroxyl groups on the biphenyl skeleton of honokiol could be subjected to metabolic oxidation by Phase I enzymes, thus diminishing its efficacy.

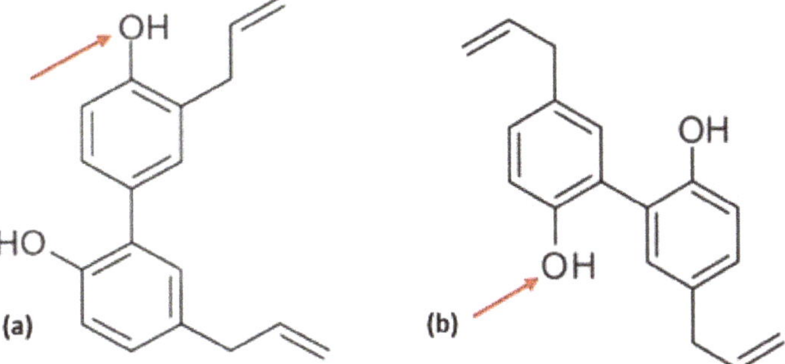

Figure 1. (**a**): The structure of honokiol [24]; (**b**): The structure of magnolol [25]. Arrow indicates the difference in the position of hydroxyl group between honokiol and magnolol.

Through the alteration of the top and bottom rings by changing the substitution pattern at its bottom ring and replacing the hydroxyl group in the top ring with a methoxy group, six different analogues were produced, as shown in Figure 2. A structure–activity relationship (SAR) study was conducted and it was found that replacing the hydroxyl group in the top ring of honokiol with a methoxy group greatly improved its cytotoxicity against lung, melanoma, and colon cancer cells. The two hydroxyl group substituted analogues (3′-Bromo-3,5′-dy-allyl-2′hydroxyl-4-methoxy-1,1′-biphenyl and 3,3′-Diallyl-4-methoxy-4′-hydroxy-1,1′-biphenyl) have induced G0/G1 phase cell cycle arrest and a swift decrement in Cdk1 and cyclin B1 protein levels, similarly to the parental honokiol compound [24]. Overall, obstruction of the potential oxidation of the phenolic hydroxyl group in the biphenyl group skeleton of honokiol improved its anti-cancer effect.

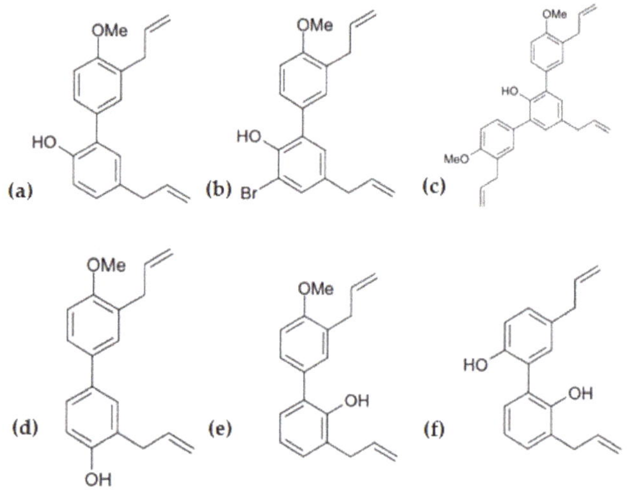

Figure 2. The structure of honokiol analogues. (**a**): 3,5′-Diallyl-2′-hydroxyl-4-methoxy-1,1′-biphenyl; (**b**): 3′-Bromo-3,5′-di-allyl-2′-hydroxyl-4-methoxy-1,1′-biphenyl; (**c**): 2,6-Di-(4′-methoxy-3′-allylphenyl)-1-phenol; (**d**): 3,3′-Diallyl-4-methoxy-4′-hydroxy-1,1′-biphenyl; (**e**): 3,3′Diallyl-2′-hydroxyl-4-methoxy-1,1′-biphenyl; (**f**): 3′,5-Diallyl-2,2′-di-hydroxy-1,1′-biphenyl [24].

4. Anticancer Properties of Honokiol

4.1. In Vitro Studies

Honokiol has been shown to exhibit antiproliferation effects against numerous cancer cells, including bone, bladder, brain, breast, blood, and colon, as shown in Table 1. Generally, the concentrations used for the in vitro studies are between 0–150 µM, which majority of these concentration ranges have been shown to significantly inhibit cell proliferation or cell viability of various cancer cell lines. The trend for the IC_{50} values of numerous cancer cell lines were time-dependent, whereby the IC_{50} values decreases as duration of the experiment increases. As seen in Table 1, human blood cancer Raji cells were highly susceptible to honokiol treatment (IC_{50} = 0.092) compared to highly resistant human nasopharyngeal cancer HNE-1 cells (IC_{50} = 144.71 µM). Interestingly, honokiol has been shown to exhibit minimal cytotoxicity against on normal cell lines, including human fibroblast FB-1, FB-2, Hs68, and NIH-3T3 cells [25–28]. The low cytotoxicity of honokiol treatment against normal cell lines should be emphasised as current chemotherapeutic regimens have a considerable amount of side effects that harm cancer patients.

Many chemotherapeutic agents have been shown to induce severe systemic toxicity and several side effects due to their deficient pharmacokinetic profiles and non-specific distribution in the body [29]. In Yang et al.'s study [30], they have encapsulated honokiol into nanopolymers to enhance its permeability and specificity against cancer cells. They utilised the active targeting nanoparticles-loaded honokiol (ANTH) in their in vitro studies against human nasopharyngeal cancer HNE-1 cells, and this incorporation exhibited significantly lower IC_{50} values compared to free honokiol treatment. As a result, the incorporation or encapsulation of honokiol in transporting vehicles can improve the anticancer effects and concurrently overcome the water solubility issue of honokiol itself. This has shown to be a promising regimen for anticancer treatment in the future.

Furthermore, it is worthy to note that honokiol can enhance the antineoplastic effects of several chemotherapeutic agents when cells are treated in combination treatment of both honokiol and the chemotherapeutic agent. In Wang et al.'s study [31], they have shown that honokiol has enhanced the in vitro cytotoxicity of paclitaxel against human cervix cancer cell lines. The combination treatment has resulted in approximately 10–60% increase of apoptotic cells and inhibition of cell viability when compared to honokiol treatment alone [31]. In another study, honokiol potentiated the apoptotic effect of both doxorubicin and paclitaxel against human liver cancer HepG2 cells. Honokiol enhanced the apoptotic effects of paclitaxel and doxorubicin by 22% and 24% respectively [32].

Table 1. The anticancer effects of honokiol against cancer cells in in vitro experiments.

	Cell Lines	Mechanism of Action	Concentration Used	Efficacy/IC$_{50}$ (Exposure Time)	References
Colorectal cancer	RKO	Inhibit cell proliferation Induce G1 phase cell cycle arrest Induce apoptosis ↓ Bcl-xL; ↑ Caspase-3 & caspase-9	0–150 μM	46.76 μM (68 h)	[33]
	HCT116, HCT116-CH2, HCT116-CH3	Inhibit cell proliferation Induce G0/G1 & G2/M phase cell cycle arrest: ↓ cyclin D1 & A1; ↑ p53 phosphorylation Induce apoptosis: ↓ Caspase-3; ↓ Bcl-2; ↑ Bax protein	25 μM Honokiol with 2.5 or 5.0 Gy IR	N/A	[34]
	HT-29	Inhibit cell growth & proliferation Induce G1 phase cell cycle arrest: ↓ Cdk1 & cyclin B1	0–50 μM followed by 0–5 Gy IR	23.05 μM (24 h) 13.24 μM (72 h)	[24]
	HCT116 & SW480	Inhibit cell proliferation via Inhibition of Notch signalling: ↓ Notch1 & Jagged-1; ↓ Hey-1 & Hes1; ↓ γ-secretase complex; ↓ Skip1 Induce apoptosis: ↑ caspase-3/-7 activity; ↓ Bcl-2 & Bcl-xL; ↑ Bax protein; ↓ cyclin D1 & c-Myc; ↑ p21^{WAF1} protein Inhibit primary and secondary colonosphere formation	0–50 μM	N/A	[35]
	RKO & HCT116	Inhibit cell viability Induce apoptosis: ↑ caspase-3, caspase-8 & caspase-9 activation; ↑ DR5 & cleaved PARP proteins; ↑ survivin protein; ↑ phosphorylated p53 & p53 proteins; ↓ PUMA protein	0–60 μM	RKO: 38.25 μM (24 h) HCT116: 39.64 μM (24 h)	[36]
Blood cancer	B-CLL	Inhibit cell viability Induce apoptosis: ↑ caspase-3 activity; ↑ caspase-8 & caspase-9 activation; ↓ caspase-9; ↑ Bax protein; ↓ Mcl-1 protein	0–100 μM	49 μM (6 h) 38 μM (24 h)	[37]
	Raji, Molt-4	Inhibit cell growth: ↓ p65; ↓ NF-κB Induce apoptosis: ↑ JNK activation Increase ROS activity: ↑ Nrf2 & c-Jun protein activation	0–2.5 μM	Raji: 3.500 μM (24 h) 0.092 μM (72 h) Molt-4: 0.521 μM (24 h)	[38]
Breast cancer	MCF-7, MDA-MB-231, SKBR-3, ZR-75-1, BT-474	Inhibit cell viability and growth: ↓ EFGR; ↓ MAPK/PI3K pathway activity Induce apoptosis: ↑ PARP protein degradation; ↓ caspase-8; ↑ Bax proteins Induce G1 phase cell cycle arrest: ↓ cyclin D1; ↑ p21 & p27	0–100 μM	MCF-7: 40 μM (24 h) MDA-MB-231: 33 μM (24 h) SKBR-3: 29 μM (24 h) ZR-75-1: 39 μM (24 h) BT-474: 50 μM (24 h)	[39]
	MCF-7, MDA-MB-231	Inhibit cell clonogenicity Inhibit cell anchorage-dependent colony formation Inhibit cell growth, migration & invasion: ↓ pS6K & 4EBP1 phosphorylation; ↑ AMPK activation; ↓ mTORC1 function; ↑ LKB1 & cytosolic localisation	1–25 μM	N/A	[40]
	MCF-7, MDA-MB-231, SUM149, SUM159	Inhibit cell migration & invasion: ↑ AMPK phosphorylation; ↑ LKB1 Inhibit stem-like characteristics: ↓ Oct4, Nanog & Sox4 protein; ↓ STAT3; ↓ iPSC inducer mRNA	5 μM	N/A	[41]

Table 1. *Cont.*

Cell Lines	Mechanism of Action	Concentration Used	Efficacy/IC$_{50}$ (Exposure Time)	References
MCF-7, MDA-MB-231, T47D, SKBR-3, Zr-75, BT-474	Inhibit cell growth: ↓ PI3K/Akt/mTOR signalling Inhibit cell invasion Induce G0/G1 phase cell cycle arrest: ↓ cyclin D1 & cyclin E; ↓ Cdk2 & c-myc; ↑ PTEN Induce apoptosis: ↑ caspase-3, caspase-6 & caspase-9 activation	0–40 μM	MCF7: 34.9 μM (24 h) 13.7 μM (48 h) 13.5 μM (72 h) 10.5 μM (96 h) MDA-MB-231: 56.9 μM (24 h) 44.4 μM (48 h) 16.0 μM (72 h) 12.0 μM (96 h) T47D: 47.7 μM (24 h) 41.6 μM (48 h) 17.6 μM (72 h) 7.1 μM (96 h) SKBR-3: 76.1 μM (24 h) 68.1 μM (48 h) 62.7 μM (72 h) 15.7 μM (96 h) ZR-75: 71.1 μM (24 h) 58.1 μM (48 h) 28.7 μM (72 h) 14.5 μM (96 h) BT-474: 80.2 μM (24 h) 65.6 μM (48 h) 39.5 μM (72 h) 15.1 μM (96 h)	[42]
MDA-MB-231	Inhibit cell proliferation: ↓ c-Src/EGFR-mediated signalling pathway; ↓ c-Myc protein Induce G0/G1 phase cell cycle arrest: ↓ cyclin A, cyclin D1 & cyclin E; ↓ Cdk2, Cdk4 & p-pRb^{Ser780}; ↑ p27^{Kip-1} Induce apoptosis: ↑ caspase-3, caspase-8 & caspase-9 cascade; ↓ Bcl-2 & Bid protein; ↑ PARP cleavage	0–100 μM	59.5 μM (72 h)	[43]

Table 1. *Cont.*

	Cell Lines	Mechanism of Action	Concentration Used	Efficacy/IC$_{50}$ (Exposure Time)	References
Lung cancer	A549	Inhibit cell growth & proliferation Induce G0/G1 phase cell cycle arrest: ↓ Cdk1 & cyclin B1	0–50 μM	12.51 μM (24 h) 7.75 μM (72 h)	[24]
	A549, H460, H226, H1299	Reduce invasive potential Inhibit PGE$_2$-induced cell migration: ↓ PGE$_2$ production ↓ COX-2 ↑ β-catenin degradation ↓ NF-κB/p65 activity ↓ IKKα	0–20 μM	N/A	[44]
	A549, H1299	Inhibit cell viability and growth: ↓ class I HDAC proteins; ↓ HDAC activity; ↑ histone acetyltransferase (HAT) activity; ↑ histone H3 & H4 Induce G1 phase cell cycle arrest: ↓ cyclin D1 & cyclin D2; ↓ Cdk2, Cdk4 & Cdk6	0–60 μM	N/A	[45]
	H460 & A549	Inhibit cell proliferation Induce apoptosis: ↑ cathepsin D; ↑ cleaved PARP; ↑ caspase-3 Inhibit autophagy: ↑ p62; ↑ LC3-II	0–60 μM	H460: ~30 μM (48 h) A549: ~40 μM (48 h)	[46]
	Pc9-BrM3 & H2030-BrM3 (brain metastatic)	Inhibit cell proliferation and cell invasion: ↓ STAT3 protein phosphorylation; ↓ STAT-3 mediated mitochondrial respiratory function	0–50 μM	PC9-BrM3: 28.4 μM (48 h) H2030-BrM3: 25.7 μM (48 h)	[47]
	H23, A549 & HCC827	Inhibit cell growth Induce G1 phase cell cycle arrest: ↓EGFR; ↓ class I HDAC; ↓ class IIb HDAC6 activity; ↑ Hsp90 acetylation & EGFR degradation	0–40 μM	A549: 23.55 μM (24h)	[48]
	H460, A549, H358	Inhibit cell growth: ↓ c-RAF, ERK & AKT phosphorylation Inhibit colony formation capacity Induce apoptosis: ↑ Bax protein; ↓ Bcl-2 protein; ↑ PARP cleavage Induce G1 phase cell cycle arrest: ↓ cyclin D1; ↑ p21 & p27; ↓ P70S6k kinase activity Induce autophagy: ↑ LC3-I conversion to LC3-II; ↑ Sirt3 mRNA & protein; ↓ Hif-1α protein	0–80 μM	H460: 30.42 μM (72 h) A549: 50.58 μM (72 h) H358: 59.38 μM (72 h)	[49]
	A549 & 95-D	Inhibit cell viability Induce apoptosis: ↑ ER stress signalling pathway activation; ↑ GRP78, phosphorylation PERK & phosphorylated IRE1α; ↑ cleaved caspase-9 & CHOP; ↓ Bcl-2 protein; ↑ Bax, caspase-3 & caspase-9 Inhibit cell migration	0–60 μM	N/A	[50]
	CH27, H460 & H1299	Inhibit cell growth Induce apoptosis: ↓ Bcl-XL; ↑ mitochondrial cytochrome c release; ↑ BAD protein; ↑ caspase-1, caspase-2, caspase-3, caspase-6, caspase-8 & caspase-9 activity; ↑ PARP cleavage	0–100 μM	CH27: 40.9 μM (24 h) H460: 41.4 μM (24 h) H1299: 34.7 μM (24 h)	[25]
	MSTO-211H	Inhibit cell viability Induce apoptosis: ↑ PARP cleavage; ↑ caspase-3 activation; ↓ Bid & Bcl-xL protein; ↑ Bax protein; ↓ Mcl-1 & survivin protein; ↓ Sp1 Induce G1 phase cell cycle arrest: ↓ cyclin D1	0–22.5 μM	N/A	[51]

Table 1. *Cont.*

	Cell Lines	Mechanism of Action	Concentration Used	Efficacy/IC$_{50}$ (Exposure Time)	References
Skin cancer	SK-MEL2 & MeWo	Inhibit cell growth & cell proliferation Induce apoptosis via DNA degradation Induce cell death via mitochondrial depolarization	0–100 μM	N/A	[52]
	A431	Inhibit cell viability & proliferation Induce G0/G1 phase cell cycle arrest: ↓ cyclin A, cyclin D1, cyclin D2 & cyclin E; ↓ Cdk2, Cdk4 & Cdk6; ↑ p21 & p27 Induce cell apoptosis: ↑ PARP	0–75 μM	N/A	[53]
	B16-F10	Inhibit cell proliferation Induce cell death: ↑ Autophagosome (vacuoles) formation; ↓ cyclin D1; ↓ AKT/mTOR & Notch signalling	0–50 μM	N/A	[54]
	B16/F-10 & SKMEL-28	Inhibit cell proliferation & viability: ↓ Notch signalling; ↓ TACE & γ-secretase complex proteins Inhibit clonogenicity Induce G0/G1 phase cell cycle arrest Induce autophagy: ↑ autophagosome formation; ↑ LC3B cleavage Inhibit cell stemness: ↓ CD271, CD166, Jarid1B & ABCB5	0–60 μM	N/A	[55]
	UACC903	Inhibit cell growth & proliferation	0–50 μM	7.45 μM (24 h) 5.10 μM (72 h)	[24]
	SKMEL-2	Inhibit cell proliferation & viability Induce apoptotic death: ↑ caspase-3, caspase-6, caspase-8 & caspase-9; ↑ PARP cleavage; ↓ procaspase-3, procaspase-8 & procaspase-9 Induce G2/M phase cell cycle arrest: ↓ cyclin B1, cyc in D1, cyclin D2 & PCNA; ↓ Cdk2 & Cdk4; ↑ p21 & p53	0–100 μM	N/A	[56]
	UACC-62	Inhibit cell proliferation & viability Induce apoptotic death: ↑ caspase-3, caspase-6, caspase-8 & caspase-9; ↑ cleaved PARP; ↓ procaspase-3, procaspase-8 & procaspase-9 Induce G0/G1 phase cell cycle arrest: ↓ cyclin B1, cyclin D1 & cyclin D2; ↓ Cdk2, Cdk4 & Cdc2p34; ↓ p21 & p27	0–100 μM	N/A	[56]
Renal cancer	A498	Inhibit cell proliferation Inhibit colony formation capability Inhibit cell migration and invasion: ↓ Epithelial-mesenchymal transition (EMT); ↓ cancer stem cells (CSC) properties; ↑ miR-141; ↓ ZEB2 Inhibit tumoursphere formation	0–80 μM	~12 μM (72 h)	[57]
Cervix cancer	KB-3-1, KB-8-5, KB-C1, KB-V1	Inhibit cell viability: ↓ EGFR-STAT3 signalling Induce mitochondria-dependent & death receptor-dependent apoptosis: ↓ Bcl-2, Mcl-1 & survivin; ↑ PARP & caspase-3 cleavage; ↑ mitochondrial release of cytochrome c; ↑ DR5 Enhances in vitro cytotoxicity of Paclitaxel	0–75 μM	KB-3-1: 12.56 μM (72 h) KB-8-5: 12.08 μM (72 h) KB-C1: 11.40 μM (72 h) KB-V1: 10.39 μM (72 h)	[31]

Table 1. *Cont.*

	Cell Lines	Mechanism of Action	Concentration Used	Efficacy/IC$_{50}$ (Exposure Time)	References
Pancreatic cancer	MiaPaCa & Colo-357	Suppress plating efficiency of cells Reduce anchorage-independent clonogenicity growth Suppress migration and invasion ability	0–5 µM	N/A	[58]
	MiaPaCa & Panc1	Inhibit cell growth Induce G1 phase cell cycle arrest: ↓ cyclin D1 & cyclin E; ↓ Cdk2 & Cdk4; ↑ p21 & p27 Induce apoptosis: ↓ Bcl-2 & Bcl-xL proteins; ↑ Bax protein; ↓ IKB-α phosphorylation; ↓ NF-κB constitutive activation	0–60 µM	MiaPaCa: 43.25 µM (24 h) 31.08 µM (48 h) 18.54 µM (72 h) Panc1: 47.44 µM (24 h) 34.17 µM (48 h) 21.86 µM (72 h)	[59]
Thyroid cancer	ARO, WRO	Inhibit cell growth & proliferation: ↓ ERK, JNK & p37 activation and expression; ↓ mTOR & p70S6K Inhibit colony formation Induce apoptosis: ↑ PARP cleavage; ↑ caspase-3, caspase-8 & PARP activation; ↓ PI3K/AKT & MAPK pathways Induce G0/G1 cell cycle arrest: ↓ cyclin D1; ↓ Cdk2 & Cdk4; ↑ p21 & p27 Induce autophagy & autophagy flux: ↑ LC3-II	ARO & WRO: 0–60 µM SW579: 0–40 µM	ARO: 36.3 µM (24 h) 40.1 µM (48 h) 44.8 µM (72 h) WRO: 37.7 µM (24 h) 31.8 µM (48 h) 30.7 µM (72 h) SW579: 19.9 µM (24 h) 10.5 µM (48 h) 8.8 µM (72 h)	[60]
Nasopharyngeal cancer	HNE-1	Inhibit cell growth Induce apoptosis Induce G1 phase cell cycle arrest	0–150 µM (Honokiol & ATNH—Active targeting nanoparticles-loaded honokiol)	Honokiol: 144.71 µM (24 h) ATNH: 69.04 µM (24 h)	[30]
Brain cancer	U251	Inhibit cell growth Inhibit cell proliferation Induce apoptosis	0–120 µM	61.43 µM (24 h)	[61]
	T98G	Inhibit cell viability Inhibit cell invasion Induce cell apoptosis: ↑ Bax protein; ↓ Bcl-2; ↑ Bax/Bcl-2 ratio	0–50 µM	N/A	[62]
	GBM8401 (Parental) & GBM8401 SP	Inhibit cell proliferation & viability Induce sub-G1 phase cell cycle arrest Induce apoptosis: ↓ Notch3/Hes1 pathway	0–20 µM	GBM8401 (Parental): 5.30 µM (48 h) GBM8401 SP: 11.20 µM (48 h)	[36]
	U251 & U-87 MG	Inhibit cell viability & proliferation: ↓ PI3K/Akt & MAPK/Erk signalling pathways Inhibit cell invasion & migration: ↓ MMP2 & MMP9; ↓ NF-κB-mediated E-cadherin pathway Inhibit colony formation Induce apoptosis: ↓ Bcl-2, p-AKT & p-ERK; ↑ Bax protein; ↑ caspase-3 cleavage; ↓ EGFR-STAT3 signalling Reduce spheroid formation: ↓ CD133 & Nestin protein	0–60 µM	U251: 54.00 µM (24 h) U-87 MG: 62.50 µM (24 h)	[63]

Table 1. *Cont.*

	Cell Lines	Mechanism of Action	Concentration Used	Efficacy/IC$_{50}$ (Exposure Time)	References
	DBTRG-05MG	Inhibit cell growth Induce apoptosis: ↓ Rb protein; ↑ PARP & Bcl-x(S/L) cleavage Induce autophagy: ↑ Beclin-1 & LC3-II	0–50 μM	~30 μM	[64]
	U87 MG (Human) BMEC (Mouse)	Inhibit cell viability Inhibit epithelial-mesenchymal transition (EMT): ↓ Snail, β-catenin & N-cadherin; ↑ E-cadherin Inhibit cell adhesion & invasion: ↓ VCAM-1; ↓ phosphor-VE-cadherin-mediated BMEC permeability	0–20 μM	U87MG: 22.66 μM (24 h) BMEC: 13.09 μM (24 h)	[65]
	U87 MG	Inhibit cell viability Induce G1 phase cell cycle arrest: ↑ p21 & p53; ↓ cyclin D1; ↓ Cdk4 & Cdk6; ↓ p-Rb protein; ↓ E2F1 Induce apoptosis: ↓ procaspase-3; ↑ caspase-8 & caspase-9 activity	0–100 μM	52.70 μM	[66]
	HOS & U20S	Inhibit cell proliferation Inhibit colony formation Induce G0/G1 phase cell cycle arrest: ↓ cyclin D1 & cyclin E; ↓ Cdk4 Induce mitochondria-mediated apoptosis: ↑ caspase-3 & caspase-9 activation; ↑ PARP cleavage; ↓ Bcl-2, Bcl-xL & survivin; ↑ ERK activation; ↓ proteasome activity; ↑ ER stress and subsequent ROS overgeneration; ↑ GRP78 Induce autophagy: ↑ Atg7 protein activation; ↑ Atg5; ↑ LC3B-II	0–30 μM	HOS: 17.70 μM (24 h) U20S: 21.50 μM (24 h)	[67]
Bone cancer	SAOS-2, HOS, 143B, MG-63 M8, HU09, HU09 M132 Dunn, LM5, LM8 & LM8-LacZ (Mouse)	Inhibit cell metabolic activity Inhibit cell proliferation Inhibit cell migration Induce rapid cell death via Honokiol-provoked vacuolation	0–150 μM	(72 h) SAOS-2: 48.38 μM HOS: 51.38 μM 143B: 41.63 μM MG-63M8: 34.88 μM HU09: 59.25 μM HU09M132: 31.88 μM (72 h) Dunn: 36.00 μM LM5: 30.00 μM LM8: 31.13 μM	[68]
	Saos-2 & MG-63	Inhibit cell viability Induce apoptosis: ↑ caspase-3 & PARP cleavage; ↑ Bax protein; ↓ Bcl-2; ↓ PI3K/AKT signalling pathway; ↓ miR-21	0–100 μM	Saos-2: 37.85 μM (24 h) MG-63: 38.24 μM (24h)	[69]

Table 1. *Cont.*

	Cell Lines	Mechanism of Action	Concentration Used	Efficacy/IC$_{50}$ (Exposure Time)	References
Oral cancer	OC2 & OCSL	Inhibit cell growth Induce G0/G1 phase cell cycle arrest: ↑ cyclin E accumulation; ↑ p21 & p27; ↓ cyclin D1, ↓ Cdk2 & Cdk4 Induce apoptosis: ↓ caspase-8 & caspase-9; ↑ caspase-3 cleavage; ↓ Bid protein Induce autophagy and autophagic flux: ↑ LC3–II; ↓ Akt/mTORC1 pathway; ↑ AMPK signalling pathway; ↑ p62	0–60 µM	OC2: 35.00 µM (24 h) 22.00 µM (48 h) OCSL: 33 µM (24 h) 13 µM (48 h)	[26]
	HN-22 & HSC-4	Inhibit cell viability Induce apoptosis: ↓ Sp1 protein; ↑ p21 & p27; ↑ PARP & caspase-3 activation; ↓ Mcl-1 & survivin protein Induce G1 phase cell cycle arrest: ↓ cyclin D1	0–37.5 µM	HN-22: 26.63 µM (48 h) HSC-4: 30.00 µM (48 h)	[70]
Liver cancer	HepG2	Inhibit cell growth & proliferation: ↓ β-catenin protein Induce apoptosis: ↑ BAD protein; ↓ Bcl-2 protein Upregulation of BAD protein expression Downregulation of Bcl-2 protein level	0–2 µM	N/A	[71]
	SMMC-7721	Inhibit cell growth Induce G0/G1 phase cell cycle arrest Induce apoptosis: ↓ mitochondrial potential; ↑ ROS production; ↓ Bcl-2 protein; ↑ Bax protein	0–37.5 µM	N/A	[72]
	HepG2, HUH7, PLC/PRF5, Hep3B	Inhibit cell proliferation: ↓ STAT3 activation; ↓ IL-induced Akt phosphorylation; ↓ c-Src activation; ↓ JAK1 & JAK2; ↑ SHP-1 protein Induce sub-G1 phase cell cycle arrest: ↓ cyclin D1 Downregulation of cyclin D1 level Induce apoptosis: ↓ Bcl-2 & Bcl-xL; ↓ survivin & Mcl-1 protein; ↑ caspase-3 activation; ↑ PARP cleavage Enhance apoptotic effect of doxorubicin & paclitaxel	0–100 µM	N/A	[32]
Ovarian cancer	A2780s & A2780cp	Inhibit cell growth Induce apoptosis	0–100 µM	A2780s: 36.00 µM (48 h) A2780cp: 34.70 µM (48 h)	[73]
	SKOV3 & Caov-3	Inhibit cell proliferation and growth Inhibit colony formation Induce apoptosis: ↑ AMPK pathway activation; ↑ caspase-3, caspase-7 & caspase-9 activation; ↑ PARP cleavage Induce G0/G1 phase cell cycle arrest Inhibit cell migration and invasion	0–100 µM	SKOV: 48.71 µM (24 h) Caov-3: 46.42 µM (24 h)	[28]
	SKOV3, COC1, Angelen & A2780	Inhibit cell proliferation Induce cell apoptosis: ↓ Bcl-xL; ↑ BAD protein; ↑ caspase-3 activation Induce G1 phase cell cycle arrest	0–93.75 µM	SKOV3: 62.63 µM (24 h) COC1: 73.50 µM (24 h) Angelen: 61.50 µM (24 h) A2780: 55.85 µM (24 h)	[74]

Table 1. *Cont.*

	Cell Lines	Mechanism of Action	Concentration Used	Efficacy/IC$_{50}$ (Exposure Time)	References
Prostate cancer	PC-3 & LNCaP	Inhibit cell viability Induce G0/G1 phase cell cycle arrest: ↓ cyclin D1 & cyclin E; ↓ Cdk2, Cdk4 & Cdk6; ↑ p21 & p53; ↓ Rb & E2F1 proteins; ↓ Rb phosphorylation at Ser$^{807/811}$; ↑ ROS generation	0–60 μM	N/A	[75]
	PC-3, LNCaP & C4-2	Inhibit cell growth Induce apoptosis: ↑ caspase-3, caspase-8 & caspase-9 activation; ↑ PARP cleavage Induce apoptosis via DNA fragmentation: ↑ Bax & Bak proteins; ↓ Mcl-1 protein	0–75 μM	18.75–37.50 μM (24 h)	[76]
	PC-3, LNCaP	Inhibit cell viability Induce autophagy: ↑ LC3-BII protein; ↓ mTOR pathway Induce apoptosis via DNA fragmentation: ↑ ROS generation	0–40 μM	N/A	[77]
Head & neck squamous cancer	Cal-33 & MD-1483	Inhibit cell growth Induce cell apoptosis and cell cycle arrest: ↓ EGFR signalling pathway; ↓ STAT3 signalling pathway; ↓ Bcl-xL & cyclin D1; ↓ phosphorylation p42/p44 MAPK & phosphorylated Akt	0–100 μM	Cal-33: 3.80 μM (72 h) 1483: 7.44 μM (72 h)	[78]
Neuroblastoma	Neuro-2a	Induce apoptosis via DNA fragmentation: ↑ caspase-3, caspase-6 & caspase-9 activation; ↑ Bax protein; ↓ mitochondrial membrane potential; ↑ cytochrome c releaseInduce sub-G1 phase cell cycle arrest	0–100 μM	63.3 μM (72 h)	[79]
	Neuro-2a & NB41A3	Inhibit cell viability Induce autophagy: ↑ LC3-II; ↑ PI3K/Akt/mTOR signalling pathway; ↑ Grp78; ↑ ROS generation; ↑ ERK1/2; ↑ p-ERK1Induce apoptosis via DNA fragmentation Inhibit cell migration	0–100 μM	Neuro-2a: ~50 μM (72 h)	[80]
Bladder cancer	T24 & 5637	Inhibit cell viability and induce apoptosis: ↑ Bax protein; ↑ PARP cleavage; ↓ Bcl-2 protein Inhibit clonogenicity Induce G1 phase cell cycle arrest ↓ cyclin D1; ↑ p21 & p27 Inhibit sphere formation capacity Inhibit cell migration & invasion: ↓ EZH2 gene expression; ↓ MMP9 Inhibit cell stemness: ↓ EZH2 gene expression; ↓ CD44 & Sox2; ↑ miR-143 overexpression	0–72 μM	N/A	[81]

4.2. In Vivo Studies

Based on the in vivo studies, honokiol possessed the capability to inhibit tumour growth, metastasis, and angiogenesis using different animal models, as shown in Table 2. The degree of tumour inhibition was shown to be significantly effective against each distinct cancer cell line, ranging from 0–150 mg/kg via various delivery methods of honokiol between oral gavage or injection (intraperitoneal, caudal vein, or intravenous). Honokiol was shown to downregulate the expression of Oct4, Nanog, and Sox2 which were known to be expressed in osteosarcoma, breast carcinoma and germ cell tumours [41]. According to Wang et al.'s study, they have found that the average tumour size was significantly lower than the control group without affecting their body weight, suggesting inconsequential toxicity under tested conditions when treated with a combination of honokiol and paclitaxel [31]. Indisputably, honokiol was once again proven to exhibit minor to no toxicity against normal cells.

Over the years, the development of chemo-resistance in ovarian cancer cells has hindered the outcome of treatment regimen towards ovarian cancer [82]. Despite the effectiveness of honokiol to inhibit cancer cell proliferation, delivering effective concentration towards the tumour site was deemed challenging due to its water insolubility [73]. The encapsulation of honokiol in liposome, namely Lipo-HNK by Luo and his team has displayed substantial efficacy against cisplatin-resistance ovarian cancer cell line A2780cp. The tumour volume for Lipo-HNK treated mice was 408 ± 165 mm^3 compared to liposome-treated mice and control mice were 2575 ± 701 mm^3 and 2828 ± 796 mm^3 respectively after 21 days [73]. In addition, Lipo-HNK was also shown to prolong survival and induce intra-tumoral apoptosis in vivo. The promising in vivo properties of honokiol should consolidate its importance as a potential anticancer agent for future researches.

Zebrafish (*Danio rerio*) model has emerged as a newly important cancer model that complements against traditional cell culture assays and mice model due to its small size, heavy brood, and rapid maturation time. Importantly, its transparent body wall enables visibility of tumour progression and the ease of experimentation [83,84]. It was known that juvenile zebrafish (*Danio rerio*) or zebrafish embryos have the competency to study cancer cell invasion, metastasis, tumour-induced angiogenesis. Honokiol reduced U-87 MG human glioma/glioblastoma cell proliferation and migration in zebrafish yolk sac and in vivo xenograft nude mouse model [63]. These observations are associated with a reduction in EGFR, phosphorylated STAT3, CD133 and Nestin levels, thus highlighting the regulation of honokiol in EGFR-mediated STAT3/JAK signalling pathway to induce anti-tumour and anti-metastasis.

The subsections below will further discuss the mechanism of anticancer actions of honokiol including the induction of cancer cell death, inhibition of cell cycle progression, induction of autophagy, prevention of epithelial–mesenchymal transition (EMT), as well as the suppression of migration, invasion, and angiogenesis of cancer cells.

Table 2. The antitumour effect of honokiol in in vivo tumour bearing animal models.

Cancer Cell Line	Animal Model & Site of Tumour Xenograft	Dose, Duration & Route of Administration	Observation & Mechanism of Action	Efficacy on Tumour Inhibition	References
			Breast cancer		
MDA-MB-231 cells	Both flanks of athymic nude mice	100 mg/kg/day 28 days IP	Induce tumour growth arrest	Complete arrest of tumour growth from week 2 onwards	[39]
MDA-MB-231 cells	Right gluteal region of athymic nude mice	3 mg/mouse/day Three times a week 28 days IP	Inhibit tumour progression: ↓ Ki-67; ↑ LKB1 & pAMPK; ↑ ACC phosphorylation, ↓ pS6K & 4EBP1 phosphorylation	Tumour weight of honokiol-treated group was 0.22 g compared to control group which was 1.58 g	[40]
MDA-MB-231-pLKO.1 & MDA-MB-231-LKB1[shRNA] cells	Right gluteal region of athymic nude mice	3 mg/mouse/day Three times a week 42 days Oral gavage	Inhibit cell stemness: ↓ Oct4, Nanog & Sox2; ↓ pSTAT3 & Ki-67 Inhibit mammosphere formation	Decreased expression of Oct4, Nanog, Sox2 Reduce number of tumour cells showing Ki-67 & pStat3 expression	[41]
			Colorectal cancer		
RKO cells	Axilla of BALB/c nude mice	80 mg/kg/day Treatment on days 8–11, 14–17, 21–24, 28–31 51 days IP	Inhibit tumour growth Prolong survival of mice	709.9% increase in tumour growth rate in honokiol-treated group compared to 1627.6% and 1408.2% in control and vehicle groups respectively	[33]
HCT116 cells	Flank of athymic nude mice	200 µg/kg/day + 5 Gy irradiation Once a week 21 days IP	Inhibit tumour growth: ↓ CSC proteins → ↓ DCLK1, Sox-9, CD133 & CD44	Significantly lower tumour weight (<800 mg) in honokiol-IR combination, (~1500 mg) in honokiol treatment group compared to (~3300 mg) in control group	[35]
			Cervical cancer		
KB-8-5 cells	Athymic nu/nu nude mice (site of xenograft not stated)	50 mg/kg Honokiol Three times a week + 20 mg/kg Paclitaxel Once a week 23 days IP (honokiol) Tail vein injection (paclitaxel)	Suppress tumour growth: ↓ Ki-67 tissue level Induce apoptosis	Significantly lower average tumour volume for honokiol-paclitaxel combination treatment (573.9 mm^3) compared to control (2585.4 mm^3)	[31]
			Lung cancer		
H2030-BrM3 cells	Left ventricle of NOD/SCID mice	2 or 10 mg/kg/day 23 days Oral gavage	Prevent metastasis of lung cancer cells to brain	10 mg/kg: Decrease brain metastasis for >70%	[47]
H2030-BrM3 cells	Left lung via left ribcage of athymic nude mice	2 or 10 mg/kg/day Five days a week 28 days Oral gavage	Decrease lung tumour growth Inhibit metastasis to lymph node	10 mg/kg: Significantly reduce incidence of mediastinal adenopathy, decrement of weight of mediastinal lymph node for >80%, only 2/6 mice have lymphatic metastasis	[47]

Table 2. *Cont.*

Cancer Cell Line	Animal Model & Site of Tumour Xenograft	Dose, Duration & Route of Administration	Observation & Mechanism of Action	Efficacy on Tumour Inhibition	References
			Blood cancer		
Raji cells	Back of BALB/c nude mice	5 mg/20 g & 10 mg/20 g Treatment on days 8–12 & 15–19 20 days (Route of administration not specified)	Inhibit cell proliferation Inhibit tumour growth	Tumour growth of honokiol-treated mice was significantly lower (~90 cm^3) compared to control mice (~270 cm^3)	[38]
HL60 cells	Inoculated intraperitoneally into SCID mice	100 mg/kg/day Treatment on Day 1–6 47 days IP	Prolong survival of mice	Median survival time of honokiol-treated mice are longer (37.5 days) compared to vehicle-treated mice (24.5 days)	[85]
			Pancreatic cancer		
MiaPaCa cells	Pancreas of immunocompromised mice	150 mg/kg/day 28 days IP	Suppress tumour growth Inhibit metastasis: ↓ CXCR & SHH; ↓ NF-κB & downstream pathway Inhibit desmoplastic reaction: ↓ ECM protein; ↓ collagen I	Significant decrease in tumour growth for honokiol-treated mice (99.6 mm^3) compared to vehicle-treated mice (1361.0 mm^3)	[58]
			Skin cancer		
SKMEL-2 or UACC-62 cells	Right flank of athymic nude mice	50 mg/kg Three times a week 14–54 days IP	Decrease tumour growth	SKMEL-2: 40% reduction in tumour volume UACC-62: 50% reduction in tumour volume	[56]
			Thyroid cancer		
ARO cells	BALB/cAnN.Cg-Foxn1nu/CrlNarl mice (site of xenograft not stated)	5 or 15 mg/kg/mouse Every three days 21 days Oral gavage	Decrease tumour volume & tumour weight Induce apoptosis & autophagy	Control: ~1000 mm^3; 700 mg 5 mg/kg Honokiol: ~600 mm^3; 400 mg 15 mg/kg Honokiol: ~400 mm^3; 200 mg	[60]
			Nasopharyngeal cancer		
HNE-1 cells	Right dorsal aspect of right foot of BALB/c athymic nude mice	Active-targeting nanoparticles-loaded HK (ATNH), Non-active-targeting nanoparticles-loaded HK (NATNH), Free Honokiol (HK) 3 mg/mouse/day Every three days Euthanise 50% mice after 12 days, rest are left to observe tumour growth & survival time up to 60 days; IV	Inhibit tumour progression, Induce apoptosis Potential inhibitor of angiogenesis & proliferation	Efficiency in tumour delay: ATNH > NATNH > Free HK Median survival time: Control: 28.5 days Free HK: 34 days NATNH: 42.5 days ATNH: 57.5 days	[30]

Table 2. *Cont.*

Cancer Cell Line	Animal Model & Site of Tumour Xenograft	Dose, Duration & Route of Administration	Observation & Mechanism of Action	Efficacy on Tumour Inhibition	References
			Brain cancer		
U21 cells	Right flank of athymic nude mice	20 mg/kg Twice a week 27 days Caudal vein injection	Inhibit tumour growth Inhibit angiogenesis	Honokiol-treated mice have significant inhibition of tumour volume by 50.21% compared to vehicle Significantly lower microvessel present in honokiol-treated cells	[61]
U-87 MG cell suspension pre-treated with honokiol or vehicle for 48h	Yolk sac of Zebrafish larvae	(Concentration N/A) 3 days Injection of cells into zebrafish	Inhibit cell proliferation Inhibit cell migration	Reduced number of cell mass compared to vehicle-treated cells	[63]
U-87 MG cells	Right flank near upper extremity of nude mice	100 mg/kg/day Treatment at days 1–7 21 days IP	Reduce tumour growth: ↓ EGFR, pSTAT3, CD133 & Nestin	Increased number of apoptotic cells in honokiol-treated tissue, Significantly lower tumour volume & tumour weight in honokiol-treated mice	[63]
			Bone cancer		
HOS cells	Dorsal area of BALB/c-nu mice	40 mg/kg/day 7 days IP	Reduce tumour growth Induce apoptosis & autophagy: ↑ cleaved caspase-3; ↑ LC3B-II & phosphor-ERK (ROS/ERK1/2 signalling pathway)	Significant decrease in tumour volume & weight of honokiol-treated mice (200 mm^3; 0.2 g) compared to control group (~500 mm^3; 0.5 g) Increased number of TUNEL-positive cells	[26]
LM8-LacZ cells	Left flank of C3H/HeNCrl mice	150 mg/kg/day 25 days; IP	Inhibit metastasis	Mean number of micrometastases decreased significantly by 41.4% in honokiol-treated mice compared to control mice	[68]
			Oral cancer		
SAS cells	Right flank of BALB/cAnN.Cg-Foxn1nu.Crl·Narl nude mice	5 mg/kg or 15 mg/kg Treatment on day 1, 4, 7, 10, 13, 16, 19, 22 35 days Oral	Reduce tumour growth & volume	Significantly reduction in tumour growth in honokiol-treated mice 29% reduction (5 mg/kg; 21 days), 40% reduction (15 mg/kg; 21 days) 41% reduction (5 mg/kg; 35 days), 56% reduction (15 mg/kg; 35 days)	[26]

Table 2. *Cont.*

Cancer Cell Line	Animal Model & Site of Tumour Xenograft	Dose, Duration & Route of Administration	Observation & Mechanism of Action	Efficacy on Tumour Inhibition	References
		Prostate cancer			
C4-2 cells	Bilateral tibia of BALB/c nu/nu athymic nude mice	100 mg/kg/day 42 days IP	Inhibit cell proliferation: ↑ Ki-67 Induce apoptosis: ↑ M-31 Inhibit angiogenesis: ↓ CD-31	Lower PSA value in honokiol-treated mice compared to control group	[76]
PC-3 cells	Left & right flanks above hind limb of nude mice	1 or 2 mg/mice Monday, Wednesday & Friday two weeks before tumour implantation and duration of experiment after implantation 77 days Oral gavage	Inhibit tumour growth Inhibit cell proliferation Inhibit neovascularisation Induce apoptosis	Tumour volume of honokiol-treated mice are significantly lower (~330 mm^3; 1 mg), (~50 mm^3; 2 mg) compared to control (~400 mm^3)	[18]
		Gastric cancer			
MKN45 cells	Dorsal side of BALB/c nude mice (nu/nu)	0.5 mg/kg/day & 1.5 mg/kg/day 10 days Injection (route not stated)	Inhibit tumour growth: ↓ GRP94 overexpression	30% reduction in tumour volume (0.5 mg/kg) 60% reduction in tumour volume (1.5 mg/kg) Decreased accumulation of GRP94	[86]
MKN45 & SCM-1 cells	Peritoneal cavity of BALB/c nude mice	5 mg/kg Twice a week 28 days IP	Inhibit metastasis Inhibit angiogenesis	Honokiol inhibited STAT-3 signalling and VEGF signalling induced by calpain/SHP-1	[87]
		Ovarian cancer			
SKOV3 cells	Right axilla of BALB/c nude mice	1 mg liposome-encapsulated honokiol/day 48 days IP	Inhibit tumour growth Inhibit angiogenesis	Reduction in tumour growth rate in liposome-encapsulated honokiol-treated mice by 67–70% compared to control	[73,88]
A2780s cells	Right flank of athymic BALB/c nude mice	10 mg/kg Lipo-Honokiol Twice a week 21 days IV	Inhibit cancer growth Prolong survival of mice Increase intra-tumoural apoptosis Inhibit intra-tumoural angiogenesis	Lipo-HNK treated mice have significantly smaller tumour volume (222 ± 71 mm^3) compared to liposome-treated mice (1823 ± 606 mm^3) and control mice (3921 ± 235 mm^3)	[73]
A2780cp cells	Right flank of athymic BALB/c nude mice	10 mg/kg Lipo-Honokiol Twice a week 21 days IV	Inhibit cancer growth Prolong survival Increase intra-tumoural apoptosis Inhibit intra-tumoural angiogenesis	Lipo-HNK treated mice have significantly smaller tumour volume (408 ± 165 mm^3) compared to liposome-treated mice (2575 ± 701 mm^3) and control mice (2828 ± 796 mm^3)	[73]

5. Mechanism of Action of Honokiol

5.1. Dual Induction of Apoptotic and Necrotic Cell Death

Apoptosis is a normal physiological process that maintains the homeostatic cellular balance in multicellular organisms [89]. Generally, apoptosis can be classified into two central pathways, namely the intrinsic pathway (mitochondrial-mediated pathway) and extrinsic pathway (death receptor-mediated pathway) [90]. The intrinsic pathway is associated with changes in mitochondrial membrane permeability that lead to imbalance in Bax/Bak ratio and release of cytochrome *c* and other mitochondrial proteins into cytosol [89,90]. The released cytochrome *c* interacts with apoptosis protease-activating factor 1 (Apaf1) and forms an apoptosome complex [91], which promotes the activation of caspase-9 and later caspase-3, initiating the caspase cascade, which executes cell death in a coordinated way [91]. For the extrinsic pathway, the binding of ligands such as tumour necrosis factor (TNF), Fas ligand (Fas-L), and TNF-related apoptosis-inducing ligand (TRAIL) to their respective death receptors (type 1 TNF receptor (TNFR1), Fas (also called CD95/Apo-1) and TRAIL receptors will turn procaspase-8 into active caspase-8 to induce apoptosis [91–93].

Honokiol has been shown tp initiate caspase-dependent apoptotic pathways in different types of cancer (Table 1). Chen et al. [14] found that JJ012 human chondrosarcoma cells lose their mitochondrial membrane potential when treated at 10 µM of honokiol, thus leading to apoptosis. Other studies have also shown that honokiol markedly disrupted the balance of Bax/Bcl-2 ratio [13,18,34,63,94–97]. The increasing ratio of proapoptotic to antiapoptotic Bcl-2 family proteins (Bax/Bcl-2) will induce the release of cytochrome *c* and other apoptogenic proteins through the mitochondrial membrane to the cytosol, subsequently leading to the activation of caspase cascade and apoptosis [34]. Furthermore, honokiol downregulated the expression of several other anti-apoptosis mRNA and proteins such as Bcl-xL [13,18,25,64], survivin [67,98], and MCL-1 [18], as well as upregulated other pro-apoptotic proteins such as BAD, BAX, and BAK proteins [18,25].

Moreover, honokiol has been shown to effectively induce apoptosis in p53-deficient cancer cells, such as MDA-MD-231 breast cancer cells, as well as lung and bladder cancer cell lines by inhibiting the activation of ras-phospholipase D [39,99,100]. Besides p53, other tumour suppressor genes that will be activated in honokiol treatment include p21 [53], p21/waf1 [101], p27 [53], p38 MAPK [102,103], and p62 [26,46].

Besides the intrinsic pathway, honokiol is capable of targeting death receptors TNF-related apoptosis-inducing ligand (TRAIL) receptors and tumour necrosis factor receptors (TNFR) resulting in a sequential activation of caspase-8 and -3, which cleaves target proteins and then leads to apoptosis [104–106]. Activation of the death receptor mediated apoptotic pathway is primarily inhibited by cellular-caspase-8/FADD-like IL-1β-converting enzyme (FLICE) inhibitory protein (c-FLIP), which inhibits caspase-8 activation by preventing the recruitment of caspase-8 to the death inducing signalling complex [106]. However, honokiol was able to downregulate c-FLIP through the ubiquitin/proteasome-mediated mechanism, resulting in the sensitisation of non-small cell lung cancer cells to TRAIL-mediated apoptosis [107,108].

Other than intrinsic and extrinsic pathways, honokiol can also induce apoptosis by the endoplasmic reticulum (ER) stress-induced mechanism. A variety of ER stresses result in unfolded protein accumulation responses [109,110]. For survival, the cells induce ER chaperone proteins to increase protein aggregation, temporarily halt translation, and activate the proteasome machinery to degrade misfolded proteins. However, under severe and prolonged ER stress, an unfolded protein response activates unique pathways that lead to cell death through apoptosis [111]. According to a study by Zhu et al. [50], honokiol can upregulate the expressions of ER stress-induced apoptotic signalling molecules such as GRP78, phosphorylated PERK, phosphorylated eIF2α, CHOP, Bcl-2, Bax, and cleaved caspase-9 in human lung cancer cells. Chiu et al. [112] found that honokiol also led to an increase in ER stress activity in melanoma cell lines B16F10 (mouse), human malignant melanoma, and human metastatic melanoma. Honokiol activated ER stress and down-regulated peroxisome proliferator-activated receptor-γ (PPARγ) activity resulting in PPARγ and CRT degradation through calpain-II activity in human gastric cancer cell lines [86,113,114] and human chondrosarcoma cells [14]. This was due to the ability of honokiol to upregulate and bind effectively to the glucose regulated protein 78 (GRP78) to activate apoptosis [14,115]. However, this was opposed by another study where treatment of various human gastric cancer cells with honokiol led to the induction of GRP94 cleavage but did not affect GRP78 [86].

Necrosis is known as unprogrammed cell death whereby cell swelling and destabilisation of the cell membrane results in the leakage of cellular cytoplasmic contents into the extracellular space, thus causing inflammation [116]. Besides apoptosis, honokiol has also been found to induce necrotic cell death in MCF-7 (40 µg/mL honokiol) [117], human oesophageal adenocarcinoma cells CP-A and CP-C [118], and primary human acute myelogenous leukemia HL60 [85] via p16ink4a pathway by targeting cyclophilin D to affect several downstream mechanisms. This phenomenon was also observed in transformed Barrett's and oesophageal adenocarcinoma cells when treated with honokiol (<40 µM) by targeting their STAT3 signalling pathway, thus resulting in a decrease of Ras activity and phosphorylated ERK1/2 expression [119]. The phosphorylation of Ser727 STAT3 induces translocation towards the mitochondria followed by ROS production, ultimately leading to the induction of necrosis [120]. Taken together, honokiol demonstrates the dual induction of apoptotic and necrotic cell death.

5.2. Cell Cycle Arrest

Cancer is attributed to uncontrolled proliferation resulting from abnormal activity of different cell cycle proteins. Therefore, cell cycle regulators are becoming attractive targets in cancer therapy. Honokiol can induce cell cycle arrest in several types of cancer cells, such as in lung squamous cell carcinoma [121], prostate cancer cells [75,122], oral squamous cancer [70], UVB-induced skin cancer [123], and more as listed in Table 1, by generally inducing G0/G1 and G2/M arrest. This arrest is associated with the suppression of cyclin-B1, CDC2, and cdc25C in honokiol-treated human gastric carcinoma and human neuroglioma cells [97,124,125], downregulation of cyclin dependent kinase (CDK)-2 and CDK-4, and the upregulation of cell cycle suppressors p21 and p27 in human oral squamous cell carcinoma (OSCC) cells [26,97]. In addition, the downregulation of c-Myc and class I histone deacetylases was also identified as other contributors to cell cycle arrest at the G0/G1 phase in prostate cancer cells [97,122] and acute myeloid leukemia respectively [44,101,108].

5.3. Autophagy

Autophagy is an evolutionary conserved catabolic process that involves the delivery of dysfunctional cytoplasmic components for lysosomal degradation [126,127]. The activation of autophagy promotes cell survival and regulates cell growth during harsh and stressful conditions via a reduction of cellular energy requirements by breaking down unnecessary components [82,127]. In cancer cells, autophagy facilitates both tumour suppression and tumourigenesis by the induction of cell death and tumour growth promotion, respectively [128,129]. The regulation of mTORC complexes mTORC1 and mTORC2 is involved in controlling the autophagic process. The activation of mTORC1 plays an important role in phosphorylation of autophagy-related protein (ATG) and subsequently inhibiting autophagy, whereas the inhibition of mTORC1 complements the autophagic process [130,131]. The inhibition of mTORC1 complex will concurrently activate Unc-51-like autophagy-activating kinase (ULK) complex, inducing localisation to the phagophore and followed by class III PI3K activation [132,133]. Beclin-1 was known to play a role in tumour suppression by recruiting several proteins associated with autophagosome elongation and maturation [134]. ATGs regulate the autophagosome elongation. For instance, ATG5-ATG12/ATG16L complexes recruit microtubule-associated protein 1 light chain 3 (LC3), followed by conversion of pro-LC3 to active cytosolic isoform LC3 I by ATG4B [135,136]. Thereafter, the interaction with ATG3, ATG7, and phosphatidylethanolamine (PE) converts LC3 I to LC3 II. The LC3 II enables the autophagosome to bind to degraded substrates and mature autophagosomes are capable of fusing with lysosomes to selectively remove damaged organelles via autophagy [137].

Generally, there are two modes of autophagy known as conventional and alternative autophagy. Conventional autophagy (also known as Atg5/Atg7-dependent pathway) involves the activation of Atg5 and Atg7 which are core regulators of autophagy, and then leads to microtubule-associated protein 1A/1B light chain 3 (LC3) modification and translocation from cytosol to the isolation membrane. This LC3 translocation was considered as a reliable hallmark of autophagy. Contradictorily, alternative autophagy occurs independently without involving Atg5 and Atg7, as well as LC3 modification [128,129,137].

The regulation of autophagy in cancer remains controversial as it plays dual roles in tumour suppression and promotion. Autophagy is believed to contribute to the properties of cancer cells stemness, induction of recurrence, and the development of anticancer drugs. However, the actual mechanism of autophagy in cancer remains unclear. Several studies have highlighted the potential of honokiol to induce cell death via autophagy in human prostate cancer cells [77], human glioma cells [138], NSCLC cells [30], and human thyroid cancer cells [60].

The activation of Atg5/Atg7-dependent pathways through the upregulation of LC3B-II, Atg5, and Atg7 levels was observed in honokiol-treated osteosarcoma HOS and U2OS cells and leads to the accumulation of autophagic vacuoles [26]. According to a study by Chang et al. [64], the expression of two critical autophagic proteins, Beclin-1 and LC3, were found to have increased in the honokiol-treated glioblastoma multiforme cells (DBTRG-05MG cell line). Similarly, the expression of autophagosomal marker LC3-II was also increased in Kirsten rat sarcoma viral oncogene homolog (KRAS) mutated cell lines of non-small cell lung cancer (NSCLC).

Other signalling pathways are also found to be involved in honokiol-induced autophagy including the involvement of AMPK-mTOR signalling pathway which leads to autophagocytosis through the coordinated phosphorylation of Ulk1 in Kirsten rat sarcoma viral oncogene homolog (KRAS) mutant lung cancer and melanoma cells [55,60,66,97]. Besides this, the ROS/ERK1/2 signalling pathway is also believed to play a certain role in honokiol-induced autophagy though ERK activation and the generation of ROS in treated osteosarcoma cells [67,77,97]. All these recent studies have further supported the potential of honokiol in the induction of autophagy in cancer cells.

5.4. Epithelial-Mesenchymal Transition (EMT)

Migratory mesenchymal-like cells are involved in embryonic development, tissue repair, and regeneration, as well as several pathological processes like tissue fibrosis, tumour invasiveness, and metastasis [139,140]. These migratory mesenchymal cells originate from the conversion of the epithelial cells, and this process is known as epithelial-mesenchymal transition (EMT). This plasticity of cellular phenotypes provides a new insight into possible therapeutic interventions in cancer [140].

EMT is characterised by the loss of epithelial markers such as cytokeratins and E-cadherin, followed by an increase in mesenchymal markers such as N-cadherin and vimentin [141]. The cellular processes of EMT are composed of several key transcription factors (such as TWIST, SNAI1, SNAI2, ZEB1/2) that act in concert with epigenetic mechanisms and post-translational protein modifications to coordinate cellular alterations [139,142]. The application of gene expression signatures combining multiple EMT-linked genes has proven useful to evaluate EMT as a contributing factor in tumour development in human cancers. However, the EMT process was shown to be incomplete in tumours, venturing in between multiple translational states and expressing a mixture of both epithelial and mesenchymal genes. This hybrid in partial EMT can be more aggressive than tumour cells with a complete EMT phenotype [141]. In addition, EMT contributes to tumour metastatic progression and resistance towards cancer treatment, resulting in poor clinical outcomes [140,141].

Honokiol has been shown to block and inhibit EMT in many cancer cells such as breast cancer, melanoma, bladder cancer, human non-small cell lung cancer, and gastric cancer (Table 1). Honokiol reduced steroid receptor coactivator-3 (SRC-3), matrix metalloproteinase (MMP)-2, and Twist1, preventing the invasion of urinary bladder cancer cells [108,143]. In addition, honokiol was also capable of inducing E-cadherin and repressing N-cadherin expression, thus inhibiting the EMT process in J82 bladder cancer cells [108,143]. In breast cancer cells, honokiol inhibits the recruitment of Stat3 on mesenchymal transcription factor Zeb1 promoter, resulting in decreased Zeb1 expression and nuclear translocation [144]. In addition, honokiol increases E-cadherin expression via the Stat3-mediated release of Zeb1 from E-cadherin promoter [144]. Collectively, many studies have reported that honokiol effectively inhibits EMT in breast cancer cells, evidence has been found to support a cross-talk between honokiol and Stat3/Zeb1/E-cadherin axis [144]. On the other hand, EMT is inhibited by modulating the miR-141/ZEB2 signalling in renal cell carcinoma (A-498) [57].

Honokiol inhibited the EMT-driven migration of human NSCLC cells in vitro by targeting c-FLIP through N-cadherin/snail signalling as N-cadherin and snail are downstream targets of c-FLIP [145]. Twist1, a basic helix-loop-helix domain-containing transcription factor, promotes tumour metastasis by inducing EMT, and can be upregulated by multiple factors, including SRC-1, STAT3, MSX2, HIF-1α, integrin-linked kinase, and NF-κB. The capability of honokiol in targeting Twist1 can be regarded as a promising therapy for metastatic cancer [108,146].

Honokiol was found to inhibit breast cancer cell metastasis and eliminate human oral squamous cell carcinoma cell by blocking EMT through the modulation of Snail/Slug protein translation [147,148]. Honokiol markedly downregulated endogenous Snail, Slug, and vimentin expression and upregulated E-cadherin expression in MDA-MB-231, MCF7, and 4T1 breast cancer cells [148]. As primary EMT inducers, Snail and Slug dictate the induction of EMT by targeting E-cadherin and vimentin [144,148]. Furthermore, when cells were treated with honokiol, Snail and Slug expression levels were decreased from 12 h to 24 h in a time-dependent manner, suggesting that honokiol can reverse the EMT process via the downregulation of Snail and Slug in breast cancer cell lines [148]. Besides that, EMT was inhibited in human oral squamous cell carcinoma cell via the disruption of Wnt/β-catenin signalling pathway [147]. It was reported that the protein levels of mesenchymal markers such as Slug and Snail were markedly suppressed, while β-catenin and its downstream Cyclin D1 were inhibited [147]. It is known that β-catenin could mediate EMT [147,149], which plays a crucial role in cancer invasion and metastasis. The EMT markers such as Snail and Slug are also the target genes of β-catenin [150]. Therefore, the suppression of Snail and Slug in honokiol treated human oral squamous cell carcinoma cells was believed to be due to the inhibition of Wnt/β-catenin signalling pathway [147]. Similarly, in U87MG human glioblastoma cell and melanoma cells, Snail, N-cadherin and β-catenin expression levels were decreased, whereas E-cadherin expression was increased after honokiol treatment [65,112].

5.5. Suppression of Migration, Invasion and Angiogenesis of Cancer Cells

Metastasis is known to be the major cause of death in cancer patients [151]. It involves the migration and invasion of tumour cells into neighbouring tissues and distant organs via intravasation into blood or lymphatic system [152,153]. The formation of invadopodium was stimulated by epidermal growth factor (EGF) and is crucial for the degradation of the extracellular matrix and remodelling membrane proteins, promoting metastasis [151]. Therefore, one of the important steps in cancer management is to control tumour cell metastasis, especially for early-stage cancer patients [153]. Various studies have reported that honokiol has the capability to suppress tumour metastasis in different types of cancer including breast cancer [40,148,154], non-small cell lung cancer [44,155] ovarian carcinoma cells [28], lung cancer [50], U251 human glioma, as well as U-87MG and T98G human glioblastoma cell [63,65,94], oral squamous cell carcinoma (OSCC) [26], bladder cancer cell [143], pancreatic cancer [58], renal cell carcinoma [156,157], and gastric cancer cells [113]. For instance, the percentage of invading urinary bladder cancer (UBC) cells was significantly reduced by 67% and 92% upon 2.4 μg/mL and 4.8 μg/mL of honokiol treatment, respectively [143]. Similarly, tumour cell migration was inhibited by 38–66% in A549 cells, by 37–62% in H1299 cells, 12% to 58% in H460 cells and 32% to 69% in H226 cells, in a concentration-dependent manner after treatment with honokiol [44].

Furthermore, honokiol also demonstrated an inhibitory effect on the expression of matrix metalloproteinases (MMPs) such as MMP-2 and MMP-9 proteins, which play an essential role in the metastatic process of tumour cells, as well as the regulation of angiogenesis in the maintenance of tumour cell survivability [44,63,143]. MMPs are a group of extracellular matrix degrading enzymes that control various normal cellular processes, such as cell growth, differentiation, apoptosis, and migration [153]. However, MMP activity was increased in many tumour cells. The overexpression of MMP-2 and MMP-9 are associated with pro-oncogenic events such as neovascularisation, tumour cell proliferation, and metastasis because it can degrade the extracellular matrix, basement membranes, and adhesion molecules (intercellular adhesion molecule, ICAM, and vascular cell adhesion molecule) and become invasive [58,153,158].

The transition from an epithelial-to-mesenchymal (EMT) phenotype facilitates the breakdown of extracellular matrix followed by the subsequent invasion of the surrounding tissues in order to enter the bloodstream and/or lymph nodes, and travel to distant organ sites. Once cells have reached the distant organ sites, they undergo mesenchymal-to-epithelial transition and begin the establishment of distal metastasis by the surviving cancer cells followed by the outgrowth of secondary tumours [58,159]. Honokiol has been shown to inhibit the invasion of HT-1080 human fibrosarcoma cells and U937 leukemic cells by inhibiting MMP-9 [160]. In addition, honokiol also reduced the protein levels of MMP2 and MMP9 in U251 human glioma and U-87 MG human glioblastoma cell lines in a dose-dependent manner [63]. The expression of MMP-2 and MMP-9 were also found to be decreased in both honokiol-treated A549 and H1299 cells (NSCLC cell lines), consistent with the decreased nuclear accumulation of β-catenin as both MMP-2 and MMP-9 are the downstream targets of β-catenin [44,161,162]. In the J82 bladder cancer cell, honokiol repressed the expression of SRC-3, MMP-2, and Twist1 genes which were involved in cancer cell invasion [143].

Another proposed mechanism for the inhibitory effects of honokiol on invasion and metastasis is through the liver kinase B1 (LKB1)/adenine monophosphate-activated protein kinase (AMPK) axis. Honokiol treatment increased the expression and cytoplasmic translocation of tumour-suppressor LKB1 in breast cancer cells, which led to the phosphorylation and functional activation of AMPK and resulted in the inhibition of cell invasion and metastasis [40,58]. The activation of AMPK suppresses mTOR signalling, decreasing the phosphorylation of p70 kDA ribosomal protein S6 kinase 1 (p70S6K1) and eukaryotic translation initiation factor 4E (eIF4E)-binding protein (4EBP1). This will ultimately inhibit the reorganisation of the actin cytoskeleton in cells, subsequently inhibiting cell migration [40].

In human renal carcinoma cell (RCC) 786-0, honokiol significantly upregulated the expression of metastasis suppressor gene (KISS-1), genes encoding TIMP metalloproteinase inhibitor 4 (TIMP4), and KISS-1 receptor (KISS-1R). In addition, honokiol markedly suppressed the expression of genes encoding chemokine (C-X-C motif) ligand 12 (CXCL12), chemokine (C-C motif) ligand 7 (CCL7), interleukin-18 (IL18) and matrix metalloproteinase 7 (MMP7). It was proven that honokiol significantly upregulated KISS1 and KISS1R in the 786-0 cells when treated with honokiol since recent studies showed that the activation of KISS1/KISS1R signalling by kisspeptin treatment decreases the motility and invasive capacity of conventional RCC, and overexpression of KISS1 inhibits the invasion of RCC cells Caki-1 [14,163]. In short, the activation of KISS1/KISS1R signalling by honokiol suppresses the multistep process of metastasis, including invasion and colony formation, in RCC cells 786-0 [163].

Angiogenesis is the formation of new blood vessels for supplying nutrients and oxygen to tissues and cells. In tumourigenesis, angiogenesis is important for the development and progression of malignant tumours [164]. The endothelial cells in growing cancer are active due to the release of cell growth and motility promoting proteins, creating a network of blood vessels to overcome its oxygen tension [165]. Vascular endothelial growth factor (VEGF) and fibroblast growth factor-2 (FGF2) are among the factors that play an important role in tumour angiogenesis [153]. In human renal cancer cell lines (786-0 and Caki-1), honokiol induced down-regulation of the expression of VEGF and heme oxygenase-1 (HO-1) via the Ras signalling pathway thus inhibit angiogenesis [166,167].

In retinal pigment epithelial (RPE) cell lines, honokiol inhibited the binding of hypoxia-inducible-factor (HIF) to hypoxia-response elements present on the VEGF promoter, thereby inhibiting the secretion of VEGF protein [168,169]. This decrement of VEGF levels resulted in reduced proliferation of human retinal microvascular endothelial cells (hRMVECs) [168]. Therefore, honokiol is said to possess both anti-HIF and anti-angiogenic properties.

In the overexpression of VEGF-D Lewis lung carcinoma cell-induced tumours in C57BL/6 mice, honokiol was shown to significantly inhibit tumour-associated lymphangiogenesis and metastasis. Furthermore, a remarkable delay in tumour growth and prolonged life span in honokiol-treated mice were also observed [170]. In another study, honokiol inhibited VEGF-D-induced survival, proliferation, and microcapillary tube formation in both human umbilical vein endothelial cells (HUVECs) and lymphatic vascular endothelial cells (HLECs). These observations are believed to be due to the inhibition in Akt and MAPK phosphorylation and downregulation of VEGFR-2 expressions in HUVECs as well as VEGFR-3 of HLECs [101,160,171]. Collectively, honokiol has been shown to exert direct and indirect effects on tumour suppression via anti-metastasis, anti-angiogenesis, and anti-lymphangiogenesis by mainly affecting HIF- and VEGF/VEGFR- dependent pathways. However, an in-depth mechanism of honokiol on the inhibition of metastatic progression and spread should be further explored in the future.

6. Effect of Honokiol on Various Signalling Pathways

6.1. Nuclear Factor Kappa B (NF-κB)

The nuclear factor kappa B (NFκB) family comprises of five DNA-binding proteins (p50, p52, p65, cRel, and RelB) that differentially modulate the transcription of genes that are involved in various cellular processes such as inflammation, migration, invasion, angiogenesis, proliferation, and apoptosis [172,173]. The continuous activation of NFκB has been reported in different types of cancers. Honokiol affects the constitutive activation of NFκB and expression of NFκB-regulated gene products involved in apoptosis (survivin, Bcl-2, Bcl-xL, IAP1, IAP2, cFLIP and TRAF1), inflammation (cyclooxygenase-2, COX-2), proliferation (cyclin D1 and c-myc), invasion (ICAM-1 and MMP-9), and angiogenesis (VEGF), thereby enhancing apoptosis and suppressing cancer progression [58,174]. Several studies support the inhibitory activity of honokiol against NFκB in different types of cancer cells, including breast cancer [42,117,175], head and neck squamous cell carcinoma (HNSCC) [176], colon cancer cells [177], non-small cell lung cancer (NSCLC) cells [44], pancreatic cancer cells [13], human leukemic cell [104], embryonic kidney cells, T-cell leukemia, multiple myeloma, lung adenocarcinoma, and squamous cell carcinoma [174].

Honokiol was found to repress the transcriptional activity of NFκB in both pancreatic MiaPaCa and Panc1 cancer cells. It was found that honokiol treatment significantly reduced nuclear NFκB levels with an increase of cytoplasmic NFκB fraction in MiaPaCa and Panc1 cells, in a dose-dependent manner [13]. The cellular distribution of NFκB is controlled by the relative expression of its biological inhibitor IκB, which keeps NFκB sequestered in the cytoplasm in an inactive complex [172]. Upon honokiol treatment, IκB-α levels were increased due to the stabilisation of IκB-α post-treatment, concurrently inducing the downregulation of IκB-α phosphorylation [13]. Furthermore, honokiol has also been shown to inhibit the TNF-α-induced phosphorylation and degradation of the cytosolic NFκB inhibitor IκBa and suppression of IKK activation [104,174,178]. In addition, honokiol was also found to inhibit the nuclear translocation and phosphorylation of p65 subunit of NFκB [44,104]. Honokiol suppressed NF-κB-regulated gene products including MMP-9, TNF-α, IL-8, ICAM-1, and MCP-1 [66].

6.2. Signal Transducers and Activators of Transcription (STATs)

Signal transducers and activators of transcription (STATs) is a well-known oncogene that is regulated by receptor tyrosine kinases, G-protein-coupled receptors, and interleukin families [179,180]. STAT3 are a group of transcription factors that upon phosphorylation will undergo dimerization and translocation to either the nucleus or mitochondria to control cell survival, cell cycle, cellular growth, and angiogenesis. STATs are aberrantly activated in several types of malignancies due to functional loss of their negative regulators, or the overexpression of upstream tyrosine kinases [179].

STAT3 can also localise into the mitochondria and mediate mitochondrial biogenesis. Honokiol has been shown to target STAT3 to reduce its expression and phosphorylation in many cancer cells such as human glioblastoma [47,63,100], lung cancer [47,181], oral squamous cell carcinoma (OSCC) [95], breast cancer [41,144], human epidermoid carcinoma [31], colorectal cancers [182], gastric cancer [87], and esophageal adenocarcinoma [119].

Honokiol was found to inhibit EGFR expression and down-regulate STAT3 phosphorylation by reducing the CD133 and Nestin levels [63]. Similarly, honokiol also induces apoptosis through the suppression of JAK2/STAT3, Akt and Erk signalling pathways in human oral squamous cell carcinoma (SAS and OCEM-1) cell lines [95]. Similar effect was observed in oral cancer cells where honokiol suppressed JAK2/STAT3 activation and, inhibited IL-6-mediated cell migration [95,183]. Furthermore, another study indicated that honokiol induces apoptosis in human glioblastoma cell line U87 through suppressing the phosphorylation of STAT3 (Tyr705), down-regulating survivin, and upregulating cleaved caspase-3 expression [98].

Moreover, honokiol inhibited STAT3-phosphorylation/activation in an LKB1-dependent manner, preventing its recruitment to canonical binding-sites in the promoters of Nanog, Oct4, and Sox2 [41]. Thus, the inhibition of the coactivation function of STAT3 resulted in the suppression of expression of pluripotency factors in MCF7, MDA-MB-231, SUM149, and SUM159 breast cancer cells [41]. Furthermore, honokiol inhibited breast tumorigenesis in mice in an LKB1-dependent manner [41]. This showed that honokiol can support crosstalk between LKB1, STAT3, and pluripotency factors in breast cancer and effective anticancer modulation of this axis with honokiol treatment in both in vitro and in vivo [41]. Apart from that, honokiol suppressed metastasis and proliferation in both brain metastatic lung cancer cell lines PC9-BrM3 and H2030- BrM3 by inhibiting STAT3 phosphorylation [47].

In other studies, honokiol is proven to be an effective chemotherapeutic agent that exert its antitumour function by inhibiting the STAT3 signalling pathway. Honokiol can induce cell cycle arrest and apoptosis via the inhibition of survival signals in adult T-cell leukemia by suppressing the phosphorylation and DNA binding of different oncogene factors, such as NF-κB, activator protein 1, STAT3, and STAT5 [184]. Besides that, honokiol can induce necrosis and apoptosis in transformed Barrett's and oesophageal adenocarcinoma cells through the inhibition of the STAT3 signalling pathway [119]. Honokiol can inhibit the growth and peritoneal metastasis of gastric cancer in nude mice, which was correlated with the inhibition of STAT3 signalling via the upregulation of Src homology 2 (SH2)-containing tyrosine phosphatase 1 [87].

6.3. Epidermal Growth Factor Receptor (EGFR)

EGFR is a group of transmembrane receptor tyrosine kinases (RTKs) that are normally deregulated in various cancers [185,186]. The overexpression or activating mutations in EGFR results in increased cell proliferation, abnormal metabolism, and cell survival through the activation of the downstream mitogen-activated protein kinase (MAPK) and v-akt murine thymoma viral oncogene homolog 1 (AKT) signalling pathways, as well as phosphatidyl-inositol 3-kinase (PI3K)/Akt, and STAT3 signalling pathways [13,58]. EGFR activation occurs upon binding to its ligands, which then leads to its homo- or heterodimerization with other members of the ErbB family, and subsequent activation of downstream signalling cascades in many cancer cell types, including breast cancer and head and neck squamous cell carcinoma (HNSCC) [187,188].

Honokiol has been shown to inhibit EGFR signalling pathway through either inhibition of EGFR expression or inhibition of EGFR phosphorylation [78,189,190]. Honokiol (60 μM) was found to inhibit EGFR expression and down-regulate STAT3 phosphorylation in U251 and U-87 MG human glioma/glioblastoma cells via JAK-STAT3 signalling [63]. In another study, honokiol (2.5–7.5 μM) differentially suppressed proliferation (up to 93%) and induced the apoptosis (up to 61%) of EGFR overexpressing tumourigenic bronchial cells. These effects were observed in parallel with the downregulation of phospho-EGFR, phospho-Akt, phospho- STAT3, and cell cycle-related proteins [189]. Furthermore, in a mouse lung tumour bioassay, intranasal instillation of liposomal honokiol (5 mg/kg) for 14 weeks reduced the size and multiplicity (49%) of lung tumours and the level of total- and phospho-EGFR, phospho-Akt, and phospho-STAT3 [189]. Overall, honokiol has been proven to be a promising candidate to suppress the development and progression of lung tumours driven by EGFR deregulation. Moreover, honokiol induced mitochondria-dependent and death receptor-mediated apoptosis in multi-drug resistant (MDR) KB cells, which was associated with inhibition of EGFR-STAT3 signalling and downregulation of STAT3 target genes [31].

Furthermore, the downregulation of c-Src/EGFR-mediated signaling is involved in honokiol-induced cell cycle arrest and apoptosis in MDA-MB-231 human breast cancer cells. EGFR can also be activated in a ligand-independent manner by cellular Src (c-Src), a non-receptor tyrosine kinase. The tyrosine kinase c-Src is also upregulated in many human malignancies and promotes the activation of mitogenic signalling through EGFR [13,191]. In MDA-MB-231 human breast cancer cells, honokiol downregulated the expression and phosphorylation of c-Src, epidermal growth factor receptor (EGFR), and Akt, and consequently led to the inactivation of mTOR and its downstream signal molecules including 4E-binding protein (4E-BP) and p70 S6 kinase [43]. Besides that, inhibition of HER-2 signalling by specific human epidermal growth receptor 1/HER-2 (EGFR/HER-2) kinase inhibitor lapatinib synergistically enhanced the anti-cancer effects of honokiol in HER-2 over-expressed breast cancer cells [42].

The treatment of HNSCC cells with honokiol also decreased the expression of total EGFR as well as p-EGFR and its downstream target, mTOR. Since the activation of mTOR has been shown to contribute to tumour progression, it can be speculated that the honokiol-induced inhibition of cell proliferation in HNSCC cells is mediated through the downregulation of EGFR/mTOR signalling pathway [176,192]. These observations are consistent with the evidence that honokiol inhibits the growth of cancer cells by targeting EGFR and its downstream molecular targets and suggest that these mechanisms are in play in HNSCC.

6.4. Mammalian Target of Rapamycin (mTOR)

The mammalian target of rapamycin (mTOR) is a type of protein kinase which regulates cell metabolism, proliferation, and growth. The activation of PI3K/Akt pathway results in the aberrant activation of mTOR in most cancer cells [97,193,194]. It is known that mTOR controls the expression of many survival proteins via activating p70 S6 kinase (S6K) and inhibition of eIF4E inhibitor 4E-BP1 [193]. The mTOR signalling pathway is dysregulated in premalignant or early malignant human tissues and is highly implicated in the carcinogenic process. Honokiol suppresses the activation of mTOR and its signalling mediators (4E-BP1 and p70 S6 kinase) by inhibiting ERK and Akt pathways [43] or upregulating PTEN (Phosphatase and Tensin homolog) expression [42,157].

Honokiol was found to induce apoptosis and suppress migration and invasion in ovarian carcinoma cells (SKOV3 and Caov-3) via TSC1/TSC2 complex/AMPK/mTOR signalling pathway [28]. This is mediated via the regulation of the tumour suppressors p27, p53, and MMP-9 [28]. Furthermore, it was proven that honokiol was able to attenuate PI3K/Akt/mTOR signalling via the down-regulation of Akt phosphorylation and upregulation of PTEN expression in breast cancer cells (MCF-7, MCF-7/adr, and BT-474 cell lines) [42]. A combination of honokiol with the mTOR inhibitor rapamycin presented synergistic effects to induce apoptosis in breast cancer cells where the inhibition of PI3K/Akt/mTOR signalling by the mTOR inhibitor further sensitizes breast cancer cells to honokiol [42]. Other studies have also shown that honokiol induces autophagy in PC-3 and LNCaP prostate cancer cells via the suppression of mTOR and Akt phosphorylation [77]. Another study revealed that the treatment of neuroblastoma cells with honokiol caused significant downregulation of mTOR phosphorylation, which leads to the induction of autophagy of neuroblastoma cells (neuro-2a cells) through the PI3K/Akt/mTOR signalling pathways [96,195].

6.5. Hypoxia-Inducible-Factor (HIF) Pathway

The master regulator of neovascularisation, HIF, is a transcription factor that that plays an integral role in the body's response to low oxygen concentrations (i.e., hypoxia) [196,197]. Active HIF is composed of of two subunits: HIF-α and HIF-1/ARNT. Transcriptional regulation by oxygen is mediated by the HIF-α isoforms. In humans, three isoforms of α-subunit (HIF-1α, HIF-2α, and HIF-3α) have been identified. Recent studies suggest that transcriptional adaptation to hypoxia involves epigenetic changes in histone methylation. Strong evidence has established that the expression of pro-angiogenic factors (VEGF), which play a critical role in pathological neovascularisation in cancer, is elevated due to the activation of HIF pathway under hypoxia conditions [198].

An activation of the HIF pathway leading to hypoxia-induced neovascularisation is the central cause of pathogenesis in almost all solid tumours and ischemic retinal diseases [198,199]. There are studies reporting the capability of honokiol to inhibit HIF isoforms and the expression of hypoxic markers, as well as the binding of HIF to hypoxia-response elements present on VEGF promoter in D407 cells (human retinal pigment epithelial cells) [168]. In KRAS mutant lung cancer cells, it was discovered that Sirt3 was significantly up-regulated in honokiol-treated KRAS mutant lung cancer cells, leading to the destabilisation of its target gene Hif-1α and induction of G1 arrest and apoptosis. This suggests that the anticancer property of honokiol is regulated via a novel mechanism associated with the Sirt3/Hif-1α [49].

6.6. Notch Signalling Pathway

Notch signalling has been implicated in maintaining tissue homeostasis, including the regulation of self-renewal in adult stem cells, organ development, and embryonic development [200–202]. In mammals, the Notch receptor family comprises of four receptors (Notch-1, Notch-2, Notch-3, and Notch-4) and five ligands (Delta-like-1, Delta-like-3, Delta-like-4, Jagged-1, and Jagged-2). Each Notch receptor is activated through cell membrane-associated ligands. A series of proteolytic cleavage processes lead to the maturation and activation of Notch receptors. The first cleavage was catalysed by ADAM-family metalloprotease TACE, followed by the second cleavage mediated by γ-secretase, an enzyme complex that contains presenilin, nicastrin, presenilin enhancer 2 (PEN2), and anterior pharynx-defective 1 (APH1). The series of cleavages will lead to the release and translocation of Notch intracellular domain (NICD) into the nucleus [202]. Activated NICD is able to bind to activator proteins, including mastermind-like proteins (MAML) and recombination signalling binding protein-J (RBPJ) to form a nuclear transcriptional activator complex to regulate the transcription of downstream target genes, such as the hairy and enhancer of split (Hes) gene, Hey family genes, c-myc, cyclin D1, and p21/Waf1 [200]. The Notch pathway plays a complex role in the tumourigenesis of both hematologic and solid tissues. In fact, Notch signalling plays a vital role in regulating cellular differentiation, angiogenesis, proliferation, and apoptosis [201].

It has been shown that honokiol can eliminate cancer stem-like cells and potentiation of temozolomide (TMZ) sensitivity in glioblastoma multiforme (GBM) cells [36]. It was shown that honokiol enhanced the sensitization of GBM cells to MGMT inhibitor O6 benzylguanine (O6-BG) through the downregulation of Notch3 as well as the expression of its downstream target, Hes1 [36]. Furthermore, honokiol has been shown to inhibit B16/F-10, SKMEL-28 melanoma cell lines and SW480 colon cancer cells by targeting Notch signalling pathways [203,204]. Honokiol treatment resulted in reduced levels of cleaved Notch, particularly the Notch-2 receptor, along with a decrease in the expression of downstream target proteins, including Hes-1, cyclin D1, as well as TACE and γ-secretase complex proteins in melanoma cells [55].

Apart from that, honokiol in combination with radiation treatment reduced the number of DCLK1+ (cancer stem cell marker protein) colon cancer cells, which was accompanied by reduced levels of activated Notch-1, its ligand Jagged-1, and the downstream target gene Hes-1 [35,204]. Furthermore, the expression of components of the Notch-1 activating γ-secretase complex, presenilin 1, nicastrin, Pen2, and APH-1 were also suppressed [35]. To determine the effect of a honokiol–IR combination on tumour growth in vivo, nude mice tumour xenografts were administered honokiol intraperitoneally and exposed to IR. The honokiol–IR combination significantly inhibited tumour xenograft growth [35]. In addition, there were reduced levels of DCLK1 and the Notch signalling–related proteins in the xenograft tissues. Together, these data suggest that honokiol is a potent inhibitor of colon cancer growth that targets the stem cells by inhibiting the γ-secretase complex and the Notch signalling pathway [35,204].

6.7. Downregulation of P-Glycoprotein

The principal mechanism of multidrug resistance is due to the active transport of drugs out of cells [205]. Among the efflux transporters, P-glycoprotein (P-gp, gene symbol ABCB1) plays an important role in the resistance of cancer cells to a variety of chemotherapeutic treatments [205,206]. Furthermore, P-gp is distributed throughout the body where it interacts with various drugs of different structures to limit their bioavailability [207]. Therefore, the development of effective inhibitors of P-gp expression and/or functional activity should reverse drug resistance and enhance the bioavailability of P-gp substrates. One of the effective ways to overcome P-gp mediated drug resistance is either to block its drug-pump function or to inhibit its expression. To date, there are a total of three generations of P-gp inhibitors that have been discovered [207,208]. However, these compounds were not used widely due to toxicity at the doses required for attenuating P-gp activity, poor specificity, or unpredictable pharmacokinetic interactions. Honokiol was shown to downregulate the expression of P-gp at mRNA and protein levels in MCF-7/ADR, a human breast MDR cancer cell line [209,210]. The downregulation of P-gp was accompanied by a partial recovery of intracellular drug accumulation [210]. In MDR ovarian cancer cells (NCI/ADR-RES), honokiol has also been shown to downregulate the expression of P-gp in a concentration- and time-dependent manner [208].

7. Metabolism, Bioavailability, and Pharmacological Relevance of Honokiol

Pharmacokinetics involves the study of drug movement within the body, which includes the time course of absorption, distribution, metabolism, and excretion (ADME). Honokiol is mainly metabolized in the liver and undergoes in vivo biotransformation, whereby glucuronidation and sulfation are the main metabolic pathways to convert honokiol into mono-glucuronide honokiol and sulphated mono-hydroxyhonokiol before elimination [23]. This extensive biotransformation of honokiol may contribute to its low bioavailability. Currently, studies are being conducted to determine whether the metabolites of honokiol possess any biological activities that can extend the half-life of honokiol while maintaining its biological properties.

Most of the studies have reported that honokiol undergoes a rapid distribution and absorption, but slow elimination after intravenous (i.v.) administration [13,58,211,212]. For i.v. administration, it has been found that there was a rapid rate of distribution followed by a slower rate of elimination (elimination half-life $t_{1/2}$ = 49.22 min and 56.2 min for 5 mg or 10 mg of honokiol, respectively) observed in Sprague Dawley rats [213]. In another study, Liang et al. [214] investigated the pharmacokinetic properties of honokiol in beagle dogs after intravenous guttae, whereby the blood plasma of both male and female dogs was assessed. The elimination half-life ($t_{1/2}$ in hours) was found to be 20.13 (female), 9.27 (female), 7.06 (male), 4.70 (male), and 1.89 (male) after administration of doses of 8.8, 19.8, 3.9, 44.4, and 66.7 mg/kg, respectively. The $t_{1/2}$ decreases with an increase in the dose and length of infusion [214]. In another study, Wang et al. [61] discovered for the first time that honokiol is able to cross the blood–brain barrier (BBB) and blood–cerebrospinal fluid barrier (BCSFB) after i.v. administration when tested on intracerebral gliosarcoma model in Fisher 344 rats and human U251 xenograft glioma model in nude mice. It was also reported that the honokiol was distributed in the order of: lungs > plasma > liver > brain > kidney > heart > spleen after i.v. administration [61].

Furthermore, honokiol has been studied via an intraperitoneal route of administration. Chen et al. [33] reported a maximum plasma concentration of honokiol at 27.179 ± 6.252 min, with the $t_{1/2}$ of 312.08 ± 51.66 min after intraperitoneal injection of 250 mg/kg in BALB/c mice. On another note, studies have also shown that the presence of rhubarb and immature orange fruit extract in the decoction influenced the pharmacokinetics of honokiol, where a single oral dose of honokiol in Houpu decoction (a compound prescription of honokiol; 5 g/kg body weight) in Wistar rats demonstrated an elimination $t_{1/2}$ of 526.6 min [215]. Honokiol has a rapid absorption (Tmax = 20 min) and slow elimination ($t_{1/2}z$ = 290 min) after a single dose of oral gavage at 40 mg/kg in healthy rats [216]. In another study, honokiol showed a peak plasma concentration at 72 min, and $t_{1/2}$ of 186 min, and the absolute bioavailability for honokiol was found to be 5.3% when rats underwent oral administration of Magnolol/Honokiol emulsion (4:1) at 50 mg/kg [217]. After the rats were administered with honokiol orally, the honokiol was distributed rapidly to all parts of organs with the highest concentration being accumulated in the liver, followed by the brain and kidneys [216]. This was opposed to their discovery in tumour-bearing mice, where the highest concentration was found in the liver, followed by the kidneys and lungs [218]. This may be due to the different types of species being used as well as the tumor-burdened mice possibly affecting drug distribution [47]. With the rectal administration of Houpo extract at a dose of 245 mg/kg (equivalent to 13.5 mg/kg of honokiol) in Wistar rats, the maximal plasma concentration of honokiol found was approximately six times to that administered orally at an identical dose, indicating that rectal dosing avoids first-pass metabolism to some extent [219].

Meanwhile, the topical application of honokiol on UVB-induced contact hypersensitivity (CHS) as a model in C3H/HeN mice was also evaluated [68,220]. The topical application of honokiol (0.5 and 1.0 mg/cm^2 skin area) had a significant preventive effect on the UVB-induced suppression of the CHS response. The inflammatory mediators COX-2 and PGE$_2$ played a key role in this effect, as indicated by the honokiol-mediated inhibition of cyclooxygenase-2 (COX-2) expression and PGE$_2$ production in the UVB-exposed skin. Besides that, both topical application and oral administration of honokiol significantly inhibited (38% to 46%, $p < 0.001$) UVB-induced suppression of CHS in mice compared with the mice that were not treated with honokiol but exposed to UVB radiation. Prominently, the level of inhibition of CHS was not significantly different between the two modes of administration of honokiol [220].

Apart from that, Gao et al. [221] investigated the enhancement in the transdermal and localised delivery of honokiol through breast tissue. It was reported that microneedle-porated dermatome significantly increased the delivery of honokiol by nearly three-fold (97.81 ± 18.96 μg/cm^2) compared with passive delivery (32.56 ± 5.67 μg/cm^2). Oleic acid was found to be the best chemical penetration enhancer, increasing the delivery almost 27-fold (868.06 ± 100.91 μg/cm^2). The addition of oleic acid also resulted in a better retention of drugs in porcine mammary papilla (965.41 ± 80.26 μg/cm^2) compared with breast skin (294.16 ± 8.49 μg/cm^2) [221]. In summary, both microneedles and chemical enhancers can improve the absorption of honokiol through the skin. Directly applying honokiol on mammary papilla is a potential administration route which can increase localized delivery into breast tissue [183].

On another note, some studies have addressed the poor solubility of honokiol in hydrophilic environment. Wang et al. [222] developed polyethylene glycol-coated (PEGylated) liposomal honokiol to improve its solubility compared to free honokiol. PEGylated (polyethylene glycol coated) liposomal honokiol was shown to enhance the serum honokiol concentration and decrease clearance. The pharmacokinetic analysis of PEGylated liposomal honokiol showed a two-fold increase in elimination $t_{1/2}$ value as compared to that of free honokiol when being injected through the i.v. route (20 mg/kg body weight) in Balb/c mice (from 26 min in PEGylated liposomal honokiol to 13 min in free honokiol) [222]. Moreover, the AUC$_{0\rightarrow\infty}$ (mean concentration of drug in plasma) of PEGylated liposomal honokiol was about 1.85-fold higher than free honokiol. The protein-binding ability of honokiol in plasma was reported to be between 60% and 65% as revealed by equilibrium dialysis [222]. In another study, plasma honokiol concentrations were maintained above 30 and 10 μg/mL for 24 and 48 h, respectively, in liposomal honokiol-treated mice. However, it was reduced rapidly (<5 μg/mL) by 12 h in free honokiol-treated mice bearing A549 xenograft tumors, suggesting that liposomal honokiol extended blood circulation times in tumor-bearing mice compared to free honokiol [223].

8. Potential Drug Delivery of Honokiol

Due to the low water solubility and bioavailability of honokiol, multiple studies have been performed to develop proper honokiol delivery systems to improve its pharmacological effectiveness. A few studies have been performed to develop efficient drug carriers to deliver honokiol to its respective target, including the development of nanoparticles [224–226], micelles [227–229], and liposomes [73,171,223].

For honokiol delivery in the form of nanoparticles, Zheng et al. [230] developed monomethoxy poly(ethylene glycol)–poly(lactic acid) (MPEG–PLA) via ring opening polymerisation and then processed into nanoparticle for honokiol delivery. The honokiol-loaded MPEG–PLA nanoparticles were mono-dispersed and stable in the aqueous solution [230]. It was found that only 53% of honokiol was released from the nanoparticles within 24 h, while 100% of free honokiol was released into the outside media, suggesting that the honokiol loaded MPEG–PLA nanoparticle is a novel honokiol formulation which could meet the requirement of intravenous injection. In comparison, honokiol loaded MPEG-PLA nanoparticles significantly decreased the viability of A2780s cells (human ovarian cancer cells) than free honokiol, indicating that honokiol loaded MPEG–PLA nanoparticles might possess great potential applications for anticancer effect on cisplatin-sensitive A2780s cells in vitro [230]. In addition, the incorporation of both honokiol and doxorubicin in MPEG-PLA nanoparticles exhibited stronger anticancer activity than its individual form against A2780s cells [231].

In another study, emulsion solvent evaporation was used to develop the active targeting nanoparticle-loaded honokiol (ATNH) using copolymerpoly (ε-caprolactone)-poly (ethylene glycol)-poly (ε-caprolactone) (PCEC), which was modified with folate (FA) by introducing polyethylenimine (PEI) [30]. It was reported that ATNH showed a suitable size distribution, high encapsulation efficiency, gradual release, and targeting uptake by human nasopharynx carcinoma cells (HNE-1). Moreover, ATNH significantly inhibited tumour growth, metabolism, proliferation, micro-vessel generation, and caused cell-cycle arrest at the G1 phase [30]. Apart from that, epigallocatechin-3-gallate functionalized chitin loaded with honokiol nanoparticles (CE-HK NP), developed by Tang et al. [224], inhibit HepG2 cell growth and induce apoptosis through the suppression of mitochondrial membrane potential. Furthermore, CE-HK NPs (40 mg/kg) inhibited tumour growth by 83.55% ($p < 0.05$), which was far higher than the 30.15% inhibition of free honokiol (40 mg/kg). The proposed delivery system exhibits better tumour selectivity and growth reduction in both in vitro and in vivo models (male BALB/c nude mice treated with honokiol administrated by intertumoral injection) and did not induce any side effects [224]. Therefore, the CE-HK NPs may act as an effective delivery system for liver cancer. Recently, Yu et al. [232] further improved the design of nanoparticles for targeted delivery in breast cancer by surface modifying the honokiol nanoparticles through conjugation with folic acid to the surface of honokiol nanoparticles coated with polydopamine (HK-PDA-FA-NPs) as a pH-sensitive targeting anchor for nanoparticles. The targeted nanoparticles (HK-PDA-FA-NPs) can be stably present in various physiological media and exhibit pH sensitivity during drug release in vitro. HK-PDA-FA-NPs have better targeting ability to 4T1 cells than normal HK-NPs. Targeted nanoparticles have a tumour inhibition rate of greater than 80% in vivo (female Balb/c mice injected intraperitoneally with 40 mg/kg HK-PDA-FA-NPs), which is significantly higher than conventional HK-NPs [232].

For honokiol delivery in the form of micelles, researchers developed poly(ethylene glycol)-poly(ε-caprolactone)-poly(ethylene glycol) (PECE) micelle loaded with honokiol [229]. The cytotoxicity results showed that the composite drug delivery system is a safe carrier and the encapsulated honokiol retained its potent antitumor effect when tested against murine melanoma cell line B16 [233]. The IC_{50} values of free honokiol, honokiol nanoparticles, and honokiol micelles were 5.357, 6.274, and 6.746 μg/mL, respectively. The result indicated that the cytotoxicity of the honokiol micelles was lower than that of free honokiol, which was attributed to the sustained release behaviour of honokiol from honokiol micelles [233]. Further, comparing with honokiol nanoparticles, the cytotoxicity of honokiol micelles was a little lower, which might be due to the absence of organic solvent and surfactant in the honokiol micelles [233]. To increase the hydrophilicity of honokiol, Qiu et al. [234] developed an amphiphilic polymer–drug conjugate via the condensation of low molecular weight monomethoxy-poly(ethylene glycol) (MPEG)-2000 with honokiol through an ester linkage. The MPEG–honokiol (MPEG–HK) conjugate prepared formed nano-sized micelles, with a mean particle size of less than 20 nm (MPEG–HK, 360 $\mu g \cdot mL^{-1}$) in water, in which they could be well dispersed, and the results showed that only 20% of the conjugated honokiol was released in 2 h in beagle dog plasma, while in phosphate-buffered saline, the time required to reach 20% of honokiol release was >200 h [234]. Meanwhile, the inhibitory activity of the honokiol conjugate was found to be retained in vitro against LL/2 cell lines with an IC_{50} value of 10.7 µg/mL [234]. These results suggest that the polymer–drug conjugate provides a potential new approach to hydrophobic drugs, such as honokiol, in formulation design. In another study, nanomicellar honokiol (HNK-NM) with the size range of 20–40 nm was developed and compared against honokiol free drug (HNK-FD) [212]. Compared to HNK-FD, HNK-NM resulted in a significant increase in oral bioavailability. Cmax (4.06 and 3.60-fold) and AUC (6.26 and 5.83-fold) were significantly increased in comparison to oral 40 and 80 mg/kg HNK-FD, respectively, when tested in triple negative breast cancer cell lines (MDA-MB-231, MDA-MB-453, and MDA-MB-468). The anticancer effects of these formulations were also studied in BALB/c nude mice transplanted with orthotopic MDA-MB-231 cell induced xenografts [212]. After four weeks of daily oral administration of HNK-NM formulation, a significant reduction in the tumour volumes and weights compared to free drug ($p < 0.001$) treated groups was observed. Furthermore, in 25% of the mice, the treatment resulted in a complete eradication of tumours. Increased apoptosis and antiangiogenic effects were observed in HNK-NM groups compared to HNK-FD and untreated control mice [212].

Wang et al. [228] prepared paclitaxel (PTX) and honokiol (HK) combination methoxy poly(ethylene glycol) poly(caprolactone) micelles (P–H/M) via the solid dispersion method against breast cancer (4T1). The particle size of P–H/M was 28.7 ± 2.5 nm and spherical in shape. Both the cytotoxicity and the cellular uptake of P–H/M were increased in 4T1 cells, and P–H/M induced more apoptosis than PTX-loaded micelles or HK-loaded micelles. Furthermore, the antitumor effect of P–H/M was significantly improved compared with PTX-loaded micelles or HK-loaded micelles in vivo (Female Balb/c mice and female Balb/c nude mice treated with intravenous injection) [31,228]. P–H/M were more effective in inhibiting tumour proliferation, inducing tumour apoptosis, and decreasing the density of microvasculature accumulated more in tumour tissues compared to the free drug. After that, Wang et al. [235] developed paclitaxel (PTX) and honokiol (HNK) which are co-encapsulated into pH-sensitive polymeric micelles based on poly(2-ethyl-2-oxazoline)-poly(D,L-lactide) (PEOz-PLA). Results showed efficient inhibition of tumour metastasis by dual drug-loaded PEOz-PLA micelles in vitro anti-invasion and anti-migration assessment in MDA-MB-231 cells and in vivo in nude mice [235]. The suppression of MDR and metastasis by the micelles was assigned to the synergistic effects of pH-triggered drug release and HNK/PEOz-PLA-aroused P-gp inhibition, and pH-triggered drug release and PTX/HNK-aroused MMPs inhibition, respectively. After that, Wang et al. [236] proceeded to modify the paclitaxel plus honokiol micelles with dequalinium and tested it in non-small-cell lung cancer. When tested on Lewis lung tumour (LLT) cells, the polymeric micelles show powerful cytotoxicity, effective suppression on vasculogenic mimicry (VM) channels and tumour metastasis, as well as the activation of apoptotic enzymes caspase-3 and caspase-9, and down-regulation of FAK, PI3K, MMP-2, and MMP-9 [236]. In vivo assays (C57BL/6 mice treated through intravenous injection) indicated that polymeric micelles could increase the selective accumulation of chemotherapeutic drugs at tumour sites and showed a conspicuous anti-tumour efficacy [236].

For liposomes loaded with honokiol, Luo et al. [73] created liposomal honokiol and tested it on cisplatin-sensitive (A2780s) and -resistant (A2780cp) human ovarian cancer models. The administration of liposomal honokiol resulted in significant inhibition (84–88% maximum inhibition relative to controls) in the growth of A2780s and A2780cp tumour xenografts and prolonged the survival of the treated mice (treated twice weekly with intravenous administration) [73]. These anti-tumour responses were associated with marked increases in tumour apoptosis, and reductions in intratumoural microvessel density. Jiang et al. [223] incorporated honokiol in combination with cisplatin in the liposomes and tested it in A549 lung cancer xenograft nude mice model through intraperitoneal administration. This combination effectively suppressed tumour growth and significantly increased life span of treated mice compared to liposomal honokiol alone [223]. A similar result was observed in murine CT26 colon cancer models, where the systemic administration of liposomal honokiol with cisplatin resulted in the inhibition of subcutaneous tumour growth beyond the effects observed with either liposomal honokiol or cisplatin alone due to elevated levels of apoptosis and reduced endothelial cell density significantly [237]. In a recent study, hyaluronic acid (HA) modified daunorubicin plus honokiol cationic liposomes were prepared and characterised for the treatment of breast cancer by eliminating vasculogenic mimicry (VM) [238]. Studies found that the HA modified daunorubicin plus honokiol cationic liposomes enhanced the cellular uptake and destroyed VM channels. In addition, these liposomes prolonged their circulation time in the blood, and significantly accumulated at the tumour site to maximise its anticancer efficacy.

9. Future Perspective

Up to date, many in vitro and in vivo studies have identified the protective effects of honokiol in various types of cancers. However, the exact anticancer mechanism of honokiol is still insufficiently elucidated, especially its application in treating human cancer clinically. Since honokiol is being extensively metabolised in the body into different metabolites, it is vital to recognise the different types of metabolites circulating in the body in order to gain a better insight into the fate of honokiol after administration. The characterisation of honokiol metabolites would enable a better understanding of the overall bioactivity of honokiol as well as to determine the relationship between the bioactivity of the core molecule and its metabolites circulating within the target tissue. Moreover, future studies could focus on improving the methods used for in vitro studies to mimic more favourable in vivo conditions by considering the actual metabolites detected and concentrations found in the respective cancer tissues in order to better understand the mode of action of honokiol in cancer. Apart from that, it is essential to study the anticancer properties of the derivatives of honokiol as very few studies have been performed on the derivatives. It is important to study its derivatives as they might have improved and enhanced anticancer properties due to the change in structures and functional groups.

In short, more research can be done to confirm the anticancer properties of honokiol in more detail in order to come up with a safe and effective dosage to be used in chemoprevention and chemotherapy. Furthermore, more research can be done on the metabolism of honokiol via different routes of administration to find out the most effective route of administration for different types of cancer. The pre-formulation as well as formulation of honokiol can also be developed to prepare the transition of honokiol from pre-clinical to clinical studies in the future.

10. Conclusions

For centuries, researchers have been searching for strategies to control cancer progression through different approaches. Honokiol is a potential natural compound that exerts multiple effects on different cellular processes in various cancer models. Honokiol has been shown to regulate cell cycle arrest, induction of apoptosis, necrosis, and autophagy, as well as the inhibition of metastasis and angiogenesis through various signalling pathways. In addition, its effects are also validated in several in vivo studies with promising results where it can inhibit tumour growth and prolong survival in mouse cancer models. Current efforts are focusing on developing numerous drug delivery systems to improve the pharmacological, pharmacokinetics, and pharmacodynamic properties of honokiol. This review concludes that honokiol may be considered as a potential candidate for anticancer drug development.

Author Contributions: The literature searches and data collection were performed by W.L.L., C.P.O. and Y.Q.T. The manuscript was written by W.L.L., C.P.O., Y.Q.T. and W.H.Y. The manuscript was critically reviewed and edited by Y.Q.T. and W.H.Y. The project was conceptualized by W.H.Y. All authors have read and agreed to the published version of the manuscript.

Funding: This work was supported by the Ministry of Education (MOE) Fundamental Research Grant Scheme (FRGS/1/2019/SKK08/TAYLOR/02/2) awarded to W.H.Y. and Taylor's Internal Research Grant Scheme—Emerging Research Funding Scheme (TRGS/ERFS/1/2018/SBS/035) awarded to Y.Q.T.

Conflicts of Interest: The authors declare no conflict of interest.

References

1. Foster, I. Cancer: A cell cycle defect. *Radiography* **2008**, *14*, 144–149. [CrossRef]
2. Cabral, C.; Efferth, T.; Pires, I.M.; Severino, P.; Lemos, M.F.L. Natural Products as a Source for New Leads in Cancer Research and Treatment. *Evid.-Based Complement. Altern. Med.* **2018**, *2018*, 8243680. [CrossRef] [PubMed]
3. Wu, S.; Zhu, W.; Thompson, P.; Hannun, Y.A. Evaluating intrinsic and non-intrinsic cancer risk factors. *Nat. Commun.* **2018**, *9*, 3490. [CrossRef] [PubMed]
4. Siegel, R.L.; Miller, K.D.; Jemal, A. Cancer statistics, 2019. *CA Cancer J. Clin.* **2019**, *69*, 7–34. [CrossRef]
5. Bray, F.; Ferlay, J.; Soerjomataram, I.; Siegel, R.L.; Torre, L.A.; Jemal, A. Global cancer statistics 2018: GLOBOCAN estimates of incidence and mortality worldwide for 36 cancers in 185 countries. *CA Cancer J. Clin.* **2018**, *68*, 394–424. [CrossRef]
6. DeVita, V.T.; Canellos, G.P. New therapies and standard of care in oncology. *Nat. Rev. Clin. Oncol.* **2011**, *8*, 67–68. [CrossRef]
7. Marqus, S.; Pirogova, E.; Piva, T.J. Evaluation of the use of therapeutic peptides for cancer treatment. *J. Biomed. Sci.* **2017**, *24*, 21. [CrossRef]
8. Mitra, S.; Dash, R. Natural Products for the Management and Prevention of Breast Cancer. *Evid.-Based Complement. Altern. Med.* **2018**, *2018*, 8324696. [CrossRef]
9. Bayat Mokhtari, R.; Homayouni, T.S.; Baluch, N.; Morgatskaya, E.; Kumar, S.; Das, B.; Yeger, H. Combination therapy in combating cancer. *Oncotarget* **2017**, *8*, 38022–38043. [CrossRef]
10. Robinson, M.M.Z.; Zhang, X. *The World Medicines Situation 2011. Traditional Medicines: Global Situation, Issues and Challenges*; World Health Organization: Geneva, Switzerland, 2011.
11. Seelinger, M.; Popescu, R.; Giessrigl, B.; Jarukamjorn, K.; Unger, C.; Wallnöfer, B.; Fritzer-Szekeres, M.; Szekeres, T.; Diaz, R.; Jäger, W.; et al. Methanol extract of the ethnopharmaceutical remedy Smilax spinosa exhibits anti-neoplastic activity. *Int. J. Oncol.* **2012**, *41*, 1164–1172. [CrossRef]
12. Amaral, R.G.; dos Santos, S.A.; Andrade, L.N.; Severino, P.; Carvalho, A.A. Natural Products as Treatment against Cancer: A Historical and Current Vision. *Clin. Oncol.* **2019**, *4*, 1562.
13. Arora, S.; Singh, S.; Piazza, G.A.; Contreras, C.M.; Panyam, J.; Singh, A.P. Honokiol: A novel natural agent for cancer prevention and therapy. *Curr. Mol. Med.* **2012**, *12*, 1244–1252. [CrossRef] [PubMed]
14. Chen, Y.J.; Wu, C.L.; Liu, J.F.; Fong, Y.C.; Hsu, S.F.; Li, T.M.; Su, Y.C.; Liu, S.H.; Tang, C.H. Honokiol induces cell apoptosis in human chondrosarcoma cells through mitochondrial dysfunction and endoplasmic reticulum stress. *Cancer Lett.* **2010**, *291*, 20–30. [CrossRef] [PubMed]
15. Lee, J.D.; Lee, J.Y.; Baek, B.J.; Lee, B.D.; Koh, Y.W.; Lee, W.-S.; Lee, Y.-J.; Kwon, B.-M. The inhibitory effect of honokiol, a natural plant product, on vestibular schwannoma cells. *Laryngoscope* **2012**, *122*, 162–166. [CrossRef]
16. Amblard, F.; Delinsky, D.; Arbiser, J.L.; Schinazi, R.F. Facile purification of honokiol and its antiviral and cytotoxic properties. *J. Med. Chem.* **2006**, *49*, 3426–3427. [CrossRef]
17. Woodbury, A.; Yu, S.P.; Wei, L.; García, P. Neuro-modulating effects of honokiol: A review. *Front. Neurol.* **2013**, *4*, 130. [CrossRef]
18. Hahm, E.R.; Arlotti, J.A.; Marynowski, S.W.; Singh, S.V. Honokiol, a constituent of oriental medicinal herb magnolia officinalis, inhibits growth of PC-3 xenografts in vivo in association with apoptosis induction. *Clin. Cancer Res.* **2008**, *14*, 1248–1257. [CrossRef]
19. Chen, L. Rapid purification and scale-up of honokiol and magnolol using high-capacity high-speed counter-current chromatography A. *J. Chromatogr.* **2007**, *1142*, 115–122. [CrossRef]

20. Chen, C.-M.; Liu, Y.-C. ChemInform Abstract: A Concise Synthesis of Honokiol. *Tetrahedron Lett.* **2009**, *50*, 1151–1152. [CrossRef]
21. Gupta, M. Pharmacological Properties and Traditional Therapeutic Uses of Important Indian Spices: A Review. *Int. J. Food Prop.* **2010**, *13*, 1092–1116. [CrossRef]
22. Anand, K.W.; Wakode, S. Development of drugs based on Benzimidazole Heterocycle: Recent advancement and insights. *Int. J. Chem. Stud.* **2017**, *5*, 350–362.
23. Bohmdorfer, M.; Maier-Salamon, A.; Taferner, B.; Reznicek, G.; Thalhammer, T.; Hering, S.; Hufner, A.; Schuhly, W.; Jager, W. In vitro metabolism and disposition of honokiol in rat and human livers. *J. Pharm. Sci.* **2011**, *100*, 3506–3516. [CrossRef] [PubMed]
24. Lin, J.M.; Prakasha Gowda, A.S.; Sharma, A.K.; Amin, S. In vitro growth inhibition of human cancer cells by novel honokiol analogs. *Bioorg. Med. Chem.* **2012**, *20*, 3202–3211. [CrossRef] [PubMed]
25. Yang, S.E.; Hsieh, M.T.; Tsai, T.H.; Hsu, S.L. Down-modulation of Bcl-XL, release of cytochrome c and sequential activation of caspases during honokiol-induced apoptosis in human squamous lung cancer CH27 cells. *Biochem. Pharmacol.* **2002**, *63*, 1641–1651. [CrossRef]
26. Huang, K.J.; Kuo, C.H.; Chen, S.H.; Lin, C.Y.; Lee, Y.R. Honokiol inhibits in vitro and in vivo growth of oral squamous cell carcinoma through induction of apoptosis, cell cycle arrest and autophagy. *J. Cell. Mol. Med.* **2018**, *22*, 1894–1908. [CrossRef] [PubMed]
27. Huang, L.; Zhang, K.; Guo, Y.; Huang, F.; Yang, K.; Chen, L.; Huang, K.; Zhang, F.; Long, Q. Honokiol protects against doxorubicin cardiotoxicity via improving mitochondrial function in mouse hearts. *Sci. Rep.* **2017**, *7*, 11989. [CrossRef] [PubMed]
28. Lee, J.S.; Sul, J.Y.; Park, J.B.; Lee, M.S.; Cha, E.Y.; Ko, Y.B. Honokiol induces apoptosis and suppresses migration and invasion of ovarian carcinoma cells via AMPK/mTOR signaling pathway. *Int. J. Mol. Med.* **2019**, *43*, 1969–1978. [CrossRef]
29. Olusanya, T.O.B.; Haj Ahmad, R.R.; Ibegbu, D.M.; Smith, J.R.; Elkordy, A.A. Liposomal Drug Delivery Systems and Anticancer Drugs. *Molecules* **2018**, *23*, 907. [CrossRef]
30. Yang, B.; Ni, X.; Chen, L.; Zhang, H.; Ren, P.; Feng, Y.; Chen, Y.; Fu, S.; Wu, J. Honokiol-loaded polymeric nanoparticles: An active targeting drug delivery system for the treatment of nasopharyngeal carcinoma. *Drug Deliv.* **2017**, *24*, 660–669. [CrossRef]
31. Wang, X.; Beitler, J.J.; Wang, H.; Lee, M.J.; Huang, W.; Koenig, L.; Nannapaneni, S.; Amin, A.R.M.R.; Bonner, M.; Shin, H.J.C.; et al. Honokiol Enhances Paclitaxel Efficacy in Multi-Drug Resistant Human Cancer Model through the Induction of Apoptosis. *PLoS ONE* **2014**, *9*, e86369. [CrossRef]
32. Rajendran, P.; Li, F.; Shanmugam, M.K.; Vali, S.; Abbasi, T.; Kapoor, S.; Ahn, K.S.; Kumar, A.P.; Sethi, G. Honokiol inhibits signal transducer and activator of transcription-3 signaling, proliferation, and survival of hepatocellular carcinoma cells via the protein tyrosine phosphatase SHP-1. *J. Cell. Physiol.* **2012**, *227*, 2184–2195. [CrossRef] [PubMed]
33. Chen, F.; Wang, T.; Wu, Y.-F.; Gu, Y.; Xu, X.-L.; Zheng, S.; Hu, X. Honokiol: A potent chemotherapy candidate for human colorectal carcinoma. *World J. Gastroenterol.* **2004**, *10*, 3459–3463. [CrossRef] [PubMed]
34. He, Z.; Subramaniam, D.; Ramalingam, S.; Dhar, A.; Postier, R.G.; Umar, S.; Zhang, Y.; Anant, S. Honokiol radiosensitizes colorectal cancer cells: Enhanced activity in cells with mismatch repair defects. *Am. J. Physiol.-Gastrointest. Liver Physiol.* **2011**, *301*, G929–G937. [CrossRef] [PubMed]
35. Ponnurangam, S.; Mammen, J.M.; Ramalingam, S.; He, Z.; Zhang, Y.; Umar, S.; Subramaniam, D.; Anant, S. Honokiol in combination with radiation targets notch signaling to inhibit colon cancer stem cells. *Mol. Cancer Ther.* **2012**, *11*, 963–972. [CrossRef] [PubMed]
36. Lai, I.C.; Shih, P.-H.; Yao, C.-J.; Yeh, C.-T.; Wang-Peng, J.; Lui, T.-N.; Chuang, S.-E.; Hu, T.-S.; Lai, T.-Y.; Lai, G.-M. Elimination of Cancer Stem-Like Cells and Potentiation of Temozolomide Sensitivity by Honokiol in Glioblastoma Multiforme Cells. *PLoS ONE* **2015**, *10*, e0114830. [CrossRef] [PubMed]
37. Battle, T.E.; Arbiser, J.; Frank, D.A. The natural product honokiol induces caspase-dependent apoptosis in B-cell chronic lymphocytic leukemia (B-CLL) cells. *Blood* **2005**, *106*, 690–697. [CrossRef]
38. Gao, D.Q.; Qian, S.; Ju, T. Anticancer activity of Honokiol against lymphoid malignant cells via activation of ROS-JNK and attenuation of Nrf2 and NF-kappaB. *J. BUON* **2016**, *21*, 673–679.
39. Wolf, I.; O'Kelly, J.; Wakimoto, N.; Nguyen, A.; Amblard, F.; Karlan, B.Y.; Arbiser, J.L.; Koeffler, H.P. Honokiol, a natural biphenyl, inhibits in vitro and in vivo growth of breast cancer through induction of apoptosis and cell cycle arrest. *Int. J. Oncol.* **2007**, *30*, 1529–1537. [CrossRef]

40. Nagalingam, A.; Arbiser, J.L.; Bonner, M.Y.; Saxena, N.K.; Sharma, D. Honokiol activates AMP-activated protein kinase in breast cancer cells via an LKB1-dependent pathway and inhibits breast carcinogenesis. *Breast Cancer Res.* **2012**, *14*, R35. [CrossRef]

41. Sengupta, S.; Nagalingam, A.; Muniraj, N.; Bonner, M.Y.; Mistriotis, P.; Afthinos, A.; Kuppusamy, P.; Lanoue, D.; Cho, S.; Korangath, P.; et al. Activation of tumor suppressor LKB1 by honokiol abrogates cancer stem-like phenotype in breast cancer via inhibition of oncogenic Stat3. *Oncogene* **2017**, *36*, 5709–5721. [CrossRef]

42. Liu, H.; Zang, C.; Emde, A.; Planas-Silva, M.D.; Rosche, M.; Kuhnl, A.; Schulz, C.O.; Elstner, E.; Possinger, K.; Eucker, J. Anti-tumor effect of honokiol alone and in combination with other anti-cancer agents in breast cancer. *Eur. J. Pharmacol.* **2008**, *591*, 43–51. [CrossRef] [PubMed]

43. Park, E.J.; Min, H.Y.; Chung, H.J.; Hong, J.Y.; Kang, Y.J.; Hung, T.M.; Youn, U.J.; Kim, Y.S.; Bae, K.; Kang, S.S.; et al. Down-regulation of c-Src/EGFR-mediated signaling activation is involved in the honokiol-induced cell cycle arrest and apoptosis in MDA-MB-231 human breast cancer cells. *Cancer Lett.* **2009**, *277*, 133–140. [CrossRef] [PubMed]

44. Singh, T.; Katiyar, S.K. Honokiol Inhibits Non-Small Cell Lung Cancer Cell Migration by Targeting PGE2-Mediated Activation of β-Catenin Signaling. *PLoS ONE* **2013**, *8*, e60749. [CrossRef] [PubMed]

45. Singh, T.; Prasad, R.; Katiyar, S.K. Inhibition of class I histone deacetylases in non-small cell lung cancer by honokiol leads to suppression of cancer cell growth and induction of cell death in vitro and in vivo. *Epigenetics* **2013**, *8*, 54–65. [CrossRef]

46. Lv, X.; Liu, F.; Shang, Y.; Chen, S.Z. Honokiol exhibits enhanced antitumor effects with chloroquine by inducing cell death and inhibiting autophagy in human non-small cell lung cancer cells. *Oncol. Rep.* **2015**, *34*, 1289–1300. [CrossRef]

47. Pan, J.; Lee, Y.; Zhang, Q.; Xiong, D.; Wan, T.C.; Wang, Y.; You, M. Honokiol Decreases Lung Cancer Metastasis through Inhibition of the STAT3 Signaling Pathway. *Cancer Prev. Res.* **2017**, *10*, 133–141. [CrossRef]

48. Liou, S.-F.; Hua, K.-T.; Hsu, C.-Y.; Weng, M.-S. Honokiol from Magnolia spp. induces G1 arrest via disruption of EGFR stability through repressing HDAC6 deacetylated Hsp90 function in lung cancer cells. *J. Funct. Foods* **2015**, *15*, 84–96. [CrossRef]

49. Luo, L.-X.; Li, Y.; Liu, Z.-Q.; Fan, X.-X.; Duan, F.-G.; Li, R.-Z.; Yao, X.-J.; Leung, E.L.-H.; Liu, L. Honokiol Induces Apoptosis, G1 Arrest, and Autophagy in KRAS Mutant Lung Cancer Cells. *Front. Pharmacol.* **2017**, *8*, 199. [CrossRef]

50. Zhu, J.; Xu, S.; Gao, W.; Feng, J.; Zhao, G. Honokiol induces endoplasmic reticulum stress-mediated apoptosis in human lung cancer cells. *Life Sci.* **2019**, *221*, 204–211. [CrossRef]

51. Chae, J.I.; Jeon, Y.J.; Shim, J.H. Downregulation of Sp1 is involved in honokiol-induced cell cycle arrest and apoptosis in human malignant pleural mesothelioma cells. *Oncol. Rep.* **2013**, *29*, 2318–2324. [CrossRef]

52. Mannal, P.W.; Schneider, J.; Tangada, A.; McDonald, D.; McFadden, D.W. Honokiol produces anti-neoplastic effects on melanoma cells in vitro. *J. Surg. Oncol.* **2011**, *104*, 260–264. [CrossRef] [PubMed]

53. Chilampalli, C.; Guillermo, R.; Kaushik, R.S.; Young, A.; Chandrasekher, G.; Fahmy, H.; Dwivedi, C. Honokiol, a chemopreventive agent against skin cancer, induces cell cycle arrest and apoptosis in human epidermoid A431 cells. *Exp. Biol. Med.* **2011**, *236*, 1351–1359. [CrossRef] [PubMed]

54. Kaushik, G.; Ramalingam, S.; Subramaniam, D.; Rangarajan, P.; Protti, P.; Rammamoorthy, P.; Anant, S.; Mammen, J.M. Honokiol induces cytotoxic and cytostatic effects in malignant melanoma cancer cells. *Am. J. Surg.* **2012**, *204*, 868–873. [CrossRef] [PubMed]

55. Kaushik, G.; Kwatra, D.; Subramaniam, D.; Jensen, R.A.; Anant, S.; Mammen, J.M.V. Honokiol affects melanoma cell growth by targeting the AMP-activated protein kinase signaling pathway. *Am. J. Surg.* **2014**, *208*, 995–1002. [CrossRef] [PubMed]

56. Guillermo-Lagae, R.; Santha, S.; Thomas, M.; Zoelle, E.; Stevens, J.; Kaushik, R.S. Antineoplastic Effects of Honokiol on Melanoma. *Biomed Res. Int.* **2017**, *2017*, 5496398. [CrossRef] [PubMed]

57. Li, W.; Wang, Q.; Su, Q.; Ma, D.; An, C.; Ma, L.; Liang, H. Honokiol suppresses renal cancer cells' metastasis via dual-blocking epithelial-mesenchymal transition and cancer stem cell properties through modulating miR-141/ZEB2 signaling. *Mol. Cells* **2014**, *37*, 383–388. [CrossRef] [PubMed]

58. Averett, C.; Bhardwaj, A.; Arora, S.; Srivastava, S.K.; Khan, M.A.; Ahmad, A.; Singh, S., Carter, J.E.; Khushman, M.D.; Singh, A.P. Honokiol suppresses pancreatic tumor growth, metastasis and desmoplasia by interfering with tumor–stromal cross-talk. *Carcinogenesis* **2016**, *37*, 1052–1061. [CrossRef] [PubMed]

59. Arora, S.; Bhardwaj, A.; Srivastava, S.K.; Singh, S.; McClellan, S.; Wang, B.; Singh, A.P. Honokiol arrests cell cycle, induces apoptosis, and potentiates the cytotoxic effect of gemcitabine in human pancreatic cancer cells. *PLoS ONE* **2011**, *6*, e21573. [CrossRef] [PubMed]

60. Lu, C.H.; Chen, S.H.; Chang, Y.S.; Liu, Y.W.; Wu, J.Y.; Lim, Y.P.; Yu, H.I.; Lee, Y.R. Honokiol, a potential therapeutic agent, induces cell cycle arrest and program cell death in vitro and in vivo in human thyroid cancer cells. *Pharmacol. Res.* **2017**, *115*, 288–298. [CrossRef]

61. Wang, X.; Duan, X.; Yang, G.; Zhang, X.; Deng, L.; Zheng, H.; Deng, C.; Wen, J.; Wang, N.; Peng, C.; et al. Honokiol crosses BBB and BCSFB, and inhibits brain tumor growth in rat 9L intracerebral gliosarcoma model and human U251 xenograft glioma model. *PLoS ONE* **2011**, *6*, e18490. [CrossRef]

62. Jeong, J.J.; Lee, J.H.; Chang, K.C.; Kim, H.J. Honokiol exerts an anticancer effect in T98G human glioblastoma cells through the induction of apoptosis and the regulation of adhesion molecules. *Int. J. Oncol.* **2012**, *41*, 1358–1364. [CrossRef] [PubMed]

63. Fan, Y.; Xue, W.; Schachner, M.; Zhao, W. Honokiol Eliminates Glioma/Glioblastoma Stem Cell-Like Cells Via JAK-STAT3 Signaling and Inhibits Tumor Progression by Targeting Epidermal Growth Factor Receptor. *Cancers* **2018**, *11*, 22. [CrossRef] [PubMed]

64. Chang, K.H.; Yan, M.D.; Yao, C.J.; Lin, P.C.; Lai, G.M. Honokiol-induced apoptosis and autophagy in glioblastoma multiforme cells. *Oncol. Lett.* **2013**, *6*, 1435–1438. [CrossRef] [PubMed]

65. Joo, Y.N.; Eun, S.Y.; Park, S.W.; Lee, J.H.; Chang, K.C.; Kim, H.J. Honokiol inhibits U87MG human glioblastoma cell invasion through endothelial cells by regulating membrane permeability and the epithelial-mesenchymal transition. *Int. J. Oncol.* **2014**, *44*, 187–194. [CrossRef]

66. Lin, C.J.; Chen, T.L.; Tseng, Y.Y.; Wu, G.J.; Hsieh, M.H.; Lin, Y.W.; Chen, R.M. Honokiol induces autophagic cell death in malignant glioma through reactive oxygen species-mediated regulation of the p53/PI3K/Akt/mTOR signaling pathway. *Toxicol. Appl. Pharmacol.* **2016**, *304*, 59–69. [CrossRef]

67. Huang, K.; Chen, Y.; Zhang, R.; Wu, Y.; Ma, Y.; Fang, X.; Shen, S. Honokiol induces apoptosis and autophagy via the ROS/ERK1/2 signaling pathway in human osteosarcoma cells in vitro and in vivo. *Cell Death Dis.* **2018**, *9*, 157. [CrossRef]

68. Steinmann, P.; Walters, D.K.; Arlt, M.J.; Banke, I.J.; Ziegler, U.; Langsam, B.; Arbiser, J.; Muff, R.; Born, W.; Fuchs, B. Antimetastatic activity of honokiol in osteosarcoma. *Cancer* **2012**, *118*, 2117–2127. [CrossRef]

69. Yang, J.; Zou, Y.; Jiang, D. Honokiol suppresses proliferation and induces apoptosis via regulation of the miR21/PTEN/PI3K/AKT signaling pathway in human osteosarcoma cells. *Int. J. Mol. Med.* **2018**, *41*, 1845–1854. [CrossRef]

70. Kim, D.W.; Ko, S.M.; Jeon, Y.J.; Noh, Y.W.; Choi, N.J.; Cho, S.D.; Moon, H.S.; Cho, Y.S.; Shin, J.C.; Park, S.M.; et al. Anti-proliferative effect of honokiol in oral squamous cancer through the regulation of specificity protein 1. *Int. J. Oncol.* **2013**, *43*, 1103–1110. [CrossRef]

71. Xu, Q.; Tong, F.; He, C.; Song, P.; Xu, Q.; Chen, Z. The inhibition effect of Honokiol in liver cancer. *Int. J. Clin. Exp. Med.* **2018**, *11*, 10673–10678.

72. Han, L.L.; Xie, L.P.; Li, L.H.; Zhang, X.W.; Zhang, R.Q.; Wang, H.Z. Reactive oxygen species production and Bax/Bcl-2 regulation in honokiol-induced apoptosis in human hepatocellular carcinoma SMMC-7721 cells. *Environ. Toxicol. Pharmacol.* **2009**, *28*, 97–103. [CrossRef] [PubMed]

73. Luo, H.; Zhong, Q.; Chen, L.J.; Qi, X.R.; Fu, A.F.; Yang, H.S.; Yang, F.; Lin, H.G.; Wei, Y.Q.; Zhao, X. Liposomal honokiol, a promising agent for treatment of cisplatin-resistant human ovarian cancer. *J. Cancer Res. Clin. Oncol.* **2008**, *134*, 937–945. [CrossRef] [PubMed]

74. Li, Z.; Liu, Y.; Zhao, X.; Pan, X.; Yin, R.; Huang, C.; Chen, L.; Wei, Y. Honokiol, a natural therapeutic candidate, induces apoptosis and inhibits angiogenesis of ovarian tumor cells. *Eur. J. Obstet. Gynecol. Reprod. Biol.* **2008**, *140*, 95–102. [CrossRef] [PubMed]

75. Hahm, E.R.; Singh, S.V. Honokiol causes G0-G1 phase cell cycle arrest in human prostate cancer cells in association with suppression of retinoblastoma protein level/phosphorylation and inhibition of E2F1 transcriptional activity. *Mol. Cancer Ther.* **2007**, *6*, 2686–2695. [CrossRef] [PubMed]

76. Shigemura, K.; Arbiser, J.L.; Sun, S.Y.; Zayzafoon, M.; Johnstone, P.A.; Fujisawa, M.; Gotoh, A.; Weksler, B.; Zhau, H.E.; Chung, L.W. Honokiol, a natural plant product, inhibits the bone metastatic growth of human prostate cancer cells. *Cancer* **2007**, *109*, 1279–1289. [CrossRef] [PubMed]

77. Hahm, E.R.; Sakao, K.; Singh, S.V. Honokiol activates reactive oxygen species-mediated cytoprotective autophagy in human prostate cancer cells. *Prostate* **2014**, *74*, 1209–1221. [CrossRef]

78. Leeman-Neill, R.J.; Cai, Q.; Joyce, S.C.; Thomas, S.M.; Bhola, N.E.; Neill, D.B.; Arbiser, J.L.; Grandis, J.R. Honokiol inhibits epidermal growth factor receptor signaling and enhances the antitumor effects of epidermal growth factor receptor inhibitors. *Clin. Cancer Res.* **2010**, *16*, 2571–2579. [CrossRef]

79. Lin, J.W.; Chen, J.T.; Hong, C.Y.; Lin, Y.L.; Wang, K.T.; Yao, C.J.; Lai, G.M.; Chen, R.M. Honokiol traverses the blood-brain barrier and induces apoptosis of neuroblastoma cells via an intrinsic bax-mitochondrion-cytochrome c-caspase protease pathway. *Neuro-Oncology* **2012**, *14*, 302–314. [CrossRef]

80. Yeh, P.S.; Wang, W.; Chang, Y.A.; Lin, C.J.; Wang, J.J.; Chen, R.M. Honokiol induces autophagy of neuroblastoma cells through activating the PI3K/Akt/mTOR and endoplasmic reticular stress/ERK1/2 signaling pathways and suppressing cell migration. *Cancer Lett.* **2016**, *370*, 66–77. [CrossRef]

81. Zhang, Q.; Zhao, W.; Ye, C.; Zhuang, J.; Chang, C.; Li, Y.; Huang, X.; Shen, L.; Li, Y.; Cui, Y.; et al. Honokiol inhibits bladder tumor growth by suppressing EZH2/miR-143 axis. *Oncotarget* **2015**, *6*, 37335–37348. [CrossRef]

82. Bao, L.; Jaramillo, M.C.; Zhang, Z.; Zheng, Y.; Yao, M.; Zhang, D.D.; Yi, X. Induction of autophagy contributes to cisplatin resistance in human ovarian cancer cells. *Mol. Med. Rep.* **2015**, *11*, 91–98. [CrossRef] [PubMed]

83. Zhang, Q.; Cheng, J.; Xin, Q. Effects of tetracycline on developmental toxicity and molecular responses in zebrafish (*Danio rerio*) embryos. *Ecotoxicology* **2015**, *24*, 707–719. [CrossRef] [PubMed]

84. Hill, D.; Chen, L.; Snaar-Jagalska, E.; Chaudhry, B. Embryonic zebrafish xenograft assay of human cancer metastasis. *F1000Research* **2018**, *7*, 1682. [CrossRef] [PubMed]

85. Li, L.; Han, W.; Gu, Y.; Qiu, S.; Lu, Q.; Jin, J.; Luo, J.; Hu, X. Honokiol Induces a Necrotic Cell Death through the Mitochondrial Permeability Transition Pore. *Cancer Res.* **2007**, *67*, 4894. [CrossRef] [PubMed]

86. Sheu, M.L.; Liu, S.H.; Lan, K.H. Honokiol induces calpain-mediated glucose-regulated protein-94 cleavage and apoptosis in human gastric cancer cells and reduces tumor growth. *PLoS ONE* **2007**, *2*, e1096. [CrossRef]

87. Liu, S.H.; Wang, K.B.; Lan, K.H.; Lee, W.J.; Pan, H.C.; Wu, S.M.; Peng, Y.C.; Chen, Y.C.; Shen, C.C.; Cheng, H.C.; et al. Calpain/SHP-1 Interaction by Honokiol Dampening Peritoneal Dissemination of Gastric Cancer in nu/nu Mice. *PLoS ONE* **2012**, *7*, e43711. [CrossRef]

88. Liu, Y.; Chen, L.; He, X.; Fan, L.; Yang, G.; Chen, X.; Lin, X.; Du, L.; Li, Z.; Ye, H.; et al. Enhancement of therapeutic effectiveness by combining liposomal honokiol with cisplatin in ovarian carcinoma. *Int. J. Gynecol. Cancer* **2008**, *18*, 652–659. [CrossRef]

89. Goldar, S.; Khaniani, M.S.; Derakhshan, S.M.; Baradaran, B. Molecular mechanisms of apoptosis and roles in cancer development and treatment. *Asian Pac. J. Cancer Prev.* **2015**, *16*, 2129–2144. [CrossRef]

90. Hassan, M.; Watari, H.; AbuAlmaaty, A.; Ohba, Y.; Sakuragi, N. Apoptosis and molecular targeting therapy in cancer. *Biomed Res. Int.* **2014**, *2014*, 150845. [CrossRef]

91. Wong, R.S.Y. Apoptosis in cancer: From pathogenesis to treatment. *J. Exp. Clin. Cancer Res.* **2011**, *30*, 87. [CrossRef]

92. Gerl, R.; Vaux, D.L. Apoptosis in the development and treatment of cancer. *Carcinogenesis* **2005**, *26*, 263–270. [CrossRef] [PubMed]

93. Letai, A. Apoptosis and Cancer. *Annu. Rev. Cancer Biol.* **2017**, *1*, 275–294. [CrossRef]

94. Jeong, Y.-H.; Hur, J.H.; Jeon, E.-J.; Park, S.-J.; Hwang, T.J.; Lee, S.A.; Lee, W.K.; Sung, J.M. Honokiol Improves Liver Steatosis in Ovariectomized Mice. *Molecules* **2018**, *23*, 194. [CrossRef] [PubMed]

95. Huang, J.-S.; Yao, C.-J.; Chuang, S.-E.; Yeh, C.-T.; Lee, L.-M.; Chen, R.-M.; Chao, W.-J.; Whang-Peng, J.; Lai, G.-M. Honokiol inhibits sphere formation and xenograft growth of oral cancer side population cells accompanied with JAK/STAT signaling pathway suppression and apoptosis induction. *BMC Cancer* **2016**, *16*, 245. [CrossRef] [PubMed]

96. Prasad, R.; Katiyar, S.K. Honokiol, an Active Compound of Magnolia Plant, Inhibits Growth, and Progression of Cancers of Different Organs. *Adv. Exp. Med. Biol.* **2016**, *928*, 245–265. [CrossRef] [PubMed]

97. Banik, K.; Ranaware, A.M.; Deshpande, V.; Nalawade, S.P.; Padmavathi, G.; Bordoloi, D.; Sailo, B.L.; Shanmugam, M.K.; Fan, L.; Arfuso, F.; et al. Honokiol for cancer therapeutics: A traditional medicine that can modulate multiple oncogenic targets. *Pharmacol. Res.* **2019**, *144*, 192–209. [CrossRef] [PubMed]

98. Wang, X.; Beitler, J.J.; Huang, W.; Chen, G.; Qian, G.; Magliocca, K.; Patel, M.R.; Chen, A.Y.; Zhang, J.; Nannapaneni, S.; et al. Honokiol Radiosensitizes Squamous Cell Carcinoma of the Head and Neck by Downregulation of Survivin. *Clin. Cancer Res.* **2018**, *24*, 858–869. [CrossRef]

99. Garcia, A.; Zheng, Y.; Zhao, C.; Toschi, A.; Fan, J.; Shraibman, N.; Brown, H.A.; Bar-Sagi, D.; Foster, D.A.; Arbiser, J.L. Honokiol Suppresses Survival Signals Mediated by Ras-Dependent Phospholipase D Activity in Human Cancer Cells. *Clin. Cancer Res.* **2008**, *14*, 4267. [CrossRef]

100. Fried, L.E.; Arbiser, J.L. Honokiol, a multifunctional antiangiogenic and antitumor agent. *Antioxid. Redox Signal.* **2009**, *11*, 1139–1148. [CrossRef]

101. Li, H.Y.; Ye, H.G.; Chen, C.Q.; Yin, L.H.; Wu, J.B.; He, L.C.; Gao, S.M. Honokiol induces cell cycle arrest and apoptosis via inhibiting class I histone deacetylases in acute myeloid leukemia. *J. Cell. Biochem.* **2015**, *116*, 287–298. [CrossRef]

102. Deng, J.; Qian, Y.; Geng, L.; Chen, J.; Wang, X.; Xie, H.; Yan, S.; Jiang, G.; Zhou, L.; Zheng, S. Involvement of p38 mitogen-activated protein kinase pathway in honokiol-induced apoptosis in a human hepatoma cell line (hepG2). *Liver Int.* **2008**, *28*, 1458–1464. [CrossRef] [PubMed]

103. Hasegawa, S.; Yonezawa, T.; Ahn, J.Y.; Cha, B.Y.; Teruya, T.; Takami, M.; Yagasaki, K.; Nagai, K.; Woo, J.T. Honokiol inhibits osteoclast differentiation and function in vitro. *Biol. Pharm. Bull.* **2010**, *33*, 487–492. [CrossRef] [PubMed]

104. Tse, A.K.; Wan, C.K.; Shen, X.L.; Yang, M.; Fong, W.F. Honokiol inhibits TNF-alpha-stimulated NF-kappaB activation and NF-kappaB-regulated gene expression through suppression of IKK activation. *Biochem. Pharmacol.* **2005**, *70*, 1443–1457. [CrossRef] [PubMed]

105. Li, J.; Shao, X.; Wu, L.; Feng, T.; Jin, C.; Fang, M.; Wu, N.; Yao, H. Honokiol: An effective inhibitor of tumor necrosis factor-alpha-induced up-regulation of inflammatory cytokine and chemokine production in human synovial fibroblasts. *Acta Biochim. Biophys. Sin.* **2011**, *43*, 380–386. [CrossRef]

106. Xu, H.L.; Tang, W.; Du, G.H.; Kokudo, N. Targeting apoptosis pathways in cancer with magnolol and honokiol, bioactive constituents of the bark of *Magnolia officinalis. Drug Discov. Ther.* **2011**, *5*, 202–210. [CrossRef]

107. Raja, S.M.; Chen, S.; Yue, P.; Acker, T.M.; Lefkove, B.; Arbiser, J.L.; Khuri, F.R.; Sun, S.Y. The natural product honokiol preferentially inhibits cellular FLICE-inhibitory protein and augments death receptor-induced apoptosis. *Mol. Cancer Ther.* **2008**, *7*, 2212–2223. [CrossRef]

108. Rauf, A.; Patel, S.; Imran, M.; Maalik, A.; Arshad, M.U.; Saeed, F.; Mabkhot, Y.N.; Al-Showiman, S.S.; Ahmad, N.; Elsharkawy, E. Honokiol: An anticancer lignan. *Biomed. Pharmacother.* **2018**, *107*, 555–562. [CrossRef]

109. Schroder, M.; Kaufman, R.J. ER stress and the unfolded protein response. *Mutat. Res.* **2005**, *569*, 29–63. [CrossRef]

110. Cao, S.S.; Kaufman, R.J. Endoplasmic reticulum stress and oxidative stress in cell fate decision and human disease. *Antioxid. Redox Signal.* **2014**, *21*, 396–413. [CrossRef]

111. Ferri, K.F.; Kroemer, G. Organelle-specific initiation of cell death pathways. *Nat. Cell Biol.* **2001**, *3*, E255–E263. [CrossRef]

112. Chiu, C.-S.; Tsai, C.-H.; Hsieh, M.-S.; Tsai, S.-C.; Jan, Y.-J.; Lin, W.-Y.; Lai, D.-W.; Wu, S.-M.; Hsing, H.-Y.; Arbiser, J.L.; et al. Exploiting Honokiol-induced ER stress CHOP activation inhibits the growth and metastasis of melanoma by suppressing the MITF and β-catenin pathways. *Cancer Lett.* **2019**, *442*, 113–125. [CrossRef] [PubMed]

113. Liu, S.H.; Shen, C.C.; Yi, Y.C.; Tsai, J.J.; Wang, C.C.; Chueh, J.T.; Lin, K.L.; Lee, T.C.; Pan, H.C.; Sheu, M.L. Honokiol inhibits gastric tumourigenesis by activation of 15-lipoxygenase-1 and consequent inhibition of peroxisome proliferator-activated receptor-γ and COX-2-dependent signals. *Br. J. Pharmacol.* **2010**, *160*, 1963–1972. [CrossRef] [PubMed]

114. Liu, S.H.; Lee, W.J.; Lai, D.W.; Wu, S.M.; Liu, C.Y.; Tien, H.R.; Chiu, C.S.; Peng, Y.C.; Jan, Y.J.; Chao, T.H.; et al. Honokiol confers immunogenicity by dictating calreticulin exposure, activating ER stress and inhibiting epithelial-to-mesenchymal transition. *Mol. Oncol.* **2015**, *9*, 834–849. [CrossRef] [PubMed]

115. Martin, S.; Lamb, H.K.; Brady, C.; Lefkove, B.; Bonner, M.Y.; Thompson, P.; Lovat, P.E.; Arbiser, J.L.; Hawkins, A.R.; Redfern, C.P. Inducing apoptosis of cancer cells using small-molecule plant compounds that bind to GRP78. *Br. J. Cancer* **2013**, *109*, 433–443. [CrossRef] [PubMed]

116. Lee, S.Y.; Ju, M.K.; Jeon, H.M.; Jeong, E.K.; Lee, Y.J.; Kim, C.H.; Park, H.G.; Han, S.I.; Kang, H.S. Regulation of Tumor Progression by Programmed Necrosis. *Oxidative Med. Cell. Longev.* **2018**, *2018*, 3537471. [CrossRef]

117. Tian, W.; Xu, D.; Deng, Y.-C. Honokiol, a multifunctional tumor cell death inducer. *Die Pharm. Int. J. Pharm. Sci.* **2012**, *67*, 811–816. [CrossRef]

118. Chen, G.; Izzo, J.; Demizu, Y.; Wang, F.; Guha, S.; Wu, X.; Hung, M.-C.; Ajani, J.A.; Huang, P. Different redox states in malignant and nonmalignant esophageal epithelial cells and differential cytotoxic responses to bile acid and honokiol. *Antioxid. Redox Signal.* **2009**, *11*, 1083–1095. [CrossRef]

119. Yu, C.; Zhang, Q.; Zhang, H.Y.; Zhang, X.; Huo, X.; Cheng, E.; Wang, D.H.; Arbiser, J.L.; Spechler, S.J.; Souza, R.F. Targeting the intrinsic inflammatory pathway: Honokiol exerts proapoptotic effects through STAT3 inhibition in transformed Barrett's cells. *Am. J. Physiol. Gastrointest. Liver Physiol.* **2012**, *303*, G561–G569. [CrossRef]

120. Meier, J.A.; Hyun, M. Stress-induced dynamic regulation of mitochondrial STAT3 and its association with cyclophilin D reduce mitochondrial ROS production. *Sci. Signal.* **2017**, *10*. [CrossRef]

121. Cen, M.; Yao, Y.; Cui, L.; Yang, G.; Lu, G.; Fang, L.; Bao, Z.; Zhou, J. Honokiol induces apoptosis of lung squamous cell carcinoma by targeting FGF2-FGFR1 autocrine loop. *Cancer Med.* **2018**, *7*, 6205–6218. [CrossRef]

122. Hahm, E.R.; Singh, K.B.; Singh, S.V. c-Myc is a novel target of cell cycle arrest by honokiol in prostate cancer cells. *Cell Cycle* **2016**, *15*, 2309–2320. [CrossRef] [PubMed]

123. Vaid, M.; Sharma, S.D.; Katiyar, S.K. Honokiol, a phytochemical from the Magnolia plant, inhibits photocarcinogenesis by targeting UVB-induced inflammatory mediators and cell cycle regulators: Development of topical formulation. *Carcinogenesis* **2010**, *31*, 2004–2011. [CrossRef] [PubMed]

124. Guo, C.; Ma, L.; Zhao, Y.; Peng, A.; Cheng, B.; Zhou, Q.; Zheng, L.; Huang, K. Inhibitory effects of magnolol and honokiol on human calcitonin aggregation. *Sci. Rep.* **2015**, *5*, 13556. [CrossRef] [PubMed]

125. Yan, B.; Peng, Z.Y. Honokiol induces cell cycle arrest and apoptosis in human gastric carcinoma MGC-803 cell line. *Int. J. Clin. Exp. Med.* **2015**, *8*, 5454–5461. [PubMed]

126. Grimmel, M.; Backhaus, C.; Proikas-Cezanne, T. WIPI-Mediated Autophagy and Longevity. *Cells* **2015**, *4*, 202–217. [CrossRef] [PubMed]

127. Yun, C.W.; Lee, S.H. The Roles of Autophagy in Cancer. *Int. J. Mol. Sci.* **2018**, *19*, 3466. [CrossRef]

128. Galluzzi, L.; Pietrocola, F.; Bravo-San Pedro, J.M.; Amaravadi, R.K.; Baehrecke, E.H.; Cecconi, F.; Codogno, P.; Debnath, J.; Gewirtz, D.A.; Karantza, V.; et al. Autophagy in malignant transformation and cancer progression. *EMBO J.* **2015**, *34*, 856–880. [CrossRef]

129. Lin, L.; Baehrecke, E.H. Autophagy, cell death, and cancer. *Mol. Cell Oncol.* **2015**, *2*, e985913. [CrossRef]

130. Paquette, M.; El-Houjeiri, L.; Pause, A. mTOR Pathways in Cancer and Autophagy. *Cancers* **2018**, *10*, 18. [CrossRef]

131. Murray, J.T.; Tee, A.R. Mechanistic Target of Rapamycin (mTOR) in the Cancer Setting. *Cancers* **2018**, *10*, 168. [CrossRef]

132. Itakura, E.; Mizushima, N. Atg14 and UVRAG: Mutually exclusive subunits of mammalian Beclin 1-PI3K complexes. *Autophagy* **2009**, *5*, 534–536. [CrossRef] [PubMed]

133. Torii, S.; Yoshida, T.; Arakawa, S.; Honda, S.; Nakanishi, A.; Shimizu, S. Identification of PPM1D as an essential Ulk1 phosphatase for genotoxic stress-induced autophagy. *EMBO Rep.* **2016**, *17*, 1552–1564. [CrossRef] [PubMed]

134. Maiuri, M.C.; Criollo, A.; Kroemer, G. Crosstalk between apoptosis and autophagy within the Beclin 1 interactome. *EMBO J.* **2010**, *29*, 515–516. [CrossRef] [PubMed]

135. Carlsson, S.R.; Simonsen, A. Membrane dynamics in autophagosome biogenesis. *J. Cell Sci.* **2015**, *128*, 193. [CrossRef] [PubMed]

136. Mochida, K.; Oikawa, Y.; Kimura, Y.; Kirisako, H.; Hirano, H.; Ohsumi, Y.; Nakatogawa, H. Receptor-mediated selective autophagy degrades the endoplasmic reticulum and the nucleus. *Nature* **2015**, *522*, 359–362. [CrossRef]

137. Wrighton, K.H. Selecting ER for eating. *Nat. Rev. Mol. Cell Biol.* **2015**, *16*, 389. [CrossRef]

138. Chio, C.C.; Chen, K.Y.; Chang, C.K.; Chuang, J.Y.; Liu, C.C.; Liu, S.H.; Chen, R.M. Improved effects of honokiol on temozolomide-induced autophagy and apoptosis of drug-sensitive and -tolerant glioma cells. *BMC Cancer* **2018**, *18*, 379. [CrossRef]

139. Nieto, M.A.; Huang, R.Y.; Jackson, R.A.; Thiery, J.P. EMT: 2016. *Cell* **2016**, *166*, 21–45. [CrossRef]

140. Brabletz, T.; Kalluri, R.; Nieto, M.A.; Weinberg, R.A. EMT in cancer. *Nat. Rev. Cancer* **2018**, *18*, 128–134. [CrossRef]

141. Roche, J. The Epithelial-to-Mesenchymal Transition in Cancer. *Cancers* **2018**, *10*, 52. [CrossRef]

142. Røsland, G.V.; Dyrstad, S.E.; Tusubira, D.; Helwa, R.; Tan, T.Z.; Lotsberg, M.L.; Pettersen, I.K.N.; Berg, A.; Kindt, C.; Hoel, F.; et al. Epithelial to mesenchymal transition (EMT) is associated with attenuation of succinate dehydrogenase (SDH) in breast cancer through reduced expression of SDHC. *Cancer Metab.* **2019**, *7*, 6. [CrossRef] [PubMed]

143. Shen, L.; Zhang, F.; Huang, R.; Yan, J.; Shen, B. Honokiol inhibits bladder cancer cell invasion through repressing SRC-3 expression and epithelial-mesenchymal transition. *Oncol. Lett.* **2017**, *14*, 4294–4300. [CrossRef] [PubMed]

144. Avtanski, D.B.; Nagalingam, A.; Bonner, M.Y.; Arbiser, J.L.; Saxena, N.K.; Sharma, D. Honokiol inhibits epithelial-mesenchymal transition in breast cancer cells by targeting signal transducer and activator of transcription 3/Zeb1/E-cadherin axis. *Mol. Oncol.* **2014**, *8*, 565–580. [CrossRef] [PubMed]

145. Lv, X.-Q.; Qiao, X.-R.; Su, L.; Chen, S.-Z. Honokiol inhibits EMT-mediated motility and migration of human non-small cell lung cancer cells in vitro by targeting c-FLIP. *Acta Pharmacol. Sin.* **2016**, *37*, 1574–1586. [CrossRef]

146. Qin, L.; Liu, Z.; Chen, H.; Xu, J. The steroid receptor coactivator-1 regulates twist expression and promotes breast cancer metastasis. *Cancer Res.* **2009**, *69*, 3819–3827. [CrossRef]

147. Yao, C.-J.; Lai, G.-M.; Yeh, C.-T.; Lai, M.-T.; Shih, P.-H.; Chao, W.-J.; Whang-Peng, J.; Chuang, S.-E.; Lai, T.-Y. Honokiol Eliminates Human Oral Cancer Stem-Like Cells Accompanied with Suppression of Wnt/β-Catenin Signaling and Apoptosis Induction. *Evid.-Based Complement. Altern. Med.* **2013**, *2013*, 146136. [CrossRef]

148. Wang, W.D.; Shang, Y.; Li, Y.; Chen, S.Z. Honokiol inhibits breast cancer cell metastasis by blocking EMT through modulation of Snail/Slug protein translation. *Acta Pharm. Sin.* **2019**, *40*, 1219–1227. [CrossRef]

149. Galichon, P.; Hertig, A. Epithelial to mesenchymal transition as a biomarker in renal fibrosis: Are we ready for the bedside? *Fibrogenesis Tissue Repair* **2011**, *4*, 11. [CrossRef]

150. Conacci-Sorrell, M.; Ngouenet, C.; Anderson, S.; Brabletz, T.; Eisenman, R.N. Stress-induced cleavage of Myc promotes cancer cell survival. *Genes Dev.* **2014**, *28*, 689–707. [CrossRef]

151. Yamaguchi, H.; Wyckoff, J.; Condeelis, J. Cell migration in tumors. *Curr. Opin. Cell Biol.* **2005**, *17*, 559–564. [CrossRef]

152. Seyfried, T.N.; Huysentruyt, L.C. On the origin of cancer metastasis. *Crit. Rev. Oncog.* **2013**, *18*, 43–73. [CrossRef] [PubMed]

153. Tay, R.Y.; Fernández-Gutiérrez, F.; Foy, V.; Burns, K.; Pierce, J.; Morris, K.; Priest, L.; Tugwood, J.; Ashcroft, L.; Lindsay, C.R.; et al. Prognostic value of circulating tumour cells in limited-stage small-cell lung cancer: Analysis of the concurrent once-daily versus twice-daily radiotherapy (CONVERT) randomised controlled trial. *Ann. Oncol.* **2019**, *30*, 1114–1120. [CrossRef] [PubMed]

154. Singh, T.; Katiyar, S.K. Honokiol, a phytochemical from *Magnolia* spp., inhibits breast cancer cell migration by targeting nitric oxide and cyclooxygenase-2. *Int. J. Oncol.* **2011**, *38*, 769–776. [CrossRef]

155. Zhang, J.; Zhang, Y.; Shen, W.; Fu, R.; Ding, Z.; Zhen, Y.; Wan, Y. Cytological effects of honokiol treatment and its potential mechanism of action in non-small cell lung cancer. *Biomed. Pharmacother.* **2019**, *117*, 109058. [CrossRef] [PubMed]

156. Cheng, S.; Castillo, V.; Welty, M.; Eliaz, I.; Sliva, D. Honokiol inhibits migration of renal cell carcinoma through activation of RhoA/ROCK/MLC signaling pathway. *Int. J. Oncol.* **2016**, *49*, 1525–1530. [CrossRef]

157. Balan, M.; Chakraborty, S.; Flynn, E.; Zurakowski, D.; Pal, S. Honokiol inhibits c-Met-HO-1 tumor-promoting pathway and its cross-talk with calcineurin inhibitor-mediated renal cancer growth. *Sci. Rep.* **2017**, *7*, 5900. [CrossRef]

158. Alizadeh, A.M.; Shiri, S.; Farsinejad, S. Metastasis review: From bench to bedside. *Tumor Biol.* **2014**, *35*, 8483–8523. [CrossRef]

159. Klein, C.A. Cancer. The metastasis cascade. *Science* **2008**, *321*, 1785–1787. [CrossRef]

160. Bai, X.; Cerimele, F.; Ushio-Fukai, M.; Waqas, M.; Campbell, P.M.; Govindarajan, B.; Der, C.J.; Battle, T.; Frank, D.A.; Ye, K.; et al. Honokiol, a small molecular weight natural product, inhibits angiogenesis in vitro and tumor growth in vivo. *J. Biol. Chem.* **2003**, *278*, 35501–35507. [CrossRef]

161. Kolligs, F.T.; Bommer, G.; Goke, B. Wnt/beta-catenin/tcf signaling: A critical pathway in gastrointestinal tumorigenesis. *Digestion* **2002**, *66*, 131–144. [CrossRef]

162. Hlubek, F.; Spaderna, S.; Jung, A.; Kirchner, T.; Brabletz, T. Beta-catenin activates a coordinated expression of the proinvasive factors laminin-5 gamma2 chain and MT1-MMP in colorectal carcinomas. *Int. J. Cancer* **2004**, *108*, 321–326. [CrossRef] [PubMed]

163. Cheng, S.; Castillo, V.; Eliaz, I.; Sliva, D. Honokiol suppresses metastasis of renal cell carcinoma by targeting KISS1/KISS1R signaling. *Int. J. Oncol.* **2015**, *46*, 2293–2298. [CrossRef] [PubMed]

164. Nishida, N.; Yano, H.; Nishida, T.; Kamura, T.; Kojiro, M. Angiogenesis in cancer. *Vasc. Health Risk Manag.* **2006**, *2*, 213–219. [CrossRef] [PubMed]

165. Rajabi, M.; Mousa, S.A. The Role of Angiogenesis in Cancer Treatment. *Biomedicines* **2017**, *5*, 34. [CrossRef] [PubMed]

166. Tonini, T.; Rossi, F.; Claudio, P.P. Molecular basis of angiogenesis and cancer. *Oncogene* **2003**, *22*, 6549–6556. [CrossRef] [PubMed]

167. Banerjee, P.; Basu, A.; Arbiser, J.L.; Pal, S. The natural product honokiol inhibits calcineurin inhibitor-induced and Ras-mediated tumor promoting pathways. *Cancer Lett.* **2013**, *338*, 292–299. [CrossRef]

168. Vavilala, D.T.; Ponnaluri, V.K.C.; Kanjilal, D.; Mukherji, M. Evaluation of anti-HIF and anti-angiogenic properties of honokiol for the treatment of ocular neovascular diseases. *PLoS ONE* **2014**, *9*, e113717. [CrossRef]

169. Vavilala, D.T.; O'Bryhim, B.E.; Ponnaluri, V.K.; White, R.S.; Radel, J.; Symons, R.C.; Mukherji, M. Honokiol inhibits pathological retinal neovascularization in oxygen-induced retinopathy mouse model. *Biochem. Biophys. Res. Commun.* **2013**, *438*, 697–702. [CrossRef]

170. Wen, J.; Wang, X.; Pei, H.; Xie, C.; Qiu, N.; Li, S.; Wang, W.; Cheng, X.; Chen, L. Anti-psoriatic effects of Honokiol through the inhibition of NF-kappaB and VEGFR-2 in animal model of K14-VEGF transgenic mouse. *J. Pharmacol. Sci.* **2015**, *128*, 116–124. [CrossRef]

171. Hu, J.; Chen, L.J.; Liu, L.; Chen, X.; Chen, P.L.; Yang, G.; Hou, W.L.; Tang, M.H.; Zhang, F.; Wang, X.H.; et al. Liposomal honokiol, a potent anti-angiogenesis agent, in combination with radiotherapy produces a synergistic antitumor efficacy without increasing toxicity. *Exp. Mol. Med.* **2008**, *40*, 617–628. [CrossRef]

172. Xia, Y.; Shen, S.; Verma, I.M. NF-κB, an active player in human cancers. *Cancer Immunol. Res.* **2014**, *2*, 823–830. [CrossRef] [PubMed]

173. Xia, L.; Tan, S.; Zhou, Y.; Lin, J.; Wang, H.; Oyang, L.; Tian, Y.; Liu, L.; Su, M.; Wang, H.; et al. Role of the NFκB-signaling pathway in cancer. *Onco Targets Ther.* **2018**, *11*, 2063–2073. [CrossRef] [PubMed]

174. Ahn, K.S.; Sethi, G.; Shishodia, S.; Sung, B.; Arbiser, J.L.; Aggarwal, B.B. Honokiol Potentiates Apoptosis, Suppresses Osteoclastogenesis, and Inhibits Invasion through Modulation of Nuclear Factor-κB Activation Pathway. *Mol. Cancer Res.* **2006**, *4*, 621. [CrossRef] [PubMed]

175. Wang, Z.; Zhang, X. Chemopreventive Activity of Honokiol against 7, 12—Dimethylbenz[a]anthracene-Induced Mammary Cancer in Female Sprague Dawley Rats. *Front. Pharmacol.* **2017**, *8*, 320. [CrossRef]

176. Katiyar, S.K. Emerging Phytochemicals for the Prevention and Treatment of Head and Neck Cancer. *Molecules* **2016**, *21*, 1610. [CrossRef]

177. Hua, H.; Chen, W.; Shen, L.; Sheng, Q.; Teng, L. Honokiol augments the anti-cancer effects of oxaliplatin in colon cancer cells. *Acta Biochim. Biophys. Sin.* **2013**, *45*, 773–779. [CrossRef]

178. Nabekura, T.; Hiroi, T.; Kawasaki, T.; Uwai, Y. Effects of natural nuclear factor-kappa B inhibitors on anticancer drug efflux transporter human P-glycoprotein. *Biomed. Pharmacother.* **2015**, *70*, 140–145. [CrossRef]

179. Furqan, M.; Akinleye, A.; Mukhi, N.; Mittal, V.; Chen, Y.; Liu, D. STAT inhibitors for cancer therapy. *J. Hematol. Oncol.* **2013**, *6*, 90. [CrossRef]

180. Yu, H.; Pardoll, D.; Jove, R. STATs in cancer inflammation and immunity: A leading role for STAT3. *Nat. Rev. Cancer* **2009**, *9*, 798–809. [CrossRef]

181. Pan, J.; Lee, Y.; Cheng, G.; Zielonka, J.; Zhang, Q.; Bajzikova, M.; Xiong, D.; Tsaih, S.-W.; Hardy, M.; Flister, M.; et al. Mitochondria-Targeted Honokiol Confers a Striking Inhibitory Effect on Lung Cancer via Inhibiting Complex I Activity. *iScience* **2018**, *3*, 192–207. [CrossRef]

182. He, Z.; Subramaniam, D.; Zhang, Z.; Zhang, Y.; Anant, S. Honokiol as a Radiosensitizing Agent for Colorectal cancers. *Curr. Colorectal Cancer Rep.* **2013**, *9*, 358–364. [CrossRef] [PubMed]

183. Bi, L.; Yu, Z.; Wu, J.; Yu, K.; Hong, G.; Lu, Z.; Gao, S. Honokiol Inhibits Constitutive and Inducible STAT3 Signaling via PU.1-Induced SHP1 Expression in Acute Myeloid Leukemia Cells. *Tohoku J. Exp. Med.* **2015**, *237*, 163–172. [CrossRef] [PubMed]

184. Ishikawa, C.; Arbiser, J.L.; Mori, N. Honokiol induces cell cycle arrest and apoptosis via inhibition of survival signals in adult T-cell leukemia. *Biochim. Biophys. Acta* **2012**, *1820*, 879–887. [CrossRef] [PubMed]

185. Sasaki, T.; Hiroki, K.; Yamashita, Y. The role of epidermal growth factor receptor in cancer metastasis and microenvironment. *BioMed Res. Int.* **2013**, *2013*, 546318. [CrossRef] [PubMed]

186. Normanno, N.; De Luca, A.; Bianco, C.; Strizzi, L.; Mancino, M.; Maiello, M.R.; Carotenuto, A.; De Feo, G.; Caponigro, F.; Salomon, D.S. Epidermal growth factor receptor (EGFR) signaling in cancer. *Gene* **2006**, *366*, 2–16. [CrossRef] [PubMed]

187. Kari, C.; Chan, T.O.; Rocha de Quadros, M.; Rodeck, U. Targeting the Epidermal Growth Factor Receptor in Cancer. *Cancer Res.* **2003**, *63*, 1. [PubMed]

188. Yewale, C.; Baradia, D.; Vhora, I.; Patil, S.; Misra, A. Epidermal growth factor receptor targeting in cancer: A review of trends and strategies. *Biomaterials* **2013**, *34*, 8690–8707. [CrossRef]

189. Song, J.M.; Anandharaj, A.; Upadhyaya, P.; Kirtane, A.R.; Kim, J.-H.; Hong, K.H.; Panyam, J.; Kassie, F. Honokiol suppresses lung tumorigenesis by targeting EGFR and its downstream effectors. *Oncotarget* **2016**, *7*, 57752–57769. [CrossRef]

190. Dai, X.; Li, R.-Z.; Jiang, Z.-B.; Wei, C.-L.; Luo, L.-X.; Yao, X.-J.; Li, G.-P.; Leung, E.L.-H. Honokiol Inhibits Proliferation, Invasion and Induces Apoptosis Through Targeting Lyn Kinase in Human Lung Adenocarcinoma Cells. *Front. Pharmacol.* **2018**, *9*. [CrossRef]

191. Biscardi, J.S.; Ishizawar, R.C.; Silva, C.M.; Parsons, S.J. Tyrosine kinase signalling in breast cancer: Epidermal growth factor receptor and c-Src interactions in breast cancer. *Breast Cancer Res.* **2000**, *2*, 203–210. [CrossRef]

192. Singh, T.; Gupta, N.A.; Xu, S.; Prasad, R.; Velu, S.E.; Katiyar, S.K. Honokiol inhibits the growth of head and neck squamous cell carcinoma by targeting epidermal growth factor receptor. *Oncotarget* **2015**, *6*, 21268–21282. [CrossRef] [PubMed]

193. Dufour, M.; Dormond-Meuwly, A.; Demartines, N.; Dormond, O. Targeting the Mammalian Target of Rapamycin (mTOR) in Cancer Therapy: Lessons from Past and Future Perspectives. *Cancers* **2011**, *3*, 2478–2500. [CrossRef] [PubMed]

194. Hua, H.; Kong, Q.; Zhang, H.; Wang, J.; Luo, T.; Jiang, Y. Targeting mTOR for cancer therapy. *J. Hematol. Oncol.* **2019**, *12*, 71. [CrossRef] [PubMed]

195. Li, Z.; Dong, H.; Li, M.; Wu, Y.; Liu, Y.; Zhao, Y.; Chen, X.; Ma, M. Honokiol induces autophagy and apoptosis of osteosarcoma through PI3K/Akt/mTOR signaling pathway. *Mol. Med. Rep.* **2018**, *17*, 2719–2723. [CrossRef]

196. Pezzuto, A.; Carico, E. Role of HIF-1 in Cancer Progression: Novel Insights. A Review. *Curr. Mol. Med.* **2018**, *18*, 343–351. [CrossRef]

197. Masoud, G.N.; Li, W. HIF-1α pathway: Role, regulation and intervention for cancer therapy. *Acta Pharm. Sin. B* **2015**, *5*, 378–389. [CrossRef]

198. Semenza, G.L. Targeting HIF-1 for cancer therapy. *Nat. Rev. Cancer* **2003**, *3*, 721–732. [CrossRef]

199. Soni, S.; Padwad, Y.S. HIF-1 in cancer therapy: Two decade long story of a transcription factor. *Acta Oncol.* **2017**, *56*, 503–515. [CrossRef]

200. Aster, J.C.; Pear, W.S.; Blacklow, S.C. The Varied Roles of Notch in Cancer. *Annu. Rev. Pathol.* **2017**, *12*, 245–275. [CrossRef]

201. Venkatesh, V.; Nataraj, R.; Thangaraj, G.S.; Karthikeyan, M.; Gnanasekaran, A.; Kaginelli, S.B.; Kuppanna, G.; Kallappa, C.G.; Basalingappa, K.M. Targeting Notch signalling pathway of cancer stem cells. *Stem Cell Investig.* **2018**, *5*, 5. [CrossRef]

202. Nowell, C.S.; Radtke, F. Notch as a tumour suppressor. *Nat. Rev. Cancer* **2017**, *17*, 145–159. [CrossRef] [PubMed]

203. Kaushik, G.; Venugopal, A.; Ramamoorthy, P.; Standing, D.; Subramaniam, D.; Umar, S.; Jensen, R.A.; Anant, S.; Mammen, J.M. Honokiol inhibits melanoma stem cells by targeting notch signaling. *Mol. Carcinog.* **2015**, *54*, 1710–1721. [CrossRef] [PubMed]

204. Wynn, M.L.; Consul, N.; Merajver, S.D.; Schnell, S. Inferring the Effects of Honokiol on the Notch Signaling Pathway in SW480 Colon Cancer Cells. *Cancer Inform.* **2014**, *13*, 1–12. [CrossRef] [PubMed]

205. Callaghan, R.; Luk, F.; Bebawy, M. Inhibition of the multidrug resistance P-glycoprotein: Time for a change of strategy? *Drug Metab. Dispos.* **2014**, *42*, 623–631. [CrossRef]

206. Waghray, D.; Zhang, Q. Inhibit or Evade Multidrug Resistance P-Glycoprotein in Cancer Treatment. *J. Med. Chem.* **2018**, *61*, 5108–5121. [CrossRef]

207. Nanayakkara, A.K.; Follit, C.A.; Chen, G.; Williams, N.S.; Vogel, P.D.; Wise, J.G. Targeted inhibitors of P-glycoprotein increase chemotherapeutic-induced mortality of multidrug resistant tumor cells. *Sci. Rep.* **2018**, *8*, 967. [CrossRef]

208. Han, H.K.; Van Anh, L.T. Modulation of P-glycoprotein expression by honokiol, magnolol and 4-O-methylhonokiol, the bioactive components of Magnolia officinalis. *Anticancer Res.* **2012**, *32*, 4445–4452.

209. Wang, X.; Cho, K.; Chen, Z.; Arbiser, J.; Shin, D. Honokiol reduces drug resistance by inhibition of P-glycoprotein expression in multidrug resistant (MDR) squamous cell carcinoma of the head and neck (SCCHN). *Cancer Res.* **2007**, *67*, 2776.

210. Xu, D.; Lu, Q.; Hu, X. Down-regulation of P-glycoprotein expression in MDR breast cancer cell MCF-7/ADR by honokiol. *Cancer Lett.* **2006**, *243*, 274–280. [CrossRef]

211. Wang, H.; Liao, Z.; Sun, X.; Shi, Q.; Huo, G.; Xie, Y.; Tang, X.; Zhi, X.; Tang, Z. Intravenous administration of Honokiol provides neuroprotection and improves functional recovery after traumatic brain injury through cell cycle inhibition. *Neuropharmacology* **2014**, *86*, 9–21. [CrossRef]

212. Godugu, C.; Doddapaneni, R.; Singh, M. Honokiol nanomicellar formulation produced increased oral bioavailability and anticancer effects in triple negative breast cancer (TNBC). *Colloids Surf. B* **2017**, *153*, 208–219. [CrossRef] [PubMed]

213. Tsai, T.H.; Chou, C.J.; Cheng, F.C.; Chen, C.F. Pharmacokinetics of honokiol after intravenous administration in rats assessed using high-performance liquid chromatography. *J. Chromatogr. B* **1994**, *655*, 41–45. [CrossRef]

214. Liang, Y.; Cui, G.; Wang, X.; Zhang, W.; An, Q.; Lin, Z.; Wang, H.; Chen, S. Pharmacokinetics of honokiol after intravenous guttae in beagle dogs assessed using ultra-performance liquid chromatography-tandem mass spectrometry. *Biomed. Chromatogr.* **2014**, *28*, 1378–1383. [CrossRef] [PubMed]

215. Su, W.J.; Huang, X.; Qin, E.; Jiang, L.; Ren, P. Pharmacokinetics of honokiol in rat after oral administration of Cortex of Magnolia officinalis and its compound preparation Houpu Sanwu Decoction. *J. Chin. Med. Mater.* **2008**, *31*, 255–258.

216. Wang, J.; Miao, X.L.; Chen, J.Y.; Chen, Y. The Pharmacokinetics and Tissue Distribution of Honokiol and its Metabolites in Rats. *Eur. J. Drug Metab. Pharmacokinet.* **2016**, *41*, 587–594. [CrossRef]

217. Sheng, Y.L.; Xu, J.H.; Shi, C.H.; Li, W.; Xu, H.Y.; Li, N.; Zhao, Y.Q.; Zhang, X.R. UPLC-MS/MS-ESI assay for simultaneous determination of magnolol and honokiol in rat plasma: Application to pharmacokinetic study after administration emulsion of the isomer. *J. Ethnopharmacol.* **2014**, *155*, 1568–1574. [CrossRef]

218. Han, M.; Yu, X.; Guo, Y.; Wang, Y.; Kuang, H.; Wang, X. Honokiol nanosuspensions: Preparation, increased oral bioavailability and dramatically enhanced biodistribution in the cardio-cerebro-vascular system. *Colloids Surf. B* **2014**, *116*, 114–120. [CrossRef]

219. Wu, X.; Chen, X.; Hu, Z. High-performance liquid chromatographic method for simultaneous determination of honokiol and magnolol in rat plasma. *Talanta* **2003**, *59*, 115–121. [CrossRef]

220. Prasad, R.; Singh, T.; Katiyar, S.K. Honokiol inhibits ultraviolet radiation-induced immunosuppression through inhibition of ultraviolet-induced inflammation and DNA hypermethylation in mouse skin. *Sci. Rep.* **2017**, *7*, 1657 [CrossRef]

221. Gao, X.; Patel, M.G.; Bakshi, P.; Sharma, D.; Banga, A.K. Enhancement in the Transdermal and Localized Delivery of Honokiol Through Breast Tissue. *AAPS PharmSciTech* **2018**, *19*, 3501–3511. [CrossRef]

222. Wang, X.H.; Cai, L.L.; Zhang, X.Y.; Deng, L.Y.; Zheng, H.; Deng, C.Y.; Wen, J.L.; Zhao, X.; Wei, Y.Q.; Chen, L.J. Improved solubility and pharmacokinetics of PEGylated liposomal honokiol and human plasma protein binding ability of honokiol. *Int. J. Pharm.* **2011**, *410*, 169–174. [CrossRef] [PubMed]

223. Jiang, Q.Q.; Fan, L.Y.; Yang, G.L.; Guo, W.H.; Hou, W.L.; Chen, L.J.; Wei, Y.Q. Improved therapeutic effectiveness by combining liposomal honokiol with cisplatin in lung cancer model. *BMC Cancer* **2008**, *8*, 242. [CrossRef] [PubMed]

224. Tang, P.; Sun, Q.; Yang, H.; Tang, B.; Pu, H.; Li, H. Honokiol nanoparticles based on epigallocatechin gallate functionalized chitin to enhance therapeutic effects against liver cancer. *Int. J. Pharm.* **2018**, *545*, 74–83. [CrossRef] [PubMed]

225. Wu, W.; Wang, L.; Wang, L.; Zu, Y.; Wang, S.; Liu, P.; Zhao, X. Preparation of honokiol nanoparticles by liquid antisolvent precipitation technique, characterization, pharmacokinetics, and evaluation of inhibitory effect on HepG2 cells. *Int. J. Nanomed.* **2018**, *13*, 5469–5483. [CrossRef] [PubMed]

226. Guo, Y.; Zhao, Y.; Wang, T.; Zhao, S.; Qiu, H.; Han, M.; Wang, X. Honokiol nanoparticles stabilized by oligoethylene glycols codendrimer: In vitro and in vivo investigations. *J. Mater. Chem. B* **2017**, *5*, 697–706. [CrossRef]

227. Li, X.; Hou, X.; Ding, W.; Cong, S.; Zhang, Y.; Chen, M.; Meng, Y.; Lei, J.; Liu, Y.; Li, G. Sirolimus-loaded polymeric micelles with honokiol for oral delivery. *J. Pharm. Pharmacol.* **2015**, *67*, 1663–1672. [CrossRef] [PubMed]

228. Wang, N.; Wang, Z.; Nie, S.; Song, L.; He, T.; Yang, S.; Yang, X.; Yi, C.; Wu, Q.; Gong, C. Biodegradable polymeric micelles coencapsulating paclitaxel and honokiol: A strategy for breast cancer therapy in vitro and in vivo. *Int. J. Nanomed.* **2017**, *12*, 1499–1514. [CrossRef]

229. Gong, C.; Wei, X.; Wang, X.; Wang, Y.; Guo, G.; Mao, Y.; Luo, F.; Qian, Z. Biodegradable self-assembled PEG–PCL–PEG micelles for hydrophobic honokiol delivery: I. Preparation and characterization. *Nanotechnology* **2010**, *21*, 215103. [CrossRef]

230. Zheng, X.; Kan, B.; Gou, M.; Fu, S.; Zhang, J.; Men, K.; Chen, L.; Luo, F.; Zhao, Y.; Zhao, X.; et al. Preparation of MPEG–PLA nanoparticle for honokiol delivery in vitro. *Int. J. Pharm.* **2010**, *386*, 262–267. [CrossRef]

231. Wang, B.; Gou, M.; Zheng, X.; Wei, X.; Gong, C.; Wang, X.; Zhao, Y.; Luo, F.; Chen, L.; Qian, Z.; et al. Co-delivery honokiol and doxorubicin in MPEG-PLA nanoparticles. *J. Nanosci. Nanotechnol.* **2010**, *10*, 4166–4172. [CrossRef]

232. Yu, R.; Zou, Y.; Liu, B.; Guo, Y.; Wang, X.; Han, M. Surface modification of pH-sensitive honokiol nanoparticles based on dopamine coating for targeted therapy of breast cancer. *Colloids Surf. B* **2019**, *177*, 1–10. [CrossRef] [PubMed]

233. Gong, C.; Shi, S.; Wang, X.; Wang, Y.; Fu, S.; Dong, P.; Chen, L.; Zhao, X.; Wei, Y.; Qian, Z. Novel composite drug delivery system for honokiol delivery: Self-assembled poly(ethylene glycol)-poly(epsilon-caprolactone)-poly(ethylene glycol) micelles in thermosensitive poly(ethylene glycol)-poly(epsilon-caprolactone)-poly(ethylene glycol) hydrogel. *J. Phys. Chem. B* **2009**, *113*, 10183–10188. [CrossRef] [PubMed]

234. Qiu, N.; Cai, L.L.; Xie, D.; Wang, G.; Wu, W.; Zhang, Y.; Song, H.; Yin, H.; Chen, L. Synthesis, structural and in vitro studies of well-dispersed monomethoxy-poly(ethylene glycol)-honokiol conjugate micelles. *Biomed. Mater.* **2010**, *5*, 065006. [CrossRef] [PubMed]

235. Wang, Z.; Li, X.; Wang, D.; Zou, Y.; Qu, X.; He, C.; Deng, Y.; Jin, Y.; Zhou, Y.; Zhou, Y.; et al. Concurrently suppressing multidrug resistance and metastasis of breast cancer by co-delivery of paclitaxel and honokiol with pH-sensitive polymeric micelles. *Acta Biomater.* **2017**, *62*, 144–156. [CrossRef] [PubMed]

236. Wang, X.; Cheng, L.; Xie, H.-J.; Ju, R.-J.; Xiao, Y.; Fu, M.; Liu, J.-J.; Li, X.-T. Functional paclitaxel plus honokiol micelles destroying tumour metastasis in treatment of non-small-cell lung cancer. *Artif. Cells Nanomed. Biotechnol.* **2018**, *46*, 1154–1169. [CrossRef] [PubMed]

237. Cheng, N.; Xia, T.; Han, Y.; He, Q.J.; Zhao, R.; Ma, J.R. Synergistic antitumor effects of liposomal honokiol combined with cisplatin in colon cancer models. *Oncol. Lett.* **2011**, *2*, 957–962. [CrossRef] [PubMed]

238. Ju, R.J.; Cheng, L.; Qiu, X.; Liu, S.; Song, X.L.; Peng, X.M.; Wang, T.; Li, C.Q.; Li, X.T. Hyaluronic acid modified daunorubicin plus honokiol cationic liposomes for the treatment of breast cancer along with the elimination vasculogenic mimicry channels. *J. Drug Target.* **2018**, *26*, 793–805. [CrossRef]

Review

Molecular Insights into Potential Contributions of Natural Polyphenols to Lung Cancer Treatment

Qingyu Zhou [1,*], Hua Pan [2] and Jing Li [3]

[1]   Department of Pharmaceutical Sciences, Taneja College of Pharmacy, University of South Florida, Tampa, FL 33612, USA
[2]   Department of Cardiovascular Sciences, Mossani College of Medicine, University of South Florida, Tampa, FL 33612, USA; huapan@health.usf.edu
[3]   Karmanos Cancer Institute, Wayne State University School of Medicine, Detroit, MI 48201, USA; bb8374@wayne.edu
*   Correspondence: qzhou1@usf.edu; Tel.: +1-813-974-7081

Received: 16 September 2019; Accepted: 13 October 2019; Published: 15 October 2019

Abstract: Naturally occurring polyphenols are believed to have beneficial effects in the prevention and treatment of a myriad of disorders due to their anti-inflammatory, antioxidant, antineoplastic, cytotoxic, and immunomodulatory activities documented in a large body of literature. In the era of molecular medicine and targeted therapy, there is a growing interest in characterizing the molecular mechanisms by which polyphenol compounds interact with multiple protein targets and signaling pathways that regulate key cellular processes under both normal and pathological conditions. Numerous studies suggest that natural polyphenols have chemopreventive and/or chemotherapeutic properties against different types of cancer by acting through different molecular mechanisms. The present review summarizes recent preclinical studies on the applications of bioactive polyphenols in lung cancer therapy, with an emphasis on the molecular mechanisms that underlie the therapeutic effects of major polyphenols on lung cancer. We also discuss the potential of the polyphenol-based combination therapy as an attractive therapeutic strategy against lung cancer.

Keywords: lung cancer; natural polyphenols; anticancer activities; molecular mechanisms

---

1. Introduction

Lung cancer is the second most common cancer in both men and women and is the leading cause of cancer mortality in the United States [1], with an overall five-year survival rate of 19.4% (https://seer.cancer.gov/statfacts/html/lungb.html). The two main types of lung cancer are small cell lung cancer (SCLC) and non-small cell lung cancer (NSCLC), which account for about 15% and 85% of all lung cancer cases, respectively [2]. Tobacco smoking is the principal cause of lung cancer [3]. Secondhand smoking, chronic exposure to various occupational and environmental lung carcinogens, and previous lung diseases can also increase the risk of lung cancer [4–9]. Historically, intakes of fruits, non-starchy vegetables, whole grains, and herbs abundant in certain phytochemicals are thought to be protective against lung cancer. Epidemiological and experimental evidence suggests that natural bioactive compounds can act as chemopreventive agents to delay, suppress or reverse carcinogenic progression to advanced lung cancer through various mechanisms, including antioxidant/anti-inflammatory activities, modulation of biotransformation enzymes, anti-proliferative effect, and modulation of the immune system [10–18]. In addition, natural products that are derived from a variety of sources, including phytochemicals [19–21], hormones [22,23], and nutrients [24–26], have been shown to reduce the side effects and toxicities that are associated with chemotherapy and radiation therapy for lung cancer.

Although the consumption of fruits, vegetables, and natural products with a high content in anticancer natural compounds are generally considered to be beneficial in preventing and combating lung cancer, bioactive compounds from foods, or natural product extracts are not ready for uniform adoption into complementary and integrative therapy use due to inconclusive or negative results from clinical trials [27–33]. Over the past two decades, the rapid evolution of medical research and technologies has led to significant breakthroughs in our understanding of the molecular and genetic alterations that drive cancer development and progression. Consequently, the undergoing transition from the empirical trial-and-error medicine to precision medicine according to the unique genetic mapping of individual patients has stimulated considerable research activities in the area of natural bioactive compounds. With the advent of improved cellular and molecular experimental systems, significant progression has been made in unraveling the molecular mechanisms that underlie the antitumor properties of individual natural compounds. In this review, we focus on the compelling evidence from in vitro and in vivo studies demonstrating the ability of natural polyphenols, one of the most important groups of phytochemicals, to suppress lung cancer progression through the induction of tumor cell death and inhibition of aberrantly activated pro-proliferative and pro-survival signaling pathways. Our purpose is to highlight the potential of natural polyphenols to be developed and integrated into standard therapeutic strategies to improve clinical outcomes for patients with advanced lung cancer.

2. Mutations and Dysregulated Signaling Pathways in Lung Cancer

Lung cancer arises through a multistep process that is driven by the sequential accumulation of genetic mutations and epigenetic modifications, which leads to uncontrolled cell proliferation and inactivation of programmed cell death (apoptosis) [34–36]. As a lung tumor grows in size, the angiogenic switch is triggered so that the existing vascular network expands to form new blood vessels (angiogenesis) that are intended to sustain the nutrient and oxygen supply in the growing tumor [37–39]. Subsequently, lung tumor cells can acquire the ability to invade the surrounding tissues, intravasate into the circulatory system, travel to distant tissue sites, extravasate, and develop a secondary tumor at distant sites (metastasis) [40,41]. A typical lung cancer contains approximately 200 nonsynonymous mutations, some of which are considered "driver mutations" that play a dominant-acting role in tumor growth and progression [42]. For example, activating mutations in the Kirsten rat sarcoma viral oncogene homolog (KRAS) and epidermal growth factor receptor (EGFR) genes and rearrangements in the anaplastic lymphoma kinase (ALK) or ROS proto-oncogene 1 receptor tyrosine kinase (ROS1) genes are identified as oncogenic drivers in certain subtypes of NSCLC [43–47], so are the inactivating mutations in tumor protein p53 (TP53) and retinoblastoma 1 (RB1) genes in SCLC [48,49]. Most driver mutations are gain-of-function mutations that result in the overexpression of oncogenes or mutant proteins with dysregulated activities. For example, constitutive activation of EGFR or K-RAS that is due to mutation subsequently upregulates the Mitogen-activated protein kinase kinase (MEK)/extracellular signal-regulated kinase (ERK) and phosphoinositide 3-kinase (PI3K)/v-akt murine thymomaviral oncogene (AKT) signaling pathways, which triggers a cascade of downstream effectors promoting tumor growth, angiogenesis, and metastasis [50,51]. The p53 tumor suppressor gene is a transcription factor that activates a number of target genes to restrict aberrant cell growth through the induction of senescence, cell cycle arrest, or apoptosis [52,53]. The high expression of p53 protein is a favorable prognostic factor in a subset of patients with NSCLC [54], whereas the exon 8 mutation of p53 gene reduces the responsiveness to tyrosine kinase inhibitors (TKIs) and worsens prognosis in EGFR-mutant NSCLC patients [55]. It is well accepted that the inactivation of apoptosis plays an important role in lung carcinogenesis and resistance to treatment [56,57]. There are two main apoptotic pathways [58]. The extrinsic apoptotic pathway is initiated by the activation of multiple death receptors, such as Fas, tumor necrosis factor receptors (TNFRs), and TNF-related apoptosis-inducing ligand receptors (TRAILRs), and it proceeds through caspase-8 [59], while the intrinsic pathway is activated by mitochondrial outer membrane permeabilization (MOMP), followed by the release of cytochrome c to the cytoplasm and recruitment of pro-caspase 9 to cytochrome c [60]. The Bcl-2 family proteins, including pro-apoptotic members (such as Bad, Bak, Bax, Bcl-Xs, BID, Bik, Bim, HRK, Noxa, and PUMA) and anti-apoptotic members (such as Bcl-2, Bcl-W, Bcl-Xl, Bfl-1, and MCL-1) also regulate the intrinsic pathway [61]. The extrinsic and intrinsic pathways converge on the effector caspases (e.g., Caspases 3, 6, and 7), which are capable of cleaving hundreds of substrates, including nuclear proteins, plasma membrane proteins, and mitochondrial proteins, to trigger cell death [62]. Given that the evasion of apoptosis is one of the prominent hallmarks of cancer [63], an ideal therapeutic strategy to effectively induce apoptosis and avoid the "death by a thousand cuts" in lung cancer would be to restore p53 function and facilitate caspase activation [64,65].

3. Classification and Structures of Natural Polyphenols

Polyphenols are a large group of nature compounds present in plant-based foods and beverages, including fruits, vegetables, whole grains, tea, and wine. So far, more than 10,000 polyphenolic compounds have been identified [66]. Polyphenols can be classified into four main groups, including phenolic acids, flavonoids, stilbenes, and lignans, based on the number of aromatic rings, the structural elements connecting these rings to one another, and the substituents that are bound to the rings (Figure 1) [67,68]. Phenolic acids contain a single benzene rings and they can be further divided into two main subclasses, hydroxybenzoic acid and hydroxycinnamic acid derivatives that are based on the C6–C1 and C6–C3 backbones, respectively [69]. Flavonoids share a common structure with two aromatic rings (A and B) that are connected by three carbon atoms that form an oxygenated heterocycle (ring C), which is also known as the flavan nucleus or 1,3-diphenylpropane structure (C6-C3-C6 carbon skeleton) [69]. Flavonoids can be divided into six major subclasses, including flavonols, flavones, flavonones or dihydroflavones, isoflavones, anthocyanidins, flavanols, or catechins based on the variations in hydroxyl/methoxy group placements on the ring structures [68,70]. Natural stilbenes are structurally characterized by the presence of 1,2-diphenylethylene nucleus (C6–C2–C6 carbon skeleton) [71]. Lignans are a diverse group of optically active phenylpropanoid dimers, in which the two phenylpropane units are connected by the center carbon (C-8/C-8′) of their side chains [72].

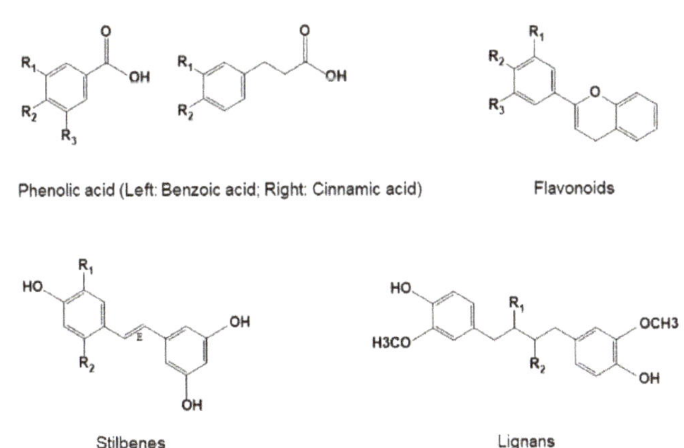

Figure 1. Chemical structures of different classes of natural polyphenols.

The structure-activity relationships of many natural polyphenols in terms of their anticancer potential have been documented. The basic chemical structure of polyphenols contains one or more aromatic rings with one or more hydroxyl groups attached. The presence of functional phenolic hydroxyl groups makes polyphenols excellent hydrogen bond donors that confer high affinities for proteins and nuclear acids [73]. Therefore, the number and position of hydroxyl groups can have a decisive impact on the cellular bioactivities of polyphenols, which are central to their antitumoral, antimutagenic, pro-apoptotic, and antioxidant effects [74–76]. Considerable efforts have been devoted to the characterization of structure-activity relationships that provide the basis of rational design of polyphenol analogs with improved anticancer effect, given that the polyphenol core represents an attractive chemical structure towards new anticancer agents. For example, the cytotoxic activity is enhanced in phenols with low bond dissociation energy (BDE) values or large negative σ+ values since the inhibitory effect of simple phenols on fast-growing murine leukemia cells is related to the O–H BDE that is required to form a phenoxy radical and the brown variation of the Hammett electronic parameter (σ+) [77,78].

Structure-activity relationship (SAR), quantitative structure-activity relationship (QSAR), and docking approaches have been used to delineate the structural mechanisms that underlie the correlation between the binding affinity of polyphenol compounds for a specific oncogenic molecule and the expected anticancer activities [76]. For example, gossypol, a polyphenol that is derived from cotton seeds, is an effective inhibitor of Bcl-2, Bcl-xL, and MCL-1 with the inhibitor constant (Ki) values at sub-micromolar levels [79,80]. The elimination of the two reactive aldehydes from gossypol based on a model of the docked structure of the compound into Bcl-xL resulted in a semisynthetic analog of gossypol, namely apogossypol, which showed superior efficacy and markedly reduced toxicity in Bcl-2-transgenic mice as compared with gossypol [81,82]. A couple of studies by Wang's group have demonstrated that gossypol forms a hydrogen bonding network with residues Arg146 and Asn143 in Bcl-2 through its aldehyde group and the adjacent hydroxyl groups on one of its naphthalene rings, while the isopropyl group on the same naphthalene ring is inserted into a hydrophobic pocket in Bcl-2 [83,84]. Based on the predicted binding model, a simplified pyrogallol-based analogue of gossypol was designed to mimic the hydrogen bonding and part of the hydrophobic interactions between gossypol and Bcl-2. Moreover, modification to the isopropyl group of pyrogallol resulted in not only improved binding affinities for Bcl-2 and Mcl-1, but also an increased cytotoxic effect on the MDA-MB-231 and PC-3 cancer cell lines with the inhibitory concentration IC50 values at sub-nanomolar levels [85].

A molecular docking study on the interaction between tea catechins and hepatocyte growth factor receptor (Met) has revealed that the gallate-containing catechins, including (−)-epicatechin gallate (ECG), (−)-epigallocatechin gallate (EGCG), and gallocatechin gallate (GCG), favorably fit into the Met binding site with hydrogen bonding being established between the aromatic hydroxyl groups of the gallate moiety and the backbone –NH of two Met kinase active sites, i.e., Met1160 and Pro1158, whereas tea catechins without the gallate group, including (−)-catechin (CAT), (−)-epicatechin (EC), and (−)-epigallocatechin (EGC), did not interact with Met1160, but exhibit affinity for the backbone –NH of Asp1222, which suggests that the gallate group is a key structure feature for binding of tea catechins to the active sites of Met kinase domain [86] The findings of the molecular docking study were confirmed by the Met kinase activity assay, which showed that ECG, EGCG, and GCG had inhibitory effects on Met kinase activity, while the other tea catechins had little or no effect [86]. Another potential molecular target of tea catechins is the proteasome, which is known to mediate the degradation of many intracellular proteins that are involved in carcinogenesis and tumor progression [87]. The SAR studies by Dou's group showed that the ester bond-containing tea polyphenols, such as ECG, GCG, EGCG, and catechin-3-gallate (CG), were strong inhibitors for the chymotrypsin-like activity of the purified 20S proteasome with IC50 values at nanomolar levels, whereas the tea polyphenols without the gallate ester function, such as EGC, EC, gallocatechin (GC), and CAT, were unable to inhibit the proteasomal chymotrypsin-like activity [88,89].

4. Molecular Underpinnings of Polyphenols in Lung Cancer Treatment

The use of bioactive natural polyphenols for therapeutic prevention and intervention is an evolving strategy in the management of cancer. A thorough understanding of the mechanisms of action is imperative for integrating those polyphenols into standard oncology care. In this section, we primarily focus on the recent advances in understanding the antitumor actions of natural polyphenols in lung cancer (Figure 2). The preclinical studies were identified through a literature review that was conducted on PubMed using the key terms polyphenol and lung cancer (Table 1). Only original studies credibly investigating molecular mechanisms underlying the antitumor potential of natural polyphenols and their analogues were included.

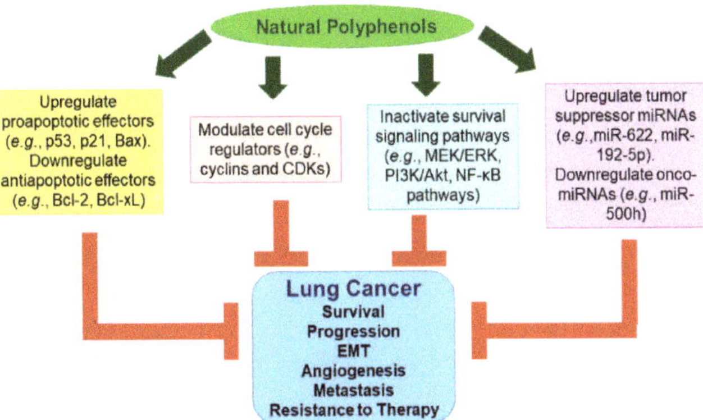

Figure 2. The role of bioactive natural polyphenols in lung cancer therapy.

Table 1. Preclinical Studies on the Mechanisms Underlying the Antitumor Activities of Natural Polyphenols.

Polyphenol Compounds or Extracts	Mechanisms	In Vitro and/or In Vivo Models	References
Resveratrol	Induction of apoptosis by up-regulation of p53 and p21, activation of the caspases and disruption of the mitochondrial membrane complex. Cell cycle arrest at the G_1 phase. Alterations in expressions of cyclin A, chk1, CDC27, and Eg5. Anti-tumor effect mediated by transforming growth factor-β pathway, particularly through the Smad proteins. i.e., down-regulation of the Smad activators 2 and 4 and up-regulation of the repressor Smad 7	A549 human NSCLC cell line	[90]
Resveratrol	Induction of apoptosis as a result of mitochondrial depolarization, release of cytochrome c from the mitochondrial compartment to the cytoplasm, apoptosis-inducing factor translocation from the mitochondrial compartment to the nucleus, and altered protein levels of Bcl-2, Bcl-xL and Bax	H446 human SCLC cells	[91]
Resveratrol	Induction of TRAIL-mediated apoptosis through suppression of NF-κB and downregulation of anti-apoptotic factors Bcl-2 and Bcl-xl	A549 and HCC15 human NSCLC cells	[92]
Resveratrol	Suppressed M2-like polarization of tumor associated macrophages and inhibited STAT3 activity	A549 and H1299 human NSCLC cells. Lewis lung cancer (LLC) s.c. xenograft model (Intraperitoneal (i.p.) administration) [a]	[93]
Resveratrol	Resveratrol enhancing the effects of cisplatin on inhibition of cancer cell proliferation, induction of cell apoptosis, depolarization of mitochondrial membrane potential, release of cytochrome c, upregulation of Bax, downregulation of Bcl-2	H520 and H838 human NSCLC cell lines	[94]
Resveratrol	Upregulation of p21 and TRAIL receptor 1 and 2 expression, and downregulation of Bcl2, cyclin D, NF-κB and IKK1 expression	A549 human NSCLC cell line	[95]
Resveratrol	Increase in production of hydrogen peroxide (H_2O_2), activation of Bid, PARP and caspase 8, and downregulation of pEGFR, pAkt, c-FLIP and NF-κB protein expression	H460 human NSCLC cells	[96]
Resveratrol	Suppress of tumor cell growth via an apoptosis-independent mechanism involving induction of premature senescence by increasing P53 and p21 expression and ROS production and decreasing EF1A expression	A549 and H460 human NSCLC cell lines	[97]
Resveratrol	Enhancing ionizing radiation through increased production of ROS, and induction of DNA double-strand breaks and senescence induction	A549 and H460 human NSCLC cell lines	[98]
Resveratrol	Resveratrol overcoming gefitinib resistance by increasing the intracellular gefitinib concentration through inhibition of CYP1A1 and ABCG2 and by inducing cell apoptosis, autophagy, cell cycle arrest and senescence through increase in expression of cleaved caspase-3, LC3B-II, p53 and p21	PC9/G human NSCLC cells	[99]
Resveratrol	Resveratrol-enhanced erlotinib-mediated apoptosis through decreasing survivin expression and induction of PUMA expression	H460, A549, PC-9 and H1975 human NSCLC cell lines	[100]

Table 1. *Cont.*

Polyphenol Compounds or Extracts	Mechanisms	In Vitro and/or In Vivo Models	References
Resveratrol	Resveratrol-enhanced etoposide-Induced cytotoxicity through down-regulating ERK1/2 and AKT-mediated X-ray repair cross-complement group 1 (XRCC1) protein expression	H1703 and H1975 human NSCLC cell lines	[101]
Resveratrol	Modulation of the expression of specific miRNAs with potential target genes involved in apoptosis, cell cycle regulation, cell proliferation, and differentiation	A549 human NSCLC cell line	[102]
Resveratrol	Upregulation of miR-622 leading to suppression of K-Ras mRNA translation without affecting its accumulation levels	16HBE-T human bronchial epithelial cell line and H460 human NSCLC cell line.	[103]
Resveratrol	Inhibition of lung cancer progression by downregulating miR-500h, which subsequently leads to downregulation of PP2A expression, inactivation of AKT/NF-kB and downregulation of FOXC2	CL1-5, A549, H322, H520 and H1435 human NSCLC cell lines	[104]
Resveratrol	Induction of G2/M cell cycle arrest through downregulation of checkpoint protein cyclin B1. Induction of apoptosis by increasing p53 and p21 expression and the release of cytochrome c in the cytosol	A549 human NSCLC cell line	[105]
Resveratrol	Inhibition of A549 cell proliferation through the reduction of the ratio of Bcl-2/Bax through activation of p53, thus activating the caspase-3- dependent apoptotic cascade and induces apoptosis	A549 human NSCLC cells.	[106]
Resveratrol	Attenuated A549 cell-induced platelet secretion and angiogenic responses in vitro and suppressed A549 lung cancer metastasis and angiogenesis in vivo through inhibition of platelets-mediated angiogenic responses induced by [106] adenosine diphosphate (ADP) through increased cGMP generation and cGMP-mediated vasodilator-stimulated phosphoprotein phosphorylation along with reduced intracellular Ca^{2+} mobilization	A549 human NSCLC cells, and A549 subcutaneous (s.c.) xenograft tumors in nude mice (i.p. administration) [a]	[107]
Resveratrol	Anticancer effects attributable to inhibition of STAT-3 Signaling	A549 human NSCLC cells	[108]
Resveratrol	Inhibition of anchorage-dependent and -independent growth of NSCLC cells by decreasing EGFR and downstream kinases Akt and ERK1/2 activation, and subsequent impairment of hexokinase II (HK2)-mediated glycolysis by inhibiting HK2 expression mediated by the Akt signaling pathway	H460, H1650 and HCC827 human NSCLC cells. H460 s.c. xenograft model (i.p. administration) [a]	[109]
Resveratrol	Synergism between Resveratrol and Metformin attributable to the suppression of DNA damage based on the downregulation of γH2AX/p53/p-chk2, inhibition of cell cycle progression via modulation of cyclin E/cdk2, Rb, p21 cyclin B1/cdk1 and plk1/cdc25c and enhancement of DNA repair indicated by the upregulation of p53R2	A549 human NSCLC cells	[110]
Resveratrol	Inhibition of the release of IL-6 and VEGF for co-cultured A549 lung cancer cells and adipose-derived mesenchymal stem cells	Co-cultured A549 human lung cancer cells and adipose-derived mesenchymal stem cells	[111]
Resveratrol	Induction of cell cycle arrest in the G0/G1 phase by downregulating the expression levels of cyclin D1, cyclin-dependent kinase (CDK)4 and CDK6, and upregulating the expression levels of the CDK inhibitors, p21 and p27	A549 human NSCLC cell line	[112]

Table 1. *Cont.*

Polyphenol Compounds or Extracts	Mechanisms	In Vitro and/or In Vivo Models	References
Resveratrol	miR-200c sensitized tumor cell response to resveratrol by targeting reversion-inducing cysteine-rich protein with Kazal motifs (RECK), followed by activation of the JNK signaling pathway and ER stress	H460 human NSCLC cell line	[113]
Resveratrol	Suppression of invasion and metastasis through reversal of TGF-β1-induced EMT through increasing E-cadherin expression and repressing Fibronectin, Vimentin, Snail1 and Slug expression	A549 human NSCLC cell line	[114]
Resveratrol	Enhancing the radiosensitivity through NF-κB inhibition and S-phase arrest	NCI-H838 human NSCLC cell line	[115]
Resveratrol	Anti-metastasis effect attributable to the inhibition of expression of MMP-9/MMP-2 by suppression of HO-1, which in part results from the suppression of NF-κB-dependent signaling pathway	A549 human NSCLC cell line	[116]
Resveratrol	Inhibition of the proliferation of SPC-A-1/CDDP cells, induction of apoptosis and cell cycle arrest at phase between G0-G1 and S phase or at the G2/M phase by downregulating survivin	Human multidrug-resistant SPC-A-1/CDDP cells	[117]
Resveratrol	Anti-proliferative effect associated with inhibition of the phosphorylation of the retinoblastoma protein (pRB) and induction of cyclin-dependent kinase (Cdk) inhibitor p21WAF1/CIP. Induction of apoptosis associated with activation of caspase-3, shift in Bax/Bcl-xL ratio and inhibition of transcriptional activity of NF-κB	A549 human NSCLC cell line	[118]
Resveratrol loaded gelatin nanoparticles	Induction of cell death through inhibition of cell cycle progression and constitutive NF-κB activation by altering the expression of p53, p21, caspase-3, Bax, Bcl-2 and NF-κB	H460 human NSCLC cell line	[119]
SS28 (a synthetic Resveratrol analog)	Inhibition of Tubulin polymerization during cell division to cause cell cycle arrest at G2/M phase of the cell cycle	A549 human NSCLC cell line	[120]
4,4′-Dihydroxy-*trans*-stilbene (DHS) (a resveratrol analog)	Inhibition on anchorage-dependent or -independent cell growth, leading to impairment of the cell cycle progression with reduction of cell numbers arresting at the G1 phase	Murine Lewis lung carcinoma (LLC) cell line	[121]
Curcumin and resveratrol alone or in combination	Improvement of lung histoarchitecture and ultrahistoarchitecture during benzopyrene-induced lung carcinogenesis in mice	3,4-Benzopyrene-induced mouse lung carcinoma model (Oral (p.o.) administration) [a]	[122]
Resveratrol and dibenzoylmethane	Induction of apoptosis through activation of caspase-9 and caspase-3 and subsequent cleavage of PARP	A549 and CH27 human NSCLC cell lines	[123]
Heyneanol A (HA) (A tetramer of resveratrol)	Induction of caspase-mediated cancer cell apoptosis by inducing cleavage of caspase-9 and caspase-3 and suppression of basic fibroblast growth factor (bFGF)-induced tumor angiogenesis.	In vivo Lewis lung tumor model (i.p. administration) [a]	[124]
EGCG, ECG, EGC and EC	Induction of apoptosis through a p53-dependent pathway.	A549 human NSCLC cell line	[125]
EGCG	Induction of G2-M arrest. Incorporation into cytosol and nuclei	PC-9 human NSCLC cell line	[126]

Table 1. *Cont.*

Polyphenol Compounds or Extracts	Mechanisms	In Vitro and/or In Vivo Models	References
EGCG	EGCG inhibited cell growth through decreasing the phosphorylation of Akt and ERK irrespective of EGFR-, ALK- or ROS1-dependency. The antiangiogenic effect of EGCG might be attributable to the inhibition of HIF-1α	PC-9, RPC-9, H1975, H2228 and HCC78 human NSCLC cell lines (EGFR- or fusion gene-driven tumor cells) and xenograft models (p.o. administration) [a]	[127]
EGCG	Downregulation of gene expression of NF-κB inducing kinase (NIK), death-associated protein kinase 1 (DAPK 1), RhoB and tyrosine-protein kinase (SKY), and upregulation of the retinoic acid receptor alpha1 gene expression	PC-9 human NSCLC cell line	[128]
EGCG	Induction of miRNA profile changes, which modulate several regulatory networks associated to AKT, NF-κB, MAP kinases, and cell cycle	4-(Methylnitrosamino)-1-(3-pyridyl)-1-butanone (NNK) induced mouse lung cancer (p.o. administration) [a]	[129]
EGCG	Co-treatment with celecoxib synergistically inducing apoptosis by upregulation of growth arrest and DNA damage-inducible 153 (GADD153) through the ERK signaling pathway	A549, ChaGo K-1 and PC-9 human NSCLC cell lines	[130]
EGCG	Co-treatment of EGCG with cisplatin resulting in proliferation inhibition, cell cycle arrest in G1 phase, increase in apoptosis along with inhibition of DNA methyltransferase (DNMT) activity and histone deacetylase (HDAC) activity, reversal of hypermethylated status and downregulated expression of *GAS1*, *TIMP4*, *ICAM1* and *WISP2* genes	Cisplatin-resistant A549 (A549/DDP) human NSCLC cell line and A549/DDP xenograft tumor model (i.p. administration)	[131]
EGCG and Luteolin	Enhanced antitumor effect attributable to ATM (ataxia telangiectasia mutated) kinase-dependent Ser15 phosphorylation of p53 as a consequence of DNA double strand break	H292, A549 and H460 human NSCLC cell lines expressing wild-type p53; A549 xenograft tumor model (p.o. administration) [a]	[132]
Tea polyphenols	Upregulation of p53 expression and downregulation of Bcl-2 expression with no influence on H-Ras and c-Myc expressions	NCI-H460 human NSCLC cell line	[133]
Black tea polyphenols	Inhibition of Cox-1 and induction of caspase-3 and caspase-7 expression	3,4-Benzopyrene induced mouse lung tumor model	[134]
Black tea polyphenols	Suppressing cell proliferation and inducing apoptosis	3,4-Benzopyrene induced mouse lung tumor model	[135]
Green tea polyphenols	Preventive effect against lung cancer by upregulating p53 and downregulating Bcl-2	3,4-Benzopyrene induced rat lung tumor model	[136]
Tea polyphenols	Increase in p53 expression and decrease in Bcl-2 expression	3,4-Benzopyrene-induced rat lung carcinoma model	[137]

Table 1. *Cont.*

Polyphenol Compounds or Extracts	Mechanisms	In Vitro and/or In Vivo Models	References
Tea polyphenols	Inhibition of Akt and cyclooxygenase-2 expression, and inactivation of nuclear factor-kappa B via blocking phosphorylation and subsequent degradation of IkappaB alpha	Diethylnitrosoamine- induced mouse lung tumor model (p.o. administration) [a]	[138]
Green tea polyphenols	TAM67-mediated changes in gene expression involving the downregulation of activator protein-1 (AP-1)	H1299 human NSCLC cell line and SPON 10 mouse lung tumor cell line	[139]
Green tea extracts	Modulation of the expression of 14 proteins involved in calcium-binding, cytoskeleton and motility, metabolism, detoxification, or gene regulation	A549 human NSCLC cell line	[140]
Green tea polyphenols	Synergistic antitumor effect with atorvastatin attributable to increased apoptosis, reduced Mcl-1 level and increased cleaved caspase-3 and cleaved poly(ADP)-ribose polymerase (PARP)	H1299 and H460 human NSCLC cell lines. 4-(Methylnitrosaminao)-1-(3-pyridyl)-1-butanone induced mouse lung tumor model (p.o. administration) [a]	[141]
Green tea extract	Induction of protective autophagy	A549 human NSCLC cell line	[142]
Thymoquinone (TQ)	Upregulation of Bax and downregulation of Bcl-2 expression and increase in the Bax/Bcl-2 ratio. Decrease in the expression of cyclin D, NF-κB and IKK1 and increase in the expression of p21 and TRAIL receptor 1 and 2 expression	A549 human NSCLC cell line	[95]
Curcumin	Induction of apoptosis through p53-independent pathway by downregulation of Bcl-2 and Bcl-xL expression	A549 and H1299 human NSCLC cell lines	[143]
Curcumin	Induction of cell cycle arrest at the G1/S phase and apoptosis through up-regulation of *GADD45* and *GADD153*	PC-9 human NSCLC cell line	[144]
Curcumin	Induction of cell cycle arrest at the G2/M phase and apoptosis through upregulation of Bax and Bad expression, downregulation of Bcl-2, Bcl-xL and XIAP expression, increase in ROS, intracellular Ca^{2+} and endoplasmic reticulum stress, activation of GRP78 and GADD153 proteins and FAS/caspase-8 pathway	H460 human NSCLC cell line	[145]
Curcumin	Induction of apoptosis through a mitochondria-dependent mechanism as manifested by the decrease in the mitochondrial membrane potential, releasing cytochrome c from mitochondria to cytoplasm	A549 human NSCLC cell line	[146]
Curcumin	Induction of apoptosis via the ROS-mediated mitochondrial pathway accompanied by increased Bax expression and decreased expression of Bcl-2 and Bcl-xL	H446 human SCLC cell line	[147]
Curcumin	Inhibition of tumor cell proliferation and induction of apoptosis through upregulation of miR-192-5p and suppression of the PI3K/Akt signaling pathway	A549 human NSCLC cell line	[148]
Curcumin	Enhancing autophagy and apoptosis through inaction of PI3K/mTOR signaling pathway	A549 and H1299 human NSCLC cell lines	[149]

Table 1. *Cont.*

Polyphenol Compounds or Extracts	Mechanisms	In Vitro and/or In Vivo Models	References
Curcumin	Induction of autophagy leading to suppression of tumor cell proliferation.	A549 human NSCLC cell line	[150]
Curcumin	Inhibition of tumor cell invasion and metastasis through attenuating GLUT1/MT1-MMP/MMP2 pathway	A549 human NSCLC cell line and xenograft tumor model (i.p. administration) [a]	[151]
Curcumin	Inhibition of tumor cell metastasis through inhibition of the adiponectin/NF-κB/MMPs signaling pathway	A549 human NSCLC cell line and xenograft tumor model (i.p. administration) [a]	[152]
A501 (Curcumin analogue)	Induction of cell cycle arrest at the G2/M phase and apoptosis through decreasing the expression of cyclinB1, cdc-2, Bcl-2, while increasing the expression of p53, cleaved caspase-3 and Bax	A549 and H460 human NSCLC cell lines	[153]
Curcumin and gefitinib	Potentiating the antitumor effect of gefitinib in gefitinib-resistant tumor cells through induction of endogenous EGFR protein degradation and downregulation of EGFR and AKT protein expression. Reduction of the gefitinib-induced villi damage and apoptosis in mouse intestine through attenuating gefitinib-induced p38 activation	CL1-5, A549 and H1975 human NSCLC cell lines and xenograft models (p.o. administration) [a]	[154]
Curcumin and carboplatin	Synergistic antitumor activity mediated by multiple mechanisms involving suppression of NF-kB via inhibition of the Akt/IKKα pathway, enhancement of ERK1/2 activity and downregulation of MMP-2 and MMP-9 expression	A549 human NSCLC cell line	[155]
Quercetin	Induction of cell cycle arrest at the G2/M phase and apoptosis through increased expression of cyclin B1 and phosph-cdc2 (T161), survivin, total p53, phosphor-p53 (S15) and p21 proteins	A549 and H1299 human NSCLC cell lines	[156]
Quercetin	Induction of apoptosis through activation of MEK-ERK pathway, inactivation of Akt and alteration in the expression of Bcl-2 family	A549 human NSCLC cell line	[157]
Quercetin	Proapoptosis activity through multiple mechanisms including upregulated the expression of genes associated with the death pathway, the JNK pathway, the IL1 receptor pathway, the caspase cascade, the NF-κB pathway and cell cycle arrest, and downregulated the expression of genes related to cell proliferation	H460 human NSCLC cell line	[158]
Quercetin	Anti-invasion activity through inhibition of monocarboxylate transporter 1	A110L human lung cancer cell line	[159]
Quercetin	Targeting aurora B kinase	A549 human NSCLC cell line and xenograft model (i.p. administration) [a]	[160]
Quercetin	Trigger Bcl-2/Bax-mediated apoptosis, necrosis and mitotic catastrophe. Inhibition of cell migration through disassembly of microfilaments, microtubules and vimentin filaments and inhibition of vimentin and N-cadherin expression	A549 human NSCLC cell line	[161]

Table 1. *Cont.*

Polyphenol Compounds or Extracts	Mechanisms	In Vitro and/or In Vivo Models	References
Quercetin	Suppression of in vitro cell migration/invasion and in vivo bone metastasis through inhibition of Snail-dependent Akt activation and Snail-independent ADAM9 expression pathways	A549 and HCC827 human NSCLC cell lines and A549 xenograft model (i.p. administration) [a]	[162]
Quercetin and chrysin	Suppressed the secretion of cytokines, IL-1β, IL-6, TNF-α and IL-10, and decreased the phosphorylation of IKKβ and IκB, the nuclear level of p65 (NF-κB) as well as the expression of MMP-9 in A549cells exposed to nickel	A549 human NSCLC cell line	[163]
Quercetin and trichostatin A	Enhancing the antitumor activity of trichostatin A through upregulation of p53 expression	A549 and H1299 human NSCLC cell lines and A549 xenograft model (i.p. administration) [a]	[164]
Quercetin and gemcitabine	Promoting apoptosis and sensitizing tumor response to gemcitabine via inhibition of HSP70 expression	A549 and H4650 human NSCLC cell lines	[165]
Caffeic acid phenethyl ester (CAPE)	Upregulation of Bax, p21 and TRAIL receptor 1 and 2 expression, and downregulation of cyclin D expression	A549 human NSCLC cell line	[95]
Pterostilbene	Exhibition of p53-dependent chemotherapeutic effects through the ATM/CHK/p53 tumor suppressive pathway leading to cell senescence	Precancerous human bronchial epithelial cell lines, HBECR and HBECR/p53i, with normal p53 and suppressed p53 expression, respectively	[166]
Bakuchiol	Increase in reactive oxygen species production, decrease in mitochondrial membrane potential (ΔΨm), cell cycle arrest at S phase, caspase 9/3 activation, p53 and Bax up-regulation, and Bcl-2 downregulation	A549 human NSCLC cell line	[167]
Chlorogenic acid (CGA)	Decrease in hypoxia-induced HIF-1α protein level and suppression of the transcriptional activity of HIF-1α under hypoxic conditions, leading to antiangiogenic activity through inhibition of HIF-1α/AKT pathway and decrease in VEGF expression	A549 human NSCLC cell line	[168]
Fisetin	Inhibition of cell growth through concomitant suppression of PI3K/Akt and mTOR signaling	A549 and H1792 human NSCLC cell lines	[169]
Fisetin	Enhancing cisplatin cytotoxicity in cisplatin-resistant cells by modulation of the MAPK/survivin/caspase pathway	A549 human NSCLC cell line	[170]
Fisetin	Synergistic interaction between paclitaxel and fisetin due to the induction of mitotic catastrophe probably through the promotion of multipolar spindle formation. Mitotic catastrophe induced protective autophagy against apoptosis, which then switched to the autophagic cell death	A549 human NSCLC cell line	[171]
Liquiritin, isoliquiritin and isoliquirigenin	Induction of apoptosis and cell cycle at the G2/M phase by increasing p53, p21 and BAX expression and decreasing PCNA, MDM2, p-GSK-3β, p-Akt, p-c-Raf, p-PTEN, caspase-3, pro-caspase-8, pro-caspase-9, PARP and Bcl-2 expression	A549 human NSCLC cell line	[172]

Table 1. *Cont.*

Polyphenol Compounds or Extracts	Mechanisms	In Vitro and/or In Vivo Models	References
Eucalyptus globulus Labill	Cell cycle arrest in the G0/G1 phase. Increase in the expression of p53, p21 and cyclin D1 proteins	NCI-H460 human NSCLC cell line	[173]
Polyphenols isolated from	Suppressed cell migration by targeting MMP-9. Induced cell apoptosis through intrinsic apoptosis pathways, accompanied by increasing the expression of Bax and caspase-3	A549 human NSCLC cell line	[174]
Rosemary extract	Reduced total and phosphorylated/activated Akt, mTOR and p70S6K levels	A549 human NSCLC cell line	[175]
Salvianolic acid A	Salvianolic acid A enhanced sensitivity to cisplatin through suppression of the c-met/AKT/mTOR signaling pathway	A549 human lung cancer cisplatin resistance cell line (A549/DDP)	[176]
Red wine	Inhibition of basal and EGF-stimulated Akt and Erk signals and enhancement of total and phosphorylated levels of p53, leading to inhibition of A549 cell proliferation and clonogenic survival	A549 human NSCLC cell line	[15]
Magnolol and polyphenol mixture (PM) derived from Magnolia officinalis	Induction of cell apoptosis by arresting the cell cycle in the G0/G1 phase while simultaneously activating various pro-apoptotic signals, including TRAIL-R2 (DR5), Bax, caspase 3, cleaved caspase 3, and cleaved PARP	A549 and H1299 human NSCLC cell lines	[177]
Pomegranate juice	Attenuated the formation of lung nodules and reduced PHH3 (marker of mitotic activity) and HIF-1α expression	Two-month-old adult male AJ mice (p.o. administration) [a]	[178]
Polyphenolic compounds of Achyranthes aspera (FCA) extract	Downregulation of the expression of pro-inflammatory cytokines IL-1β, IL-6 and TNF-α, TFs, NF-κB and Stat3, and upregulation of the expression of pro-apoptotic proteins Bax and p53. Increase in the activities and expression of antioxidant enzymes GST, GR, CAT, and SOD. Decrease in the activity and expression of LDH enzymes	Urethane-induced mouse lung cancer in vivo model (p.o. administration) [a]	[179]
Bilberry extract (BE), geristein (GEN), delphinidin-3-O-glucoside (D3G), delphinidin (DE), gallic acid (GA), and phloroglucinol aldehyde (PGA)	Antagonistic interactions of BE, D3G, DEL and GEN with erlotinib. No effect of GA on erlotinib, while synergistic interaction of PGA with erlotinib. Mechanism unknown	A431 human epithelial cell line	[180]

Note: a. The route of administration for individual natural polyphenols evaluated in in vivo studies.

4.1. Resveratrol

Resveratrol (3,5,4′-trihydroxy-*trans*-stilbene) is a naturally occurring stilbene phytoalexin that was first isolated from the white hellebore *Veratrum grandiflorum* in the 1940's [181]. It has a wide spectrum of biological activities that confer various health-promoting effects, such as antioxidant, anti-inflammatory, antidegenerative, cardioprotective, and anticarcinogenic properties [182,183]. The anticancer activities of resveratrol are often associated with modulating enzymes that are responsible for metabolism of carcinogens, activating or inhibiting molecular targets, and signaling pathways that control cancer development and progression [184–186]. In lung cancer, considerable progress has been made in understanding the mechanisms by which resveratrol inhibits cell proliferation, induces apoptosis and cell cycle arrest, and suppresses invasion and metastasis (Figure 3), which highlights the potential of resveratrol to be used as a complementary treatment to augment the efficacy of existing therapies, and providing the insight into the development of novel synthetic resveratrol analogues with improved therapeutic efficacy and reduced side-effects.

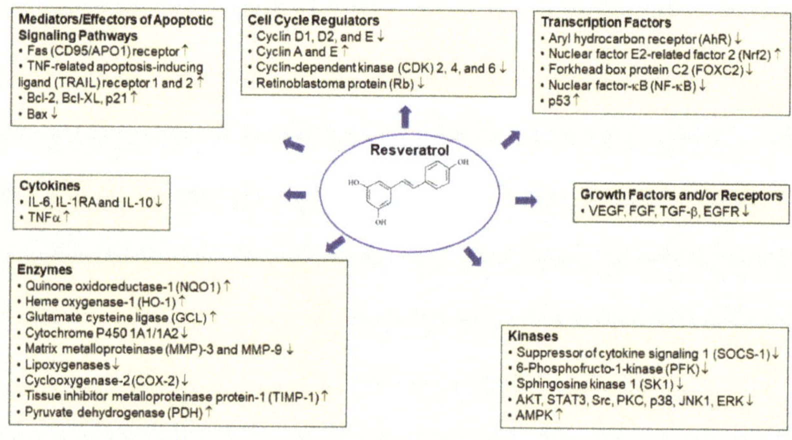

Figure 3. Molecular Mechanisms of Antitumor Activities of Resveratrol in Lung Cancer.

The in vitro anti-proliferative effect of resveratrol is often associated with the induction of cell cycle arrest and apoptosis although the molecular underpinnings may vary among individual lung cancer cell lines. The results of the high-throughput immunoblotting (PowerBlot) and microarray gene expression profiling have revealed that the growth inhibitory effect of resveratrol was mediated by the transforming growth factor-β (TGF-β)/Smad pathway through the downregulation of the TGF-β pathway activators, Smad 2 and Smad 4, and the upregulation of the repressor Smad 7. Moreover, resveratrol-induced apoptosis and G1 phase cell cycle arrest was attributable to the activation of the caspases, the loss of mitochondrial permeability transition, and the increase in the expression of pro-apoptotic tumor suppressor p53 and cyclin-dependent kinase inhibitors p21 and p27 at both the gene and protein levels [90]. Similar results have been documented by others, which indicate that resveratrol-induced apoptosis is associated with increased expression of p53, Bax, and cleaved caspase-3 and decreased expression of Bcl-2 [91–95]. Resveratrol has also been shown to induce apoptosis through the downregulation of cellular FLICE (FADD-like interleukin-1 beta-converting enzyme) inhibitory protein (c-FLIP), which leads to a decrease in phospho-Akt, phospho-EGFR and NF-κB protein expression and an increase in the cleavage and upregulation of Bid, PARP, and caspase-8 and the production of hydrogen peroxide (H_2O_2) [96]. Besides the induction of apoptosis, resveratrol-induced premature senescence is another mechanism that is associated with its anticancer activities [97,98]. Resveratrol-induced premature senescence is correlated with increased DNA double strand breaks (DSBs) and reactive oxygen species (ROS) production in lung cancer cells [97]. When A549 and H460 cells underwent ionizing radiation and resveratrol co-treatment, resveratrol enhanced ionizing radiation-induced premature senescence and increased the ROS production in those cells [98].

The ability of resveratrol to induce apoptosis and cell cycle arrest in lung cancer cells renders it an ideal candidate for combination cancer therapy with the potential to provide additive anticancer efficacy and counteract the onset of acquired drug resistance in the treatment of lung cancer. Resveratrol has been shown to potentiate the growth inhibitory effect of cisplatin through induction of cell apoptosis, which was preceded by the depolarization of mitochondrial membrane potential, opening of the mitochondrial permeability transition pore, release of cytochrome *c*, upregulation of Bax expression, and downregulation of Bcl-2 expression [94]. Preclinical evaluation of the synergistic effect of resveratrol and the EGFR inhibitor gefitinib in a panel of human NSCLC cell lines demonstrated that resveratrol increased the sensitivity to gefitinib in all the cell lines tested, regardless of their EGFR mutation status. Moreover, resveratrol enhanced the inhibitory effect of gefitinib on EGFR phosphorylation in gefitinib-resistant PC-9 (PC9/G) human NSCLC cells by inhibiting CYP1A1 and ABCG2 protein expression, thereby increasing intracellular gefitinib accumulation [99]. Furthermore, among resveratrol and gefitinib single agent and combination treatment groups, the combination of resveratrol and gefitinib showed the highest increase in the fluorescence intensity of monodansylcadaverine (MDC), which is a marker of autophagic vacuoles, and in the number of MDC-labelled tumor cells, and in LC3B II protein expression, suggesting that the antiproliferative effect of combined resveratrol and gefitinib is, in part, attributable to the increased autophagy [99]. The molecular mechanism that underlies the synergistic effect of resveratrol and the EGFR inhibitor erlotinib appears to be different from that of resveratrol and gefitinib. Resveratrol potentiated the cytotoxic effect of erlotinib and enhanced erlotinib-induced apoptosis by repressing survivin and Mcl-1 expression, which inhibits the AKT/mTOR/S6 kinase pathway and increasing p53 and PUMA expression and caspase 3 activity [100]. Resveratrol has been shown to enhance tumor TRAIL-mediated apoptosis through a p53-independent mechanism by which resveratrol decreased the expression of phosphorylated Akt and subsequently suppressed the expression of NF-κB (p65), which leads to mitochondrial dysfunction and cytochrome c translocation [92]. When used in combination with etoposide, resveratrol counteracted etoposide-induced the upregulation of X-ray repair cross-complementing group 1 (XRCC1) expression that led to activation of Akt and ERK1/2, thereby restoring tumor cell sensitivity to etoposide [101].

Several studies have identified modulation of microRNAs (miRNAs) as one of the key mechanisms by which resveratrol exerts its antitumor activities in lung cancer [102–104]. Bae and coworkers identified 71 miRNAs with considerable changes in their expression levels in resveratrol-treated A549 cells while using microarray analysis [102]. Further analysis revealed that 25 of the 71 miRNAs target genes possessing experimentally confirmed function in apoptosis (97 genes), cell cycle regulation (20 genes), cell proliferation and differentiation (28 genes) [102]. Several recent studies have demonstrated the cell line-dependent functional link between the antitumor activities of resveratrol and resveratrol-regulated miRNA expression. In a study by Han et al., resveratrol treatment resulted in the upregulation of miR-622 in 16HBE-T human bronchial epithelial cells and H460 cells [103]. miR-622 was considered to be a tumor suppressor, as an increase in the expression level of miR-622 inhibited the cell proliferation and colony formation, induced cell cycle arrest at G0 phase, and delayed tumor growth in nude mice. Moreover, increase in miR-622 expression reduced K-Ras protein expression levels but had no effect on K-Ras mRNA level, suggesting miR-622 exerts its antitumor activity via targeting K-Ras [103]. In another study, Yu and coworkers examined the role of miR-520h in mediating the antitumor effect of resveratrol in CL1-4 and A549 lung cancer cells. Resveratrol was shown to induce the mesenchymal-epithelial transition (MET) by increasing the expression of protein phosphatase 2A catalytic subunit (PP2A/C) and reducing the expression of FOXC2, phospho-Akt, and p65. As the increased expression of PP2A/C was associated with downregulation of miR-520h [187], and treatment with resveratrol decreased miR-520h expression in A549 cells, it is suggested that resveratrol-induced MET and its inhibitory effect on lung cancer cell migration and invasion are attributable to its ability to inhibit miR-520h expression and activate PP2A/C, which in turn suppresses Akt-mediated activation of the NF-κB pathway, which promotes the malignant behaviors of lung cancer cells [104].

Based on the identified structure-activity relationship, analogues of resveratrol have been synthesized and tested for their antitumor activities in lung cancer cell lines. A synthetic resveratrol named BCS (3,4,5-trimethoxy-4V-bromo-*cis*-stilbene), in which the hydroxyl group of resveratrol is substituted by the methoxy group, was about 1100 times more potent than resveratrol in the growth inhibition of A549 cells (IC_{50}; 0.03 μM vs. 33 μM) [105]. The anti-proliferative effect of BCS was highly associated with cell cycle arrest at G2/M phase and the induction of apoptosis possibly through a mitochondrial-mediated pathway, as manifested by the elevated expression levels of p53 and p21, and the release of cytochrome c in the cytosol [105]. SS28 ((E)-1,2,3-trimethoxy-5-(4-methylstyryl)benzene (6 h)) is a resveratrol-based tubulin inhibitor that exerts its antiproliferative activity by binding to its cellular target tubulin to disrupt the microtubule dynamics [120]. SS28 treatment induces G2/M cell cycle arrest by inhibiting tubulin polymerization during cell division and it leads to apoptosis via the intrinsic (mitochondrial) pathway, as indicated by the loss of mitochondrial membrane potential and activation of Caspase 9 and Caspase 3 [120]. Another resveratrol analogue, 4,4′-dihydroxy-*trans*-stilbene (DHS), significantly suppressed tumor growth and angiogenesis in C57BL/6 mice bearing Lewis lung carcinoma (LLC) and inhibited the anchorage-dependent or -independent LLC cell growth in both mouse and zebrafish lung cancer invasion models [121]. In addition, the results of the in vitro study showed that DHS inhibited LLC cell proliferation, migration, and invasion, and induced the accumulation of hypodiploid cells in the sub-G1 phase, which suggests that the antitumor effect of DHS is via inhibiting DNA synthesis and driving cells towards the apoptotic pathway [121].

4.2. Tea Catechins

Catechins belong to a family of flavonoids and they are the main component of green tea in which catechins comprise 80–90% of the flavonoids, with EGCG being the most abundant catechin (up to 60%) and EGC being the second most abundant (up to 20%), followed by ECG (up to 14%) and EC (about 6%) [188,189]. With a structure of two benzene rings (the A- and B-rings) and a dihydropyran heterocycle (the C-ring) with a hydroxyl or galloyl group over carbon 3, catechins have four possible diasteroisomers. Two isomers with *trans* configuration are called catechin ((+)-catechin and (−)-catechin), and two with *cis* configuration, called epicatechin ((+)-epicatechin and (−)-epicatechin) [190]. The number and configuration of hydroxyl groups on the B ring are the most important determinants of the antioxidant ability of catechins, while the presence of the galloyl group might further increase the antioxidant action [191–193].

The mechanism underlying the inhibitory effects of tea catechins, especially EGCG, on lung cancer progression have been extensively investigated [194–197]. In general, the growth inhibition effect of EGCG is superior to EGC, ECG, and EC [125], and it is associated with the induction of G_2-M arrest [126] and activation of p53-dependent transcription [125]. EGCG treatment has been shown to effectively inhibit the in vitro and in vivo growth of fusion gene- or EGFR-driven lung cancer cells such as H2228 and HCC78 cells that harbor the EML4-ALK fusion gene and SLC34A2-ROS1 fusion gene, respectively, and PC-9, RPC-9, and H1975 cells that harbor EGFR[19DEL], EGFR[19DEL + T790M], and EGFR[L858R + T790M] mutations, respectively [127], which suggests that EGCG has a broad growth inhibitory effect independent of the EGFR mutation status and the ALK or ROS1 fusion status. Although the results of the in vitro study showed that the anti-proliferative activity of EGCG was attributable to the suppressed phosphorylation of EGFR, ALK, and ROS1, and their downstream proteins, Akt and ERK, the in vivo growth inhibitory effect of EGCG in xenograft tumors was associated with the inhibition of HIF-1α expression and reduction tumor angiogenesis, which suggests that tumor response to EGCG is influenced by the tumor microenvironment [127].

The results from a human cancer cDNA expression array study showed that EGCG downregulated the expression of 12 genes and upregulated the expression of four genes out of the 163 genes examined [128]. Among the 12 downregulated genes, two genes (NF-κB inducing kinase (NIK) and death-associated protein kinase 1 (DAPK1)) are associated with apoptosis, two genes (MAP kinase p38γ and CDC 25B/M-phase inducer phosphatase 2) associated with cell cycle, two genes (envoplakin and synapse-associated protein 102 (SAP102)) related to cell-cell interaction, three genes (Rho B, T-lymphoma invasion and metastasis inducing protein 1 (TIAM1) and Cdc42 GTPase-activating protein (Cdc42GAP)) related to the Rho family of small GTPase and regulator, tyrosine–protein kinase (SKY) gene, dishevelled 1 gene, and *EGFR* gene [128]. The four EGCG-upregulated genes included retinoblastoma binding protein (RBQ1), *VEGF*, retinoic acid receptor α1 (RAR-α1), and insulin-like growth factor-binding protein 3 (IGFBP 3) genes [128]. It is noteworthy that high levels of IGFBP 3 in plasma are associated with reduced lung cancer risk [198].

A line of evidence has shown changes in miRNA expression in response to EGCG treatment in lung tumor cell lines and 4-(methylnitrosamino)-1-(3-pyridyl)-1-butanone (NNK)-induced mouse lung tumors [129,199]. EGCG treatment resulted in increased miR-210 expression, leading to reduced proliferation and anchorage-independent growth in CL13 mouse lung adenocarcinoma cells and H460 and H1299 human NSCLC cells [199]. Moreover, EGCG increased the activity of both mouse and human miR-210 gene promoters in H1299 and H460 cells that were transfected with the 2 kb of mouse and 600 bp of human miR-210 gene promoter driven luciferase reporters [199]. Furthermore, the activity of HRE-luciferase in response to EGCG was increased from 2000–4500 U to 8000–40,000 U with the addition of HIF-1a expression vector, and an increased HIF-1α protein expression was observed in EGCG-treated lung cancer cells, which suggests that the upregulation of miR-210 by EGCG is mediated through the HRE in the promoter of miR-210 and stabilization of HIF1α [199]. The involvement of miRNA-mediated gene regulation in antitumor activities of EGCG has also been demonstrated to modulate miRNA expression in 4-(methylnitrosamino)-1-(3-pyridyl)-1-butanone (NNK)-induced mouse lung tumor model [200]. The results of the miRNA microarray showed that 12 miRNAs were upregulated in response to the EGCG treatment, while 9 miRNAs were downregulated. ECGC treatment was found to induce changes in the expression of 21 mRNAs in NNK-induced mouse lung tumors. Moreover, a group of 26 genes were identified as potential targets of the EGCG-regulated miRNAs. Changes in the expression levels of those genes were inversely correlated to changes in the expression levels of the corresponding miRNAs [200]. Further analysis of the role of the 26 miRNA targeted genes revealed an interaction network that is centralized by IGFBP5 and is involved in the regulation of Akt, MAP kinases, NF-κB, and cell cycle [200], which is consistent with the documented mechanisms of inhibitory effects of EGCG on cell cycle and inflammation [194,201]. It was noted that EGCG-induced the upregulation of miR-210 in cultured lung tumor cells was not one of the 26 miRNAs which expression levels were significantly altered in response to EGCG treatment in vivo. This discrepancy could be due to differences in the oxidative stress levels and EGCG-binding proteins between in vitro cultured cells and primary tumors, and in EGCG bioavailability and the elimination half-life between the in vitro and in vivo systems [200].

EGCG has been proved to be beneficial in combination with cancer preventive and chemotherapeutic agents [130,131]. Cotreatment of EGCG with celecoxib, a cyclooxygenase-2 selective inhibitor, synergistically induced apoptosis through the upregulation of *GADD153* gene expression and activation of the mitogen-activated protein kinase (MAPK) signaling pathway [130]. EGCG in combination with cisplatin significantly inhibited cell proliferation and induced cell cycle arrest in G1 phase and apoptosis in cisplatin-resistant A549 cells and suppressed the growth of cisplatin-resistant A549 xenograft tumors [131]. The mechanism of resensitization of tumor cells to cisplatin by EGCG is linked to the inhibition of DNA methyltransferase (DNMT) activity and histone deacetylase (HDAC) activity, reversal of hypermethylated status, and downregulated expression of the GAS1, TIMP4, ICAM1, and WISP2 gene [131]. The combined treatment of EGCG and another dietary polyphenol, luteolin, resulted in synergistic/additive apoptotic and growth inhibitory effects in both in vitro and in vivo lung tumor models. It was noted that p53 wildtype lung cancer cell lines showed greater sensitivity to co-treatment with EGCG and luteolin than p53-mutant or p53-null cell lines, and the combination effectively increased stabilization and ATM-dependent S15 phosphorylation of p53 and mitochondrial translocation of p53. Those results suggest that p53 is required for apoptosis that is induced by the combination of EGCG and luteolin [132].

Similar to EGCG, green tea extract has been demonstrated to exert anticancer activities across a spectrum of lung cancer cell lines and in vivo tumor models through different mechanisms, including the induction of apoptosis through upregulation of p53 expression and downregulation of Bcl-2 expression [133–137], and the inhibition of tumorigenesis through inhibition of cyclooxygenase-2, inactivation of Akt and NF-κB and degradation of IκBα [138], and through the induction of dominant-negative activator protein 1 (*TAM67*) and inhibition of activator protein-1 (AP-1) pathway [139]. The results of the proteomic analysis of A549 cells treated with green tea exact reveals 14 proteins with a ≥2-fold change in the expression level are involved in calcium-binding, cytoskeleton and motility, metabolism, detoxification, or gene regulation [140]. In particular, green tea extract was found to upregulate the expression of lamin A/C, which regulates actin polymerization in nucleus, which leads to decreased cell motility and growth and increased apoptosis [140]. Combination of polyphenon E (a standardized green tea polyphenol preparation) and atorvastatin, an inhibitor of 3-hydroxy-3-methylglutaryl CoA reductase that is commonly used for the treatment of hypercholesterolemia, synergistically inhibited 4-(methylnitrosaminao)-1-(3-pyridyl)-1-butanone induced lung tumorigenesis in mice and the tumor cell proliferation through enhanced apoptosis, which implicates that the combined use of green tea polyphenols and atorvastatin might be beneficial in lung cancer prevention and therapy [141].

4.3. Curcumin

Curcumin is a bioactive phytochemical in the dietary spice turmeric and it has potential anticancer activity against various types of cancer, including lung cancer [202,203]. The results from the in vitro studies have demonstrated that curcumin treatment inhibits tumor cell growth by inducing apoptosis through a variety of p53-independent and mitochondria-dependent pathways [143–147]. In a study by Wu et al., curcumin treatment in cultured H460 cells resulted in cell cycle arrest at the G2/M phase, initial upregulation, followed by the downregulation of cell cycle regulator cyclin D and E, upregulation of Bax, Bad and FAS/CD95 and downregulation of Bcl-2, Bcl-xL, and XIAP protein expression, increase in ROS, intracellular Ca^{2+} and endoplasmic reticulum stress, which led to a loss of mitochondrial membrane potential ($\Delta\Psi_m$) and activation of caspase-3, release of growth arrest and DNA damage inducible gene 153 (*GADD153*) and glucose-regulated protein 78 (*GRP78*) from mitochondria to cytosol and nuclei, and decreased CDK1, CDK2, CDK4, and CDK6 protein expression and increased caspase 8 and *Endo G* mRNA expression [145]. Similar results elucidating the mechanism underlying the apoptotic activity of curcumin have been reported by other research groups using different tumor cell lines. For example, curcumin inhibited the proliferation of A549 cells through upregulation of Bax and downregulation of Bcl-2 expression, and activation of the mitochondrial apoptosis pathway, as manifested by the decreased mitochondrial membrane potential and increased release of cytochrome C from mitochondria to cytoplasm [146]. The effect of curcumin on inhibiting cell growth and inducing cell cycle arrest at the G1/S phase and apoptosis in PC-9 cells has been associated with the upregulation of the expression of *GADD45*, *GADD153*, CDK inhibitors *p21* and *p27* genes, and downregulation of the expression of *cyclin D1*, *CDK2*, *CDK4*, and *CDK6* genes [144]. A501, a synthetic analogue of curcumin with improved anticancer activities, induced cell cycle arrest at the G2/M phase by decreasing the expression of cyclin B1 and cdc-2, and promoted apoptosis by increasing the expression of p53 and Bax and decreasing the expression of Bcl-2 [153].

Although apoptosis induction appears to be the main mechanism underlying the antitumor activities of curcumin in lung cancer, there is evidence of other mechanisms being involved in the inhibitory effect of curcumin on lung tumor survival and progression. The anti-proliferative effect of curcumin has been associated with the inactivation of the PI3K/Akt/mTOR signaling pathway [148,149], upregulation of miR-192-5p [148], and induction of autophagy [149,150]. In a study by Liao et al., the ability of curcumin to suppress the proliferation, invasion, and metastasis of A549 cells was attributable to its inhibitory effect on the expression of GLUT1, MT1-MMP, and MMP2 in A549 cells [151]. Targeting GLUT1 has been sought as an attractive approach for cancer therapy, as upregulation of GLUT1 expression in malignant tumor cells is known to be responsible for the increased glucose uptake needed to drive ATP production through aerobic glycolysis, also known as the "Warburg effect" [204–206]. However, the overexpression of GLUT1 in A549 cells was found to attenuate the inhibitory effect of curcumin against tumor cell invasion in vitro and metastasis in vivo and increase the intracellular expression levels of MT1-MMP and MMP2, implicating that curcumin inhibits lung tumor growth and metastasis through its modulatory effect on the GLUT1/MT1-MMP/MMP2 pathway, but not by targeting GLUT1. In addition, it is suggested that GLUT1 overexpression might potentially confer resistance to curcumin treatment in lung cancer [151]. In another study by Tsai et al., the anti-migratory and anti-invasive effect of curcumin was attributable to the inhibited adiponectin expression via blockage of the adiponectin receptor 1 expression, the inactivated p38 and ERK pathways, and the downregulated expression levels of p65, MMP-2, -9, -3, -13, and -14 [152]. Given the additional evidence indicating that adiponectin regulated NF-κB expression through the Akt pathway, it was concluded that curcumin inhibited lung cancer metastasis through the adiponectin/NF-κB/MMP signaling pathway [152].

The potential of curcumin in combination with other anticancer drugs for lung cancer treatment has been documented. Curcumin potentiated the anti-proliferative effect of gefitinib in three gefitinib-resistant NSCLC cell lines, including CL1-5 (EGFR^{wt}), A549 (EGFR^{wt}), and H1975 (EGFR$^{L858R + T790M}$) cell lines through blockage of EGFR activation and induction of EGFR degradation [154]. Moreover, curcumin enhanced the antitumor effect of gefitinib in CL1-5, A549, and H1975 xenografts in vivo. Notably, curcumin alone, and in combination with gefitinib, decreased the protein expression of EGFR and Akt in CL1-5 xenografts, which was not affected by gefitinib treatment alone. In addition, co-treatment with curcumin reduced the gefitinib-induced villi damage and apoptosis in mouse intestines possibly through the modulatory effect of curcumin on gefitinib-induced p38 activation [154]. Combined treatment with curcumin and carboplatin resulted in synergistic effect on cell proliferation, apoptosis, invasion, and migration [155]. This synergism appeared to be mediated by multiple mechanisms, including efficient downregulation of MMP-2 and MMP-9, substantial suppression of NF-kB via the inhibition of the Akt/IKKa pathway and enhanced ERK1/2 activity, augmented apoptosis induction through increased upregulation of p53 and p21, and downregulation of Bcl-2 protein expression [155].

4.4. Quercetin

Quercetin, a plant pigment and the most abundant dietary flavonol, is known to possess anti-proliferative and proapoptotic effects against many human cancers, including lung cancer [207]. It has been demonstrated that quercetin induces cytotoxicity and apoptosis in human NSCLC cells through multiple mechanisms. The mechanisms by which quercetin induces cell growth inhibition, cell cycle arrest at the G2/M phase and apoptosis involve the increase in the expression levels of survivin, cyclin B1, phospho-cdc2 (threonine 161), total p53 (DO-1), phospho-p53 (serine 15) and p21 proteins, and the induction of abnormal chromosome segregation [156]. Besides inactivation of Akt, quercetin-induced cleavage of caspase-3, caspase-7 and PARP has been found to be accompanied by the increased phosphorylation of MEK, ERK, c-Jun, and JNK, which suggests that the activation of the MEK-ERK pathway plays an important role in quercetin-induced apoptosis [157]. The results of the microarray analysis of quercetin-regulated genes in H460 cells revealed that quercetin upregulated genes that are associated with cell cycle arrest (p21^{Cip1}, GADD45), the death pathway (including TRAILR, FAS, TNFR1), the JNK pathway (MEKK1, MKK4, JNK), the IL1 receptor pathway (IL1, IL1R, IRAK), the caspase cascade (caspase-10, DFF45), and the NF-κB pathway (IκBα), while it downregulated genes that are involved in cell survival (NF-κB, IKK, AKT) and proliferation (SCF, SKP2, CDKs, cyclins) [158].

Quercetin has been demonstrated to exert anti-invasive and anti-metastatic activities in lung tumor cells through the downregulation of monocarboxylate transporter 1 (MCT1) [159], inhibition of aurora B kinase activity and histone 3 phosphorylation [160], disassembly of microfilaments, microtubules, and vimentin filaments along with the inhibition of vimentin and N-cadherin expression [161]. Quercetin treatment effectively suppressed the in vitro migration/invasion and in vivo bone metastasis of NSCLC cells by increasing the expression of the epithelial marker, E-cadherin, and decreasing the expressions of the mesenchymal markers, N-cadherin, and vimentin [162]. The mechanism that is associated with quercetin inhibited cell motility involved F-actin-containing microfilament bundle rearrangement and the suppression of EMT through both Snail-dependent Akt activation and Snail-independent ADAM9 pathway [162]. A recent study on the effect of five phytochemicals, including quercetin, curcumin, chrysin, apigenin, and luteolin on $NiCl_2$ (Ni)-induced the migration and invasion of cultured lung cancer cells revealed that the most efficient phytochemical compound inhibiting cell migration and invasion was quercetin, followed by chrysin and apigenin [163]. Further investigation demonstrated that quercetin and chrysin at 2 and 5 μM significantly suppressed Ni-induced rise in Toll-like receptor 4 (TLR4) expression, nuclear p65 level, and relative phospho-IKK-β and phospho-IKK-α levels, which suggests that the anti-invasive effect of quercetin is associated with the downregulation of TLR4/NF-κB signaling pathway [163].

Quercetin treatment in combination with Trichostatin A, a histone deacetylase inhibitor, significantly increased growth arrest and apoptosis through the mitochondrial pathway in A549 cells expressing wild-type p53, but not in H1299 cells harboring a p53 null mutation [164]. Moreover, quercetin treatment enhances TSA-induced acetylation of histones H3 and H4 through the p53-independent mechanism [164]. Cotreatment of quercetin with gemcitabine, a pyrimidine nucleoside analogue that inhibits DNA synthesis, promoted apoptosis via the inhibition of heat shock protein 70 (HSP70) expression [165]. It is evident that quercetin-induced HSP70 inhibition is associated with the caspase-dependent apoptosis through intrinsic apoptotic pathway, given the fact that quercetin-induced HSP70 inhibition significantly increased the caspase-3 activity, while the combination of quercetin and gemcitabine significantly increased caspase-9 activity [165]. HSP70 is known to control proteostasis and anti-stress responses in rapidly proliferating tumor cells and thus reduce the sensitivity of tumors to conventional anti-cancer drugs [208]. The mild toxicity profile of quercetin and its potential to act as a HSP70 inhibitor render it an attractive agent for use as part of a combination regimen to improve tumor response to chemotherapy with less severe side effects.

4.5. Other Naturally Occurring Polyphenols

Thymoquinone (TQ), the predominant bioactive constituent that is present in black seed oil (Nigella sativa), and Caffeic acid phenethyl ester (CAPE), a phenolic compound that is isolated from propolis, have been shown to induce G2/M cell cycle arrest and apoptosis through mechanisms similar to those of resveratrol [95]. Notably, all three agents (i.e., CAPE, TQ, and resveratrol) decreased the expression of cyclin D and increased the expression of TRAIL receptor 1 and 2, and p21 with the highest increase in p21 expression being observed in TQ-treated A549 cells. Moreover, CAPE and TQ upregulated Bax expression, while TQ and resveratrol downregulated Bcl-2, NF-κB, and IKK1 expression in A549 cells [95]. Based on those findings, further studies are warranted to evaluate the potential benefit of using TQ and CAPE in combination with other therapeutic agents for the treatment of lung cancer.

Pterostilbene (*trans*-3,5-dimethoxy-4′-hydroxystilbene), a naturally derived phytoalexin and a demethylated analog of resveratrol, was shown to inhibit A549 cell proliferation and induce S-phase cell cycle arrest by activating the ATM/ATR-CHK1/2-p53 signaling pathway [166]. Moreover, in two precancerous human bronchial epithelial cell lines, HBECR and HBECR/p53i, which have normal and reduced p53 expression levels, respectively, low-dose pterostilbene (at 1 and 5 μM) inhibited cell growth and induced cell cycle arrest in S phase and senescence in HBECR cells more efficiently than in HBECR/p53i cells, which suggests that the chemopreventive activity of pterostilbene is p53-dependent. This finding implicates that the use of pterostilbene as a chemopreventive agent for squamous lung carcinogenesis should be initiated at the early stage before p53 mutation occurs. Another analogue of resveratrol, bakuchiol (1-(4-hydroxyphenyl)-3,7-dimethyl-3-vinyl-1,6-octadiene) that was isolated from the seeds of *Psoralea corylifolia* L. (Leguminosae), exhibits a more significant cytotoxic effect in A549 cell line than in EA.hy926 endothelial cells, HUVECs, and primary cultured mouse embryo fibroblasts. It induces apoptosis and cell cycle arrest in S phase by increasing ROS production, interrupting mitochondrial homeostasis, increasing Bax/Bcl-2 ratio, upregulating p53, and activating Caspase 9/3, which suggests that the apoptotic effect of bakuchiol is p53-dependent and involves a mitochondrial-mediated pathway [167].

Chlorogenic acid (CGA) is the ester of caffeic acid and (−)-quinic acid, one of the most abundant phenolic acid compounds found in coffee and tea [209]. A substantial body of evidence has indicated that CGA exerts antioxidant [210,211], anti-inflammatory [212], antidiabetic [209], antimicrobial [213,214], and anticancer [215,216] activities. Different mechanisms that are associated with the antitumor properties of CGA have been proposed, including enhancing the activity of aryl hydrocarbon hydroxylase, suppressing the oxidative formation of 8-hydroxy-2′-deoxyguanosine (8-OH-dG) in DNA, reducing the production of ROS, and regulating the immune system [217–219]. A study by Part et al. demonstrated that CGA significantly decreased the HIF-1α protein level without changing its mRNA level in A549 cells under hypoxic conditions and subsequently suppressed the transcriptional activity of HIF-1α, leading to decreased expression of its downstream target VEGF [168]. Moreover, CGA inhibited hypoxia-stimulated HUVEC migration, invasion, and tumor formation in vitro and VEGF-stimulated angiogenesis in Matrigel plugs in vivo through the mechanism of inhibiting the HIF-1α/AKT signaling pathway [168]. The observed antiangiogenic potential of CGA suggests that CGA could be a novel therapeutic option for the treatment of lung cancer.

Fisetin (3,3′,4′,7-tetrahydroxyflavone), a naturally occurring diet-based flavonoid, exerts anticancer activity against different cancer cell lines, including NSCLC cell lines, when used alone or in combination with other chemotherapeutic agents [169–171,220–223]. The inhibitory effect of fisetin on lung tumor cell growth is attributable to dual suppression of PI3K/Akt and mTOR signaling, as evidenced by the activation of PTEN, phospho-AMPKα, and TSC2, and the inhibition of PI3K, phospho-Akt, phospho-mTOR, and several downstream targets of mTOR [169]. Fisetin was shown not only to inhibit the growth and induce the apoptosis of A549 cells with acquired cisplatin resistant, but also enhance the cisplatin cytotoxicity in cisplatin-resistant cells through the modulation of the MAPK/survivin/caspase pathway [170]. Fisetin showed a synergistic effect with paclitaxel on growth inhibition and mitotic catastrophe induction [171]. The fisetin-enhanced paclitaxel-induced mitotic catastrophe triggered cytoprotective autophagy, subsequently changing to autophagic cell death, which led to enhanced cytotoxicity [171].

Treatment with the ethyl acetate fraction of Glycyrrhiza uralensis extract that contains liquiritin, isoliquiritin, and isoliquirigenin decreased the viability of A549 cells, induced cell cycle arrest at G2/M phase, and apoptosis [172]. The ethyl acetate fraction significantly decreased the protein expression of PCNA, MDM2, phospho-GSK-3β, phospho-Akt, phospho-c-Raf, p-PTEN, caspase-3, pro-caspase-8, pro-caspase-9, PARP, and Bcl-2, and increased the expression of p53, p21, and Bax in a concentration-dependent manner, which suggested that the antitumor effects of liquiritin, isoliquiritin, and isoliquirigenin are orchestrated by the crosstalk among p53, Bcl-2 family, caspase cascades, and the Akt pathway [172].

Experimental evidence for the protective effects of several beverages and plant extracts against lung cancer through different mechanisms has been documented [15,173–175,177–179]. For example, the decoction extract of *Eucalyptus globulus* Labill. decreased the viability of H460 cells in a concentration dependent manner, which was correlated with cell cycle arrest at the G0/G1 phase, decrease in cell proliferation, and increase in the expression of p53, p21, and cyclin D1 proteins [173].

Polyphenol compounds that were isolated from Selaginella tamariscina suppressed the migration of A549 cells by targeting matrix metalloproteinases (MMPs) [174]. The antitumor activity of the polyphenol-containing rosemary extract was associated with the inactivation of the Akt/mTOR signaling pathway [175]. The inhibitory effect of red wine on the proliferation and lonogenic survival of A549 cells was associated with the inhibition of basal and EGF-stimulated Akt and Erk phosphorylation and increased total and phosphorylated p53 levels [15]. Magnolol and polyphenol mixture derived from Magnolia officinali significantly suppressed the expression levels and function of class I histone deacetylases (HDACs) and enriched the histone acetyl mark (H3K27ac) in the promoter region of DR5, which is a key protein in the death receptor signaling pathway [177]. Pomegranate concentrate that was administered via drinking bottle to cigarette smoking (CS)-exposed mice prevented the formation of CS-induced lung nodules by reducing the mitotic activity and HIF-1αexpression in CS-exposed animals [178]. Oral administration of Achyranthes aspera (PCA) extract to urethane primed lung cancerous mice increased the expression and activities of antioxidant enzymes GST, GR, CAT, and SOD, decreased the expression and activity of LDH, downregulated the expression of pro-inflammatory cytokines IL-1β, IL-6, and TNF-α, along with TFs, NF-κB, and Stat3, and increased expression of Bax and p53 [179]. In addition, PCA was found to counteract urethane-mediated conformational changes of DNA evident by the shift in guanine and thymine bands in Fourier Trans-form Infrared (FTIR) spectroscopy, which suggests that the anticancer activity of PCA is associated with its immunomodulatory role and DNA conformation restoring effect [179].

5. Conclusions and Future Perspectives

Taken together, many years of research on the mechanisms of anticancer action of natural polyphenols have yielded an amazing amount of information. Strong lines of evidence have confirmed that certain natural polyphenols possess potential antitumor activities against lung cancer, which is the leading cause of cancer death in the United States and worldwide. Encouraging data from preclinical studies that were conducted in cell cultures and tumor models have provided much insight into a broad spectrum of molecular mechanisms underlying the anti-proliferative, anti-migratory, anti-metastasis, anti-angiogenic, and pro-apoptotic effects of various bioactive natural polyphenols in lung cancer. However, given that current chemotherapies for lung cancer have not advanced dramatically despite our increased knowledge base, much research is still needed to pave the way for the optimal integration of bioactive polyphenols with traditional chemotherapeutic regimens for lung cancer treatment, and for a full exploration of the polyphenol compounds that have the potential to form the basis for novel anticancer drugs of the future. In addition to continually refining and expanding our knowledge of the molecular mechanisms by which natural polyphenols exert their antiproliferative and proapoptotic activities against lung cancer, future research endeavors should also focus on the mechanistic understanding of bioavailability and the biodistribution process of natural polyphenols, which has been considered to be a challenging research field [224,225]. With the enhanced insight into the factors controlling tumor uptake of phenolic compounds and the advent of innovative drug delivery technologies, it is anticipated that new exploitable avenues will be opened for improved delivery of bioactive natural polyphenols to the site of action, thereby advancing their therapeutic utility in the treatment of lung cancer.

Funding: This research received no external funding.

Conflicts of Interest: The authors declare no conflicts of interest.

References

1. Jemal, A.; Siegel, R.; Ward, E.; Hao, Y.; Xu, J.; Murray, T.; Thun, M.J. Cancer statistics, 2008. *CA A Cancer J. Clin.* **2008**, *58*, 71–96. [CrossRef] [PubMed]
2. Herbst, R.S.; Heymach, J.V.; Lippman, S.M. Lung cancer. *N. Engl. J. Med.* **2008**, *359*, 1367–1380. [CrossRef] [PubMed]

3. Pesch, B.; Kendzia, B.; Gustavsson, P.; Jockel, K.H.; Johnen, G.; Pohlabeln, H.; Olsson, A.; Ahrens, W.; Gross, I.M.; Bruske, I.; et al. Cigarette smoking and lung cancer—Relative risk estimates for the major histological types from a pooled analysis of case-control studies. *Int. J. Cancer* **2012**, *131*, 1210–1219. [CrossRef]

4. Taylor, R.; Najafi, F.; Dobson, A. Meta-analysis of studies of passive smoking and lung cancer: Effects of study type and continent. *Int. J. Epidemiol.* **2007**, *36*, 1048–1059. [CrossRef]

5. Field, R.W.; Withers, B.L. Occupational and environmental causes of lung cancer. *Clin. Chest Med.* **2012**, *33*, 681–703. [CrossRef]

6. Hosgood, H.D.; Boffetta, P.; Greenland, S.; Lee, Y.C.; McLaughlin, J.; Seow, A.; Duell, E.J.; Andrew, A.S.; Zaridze, D.; Szeszenia-Dabrowska, N.; et al. In-home coal and wood use and lung cancer risk: A pooled analysis of the International Lung Cancer Consortium. *Environ. Health Perspect.* **2010**, *118*, 1743–1747. [CrossRef]

7. Chen, C.L.; Chiou, H.Y.; Hsu, L.I.; Hsueh, Y.M.; Wu, M.M.; Chen, C.J. Ingested arsenic, characteristics of well water consumption and risk of different histological types of lung cancer in northeastern Taiwan. *Environ. Res.* **2010**, *110*, 455–462. [CrossRef]

8. Brenner, D.R.; Boffetta, P.; Duell, E.J.; Bickeboller, H.; Rosenberger, A.; McCormack, V.; Muscat, J.E.; Yang, P.; Wichmann, H.E.; Brueske-Hohlfeld, I.; et al. Previous lung diseases and lung cancer risk: A pooled analysis from the International Lung Cancer Consortium. *Am. J. Epidemiol.* **2012**, *176*, 573–585. [CrossRef]

9. Littman, A.J.; Thornquist, M.D.; White, E.; Jackson, L.A.; Goodman, G.E.; Vaughan, T.L. Prior lung disease and risk of lung cancer in a large prospective study. *Cancer Causes Control* **2004**, *15*, 819–827. [CrossRef]

10. Feskanich, D.; Ziegler, R.G.; Michaud, D.S.; Giovannucci, E.L.; Speizer, F.E.; Willett, W.C.; Colditz, G.A. Prospective study of fruit and vegetable consumption and risk of lung cancer among men and women. *J. Natl. Cancer Inst.* **2000**, *92*, 1812–1823. [CrossRef]

11. Middleton, E.; Kandaswami, C.; Theoharides, T.C. The effects of plant flavonoids on mammalian cells: Implications for inflammation, heart disease, and cancer. *Pharmacol. Rev.* **2000**, *52*, 673–751. [PubMed]

12. Smith-Warner, S.A.; Spiegelman, D.; Yaun, S.S.; Albanes, D.; Beeson, W.L.; van den Brandt, P.A.; Feskanich, D.; Folsom, A.R.; Fraser, G.E.; Freudenheim, J.L.; et al. Fruits, vegetables and lung cancer: A pooled analysis of cohort studies. *Int. J. Cancer* **2003**, *107*, 1001–1011. [CrossRef] [PubMed]

13. Wright, M.E.; Park, Y.; Subar, A.F.; Freedman, N.D.; Albanes, D.; Hollenbeck, A.; Leitzmann, M.F.; Schatzkin, A. Intakes of fruit, vegetables, and specific botanical groups in relation to lung cancer risk in the NIH-AARP Diet and Health Study. *Am. J. Epidemiol.* **2008**, *168*, 1024–1034. [CrossRef] [PubMed]

14. Amararathna, M.; Johnston, M.R.; Rupasinghe, H.P. Plant Polyphenols as Chemopreventive Agents for Lung Cancer. *Int. J. Mol. Sci.* **2016**, *17*, 1352. [CrossRef]

15. Barron, C.C.; Moore, J.; Tsakiridis, T.; Pickering, G.; Tsiani, E. Inhibition of human lung cancer cell proliferation and survival by wine. *Cancer Cell Int.* **2014**, *14*, 6. [CrossRef]

16. Lim, S.L.; Goh, Y.M.; Noordin, M.M.; Rahman, H.S.; Othman, H.H.; Abu Bakar, N.A.; Mohamed, S. Morinda citrifolia edible leaf extract enhanced immune response against lung cancer. *Food Funct.* **2016**, *7*, 741–751. [CrossRef]

17. Christensen, K.Y.; Naidu, A.; Parent, M.E.; Pintos, J.; Abrahamowicz, M.; Siemiatycki, J.; Koushik, A. The risk of lung cancer related to dietary intake of flavonoids. *Nutr. Cancer* **2012**, *64*, 964–974. [CrossRef]

18. Grosso, G.; Godos, J.; Lamuela-Raventos, R.; Ray, S.; Micek, A.; Pajak, A.; Sciacca, S.; D'Orazio, N.; Del Rio, D.; Galvano, F. A comprehensive meta-analysis on dietary flavonoid and lignan intake and cancer risk: Level of evidence and limitations. *Mol. Nutr. Food Res.* **2017**, *61*, 1600930. [CrossRef]

19. Wessner, B.; Strasser, E.M.; Koitz, N.; Schmuckenschlager, C.; Unger-Manhart, N.; Roth, E. Green tea polyphenol administration partly ameliorates chemotherapy-induced side effects in the small intestine of mice. *J. Nutr.* **2007**, *137*, 634–640. [CrossRef]

20. Yao, Q.; Ye, X.; Wang, L.; Gu, J.; Fu, T.; Wang, Y.; Lai, Y.; Wang, Y.; Wang, X.; Jin, H.; et al. Protective effect of curcumin on chemotherapy-induced intestinal dysfunction. *Int. J. Clin. Exp. Pathol.* **2013**, *6*, 2342–2349.

21. Khan, R.; Khan, A.Q.; Qamar, W.; Lateef, A.; Tahir, M.; Rehman, M.U.; Ali, F.; Sultana, S. Chrysin protects against cisplatin-induced colon. toxicity via amelioration of oxidative stress and apoptosis: Probable role of p38MAPK and p53. *Toxicol. Appl. Pharmacol.* **2012**, *258*, 315–329. [CrossRef] [PubMed]

22. Lissoni, P.; Tancini, G.; Barni, S.; Paolorossi, F.; Ardizzoia, A.; Conti, A.; Maestroni, G. Treatment of cancer chemotherapy-induced toxicity with the pineal hormone melatonin. *Support Care Cancer* **1997**, *5*, 126–129. [CrossRef]

23. Ghielmini, M.; Pagani, O.; de Jong, J.; Pampallona, S.; Conti, A.; Maestroni, G.; Sessa, C.; Cavalli, F. Double-blind randomized study on the myeloprotective effect of melatonin in combination with carboplatin and etoposide in advanced lung cancer. *Br. J. Cancer* **1999**, *80*, 1058–1061. [CrossRef] [PubMed]

24. Jaakkola, K.; Lahteenmaki, P.; Laakso, J.; Harju, E.; Tykka, H.; Mahlberg, K. Treatment with antioxidant and other nutrients in combination with chemotherapy and irradiation in patients with small-cell lung cancer. *Anticancer Res.* **1992**, *12*, 599–606. [PubMed]

25. Pace, A.; Savarese, A.; Picardo, M.; Maresca, V.; Pacetti, U.; Del Monte, G.; Biroccio, A.; Leonetti, C.; Jandolo, B.; Cognetti, F.; et al. Neuroprotective effect of vitamin E supplementation in patients treated with cisplatin chemotherapy. *J. Clin. Oncol.* **2003**, *21*, 927–931. [CrossRef] [PubMed]

26. Schmidinger, M.; Budinsky, A.C.; Wenzel, C.; Piribauer, M.; Brix, R.; Kautzky, M.; Oder, W.; Locker, G.J.; Zielinski, C.C.; Steger, G.G. Glutathione in the prevention of cisplatin induced toxicities. A prospectively randomized pilot trial in patients with head and neck cancer and non small cell lung cancer. *Wien. Klin. Wochenschr.* **2000**, *112*, 617–623.

27. Ferry, D.R.; Smith, A.; Malkhandi, J.; Fyfe, D.W.; deTakats, P.G.; Anderson, D.; Baker, J.; Kerr, D.J. Phase I clinical trial of the flavonoid quercetin: Pharmacokinetics and evidence for in vivo tyrosine kinase inhibition. *Clin. Cancer Res.* **1996**, *2*, 659–668.

28. Albanes, D.; Heinonen, O.P.; Taylor, P.R.; Virtamo, J.; Edwards, B.K.; Rautalahti, M.; Hartman, A.M.; Palmgren, J.; Freedman, L.S.; Haapakoski, J.; et al. Alpha-Tocopherol and beta-carotene supplements and lung cancer incidence in the alpha-tocopherol, beta-carotene cancer prevention study: Effects of base-line characteristics and study compliance. *J. Natl. Cancer Inst.* **1996**, *88*, 1560–1570. [CrossRef]

29. Pisters, K.M.; Newman, R.A.; Coldman, B.; Shin, D.M.; Khuri, F.R.; Hong, W.K.; Glisson, B.S.; Lee, J.S. Phase I trial of oral green tea extract in adult patients with solid tumors. *J. Clin. Oncol.* **2001**, *19*, 1830–1838. [CrossRef]

30. Hakim, I.A.; Harris, R.B.; Brown, S.; Chow, H.H.; Wiseman, S.; Agarwal, S.; Talbot, W. Effect of increased tea consumption on oxidative DNA damage among smokers: A randomized controlled study. *J. Nutr.* **2003**, *133*, 3303–3309. [CrossRef]

31. Goodman, G.E.; Thornquist, M.D.; Balmes, J.; Cullen, M.R.; Meyskens, F.L.; Omenn, G.S.; Valanis, B.; Williams, J.H. The Beta-Carotene and Retinol Efficacy Trial: Incidence of lung cancer and cardiovascular disease mortality during 6-year follow-up after stopping beta-carotene and retinol supplements. *J. Natl. Cancer Inst.* **2004**, *96*, 1743–1750. [CrossRef] [PubMed]

32. Laurie, S.A.; Miller, V.A.; Grant, S.C.; Kris, M.G.; Ng, K.K. Phase I study of green tea extract in patients with advanced lung cancer. *Cancer Chemother. Pharmacol.* **2005**, *55*, 33–38. [CrossRef] [PubMed]

33. Pathak, A.K.; Bhutani, M.; Guleria, R.; Bal, S.; Mohan, A.; Mohanti, B.K.; Sharma, A.; Pathak, R.; Bhardwaj, N.K.; Prasad, K.N.; et al. Chemotherapy alone vs. chemotherapy plus high dose multiple antioxidants in patients with advanced non small cell lung cancer. *J. Am. Coll. Nutr.* **2005**, *24*, 16–21. [CrossRef]

34. Osada, H.; Takahashi, T. Genetic alterations of multiple tumor suppressors and oncogenes in the carcinogenesis and progression of lung cancer. *Oncogene* **2002**, *21*, 7421–7434. [CrossRef]

35. Blanco, D.; Vicent, S.; Fraga, M.F.; Fernandez-Garcia, I.; Freire, J.; Lujambio, A.; Esteller, M.; Ortiz-de-Solorzano, C.; Pio, R.; Lecanda, F.; et al. Molecular analysis of a multistep lung cancer model induced by chronic inflammation reveals epigenetic regulation of p16 and activation of the DNA damage response pathway. *Neoplasia* **2007**, *9*, 840–852. [CrossRef]

36. Bettio, D.; Venci, A.; Achille, V.; Alloisio, M.; Santoro, A. Lung cancer in which the hypothesis of multi-step progression is confirmed by array-CGH results: A case report. *Exp. Ther. Med.* **2016**, *11*, 98–100. [CrossRef]

37. Gazdar, A.F.; Minna, J.D. Angiogenesis and the multistage development of lung cancers. *Clin. Cancer Res.* **2000**, *6*, 1611–1612.

38. McClelland, M.R.; Carskadon, S.L.; Zhao, L.; White, E.S.; Beer, D.G.; Orringer, M.B.; Pickens, A.; Chang, A.C.; Arenberg, D.A. Diversity of the angiogenic phenotype in non-small cell lung cancer. *Am. J. Respir. Cell Mol. Biol.* **2007**, *36*, 343–350. [CrossRef]

39. Hilbe, W.; Manegold, C.; Pircher, A. Targeting angiogenesis in lung cancer—Pitfalls in drug development. *Transl. Lung Cancer Res.* **2012**, *1*, 122–128. [CrossRef]

40. Lambert, A.W.; Wong, C.K.; Ozturk, S.; Papageorgis, P.; Raghunathan, R.; Alekseyev, Y.; Gower, A.C.; Reinhard, B.M.; Abdolmaleky, H.M.; Thiagalingam, S. Tumor Cell-Derived Periostin Regulates Cytokines That Maintain Breast Cancer Stem Cells. *Mol. Cancer Res.* **2016**, *14*, 103–113. [CrossRef]

41. Popper, H.H. Progression and metastasis of lung cancer. *Cancer Metastasis Rev.* **2016**, *35*, 75–91. [CrossRef] [PubMed]

42. Vogelstein, B.; Papadopoulos, N.; Velculescu, V.E.; Zhou, S.; Diaz, L.A.; Kinzler, K.W. Cancer genome landscapes. *Science* **2013**, *339*, 1546–1558. [CrossRef] [PubMed]

43. Pao, W.; Girard, N. New driver mutations in non-small-cell lung cancer. *Lancet Oncol.* **2011**, *12*, 175–180. [CrossRef]

44. Chan, B.A.; Hughes, B.G. Targeted therapy for non-small cell lung cancer: Current standards and the promise of the future. *Transl. Lung Cancer Res.* **2015**, *4*, 36–54. [CrossRef] [PubMed]

45. Vecchiarelli, S.; Bennati, C. Oncogene addicted non-small-cell lung cancer: Current standard and hot topics. *Future Oncol.* **2018**, *14*, 3–17. [CrossRef] [PubMed]

46. Facchinetti, F.; Rossi, G.; Bria, E.; Soria, J.C.; Besse, B.; Minari, R.; Friboulet, L.; Tiseo, M. Oncogene addiction in non-small cell lung cancer: Focus on ROS1 inhibition. *Cancer Treat. Rev.* **2017**, *55*, 83–95. [CrossRef] [PubMed]

47. Tsakonas, G.; Ekman, S. Oncogene-addicted non-small cell lung cancer and immunotherapy. *J. Thorac. Dis.* **2018**, *10*, 1547–1555. [CrossRef] [PubMed]

48. Peifer, M.; Fernandez-Cuesta, L.; Sos, M.L.; George, J.; Seidel, D.; Kasper, L.H.; Plenker, D.; Leenders, F.; Sun, R.; Zander, T.; et al. Integrative genome analyses identify key somatic driver mutations of small-cell lung cancer. *Nat. Genet.* **2012**, *44*, 1104–1110. [CrossRef] [PubMed]

49. George, J.; Lim, J.S.; Jang, S.J.; Cun, Y.; Ozretic, L.; Kong, G.; Leenders, F.; Lu, X.; Fernandez-Cuesta, L.; Bosco, G.; et al. Comprehensive genomic profiles of small cell lung cancer. *Nature* **2015**, *524*, 47–53. [CrossRef]

50. Brambilla, E.; Gazdar, A. Pathogenesis of lung cancer signalling pathways: Roadmap for therapies. *Eur. Respir. J.* **2009**, *33*, 1485–1497. [CrossRef]

51. Ciuffreda, L.; Incani, U.C.; Steelman, L.S.; Abrams, S.L.; Falcone, I.; Curatolo, A.D.; Chappell, W.H.; Franklin, R.A.; Vari, S.; Cognetti, F.; et al. Signaling intermediates (MAPK and PI3K) as therapeutic targets in NSCLC. *Curr. Pharm. Des.* **2014**, *20*, 3944–3957. [CrossRef] [PubMed]

52. Giaccia, A.J.; Kastan, M.B. The complexity of p53 modulation: Emerging patterns from divergent signals. *Genes Dev.* **1998**, *12*, 2973–2983. [CrossRef] [PubMed]

53. Lohrum, M.A.; Vousden, K.H. Regulation and activation of p53 and its family members. *Cell Death Differ.* **1999**, *6*, 1162–1168. [CrossRef]

54. Lee, J.S.; Yoon, A.; Kalapurakal, S.K.; Ro, J.Y.; Lee, J.J.; Tu, N.; Hittelman, W.N.; Hong, W.K. Expression of p53 oncoprotein in non-small-cell lung cancer: A favorable prognostic factor. *J. Clin. Oncol. Off. J. Am. Soc. Clin. Oncol.* **1995**, *13*, 1893–1903. [CrossRef]

55. Canale, M.; Petracci, E.; Delmonte, A.; Chiadini, E.; Dazzi, C.; Papi, M.; Capelli, L.; Casanova, C.; De Luigi, N.; Mariotti, M.; et al. Impact of TP53 Mutations on Outcome in EGFR-Mutated Patients Treated with First-Line Tyrosine Kinase Inhibitors. *Clin. Cancer Res.* **2017**, *23*, 2195–2202. [CrossRef]

56. Shivapurkar, N.; Reddy, J.; Chaudhary, P.M.; Gazdar, A.F. Apoptosis and lung cancer: A review. *J. Cell. Biochem.* **2003**, *88*, 885–898. [CrossRef]

57. Liu, G.; Pei, F.; Yang, F.; Li, L.; Amin, A.D.; Liu, S.; Buchan, J.R.; Cho, W.C. Role of Autophagy and Apoptosis in Non-Small-Cell Lung Cancer. *Int. J. Mol. Sci.* **2017**, *18*, 367. [CrossRef]

58. Ouyang, L.; Shi, Z.; Zhao, S.; Wang, F.T.; Zhou, T.T.; Liu, B.; Bao, J.K. Programmed cell death pathways in cancer: A review of apoptosis, autophagy and programmed necrosis. *Cell Prolif.* **2012**, *45*, 487–498. [CrossRef]

59. Li, H.; Zhu, H.; Xu, C.J.; Yuan, J. Cleavage of BID by caspase 8 mediates the mitochondrial damage in the Fas pathway of apoptosis. *Cell* **1998**, *94*, 491–501. [CrossRef]

60. Zou, H.; Li, Y.; Liu, X.; Wang, X. An APAF-1.cytochrome c multimeric complex is a functional apoptosome that activates procaspase-9. *J. Biol. Chem.* **1999**, *274*, 11549–11556. [CrossRef]

61. Hata, A.N.; Engelman, J.A.; Faber, A.C. The BCL2 Family: Key Mediators of the Apoptotic Response to Targeted Anticancer Therapeutics. *Cancer Discov.* **2015**, *5*, 475–487. [CrossRef] [PubMed]

62. Luthi, A.U.; Martin, S.J. The CASBAH: A searchable database of caspase substrates. *Cell Death Differ.* **2007**, *14*, 641–650. [CrossRef] [PubMed]

63. Hanahan, D.; Weinberg, R.A. Hallmarks of cancer: The next generation. *Cell* **2011**, *144*, 646–674. [CrossRef] [PubMed]

64. Fennell, D.A. Caspase regulation in non-small cell lung cancer and its potential for therapeutic exploitation. *Clin. Cancer Res.* **2005**, *11*, 2097–2105. [CrossRef] [PubMed]

65. Choi, H.S.; Cho, M.C.; Lee, H.G.; Yoon, D.Y. Indole-3-carbinol induces apoptosis through p53 and activation of caspase-8 pathway in lung cancer A549 cells. *Food Chem. Toxicol. Int. J. Publ. Br. Ind. Biol. Res. Assoc.* **2010**, *48*, 883–890. [CrossRef]

66. Chirumbolo, S. Dietary assumption of plant polyphenols and prevention of allergy. *Curr. Pharm. Des.* **2014**, *20*, 811–839. [CrossRef]

67. Pandey, K.B.; Rizvi, S.I. Plant polyphenols as dietary antioxidants in human health and disease. *Oxid. Med. Cell Longev.* **2009**, *2*, 270–278. [CrossRef]

68. Tsao, R. Chemistry and biochemistry of dietary polyphenols. *Nutrients* **2010**, *2*, 1231–1246. [CrossRef]

69. Manach, C.; Scalbert, A.; Morand, C.; Remesy, C.; Jimenez, L. Polyphenols: Food sources and bioavailability. *Am. J. Clin. Nutr.* **2004**, *79*, 727–747. [CrossRef]

70. Panche, A.N.; Diwan, A.D.; Chandra, S.R. Flavonoids: An overview. *J. Nutr. Sci.* **2016**, *5*, e47. [CrossRef]

71. Shen, T.; Wang, X.N.; Lou, H.X. Natural stilbenes: An overview. *Nat. Prod. Rep.* **2009**, *26*, 916–935. [CrossRef]

72. Teponno, R.B.; Kusari, S.; Spiteller, M. Recent advances in research on lignans and neolignans. *Nat. Prod. Rep.* **2016**, *33*, 1044–1092. [CrossRef] [PubMed]

73. Haslam, E. Polyphenol-protein interactions. *Biochem. J.* **1974**, *139*, 285–288. [CrossRef] [PubMed]

74. Hansch, C.; Gao, H. Comparative QSAR: Radical Reactions of Benzene Derivatives in Chemistry and Biology. *Chem. Rev.* **1997**, *97*, 2995–3060. [CrossRef]

75. Nandi, S.; Vracko, M.; Bagchi, M.C. Anticancer activity of selected phenolic compounds: QSAR studies using ridge regression and neural networks. *Chem. Biol. Drug Des.* **2007**, *70*, 424–436. [CrossRef]

76. Scotti, L.; Bezerra, M.F.J.; Magalhaes, M.D.R.; da Silva, M.S.; Pitta, I.R.; Scotti, M.T. SAR, QSAR and docking of anticancer flavonoids and variants: A review. *Curr. Top. Med. Chem.* **2012**, *12*, 2785–2809. [CrossRef]

77. Selassie, C.; Shusterman, A.; Kapur, S.P.; Verma, R.; Zhang, L.; Hansch, C. On the toxicity of phenols to fast growing cells. A QSAR model for a radical-based toxicity. *J. Chem. Soc. Perkin Trans.* **1999**, *2*, 2729–2733. [CrossRef]

78. Verma, R.P.; Kapur, S.; Barberena, O.; Shusterman, A.; Hansch, C.H.; Selassie, C.D. Synthesis, cytotoxicity, and QSAR analysis of X-thiophenols in rapidly dividing cells. *Chem. Res. Toxicol.* **2003**, *16*, 276–284. [CrossRef]

79. Oliver, C.L.; Miranda, M.B.; Shangary, S.; Land, S.; Wang, S.; Johnson, D.E. (-)-Gossypol acts directly on the mitochondria to overcome Bcl-2- and Bcl-X(L)-mediated apoptosis resistance. *Mol. Cancer Ther.* **2005**, *4*, 23–31.

80. Bruncko, M.; Oost, T.K.; Belli, B.A.; Ding, H.; Joseph, M.K.; Kunzer, A.; Martineau, D.; McClellan, W.J.; Mitten, M.; Ng, S.C.; et al. Studies leading to potent, dual inhibitors of Bcl-2 and Bcl-xL. *J. Med. Chem.* **2007**, *50*, 641–662. [CrossRef]

81. Kitada, S.; Leone, M.; Sareth, S.; Zhai, D.; Reed, J.C.; Pellecchia, M. Discovery, characterization, and structure-activity relationships studies of proapoptotic polyphenols targeting B-cell lymphocyte/leukemia-2 proteins. *J. Med. Chem.* **2003**, *46*, 4259–4264. [CrossRef] [PubMed]

82. Kitada, S.; Kress, C.L.; Krajewska, M.; Jia, L.; Pellecchia, M.; Reed, J.C. Bcl-2 antagonist apogossypol (NSC736630) displays single-agent activity in Bcl-2-transgenic mice and has superior efficacy with less toxicity compared with gossypol (NSC19048). *Blood* **2008**, *111*, 3211–3219. [CrossRef] [PubMed]

83. Wang, G.; Nikolovska-Coleska, Z.; Yang, C.Y.; Wang, R.; Tang, G.; Guo, J.; Shangary, S.; Qiu, S.; Gao, W.; Yang, D.; et al. Structure-based design of potent small-molecule inhibitors of anti-apoptotic Bcl-2 proteins. *J. Med. Chem.* **2006**, *49*, 6139–6142. [CrossRef] [PubMed]

84. Tang, G.; Yang, C.Y.; Nikolovska-Coleska, Z.; Guo, J.; Qiu, S.; Wang, R.; Gao, W.; Wang, G.; Stuckey, J.; Krajewski, K.; et al. Pyrogallol-based molecules as potent inhibitors of the antiapoptotic Bcl-2 proteins. *J. Med. Chem.* **2007**, *50*, 1723–1726. [CrossRef]

85. Tang, G.; Nikolovska-Coleska, Z.; Qiu, S.; Yang, C.Y.; Guo, J.; Wang, S. Acylpyrogallols as inhibitors of antiapoptotic Bcl-2 proteins. *J. Med. Chem.* **2008**, *51*, 717–720. [CrossRef]

86. Larsen, C.A.; Bisson, W.H.; Dashwood, R.H. Tea catechins inhibit hepatocyte growth factor receptor (MET kinase) activity in human colon cancer cells: Kinetic and molecular docking studies. *J. Med. Chem.* **2009**, *52*, 6543–6545. [CrossRef]

87. Ciechanover, A.; Orian, A.; Schwartz, A.L. Ubiquitin-mediated proteolysis: Biological regulation via destruction. *Bioessays* **2000**, *22*, 442–451. [CrossRef]

88. Nam, S.; Smith, D.M.; Dou, Q.P. Ester bond-containing tea polyphenols potently inhibit proteasome activity in vitro and in vivo. *J. Biol. Chem.* **2001**, *276*, 13322–13330. [CrossRef]

89. Kazi, A.; Wang, Z.; Kumar, N.; Falsetti, S.C.; Chan, T.H.; Dou, Q.P. Structure-activity relationships of synthetic analogs of (-)-epigallocatechin-3-gallate as proteasome inhibitors. *Anticancer Res.* **2004**, *24*, 943–954.

90. Whyte, L.; Huang, Y.Y.; Torres, K.; Mehta, R.G. Molecular mechanisms of resveratrol action in lung cancer cells using dual protein and microarray analyses. *Cancer Res.* **2007**, *67*, 12007–12017. [CrossRef]

91. Li, W.; Shi, Y.; Wang, R.; Pan, L.; Ma, L.; Jin, F. Resveratrol promotes the sensitivity of small-cell lung cancer H446 cells to cisplatin by regulating intrinsic apoptosis. *Int. J. Oncol.* **2018**, *53*, 2123–2130. [CrossRef] [PubMed]

92. Rasheduzzaman, M.; Jeong, J.K.; Park, S.Y. Resveratrol sensitizes lung cancer cell to TRAIL by p53 independent and suppression of Akt/NF-kappaB signaling. *Life Sci.* **2018**, *208*, 208–220. [CrossRef] [PubMed]

93. Sun, L.; Chen, B.; Jiang, R.; Li, J.; Wang, B. Resveratrol inhibits lung cancer growth by suppressing M2-like polarization of tumor associated macrophages. *Cell Immunol.* **2017**, *311*, 86–93. [CrossRef]

94. Ma, L.; Li, W.; Wang, R.; Nan, Y.; Wang, Q.; Liu, W.; Jin, F. Resveratrol enhanced anticancer effects of cisplatin on non-small cell lung cancer cell lines by inducing mitochondrial dysfunction and cell apoptosis. *Int. J. Oncol.* **2015**, *47*, 1460–1468. [CrossRef]

95. Ulasli, S.S.; Celik, S.; Gunay, E.; Ozdemir, M.; Hazman, O.; Ozyurek, A.; Koyuncu, T.; Unlu, M. Anticancer effects of thymoquinone, caffeic acid phenethyl ester and resveratrol on A549 non-small cell lung cancer cells exposed to benzo (a) pyrene. *Asian Pac. J. Cancer Prev.* **2013**, *14*, 6159–6164. [CrossRef]

96. Wright, C.; Iyer, A.K.V.; Yakisich, J.S.; Azad, N. Anti-Tumorigenic Effects of Resveratrol in Lung Cancer Cells Through Modulation of c-FLIP. *Curr. Cancer Drug Targets* **2017**, *17*, 669–680. [CrossRef]

97. Luo, H.; Yang, A.; Schulte, B.A.; Wargovich, M.J.; Wang, G.Y. Resveratrol induces premature senescence in lung cancer cells via ROS-mediated DNA damage. *PLoS ONE* **2013**, *8*, e60065. [CrossRef]

98. Luo, H.; Wang, L.; Schulte, B.A.; Yang, A.; Tang, S.; Wang, G.Y. Resveratrol enhances ionizing radiation-induced premature senescence in lung cancer cells. *Int. J. Oncol.* **2013**, *43*, 1999–2006. [CrossRef]

99. Zhu, Y.; He, W.; Gao, X.; Li, B.; Mei, C.; Xu, R.; Chen, H. Resveratrol overcomes gefitinib resistance by increasing the intracellular gefitinib concentration and triggering apoptosis, autophagy and senescence in PC9/G NSCLC cells. *Sci. Rep.* **2015**, *5*, 17730. [CrossRef]

100. Nie, P.; Hu, W.; Zhang, T.; Yang, Y.; Hou, B.; Zou, Z. Synergistic Induction of Erlotinib-Mediated Apoptosis by Resveratrol in Human Non-Small-Cell Lung Cancer Cells by Down-Regulating Survivin and Up-Regulating PUMA. *Cell Physiol. Biochem.* **2015**, *35*, 2255–2271. [CrossRef]

101. Ko, J.C.; Syu, J.J.; Chen, J.C.; Wang, T.J.; Chang, P.Y.; Chen, C.Y.; Jian, Y.T.; Jian, Y.J.; Lin, Y.W. Resveratrol Enhances Etoposide-Induced Cytotoxicity through Down-Regulating ERK1/2 and AKT-Mediated X-ray Repair Cross-Complement Group 1 (XRCC1) Protein Expression in Human Non-Small-Cell Lung Cancer Cells. *Basic Clin. Pharm. Toxicol.* **2015**, *117*, 383–391. [CrossRef] [PubMed]

102. Bae, S.; Lee, E.M.; Cha, H.J.; Kim, K.; Yoon, Y.; Lee, H.; Kim, J.; Kim, Y.J.; Lee, H.G.; Jeung, H.K.; et al. Resveratrol alters microRNA expression profiles in A549 human non-small cell lung cancer cells. *Mol. Cells* **2011**, *32*, 243–249. [CrossRef] [PubMed]

103. Han, Z.; Yang, Q.; Liu, B.; Wu, J.; Li, Y.; Yang, C.; Jiang, Y. MicroRNA-622 functions as a tumor suppressor by targeting K-Ras and enhancing the anticarcinogenic effect of resveratrol. *Carcinogenesis* **2012**, *33*, 131–139. [CrossRef]

104. Yu, Y.H.; Chen, H.A.; Chen, P.S.; Cheng, Y.J.; Hsu, W.H.; Chang, Y.W.; Chen, Y.H.; Jan, Y.; Hsiao, M.; Chang, T.Y.; et al. MiR-520h-mediated FOXC2 regulation is critical for inhibition of lung cancer progression by resveratrol. *Oncogene* **2013**, *32*, 431–443. [CrossRef]

105. Lee, E.J.; Min, H.Y.; Joo Park, H.; Chung, H.J.; Kim, S.; Nam Han, Y.; Lee, S.K. G2/M cell cycle arrest and induction of apoptosis by a stilbenoid, 3,4,5-trimethoxy-4′-bromo-cis-stilbene, in human lung cancer cells. *Life Sci.* **2004**, *75*, 2829–2839. [CrossRef]

106. Wang, X.; Wang, D.; Zhao, Y. Effect and Mechanism of Resveratrol on the Apoptosis of Lung Adenocarcinoma Cell Line A549. *Cell Biochem. Biophys.* **2015**, *73*, 527–531. [CrossRef]

107. He, L.; Fan, F.; Hou, X.; Gao, C.; Meng, L.; Meng, S.; Huang, S.; Wu, H. Resveratrol suppresses pulmonary tumor metastasis by inhibiting platelet-mediated angiogenic responses. *J. Surg. Res.* **2017**, *217*, 113–122. [CrossRef]

108. Li, X.; Wang, D.; Zhao, Q.C.; Shi, T.; Chen, J. Resveratrol Inhibited Non-small Cell Lung Cancer Through Inhibiting STAT-3 Signaling. *Am. J. Med. Sci.* **2016**, *352*, 524–530. [CrossRef]

109. Li, W.; Ma, X.; Li, N.; Liu, H.; Dong, Q.; Zhang, J.; Yang, C.; Liu, Y.; Liang, Q.; Zhang, S.; et al. Resveratrol inhibits Hexokinases II mediated glycolysis in non-small cell lung cancer via targeting Akt signaling pathway. *Exp. Cell Res.* **2016**, *349*, 320–327. [CrossRef]

110. Lee, Y.S.; Doonan, B.B.; Wu, J.M.; Hsieh, T.C. Combined metformin and resveratrol confers protection against UVC-induced DNA damage in A549 lung cancer cells via modulation of cell cycle checkpoints and DNA repair. *Oncol. Rep.* **2016**, *35*, 3735–3741. [CrossRef]

111. Sahin, E.; Baycu, C.; Koparal, A.T.; Burukoglu Donmez, D.; Bektur, E. Resveratrol reduces IL-6 and VEGF secretion from co-cultured A549 lung cancer cells and adipose-derived mesenchymal stem cells. *Tumour Biol. J. Int. Soc. Oncodev. Biol. Med.* **2016**, *37*, 7573–7582. [CrossRef] [PubMed]

112. Yuan, L.; Zhang, Y.; Xia, J.; Liu, B.; Zhang, Q.; Liu, J.; Luo, L.; Peng, Z.; Song, Z.; Zhu, R. Resveratrol induces cell cycle arrest via a p53-independent pathway in A549 cells. *Mol. Med. Rep.* **2015**, *11*, 2459–2464. [CrossRef] [PubMed]

113. Bai, T.; Dong, D.S.; Pei, L. Synergistic antitumor activity of resveratrol and miR-200c in human lung cancer. *Oncol. Rep.* **2014**, *31*, 2293–2297. [CrossRef] [PubMed]

114. Wang, H.; Zhang, H.; Tang, L.; Chen, H.; Wu, C.; Zhao, M.; Yang, Y.; Chen, X.; Liu, G. Resveratrol inhibits TGF-beta1-induced epithelial-to-mesenchymal transition and suppresses lung cancer invasion and metastasis. *Toxicology* **2013**, *303*, 139–146. [CrossRef]

115. Liao, H.F.; Kuo, C.D.; Yang, Y.C.; Lin, C.P.; Tai, H.C.; Chen, Y.Y.; Chen, Y.J. Resveratrol enhances radiosensitivity of human non-small cell lung cancer NCI-H838 cells accompanied by inhibition of nuclear factor-kappa B activation. *J. Radiat. Res.* **2005**, *46*, 387–393. [CrossRef]

116. Liu, P.L.; Tsai, J.R.; Charles, A.L.; Hwang, J.J.; Chou, S.H.; Ping, Y.H.; Lin, F.Y.; Chen, Y.L.; Hung, C.Y.; Chen, W.C.; et al. Resveratrol inhibits human lung adenocarcinoma cell metastasis by suppressing heme oxygenase 1-mediated nuclear factor-kappaB pathway and subsequently downregulating expression of matrix met alloproteinases. *Mol. Nutr. Food Res.* **2010**, *54*, 196–204. [CrossRef]

117. Zhao, W.; Bao, P.; Qi, H.; You, H. Resveratrol down-regulates survivin and induces apoptosis in human multidrug-resistant SPC-A-1/CDDP cells. *Oncol. Rep.* **2010**, *23*, 279–286. [CrossRef]

118. Kim, Y.A.; Lee, W.H.; Choi, T.H.; Rhee, S.H.; Park, K.Y.; Choi, Y.H. Involvement of p21WAF1/CIP1, pRB, Bax and NF-kappaB in induction of growth arrest and apoptosis by resveratrol in human lung carcinoma A549 cells. *Int. J. Oncol.* **2003**, *23*, 1143–1149.

119. Karthikeyan, S.; Hoti, S.L.; Prasad, N.R. Resveratrol loaded gelatin nanoparticles synergistically inhibits cell cycle progression and constitutive NF-kappaB activation, and induces apoptosis in non-small cell lung cancer cells. *Biomed. Pharmacother.* **2015**, *70*, 274–282. [CrossRef]

120. Thomas, E.; Gopalakrishnan, V.; Hegde, M.; Kumar, S.; Karki, S.S.; Raghavan, S.C.; Choudhary, B. A Novel Resveratrol Based Tubulin Inhibitor Induces Mitotic Arrest and Activates Apoptosis in Cancer Cells. *Sci. Rep.* **2016**, *6*, 34653. [CrossRef]

121. Savio, M.; Ferraro, D.; Maccario, C.; Vaccarone, R.; Jensen, L.D.; Corana, F.; Mannucci, B.; Bianchi, L.; Cao, Y.; Stivala, L.A. Resveratrol analogue 4,4′-dihydroxy-trans-stilbene potently inhibits cancer invasion and metastasis. *Sci. Rep.* **2016**, *6*, 19973. [CrossRef] [PubMed]

122. Malhotra, A.; Nair, P.; Dhawan, D.K. Premature mitochondrial senescence and related ultrastructural changes during lung carcinogenesis modulation by curcumin and resveratrol. *Ultrastruct. Pathol.* **2012**, *36*, 179–184. [CrossRef] [PubMed]

123. Weng, C.J.; Yang, Y.T.; Ho, C.T.; Yen, G.C. Mechanisms of apoptotic effects induced by resveratrol, dibenzoylmethane, and their analogues on human lung carcinoma cells. *J. Agric. Food Chem.* **2009**, *57*, 5235–5243. [CrossRef] [PubMed]

124. Lee, E.O.; Lee, H.J.; Hwang, H.S.; Ahn, K.S.; Chae, C.; Kang, K.S.; Lu, J.; Kim, S.H. Potent inhibition of Lewis lung cancer growth by heyneanol A from the roots of Vitis amurensis through apoptotic and anti-angiogenic activities. *Carcinogenesis* **2006**, *27*, 2059–2069. [CrossRef] [PubMed]

125. Yamauchi, R.; Sasaki, K.; Yoshida, K. Identification of epigallocatechin-3-gallate in green tea polyphenols as a potent inducer of p53-dependent apoptosis in the human lung cancer cell line A549. *Toxicol. In Vitro* **2009**, *23*, 834–839. [CrossRef]

126. Okabe, S.; Suganuma, M.; Hayashi, M.; Sueoka, E.; Komori, A.; Fujiki, H. Mechanisms of growth inhibition of human lung cancer cell line, PC-9, by tea polyphenols. *Jpn. J. Cancer Res.* **1997**, *88*, 639–643. [CrossRef]

127. Honda, Y.; Takigawa, N.; Ichihara, E.; Ninomiya, T.; Kubo, T.; Ochi, N.; Yasugi, M.; Murakami, T.; Yamane, H.; Tanimoto, M.; et al. Effects of (-)-epigallocatechin-3-gallate on EGFR- or Fusion Gene-driven Lung Cancer Cells. *Acta Med. Okayama* **2017**, *71*, 505–512. [CrossRef]

128. Okabe, S.; Fujimoto, N.; Sueoka, N.; Suganuma, M.; Fujiki, H. Modulation of gene expression by (-)-epigallocatechin gallate in PC-9 cells using a cDNA expression array. *Biol. Pharm. Bull.* **2001**, *24*, 883–886. [CrossRef]

129. Zhou, H.; Chen, J.X.; Yang, C.S.; Yang, M.Q.; Deng, Y.; Wang, H. Gene regulation mediated by microRNAs in response to green tea polyphenol EGCG in mouse lung cancer. *BMC Genom.* **2014**, *15*, S3. [CrossRef]

130. Suganuma, M.; Kurusu, M.; Suzuki, K.; Tasaki, E.; Fujiki, H. Green tea polyphenol stimulates cancer preventive effects of celecoxib in human lung cancer cells by upregulation of GADD153 gene. *Int. J. Cancer* **2006**, *119*, 33–40. [CrossRef]

131. Zhang, Y.; Wang, X.; Han, L.; Zhou, Y.; Sun, S. Green tea polyphenol EGCG reverse cisplatin resistance of A549/DDP cell line through candidate genes demethylation. *Biomed. Pharmacother.* **2015**, *69*, 285–290. [CrossRef] [PubMed]

132. Amin, A.R.; Wang, D.; Zhang, H.; Peng, S.; Shin, H.J.; Brandes, J.C.; Tighiouart, M.; Khuri, F.R.; Chen, Z.G.; Shin, D.M. Enhanced anti-tumor activity by the combination of the natural compounds (-)-epigallocatechin-3-gallate and luteolin: Potential role of p53. *J. Biol. Chem.* **2010**, *285*, 34557–34565. [CrossRef] [PubMed]

133. Ganguly, C.; Saha, P.; Panda, C.K.; Das, S. Inhibition of growth, induction of apoptosis and alteration of gene expression by tea polyphenols in the highly metastatic human lung cancer cell line NCI-H460. *Asian Pac. J. Cancer Prev.* **2005**, *6*, 326–331. [PubMed]

134. Banerjee, S.; Manna, S.; Mukherjee, S.; Pal, D.; Panda, C.K.; Das, S. Black tea polyphenols restrict benzopyrene-induced mouse lung cancer progression through inhibition of Cox-2 and induction of caspase-3 expression. *Asian Pac. J. Cancer Prev.* **2006**, *7*, 661–666. [PubMed]

135. Banerjee, S.; Manna, S.; Saha, P.; Panda, C.K.; Das, S. Black tea polyphenols suppress cell proliferation and induce apoptosis during benzo(a)pyrene-induced lung carcinogenesis. *Eur. J. Cancer Prev.* **2005**, *14*, 215–221. [CrossRef]

136. Gu, Q.; Hu, C.; Chen, Q.; Xia, Y.; Feng, J.; Yang, H. Development of a rat model by 3,4-benzopyrene intra-pulmonary injection and evaluation of the effect of green tea drinking on p53 and bcl-2 expression in lung carcinoma. *Cancer Detect. Prev.* **2009**, *32*, 444–451. [CrossRef]

137. Gu, Q.; Hu, C.; Chen, Q.; Xia, Y. Tea polyphenols prevent lung from preneoplastic lesions and effect p53 and bcl-2 gene expression in rat lung tissues. *Int. J. Clin. Exp. Pathol.* **2013**, *6*, 1523–1531.

138. Roy, P.; Nigam, N.; Singh, M.; George, J.; Srivastava, S.; Naqvi, H.; Shukla, Y. Tea polyphenols inhibit cyclooxygenase-2 expression and block activation of nuclear factor-kappa B and Akt in diethylnitrosoamine induced lung tumors in Swiss mice. *Investig. New Drugs* **2010**, *28*, 466–471. [CrossRef]

139. Pan, J.; Zhang, Q.; Xiong, D.; Vedell, P.; Yan, Y.; Jiang, H.; Cui, P.; Ding, F.; Tichelaar, J.W.; Wang, Y.; et al. Transcriptomic analysis by RNA-seq reveals AP-1 pathway as key regulator that green tea may rely on to inhibit lung tumorigenesis. *Mol. Carcinog.* **2014**, *53*, 19–29. [CrossRef]

140. Lu, Q.Y.; Yang, Y.; Jin, Y.S.; Zhang, Z.F.; Heber, D.; Li, F.P.; Dubinett, S.M.; Sondej, M.A.; Loo, J.A.; Rao, J.Y. Effects of green tea extract on lung cancer A549 cells: Proteomic identification of proteins associated with cell migration. *Proteomics* **2009**, *9*, 757–767. [CrossRef]

141. Lu, G.; Xiao, H.; You, H.; Lin, Y.; Jin, H.; Snagaski, B.; Yang, C.S. Synergistic inhibition of lung tumorigenesis by a combination of green tea polyphenols and atorvastatin. *Clin. Cancer Res.* **2008**, *14*, 4981–4988. [CrossRef] [PubMed]

142. Izdebska, M.; Klimaszewska-Wisniewska, A.; Halas, M.; Gagat, M.; Grzanka, A. Green tea extract induces protective autophagy in A549 non-small lung cancer cell line. *Postepy. Hig. Med. Dosw.* **2015**, *69*, 1478–1484.

143. Radhakrishna, P.G.; Srivastava, A.S.; Hassanein, T.I.; Chauhan, D.P.; Carrier, E. Induction of apoptosis in human lung cancer cells by curcumin. *Cancer Lett.* **2004**, *208*, 163–170. [CrossRef] [PubMed]

144. Saha, A.; Kuzuhara, T.; Echigo, N.; Fujii, A.; Suganuma, M.; Fujiki, H. Apoptosis of human lung cancer cells by curcumin mediated through up-regulation of "growth arrest and DNA damage inducible genes 45 and 153". *Biol. Pharm. Bull.* **2010**, *33*, 1291–1299. [CrossRef]

145. Wu, S.H.; Hang, L.W.; Yang, J.S.; Chen, H.Y.; Lin, H.Y.; Chiang, J.H.; Lu, C.C.; Yang, J.L.; Lai, T.Y.; Ko, Y.C.; et al. Curcumin induces apoptosis in human non-small cell lung cancer NCI-H460 cells through ER stress and caspase cascade- and mitochondria-dependent pathways. *Anticancer Res.* **2010**, *30*, 2125–2133.

146. Li, Y.; Zhang, S.; Geng, J.X.; Hu, X.Y. Curcumin inhibits human non-small cell lung cancer A549 cell proliferation through regulation of Bcl-2/Bax and cytochrome C. *Asian Pac. J. Cancer Prev.* **2013**, *14*, 4599–4602. [CrossRef]

147. Yang, C.L.; Ma, Y.G.; Xue, Y.X.; Liu, Y.Y.; Xie, H.; Qiu, G.R. Curcumin induces small cell lung cancer NCI-H446 cell apoptosis via the reactive oxygen species-mediated mitochondrial pathway and not the cell death receptor pathway. *DNA Cell Biol.* **2012**, *31*, 139–150. [CrossRef]

148. Jin, H.; Qiao, F.; Wang, Y.; Xu, Y.; Shang, Y. Curcumin inhibits cell proliferation and induces apoptosis of human non-small cell lung cancer cells through the upregulation of miR-192-5p and suppression of PI3K/Akt signaling pathway. *Oncol. Rep.* **2015**, *34*, 2782–2789. [CrossRef]

149. Wang, A.; Wang, J.; Zhang, S.; Zhang, H.; Xu, Z.; Li, X. Curcumin inhibits the development of non-small cell lung cancer by inhibiting autophagy and apoptosis. *Exp. Ther. Med.* **2017**, *14*, 5075–5080. [CrossRef]

150. Liu, F.; Gao, S.; Yang, Y.; Zhao, X.; Fan, Y.; Ma, W.; Yang, D.; Yang, A.; Yu, Y. Curcumin induced autophagy anticancer effects on human lung adenocarcinoma cell line A549. *Oncol. Lett.* **2017**, *14*, 2775–2782. [CrossRef]

151. Liao, H.; Wang, Z.; Deng, Z.; Ren, H.; Li, X. Curcumin inhibits lung cancer invasion and metastasis by attenuating GLUT1/MT1-MMP/MMP2 pathway. *Int. J. Clin. Exp. Med.* **2015**, *8*, 8948–8957. [PubMed]

152. Tsai, J.R.; Liu, P.L.; Chen, Y.H.; Chou, S.H.; Cheng, Y.J.; Hwang, J.J.; Chong, I.W. Curcumin Inhibits Non-Small Cell Lung Cancer Cells Metastasis through the Adiponectin/NF-kappab/MMPs Signaling Pathway. *PLoS ONE* **2015**, *10*, e0144462. [CrossRef] [PubMed]

153. Xia, Y.Q.; Wei, X.Y.; Li, W.L.; Kanchana, K.; Xu, C.C.; Chen, D.H.; Chou, P.H.; Jin, R.; Wu, J.Z.; Liang, G. Curcumin analogue A501 induces G2/M arrest and apoptosis in non-small cell lung cancer cells. *Asian Pac. J. Cancer Prev.* **2014**, *15*, 6893–6898. [CrossRef] [PubMed]

154. Lee, J.Y.; Lee, Y.M.; Chang, G.C.; Yu, S.L.; Hsieh, W.Y.; Chen, J.J.; Chen, H.W.; Yang, P.C. Curcumin induces EGFR degradation in lung adenocarcinoma and modulates p38 activation in intestine: The versatile adjuvant for gefitinib therapy. *PLoS ONE* **2011**, *6*, e23756. [CrossRef] [PubMed]

155. Kang, J.H.; Kang, H.S.; Kim, I.K.; Lee, H.Y.; Ha, J.H.; Yeo, C.D.; Kang, H.H.; Moon, H.S.; Lee, S.H. Curcumin sensitizes human lung cancer cells to apoptosis and metastasis synergistically combined with carboplatin. *Exp. Biol. Med.* **2015**, *240*, 1416–1425. [CrossRef] [PubMed]

156. Kuo, P.C.; Liu, H.F.; Chao, J.I. Survivin and p53 modulate quercetin-induced cell growth inhibition and apoptosis in human lung carcinoma cells. *J. Biol. Chem.* **2004**, *279*, 55875–55885. [CrossRef] [PubMed]

157. Nguyen, T.T.; Tran, E.; Nguyen, T.H.; Do, P.T.; Huynh, T.H.; Huynh, H. The role of activated MEK-ERK pathway in quercetin-induced growth inhibition and apoptosis in A549 lung cancer cells. *Carcinogenesis* **2004**, *25*, 647–659. [CrossRef] [PubMed]

158. Youn, H.; Jeong, J.C.; Jeong, Y.S.; Kim, E.J.; Um, S.J. Quercetin potentiates apoptosis by inhibiting nuclear factor-kappaB signaling in H460 lung cancer cells. *Biol. Pharm. Bull.* **2013**, *36*, 944–951. [CrossRef]

159. Izumi, H.; Takahashi, M.; Uramoto, H.; Nakayama, Y.; Oyama, T.; Wang, K.Y.; Sasaguri, Y.; Nishizawa, S.; Kohno, K. Monocarboxylate transporters 1 and 4 are involved in the invasion activity of human lung cancer cells. *Cancer Sci.* **2011**, *102*, 1007–1013. [CrossRef]

160. Zhu, X.; Ma, P.; Peng, D.; Wang, Y.; Wang, D.; Chen, X.; Zhang, X.; Song, Y. Quercetin suppresses lung cancer growth by targeting Aurora B kinase. *Cancer Med.* **2016**, *5*, 3156–3165. [CrossRef]

161. Klimaszewska-Wisniewska, A.; Halas-Wisniewska, M.; Izdebska, M.; Gagat, M.; Grzanka, A.; Grzanka, D. Antiproliferative and antimetastatic action of quercetin on A549 non-small cell lung cancer cells through its effect on the cytoskeleton. *Acta Histochem.* **2017**, *119*, 99–112. [CrossRef] [PubMed]

162. Chang, J.H.; Lai, S.L.; Chen, W.S.; Hung, W.Y.; Chow, J.M.; Hsiao, M.; Lee, W.J.; Chien, M.H. Quercetin suppresses the metastatic ability of lung cancer through inhibiting Snail-dependent Akt activation and Snail-independent ADAM9 expression pathways. *Biochim. Biophys. Acta Mol. Cell Res.* **2017**, *1864*, 1746–1758. [CrossRef] [PubMed]

163. Wu, T.C.; Chan, S.T.; Chang, C.N.; Yu, P.S.; Chuang, C.H.; Yeh, S.L. Quercetin and chrysin inhibit nickel-induced invasion and migration by downregulation of TLR4/NF-kappaB signaling in A549 cells. *Chem. Biol. Interact.* **2018**, *292*, 101–109. [CrossRef]

164. Chan, S.T.; Yang, N.C.; Huang, C.S.; Liao, J.W.; Yeh, S.L. Quercetin enhances the antitumor activity of trichostatin A through upregulation of p53 protein expression in vitro and in vivo. *PLoS ONE* **2013**, *8*, e54255. [CrossRef]

165. Lee, S.H.; Lee, E.J.; Min, K.H.; Hur, G.Y.; Lee, S.H.; Lee, S.Y.; Kim, J.H.; Shin, C.; Shim, J.J.; In, K.H.; et al. Quercetin Enhances Chemosensitivity to Gemcitabine in Lung Cancer Cells by Inhibiting Heat Shock Protein 70 Expression. *Clin. Lung Cancer* **2015**, *16*, 235–243. [CrossRef]

166. Lee, H.; Kim, Y.; Jeong, J.H.; Ryu, J.H.; Kim, W.Y. ATM/CHK/p53 Pathway Dependent Chemopreventive and Therapeutic Activity on Lung Cancer by Pterostilbene. *PLoS ONE* **2016**, *11*, e0162335. [CrossRef]

167. Chen, Z.; Jin, K.; Gao, L.; Lou, G.; Jin, Y.; Yu, Y.; Lou, Y. Anti-tumor effects of bakuchiol, an analogue of resveratrol, on human lung adenocarcinoma A549 cell line. *Eur. J. Pharmacol.* **2010**, *643*, 170–179. [CrossRef]

168. Park, J.J.; Hwang, S.J.; Park, J.H.; Lee, H.J. Chlorogenic acid inhibits hypoxia-induced angiogenesis via down-regulation of the HIF-1alpha/AKT pathway. *Cell. Oncol.* **2015**, *38*, 111–118. [CrossRef]

169. Khan, N.; Afaq, F.; Khusro, F.H.; Mustafa Adhami, V.; Suh, Y.; Mukhtar, H. Dual inhibition of phosphatidylinositol 3-kinase/Akt and mammalian target of rapamycin signaling in human nonsmall cell lung cancer cells by a dietary flavonoid fisetin. *Int. J. Cancer* **2012**, *130*, 1695–1705. [CrossRef]

170. Zhuo, W.; Zhang, L.; Zhu, Y.; Zhu, B.; Chen, Z. Fisetin, a dietary bioflavonoid, reverses acquired Cisplatin-resistance of lung adenocarcinoma cells through MAPK/Survivin/Caspase pathway. *Am. J. Transl. Res.* **2015**, *7*, 2045–2052.

171. Klimaszewska-Wisniewska, A.; Halas-Wisniewska, M.; Tadrowski, T.; Gagat, M.; Grzanka, D.; Grzanka, A. Paclitaxel and the dietary flavonoid fisetin: A synergistic combination that induces mitotic catastrophe and autophagic cell death in A549 non-small cell lung cancer cells. *Cancer Cell Int.* **2016**, *16*, 10. [CrossRef] [PubMed]

172. Zhou, Y.; Ho, W.S. Combination of liquiritin, isoliquiritin and isoliquirigenin induce apoptotic cell death through upregulating p53 and p21 in the A549 non-small cell lung cancer cells. *Oncol. Rep.* **2014**, *31*, 298–304. [CrossRef] [PubMed]

173. Teixeira, A.; DaCunha, D.C.; Barros, L.; Caires, H.R.; Xavier, C.P.R.; Ferreira, I.; Vasconcelos, M.H. Eucalyptus globulus Labill. decoction extract inhibits the growth of NCI-H460 cells by increasing the p53 levels and altering the cell cycle profile. *Food Funct.* **2019**, *6*. [CrossRef] [PubMed]

174. Wang, C.G.; Yao, W.N.; Zhang, B.; Hua, J.; Liang, D.; Wang, H.S. Lung cancer and matrix met alloproteinases inhibitors of polyphenols from Selaginella tamariscina with suppression activity of migration. *Bioorg. Med. Chem. Lett.* **2018**, *28*, 2413–2417. [CrossRef] [PubMed]

175. Moore, J.; Megaly, M.; MacNeil, A.J.; Klentrou, P.; Tsiani, E. Rosemary extract reduces Akt/mTOR/p70S6K activation and inhibits proliferation and survival of A549 human lung cancer cells. *Biomed. Pharmacother.* **2016**, *83*, 725–732. [CrossRef]

176. Tang, X.L.; Yan, L.; Zhu, L.; Jiao, D.M.; Chen, J.; Chen, Q.Y. Salvianolic acid A reverses cisplatin resistance in lung cancer A549 cells by targeting c-met and attenuating Akt/mTOR pathway. *J. Pharm. Sci.* **2017**, *135*, 1–7. [CrossRef]

177. Liu, Y.; Tong, Y.; Yang, X.; Li, F.; Zheng, L.; Liu, W.; Wu, J.; Ou, R.; Zhang, G.; Hu, M.; et al. Novel histone deacetylase inhibitors derived from Magnolia officinalis significantly enhance TRAIL-induced apoptosis in non-small cell lung cancer. *Pharmacol. Res.* **2016**, *111*, 113–125. [CrossRef]

178. Husari, A.; Hashem, Y.; Zaatari, G.; El Sabban, M. Pomegranate Juice Prevents the Formation of Lung Nodules Secondary to Chronic Cigarette Smoke Exposure in an Animal Model. *Oxid. Med. Cell Longev.* **2017**, *2017*, 6063201. [CrossRef]

179. Narayan, C.; Kumar, A. Antineoplastic and immunomodulatory effect of polyphenolic components of Achyranthes aspera (PCA) extract on urethane induced lung cancer in vivo. *Mol. Biol. Rep.* **2014**, *41*, 179–191. [CrossRef]

180. Aichinger, G.; Pahlke, G.; Nagel, L.J.; Berger, W.; Marko, D. Bilberry extract, its major polyphenolic compounds, and the soy isoflavone genistein antagonize the cytostatic drug erlotinib in human epithelial cells. *Food Funct.* **2016**, *7*, 3628–3636. [CrossRef]

181. Aggarwal, B.B.; Bhardwaj, A.; Aggarwal, R.S.; Seeram, N.P.; Shishodia, S.; Takada, Y. Role of resveratrol in prevention and therapy of cancer: Preclinical and clinical studies. *Anticancer Res.* **2004**, *24*, 2783–2840. [PubMed]

182. Salehi, B.; Mishra, A.P.; Nigam, M.; Sener, B.; Kilic, M.; Sharifi-Rad, M.; Fokou, P.V.T.; Martins, N.; Sharifi-Rad, J. Resveratrol: A Double-Edged Sword in Health Benefits. *Biomedicines* **2018**, *6*, 91. [CrossRef]

183. Poulose, S.M.; Thangthaeng, N.; Miller, M.G.; Shukitt-Hale, B. Effects of pterostilbene and resveratrol on brain and behavior. *Neurochem. Int.* **2015**, *89*, 227–233. [CrossRef] [PubMed]

184. Ko, J.H.; Sethi, G.; Um, J.Y.; Shanmugam, M.K.; Arfuso, F.; Kumar, A.P.; Bishayee, A.; Ahn, K.S. The Role of Resveratrol in Cancer Therapy. *Int. J. Mol. Sci.* **2017**, *18*, 2589. [CrossRef]

185. Whitlock, N.C.; Baek, S.J. The anticancer effects of resveratrol: Modulation of transcription factors. *Nutr. Cancer* **2012**, *64*, 493–502. [CrossRef]

186. Bergman, M.; Levin, G.S.; Bessler, H.; Djaldetti, M.; Salman, H. Resveratrol affects the cross talk between immune and colon cancer cells. *Biomed. Pharmacother.* **2013**, *67*, 43–47. [CrossRef]

187. Su, J.L.; Chen, P.B.; Chen, Y.H.; Chen, S.C.; Chang, Y.W.; Jan, Y.H.; Cheng, X.; Hsiao, M.; Hung, M.C. Downregulation of microRNA miR-520h by E1A contributes to anticancer activity. *Cancer Res.* **2010**, *70*, 5096–5108. [CrossRef]

188. Graham, H.N. Green tea composition, consumption, and polyphenol chemistry. *Prev. Med.* **1992**, *21*, 334–350. [CrossRef]

189. Balentine, D.A.; Wiseman, S.A.; Bouwens, L.C. The chemistry of tea flavonoids. *Crit. Rev. Food Sci. Nutr.* **1997**, *37*, 693–704. [CrossRef]

190. Colomer, R.; Sarrats, A.; Lupu, R.; Puig, T. Natural Polyphenols and their Synthetic Analogs as Emerging Anticancer Agents. *Curr. Drug Targets* **2017**, *18*, 147–159. [CrossRef]

191. Sekher Pannala, A.; Chan, T.S.; O'Brien, P.J.; Rice-Evans, C.A. Flavonoid B-ring chemistry and antioxidant activity: Fast reaction kinetics. *Biochem. Biophys. Res. Commun.* **2001**, *282*, 1161–1168. [CrossRef] [PubMed]

192. Burda, S.; Oleszek, W. Antioxidant and antiradical activities of flavonoids. *J. Agric. Food Chem.* **2001**, *49*, 2774–2779. [CrossRef] [PubMed]

193. Sang, S.; Lambert, J.D.; Ho, C.T.; Yang, C.S. The chemistry and biotransformation of tea constituents. *Pharmacol. Res.* **2011**, *64*, 87–99. [CrossRef] [PubMed]

194. Yang, C.S.; Wang, X.; Lu, G.; Picinich, S.C. Cancer prevention by tea: Animal studies, molecular mechanisms and human relevance. *Nat. Rev. Cancer* **2009**, *9*, 429–439. [CrossRef] [PubMed]

195. Yang, C.S.; Wang, H.; Li, G.X.; Yang, Z.; Guan, F.; Jin, H. Cancer prevention by tea: Evidence from laboratory studies. *Pharmacol. Res.* **2011**, *64*, 113–122. [CrossRef] [PubMed]

196. Yang, C.S.; Wang, H.; Chen, J.X.; Zhang, J. Effects of Tea Catechins on Cancer Signaling Pathways. *Enzymes* **2014**, *36*, 195–221. [CrossRef] [PubMed]

197. Miyata, Y.; Matsuo, T.; Araki, K.; Nakamura, Y.; Sagara, Y.; Ohba, K.; Sakai, H. Anticancer Effects of Green Tea and the Underlying Molecular Mechanisms in Bladder Cancer. *Medicines* **2018**, *5*, 87. [CrossRef]

198. Wu, X.; Yu, H.; Amos, C.I.; Hong, W.K.; Spitz, M.R. Joint effect of insulin-like growth factors and mutagen sensitivity in lung cancer risk. *J. Natl. Cancer Inst.* **2000**, *92*, 737–743. [CrossRef]

199. Wang, H.; Bian, S.; Yang, C.S. Green tea polyphenol EGCG suppresses lung cancer cell growth through upregulating miR-210 expression caused by stabilizing HIF-1alpha. *Carcinogenesis* **2011**, *32*, 1881–1889. [CrossRef]

200. Liu, F.; Cao, X.; Liu, Z.; Guo, H.; Ren, K.; Quan, M.; Zhou, Y.; Xiang, H.; Cao, J. Casticin suppresses self-renewal and invasion of lung cancer stem-like cells from A549 cells through down-regulation of pAkt. *Acta Biochim. Biophys. Sin.* **2014**, *46*, 15–21. [CrossRef]

201. Singh, B.N.; Shankar, S.; Srivastava, R.K. Green tea catechin, epigallocatechin-3-gallate (EGCG): Mechanisms, perspectives and clinical applications. *Biochem. Pharmacol.* **2011**, *82*, 1807–1821. [CrossRef] [PubMed]

202. Vallianou, N.G.; Evangelopoulos, A.; Schizas, N.; Kazazis, C. Potential anticancer properties and mechanisms of action of curcumin. *Anticancer Res.* **2015**, *35*, 645–651. [PubMed]

203. Tomeh, M.A.; Hadianamrei, R.; Zhao, X. A Review of Curcumin and Its Derivatives as Anticancer Agents. *Int. J. Mol. Sci.* **2019**, *20*, 1033. [CrossRef]

204. Adekola, K.; Rosen, S.T.; Shanmugam, M. Glucose transporters in cancer metabolism. *Curr. Opin. Oncol.* **2012**, *24*, 650–654. [CrossRef]

205. Liu, Y.; Cao, Y.; Zhang, W.; Bergmeier, S.; Qian, Y.; Akbar, H.; Colvin, R.; Ding, J.; Tong, L.; Wu, S.; et al. A small-molecule inhibitor of glucose transporter 1 downregulates glycolysis, induces cell-cycle arrest, and inhibits cancer cell growth in vitro and in vivo. *Mol. Cancer Ther.* **2012**, *11*, 1672–1682. [CrossRef]

206. Meng, Y.; Xu, X.; Luan, H.; Li, L.; Dai, W.; Li, Z.; Bian, J. The progress and development of GLUT1 inhibitors targeting cancer energy metabolism. *Future Med. Chem.* **2019**, *11*, 2333–2352. [CrossRef]

207. Rauf, A.; Imran, M.; Khan, I.A.; Ur-Rehman, M.; Gilani, S.A.; Mehmood, Z.; Mubarak, M.S. Anticancer potential of quercetin: A comprehensive review. *Phytother. Res.* **2018**, *32*, 2109–2130. [CrossRef]

208. Garrido, C.; Schmitt, E.; Cande, C.; Vahsen, N.; Parcellier, A.; Kroemer, G. HSP27 and HSP70: Potentially oncogenic apoptosis inhibitors. *Cell Cycle* **2003**, *2*, 579–584. [CrossRef]

209. Meng, S.; Cao, J.; Feng, Q.; Peng, J.; Hu, Y. Roles of chlorogenic Acid on regulating glucose and lipids metabolism: A review. *Evid. Based Complementary Altern. Med.* **2013**, *2013*, 801457. [CrossRef]

210. Santana-Galvez, J.; Cisneros-Zevallos, L.; Jacobo-Velazquez, D.A. Chlorogenic Acid: Recent Advances on Its Dual Role as a Food Additive and a Nutraceutical against Metabolic Syndrome. *Molecules* **2017**, *22*, 358. [CrossRef]

211. Naveed, M.; Hejazi, V.; Abbas, M.; Kamboh, A.A.; Khan, G.J.; Shumzaid, M.; Ahmad, F.; Babazadeh, D.; FangFang, X.; Modarresi-Ghazani, F.; et al. Chlorogenic acid (CGA): A pharmacological review and call for further research. *Biomed. Pharmacother.* **2018**, *97*, 67–74. [CrossRef] [PubMed]

212. Liu, C.C.; Zhang, Y.; Dai, B.L.; Ma, Y.J.; Zhang, Q.; Wang, Y.; Yang, H. Chlorogenic acid prevents inflammatory responses in IL1betastimulated human SW1353 chondrocytes, a model for osteoarthritis. *Mol. Med. Rep.* **2017**, *16*, 1369–1375. [CrossRef] [PubMed]

213. Karunanidhi, A.; Thomas, R.; van Belkum, A.; Neela, V. In vitro antibacterial and antibiofilm activities of chlorogenic acid against clinical isolates of Stenotrophomonas maltophilia including the trimethoprim/sulfamethoxazole resistant strain. *Biomed. Res. Int.* **2013**, *2013*, 392058. [CrossRef] [PubMed]

214. Sung, W.S.; Lee, D.G. Antifungal action of chlorogenic acid against pathogenic fungi, mediated by membrane disruption. *Pure Appl. Chem.* **2010**, *82*, 219–226. [CrossRef]

215. Shimizu, M.; Yoshimi, N.; Yamada, Y.; Matsunaga, K.; Kawabata, K.; Hara, A.; Moriwaki, H.; Mori, H. Suppressive effects of chlorogenic acid on N-methyl-N-nitrosourea-induced glandular stomach carcinogenesis in male F344 rats. *J. Toxicol. Sci.* **1999**, *24*, 433–439. [CrossRef]

216. Matsunaga, K.; Katayama, M.; Sakata, K.; Kuno, T.; Yoshida, K.; Yamada, Y.; Hirose, Y.; Yoshimi, N.; Mori, H. Inhibitory Effects of Chlorogenic Acid on Azoxymethane-induced Colon Carcinogenesis in Male F344 Rats. *Asian Pac. J. Cancer Prev.* **2002**, *3*, 163–166.

217. Kasai, H.; Fukada, S.; Yamaizumi, Z.; Sugie, S.; Mori, H. Action of chlorogenic acid in vegetables and fruits as an inhibitor of 8-hydroxydeoxyguanosine formation in vitro and in a rat carcinogenesis model. *Food Chem. Toxicol. Int. J. Publ. Br. Ind. Biol. Res. Assoc.* **2000**, *38*, 467–471. [CrossRef]

218. Ignatowicz, E.; Balana, B.; Vulimiri, S.V.; Szaefer, H.; Baer-Dubowska, W. The effect of plant phenolics on the formation of 7,12-dimethylbenz[a]anthracene-DNA adducts and TPA-stimulated polymorphonuclear neutrophils chemiluminescence in vitro. *Toxicology* **2003**, *189*, 199–209. [CrossRef]

219. Kang, T.Y.; Yang, H.R.; Zhang, J.; Li, D.; Lin, J.; Wang, L.; Xu, X. The studies of chlorogenic Acid antitumor mechanism by gene chip detection: The immune pathway gene expression. *J. Anal. Methods Chem.* **2013**, *2013*, 617243. [CrossRef]

220. Khan, N.; Afaq, F.; Syed, D.N.; Mukhtar, H. Fisetin, a novel dietary flavonoid, causes apoptosis and cell cycle arrest in human prostate cancer LNCaP cells. *Carcinogenesis* **2008**, *29*, 1049–1056. [CrossRef]

221. Syed, D.N.; Adhami, V.M.; Khan, M.I.; Mukhtar, H. Inhibition of Akt/mTOR signaling by the dietary flavonoid fisetin. *Anticancer Agents Med. Chem.* **2013**, *13*, 995–1001. [CrossRef] [PubMed]

222. Yang, P.M.; Tseng, H.H.; Peng, C.W.; Chen, W.S.; Chiu, S.J. Dietary flavonoid fisetin targets caspase-3-deficient human breast cancer MCF-7 cells by induction of caspase-7-associated apoptosis and inhibition of autophagy. *Int. J. Oncol.* **2012**, *40*, 469–478. [CrossRef] [PubMed]

223. Kashyap, D.; Garg, V.K.; Tuli, H.S.; Yerer, M.B.; Sak, K.; Sharma, A.K.; Kumar, M.; Aggarwal, V.; Sandhu, S.S. Fisetin and Quercetin: Promising Flavonoids with Chemopreventive Potential. *Biomolecules* **2019**, *9*, 174. [CrossRef]

224. D'Archivio, M.; Filesi, C.; Vari, R.; Scazzocchio, B.; Masella, R. Bioavailability of the polyphenols: Status and controversies. *Int. J. Mol. Sci.* **2010**, *11*, 1321–1342. [CrossRef]

225. Teng, H.; Chen, L. Polyphenols and bioavailability: An update. *Crit. Rev. Food Sci. Nutr.* **2019**, *59*, 2040–2051. [CrossRef]

Review

Nobiletin and Derivatives: Functional Compounds from Citrus Fruit Peel for Colon Cancer Chemoprevention

Joanna Xuan Hui Goh [1], Loh Teng-Hern Tan [2,3], Joo Kheng Goh [4], Kok Gan Chan [5,6,*], Priyia Pusparajah [7], Learn-Han Lee [2,8] and Bey-Hing Goh [1,8,*]

1   Biofunctional Molecule Exploratory (BMEX) Research Group, School of Pharmacy, Monash University Malaysia, Bandar Sunway 47500, Selangor Darul Ehsan, Malaysia; jxgoh2@student.monash.edu
2   Novel Bacteria and Drug Discovery (NBDD) Research Group, Microbiome and Bioresource Research Strength, Jeffrey Cheah School of Medicine and Health Sciences, Monash University Malaysia, Bandar Sunway 47500, Selangor Darul Ehsan, Malaysia; Loh.Tan@monash.edu (L.T.-H.T.); lee.learn.han@monash.edu (L.-H.L.)
3   Institute of Biomedical and Pharmaceutical Sciences, Guangdong University of Technology, Guangzhou 510006, China
4   School of Science, Monash University Malaysia, Jalan Lagoon Selatan, Bandar Sunway 47500, Selangor Darul Ehsan, Malaysia; goh.joo.kheng@monash.edu
5   Division of Genetics and Molecular Biology, Institute of Biological Sciences, Faculty of Science, University of Malaya, Kuala Lumpur 50603, Malaysia
6   International Genome Centre, Jiangsu University, Zhenjiang 212013, China
7   Medical Health and Translational Research Group, Jeffrey Cheah School of Medicine and Health Sciences, Monash University Malaysia, Bandar Sunway 47500, Selangor, Malaysia; priyia.pusparajah@monash.edu
8   Asian Centre for Evidence Synthesis in Population, Implementation and Clinical Outcomes (PICO), Health and Well-being Cluster, Global Asia in the 21st Century (GA21) Platform, Monash University Malaysia, Bandar Sunway, 47500, Malaysia
*   Correspondence: kokgan@um.edu.my (K.G.C.); goh.bey.hing@monash.edu (B.-H.G.)

Received: 28 May 2019; Accepted: 19 June 2019; Published: 21 June 2019

Abstract: The search for effective methods of cancer treatment and prevention has been a continuous effort since the disease was discovered. Recently, there has been increasing interest in exploring plants and fruits for molecules that may have potential as either adjuvants or as chemopreventive agents against cancer. One of the promising compounds under extensive research is nobiletin (NOB), a polymethoxyflavone (PMF) extracted exclusively from citrus peel. Not only does nobiletin itself exhibit anti-cancer properties, but its derivatives are also promising chemopreventive agents; examples of derivatives with anti-cancer activity include 3′-demethylnobiletin (3′-DMN), 4′-demethylnobiletin (4′-DMN), 3′,4′-didemethylnobiletin (3′,4′-DMN) and 5-demethylnobiletin (5-DMN). In vitro studies have demonstrated differential efficacies and mechanisms of NOB and its derivatives in inhibiting and killing of colon cancer cells. The chemopreventive potential of NOB has also been well demonstrated in several in vivo colon carcinogenesis animal models. NOB and its derivatives target multiple pathways in cancer progression and inhibit several of the hallmark features of colorectal cancer (CRC) pathophysiology, including arresting the cell cycle, inhibiting cell proliferation, inducing apoptosis, preventing tumour formation, reducing inflammatory effects and limiting angiogenesis. However, these substances have low oral bioavailability that limits their clinical utility, hence there have been numerous efforts exploring better drug delivery strategies for NOB and these are part of this review. We also reviewed data related to patents involving NOB to illustrate the extensiveness of each research area and its direction of commercialisation. Furthermore, this review also provides suggested directions for future research to advance NOB as the next promising candidate in CRC chemoprevention.

Keywords: nobiletin; colorectal cancer; chemoprevention; bioactivities

---

1. Introduction

Colorectal cancer (CRC) is the third most prevalent cancer reported in both men and women, ranking just after prostate or breast cancer and lung cancer [1]. Although in many cases there is no readily apparent cause of CRC, a number of factors have been found to be closely associated with this malignancy including gender, age, genetic predisposition, lifestyle, diet or as a complication from other diseases such as inflammatory bowel disease (IBD). Statistics showed that the death rate from CRC is 40% higher in males as compared to females, with the prevalence increasing with age, especially above 50 years old; however, there is a new and worrying trend of increasing incidence of colorectal cancer in the age group younger than 50, which, while slight, is still worrying [2,3].

The incidence of CRC has reduced as modern screening strategies have enabled much earlier detection of potentially malignant lesions, allowing for early intervention such as surgical excision of adenoma before it undergoes malignant transformation [4,5]. Although there has been a reduction, the high number of cases remains a major concern and the search for new and better treatments for CRC has been a key focus in pharmacological research. Standard therapy for cancer typically involves the triple regimen of surgery, chemotherapy, and radiation treatment. Efforts in exploring and developing new treatments are very much needed due to the limitations of the current treatment regimen—ranging from side effects, to complications and the development of drug resistance.

Researchers are attempting to explore multiple avenues for novel leads as anti-cancer agents with an increasing trend to focus on natural sources like plants and fruits [6–8]. However, while it is key to find new treatments to existing cancers, a crucial aspect that is also being explored is prevention of cancerous growths; in particular, this would be of benefit for those at risk due to the various factors outlined earlier. One of the effective strategies to control cancer is chemoprevention, which is defined as the use of a natural or synthetic agent to reverse, inhibit, or prevent the progression of cancer [9].

Plants and fruits are often part of a diet recommended to prevent various illnesses including cancer [10]. These beneficial properties may be from the chemicals they contain as well as their metabolites which enter our alimentary canal and eventually end up in our colon and rectum. If the compounds responsible can be isolated and purified for use as a treatment, this may be a milestone in new cancer therapies and prophylaxis. While an extensive review of polyphenols like apigenin and luteolin on anti-colorectal cancer effect can readily be found [11], this study highlights the potential chemopreventive effect on CRC of another flavonoid, namely nobiletin (NOB).

NOB, a polymethoxyflavone (PMF), is likely named after *Citrus nobilis*. This compound is one of the most ubiquitous flavones that can be isolated exclusively from the peel of citrus fruits [12]. Besides CRC, there is concurrently ongoing research looking into the effect of NOB on other types of cancers such as breast cancer [13,14], ovarian cancer [15], gastric cancer [16,17], lung cancer [18,19], liver cancer [20] and bone cancer [21]. There are also recent studies attesting to the benefits of NOB in anti-neurodegeneration [22,23], anti-diabetes [24], anti-obesity [25–27], antimicrobial [28], anti-allergy [29] and anti-inflammatory effects [30,31]. There are also a number of articles that support claims purporting to the role of NOB in reducing the risk of cardiovascular diseases [32,33] and osteoporosis [34,35].

Interestingly, this compound can be metabolised into a number of metabolites which also show significant anti-cancer effects. There are several recent reviews on the bioactivities of these citrus PMF [36] as well as the potential chemopreventive abilities of these PMFs toward cancers in general [37]. This review paper aims to gather the results of the in vivo and in vitro studies done in recent years and compile various molecular pathways by which the compound NOB and its derivatives act in CRC prevention which will in turn help to facilitate future research that targets these specific mechanisms.

2. Research Methodology

The main focus was to search for all relevant primary research papers published which looked into the use of NOB and its derivatives as a chemopreventive agent for CRC. A systematic search was performed to identify published literature on the chemopreventive potentials of NOB against CRC using Google Scholar. For studies published in foreign languages like Chinese, Japanese or Korean, an attempt was made to locate the translated version. The search strategy was performed using keywords 'nobiletin' and 'colorectal cancer' to locate relevant papers. This was also supplemented with keyword search of the terms 'cancer statistics', 'colorectal cancer', 'colon cancer', 'metabolites', 'synergism', 'biotransformation', 'mechanism', 'apoptosis', 'anti-inflammatory', 'inflammation', 'cell proliferation', 'cell cycle arrest', 'metastasis', 'tumour', 'angiogenesis', 'absorption', 'metabolism', 'toxicity', 'distribution', 'elimination', 'solubility' and 'delivery' combined using Boolean operators with 'nobiletin'. PubChem and EMBASE were used as alternatives to ensure inclusion of all relevant papers while SciFinder database was mainly used to locate patents related to NOB. The reference lists of relevant articles collected were also screened for additional studies to be included in the review.

3. Nobiletin and Its Derivatives

The compound nobiletin (NOB) can be extracted exclusively from citrus fruits, namely mandarin oranges (*Citrus reticulate*), sweet oranges or Valencia oranges (*Citrus sinesis*), Miaray mandarins (*Citrus miaray*) [38], flat lemons or Hayata (*Citrus depressa*) [39,40], tangerines (*Citrus tangerine*), bitter oranges (*Citrus aurantium*) [12], Unshu Mikans or Satsuma mandarins (*Citrus Unshiu arnicia indica*) [41,42], Cleopatra mandarins (*Citrus reshni*) [43], mandarin oranges (*Citrus tachibana*), Koji Oranges (*Citrus leiocarpa*), Natsu Mikans (*Citrus tardiva*), Jimikan (*Citrus succosa*), Kinokuni Mandarins (*Citrus Kinokuni*), Fukushu (*Citrus erythrosa*), Sunkat (*Citrus sunki*) and hybrids of the mandarin orange with pomelo (*Citrus deliciosa*) [44]. *Citrus tangerine* was reported to contain the highest content of NOB, approximately five times of that in *Citrus sinesis* [45].

PMF can be isolated from orange peel through different types of chemical extraction processes, for example, the supercritical fluid extraction, microwave assisted extraction [46] and Soxhlet method, which is capable of extracting large sample volumes [43]. Through the supercritical fluid extraction process, the supercritical fluid extractor is used to process the orange peel grinds that have been freeze-dried. Then, the extract is further treated with carbon dioxide and ethanol to concentrate the bioactive compound [47]. A special method to improve NOB yield through the supercritical fluid extraction method is currently patented in Korea [48]. It was found that the maximal yield of NOB occurs at a temperature of 80 °C and pressure of 30 MPa with an optimum sample particle size of 375 μm [40].

NOB is a PMF classified under the flavonoid family of polyphenols. The International Union of Pure and Applied Chemistry (IUPAC) nomenclature is 2-(3,4-dimethoxyphenyl)-5,6,7,8-tetramethoxychromen-4-one. It is also known as 5,6,7,8,3′,4′-hexamethoxyflavone or 2-(3,4-dimethoxyphenyl)-5,6,7,8-tetramethoxy-4H-1-benzopyran-4-one [12]. NOB has a molecular formula of $C_{21}H_{22}O_8$ and a molecular weight of 402.399 g/mol. The chemical structure of NOB is illustrated in Figure 1. This flavone has a distinct three aromatic ring structure (labelled A, B and C in Figure 1), with the ketone and ether group in ring C along with four methoxy groups at the 5, 6, 7 and 8 positions of ring A and 2 methoxy groups at the 3 and 4 positions of ring B. Under long-term storage, NOB can degrade into 5-demethylnobiletin (5-DMN), IUPAC name 5-hydroxy- 6,7,8,3′,4′-pentamethoxyflavone (structure illustrated in Figure 1), through the process of autohydrolysis [49]. It has also been proposed that 5-DMN could be formed through conversion of nobiletin by gastric acid after oral consumption [50].

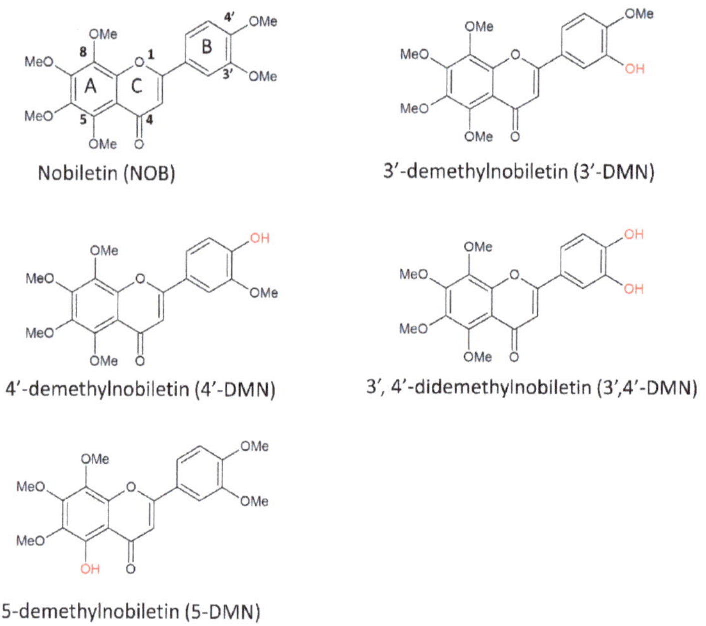

Nobiletin (NOB)
3'-demethylnobiletin (3'-DMN)
4'-demethylnobiletin (4'-DMN)
3', 4'-didemethylnobiletin (3',4'-DMN)
5-demethylnobiletin (5-DMN)

Figure 1. Chemical structures of nobiletin and its derivatives.

Both the NOB and 5-DMN undergo further transformation to form a number of metabolites in the body after ingestion [50,51]. More than 20 metabolites have been identified and the types vary significantly according to the species of citrus plants [12]. The three common phase I metabolites of NOB identified in urine after administration to rodents are 3'-DMN, 4'-DMN and 3',4'-DMN [52,53]. Wu et al. successfully quantitated the amount of NOB, 3'-DMN, 4'-DMN and 3',4'-DMN at 2.03, 3.28, 24.13 and 12.03 nmol/(gram of tissue of colonic mucosa) at the end of 20 weeks daily feeding of 500 ppm NOB to CD-1 mice [54].

After absorption, NOB generally undergoes Phase I and Phase II metabolism. In vivo tests show the Phase I demethylation of NOB is likely caused by the action of cytochrome P450 [55]. Koga et al. researched the enzymes involved in NOB metabolism and confirmed that CYP1A1, CYP1A2, CYP1B1 and CYP3A5 are involved in the conversion of NOB to 3'-DMN; further action from CYP1A1 and CYP1A2 is required to convert 3'-DMN to 3',4'-DMN [56]. NOB was also found to undergo extensive Phase II metabolism in the small intestine involving glucuronides or sulphates. [57] Four phase II metabolites of NOB have been identified in rodent serum, bile and urine. These Phase II metabolites are formed from the glucuronidation/sulphation of the Phase I products, namely 4'-DMN and 3',4'-DMN [58]. However, research on these Phase II metabolites are limited likely due to the fact that existing literature suggests a high likelihood that these substances have decreased activity. For example, Manthey et al. showed there was a reduced anti-inflammatory effect of the compound after glucuronidation [59].

In contrast to the dominant Phase II metabolites in the small intestines, the majority of the metabolites in the large intestine undergo deconjugation mainly through the action of the microflora in the gut. The microbiome produces enzymes such as C-deglycosidases, O-deglycosidases and hydrolases that break down the unabsorbed compounds from the small intestine. The microbiome also releases enzymes such as glucuronidases and sulphatases that hydrolyze the conjugate bonds, resulting in the reformation of free molecules that either undergo reuptake into the colonocytes or enter into the blood stream [60]. At present, only a limited species of the microbiome have been identified and further research is crucial to understand the in vivo biotransformation of the NOB compound resulting in the generation of multiple metabolites with different activities [61]. It is likely that the subtle variances in the gut microbiome in different individuals may result in different pharmacodynamic effects after administration of NOB. For instance, 4′-DMN and 3′,4′-DMN have been shown to exhibit higher anti-cancer and anti-inflammatory effects than NOB itself, but the rate of conversion from NOB to these metabolites may vary from one person to another [54,62]. The mechanisms of NOB in chemoprevention are elaborated under 'Section 5—Chemopreventive effects of NOB, 5-DMN and NOB-metabolites'.

Early in vitro studies using rat liver S9 extracts reveals 3′-DMN as the main metabolite of NOB after 24 h of treatment [42]. However, further High-Performance Liquid Chromatography (HPLC) analysis on in vivo experiments showed that the concentration of nobiletin and its metabolites differ in the colonic mucosa—the concentration of 3′-DMN is almost equal to NOB, while 3′,4′-DMN is about 5.9-fold more than NOB, and 4′-DMN being the most concentrated, at 11.9 times the concentration of NOB. Integrating these values, the concentration of NOB is actually 20 times significantly lower in the colon when compared to the total concentration of its metabolites [54]. Convincing evidence has shown that these metabolites generated in vivo following oral administration of NOB result in significant accumulation in colonic tissues which is associated with the chemopreventive effect for CRC.

Interestingly, growing evidence suggests that the metabolites have more potent anti-cancer activity than their parent compounds, and the high concentration of the metabolites of NOB found in the colon may indicate that anti-cancer effect of NOB is largely conferred by its metabolites. This is consistent with the findings of Wu et al. who discovered that by treating HCT116 cell lines with NOB and its metabolites results in a 3.3 to 7.6-fold increase in apoptotic cells [54]. A recent study by Chiou et al. also shows that the hydroxylated PMF, 5-DMN is more potent than NOB in terms of its chemopreventive effect on colon malignancy for both in vivo studies using xenograft mice and in vivo studies using three different colon cell lines. Chiou and colleagues reported that 5-DMN shows different levels of inhibition in different types of cell lines, with the highest efficacies in COLO205 cell lines, followed by HCT116 and HT-29 [49]. This is consistent with the findings of Qiu et al. stating that the half maximal inhibitory concentration (IC_{50}) required for 5-DMN to exert an inhibitory effect on the growth of HCT116 cells is 8.4 µM as compared to the notably higher value of 37 µM for NOB. Similarly, the IC_{50} required for 5-DMN against HT-29 cells is 22 µM as compared to the higher IC_{50} of 46.2 µM for NOB [63]. This may suggest that the hydroxyl group at the 5th position on the A ring is an important functional group involved in the molecular interactions [49].

4. Pathogenesis of Colorectal Cancer

The mechanisms leading to CRC development are part of a rather complex process. The pathogenesis of CRC is arbitrarily subdivided into three stages here: initiation, progression and metastasis. Each pathway is known to be regulated by chemical signals, called cytokines, which allow the progression from one stage to the next, whilst inflammation is the underlying result of each stage [11].

Cancer generally starts with a mutated cell which deviates from the normal cell growth cycle and progresses through the cell cycle rapidly with no differentiation of structure or function. It can be attributed to the down regulation of the regulatory genes or up-regulation of oncogenes. This gradually leads to the formation of a mass of undifferentiated cells called an adenoma. This lump of cells does not perform any specific function but competes with the surrounding normal cells for nutrients. In more than 60% of colorectal adenomas, the dysregulation of adenomatous polyposis coli gene resulting from the Wnt/β-catenin pathway is the major culprit in triggering this process [64].

The initial adenoma progresses on to an intermediate adenoma when the epidermal growth factor receptor (EGFR) is activated, which in turn triggers the phosphatidylinositol-3-kinase pathway and results in tumour formation [65]. Also, the inactivation of transforming growth factor-β (TGF-β) and the loss of function of p53 further aggravates tumour growth by preventing apoptosis [66,67].

Eventually, a tumour becomes malignant when angiogenesis occurs, and the cancer cells are released into the bloodstream and spread to other parts of the body through a process known as metastasis. Intercellular adhesion molecule-1 (ICAM-1) and matrix metalloproteinase (MMPs) are closely associated with the promotion of angiogenesis and metastasis. To illustrate, the MMP disrupts the integrity of the basal membrane allowing the cancer cells to enter the surrounding blood vessels and thus the blood stream through a process known as intravasation [68,69].

5. Chemopreventive Effects of Nobiletin, 5-DMN and NOB-Metabolites

In one of earliest in vitro studies, the antiproliferative effect of NOB was evaluated against HT-29 colon cancer cells [70]. The study determined that the IC_{50} and IC_{90} of NOB against HT-29 cell were 4.7 µM and 13.9 µM, respectively, via the 3H-thymidine uptake assay [70]. As a product of autohydrolysis of NOB, 5-DMN was also evaluated for its antiproliferative effect against colon cancer cells. In the H-thymidine uptake assay, the IC_{50} and IC_{90} of 5-DMN against HT-29 was reported to be 8.5 µM and 171 µM, respectively [70]. In the following years, NOB and 5-DMN were also reported to be cytotoxic towards different colon cancer cell lines, including HCT116, HT-29, SW489, COLO320, COLO205 and Caco-2 (Table 1). Despite the stronger anti-proliferative effect of NOB observed in the earlier study [70], recent studies increasingly showed that 5-DMN exhibits stronger cytotoxic effects against different colon cancer cells as compared to NOB [49,63]. These contradictory results are potentially due to the different aspects of cancer focused in each study. Based on these in vitro studies, NOB and 5-DMN were shown to exhibit their cytotoxic effects towards colon cancer cells, predominantly via cell cycle arrest and induction of apoptosis (Table 1).

Multiple in vivo studies demonstrated that NOB offers a protective effect against several carcinogens, such as the azoxymethane (AOM) and the 2-amino-1-methyl-6-phenylimidazo[4,5-b]pyridine (PhIP) (Table 2). AOM/DSS has been used to induce colitis in mice for the purpose of creating mouse models that replicate colitis induced CRC in humans [71]; however, PhIP, a heterocyclic amine, is a food-derived carcinogen that is abundantly released in the process of cooking fish and meat [72,73]. Administration of 0.01% wt of NOB to mice for five weeks in their diet resulted in the reduction of abnormal growths induced by colonic carcinogen AOM in the colons of the mice; there was a 50% reduction as compared to the controls [41]. Another similar study to determine the anti-adenocarcinoma effects of NOB also showed positive results but with lower efficacies, whereby 34 weeks administration of 0.01% or 0.05% wt. of NOB reduced the frequency of adenocarcinoma by 12% and 32%, respectively [74]. In addition to that, Wu et al. demonstrated that NOB treatment successfully reduced the rate of cell proliferation by 69%, tumour incidence by 40%, tumour multiplicity by 71%, and downregulated TNF-α, IL-1β and IL-6 by 65%, 69% and 45% respectively in AOM/DSS treated mice [54]. Consistent with the inhibitory effect against AOM induced colon carcinogenesis, NOB also showed significant reduction in the high density of colonic aberrant crypt foci (ACF) located in the transverse colon in PhIP-induced F344 rats [75]. This shows that NOB is effective in preventing CRC triggered by different types of carcinogens.

Further support for NOB as a prospective candidate for chemoprevention is that NOB is known to inhibit different pathways leading to cancer via a number of different mechanisms which includes inhibiting cell cycle progression [54,76], limiting inflammation [76], inducing apoptosis [54], preventing angiogenesis [77] and reducing tumour formation [49,54,78]. This subsection will describe the mechanism of action of NOB, its autohydrolysis product, 5-DMN and its three common metabolites, namely 3'-DMN, 4'-DMN and 3',4'-DMN, in chemoprevention of CRC in detail.

Table 1. In vitro chemopreventive properties of NOB, 5-DMN and NOB-metabolites.

Compounds	Activities	Cell lines	Treatment/Assay (Treatment duration)	Assays/Results/Mechanisms	References
NO3 5-DMN	Anti-proliferative	HT-29	H-thymidine uptake assay	- IC_{50} of NOB = 4.7 μM - IC_{90} of NOB = 13.9 μM - IC_{50} of 5-DMN = 8.5 μM - IC_{90} of 5-DMN = 171 μM	[70]
NOB	Cytotoxicity	COLO320, SW480 and Caco-2	MTS viability assay (48 h)	- IC_{50} for COLO320 = 40.4 ± 9.1 μM - IC_{50} for SW480 = 245 ± 9.1 μM - IC_{50} for Caco-2 = 305.6 ± 41.9 μM	[79]
	Apoptosis-inducing		Apoptosis assays—DNA fragmentation - gel electrophoresis (48 h)	- DNA ladder pattern 200 μM—2-fold increase DNA fragmentation in COLO320	
	Anti-proliferative		BrdU labelling index	- 34.7 ± 4.7% BrdU-binding cells at 100 μM - 44.4 ± 6.4% BrdU-binding cells at 40 μM	
NOB	Anti-metastasis	HT-29	ELISA - proMMP-7 expression qPCR and Western blot	- At 100 μM, no detection of proMMP-7 in media, ~280 pg/mL proMMP-7 in media - >25 μM, reduced RNA and protein (both intracellular and supernatant) expression of proMMP-7	[77]
			AP-1 binding activity	- Inhibited binding activity of AP-1 (transcription factor for MMP-7 gene)	
NOB	Anti-proliferative Cell cycle arrest	HT-29	Cell counting assay	- IC_{50} of NOB ≈ 50 μM - Inhibited cell proliferation in a time- and dose-dependent manner	[14]
			Cell cycle analysis - Propidium iodide staining	- Induced G1 phase cell cycle arrest (60 and 200 μM)	
	Apoptosis-inducing		Apoptosis assay Resumption of proliferation	- No significant apoptosis detected at 60 and 100 μM - Resumed proliferation within 24 h of removal of NOB and achieve the same stage of growth as compared to control after four days of removal of NOB	
NOB 5-DMN	Cytotoxicity	HCT116, HT-29	MTT viability assay (48 h)	- IC_{50} of NOB on HCT116 = 37 μM - IC_{50} of 5-DMN on HCT116 = 8.7 μM - IC_{50} of NOB on HT-29 = 46.2 μM - IC_{50} of 5-DMN on HT-29 = 22 μM	[63]
	Cell cycle arrest		Cell cycle analysis - Propidium iodide staining (24 h) Western blot	- At 8 μM, 5-DMN induced G2/M phase arrest in HCT116 - At 36 μM, 5-DMN induced G2/M phase arrest in HT-29 - At 16 μM, NOB reduced CDK-2 expression - At 4 μM and 8 μM, 5-DMN increased p21 and Rb, while decreased CDK-2 and p-Rb.	
	Apoptosis-inducing		Apoptosis assay Annexin-V/PI (48 h) Western blot	- At 8 μM, 5-DMN increased early apoptosis by 2.2-fold in HCT116 - At 36 μM, 5-DMN increased early apoptosis by ~2-fold in HT-29 - At 16 μM, NOB did not increase apoptotic cell population in HCT116/HT-29 - At 4 μM and 8 μM, 5-DMN increased expressions of cleaved caspase 8, cleaved caspase 3 and cleaved PARP.	

Table 1. *Cont.*

Compounds	Activities	Cell lines	Treatment/Assay (Treatment Duration)	Assays/Results/Mechanisms	References
5-DMN	Apoptosis-inducing	HCT116 (p53 +/+) and HCT116 (p53 -/-);	Apoptosis assay Annexin-V/PI	- At 15 µM, 5-DMN increased late apoptotic/necrotic cell in HCT116 (p53 -/-) > HCT115 (p53 +/+), suggesting the apoptotic inducing action is independent of p53 - At 15 µM, 5-DMN increased early apoptotic cell in HCT116 (Bax +/-), but not in HCT116 (Bax -/-)	[80]
	Cell cycle arrest	HCT116 (Bax +/-) and HCT116 (Bax -/-);	Cell cycle analysis - Propidium iodide staining	- At 15 µM, 5-DMN arrested cells at G2/M and G0/G1 phases in HCT116 (p53 +/+) cells, but only caused G2/M phase arrest in HCT116 (p53 -/-) cells - G0/G1 is p53 dependent and G2/M is p53-independent	
NOB; 3'-DMN; 4'-DMN; 3',4'-DMN	Cytotoxicity	HCT116 (p53 -/-), HT-29	MTT viability assay	- At 2.03 µM and 3.28 µM, NOB and 3'-DMN, respectively showed no significant cytotoxicity against HCT116 and HT-29 - At 24.13 µM, 4'-DMN inhibited growth of HCT-116 by 45% and HT-29 by 33% - At 12.03 µM, 3',4'-DMN inhibited growth of HCT116 by 30% and HT-29 by 9% - combination of all three NOB-metabolites inhibited growth of HCT116 by 64% and HT-29 by 62% (no significant difference to three NOB-metabolites + NOB)	[54]
	Cell cycle arrest		Cell cycle analysis - Propidium iodide staining (24 h)	- NOB (40 µM) arrested cells at G0/G1 phase in both HCT-116 and HT-29 - 3'-DMN (40 µM) arrested cells at both S phase and G2/M phase in HCT-116; while arrested cells at both G0/G1 and G2/M phase in HT-29 - 4'-DMN (40 µM) induced a stronger effect than NOB in arresting cells at G0/G1 phase in HCT-116 and HT-29 - 3',4'-DMN (20 µM) arrested cells at both S phase and G2/M phase in HCT-116; while arrested cells at both G0/G1 and G2/M phase in HT-29	
	Apoptosis inducing		Western blot	- NOB and all three NOB-metabolites cause profound increase in expression of p21$^{Cip1/Waf1}$	
			Annexin-V/PI (48 h)	- NOB (40 µM) increased early apoptotic cell population by 3.3-fold, increased late apoptotic cell population by 4.2-fold in HCT116 - 3'-DMN (40 µM) increased early apoptotic cell population by 5.0-fold, increased late apoptotic cell population by 3.5-fold in HCT116 - 4'-DMN (40 µM) increased early apoptotic cell population by 4.9-fold, increased late apoptotic cell population by 7.1-fold in HCT116 - 3',4'-DMN (20 µM) increased early apoptotic cell population by 7.6-fold, increase late apoptotic cell population by 4.5-fold in HCT116 -3'-DMN (40 µM) and 4'-DMN (40 µM) did not cause significant apoptosis in HT-29 - 3',4'-DMN (20 µM) exhibits stronger apoptosis effect than NOB (40 µM) in HT-29	
			Western blot	- NOB (40 µM) only increased activation of caspase-9 and did not affect caspase-3 or PARP levels in HCT116 - NOB-metabolites increased activation of caspase-3, caspase-9 and other downstream proteins like PARP in HCT116	

Table 1. *Cont.*

Compounds	Activities	Cell lines	Treatment/Assay (Treatment duration)	Assays/Results/Mechanisms	References
NOB-Met (2.03 μM NOB: 3.28 μM 3′-DMN: 24.13 μM 4′-DMN: 12.03 μM 3′,4′-DMN	Anti-inflammatory	RAW264.7	Western Blot	- At 0.5× concentration of NOB-Met, supressed LPS-induced iNOS expression by 56.4% - At 1× and 2× concentration of NOB-Met, completely abrogated LPS-induced iNOS expression - At ×0.5, increased expression of NQO1 by 21% as compared to LPS-treated cells - At ×1, increased expression of HO-1 by 10%, increased expression of NQO1 by 34% as compared to LPS-treated cells - At ×2, increased expression of HO-1 by 37%, increased expression of NQO1 by 50% as compared to LPS-treated cells - Induced translocation of Nrf2	[76]
	Cell cycle arrest	HCT116	Cell cycle analysis - Propidium iodide staining Western blot	- At 1×, induced G0/G1 phase arrest; while at 2×, induced G0/G1 and G2/M phases arrest - Reduced expressions of CDK-2, CDK-4, CDK-6 and cyclin D, while increased expressions of p53 and p27	
NOE, 5-DMN	Cytotoxicity	HCT116, HT-29, COLO205	MTT viability assay	- At 40 μM, NOB significantly reduced viability of HCT116, HT-29 and COLO205 by ~20–30% - At >5 μM, 5-DMN significantly reduced viability of HCT116, HT-29 and COLO205	[49]
	Apoptosis inducing		Cell cycle analysis - SubG1 quantification	- At 20 μM, 5-DMN increased apoptosis ratio by ~26%, while no increased in subG1 population in NOB-treated COLO205	
		Human	Western	- At 10 and 20 μM, significantly increased expression of cleaved PARP in COLO205	
NOB	Anti-inflammatory	synovial fibroblast, mouse macrophage J774A.1	ELISA Western blot and qPCR qPCR Western blot	- At >4 μM, NOB inhibited PGE$_2$ induced by IL-1α in human synovial fibroblast - At >16 μM, NOB reduced mRNA of COX-2 induced by IL-1α in human synovial fibroblast - At 64 μM, NOB inhibited COX-2 protein expression induced by IL-1α in human synovial fibroblast - At 32 μM, NOB reduced mRNA of IL-1α, IL-1β, IL-6, TNF-α induced by LPS in J774A.1 - At >16 μM, NOB reduced proMMP-1 and proMMP-3 induced by IL-1α in human synovial fibroblast - At >16 μM, NOB enhanced TIMP-1 expression in response to IL-1α in human synovial fibroblast	[81]
NOB	Anti-inflammatory	Mouse adipocyte 3T3-L1	ELISA Western blot	- At 50 and 100 μM, NOB suppressed MCP-1 secretion induced by TNF-α IN 3T3-L1 adipocytes - At 50 and 100 μM, NOB reduced ERK phosphorylation in 3T3-L1 adipocytes treated with TNF-α	[82]

IC$_{50}$—half maximal inhibitory concentration; AP-1—activator protein-1; PI—propidium iodide; PARP—poly (ADP-ribosome) polymerase; CDK—cyclin-dependent kinase; Rb—retinoblastoma; LPS—lipopolysaccharide; iNOS—inducible NO synthase; NQO1—NAD(P)H quinone oxidoreductase 1; HO-1—heme oxygenase-1; PGE$_2$—prostaglandin E$_2$; IL—interleukin; TNF—tumor necrosis factor; TIMP-1—tissue inhibitor metalloprotease-1; MCP-1—monocyte chemoattractant protein-1.

Table 2. In vivo studies of NOB for colon cancer chemoprevention.

Animal Models	Treatment/Dosage	Mechanisms	Detailed Results	References
Colitis-associated colon carcinogenesis model - AOM (12 mg/kg i.p.)/1% DSS in drinking water treated male CD-1 mice (5-week-old)	AIN93G diet containing 0.05% wt NOB (20 weeks)	Cell cycle arrest Anti-inflammatory effects	Protein expression in colonic mucosa by Western blot - Reduced levels of CDK-2, CDK-4, CDK-6, cyclin D and cyclin E - Increased levels of p21, p27 and p53 Immunohistochemical analysis - Reduced expression of iNOS reduced by 35% when compared to the positive control Protein expression in colonic mucosa by Western blot - Increased level of HO-1 - Increased level of NQO1 - Induced translocation of level of Nrf2 transcription factor (Nuclear fraction < Cytoplasmic fraction)	[76]
Colitis-associated colon carcinogenesis model - AOM/DSS treated AOM (12 mg/kg i.p.)/1% DSS in drinking water treated male CD-1 mice (5 week old)	AIN93G diet containing 0.05% wt NOB (20 weeks)	Inhibit AOM/DSS-induced colon carcinogenesis Anti-proliferative effect Apoptosis-inducing effect Anti-inflammatory effects	- Prevented shortening of colon length, reduced the increased colon weight/length ratio - Reduced tumor incidence by 40% and tumor multiplicity by 71% - Maintained histological characteristic of normal mucosa - Reduced PCNA-positive colonocytes by 69% in mucosal crypts - Increased cleaved caspase-3 positive cells by 2.3-fold in colonic tumor - Reduced levels of proinflammatory cytokines - ELISA showed reduction of TNF-α by 51%, IL-1ß by 92% and IL-6 by 69% compared - qRT-PCR analysis showed reduction of TNF-α by 65%, IL-1ß by 69% and IL-6 by 45%	[54]
Colon carcinogenesis model - AOM (15 mg/kg i.p.) treated male *db/db* mice	Diet containing 100 ppm NOB (0.1% wt) (10 weeks)	Inhibit AOM induced colon carcinogenesis	- Reduced frequency of preneoplastic lesions (colonic aberrant crypt foci (ACF) and β-catenin-accumulated crypts (BCAC)) - Reduced incidence of ACF by 68-91% and BCAC by 64–71% - Reduced PCNA-labeling index in ACF by 21% and BCAC by 19%	[83]
Colon carcinogenesis model - AOM (10 mg/kg i.p.)/1% DSS in drinking water treated male CD-1 mice	Diet containing 100 ppm NOB (0.1% wt) (for 17 weeks)	Inhibit AOM/DSS-induced colon carcinogenesis Inhibit leptin-induced colon carcinogenesis	- Suppressed incidence of neoplasms (adenoma and adenocarcinoma), lowered multiplicity of tumor - Suppressed serum levels of leptin by 75–84%	[84]
Colon carcinogenesis model - AOM (20 mg/kg s.c.) treated male F344 rats	Diet containing NOB (0.01% wt and 0.05% wt) (34 weeks)	Inhibit AOM induced colon carcinogenesis Anti-proliferative effect Anti-inflammatory effect	- Reduced incidence and multiplicity of colonic adenocarcinoma - Increased apoptosis index of adenocarcinoma - Reduced level of PGE$_2$ in colonic adenocarcinoma and surrounding mucosa	[74]

Table 2. *Cont.*

Animal Models	Treatment/Dosage	Mechanisms	Detailed Results	References
Colon carcinogenesis model - AOM (20 mg/kg s.c.) treated male F344 rats	Diet containing NOB (0.01% 0.01% wtwt and 0.05% wt) (5 weeks)	Inhibit AOM-induced colon carcinogenesis Anti-proliferative effect Anti-inflammatory effect	- Reduced the frequency of colonic aberrant crypt foci formation - Reduced number of ACF in proximal, middle and distal colon - Reduced MIB-5 labeling index of ACF but not of normal colonic crypts - Reduced level of PGE$_2$ in colonic mucosa	[41]
Colon carcinogenesis model - PhIP hydrochloride (100 mg/kg i.g.) treated F344 male rats (twice/week for 10 weeks)	Diet containing NOB (0.05% wt.) (50 weeks)	Inhibit PhIP-induced ACF in transverse colon	- Reduced the total colonic ACF indices in transverse colon	[75]
Colorectal cancer xenograft mouse model - COLO205 cells s.c.	NOB 100 mg/kg i.p. daily for 3 weeks 5-DMN 50 mg/kg and 100 mg/kg i.p. daily for 3 weeks	Anti-tumor effect Autophagy induction Anti-inflammatory effect Anti-angiogenesis	- NOB reduced tumor size and weight but not significant as compared to control - 5-DMN reduced tumor size and weight significantly as compared to control - 5-DMN increased LC3 expression - 5-DMN increased p53 expression - 5-DMN reduced COX-2 expression - 5-DMN reduced VEGF expression	[49]

AOM—azoxymethane; DSS—dextran sulfate sodium; i.p.—intraperitoneal injection; s.c.—subcutaneous injection; i.g.—intragastric administration; PCNA—proliferating cell nuclear antigen; ACF—aberrant crypt foci; BCAC—β-catenin-accumulated crypts; PhIP—1-Methyl-6-phenyl-1H-imidazo[4,5-b]pyridin-2-amine; LC3—microtubule-associated protein light chain 3; VEGF—vascular endothelial growth factor.

5.1. Cell Cycle Arrest

Uncontrolled cell growth that arises from genomic instability is known to contribute to tumorigenesis [85]. One way to counteract CRC is to halt its cell cycle progression. The cell cycle is akin to a biological growth clock that tightly regulates each stage of cell growth, where any mutated or abnormal cells will be arrested at either the G1 or G2 checkpoints; however, this mechanism is disrupted in cancerous conditions [86]. To progress through the stages, the regulatory protein cyclin acts like a key, as it needs to phosphorylate the cyclin-dependent kinase (CDK) complexes to allow progression to the next stage [87].

5.1.1. Action of NOB and Its Metabolites Inducing Cell Arrest

Notably, different metabolites of NOB work by different mechanisms against different cells. The flow cytometry test showed NOB and 4'-DMN arrest cells at G0/G1 phase in both HCT116 and HT-29 cell lines, despite the inhibitory effect of 4'-DMN being higher than that of NOB. Both 3'-DMN and 3',4'-DMN arrest cells at both S phase and G2/M phase in HCT116 cell lines but arrest cells at both G0/G1 and G2/M phase in HT-29 cells. The inhibitory effect of 3',4'-DMN is higher than that of 3'-DMN as only half the concentration is needed to induce a similar end result [54]. The fact that the metabolites 3'-DMN and 3',4'-DMN exhibit more potent anti-cancer effects than NOB may suggest that demethylation at the 3' and 4'-position significantly enhances its inhibitory effect [51].

In vitro tests using HCT116 cells reveal NOB and all three of the common metabolites increase the expression of CDK inhibitor, p21$^{Cip1/Waf1}$ [54]. p21$^{Cip1/Waf1}$, also known as p21 or P21/CDKN1A is a negative regulator for progression of the cell cycle that is responsible for the hypo-phosphorylation of retinoblastoma (Rb) proteins, leading to cell cycle arrest at the G1/S transition [86,88,89]. Although p21 is usually associated with the degradation of cyclin D1 [86], it is interesting to note that only 4'-DMN but not other metabolites nor the NOB itself causes significant reduction in cyclin D1 level. This may partly explain the strongest cell cycle arresting effect of 4'-DMN at the G0/G1 phase as compared to the other compounds aforementioned [54].

Proliferating cell nuclear antigen (PCNA) acts as a cofactor for DNA polymerase δ. It is an important marker commonly used to detect cell proliferation due to its increased expression through the G1 phase and S phase transition of cells [90,91]. Analysis from immunohistochemical tests recorded 69% reduction of cells with PCNA compared with the untreated controls [54]. Interestingly, evidence also reveals that p21 potentially suppresses action of PCNA. Interaction with the carboxy terminal of p21 inhibits PCNA from activating DNA polymerase δ, thus blocking DNA synthesis and preventing cell proliferation [92,93]. In this light, specific research of NOB on p21 and PCNA may be required to elucidate the pathways in further details.

Wu et al. studied the combinatory effect of NOB and its metabolites at different concentrations on HCT116 cells. At half the original concentration present in the colon, there is a decreasing trend of cells in S phase and G2/M phase but an increasing trend was noted in the G0/G1 phase. The cell cycle arrest effect seems to be dose dependent as flow cytometry recorded the population of cells arrested at the G0/G1 phase to be 57.8% higher than the untreated cells and significantly increased to 91.0% when the concentration of NOB and metabolites was doubled. To validate the findings, the levels of key signalling proteins were measured. Results showed that treatment with NOB and its metabolites lowered the levels of CDK-2, CDK-4, CDK-6 and cyclin D, raised the level of p52 and p27, but did not alter levels of p21 and cyclin E. In contrast, in vivo tests in AOM/DSS induced mice solely treated with NOB did document decreased expression of cyclin E and increased expression of CDK inhibitor p21. This difference is hypothesised to be due to the cell type specific response towards NOB, which is yet to be confirmed by further research [76]. The cell cycle is tightly regulated by key signalling proteins such as cyclins and cyclin dependent kinases (CDKs). To illustrate, complexes such as cyclin D-CDK-4/6 and cyclin E-CDK-2 facilitates the transition from G1 phase to S phase, while cell transition from G1 phase to S phase can be inhibited by tumour suppressor p53 and CDK inhibitors p21 and p27 [89,91].

It is also worth mentioning here that, at the effective concentration that arrests cell cycle, NOB produces a cytostatic effect, meaning it arrests growth without killing the cell [14]. As compared to other flavonoids like tangeretin (IC_{50} = 1.6 µM) and quercetin (IC_{50} = 0.84 µM), a slightly higher IC_{50} of 4.7 µM is required for the cell proliferation inhibition action by NOB in HT-29 cell lines and IC_{50} of 8.4 µM for 5-DMN [70]. However, the inhibitory effect of NOB may only be temporary. It is demonstrated that, with the removal of NOB, the treated cells resume cell proliferation within 24 h and regain similar growth status comparable to the control within 96 h [14]. This may also imply that, in order to sustain the inhibitory effect of NOB, the treatment with NOB has to be long-term to ensure continuous cell proliferation inhibition. This is possible as NOB is considered a natural compound and has no effect on healthy cells [83]. One might argue that the effect of NOB may be problematic for naturally fast-proliferating cells like healthy non-adenomatous intestinal lining cells. However, there is reassurance based on previous research that showed NOB is 10 times more selective towards transformed cancerous cells as compared to normal healthy cells [79].

5.1.2. Action of 5-DMN Inducing Cell Cycle Arrest

Treatment with 5-DMN also shows a similar increase of Rb in a dose dependent manner. Notably, 5-DMN does not affect the level of CDK-4, but there is a significant reduction of CDK-2 levels [63], hence indicating a reduced possibility of complex formation with cyclin A or cyclin E [94]. p21 is known to play a key role in arresting the cells at the cells at the G2/M phase through the inhibition of CDK-2/Cyclin E complex formation [95,96], and 5-DMN has been found to be able to arrest cell cycles at both the G0/G1 phase and G2/M phase in HCT116 (p53 $^{+/+}$), but is only able to accumulate cells at the G2/M phase in HCT116 (p53 $^{-/-}$). This suggests that G0/G1 arrest is dependent on p53 while G2/M is independent of p53 [80]. Using HT-29 cell lines, Qiu et al. reported that 5-DMN effectively causes cell cycle arrest at the G2/M phase [63]. This effect possibly arises from the downregulation of cdc25 protein expression, which is important to activate the cyc2/cyclin B1 through the process of dephosphorylating the inactive tyrosine residues Thr-14 and Thr-15 located in cdc2 ATP binding domain [97,98].

To sum up, different derivatives of NOB potentially arrest the cell at different stages of the cell cycle, mainly through downregulating the expression of proteins or kinases such as CDKs involved in the cell proliferation pathways and preventing the formation of cyclin complexes that allow cell cycle progression.

5.2. Programmed Cell Death

As growth of a cell is tightly regulated by the cell cycle, death of a damaged or aged cell also needs to be programmed to maintain homeostasis in our body. There are three models of programmed cell death (PCD), namely apoptosis, autophagy and necrosis [99]. A tumour mass of cancerous cells is formed when the cancerous cells develop the ability to evade cell death. Not responding to the death signal, the cells continue to grow and proliferate, leading to progression of cancer. Thus, NOB, being an agent that targets the key signalling pathways of programmed cell death, may help in chemoprevention of CRC. Apoptosis will be the main focus in this subsection, while autophagy will only be discussed briefly. Necrosis, the most abrupt death of all three, will not be discussed in this section as there are no data in this area. Although necrotic death is usually associated with inflammation, this does not exclude its possibility to be exploited as a means to eliminate cancerous cells [100].

Apoptosis can be induced by two core mechanisms, namely the extrinsic and the intrinsic pathway. The cell death signals in the extrinsic pathway come from external sources such as the Fas ligands or tumour necrosis factors [101]. The intrinsic pathway generally arises from the mitochondrial intracellular protein of the Bcl-2 family. Bcl-2 is an important regulator for apoptosis, which plays a role in mitochondrial disruption that activates the caspases [102]. High levels of Bcl-2 are expressed in various types of cancer and is associated with chemoresistance. Levels of Bcl-2 need to be lowered to promote apoptosis [103,104]. As a result of reduced Bcl-2 levels, a cascade of activity is activated in the cell leading to apoptosis with caspase-9 acting as the initiator caspase in the intrinsic pathway [105] and caspase-8 in the extrinsic pathway. It may also be crucial to mention here that the procaspase-8 forms a complex called Death Inducing Signalling Complex (DISC) before it is activated to caspase-8. The downstream effect would be the activation of the executioner caspase-3, and other caspases such as caspase-1, caspase-6 and caspase-7 [106,107] which then cleaves Poly (ADP-ribosome) polymerase (PARP) [108,109]. Soon after, the cell starts to bleb and shrink while its nucleus is condensed and fragmentised, proteolysis happens and the cell loses adhesion to the extracellular matrix and neighbouring cells [110,111]. Once the cell undergoes apoptosis, its contents are taken up by the body and recycled for new cell synthesis.

Action of NOB and Metabolites Inducing Programmed Cell Death

In vitro tests using different colon cell lines such as HCT116 and HT-29 reveals that the action of NOB and its various metabolites vary in different cell types. NOB was only shown to induce apoptosis of colon cancer cells when tested at high concentration. Zheng et al. [79] demonstrated that NOB increased DNA fragmentation in COLO302 only at 200 µM. Treatment with NOB, 3′-DMN, 4′-DMN and 3′,4′-DMN in HCT116 cell lines raise the early apoptotic cell population by 3.3-fold, 5.0-fold, 4.9-fold and 7.6-fold, respectively, while also resulting in 4.2-fold, 3.5-fold 7.1-fold and 4.5-fold increments in the late apoptotic cell population, respectively. In contrast, 3′-DMN and 4′-DMN did not cause any significant changes in apoptotic cell population in HT-29 cell lines, but the pro-apoptotic effect of 3′,4′-DMN was observed to be higher than that of NOB. An in vitro test using HCT116 shows that all three metabolites of NOB are able to induce the activation of caspase-3, caspase-9 and PARP, while NOB can only induce activation of caspase-9 but not that of caspase-3 or PARP. A negative result was also reported on the apoptosis inducing effect of NOB, whereby no apoptosis was detected when NOB was tested at concentrations of up to 100 µM in HT-29 [14]. In contrast, an in vivo test in AOM/DSS treated mice revealed a 2.3-fold increase of caspase-3 levels with NOB treatment [54]. Nevertheless, we can be certain that the metabolites of NOB render a higher proapoptotic effect as compared to their parent compound NOB.

Several previous works have demonstrated that NOB and its derivatives exhibit pro-apoptosis properties. However, this effect is known to be tissue specific. To illustrate, apoptosis is observed in colon cell lines but not the HL-60, promyelocytic leukaemia cell lines [79,112]. Interestingly, apoptosis could also be related to MMP-7. It has been discovered that cells that express MMP-7 are less sensitive to the Fas ligand-induced apoptosis as MMP-7 has a higher tendency to produce the non-apoptotic form of soluble Fas ligand by releasing the ligands in the cell membranes [77]. MMP will be further discussed in Section 5.4.

In vivo tests using xenograft mice show that 5-DMN triggers apoptosis at a concentration of 40 µM [49]. 5-DMN has been reported to increase levels of caspase-8, caspase-3 and PARP in a dose dependent manner, and results in 2.2-fold increase in the population of early apoptotic cells as compared to the control. In contrast, there is no apparent increase in the apoptotic effect even after doubling the concentration of NOB, suggesting NOB is required at a significantly higher concentration to induce a pro-apoptotic effect [14,63]. However, the effect of 5-DMN was notably distinct for different types of cancer cell lines. To illustrate, 5-DMN can induce early apoptosis in HCT116 colon cancer cell lines at a concentration as low as 8 µM while 5-DMN only slightly raises the apoptotic activities at a concentration as high as 36 µM in colon HT-29 cancer cell lines [63].

Annexin-V/PI analysis reveals that 5-DMN significantly increased the Annexin-V positive cells, especially in the late apoptotic or necrotic cell population among the HCT116 (p53 $^{-/-}$) cells, suggesting that the action of 5-DMN may be independent of p53 [80]. The fact that 5-DMN induces early apoptosis in HCT116 (Bax $^{+/+}$) cells but not in HCT116 (Bax $^{-/-}$) may suggest that Bax (Bcl-2 associated X protein) is important for apoptosis to occur. In other words, absence of Bax confers resistance to apoptosis [80,113,114]. A recent study by Chiou et al. proved that 5-DMN increases the expression of p53 proteins, which not only induces apoptosis, but also triggers cell death by autophagy that contributes to the prevention of tumour growth [49]. However, the detailed mechanism of how NOB prevents CRC development through the autophagy process is yet to be elucidated. From the current stage of knowledge, autophagy is known to exhibit both the pro-tumour and anti-tumour formation effects. In response to stress, autophagy acts as a protective mechanism for cell survival, which has already been elaborated in previous literature [115]. It is likely that overactivation of autophagy contributes to suppressing tumour formation by inhibiting the anti-autophagic-related genes (ATGs) in oncogenesis and activating the pro-ATG [116]. The effect of autophagy and apoptosis may be synergistic in chemoprevention of cancer [117,118].

To conclude, NOB was shown to be less effective in inducing apoptosis of colon cancer cells. Instead, the metabolites of NOB, 3'-DMN, 4'-DMN and 3',4'-DMN were suggested to be responsible for the induction of apoptosis of colon and modulation of cancer cell growth in the colon carcinogenesis animal model [54] On top of that, the autohydrolysis product of NOB, 5-DMN could induce apoptosis in colon cancer cells at a lower concentration as compared to NOB.

5.3. Anti-Inflammation

Inflammation is a natural physiological response of our body, characterised by five main signs, namely loss of function, redness, pain, heat and swelling. Inflammation plays a significant role at times of infection and injury. It triggers the immune system to function and helps to protect our body through the release of chemical molecules called pro-inflammatory signals. However, too much inflammation is also not a good sign. There is increasing evidence narrating the interrelation between tumorigenesis and inflammation [119]. Whilst chronic inflammation is a hallmark of cancer, the inflammatory cytokines aggravate cancer progression by preventing differentiation of cells and promoting tumour formation [120]. The inflammatory cells release ROS after being activated, leading to the oxidative damage of DNA and p53 mutation [121–123]. The mechanism that triggers inflammation is a rather complex pathway and has been covered by previous literature [123,124]; only the anti-inflammatory effects mediated by NOB are discussed in this review.

Increasing evidence shows that progression of CRC can be accelerated by the upregulation of pro-inflammatory cytokines expressions—for example, TNF-α, IL-1α, IL-1β and IL-6 [81,119,125]. These proinflammatory cytokines enhance the secretion of inflammatory mediator PGE$_2$. Song et al. reported that treatment with NOB results in a noteworthy reduction of tumour size and frequency correlated with the significant lowering of IL-1, IL-6, iNOS and COX-2 levels [78]. ELISA test quantified the reduction of TNF-α, IL-1β and IL-6 at 51%, 92% and 69%, respectively, in the NOB treated group while real-time qRT-PCR quantified the reduction of the above pro-inflammatory cytokines at 65%, 69% and 45%, respectively, when compared to the control mice [54].

Anti Inflammation Effect of NOB and Its Metabolites

Besides NOB, multiple studies have shown that its metabolites, especially, 4'-DMN and 3',4'-DMN, also exhibit significant inhibitory effects towards nitric oxide production, iNOS and cyclooxygenase (COX) expressions in both in vivo and in vitro conditions [30,31,62,76,81,126]. However, the combined effect of NOB and its metabolites warrants further investigation [76]. Notably, NOB selectively inhibits COX-2 and did not affect COX-1 [81]. COX-2 is normally absent in healthy cells, but its release is triggered when the environment is inflammatory or hypoxic [127]. COX-2 is known to enhance CRC carcinogenesis, and inhibiting COX-2 also limits the production of PGE$_2$ [128], which may be associated with the inhibition of cell proliferation in colonic mucosa [41]. iNOS speeds up the conversion of L-arginine to NO through the process of oxidative deamination. NO is a potent inflammatory mediator that activates signalling molecules that trigger the process of inflammation and mutagenesis [129]. By inhibiting the iNOS and its downstream products, NOB helps in reducing the inflammation observed in chronic diseases like ulcerative colitis and CRC [130]. Introduction of NOB to colonic tissues harvested from mice treated with AOM/DSS results in a 35% reduction of cells expressing iNOS compared to the untreated tissue. This is consistent with the in vitro test. By administering NOB and its metabolites at a concentration equivalent to that found in the colons to the LPS-induced RAW 246.7

macrophages is shown to effectively and completely inhibits expression of iNOS, while, at half the concentration, the expression of iNOS was lowered by 56.4% compared to the untreated LPS-induced macrophages [76]. This shows that a similar process is likely to happen in the human body and NOB is indeed a promising anti-inflammatory agent.

Additionally, NOB also increases the release of the Nrf2-dependent enzymes which regulate Phase II enzyme production, such as heme oxygenase-1 (HO-1) and NQO1. HO-1 is an anti-oxidative enzyme that exhibits its anti-inflammatory effect by producing anti-oxidants like carbon monoxide and bilirubin. It is also important to note that HO-1 is not solely controlled by Nrf-2 [76]. NQO1 upregulation counteracts the increased expression of IL-1β and TNF-α induced by LPS [131]. This is consistent with the findings of Khor et al. reporting that a lower Nrf-2 expression greatly increases susceptibility of mice models to AOM-induced colitis [132]. The colonic mucosa of AOM/DSS-induced mice orally administrated with NOB was found to have increment of nuclear Nrf2 by 1.94-fold and reduction of cytoplasmic Nrf2 by 36% when compared to the AOM/DSS-treated mice. The upregulation and translocation of Nrf2 transcription factor is thought to be the cause for subsequent increment of HO-1 and NQO1 by 2.78-fold and 2.59-fold, respectively. This is consistent with the results from the combinatory effect of NOB and metabolites treatment on macrophages cell lines, which induces a 10% increase in the level of HO-1 and a 34% increase in the level of NQO1 when the concentration ratio of NOB and metabolites is equivalent to that in the colon [76]. In short, Nrf-2, which neutralises carcinogens and reactive oxygen species (ROS), is identified as a key signalling pathway to target in the effort to halt CRC progression [133,134].

5.4. Anti-Angiogenesis

Angiogenesis refers to the process by which new blood vessels are formed. This process is an important pathway that results in the progression of all types of cancer. This is because, when the tumour mass grows, it naturally needs more nutrients and nourishment to support its growth. To achieve this necessity, new blood vessels have to be formed surrounding the tumour mass to ensure a continuous supply of oxygen and glucose to support the growing cell mass [135,136].

Anti-Angiogenesis Effect of NOB

It is postulated that NOB prevents metastasis by inhibiting the activity of activator protein-1 (AP-1), a dimeric protein, thus preventing DNA binding [77]. Another hypothesis suggests that NOB acts via the Nuclear Factor-kappa B (NF-κB) pathway, altering the gene expression by modulating the promoter regions [126,137].

For angiogenesis to occur, the vascular endothelial growth factor (VEGF) plays a key role. VEGF not only acts as a signal to induce new blood vessel growth, but also inhibits the apoptosis induction. VEGF works by activating the mitogen activated protein kinase (MAPK). This kinase triggers the signal transduction and allows the endothelial cells to proliferate in order to form new blood vessels [138–140]. To elaborate on VEGF, it is necessary to mention leptin and insulin-like growth factor 1 (IGF-1) here. There is evidence suggesting that bidirectional cross talk exists between the leptin protein and IGF-1, a serum growth factor. Acting together, they not only catalyse the cell proliferation process, but also transactivate the epidermal growth factor receptor (EGFR), which enhances the migration and invasion power of the cancer malignancy [141].

Miyamoto et al. discovered that leptin, a protein that regulates energy balance and body mass has a positive correlation with CRC, where leptin is thought to be a mitogenic factor that leads to the development of colon cancer [84]. It induces cell proliferation by activating the nuclear factor κB, p38 MAPK and p42/44 MAPK [142]. Previous evidence demonstrates that introduction of 0.1 to 10 nM of leptin enhances the proliferation rate of HT-29 cells by 1.3 to 1.6 times [84] through c-Jun NH$_2$ terminal kinase and extracellular regulated kinase (ERK) 1/2 activation [143,144]. In other words, the chances of developing CRC can be reduced if leptin concentration is regulated. Treatment with NOB suppresses cell proliferation induced by leptin through inhibition of mitogen-activated protein/extracellular signal-regulated kinase (MEK) 1/2 [145]. Consistent with the in vitro findings, a reduction of 75% of leptin concentration by NOB, partly through the inactivation of the insulin signalling pathway, was reported at the end of the 17-week study in an in vivo model using Institute for Cancer Research (ICR) mice [83,84]. In this light, Miyamoto et al. conducted a study aiming to determine the prognosis of cancer in obese rats with flavonoid intervention. They reported that the flavonoids significantly reduced the incidence of β-catenin accumulated crypt (BCAC) by 64% to 71% and aberrant crypt foci (ACF) by 68% to 91% and proposed that this arises from the effect of NOB in downregulating the secretion of IGF-1 [83].

Metalloproteinase (MMP) plays a fundamental role in angiogenesis. MMP induces the protein that breaks down the extracellular matrix (ECM), thus making the blood vessel more permeable and allowing the cancerous cells to detach from the lump to flow, extravasate or invade the other parts of the body, causing the spread of tumour to other vital organs. This is a process called metastasis, leading to malignancy and cancer aggravation. The mortality rate greatly increases when cancerous cells metastasise to vital organs like the liver [146]. Similar to how a dexamethasone steroid acts, NOB is proven to be able to increase the expression of tissue inhibitor metalloprotease-1 (TIMP-1) in human synovial cells [81]. However, the benefit of upregulating TIMP-1 in CRC is debatable due to its bilateral role in cancer progression. Although TIMP-1 upregulation contributes to the anti-oncogenic effect, enhanced expression of it may lead to early phase tumour development via the pathways independent of MMPs. With this understanding, TIMP-1 glycosylation can function as a biomarker to aid in CRC staging [147].

There are more than 20 types of MMP involved in metastasis [148], with each one of them playing a distinct role [149]. However, whether MMP is produced by cancer cells or their surrounding stromal cells is still an ongoing debate [150]. Abnormally high levels of MMP-1, MMP-2, MMP-3, MMP-7, MMP-9 and MMP-13 have been implicated in CRC [150]. Treatment with NOB significantly inhibits release of pro-MMPs especially pro-MMP-7 (also known as metrilysin) mRNA in HT-29 cell lines. To illustrate, NOB at a concentration range of 25 μM to 100 μM, the proMMP-7 levels in the media diminishes significantly by 35% to 47% [77]. The maximal expression of MMP-7 arises from the β-catenin/TCF complex transcription factors formed in the presence of mutated APC genes [150–152]. Apart from MMP-7, the action of NOB on other MMPs in CRC is yet to be investigated. MMP-9, which is mainly secreted by inflammatory cells, is correlated with the transition phase from adenoma to adenocarcinoma, while the upregulation of MMP-3 usually suggests poor prognosis as it has a positive correlation with low microsatellite stability. On the other hand, high levels of MMP-12 reduces CRC mortality as it can potentially inhibit angiogenesis [150] by secreting angiostatin, a chemical that halts tumour progression and inhibits tumour neovascularisation [153–155]. As mentioned in the previous section, NOB suppresses MEK. This suppression of MEK then further diminishes the expression of pro-MMPs, which results in the reduction of MMP and subsequently confers anti-angiogenesis effect [144,145]. Briefly, NOB prevents angiogenesis and metastasis in CRC mainly via the inhibition of MMP, EGFR and VEGF through the regulation of leptin and IGF-1.

6. Pharmacokinetics, Bioavailability and Delivery Systems of NOB

The pharmacokinetic properties of NOB represent a key factor to be considered in an attempt to formulate it into a therapeutic product. Understanding the interactions between the compound and our body opens ways to creative strategies in solving the problem which require novel formulation in delivering NOB for chemoprevention purpose. For oral delivery, an important consideration is the bioavailability of the active compound. However, the bioavailability studies on NOB are limited [156]. Therefore, understanding the pharmacokinetic profile of NOB becomes even more crucial to assisting in the prediction of bioefficacy as the absorption, metabolism and elimination pattern indirectly affect its bioavailability.

There are many factors that affect the absorption of a compound; one important consideration is the molecular structure [157]. The proper absorption of any compound is depicted by its solubility and permeability across physiological barriers of which both properties are directly related to its molecular structure. Attributed to its unique chemical structure with multiple methoxy groups, NOB is lipophilic in nature and can easily pass through the cell membrane. Murakami et al. successfully demonstrated the relatively high permeation of NOB across differentiated Caco-2 cells which mimics the epithelial cells lining the small intestine. A significantly high 48.1% of NOB has been found to permeate through the basolateral compartment while another 39.3% remains on the apical compartment four hours after introduction of NOB in a Caco-2 monolayer trans-well permeability assay [158]. Parallel artificial membrane permeation assay (PAMPA) deciphered the permeability of NOB, 4′-DMN and 3′-DMN at 1.38×10^{-6} cm/s, 1.14×10^{-6} cm/s and 1.05×10^{-6} cm/s, respectively [159]. It was discovered that the methoxylated flavonoids show five to eight-fold higher permeability in the intestinal wall than its unmethoxylated counterparts [160]. The drawback is that PMF in general has limited solubility. Results from the high-throughput lyophilisation solubility assay (LYSA) reveal that NOB has a low solubility at 12 μg/mL, while its metabolites, 4′-DMN, 3′-DMN and 5-DMN exhibited two to three-fold higher solubilities of 22 μg/mL, 29 μg/mL and 32 μg/mL, respectively [156,161]. In general, the solubility increased with the number of hydroxyl group of the compound. This may also partly explain the higher activity of the derivatives of NOB as compared to its parent compound, which has a higher number of methoxy groups.

After absorption, NOB is found to be widely distributed throughout the body, as a significant amount of NOB could be detected in organs such as the stomach, small intestine, large intestine, brain, liver and kidney within four hours of single dose administration [162,163]. Interestingly, NOB was suggested to be absorbed through the muscularis layer of the gastrointestinal tract, especially the stomach tissue into the blood circulation given the distinctly higher concentration of NOB in the muscularis (390 ± 120 nmol/g) as compared to other organs [162]. Furthermore, it is also noteworthy that NOB was found to be distributed in the mucous membrane and muscularis from the large intestine (cecum, colon and rectum) of a rat at 4.3 ± 1.6 nmol/g after one hour of oral administration by gastric intubation [162]. More recent evidence demonstrated that the levels of NOB in the colonic mucosa of mice were 2.03 nmol/g of tissue after long-term oral administration of NOB (0.05 wt%) containing diet. Furthermore, Wu et al. [164] also suggested that the dose of NOB used (0.05 wt%) could be equivalent to approximately 100 mg/day for human oral consumption, which is achievable in humans.

After oral administration of NOB to rats, the mean plasma concentration of NOB was quantified in several pharmacokinetic studies. Wang et al. [163] reported that the plasma levels of total NOB and its metabolites could reach as high as 10 μg/mL (25 μM). Using a highly sensitive Liquid Chromatography-Mass Spectrometry/Mass Spectrometry-Electrospray ionisation (LC-MS/MS-ESI) method, the maximum concentration (C_{max}) of NOB in rat plasma was determined at 0.4 μg/mL (1 μM) after oral administration of 5 mg/kg NOB [165]. Meanwhile, a maximum concentration of 1.78 μg/mL (4.4 μM) was measured by a validated HPLC method in rat plasma after oral administration of 50 mg/kg NOB [166]. In addition, another study by Manthey et al. [167] reported that a peak of NOB serum level of 9.03 μg/mL (22.4 μM) was detected by HPLC-ESI-MS in rats after oral gavage of 50 mg/kg NOB. Nevertheless, these studies demonstrated a relatively early peak time (T_{max}) of 0.25 to one hour after oral administration of NOB in rats. This high rate of cellular uptake may be attributed to the highly hydrophobic nature of the compound rendered by the presence of six methoxy groups [162].

NOB undergoes extensive metabolism after being taken orally. As detailed in Section 3, NOB undergoes Phase I and Phase II metabolism after being absorbed in the small intestine where it may be conjugated to sulphate and glucuronide, and then again deconjugated by the microflora in the colon [60]. The three common phase I metabolites of NOB have been identified as 3′-DMN, 4′-DMN and 3′,4′-DMN [52,53]. Wang et al. found evidence of transformation of 3′-DMN and 4′-DMN into 3′,4′-DMN in the colon [163]. The liver is another important organ involved in metabolising NOB. Koga et al. identified three metabolites, demethylated at the 4, 6 or 7 positions respectively under the action of human liver microsomes when incubated aerobically with NADPH [56]. The metabolites exhibit distinct activity and distribution pattern in different areas of the body. It was found that 4′-DMN is the major metabolite present in the small intestine and liver while 3′,4′-DMN was predominantly present in the colon and spleen [163]. An in vitro test on NOB using rat liver S-9 extract shows that only 7% of NOB metabolites were detected towards the end of a 24 hour treatment, while 72.6% of NOB remains unchanged towards the end of the experiment [158]. This may be attributed to the slow rate of demethylation of NOB showed by Murakami et al. [162].

The elimination half-life of NOB from the blood plasma of a rat was reported as 1.8 h via a validated HPLC test [166] while Kumar et al. reported a terminal half-life of NOB at 4.75 ± 0.57 h following oral administration and a terminal half-life of 1.51 ± 0.61 h following parenteral administration using the LC-MS/MS-ESI method [165]. Despite the wide distribution throughout the body, concentration of NOB quickly diminishes with time and becomes undetectable in the serum, stomach, intestines, liver and kidney. Aside from the parent compound, mono-demethylated metabolites and conjugated NOB are detected in the urine, with the concentration of conjugated NOB revealing a time-dependent increment over a period of 24 h [162]. Since NOB is rapidly eliminated from the body, significant adverse effects reported after administration of NOB are rare.

Although the in vitro results were promising, most of the reported concentrations of NOB evaluated (>20 μM) were not achievable in physiological conditions as demonstrated by in vivo pharmacokinetic studies of NOB. Comparing the high experimental levels used against the relatively low peak plasma concentration—a mere 1.78 μg/mL (4.4 μM)—after one hour of oral administration of 50 mg/kg NOB [166] and the rapid elimination from the body [162] points out a limitation to utilizing NOB in its unaltered natural form as a clinical drug. In fact, the levels of NOB detected in the colonic mucosa ranged between 2 to 4 μM using the assumption that one gram of tissue is equivalent of 1 mL of volume [54,162]. However, there was a study demonstrating that NOB at lower concentration (≤5 μM) exhibited antiproliferative effects against colon cancer cells [70], perhaps indicating true promise for clinical use after all.

To further substantiate the notion of NOB for colon cancer chemoprevention, multiple in vivo studies have demonstrated that dietary treatment with NOB could inhibit colon carcinogenesis in rats [54,76]. As mentioned earlier, this may be related to the fact that, while bioavailability of NOB itself is low, much of its anti-CRC effect may be via its metabolites. Wu et al. [54] indicated that the NOB level in the colonic mucosa only accounted for <5% of the total levels of NOB and its metabolites after oral administration of NOB. The study further suggested that the NOB metabolites, which were formed as a result of phase I and II metabolism and biotransformation by gut microbiome, play an important role in colon carcinogenesis inhibition [54]. Although there is some suggestion that lower doses can have an effect on cancer, clearly, enhancement of NOB bioavailability is necessary and also represents a major challenge that needs to be addressed to achieve the desired therapeutic effect.

Given the importance of actually delivering adequate amounts of NOB to the target site to achieve chemopreventive activity, we also reviewed the delivery systems aiming to enhance the bioavailability of NOB in the gut. For chemoprevention of CRC, oral delivery represents the preferred route. There is a growing interest to formulate lipophilic natural compounds such as NOB into emulsion, as these systems not only improve the bioavailability of the active compound, but also reduce the rate of degradation during storage [168]. Yang et al. attempted to enhance the solubility of NOB by encapsulating NOB with citrus oil-based emulsion. The team discovered that dissolving NOB at a higher temperature and in an oil with log P close to NOB, such as bergamot oil, helps to increase solubility of the compound [169]. Yao et al. also experimented with the possibility of using self-microemulsifying drug delivery systems (SMEDDS) to improve the permeability of NOB in the rat intestines and reported that SMEDDS resulted in similar efficacies to micelles, but showed better absorption profile when compared to sub-microemulsions [170]. Self-assembled NOB proliposomes were also reported to improve the absorptive rate and confer longer mean residence time as compared to NOB suspension in rats [171].

Furthermore, Chen and colleagues demonstrated that, through the addition of hydroxypropyl methylcellulose (HPMC), the retention of NOB in nanoemulsion is increased by 25% [172]. Even though the fabrication of supersaturating nanoemulsion with the addition of HPMC aimed to improve the physical stability of NOB and prevent precipitation of NOB in the emulsion, the fabrication did not perform as expected at high NOB concentration where precipitation still occurred during storage and digestion process in the gut [172]. To address the issue of component precipitation in the emulsion system, a recent intervention of nanoemulsion-filled hydrogel matrix has been developed to stabilize NOB and prevent precipitation during delivery along the GI tract [173]. Interestingly, the hydrogels could provide a controlled release of NOB along the GI tract, thereby the hydrogel shrank at acidic condition pH 1.2 but swelled and burst at pH 7.4. Due to the lower bioaccessibility of NOB in hydrogel as compared to nanoemulsion during digestion, the nanoemulsion-filled hydrogel matrix could confer a sustainable absorption of NOB through a controlled release in the intestinal tract [173].

Aside from the liquid formulations, Onoue and colleagues proposed a solid formulation of NOB with the intention to further enhance the bioavailability in addition to solving the stability issues which showed a remarkable 13-fold increment in bioavailability compared to the nanosized NOB amorphous solid dispersion [174]. However, the results only quantitate the brain permeability, but the data for colon effect is still lacking. Further research is needed to establish the practicability and feasibility of each delivery method to address the bioavailability challenges before NOB can be used in aiding patients at high risk of CRC.

7. Toxicity

Although NOB is derived from a natural source, excessive intake of any substance might lead to some changes in the body. To illustrate, there are several case studies reporting that ingestion of products containing bitter orange causes adverse effects such as tachycardia and ventricular fibrillation [175]. To address this concern, a number of studies have been carried out to further evaluate this problem.

Body weight is commonly used as a surrogate marker for toxicity, whereby the weight is expected to reduce significantly if toxicity occurs. Wu et al. concluded there was no significant change in body weight, weight of liver, spleen appearance and behaviour throughout the length of their research, suggesting that the oral intake of NOB at the effective concentration 40 µM or 0.05% for a period of 3 to 20 weeks does not lead to adverse side effects [41,49,54]. This is consistent with the findings of Murakami et al. suggesting that NOB leads to no cytotoxic effects [42]. The action of NOB is more likely of a cytostatic nature rather than cytocidal as, at the concentration that inhibits cell proliferation, it does not trigger apoptosis. Furthermore, the treated cells continue to grow normally once the effect of NOB diminishes [14]. This shows that NOB could be a safer option for CRC treatment as it is less cytotoxic as compared to the available chemotherapy agents.

8. Commercial Uses

A search on Google Scholar using the keywords "nobiletin patents" gives about 1160 relevant results. This initial search was refined through the advanced search function for patents using the SciFinder database which helps to narrow down the number of patents directly and indirectly related to NOB to 300.

After a close analysis of the patents, we found that, among the 300 patents related to the concept of NOB, the largest portion of the total patents involves the usage of NOB in the medical, pharmaceutical and nutraceutical fields. The patents include both the application in traditional treatments and also western medications, where there are about 20 patents related to cardiovascular diseases [176–178], hypertension and hypercholesterolemia [179], roughly 10 patents targeting the central nervous system [180,181] or neurodegenerative disorders [182], diabetes [183,184] and obesity [185] each, and about five patents concerning body metabolism and hormonal functions, bone-related disorders [186], oral issues such as ulcer and halitosis [187,188], liver-related problems like hepatitis [189] and anti-infectives such as anti-bacterial, anti-viral [190] and vaccines, respectively. The patents also include a small number of

NOB usage in diseases like prostate disease, asthma [191,192], allergy [193], eye relief, prevention and improvements of conditions like hair fall [194], dysuria [195] and muscular atrophy [196].

A large proportion of the patents are related to anti-cancer treatments, which account for almost 13% of the total patents. The types of cancers covered are broad, ranging from the more prevalent ones like lung cancer [197] and breast cancer [198] to those lower down the prevalence indices like uterine, liver cancer, oral [199], and skin cancer [19]. Application wise, some major areas that involve the usage of NOB compounds include the synergistic effect of NOB with existing chemotherapeutic agents targeting the multidrug resistance cancer [200] which aim to increase therapeutic efficacy as well as aiming to address the side effects from conventional chemotherapeutic treatments, especially diarrhoea [201]. Some common cancer inhibition pathways leading to cancer that are targeted by the compound include anti-angiogenesis [202], anti-proliferation and anti-tumour or anti-neoplastic effects.

The usage of NOB in nutraceutical industries is also extensive; there are up to 20 patents of beverages that contain NOB, with some of the fortified drinks claiming to have pharmacological effects [203]. Next in the line, comprising one-fifth of the total patents, are the methods of extraction [204–206], purification [207,208], preparation [209,210], manufacturing [44,211,212] analysis [213], drug delivery and pharmacokinetics information such as ways to improve absorption, solubility and bioavailability.

Apart from that, the use of NOB in fields other than medicine is also very broad, which includes about 20 patents in cosmetics [214–216], another 20 patents in the food industry, which include its use as preservatives [217], flavourings or food additives [218], and four patents in the agricultural industry, of which mostly it is used as pesticide controls. Last but not least, there are also patents of NOB usage in stem cell technology and genetic analysis.

9. Future Directions

While NOB and its metabolites seem to have tremendous potential as chemopreventive agents, at this juncture of time, more intensive research is needed to resolve the challenges that arise from the limitations of this compound. As mentioned earlier, NOB showed dose dependent anti-cancer effects, but the challenge is to increase its bioavailability to enhance the chemopreventive effect. This is important as oral administration seems to be a more promising route of administration at the moment as intraperitoneal injection has been associated with severe side effects such as ischemic stroke [36]. In addition, given that the chemopreventive metabolites appear to be formed by via metabolism within the gut, the oral route seems to be a promising way of delivering drugs to the target site.

Although several effective delivery systems were developed to enhance the bioavailability of NOB, studies on targeted-delivery of NOB to the colon are still limited. Despite having delivery systems that enhance aqueous solubility and bioavailability, a colon-specific drug delivery system is highly desirable for efficient drug delivery of NOB to the colon or where the colorectal cancer reside. In 2012, a folate-modified self-microemulsifying drug delivery system (FSMEDDS) was developed with the aim to improve solubility of curcumin and specifically target colorectal cancer cells mediated by the binding of folate receptors in facilitating the endocytosis of the formulation [219]. Given the previous evidence of the successful preparation of NOB in SMEDDS [170], this intervention could be possible to be implemented to facilitate specific uptake of NOB into colorectal cancer cells via the FSMEDDS and further coated by Eudragit®S 100 (Evonik Industries AG, Essen, Germany), which prevent dissolution of the formulation under the condition of pH < 7.5 [219,220]. Recently, a dual stimuli-responsive Pickering emulsion (pH and magnetic- responsive) reported may hold immense potential for the biomedical field, particularly in the treatment of colorectal cancer [221,222]; to achieve an active targeting of specific sites, an external magnetic field could be utilized to direct the movement and accumulation of the drug carrier at the targeted sites to exert their therapeutic effects [223].

In addition to that, biotechnology can be applied to biosynthetically produce NOB in larger yields as citrus peel may only contain a limited amount of this bioactive product. In this regard, Itoh et al. successfully isolated five genes from *C. depressa* which encode the flavonoids by *O*-methyltransferases (FOMT), a precursor for a number of flavonoids. Quercetin has been synthesised via this method and it is highly likely that the same enzyme is also involved in the biosynthesis of NOB, suggesting that it may also be possible for NOB to be synthesised using this strategy [39]. However, the data on the effectiveness in this application is still lacking as there is limited research that uses this method to synthesise NOB. Apart from biotechnology, the introduction of reliable, efficient and economical validation methods such as ultraperformance liquid chromatography coupled with quadrupole time-of-flight mass spectrometry (UPLC-Q-TOP-MS) which allows high rate of separation of PMF compounds within 12 min also opens up more possibilities for NOB to be marketed [224].

In addition, another major challenge of chemotherapy that we are facing today is the development of drug resistance in cancer treatment. One possible cause that results in chemoresistance may be attributed to the cancer stem cell (CSC). CSCs are known to play a crucial role in tumour formation as they possess unique characteristics including unlimited cell renewal capacity and the ability to evade drug penetration [225,226]. Seeing the limitation of the single cell in vitro model [227], Silva et al. came up with a brilliant method of culturing cells into a three-dimensional block, which they named a 3D spheroid. At day seven, the 3D spheroids mimic the tumour lump, with the undifferentiated cells in the outer region surrounding the hypoxic inner core. Experiments showed that 2.9-fold higher concentration is needed to exhibit the same effect reported in the two-dimensional cell model [47].

Interestingly, the concomitant exposure of NOB and its metabolites gives rise to synergistic effects that are distinct from the response caused by NOB alone [228,229]. Therefore, the combinatory effect of NOB and its various metabolites should be explored in order to establish a solid foundation of understanding of the synergistic effect of NOB and its natural metabolites generated through the biotransformation process. Apart from that, compelling evidence showed that NOB produces a synergistic effect in tumour growth inhibition when co-administered with atorvastatin. When used together, only half the minimal effective concentration of each drug is required to achieve the targeted therapeutic outcome. Wu and co-authors reported a series of mechanisms by which this combination works, namely through altering important cellular signals that triggers inflammation, inhibiting cell cycle progression, inducing apoptosis and preventing angiogenesis and metastasis [164,230]. In this light, the drugs already in the market can be combined with NOB and tested for their synergistic effects in inhibiting CRC. In addition, the combinatory effect of NOB and its metabolites needs to be further elucidated to achieve a precisely targeted biological action in CRC chemoprevention. More clinical trials in human subjects with due ethical considerations are warranted as disparity will certainly exist if the data is solely extracted from in vitro or animal tests.

10. Conclusions

While there is a significant research focus on cancer, science is still at an early stage in understanding this noxious condition affecting people from every segment of society, but answers are critical as cancer's prevalence and variance are continuously on the rise. The current clinical practice in cancer treatment, which largely consists of the three broad fields, namely surgery, chemotherapy and radiotherapy, may be helpful to patients to some extent but more intensive and in-depth ongoing studies are needed in the quest for a panacea for cancer given the high mortality rates of this malady. Many more patients will be relieved from pain and suffering if scientific research can shine a light on the root causes of cancer and focus on its prevention so as to nip the problem in the bud before the need to treat it arises.

The advancement in science has allowed the discovery of numerous beneficial compounds offered by nature. It is reassuring to learn that NOB, a compound that is extracted from the ubiquitous citrus species confers a wide range of beneficial biological effects that includes cancer prevention. On top of that, the autohydrolysis product, 5-DMN and several metabolites of NOB such as 3′-DMN, 4′-DMN and 3′,4′-DMN, demonstrate more potent effects as compared to their parent compound NOB. It is apparent that NOB is indeed a prospective compound that exhibits a promising chemopreventive effect on CRC, especially for the types which are induced by carcinogens or associated with diseases such as colitis. In addition to that, this review also focuses on the underlying molecular mechanism of which NOB acts in CRC. The plus point is that NOB and its products target a number of different hallmarks of cancer. To illustrate, NOB is endowed with anti-proliferative, pro-apoptotic, anti-inflammatory and anti-angiogenesis effects, which renders it the potential to counteract the pathology of CRC in patients at various stages of cancer progression.

Besides NOB, many compounds under the polymethoxyflavones family are currently promising candidates in the field of cancer research, yet it is too early for science to conclude a best compound to formulate as the elixir. More studies, be it in vitro, in vivo or clinical studies, are needed to unravel the full potential of each possible compound. Furthermore, it would be worthwhile to explore the synergistic effect or possible interactions between NOB and well-known anti-cancer drugs by both experimental and clinical studies. The vast number of existing patents of NOB across various industries may suggest that this compound does have commercial value besides its noteworthy pharmacological benefits. Further research work needs to be intensified to overcome the current gap and limitation in formulation, for instance to increase the bioavailability and to enhance the efficacies of NOB in CRC chemoprevention. Although significant advances have been made, there is still a long way to go before NOB could truly become part of the arsenal of CRC chemoprevention.

Author Contributions: The literature searches and data collection was performed by J.X.H.G. & L.T.-H.T. The manuscript was written by J.X.H.G. and L.T.-H.T. The manuscript was critically reviewed and edited by J.X.H.G., L.T.-H.T., P.P., L.-H.L., J.K.G., K.G.C. and B.-H.G., B.-H.G. provided vital guidance and insight to the work. The project was conceptualized by B.-H.G.

Funding: This research was funded by a MOSTI eScience Fund (02-02-10-SF0215), University grants from the University of Malaya (PG136-2016A and PG135-2016A) and External Industry Grant from Biotek Abadi Sdn. Bhd. (vote no. GBA-81811A).

Acknowledgments: This work was inspired by the Monash Pharmacy Degree Course, Unit PAC3512 entitled "Current aspects of pharmaceutical research".

Conflicts of Interest: The authors declare no conflict of interest.

References

1. Siegel, R.L.; Miller, K.D.; Jemal, A. Cancer statistics, 2017. *CA Cancer J. Clin.* **2017**, *67*, 7–30. [CrossRef] [PubMed]
2. Bernstein, C.N.; Blanchard, J.F.; Kliewer, E.; Wajda, A. Cancer risk in patients with inflammatory bowel disease: A population-based study. *Cancer* **2001**, *91*, 854–862. [CrossRef]
3. Rubin, D.C.; Shaker, A.; Levin, M.S. Chronic intestinal inflammation: Inflammatory bowel disease and colitis-associated colon cancer. *Front. Immunol.* **2012**, *3*, 107. [CrossRef] [PubMed]
4. Siegel, R.L.; Miller, K.D.; Fedewa, S.A.; Ahnen, D.J.; Meester, R.G.; Barzi, A.; Jemal, A. Colorectal cancer statistics, 2017. *CA Cancer J. Clin.* **2017**, *67*, 177–193. [CrossRef] [PubMed]
5. Simon, K. Colorectal cancer development and advances in screening. *Clin. Interv. Aging* **2016**, *11*, 967–976. [PubMed]
6. Tan, L.T.H.; Lee, L.H.; Yin, W.F.; Chan, C.K.; Abdul Kadir, H.; Chan, K.G.; Goh, B.H. Traditional Uses, Phytochemistry, and Bioactivities of Cananga odorata (Ylang-Ylang). *Evid. Based Complement. Altern. Med.* **2015**, *2015*, 30. [CrossRef]
7. Chan, W.-K.; Tan, L.T.-H.; Chan, K.-G.; Lee, L.-H.; Goh, B.-H. Nerolidol: A Sesquiterpene Alcohol with Multi-Faceted Pharmacological and Biological Activities. *Molecules* **2016**, *21*, 529. [CrossRef]
8. Tan, H.-L.; Chan, K.-G.; Pusparajah, P.; Saokaew, S.; Duangjai, A.; Lee, L.-H.; Goh, B.-H. Anti-Cancer Properties of the Naturally Occurring Aphrodisiacs: Icariin and Its Derivatives. *Front. Pharmacol.* **2016**, *7*. [CrossRef]
9. Steward, W.P.; Brown, K. Cancer chemoprevention: A rapidly evolving field. *Br. J. Cancer* **2013**, *109*, 1. [CrossRef]
10. Tang, C.; Hoo, P.C.-X.; Tan, L.T.-H.; Pusparajah, P.; Khan, T.M.; Lee, L.-H.; Goh, B.-H.; Chan, K.-G. Golden Needle Mushroom: A Culinary Medicine with Evidenced-Based Biological Activities and Health Promoting Properties. *Front. Pharmacol.* **2016**, *7*. [CrossRef]
11. Alam, M.N.; Almoyad, M.; Huq, F. Polyphenols in Colorectal Cancer: Current State of Knowledge including Clinical Trials and Molecular Mechanism of Action. *BioMed Res. Int.* **2018**, *2018*. [CrossRef] [PubMed]

12. Chen, S.; Cai, D.; Pearce, K.; Sun, P.Y.; Roberts, A.C.; Glanzman, D.L. Reinstatement of long-term memory following erasure of its behavioral and synaptic expression in Aplysia. *eLife* **2014**, *3*. [CrossRef] [PubMed]

13. Surichan, S.; Arroo, R.R.; Ruparelia, K.; Tsatsakis, A.M.; Androutsopoulos, V.P. Nobiletin bioactivation in MDA-MB-468 breast cancer cells by cytochrome P450 CYP1 enzymes. *Food Chem. Toxicol.* **2018**, *113*, 228–235. [CrossRef] [PubMed]

14. Morley, K.L.; Ferguson, P.J.; Koropatnick, J. Tangeretin and nobiletin induce G1 cell cycle arrest but not apoptosis in human breast and colon cancer cells. *Cancer Lett.* **2007**, *251*, 168–178. [CrossRef] [PubMed]

15. Jiang, Y.-P.; Guo, H.; Wang, X.-B. Nobiletin (NOB) suppresses autophagic degradation via over-expressing AKT pathway and enhances apoptosis in multidrug-resistant SKOV3/TAX ovarian cancer cells. *Biomed. Pharmacother.* **2018**, *103*, 29–37. [CrossRef] [PubMed]

16. Moon, J.Y.; Cho, M.; Ahn, K.S.; Cho, S.K. Nobiletin induces apoptosis and potentiates the effects of the anticancer drug 5-fluorouracil in p53-mutated SNU-16 human gastric cancer cells. *Nutr. Cancer* **2013**, *65*, 286–295. [CrossRef] [PubMed]

17. Moon, J.Y.; Cho, S.K. Nobiletin induces protective autophagy accompanied by ER-stress mediated apoptosis in human gastric cancer SNU-16 cells. *Molecules* **2016**, *21*, 914. [CrossRef] [PubMed]

18. Uesato, S.; Yamashita, H.; Maeda, R.; Hirata, Y.; Yamamoto, M.; Matsue, S.; Nagaoka, Y.; Shibano, M.; Taniguchi, M.; Baba, K. Synergistic antitumor effect of a combination of paclitaxel and carboplatin with nobiletin from Citrus depressa on non-small-cell lung cancer cell lines. *Planta Med.* **2014**, *80*, 452–457. [CrossRef]

19. Song, M.; Wu, X.; Charoensinphon, N.; Wang, M.; Zheng, J.; Gao, Z.; Xu, F.; Li, Z.; Li, F.; Zhou, J. Dietary 5-demethylnobiletin inhibits cigarette carcinogen NNK-induced lung tumorigenesis in mice. *Food Funct.* **2017**, *8*, 954–963. [CrossRef]

20. Ma, X.; Jin, S.; Zhang, Y.; Wan, L.; Zhao, Y.; Zhou, L. Inhibitory effects of nobiletin on hepatocellular carcinoma in vitro and in vivo. *Phytother. Res.* **2014**, *28*, 560–567. [CrossRef]

21. Cheng, H.-L.; Hsieh, M.-J.; Yang, J.-S.; Lin, C.-W.; Lue, K.-H.; Lu, K.-H.; Yang, S.-F. Nobiletin inhibits human osteosarcoma cells metastasis by blocking ERK and JNK-mediated MMPs expression. *Oncotarget* **2016**, *7*, 35208–35223. [CrossRef] [PubMed]

22. Braidy, N.; Behzad, S.; Habtemariam, S.; Ahmed, T.; Daglia, M.; Mohammad Nabavi, S.; Sobarzo-Sanchez, E.; Fazel Nabavi, S. Neuroprotective effects of citrus fruit-derived flavonoids, nobiletin and tangeretin in Alzheimer's and Parkinson's disease. *CNS Neurol. Disord. Drug Targets (Former. Curr. Drug Targets CNS Neurol. Disord.)* **2017**, *16*, 387–397. [CrossRef] [PubMed]

23. Qi, G.; Guo, R.; Tian, H.; Li, L.; Liu, H.; Mi, Y.; Liu, X. Nobiletin protects against insulin resistance and disorders of lipid metabolism by reprogramming of circadian clock in hepatocytes. *Biochim. Biophys. Acta Mol. Cell Biolo. Lipids* **2018**, *1863*, 549–562. [CrossRef] [PubMed]

24. Kumar, V.; Veeranjaneyulu, A. Complications of Diabetes and Role of a Citrus Flavonoid Nobiletin in Its Treatment. In *Herbs for Diabetes and Neurological Disease Management*; Apple Academic Press: Oakville, ON, Canada, 2018; pp. 197–224.

25. Morrow, N.M.; Telford, D.E.; Sutherland, B.G.; Edwards, J.Y.; Huff, M.W. Nobiletin Corrects Intestinal Lipid Metabolism in Ldlr-/-Mice Fed a High-Fat Diet. *Atheroscler. Suppl.* **2018**, *32*, 28. [CrossRef]

26. Yuk, T.; Kim, Y.; Yang, J.; Sung, J.; Jeong, H.S.; Lee, J. Nobiletin Inhibits Hepatic Lipogenesis via Activation of AMP-Activated Protein Kinase. *Evid. Based Complement. Altern. Med.* **2018**, *2018*. [CrossRef] [PubMed]

27. Tung, Y.-C.; Li, S.; Huang, Q.; Hung, W.-L.; Ho, C.-T.; Wei, G.-J.; Pan, M.-H. 5-Demethylnobiletin and 5-Acetoxy-6, 7, 8, 3′, 4′-pentamethoxyflavone Suppress Lipid Accumulation by Activating the LKB1-AMPK Pathway in 3T3-L1 Preadipocytes and High Fat Diet-Fed C57BL/6 Mice. *J. Agric. Food Chem.* **2016**, *64*, 3196–3205. [CrossRef] [PubMed]

28. Yao, X.; Zhu, X.; Pan, S.; Fang, Y.; Jiang, F.; Phillips, G.O.; Xu, X. Antimicrobial activity of nobiletin and tangeretin against Pseudomonas. *Food Chem.* **2012**, *132*, 1883–1890. [CrossRef]

29. Onishi, S.; Nishi, K.; Yasunaga, S.; Muranaka, A.; Maeyama, K.; Kadota, A.; Sugahara, T. Nobiletin, a polymethoxy flavonoid, exerts anti-allergic effect by suppressing activation of phosphoinositide 3-kinase. *J. Funct. Foods* **2014**, *6*, 606–614. [CrossRef]

30. Lai, C.-S.; Li, S.; Chai, C.-Y.; Lo, C.-Y.; Dushenkov, S.; Ho, C.-T.; Pan, M.-H.; Wang, Y.-J. Anti-inflammatory and antitumor promotional effects of a novel urinary metabolite, 3′, 4′-didemethylnobiletin, derived from nobiletin. *Carcinogenesis* **2008**, *29*, 2415–2424. [CrossRef]

31. Narayana, J.L.; Huang, H.-N.; Wu, C.-J.; Chen, J.-Y. Epinecidin-1 antimicrobial activity: In vitro membrane lysis and In vivo efficacy against Helicobacter pylori infection in a mouse model. *Biomaterials* **2015**, *61*, 41–51. [CrossRef]

32. Eguchi, A.; Murakami, A.; Li, S.; Ho, C.T.; Ohigashi, H. Suppressive effects of demethylated metabolites of nobiletin on phorbol ester-induced expression of scavenger receptor genes in THP-1 human monocytic cells. *Biofactors* **2007**, *31*, 107–116. [CrossRef] [PubMed]

33. Eguchi, A.; Murakami, A.; Ohigashi, H. Nobiletin, a citrus flavonoid, suppresses phorbol ester-induced expression of multiple scavenger receptor genes in THP-1 human monocytic cells. *FEBS Lett.* **2006**, *580*, 3321–3328. [CrossRef] [PubMed]

34. Lee, Y.-S.; Asai, M.; Choi, S.-S.; Yonezawa, T.; Teruya, T.; Nagai, K.; Woo, J.-T.; Cha, B.-Y. Nobiletin prevents body weight gain and bone loss in ovariectomized C57BL/6J mice. *Pharmaco. Pharm.* **2014**, *5*, 959–965. [CrossRef]

35. Tominari, T.; Hirata, M.; Matsumoto, C.; Inada, M.; Miyaura, C. Polymethoxy flavonoids, nobiletin and tangeretin, prevent lipopolysaccharide-induced inflammatory bone loss in an experimental model for periodontitis. *J. Pharmacol. Sci.* **2012**, *119*, 390–394. [CrossRef] [PubMed]

36. Gao, Z.; Gao, W.; Zeng, S.-L.; Li, P.; Liu, E.-H. Chemical structures, bioactivities and molecular mechanisms of citrus polymethoxyflavones. *J. Funct. Foods* **2018**, *40*, 498–509. [CrossRef]

37. Tung, Y.-C.; Chou, Y.-C.; Hung, W.-L.; Cheng, A.-C.; Yu, R.-C.; Ho, C.-T.; Pan, M.-H. Polymethoxyflavones: Chemistry and Molecular Mechanisms for Cancer Prevention and Treatment. *Curr. Pharmacol. Rep.* **2019**, *5*, 98–113. [CrossRef]

38. Uckoo, R.M.; Jayaprakasha, G.; Vikram, A.; Patil, B.S. Polymethoxyflavones isolated from the peel of Miaray Mandarin (Citrus miaray) have biofilm inhibitory activity in Vibrio harveyi. *J. Agric. Food Chem.* **2015**, *63*, 7180–7189. [CrossRef] [PubMed]

39. Itoh, N.; Iwata, C.; Toda, H. Molecular cloning and characterization of a flavonoid-O-methyltransferase with broad substrate specificity and regioselectivity from Citrus depressa. *BMC Plant Biol.* **2016**, *16*, 180. [CrossRef]

40. Lee, Y.-H.; Charles, A.L.; Kung, H.-F.; Ho, C.-T.; Huang, T.-C. Extraction of nobiletin and tangeretin from Citrus depressa Hayata by supercritical carbon dioxide with ethanol as modifier. *Ind. Crops Prod.* **2010**, *31*, 59–64. [CrossRef]

41. Kohno, H.; Yoshitani, S.-I.; Tsukio, Y.; Murakami, A.; Koshimizu, K.; Yano, M.; Tokuda, H.; Nishino, H.; Ohigashi, H.; Tanaka, T. Dietary administration of citrus nobiletin inhibits azoxymethane-induced colonic aberrant crypt foci in rats. *Life Sci.* **2001**, *69*, 901–913. [CrossRef]

42. Murakami, A.; Nakamura, Y.; Torikai, K.; Tanaka, T.; Koshiba, T.; Koshimizu, K.; Kuwahara, S.; Takahashi, Y.; Ogawa, K.; Yano, M. Inhibitory effect of citrus nobiletin on phorbol ester-induced skin inflammation, oxidative stress, and tumor promotion in mice. *Cancer Res.* **2000**, *60*, 5059–5066. [PubMed]

43. Uckoo, R.M.; Jayaprakasha, G.K.; Patil, B.S. Rapid separation method of polymethoxyflavones from citrus using flash chromatography. *Sep. Purif. Technol.* **2011**, *81*, 151–158. [CrossRef]

44. Teruya, T.; Teruya, Y.; Sueyoshi, K.; Yamano, A.; Jitai, Y. Manufacturing method of fermentation treated products containing high-content nobiletin and tangeretin. Patent JP 2015202065, 16 November 2015.

45. Kawaii, S.; Tomono, Y.; Katase, E.; Ogawa, K.; Yano, M. HL-60 differentiating activity and flavonoid content of the readily extractable fraction prepared from Citrus juices. *J. Agric. Food Chem.* **1999**, *47*, 128–135. [CrossRef]

46. Tsukayama, M.; Ichikawa, R.; Yamamoto, K.; Sasaki, T.; Kawamura, Y. Microwave-assisted rapid extraction of polymethoxyflavones from dried peels of Citrus yuko Hort. ex Tanaka. *J. Jpn. Soc. Food Sci. Technol.* **2009**, *56*, 359–362. [CrossRef]

47. Silva, I.; Estrada, M.F.; V. Pereira, C.; da Silva, A.B.; Bronze, M.R.; Alves, P.M.; Duarte, C.M.; Brito, C.; Serra, A.T. Polymethoxylated Flavones from Orange Peels Inhibit Cell Proliferation in a 3D Cell Model of Human Colorectal Cancer. *Nutr. Cancer* **2018**, *70*, 257–266. [CrossRef] [PubMed]

48. Seo, J.W.; Jang, D.R.; Kim, Y.U. Preparation Method of Citrus Peel Extract with Increased Polymethoxyflavone Content by Supercritical Fluid Extraction. Patent KR 1838266, 14 March 2018.

49. Chiou, Y.-S.; Zheng, Y.-N.; Tsai, M.-L.; Lai, C.-S.; Ho, C.-T.; Pan, M.-H. 5-Demethylnobiletin more potently inhibits colon cancer cell growth than nobiletin in vitro and in vivo. *J. Food Bioact.* **2018**, *2*, 91–97. [CrossRef]

50. Zheng, J.; Bi, J.; Johnson, D.; Sun, Y.; Song, M.; Qiu, P.; Dong, P.; Decker, E.; Xiao, H. Analysis of 10 metabolites of polymethoxyflavones with high sensitivity by electrochemical detection in high-performance liquid chromatography. *J. Agric. Food Chem.* **2015**, *63*, 509–516. [CrossRef]

51. Zheng, J.; Song, M.; Dong, P.; Qiu, P.; Guo, S.; Zhong, Z.; Li, S.; Ho, C.T.; Xiao, H. Identification of novel bioactive metabolites of 5-demethylnobiletin in mice. *Mol. Nutr. Food Res.* **2013**, *57*, 1999–2007. [CrossRef]

52. Li, S.; Wang, Z.; Sang, S.; Huang, M.T.; Ho, C.T. Identification of nobiletin metabolites in mouse urine. *Mol. Nutr. Food Res.* **2006**, *50*, 291–299. [CrossRef]

53. Yasuda, T.; Yoshimura, Y.; Yabuki, H.; Nakazawa, T.; Ohsawa, K.; Mimaki, Y.; Sashida, Y. Urinary metabolites of nobiletin orally administered to rats. *Chem. Pharm. Bull.* **2003**, *51*, 1426–1428. [CrossRef]

54. Wu, X.; Song, M.; Wang, M.; Zheng, J.; Gao, Z.; Xu, F.; Zhang, G.; Xiao, H. Chemopreventive effects of nobiletin and its colonic metabolites on colon carcinogenesis. *Mol. Nutr. Food Res.* **2015**, *59*, 2383–2394. [CrossRef] [PubMed]

55. Wang, Z.; Li, S.; Jonca, M.; Lambros, T.; Ferguson, S.; Goodnow, R.; Ho, C.T. Comparison of supercritical fluid chromatography and liquid chromatography for the separation of urinary metabolites of nobiletin with chiral and non-chiral stationary phases. *Biomed. Chromatogr.* **2006**, *20*, 1206–1215. [CrossRef] [PubMed]

56. Koga, N.; Ohta, C.; Kato, Y.; Haraguchi, K.; Endo, T.; Ogawa, K.; Ohta, H.; Yano, M. In vitro metabolism of nobiletin, a polymethoxy-flavonoid, by human liver microsomes and cytochrome P450. *Xenobiotica* **2011**, *41*, 927–933. [CrossRef] [PubMed]

57. Wang, M. Biotransformation of Polymethoxyflavones and Its Implication on Biological Activities. Ph.D. Thesis, University of Massachusetts, Amherst, MA, USA, 2017.

58. Xu, L.; He, Y.; Guo, X.; Lu, Y.; Wang, C.; Wang, Z. Identification of metabolites of nobiletin in rats using ultra-performance liquid chromatography coupled with triple-quadrupole mass spectrometry. *Yao Xue Xue Bao (Acta Pharm. Sin.)* **2011**, *46*, 1483–1487.

59. Manthey, J.A.; Bendele, P. Anti-inflammatory activity of an orange peel polymethoxylated flavone, 3′, 4′, 3, 5, 6, 7, 8-heptamethoxyflavone, in the rat carrageenan/paw edema and mouse lipopolysaccharide-challenge assays. *J. Agric. Food Chem.* **2008**, *56*, 9399–9403. [CrossRef] [PubMed]

60. Kemperman, R.A.; Bolca, S.; Roger, L.C.; Vaughan, E.E. Novel approaches for analysing gut microbes and dietary polyphenols: Challenges and opportunities. *Microbiology* **2010**, *156*, 3224–3231. [CrossRef] [PubMed]

61. Ma, C. Biotransformation of Polymethoxyflavones by Gut Microbiome and Molecular Characterization of Polymethoxyflavones by Surface Enhanced Raman Spectroscopy. Ph.D. Thesis, University of Massachusetts, Amherst, MA, USA, 2015.

62. Li, S.; Sang, S.; Pan, M.-H.; Lai, C.-S.; Lo, C.-Y.; Yang, C.S.; Ho, C.-T. Anti-inflammatory property of the urinary metabolites of nobiletin in mouse. *Bioorg. Med. Chem. Lett.* **2007**, *17*, 5177–5181. [CrossRef] [PubMed]

63. Qiu, P.; Dong, P.; Guan, H.; Li, S.; Ho, C.T.; Pan, M.H.; McClements, D.J.; Xiao, H. Inhibitory effects of 5-hydroxy polymethoxyflavones on colon cancer cells. *Mol. Nutr. Food Res.* **2010**, *54*, S244–S252. [CrossRef] [PubMed]

64. Fearon, E.R.; Vogelstein, B. A genetic model for colorectal tumorigenesis. *Cell* **1990**, *61*, 759–767. [CrossRef]

65. Ciardiello, F.; Tortora, G. EGFR antagonists in cancer treatment. *N. Engl. J. Med.* **2008**, *358*, 1160–1174. [CrossRef] [PubMed]

66. Shi, Y.; Massagué, J. Mechanisms of TGF-β signaling from cell membrane to the nucleus. *Cell* **2003**, *113*, 685–700. [CrossRef]

67. Neuzillet, C.; Tijeras-Raballand, A.; Cohen, R.; Cros, J.; Faivre, S.; Raymond, E.; de Gramont, A. Targeting the TGFβ pathway for cancer therapy. *Pharmacol. Ther.* **2015**, *147*, 22–31. [CrossRef] [PubMed]

68. Said, A.H.; Raufman, J.-P.; Xie, G. The role of matrix metalloproteinases in colorectal cancer. *Cancers* **2014**, *6*, 366–375. [CrossRef] [PubMed]

69. Maeda, K.; Kang, S.-M.; Sawada, T.; Nishiguchi, Y.; Yashiro, M.; Ogawa, Y.; Ohira, M.; Ishikawa, T.; Hirakawa-YS Chung, K. Expression of intercellular adhesion molecule-1 and prognosis in colorectal cancer. *Oncol. Rep.* **2002**, *9*, 511–514. [CrossRef] [PubMed]

70. Manthey, J.A.; Guthrie, N. Antiproliferative activities of citrus flavonoids against six human cancer cell lines. *J. Agric. Food Chem.* **2002**, *50*, 5837–5843. [CrossRef] [PubMed]

71. Parang, B.; Barrett, C.W.; Williams, C.S. AOM/DSS Model of Colitis-Associated Cancer. In *Gastrointestinal Physiology and Diseases*; Springer: Berlin/Heidelberg, Germany, 2016; pp. 297–307.

72. Ito, N.; Hasegawa, R.; Sano, M.; Tamano, S.; Esumi, H.; Takayama, S.; Sugimura, T. A new colon and mammary carcinogen in cooked food, 2-amino-1-methyl-6-phenylimidazo [4, 5-b] pyridine (PhIP). *Carcinogenesis* **1991**, *12*, 1503–1506. [CrossRef] [PubMed]

73. Nakagama, H.; Nakanishi, M.; Ochiai, M. Modeling human colon cancer in rodents using a food-borne carcinogen, PhIP. *Cancer Sci.* **2005**, *96*, 627–636. [CrossRef]

74. Suzuki, R.; Kohno, H.; Murakami, A.; Koshimizu, K.; Ohigashi, H.; Yano, M.; Tokuda, H.; Nishino, H.; Tanaka, T. Citrus nobiletin inhibits azoxymethane-induced large bowel carcinogenesis in rats. *Biofactors* **2004**, *21*, 111–114. [CrossRef]

75. Tang, M.X.; Ogawa, K.; Asamoto, M.; Chewonarin, T.; Suzuki, S.; Tanaka, T.; Shirai, T. Effects of nobiletin on PhIP-induced prostate and colon carcinogenesis in F344 rats. *Nutr. Cancer* **2011**, *63*, 227–233. [CrossRef]

76. Wu, X.; Song, M.; Gao, Z.; Sun, Y.; Wang, M.; Li, F.; Zheng, J.; Xiao, H. Nobiletin and its colonic metabolites suppress colitis-associated colon carcinogenesis by down-regulating iNOS, inducing antioxidative enzymes and arresting cell cycle progression. *J. Nutr. Biochem.* **2017**, *42*, 17–25. [CrossRef]

77. Kawabata, K.; Murakami, A.; Ohigashi, H. Nobiletin, a citrus flavonoid, down-regulates matrix metalloproteinase-7 (matrilysin) expression in HT-29 human colorectal cancer cells. *Biosci Biotechnol. Biochem.* **2005**, *69*, 307–314. [CrossRef] [PubMed]

78. Song, M.; Wu, X.; Zheng, J.; Xiao, H. 5-Demethylnobiletin inhibits colon carcinogenesis in azoxymethane/dextran sulfate sodium-treated mice (123.3). *FASEB J.* **2014**, *28*, 123.3.

79. Zheng, Q.; Hirose, Y.; Yoshimi, N.; Murakami, A.; Koshimizu, K.; Ohigashi, H.; Sakata, K.; Matsumoto, Y.; Sayama, Y.; Mori, H. Further investigation of the modifying effect of various chemopreventive agents on apoptosis and cell proliferation in human colon cancer cells. *J. Cancer Res. Clin. Oncol.* **2002**, *128*, 539–546. [CrossRef] [PubMed]

80. Qiu, P.; Guan, H.; Dong, P.; Li, S.; Ho, C.T.; Pan, M.H.; McClements, D.J.; Xiao, H. The p53-, Bax-and p21-dependent inhibition of colon cancer cell growth by 5-hydroxy polymethoxyflavones. *Mol. Nutr. Food Res.* **2011**, *55*, 613–622. [CrossRef] [PubMed]

81. Lin, N.; Sato, T.; Takayama, Y.; Mimaki, Y.; Sashida, Y.; Yano, M.; Ito, A. Novel anti-inflammatory actions of nobiletin, a citrus polymethoxy flavonoid, on human synovial fibroblasts and mouse macrophages. *Biochem. Pharmacol.* **2003**, *65*, 2065–2071. [CrossRef]

82. Yasunaga, S.; Domen, M.; Nishi, K.; Kadota, A.; Sugahara, T. Nobiletin suppresses monocyte chemoattractant protein 1 (MCP 1) expression by regulating MAPK signaling in 3T3-L1 cells. *J. Funct. Foods* **2016**, *27*, 406–415. [CrossRef]

83. Miyamoto, S.; Yasui, Y.; Ohigashi, H.; Tanaka, T.; Murakami, A. Dietary flavonoids suppress azoxymethane-induced colonic preneoplastic lesions in male C57BL/KsJ-db/db mice. *Chem. Biol. Interact.* **2010**, *183*, 276–283. [CrossRef]

84. Miyamoto, S.; Yasui, Y.; Tanaka, T.; Ohigashi, H.; Murakami, A. Suppressive effects of nobiletin on hyperleptinemia and colitis-related colon carcinogenesis in male ICR mice. *Carcinogenesis* **2008**, *29*, 1057–1063. [CrossRef]

85. Armaghany, T.; Wilson, J.D.; Chu, Q.; Mills, G. Genetic alterations in colorectal cancer. *Gastrointest. Cancer Res.* **2012**, *5*, 19.

86. Owa, T.; Yoshino, H.; Yoshimatsu, K.; Nagasu, T. Cell cycle regulation in the G1 phase: A promising target for the development of new chemotherapeutic anticancer agents. *Curr. Med. Chem.* **2001**, *8*, 1487–1503. [CrossRef]

87. Johnson, D.; Walker, C. Cyclins and cell cycle checkpoints. *Ann. Rev. Pharmacol. Toxicol.* **1999**, *39*, 295–312. [CrossRef] [PubMed]

88. Sherr, C.J. Cancer cell cycles. *Science* **1996**, *274*, 1672–1677. [CrossRef] [PubMed]

89. Malumbres, M.; Barbacid, M. Cell cycle, CDKs and cancer: A changing paradigm. *Nat. Rev. Cancer* **2009**, *9*, 153. [CrossRef] [PubMed]

90. Kurki, P.; Vanderlaan, M.; Dolbeare, F.; Gray, J.; Tan, E. Expression of proliferating cell nuclear antigen (PCNA)/cyclin during the cell cycle. *Exp. Cell Res.* **1986**, *166*, 209–219. [CrossRef]

91. McKay, J.A.; Douglas, J.J.; Ross, V.G.; Curran, S.; Loane, J.F.; Ahmed, F.Y.; Cassidy, J.; McLeod, H.L.; Murray, G.I. Analysis of key cell-cycle checkpoint proteins in colorectal tumours. *J. Pathol. J. Pathol. Soc. Great Br. Irel.* **2002**, *196*, 386–393. [CrossRef]

92. Kroker, A.J.; Bruning, J.B. p21 exploits residue Tyr151 as a tether for high-affinity PCNA binding. *Biochemistry* **2015**, *54*, 3483–3493. [CrossRef] [PubMed]

93. Soria, G.; Gottifredi, V. PCNA-coupled p21 degradation after DNA damage: The exception that confirms the rule? *DNA Repair* **2010**, *9*, 358–364. [CrossRef]

94. Morgan, D.O. Cyclin-dependent kinases: Engines, clocks, and microprocessors. *Annu. Rev. Cell Dev. Biol.* **1997**, *13*, 261–291. [CrossRef]

95. Karimian, A.; Ahmadi, Y.; Yousefi, B. Multiple functions of p21 in cell cycle, apoptosis and transcriptional regulation after DNA damage. *DNA Repair* **2016**, *42*, 63–71. [CrossRef]

96. Bertoli, C.; Skotheim, J.M.; De Bruin, R.A. Control of cell cycle transcription during G1 and S phases. *Nat. Rev. Mol. Cell Biol.* **2013**, *14*, 518–528. [CrossRef]

97. Taylor, W.R.; Stark, G.R. Regulation of the G2/M transition by p53. *Oncogene* **2001**, *20*, 1803–1815. [CrossRef] [PubMed]

98. Borgne, A.; Meijer, L. Sequential dephosphorylation of p34cdc2 on Thr-14 and Tyr-15 at the prophase/metaphase transition. *J. Biol. Chem.* **1996**, *271*, 27847–27854. [CrossRef] [PubMed]

99. Ouyang, L.; Shi, Z.; Zhao, S.; Wang, F.T.; Zhou, T.T.; Liu, B.; Bao, J.K. Programmed cell death pathways in cancer: A review of apoptosis, autophagy and programmed necrosis. *Cell Prolif.* **2012**, *45*, 487–498. [CrossRef] [PubMed]

100. Zong, W.-X.; Ditsworth, D.; Bauer, D.E.; Wang, Z.-Q.; Thompson, C.B. Alkylating DNA damage stimulates a regulated form of necrotic cell death. *Genes Dev.* **2004**, *18*, 1272–1282. [CrossRef] [PubMed]

101. Sayers, T.J. Targeting the extrinsic apoptosis signaling pathway for cancer therapy. *Cancer Immunol. Immunother* **2011**, *60*, 1173–1180. [CrossRef]

102. Wang, X. The expanding role of mitochondria in apoptosis. *Genes Dev.* **2001**, *15*, 2922–2933. [PubMed]

103. Llambi, F.; Green, D.R. Apoptosis and oncogenesis: Give and take in the BCL-2 family. *Curr. Opin. Genet. Dev.* **2011**, *21*, 12–20. [CrossRef]

104. Engel, T.; Henshall, D.C. Apoptosis, Bcl-2 family proteins and caspases: The ABCs of seizure-damage and epileptogenesis? *Int. J. Physiol. Pathophysiol. Pharmacol.* **2009**, *1*, 97–115.

105. Chan, C.K.; Supriady, H.; Goh, B.H.; Kadir, H.A. Elephantopus scaber induces apoptosis through ROS-dependent mitochondrial signaling pathway in HCT116 human colorectal carcinoma cells. *J. Ethnopharmacol.* **2015**, *168*, 291–304. [CrossRef]

106. Nuñez, G.; Benedict, M.A.; Hu, Y.; Inohara, N. Caspases: The proteases of the apoptotic pathway. *Oncogene* **1998**, *17*, 3237–3245. [CrossRef]

107. Fernandes-Alnemri, T.; Litwack, G.; Alnemri, E.S. CPP32, a novel human apoptotic protein with homology to Caenorhabditis elegans cell death protein Ced-3 and mammalian interleukin-1 beta-converting enzyme. *J. Biol. Chem.* **1994**, *269*, 30761–30764. [PubMed]

108. Kerr, J.F.; Wyllie, A.H.; Currie, A.R. Apoptosis: A basic biological phenomenon with wideranging implications in tissue kinetics. *Br. J. Cancer* **1972**, *26*, 239. [CrossRef] [PubMed]

109. Oliver, F.J.; de la Rubia, G.; Rolli, V.; Ruiz-Ruiz, M.C.; de Murcia, G.; Ménissier-de Murcia, J. Importance of poly (ADP-ribose) polymerase and its cleavage in apoptosis Lesson from an uncleavable mutant. *J. Biol. Chem.* **1998**, *273*, 33533–33539. [CrossRef] [PubMed]

110. Saraste, A.; Pulkki, K. Morphologic and biochemical hallmarks of apoptosis. *Cardiovasc. Res.* **2000**, *45*, 528–537. [CrossRef]

111. Martin, S.J.; Green, D.R. Protease activation during apoptosis: Death by a thousand cuts? *Cell* **1995**, *82*, 349–352. [CrossRef]

112. Lee, W.-R.; Shen, S.-C.; Lin, H.-Y.; Hou, W.-C.; Yang, L.-L.; Chen, Y.-C. Wogonin and fisetin induce apoptosis in human promyeloleukemic cells, accompanied by a decrease of reactive oxygen species, and activation of caspase 3 and Ca2+-dependent endonuclease. *Biochem. Pharmacol.* **2002**, *63*, 225–236. [CrossRef]

113. Fletcher, J.I.; Huang, D.C. Controlling the cell death mediators Bax and Bak: Puzzles and conundrums. *Cell Cycle* **2008**, *7*, 39–44. [CrossRef]

114. Wei, M.C.; Zong, W.-X.; Cheng, E.H.-Y.; Lindsten, T.; Panoutsakopoulou, V.; Ross, A.J.; Roth, K.A.; MacGregor, G.R.; Thompson, C.B.; Korsmeyer, S.J. Proapoptotic BAX and BAK: A requisite gateway to mitochondrial dysfunction and death. *Science* **2001**, *292*, 727–730. [CrossRef]

115. Wang, S.-Y.; Yu, Q.-J.; Zhang, R.-D.; Liu, B. Core signaling pathways of survival/death in autophagy-related cancer networks. *Int. J. Biochem. Cell Biol.* **2011**, *43*, 1263–1266. [CrossRef]

116. Kundu, M.; Thompson, C.B. Autophagy: Basic principles and relevance to disease. *Annu. Rev. Pathmechdis Mech. Dis.* **2008**, *3*, 427–455. [CrossRef]

117. Eum, K.-H.; Lee, M. Crosstalk between autophagy and apoptosis in the regulation of paclitaxel-induced cell death in v-Ha-ras-transformed fibroblasts. *Mol. Cell. Biochem.* **2011**, *348*, 61–68. [CrossRef] [PubMed]

118. Amelio, I.; Melino, G.; Knight, R.A. Cell death pathology: Cross-talk with autophagy and its clinical implications. *Biochem. Biophys. Res. Commun.* **2011**, *414*, 277–281. [CrossRef] [PubMed]

119. Terzić, J.; Grivennikov, S.; Karin, E.; Karin, M. Inflammation and colon cancer. *Gastroenterology* **2010**, *138*, 2101–2114.e5. [CrossRef] [PubMed]

120. Klampfer, L. Cytokines, inflammation and colon cancer. *Curr. Cancer Drug Targets* **2011**, *11*, 451–464. [CrossRef] [PubMed]

121. Meira, L.B.; Bugni, J.M.; Green, S.L.; Lee, C.-W.; Pang, B.; Borenshtein, D.; Rickman, B.H.; Rogers, A.B.; Moroski-Erkul, C.A.; McFaline, J.L. DNA damage induced by chronic inflammation contributes to colon carcinogenesis in mice. *J. Clin. Investig.* **2008**, *118*, 2516–2525. [CrossRef] [PubMed]

122. Westbrook, A.M.; Wei, B.; Braun, J.; Schiestl, R.H. Intestinal mucosal inflammation leads to systemic genotoxicity in mice. *Cancer Res.* **2009**, *69*, 4827–4834. [CrossRef]

123. Kraus, S.; Arber, N. Inflammation and colorectal cancer. *Curr. Opin. Pharmacol.* **2009**, *9*, 405–410. [CrossRef]

124. Itzkowitz, S.H.; Yio, X. Inflammation and cancer IV. Colorectal cancer in inflammatory bowel disease: The role of inflammation. *Am. J. Physiol. Gastrointest. Liver Physiol.* **2004**, *287*, G7–G17. [CrossRef]

125. Takahashi, M.; Wakabayashi, K. Gene mutations and altered gene expression in azoxymethane-induced colon carcinogenesis in rodents. *Cancer Sci.* **2004**, *95*, 475–480. [CrossRef]

126. Xiong, Y.; Chen, D.; Yu, C.; Lv, B.; Peng, J.; Wang, J.; Lin, Y. Citrus nobiletin ameliorates experimental colitis by reducing inflammation and restoring impaired intestinal barrier function. *Mol. Nutr. Food Res.* **2015**, *59*, 829–842. [CrossRef]

127. Kaidi, A.; Qualtrough, D.; Williams, A.C.; Paraskeva, C. Direct transcriptional up-regulation of cyclooxygenase-2 by hypoxia-inducible factor (HIF)-1 promotes colorectal tumor cell survival and enhances HIF-1 transcriptional activity during hypoxia. *Cancer Res.* **2006**, *66*, 6683–6691. [CrossRef] [PubMed]

128. Goodwin, J. Are prostaglandins proinflammatory, antiinflammatory, both or neither? *J. Rheumatol. Suppl.* **1991**, *28*, 26–29. [PubMed]

129. Surh, Y.-J.; Chun, K.-S.; Cha, H.-H.; Han, S.S.; Keum, Y.-S.; Park, K.-K.; Lee, S.S. Molecular mechanisms underlying chemopreventive activities of anti-inflammatory phytochemicals: Down-regulation of COX-2 and iNOS through suppression of NF-κB activation. *Mutat. Res./Fundam. Mol. Mech. Mutagen.* **2001**, *480*, 243–268. [CrossRef]

130. Rao, C.V. Nitric oxide signaling in colon cancer chemoprevention. *Mutat. Res./Fundam. Mol. Mech. Mutagen.* **2004**, *555*, 107–119. [CrossRef] [PubMed]

131. Rushworth, S.A.; MacEwan, D.J.; O'Connell, M.A. Lipopolysaccharide-induced expression of NAD (P) H: Quinone oxidoreductase 1 and heme oxygenase-1 protects against excessive inflammatory responses in human monocytes. *J. Immunol.* **2008**, *181*, 6730–6737. [CrossRef]

132. Khor, T.O.; Huang, M.-T.; Kwon, K.H.; Chan, J.Y.; Reddy, B.S.; Kong, A.-N. Nrf2-deficient mice have an increased susceptibility to dextran sulfate sodium–induced colitis. *Cancer Res.* **2006**, *66*, 11580–11584. [CrossRef]

133. Klaunig, J.E.; Kamendulis, L.M.; Hocevar, B.A. Oxidative stress and oxidative damage in carcinogenesis. *Toxicol. Pathol.* **2010**, *38*, 96–109. [CrossRef]

134. Kwak, M.-K.; Kensler, T.W. Targeting NRF2 signaling for cancer chemoprevention. *Toxicol. Appl. Pharmacol.* **2010**, *244*, 66–76. [CrossRef]

135. Nishida, N.; Yano, H.; Nishida, T.; Kamura, T.; Kojiro, M. Angiogenesis in cancer. *Vasc. Health Risk Manag.* **2006**, *2*, 213–219. [CrossRef]

136. Prager, G.; Poettler, M. Angiogenesis in cancer. *Hämostaseologie* **2012**, *32*, 105–114.
137. Park, M.H.; Hong, J.T. Roles of NF-κB in cancer and inflammatory diseases and their therapeutic approaches. *Cells* **2016**, *5*, 15. [CrossRef]
138. Ferrara, N.; Gerber, H.-P.; LeCouter, J. The biology of VEGF and its receptors. *Nat. Med.* **2003**, *9*, 669. [CrossRef] [PubMed]
139. Carmeliet, P. VEGF as a key mediator of angiogenesis in cancer. *Oncology* **2005**, *69*, 4–10. [CrossRef] [PubMed]
140. Berra, E.; Pagès, G.; Pouysségur, J. MAP kinases and hypoxia in the control of VEGF expression. *Cancer Metastasis Rev.* **2000**, *19*, 139–145. [CrossRef] [PubMed]
141. Saxena, N.K.; Taliaferro-Smith, L.; Knight, B.B.; Merlin, D.; Anania, F.A.; O'Regan, R.M.; Sharma, D. Bidirectional crosstalk between leptin and insulin-like growth factor-I signaling promotes invasion and migration of breast cancer cells via transactivation of epidermal growth factor receptor. *Cancer Res.* **2008**, *68*, 9712–9722. [CrossRef] [PubMed]
142. Fenton, J.I.; Hord, N.G.; Lavigne, J.A.; Perkins, S.N.; Hursting, S.D. Leptin, insulin-like growth factor-1, and insulin-like growth factor-2 are mitogens in ApcMin/+ but not Apc+/+ colonic epithelial cell lines. *Cancer Epidemiol. Prev. Biomark.* **2005**, *14*, 1646–1652. [CrossRef] [PubMed]
143. Rouet-Benzineb, P.; Aparicio, T.; Guilmeau, S.; Pouzet, C.; Descatoire, V.; Buyse, M.; Bado, A. Leptin counteracts sodium butyrate-induced apoptosis in human colon cancer HT-29 cells via NF-κB signaling. *J. Biol. Chem.* **2004**, *279*, 16495–16502. [CrossRef] [PubMed]
144. Miyata, Y.; Sato, T.; Yano, M.; Ito, A. Activation of protein kinase C βII/ε-c-Jun NH2-terminal kinase pathway and inhibition of mitogen-activated protein/extracellular signal-regulated kinase 1/2 phosphorylation in antitumor invasive activity induced by the polymethoxy flavonoid, nobiletin. *Mol. Cancer Ther.* **2004**, *3*, 839–847. [PubMed]
145. Miyata, Y.; Sato, T.; Imada, K.; Dobashi, A.; Yano, M.; Ito, A. A citrus polymethoxyflavonoid, nobiletin, is a novel MEK inhibitor that exhibits antitumor metastasis in human fibrosarcoma HT-1080 cells. *Biochem. Biophys. Res. Commun.* **2008**, *366*, 168–173. [CrossRef]
146. Fong, Y. Surgical therapy of hepatic colorectal metastasis. *CA Cancer J. Clin.* **1999**, *49*, 231–255. [CrossRef]
147. Kim, Y.-S.; Kim, S.-H.; Kang, J.-G.; Ko, J.-H. Expression level and glycan dynamics determine the net effects of TIMP-1 on cancer progression. *BMB Rep.* **2012**, *45*, 623–628. [CrossRef] [PubMed]
148. Chambers, A.F.; Matrisian, L.M. Changing views of the role of matrix metalloproteinases in metastasis. *J. Natl. Cancer Inst.* **1997**, *89*, 1260–1270. [CrossRef] [PubMed]
149. Waas, E.; Wobbes, T.; Lomme, R.; DeGroot, J.; Ruers, T.; Hendriks, T. Matrix metalloproteinase 2 and 9 activity in patients with colorectal cancer liver metastasis. *Br. J. Surg.* **2003**, *90*, 1556–1564. [CrossRef] [PubMed]
150. Zucker, S.; Vacirca, J. Role of matrix metalloproteinases (MMPs) in colorectal cancer. *Cancer Metastasis Rev.* **2004**, *23*, 101–117. [CrossRef] [PubMed]
151. Brabletz, T.; Jung, A.; Dag, S.; Hlubek, F.; Kirchner, T. β-Catenin regulates the expression of the matrix metalloproteinase-7 in human colorectal cancer. *Am. J. Pathol.* **1999**, *155*, 1033–1038. [CrossRef]
152. Crawford, H.C.; Fingleton, B.M.; Rudolph-Owen, L.A.; Goss, K.J.H.; Rubinfeld, B.; Polakis, P.; Matrisian, L.M. The metalloproteinase matrilysin is a target of β-catenin transactivation in intestinal tumors. *Oncogene* **1999**, *18*, 2883–2891. [CrossRef]
153. Egeblad, M.; Werb, Z. New functions for the matrix metalloproteinases in cancer progression. *Nat. Rev. Cancer* **2002**, *2*, 161–174. [CrossRef]
154. Bingle, á.; Brown, N.; Lewis, C. The role of tumour-associated macrophages in tumour progression: Implications for new anticancer therapies. *J. Pathol. J. Pathol. Soc. Great Br. Irel.* **2002**, *196*, 254–265. [CrossRef]
155. Kerkelä, E.; Ala-aho, R.; Klemi, P.; Grénman, S.; Shapiro, S.D.; Kähäri, V.M.; Saarialho-Kere, U. Metalloelastase (MMP-12) expression by tumour cells in squamous cell carcinoma of the vulva correlates with invasiveness, while that by macrophages predicts better outcome. *J. Pathol.* **2002**, *198*, 258–269. [CrossRef]
156. Li, S.; Pan, M.-H.; Lo, C.-Y.; Tan, D.; Wang, Y.; Shahidi, F.; Ho, C.-T. Chemistry and health effects of polymethoxyflavones and hydroxylated polymethoxyflavones. *J. Funct. Foods* **2009**, *1*, 2–12. [CrossRef]
157. Scholz; Williamson. Interactions affecting the bioavailability of dietary polyphenols in vivo. *Int. J. Vitam. Nutr. Res.* **2007**, *77*, 224–235. [CrossRef] [PubMed]
158. Murakami, A.; Kuwahara, S.; Takahashi, Y.; Ito, C.; Furukawa, H.; Ju-ichi, M.; Koshimizu, K.; OHIGASHI, H. In vitro absorption and metabolism of nobiletin, a chemopreventive polymethoxyflavonoid in citrus fruits. *Biosci. Biotechnol. Biochem.* **2001**, *65*, 194–197. [CrossRef] [PubMed]

159. Kansy, M.; Senner, F.; Gubernator, K. Physicochemical high throughput screening: Parallel artificial membrane permeation assay in the description of passive absorption processes. *J. Med. Chem.* **1998**, *41*, 1007–1010. [CrossRef] [PubMed]

160. Wen, X.; Walle, T. Methylated flavonoids have greatly improved intestinal absorption and metabolic stability. *Drug Metab. Dispos.* **2006**, *34*, 1786–1792. [CrossRef] [PubMed]

161. Van de Waterbeemd, H. Physico-Chemical Approaches to Drug Absorption. In *Drug Bioavailability: Estimation of Solubility, Permeability, Absorption and Bioavailability*; Wiley: Hoboken, NJ, USA, 2003; pp. 3–20.

162. Murakami, A.; Koshimizu, K.; Ohigashi, H.; Kuwahara, S.; Kuki, W.; Takahashi, Y.; Hosotani, K.; Kawahara, S.; Matsuoka, Y. Characteristic rat tissue accumulation of nobiletin, a chemopreventive polymethoxyflavonoid, in comparison with luteolin. *Biofactors* **2002**, *16*, 73–82. [CrossRef] [PubMed]

163. Wang, M.; Zheng, J.; Zhong, Z.; Song, M.; Wu, X. *Tissue Distribution of Nobiletin and Its Metabolites in Mice after Oral Administration of Nobiletin*; Federation of American Societies for Experimental Biology: Bethesda, MD, USA, 2013.

164. Wu, X.; Song, M.; Qiu, P.; Rakariyatham, K.; Li, F.; Gao, Z.; Cai, X.; Wang, M.; Xu, F.; Zheng, J. Synergistic chemopreventive effects of nobiletin and atorvastatin on colon carcinogenesis. *Carcinogenesis* **2017**, *38*, 455–464. [CrossRef] [PubMed]

165. Kumar, A.; Devaraj, V.; Giri, K.C.; Giri, S.; Rajagopal, S.; Mullangi, R. Development and validation of a highly sensitive LC-MS/MS-ESI method for the determination of nobiletin in rat plasma: Application to a pharmacokinetic study. *Biomed. Chromatogr.* **2012**, *26*, 1464–1471. [CrossRef] [PubMed]

166. Singh, S.P.; Tewari, D.; Patel, K.; Jain, G.K. Permeability determination and pharmacokinetic study of nobiletin in rat plasma and brain by validated high-performance liquid chromatography method. *Fitoterapia* **2011**, *82*, 1206–1214. [CrossRef] [PubMed]

167. Manthey, J.A.; Cesar, T.B.; Jackson, E.; Mertens-Talcott, S. Pharmacokinetic Study of Nobiletin and Tangeretin in Rat Serum by High-Performance Liquid Chromatography—Electrospray Ionization—Mass Spectrometry. *J. Agric. Food Chem.* **2011**, *59*, 145–151. [CrossRef]

168. McClements, D.J. Emulsion design to improve the delivery of functional lipophilic components. *Annu. Rev. Food Sci. Technol.* **2010**, *1*, 241–269. [CrossRef] [PubMed]

169. Yang, Y.; Zhao, C.; Chen, J.; Tian, G.; McClements, D.J.; Xiao, H.; Zheng, J. Encapsulation of polymethoxyflavones in citrus oil emulsion-based delivery systems. *J. Agric. Food Chem.* **2017**, *65*, 1732–1739. [CrossRef] [PubMed]

170. Yao, J.; Lu, Y.; Zhou, J.P. Preparation of nobiletin in self-microemulsifying systems and its intestinal permeability in rats. *J. Pharm. Pharm. Sci.* **2008**, *11*, 22–29. [CrossRef] [PubMed]

171. Lin, W.; Yao, J.; Zhou, J. Preparation of self-assemble nobiletin proliposomes and its pharmacokinetics in rats. *Yao Xue Xue Bao (Acta Pharm. Sin.)* **2009**, *44*, 192–196.

172. Chen, H.; An, Y.; Yan, X.; McClements, D.J.; Li, B.; Li, Y. Designing self-nanoemulsifying delivery systems to enhance bioaccessibility of hydrophobic bioactives (nobiletin): Influence of hydroxypropyl methylcellulose and thermal processing. *Food Hydrocoll.* **2015**, *51*, 395–404. [CrossRef]

173. Lei, L.; Zhang, Y.; He, L.; Wu, S.; Li, B.; Li, Y. Fabrication of nanoemulsion-filled alginate hydrogel to control the digestion behavior of hydrophobic nobiletin. *LWT Food Sci. Technol.* **2017**, *82*, 260–267. [CrossRef]

174. Onoue, S.; Uchida, A.; Takahashi, H.; Seto, Y.; Kawabata, Y.; Ogawa, K.; Yuminoki, K.; Hashimoto, N.; Yamada, S. Development of high-energy amorphous solid dispersion of nanosized nobiletin, a citrus polymethoxylated flavone, with improved oral bioavailability. *J. Pharm. Sci.* **2011**, *100*, 3793–3801. [CrossRef] [PubMed]

175. Jordan, S.; Murty, M.; Pilon, K. Products containing bitter orange or synephrine: Suspected cardiovascular adverse reactions. *Can. Med. Assoc. J.* **2004**, *171*, 993–994.

176. Yang, G.; Li, S.; Long, T.; Yang, Y.; Li, Y. Application of Polymethoxylflavone in Preparation of Prevention Drug for Cardiovascular Inflammation. Patent CN 107281179, 24 October 2017.

177. Wu, X.; Zheng, D.; Qin, Y.; Liu, Z.; Zhu, X. Application of Nobiletin in Medicine for Preventing or Treating Heart Failure. Patent CN 106924241, 7 July 2017.

178. Morimoto, T.; Hasegawa, K.; Murakami, A.; Fukuda, H.; Takahashi, K. Cardiac Disease Treatment Agents Containing Nobiletin. Patent JP 2011037798, 24 February 2011.

179. Caramelli, G. Product with Blood Lipid-Lowering Activity. Patent IT 2008RM0232, 2 August 2008.

180. Ohizumi, Y.; Kajima, K.; Maruyama, K.; Ishibashi, M. Pharmaceutical Composition and Food Containing Citrus Butanol Extract for Preventing and/or Treating Central Nervous System Disease. Patent WO 2017208869, 7 December 2017.
181. Ohizumi, Y.; Kajima, K.; Maruyama, K. Pharmaceutical and food composition containing Anredera cordifolia and nobiletin. Patent JP 6238089, 29 November 2017.
182. Jeon, M.R.; L, S.A.; Yoon, G.J.; Park, J.H. Composition for Preventing or Treating Neurodegenerative Disease Comprising Nobiletin as Active Ingredient. Patent KR 2017090073, 25 September 2017.
183. Wu, X.; Mei, Z.; Zheng, D.; Liu, Z.; Zhu, X.; Zhou, Y.; Zeng, L.; Liang, Z. Application of Nobiletin in Preparation or Screening of Diabetic Cardiomyopathy Drug. Patent CN 108403684, 17 August 2018.
184. Guthrie, N. Compositions Comprising at Least one Polymethoxyflavone, Flavonoid, Liminoid, and/or Tocotrienol Useful in Combination Therapies for Treating Diabetes. Patent WO 2014203059, 24 December 2014.
185. Kim, T.J.; Kim, H.G.; Kwon, Y.I.; Lee, J.U. Obesity inhibiting Composition Comprising Powder of Citrus Grandis Cultivated by Eco Friendly Method as Active Ingredient. Patent KR 2016111554, 27 September 2016.
186. Miyaura, C.; Inada, M. Preventive or Therapeutic Compositions Containing Heptamethoxyflavone for Bone Diseases. Patent JP 2012232916, 22 October 2012.
187. Liao, X. Manufacture Method of Chinese Medicine Composition for Treatment of Halitosis. Patent CN 105434729, 30 March 2016.
188. Wang, L.; Tian, A.; Li, S.; Chen, J.; Li, B. Mouth Smell-Improving Agent and Its Preparation Method. Patent CN 103893334, 2 July 2014.
189. Huang, R.L.; Hsu, S.W. Polymethoxylated Flavone for Manufacturing Drugs Against Hepatitis-B with Drug Resistance. Patent TW I535439, 1 June 2016.
190. Kim, D.H.; Han, M.J.; Cho, E.H.; Kim, Y.R. Natural Products for Treating Cancer and HIV-Related diseases. Patent KR 2012011169, 7 February 2012.
191. Zhang, T.; Liao, M.; Gong, S.; Xie, X.; Sun, W.; Wang, L.; Zheng, Y. Application of Total Flavonoid Extract from Citrus Aurantium in Manufacturing Medicines for Treating Asthma. Patent CN 102935131, 20 February 2013.
192. Li, K. Application of Nobiletin in Medicine for Treating Allergic Asthma. Patent CN 102552242, 11 July 2012.
193. Sugawara, T.; Kadota, A.; Kikuchi, T. Antiallergic Oral Composition Containing β-Lactoglobulin and Nobiletin. Patent JP 2015036369, 23 February 2015.
194. Seo, J.W.; Choi, B.G.; Cheng, J.H.; Cho, M.J. Citrus Pericarp Extracts for Preventing Hair Loss and Promoting Hair Growth. Patent KR 1651833, 19 September 2016.
195. Ito, Y.; Hikiyama, E.; Yamada, S.; Woo, J.-T.; Teruya, Y.; Sugaya, K.; Nishijima, S.; Wakuda, H.; Shinozuka, K. Medicinal Composition for Preventing or Improving Dysuria, Antagonist Against Dysuria-Related Receptor, and Method for Preventing or Improving Dysuria Using Medicinal Composition or Antagonist. Patent WO 2016075960, 19 May 2016.
196. Sakata, Y.; Nakamura, H.; Oshio, K. Muscular Atrophy Preventing Agent Containing Citrus Depressa Extract. Patent WO 2013099982, 4 July 2013.
197. Li, S.; Yang, G.; Long, T. Application of (demethyl) polymethoxyflavone and taxol medicine in producing the medicine for treating non-small cell lung cancer. Patent CN 106562954, 19 April 2017.
198. Nakano, S.; Ono, M.; Hayashi, C. Agent and Method for Inhibiting Breast Cancer Cell Proliferation Comprising Nobiletin. Patent JP 2016017042, 4 February 2016.
199. Chen, G.; Wang, H. Application of Nobiletin in the Preparation of Health Products or Medicines for Preventing and/or Treating Oral Cancer. Patent CN 105030559, 11 November 2015.
200. Ma, W.-Z.; Feng, S.-L.; Yao, X.-J.; Yuan, Z.-W.; Liu, L.; Xie, Y. Use of Nobiletin in Cancer Treatment. Patent AU 2015101287, 22 October 2015.
201. Zhang, Z. Chinese Medicinal Composition Containing Extracts from Citrus and Scutellaria for Treating Cancer Chemotherapy Related Diarrhea. Patent CN 103655835, 26 March 2014.
202. Li, M.; Jin, H.; Yang, Z.; Xu, G.; Lin, Y.; Lin, Q.; Zhang, Z. Medical Application of Flavonoids of Citrus Reticulata Pericarp as Angiogenesis Inhibitor. Patent CN 101947215, 19 January 2011.
203. Zhou, H.; Xie, B.; Zang, X.; Cheng, L.; Liang, G. A Multiple Index Component content Determination, Fingerprint Construction and Preparation Method for Liver-Tonifying Eyesight-Improving Oral Liquid [Machine Translation]. Patent CN 105510452, 20 April 2016.
204. Guo, J.; Liang, L.; Song, J.; Li, H.; Yang, J.; Chen, B.; Wang, S. Method for Extracting Nobiletin and Hesperetin from Citrus. Patent CN 106632196, 10 May 2017.

205. Cao, J.; Hu, S.; Liu, X.; Cao, W.; Pang, X.; Dai, H.; Da, J. A method of Extracting Flavonoids active Ingredients in Citrus Reticulata Pericarp. Patent CN 104297026, 21 January 2015.

206. Yamaguchi, K.; Mogami, K.; Yamaguchi, Y.; Hitomi, N.; Murata, K.; Tani, Y. Manufacture of Nobiletin by Solvent Extraction and Nobiletin-Containing Extract. Patent JP 2012056938, 22 March 2012.

207. Sun, C.; Wang, Y.; Chen, K.; Li, X.; Cao, J. Process Forextn. And Purifn. of Polymethoxylated Flavonoids Compound from Fruit of Citrus Reticulate. Patent CN 107011308, 4 August 2017.

208. Li, X.; Zhang, J.; Sun, C.; Chen, K. Method for Isolating and Purifying Seven Flavonoids from Citrus Tangerina oil Cell Layer. Patent CN 103610800, 5 March 2014.

209. Liang, H.; Wu, D.; Li, B.; Li, Y.; Li, J. Stable Nobiletin liquid Preparation and Preparation Method Thereof. Patent CN 107998073, 8 May 2018.

210. Yang, W.; Song, Y.; Chen, H.; Luo, X.; Yuan, J. A Technique Based on Multi-Solvents for Preparing Nobiletin. Patent CN 105669626, 15 June 2016.

211. Iwashita, M.; Umehara, M.; Onishi, S.; Yamamoto, M.; Yamagami, K.; Ishigami, T. Method for Manufacturing Nobiletin-Containing Solid Dispersion. Patent WO 2018025871, 8 February 2018.

212. Woo, J.T.; Komaki, M. Polymethoxyflavonoid Dissolved Composition and its Manufacturing Method. Patent JP 2015221761, 10 December 2015.

213. Chen, Y.; Yu, Y.; Yang, D.; Wei, W.; He, Z.; Lin, X.; Xie, H. Measurement Method for Seventeen Kinds of Phenol Substances in Grape and Citrus Fruit Using High Performance Liquid Chromatography (HPLC). Patent CN 102706980, 3 October 2012.

214. Kusano, S.; Tamasu, S. Composition Containing 4′-Demethylnobiletin for skin Whitening Cosmetics, Medicines, Foods and Drinks. Patent JP 2017226612, 28 December 2017.

215. Choi, B.G.; Lee, D.R. Skin Moisturizers Containing Citrus Peel Extracts. Patent KR 2017000068, 6 January 2017.

216. Karabey, F. Nobiletin Molecules in Cosmetic Preparationsuse. Patent TR 2014000324, 2015.

217. Zhang, X.; Chen, S.; Wang, X.; Xie, F.; Liu, X.; Wang, J.; Yan, A.; Gao, N.; Li, F. A Snap Bean Preservative [Machine Translation]. Patent CN 106172719, 7 December 2016.

218. Krohn, M.; Seibert, S.; Kleber, A.; Wonschik, J. Sweetener and/or Sweetness Enhancer, Sweetener Composition, Methods of Making the Same and Consumables Containing the Same. Patent WO 2012107203, 16 August 2012.

219. Zhang, L.; Zhu, W.; Yang, C.; Guo, H.; Yu, A.; Ji, J.; Gao, Y.; Sun, M.; Zhai, G. A novel folate-modified self-microemulsifying drug delivery system of curcumin for colon targeting. *Int. J. Nanomed.* **2012**, *7*, 151–162. [CrossRef]

220. Bansode, S.T.; Kshirsagar, S.J.; Madgulkar, A.R.; Bhalekar, M.R.; Bandivadekar, M.M. Design and development of SMEDDS for colon specific drug delivery. *Drug Dev. Ind. Pharm.* **2016**, *42*, 611–623. [CrossRef]

221. Low, L.E.; Tan, L.T.-H.; Goh, B.-H.; Tey, B.T.; Ong, B.H.; Tang, S.Y. Magnetic cellulose nanocrystal stabilized Pickering emulsions for enhanced bioactive release and human colon cancer therapy. *Int. J. Biol. Macromol.* **2019**, *127*, 76–84. [CrossRef] [PubMed]

222. Low, L.E.; Tey, B.T.; Ong, B.H.; Chan, E.S.; Tang, S.Y. Palm olein-in-water Pickering emulsion stabilized by Fe_3O_4-cellulose nanocrystal nanocomposites and their responses to pH. *Carbohydr. Polym.* **2017**, *155*, 391–399. [CrossRef] [PubMed]

223. Vangijzegem, T.; Stanicki, D.; Laurent, S. Magnetic iron oxide nanoparticles for drug delivery: Applications and characteristics. *Exp. Opin. Drug Deliv.* **2019**, *16*, 69–78. [CrossRef] [PubMed]

224. Xing, T.T.; Zhao, X.J.; Zhang, Y.D.; Li, Y.F. Fast separation and sensitive quantitation of polymethoxylated flavonoids in the peels of citrus using UPLC-Q-TOF-MS. *J. Agric. Food Chem.* **2017**, *65*, 2615–2627. [CrossRef] [PubMed]

225. Yang, T.; Rycaj, K.; Liu, Z.-M.; Tang, D.G. *Cancer Stem Cells: Constantly Evolving and Functionally Heterogeneous Therapeutic Targets*; AACR: Philadelphia, PA, USA, 2014.

226. Chen, K.; Huang, Y.-H.; Chen, J.-L. Understanding and targeting cancer stem cells: Therapeutic implications and challenges. *Acta Pharmacol. Sin.* **2013**, *34*, 732–740. [CrossRef] [PubMed]

227. Tan, L.T.H.; Low, L.E.; Tang, S.Y.; Yap, W.H.; Chuah, L.H.; Chan, C.K.; Lee, L.H.; Goh, B.H. A reliable and affordable 3D tumor spheroid model for natural product drug discovery: A case study of curcumin. *Prog. Drug Discov. Biomed. Sci.* **2019**, *2*, 1–5.

228. DiMarco-Crook, C.; Xiao, H. Diet-based strategies for cancer chemoprevention: The role of combination regimens using dietary bioactive components. *Annu. Rev. Food Sci. Technol.* **2015**, *6*, 505–526. [CrossRef]

229. Funaro, A.; Wu, X.; Song, M.; Zheng, J.; Guo, S.; Rakariyatham, K.; Rodriguez-Estrada, M.T.; Xiao, H. Enhanced Anti-Inflammatory Activities by the Combination of Luteolin and Tangeretin. *J. Food Sci.* **2016**, *81*, H1320–H1327. [CrossRef]

230. Wu, X.; Song, M.; Qiu, P.; Li, F.; Wang, M.; Zheng, J.; Wang, Q.; Xu, F.; Xiao, H. A metabolite of nobiletin, 4′-demethylnobiletin and atorvastatin synergistically inhibits human colon cancer cell growth by inducing G0/G1 cell cycle arrest and apoptosis. *Food Funct.* **2018**, *9*, 87–95. [CrossRef]

MDPI

St. Alban-Anlage 66

4052 Basel

Switzerland

Tel. +41 61 683 77 34

Fax +41 61 302 89 18

www.mdpi.com

Cancers Editorial Office

E-mail: cancers@mdpi.com

www.mdpi.com/journal/cancers